SOME PHYSICAL CONSTANTS (see also Appendix 3)

Gravitational constant	G	6.67×10^{-11} N m²/kg²
Avogadro's number	N_0	6.02×10^{23} particles/mole
Boltzmann's constant	k	1.381×10^{-23} J/K
Gas constant	R	8.31 J/mole K
Quantum unit of charge	e	1.602×10^{-19} C
Permittivity constant (combined form)	$\frac{1}{4\pi\epsilon_0}$	8.99×10^9 ($\cong 9 \times 10^9$) N m²/C²
Permeability constant (combined form)	$\frac{\mu_0}{4\pi}$	10^{-7} N/A²
Speed of light	c	3.00×10^8 m/sec
Planck's constant	h	6.63×10^{-34} J sec
	$\hbar$	1.055×10^{-34} J sec
Mass of electron	m_e	9.11×10^{-31} kg
Mass of proton	m_p	1.673×10^{-27} kg
Bohr radius	a_0	5.29×10^{-11} m

SOME PHYSICAL DATA (see also Appendix 3)

Acceleration of gravity at earth's surface	(g)	9.8 m/sec²
Mass of earth		5.98×10^{24} kg
Radius of earth		6.37×10^6 m
Earth-moon distance (center to center)		3.84×10^8 m
Average earth-sun distance		1.50×10^{11} m
Average orbital speed of earth		2.98×10^4 m/sec
Standard conditions		1 atm = 1.013×10^5 N/m²
		0 °C = 273.15 K
Standard dry air density		1.293 kg/m³
Speed of sound in standard dry air		331 m/sec
Density of water		1.00×10^3 kg/m³

SOME CONVERSION FACTORS (see also Appendix 4)

2.54 cm/in
0.3048 m/ft
1.609 km/mile
10^{-10} m/Å
10^{-15} m/fm
3.156×10^7 sec/yr
0.447 (m/sec)/(mile/hr)
57.3 deg/radian
0.454 kg/lb
1.661×10^{-27} kg/amu

10^3 (kg/m³)/(gm/cm³)
10^5 dyne/N
10^7 erg/J
4.184 J/cal
4,184 J/kcal
1.602×10^{-19} J/eV
1.80 F°/C°
3.00×10^9 esu/C (charge)
10^4 G/T (magnetic field)

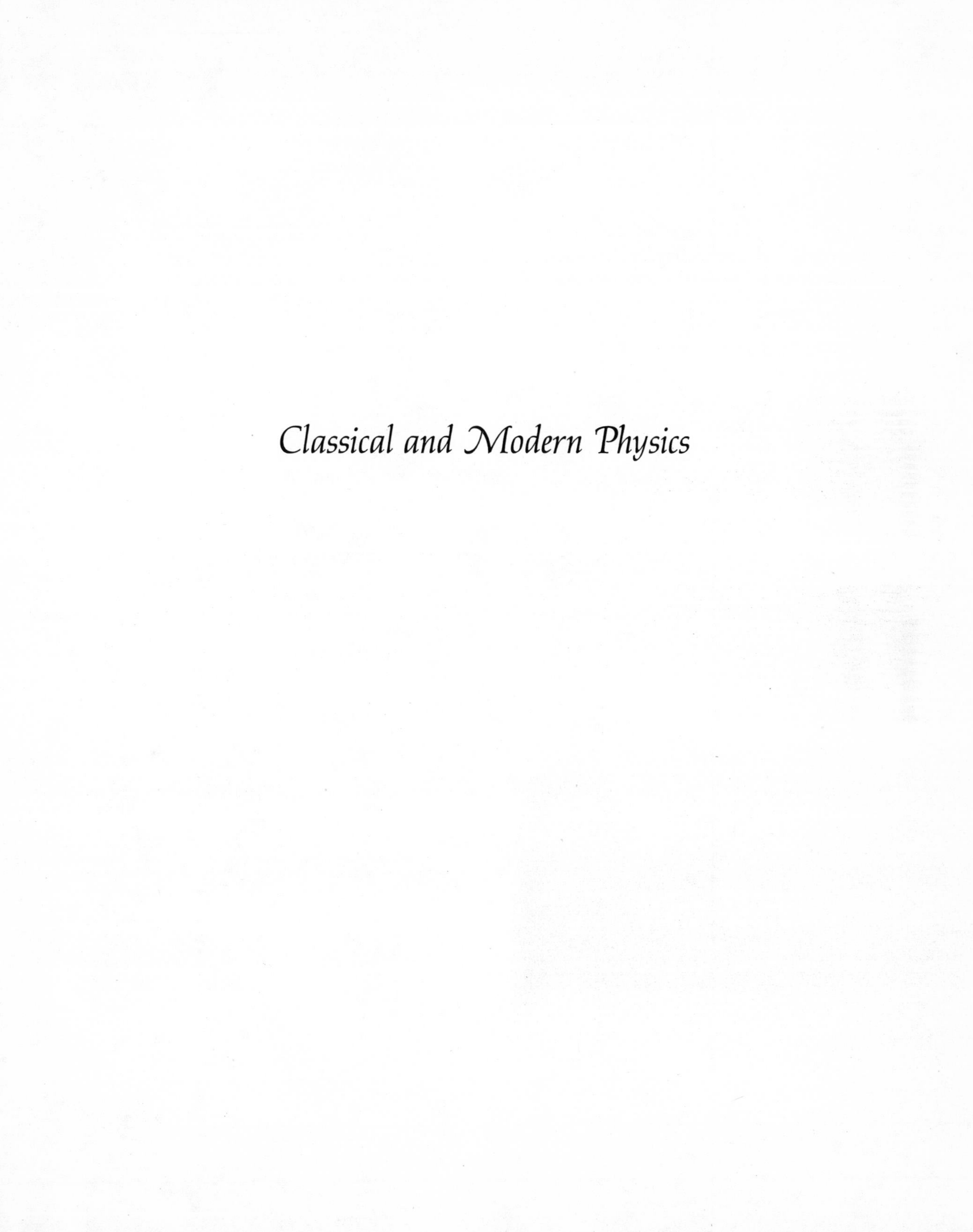

Classical and Modern Physics

THIS BOOK IS AVAILABLE IN THREE VOLUMES
AND IN A COMBINED EDITION OF VOLUMES 1 AND 2

In Volume 1

Introduction to Physics
Mathematics
Mechanics

In Volume 2

Thermodynamics
Electromagnetism

In Volume 3

Relativity
Quantum Mechanics

Kenneth W. Ford
UNIVERSITY OF MASSACHUSETTS AT BOSTON

Volume 1

Classical and Modern Physics

A TEXTBOOK FOR STUDENTS OF SCIENCE AND ENGINEERING

XEROX COLLEGE PUBLISHING *Lexington, Massachusetts | Toronto*

CONSULTING EDITOR

Brenton F. Stearns, *Hobart and William Smith Colleges*

ISB Number: 0-536-00723-3 (Domestic)
ISB Number: 0-536-00870-1 (International)
Library of Congress Catalog Card Number: 76-161385
Printed in the United States of America.

to Joanne

Preface

This is a textbook for a three-semester introductory physics course for students of science and engineering. Roughly the first two-thirds of the book (Parts 1–5) could serve as a text for a one-year course in classical physics; the last third (Parts 6 and 7) could serve as a text for a one-semester course in modern physics. With certain sections and subsections—or even whole chapters—omitted, it should also meet the needs of a one-year course that includes modern physics. Because of its substantial length and because of the several kinds of courses in which it might be used, the book is published in three volumes as well as in a combined edition of Volumes 1 and 2. The parts of the volumes are as follows:

Volume 1	1. *Introduction to Physics*
	2. *Mathematics*
	3. *Mechanics*
Volume 2	4. *Thermodynamics*
	5. *Electromagnetism*
Volume 3	6. *Relativity*
	7. *Quantum Mechanics*

The appendices and the index for the complete book appear in each volume.*

Probably every author has in mind a particular kind of student for whom he is writing. My "model student" has had high-school physics and is taking calculus concurrently with college physics. He or she is a serious but not necessarily gifted student, is interested in ideas as well as technical skills, and

* This early printing of Volume 1 contains an index for Parts 1–4 only.

learns best when mathematical derivations are supplemented by verbal explanations and physical examples. In terms of the intellectual demand placed on the student, this text is comparable to the popular text by Halliday and Resnick.* It is less demanding than the Berkeley Physics Course† or the M.I.T. Introductory Physics Series.‡

As originally conceived, this book was to be a "calculus version" of my earlier text, *Basic Physics*.§ Having passed through numerous evolutionary stages of writing and rewriting, deleting and adding, however, the book as now published is distinct from the earlier one in various ways besides its mathematical level.

Some of the principal features of this text are the following: (1) I have tried to give a unified presentation of both classical and modern physics. Although theoretical developments of relativity and quantum physics are saved for the last two parts of the book, certain ideas (mass-to-energy conversion, for instance, and nature's speed limit) are introduced early, and modern examples often serve to illustrate classical laws. (2) A series of introductory chapters give time for some maturing of the student's view of physics and his command of mathematics before the intricacies of classical mechanics are approached. (3) Ideas of calculus are introduced (in Chapter 5) somewhat more fully than in most other physics texts. (4) I have tried to steer a course through the discipline of physics that keeps the student in touch with the large view of the subject—the economy and simplicity of its concepts, the elegance of its overall structure—at the same time that he is mastering practical skills and polishing his problem-solving ability. (5) As aids to study and review, the text is divided into fairly numerous sections and subsections, marginal notes highlight key ideas and important equations, and summaries of ideas and definitions appear at the end of every chapter. (6) I have tried to bring out the excitement of physics as a living, evolving discipline, powerful yet incomplete. A limited amount of historical material is included; I have taken some care with this and hope that most of it is real history and not myth.

The "Notes on the Text" that begin on page xiii are intended as a brief guide to instructors. Students, too, can be encouraged to read these notes. As an aid to selective use of the material, some sections and subsections are marked with a star (★) to indicate that they are optional. A section or subsection may be so marked either because it is of greater than average difficulty or because it is peripheral to the main development of a chapter. Any such designation of optional material is necessarily rather arbitrary. Most instructors will have their own ideas about which material to include and which to omit; the stars provide only a first set of suggestions.

At the end of each chapter appear questions, exercises, and problems. *Questions*, with few exceptions, are to be answered in words. Many of them are intended to be thought-provoking and may have no specific right answer.

* David Halliday and Robert Resnick, *Physics* (New York: John Wiley and Sons, Inc., 1966).

† *The Berkeley Physics Course*, a five-volume series by various authors (New York: McGraw-Hill Book Co.).

‡ *The M.I.T. Introductory Physics Series*, three volumes by A. P. French in print in 1971, with three more volumes scheduled (New York: W. W. Norton and Co.).

§ Xerox College Publishing, 1968.

Some are difficult. *Exercises* are intended to be straightforward tests of understanding of the material in the chapter without special twists or subtleties. The exercises involve numerical work as well as algebra and some calculus. Often an exercise may ask for a brief explanation as well as a quantitative result. *Problems* are, in general, more challenging. They may be in the nature of difficult exercises; they may draw together material from more than one section; or they may build on material in the text but go somewhat beyond it. The number of questions, exercises, and problems is large—much larger than the number that would ordinarily be assigned in a course. This large number is provided in order to meet the needs and tastes of different instructors, to enable the student to practice on items that are not assigned, and to enable the instructor, if he wishes, to choose examination questions from the text. Because the chapters are rather long and the end-of-chapter items are numerous, marginal notes are used to classify the questions, exercises, and problems. Questions and exercises are keyed to specific sections. Problems are labeled by their subject.

I have used SI (mks) units throughout. Some special units—such as the calorie, the astronomical unit, and the electron volt—are introduced, and some exercises and problems require conversion of units. However, no effort is made to have the student develop any routine familiarity with more than one set of units. To aid the student in case he encounters Gaussian (cgs) units in another text or in a research paper, Appendix 5 contains an extensive list of the equations of electromagnetism in SI and Gaussian units. My only significant deviation from "purity" in handling units occurs in Chapters 13 and 14, where calories and kilocalories are used as often as joules and where Avogadro's number is defined as the number of molecules in 1 mole rather than the number in 1 kmole.

I want students to enjoy this book and to profit from it. I think it will serve its purpose best if students are not rushed too quickly through too much of it. Careful treatment of some material and judicious omission of other material will probably provide better preparation for further work in physics, engineering, and other sciences than will a fast trip through every section.

KENNETH W. FORD

Acknowledgments

I have had the great benefit of collaboration with Neal D. Newby, Jr., and Brenton F. Stearns on questions, exercises, and problems. Their hundreds of suggestions for end-of-chapter items were vitally important when my own imagination began to flag. In his role of Consulting Editor, Brenton Stearns has also been of inestimable value as careful reader and thoughtful critic throughout the writing and rewriting of this text. I am indebted to Russell K. Hobbie, Donald E. Schuele, and N. S. Wall, who read one draft of the manuscript, and to David J. Cowan, who read two, for their numerous helpful suggestions. So many colleagues have contributed facts, data, photographs, and suggestions that a complete list is impossible. Among them are Olexa-Myron Bilaniuk, Alfred M. Bork, George J. Igo, Henry H. Kolm, Alexander Landé, Arthur W. Martin, Edward M. Purcell, Frederick Reines, Gerald Schubert, and Barry N. Taylor.

Yale Altman and Warren Blaisdell encouraged the initiation of this project and helped to keep it going. They deserve the credit (or blame) for turning me into an author in the first place. At Xerox College Publishing, James Piles has been an agreeable and helpful mentor, and Bernice Borgeson has handled thousands of details with extraordinary dedication. I have been fortunate in having the services of two outstanding typists, Lisa Munsat and Elizabeth Higgins.

Notes on the text

Of the book's seven sections, the first two (*Introduction to Physics* and *Mathematics*) provide introductory and background material. There is great latitude in the way these two parts may be used. The remaining five parts (*Mechanics*, *Thermodynamics*, *Electromagnetism*, *Relativity*, and *Quantum Mechanics*) are devoted to specific major theories of physics. The fullest mathematical development is carried out for the theories of mechanics, electromagnetism, and special relativity. Thermodynamics and quantum mechanics are handled with somewhat more attention to phenomena and less to mathematical formalism. (Nevertheless, I have avoided more modest titles, such as *Heat* and *Atomic and Nuclear Phenomena*, because these parts do also emphasize the unity and power of physical theories, and they are by no means lacking in mathematics.) The division of the book into parts and the choice of rather long chapters in preference to more numerous shorter chapters are designed to serve the same end: to keep the overall structure of physics in view at a time when it is all too easy for the student to see the subject as a bewildering array of unrelated pieces.

PART 1: Chapter 1 provides a brief overview of physics—its relation to mathematics and technology, its division into theory and experiment, and its development over the past few centuries. I recommend that students be asked to read this short chapter and to consider some of the questions at the end even if it receives very little class time.

For students who have previously had a physics course, most of what is in Chapter 2 will be review and consolidation of material studied earlier. Students coming to physics for the first time will want to pay closer attention to this chapter. The chapter has several goals: to introduce SI units and define standards, to provide qualitative insights into the meaning of various important concepts, and to provide useful hints on dimensional consistency and units arithmetic.

Chapter 3 is an optional chapter that can be omitted without lack of continuity or be assigned to be read "for fun." There are, however, some serious reasons for putting a survey of elementary particles and submicroscopic nature near the beginning of a general physics text, and these are discussed in Section 3.1. Particles are used from time to time in later parts of the book for illustrative purposes, but these uses do not presuppose an assimilation of all the material in Chapter 3.

The seven absolute conservation laws considered in Chapter 4 introduce an important theme and reveal some common elements of classical and modern physics. Just as the concepts discussed in Chapter 2 are refined in later chapters, most of the laws discussed in this chapter are developed more elaborately later. An instructor who prefers to move rapidly toward kinematics and Newton's laws could omit this chapter or postpone it until reaching Chapters 8, 9, and 10, where the laws of conservation of momentum, angular momentum, and energy reappear.

PART 2: The consolidation of most (but not all) of the mathematical developments in the book in this early part gives the instructor flexibility. He may pause to consider mathematical topics by themselves, or he may skip some sections of Chapters 5 and 6 and refer back to them later. Although entitled "Mathematics," this part does contain considerable physics. See in particular the kinematics in Sections 5.7 and 6.10 and the discussion of experimental uncertainty in Section 5.12. Some mathematical topics that are saved for later chapters are the idea of partial differentiation (Chapter 10), the line integral (Chapter 10), and the surface integral (Chapter 15).

In Chapter 5, differentiation and integration are developed more fully than in most other physics texts. These sections cannot, of course, substitute for a mathematics course, but they can help the student to gain a more intuitive grasp of what he is learning in mathematics and see how to apply it to physical situations. Sections 5.3–5.11 are devoted to essential practical matters. Sections 5.1 and 5.2 are quite different; they are devoted to the nature of mathematics and its relation to science. Section 5.12 ought to be most effective if used in conjunction with laboratory work; it is not essential for further developments in this text.

Most of Chapter 6 does not differ greatly from many standard introductions to vectors. At the beginning, it ties vector algebra to the geometrical arithmetic that students now learn in school. Optional sections treat the transformation of components and the distinction between polar vectors and axial vectors. Some care is taken in the chapter to distinguish a physical vector quantity from a mathematical vector and to emphasize that the vector nature of a physical quantity must be decided by experiment.

PART 3: Chapters 7–10 can be thought of as an ascending staircase of increasing difficulty and sophistication. The formal development of mechanics is essentially completed in these four chapters. Chapter 11 should then come as a welcome relief to the student. Chapter 12 is an assortment of distinct supplementary topics, any combination of which could be included to round out the study of mechanics.

Newton's first and second laws appear in Chapter 7. In addition to numerous standard applications, this chapter includes discussions of frames of reference, the distinction between inertial and gravitational mass, the significance of Newton's first law, and the logical structure of mechanics—especially the question of the intermingling of laws and definitions.

Newton's third law is placed in a separate chapter (Chapter 8) to emphasize that it is quite a different sort of law from Newton's first and second laws and to bring out its special connection to the concept of momentum. Chapter 7 was devoted to particle mechanics; Chapter 8 is devoted to systems. A discussion in Section 8.5 ties together the two chapters and shows the interconnections of Newton's three laws. Section 8.12 is related to an earlier discussion of momentum conservation in Chapter 4.

In Chapter 9, angular momentum is defined for particles and is then generalized to systems. Throughout the chapters I have tried to give a balanced and unified view of angular momentum for bodies moving through space and bodies rotating about fixed axes. To this end, orbital angular momentum and spin angular momentum are introduced in Section 9.3. Preliminary to Chapter 11, the law of areas (Kepler's second law) is introduced in Section 9.9. The consideration of rotational energy is postponed to Section 10.9.

In the interest of logical development I have put a full treatment of work and kinetic energy in the first four sections of Chapter 10, even though the last part of Section 10.3 and all of Section 10.4 are more difficult than what follows in the next few sections. This optional material can be omitted or postponed. Examples in Section 10.7 are repeated in Section 10.8 in order to add the important element of the energy diagram to their analysis. The simple pendulum, treated late in the chapter (Section 10.11), should not be overlooked.

Chapter 11 is devoted to the single subject of gravitation because of the fundamental importance of gravitation in nature and because of its importance historically in the genesis of mechanics and modern science. As noted earlier, this chapter is less demanding than those that immediately precede it.

All of Chapter 12 may be considered optional. Any choice could be made from among its five nearly independent topics: surface friction, statics of rigid bodies, fluids, frictional air drag, and two-body collisions.

Contents

PART I *Introduction to Physics*

1 **THE NATURE OF PHYSICS** **3**

1.1 The hierarchy of matter 4
1.2 The great theories of physics 6
1.3 Theory and experiment in physics 7
1.4 The relation of physics to mathematics and technology 8
1.5 The evolution of physics 11

2 **CONCEPTS, UNITS, AND DIMENSIONS** **16**

2.1 Quantitative concepts 16
2.2 Units 18
2.3 Dimensions 19
2.4 Dimensional consistency and units consistency 20
2.5 Length 23
2.6 Time 24
2.7 Speed 27
2.8 Mass 28
2.9 Energy 30
2.10 Charge 32
2.11 Angular momentum 33
★ 2.12 Natural units and dimensionless physics* 35

* Sections and subsections marked with stars are optional.

3 ELEMENTARY PARTICLES 44

3.1 The submicroscopic frontier 44
3.2 The early particles 45
3.3 The pion and the muon 49
3.4 The modern particles 51
3.5 The properties of the particles 54
★ 3.6 Experimental tools 58
3.7 The significance of the particles 64

4 CONSERVATION LAWS 70

4.1 Absolute conservation laws 70
4.2 Charge conservation 71
4.3 Family-number conservation laws 72
4.4 Energy conservation 75
4.5 Momentum conservation 76
4.6 Angular-momentum conservation 78
★ 4.7 Conservation laws and symmetry principles 80
★ 4.8 The uniformity of space 82

PART II *Mathematics*

5 MATHEMATICS IN SCIENCE 92

★ 5.1 Two kinds of mathematical truth 92
★ 5.2 Mathematics and nature 95
5.3 Coordinate systems and frames of reference 97
5.4 Speed; the derivative 100
5.5 Angles and angular speed 106
5.6 Functions, tables, and graphs 109
5.7 One-dimensional kinematics 114
5.8 The indefinite integral 121
5.9 The definite integral 125
5.10 Sine and cosine functions 134
5.11 Exponential and logarithmic functions 137
★ 5.12 Probability, experimental error, and uncertainty 142

6 VECTORS 161

6.1 Scalars and numerics: geometrical arithmetic 161
6.2 Vectors: addition and subtraction 163

6.3 Multiplication of a vector by a numeric 168
6.4 The position vector 170
6.5 Components 172
6.6 Vectorial consistency 178
6.7 The scalar product 179
6.8 The vector product 180
6.9 Vectors changing in time; the derivative of a vector 184
6.10 Uniform motion in a circle 187
6.11 Vector calculus 190

PART III *Mechanics*

7 FORCE AND MOTION: NEWTON'S FIRST AND SECOND LAWS 206

7.1 Motion without force; Newton's first law 206
7.2 The concept of force 209
7.3 Newton's second law; inertial mass 212
7.4 Applications of Newton's second law 215
7.5 The harmonic oscillator 219
7.6 Motion near the earth 224
7.7 Gravitational mass 229
7.8 Motion in two dimensions 232
★ 7.9 The motion of charged particles in electric and magnetic fields 239
7.10 Frames of reference 244

8 MOMENTUM AND THE MOTION OF SYSTEMS: NEWTON'S THIRD LAW 266

8.1 Momentum and Newton's second law; impulse 267
8.2 Center of mass 270
8.3 Newton's second law for systems 275
8.4 Newton's third law 277
8.5 The cancellation of internal forces 281
8.6 The motion of systems in one dimension 284
8.7 Conservation of momentum 287
8.8 Momentum conservation in systems that are not isolated 291
8.9 Momentum conservation and center of mass 292
8.10 Rocket propulsion 295
8.11 Systems of two particles; reduced mass 299
8.12 The significance of momentum conservation 302

9 **ANGULAR MOMENTUM** **316**

★ 9.1 The concept of angular momentum 316
9.2 The angular momentum of systems 322
9.3 Orbital angular momentum and spin angular momentum 325
9.4 Moments of inertia 328
9.5 Torque and the law of angular momentum change 336
9.6 Rotation about a fixed axis 340
★ 9.7 Precession 344
9.8 The conservation of angular momentum 347
9.9 The law of areas 351
★ 9.10 The isotropy of space 357

10 **ENERGY** **376**

10.1 Work done by a constant force 376
10.2 Work and kinetic energy for one-dimensional motion 379
10.3 Work and kinetic energy for the general motion of a particle 383
★ 10.4 Work and kinetic energy for a system of particles 388
10.5 Work as a mode of energy transfer 392
10.6 Conservation of work 395
10.7 Potential energy in one dimension 398
10.8 Conservation of mechanical energy in one dimension; energy diagrams 405
10.9 Rotational energy 412
★ 10.10 Conservative forces and potential energy in general 416
10.11 The simple pendulum 422
10.12 Assessment of energy conservation 423

11 **GRAVITATION** **443**

11.1 Laws of force in mechanics 443
11.2 The law of universal gravitation 444
11.3 Gravity of the earth 448
11.4 Gravitational potential energy: escape speed 450
11.5 The shape of the earth 454
11.6 Kepler's laws 458
★ 11.7 Deduction of the law of gravitational force 464
11.8 The discovery of Neptune 472

12 **FURTHER APPLICATIONS OF MECHANICS** **489**

12.1 Surface friction 489
12.2 Statics of rigid bodies 492
12.3 Forces in fluids 496
12.4 Fluid flow 504
12.5 Frictional air drag 513
12.6 Two-body collisions 517

APPENDICES

Classical and Modern Physics

PART ONE

Introduction to Physics

The Nature of Physics

Physics, once called natural philosophy, is the discipline of science most directly concerned with the fundamental laws of nature. Other areas of science and various branches of engineering have taken on a flourishing life of their own, yet all of them build ultimately on the basic laws that make up the subject matter of physics.

Physics: a human endeavor

According to one old definition, physics is the study of matter and motion. Neither this nor any other one-sentence definition can adequately reflect the mixture of creative effort, accumulated knowledge, unifying ideas, mathematical equations, philosophical impact, and practical application that comprise physics. The modern physicist has generalized the idea of matter to include the distributed energy of wave fields and the transitory energy of unstable particles; also, as we shall frequently emphasize in this book, he is as much concerned with the unchanging aspects of nature as he is with motion and change. Yet it is true that the material world and the interaction of one part of it with another remain at the heart of physics. To encompass as much as possible of the behavior of matter with the simplest possible array of ideas and equations is the primary goal of the physicist.

Physics: the study of matter and its interactions

In some ways the beginning student of physics faces a more formidable task than does the advanced student. In surveying physics for the first time at the college level, you must do more than learn facts, laws, equations, and problem-solving techniques. You must also seek to grasp the whole of physics, appreciate its generality, see the interconnections of its parts, and perceive its boundaries. You must learn to distinguish between theory and application, between general law and specific fact, between physical ideas and mathematical tools. This book is designed to help you toward such general insights at the same time that you are gaining the power to use physics for practical purposes. Yet only through your own dedicated effort can these dual goals be achieved.

The dual goals of study: general insights and practical power

1.1 The hierarchy of matter

Much of the history of science can be characterized as a probing upward and downward away from the world of man's immediate sense experience. The frontiers of physics have left the human-sized world, which we refer to as the macroscopic world, and have reached the submicroscopic world (Figure 1.1) and the cosmological world (Figure 1.2). Between these distant limits is arrayed that part of the physical world that is understood, if only imperfectly.

From the subatomic elementary particles up to the collection of galactic clusters that is optimistically called the universe, man is now familiar with a hierarchy of objects joined together in structures of ever increasing size. Incomplete though the picture may be, it is a grand panorama spanning a factor of

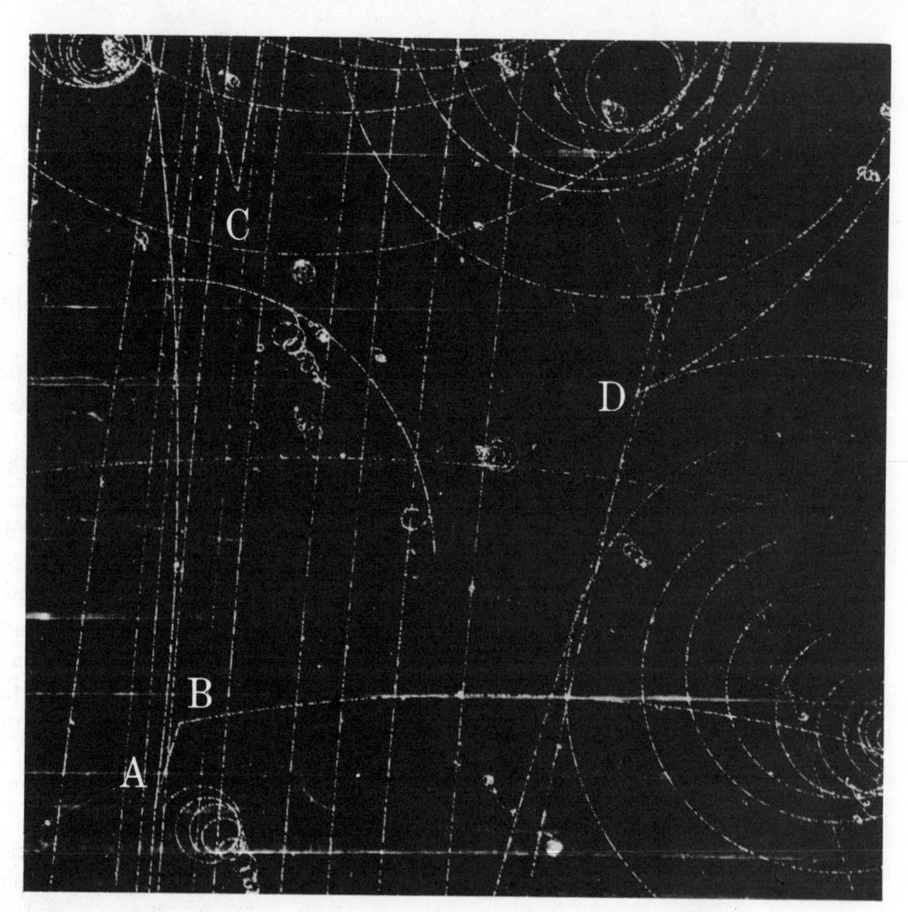

FIGURE 1.1 The frontier of the very small. Tracks of elementary particles in a bubble chamber. At the point marked A, two particles were destroyed and three others created in the submicroscopic violence of a collision extending over 10^{-15} m in space and 10^{-23} sec in time. Other particle interaction events occurred at B, C, and D. This photograph, made in 1964, helped to establish the existence of a new particle called the Ω^- (omega-minus), which left the track from A to B. The bubble chamber in which these tracks were recorded is shown in Figure 3.10(b). (Photograph courtesy of Brookhaven National Laboratory.)

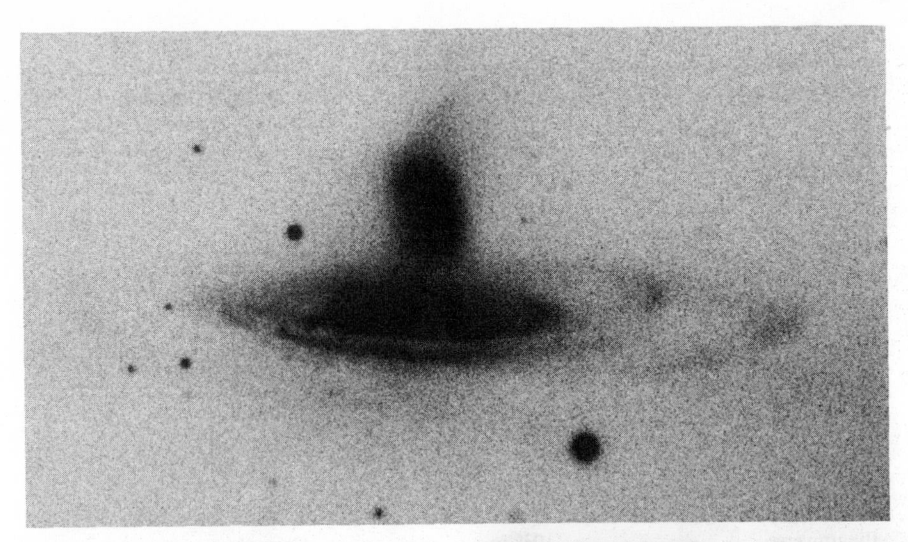

FIGURE 1.2 The frontier of the very large. (a) A pair of interacting galaxies. The smaller galaxy appears to be "raining" stars into the center of the larger spiral galaxy. Little is known about the life histories of galaxies or the reasons for their variety of form. (b) The 200-in. Hale telescope at Mt. Palomar. A half-hour exposure with this telescope was required to obtain the photograph of the interacting galaxies. [(a) Photograph courtesy of Halton Arp, Hale Observatories. (b) Photograph courtesy of Hale Observatories.]

10^{41} in scale. The names of some of the structures and of some of the special branches of science concerned with particular levels of the hierarchy are given in Table 1.1.

Maximum complexity at the level of man

As we depart in either direction from the size of man, the structures of the world grow simpler. There is no organization of constituents in either the world of the very small or the world of the very large that begins to approach the complexity and degree of organization found in living creatures. It might be argued that there exist larger and more complicated degrees of organization in the universe, which man's limited intellect is incapable of grasping, but this is an argument outside the scope of science. So far there is no evidence for any such organization and much evidence against it. The theories of the submicroscopic world have been successfully used in dealing with the structure of stars and, to some extent, galaxies.

TABLE 1.1 THE STRUCTURES OF THE WORLD

Object	Size*	Special Associated Branch of Science
Elementary particle	10^{-15} m or less	Particle physics
Atomic nucleus	10^{-14} m	Nuclear physics
Atom	10^{-10} m	Atomic physics
Molecule	10^{-9} m	Chemistry
Giant molecule	10^{-7} m	Biochemistry
Solids		Solid-state physics
Liquids		Hydrodynamics
Gases		Aerodynamics
Plants and animals	10^{-7} m to 10^{2} m	Biology
The planet Earth	10^{7} m	Geology
Star	10^{7} m to 10^{12} m	Astrophysics
Galaxy	10^{20} m	Astronomy
Galactic cluster	10^{23} m	
The known part of the universe	10^{26} m	Cosmology

* Lengths in this table are expressed in meters, usually abbreviated m. One meter is equal to 100 centimeters (cm), or 39.37 inches (in.), or 3.281 feet (ft).

1.2 The great theories of physics

As indicated in Table 1.1, special branches of science have come to be associated with particular sizes and levels of arrangement of the matter in the world. But at a deeper level science is organized around ideas that span many parts of nature. The structure of physics, for example, is built not around objects or the physical structure of the world, or even about particular phenomena in the world; it is rather a set of general theories, each of which describes a wide range of phenomena and objects. Mechanics, one such general theory, accounts for the motion of objects over almost the whole explored range of sizes, from the submicroscopic to the cosmological.

A law of physics is a statement of fact, usually about a restricted range of phenomena. The word "theory" is used more broadly to mean anything from an untested hypothesis to a firmly established set of ideas capable of accounting for many laws. In this book we are concerned mainly with the few theories that have progressed well beyond the stage of hypothesis, theories that deserve being called the great theories of physics. Each is a structure of ideas and equations thoroughly tested by experiment. Each is remarkable for its small number of basic concepts and its large number of applications. The great theories of physics are

Five great theories

1. Mechanics (sometimes called Newtonian mechanics or classical mechanics): the theory of the motion of material objects.
2. Thermodynamics: the theory of heat, temperature, and the behavior of large arrays of particles.
3. Electromagnetism: the theory of electricity, magnetism, and electromagnetic radiation.
4. Relativity: the theory of invariance in nature and the theory of high-speed motion.

5. Quantum mechanics: the theory of the mechanical behavior of the submicroscopic world.

Even these five are closely interrelated. Every one of the myriad of phenomena in the physical world that is understood can be explained in terms of one or more of these few theories. The behavior of single atoms, for example, is governed by quantum mechanics, relativity, and electromagnetism. To describe a collection of many atoms also requires the theories of mechanics and thermodynamics.

About these five theories we can say with some confidence that none will ever be completely overthrown. If history is a reliable guide, we can say with equal confidence that none will prove to be entirely correct. Mechanics is already known to be "incorrect." Relativity "overthrew" mechanics when it showed Newton's laws of mechanics to be incorrect for describing ultrahigh-speed motion. Later, quantum mechanics showed classical mechanics to be incorrect for describing the internal motions within atoms. Each of these twentieth-century developments proved mechanics to be wrong. Why then do we doggedly list mechanics as one of the great theories of physical science? Because over what is still a vast domain of sizes and speeds, mechanics is so extremely accurate that it is for all practical purposes completely correct. It is the best and simplest tool for describing nature in a certain domain. It is better to say that relativity and quantum mechanics have chipped away at the boundaries of mechanics, reducing it from an infinite to a finite domain, than it is to say that either theory has overthrown mechanics.

Mechanics has known limits of validity

A good many branches of physical science that will occur to the reader do not appear in the list of great theories. Where, for example, is the enormous subject of chemistry? Where are astronomy, geology, and aerodynamics? They are omitted because, being less fundamental, they derive their sustenance from these theories. This is to say not that they are less interesting or less complicated (they are indeed more complicated) but that they are further removed from the fundamental underlying laws of nature. Throughout this book we will be concerned primarily with the basic concepts, laws, and theories of physics.

At the limits of the very large and very small, our theoretical base is least secure. To describe nature at these limits, new general theories may be necessary. Physics is an active and still-evolving science.

The search for new theories continues

1.3 Theory and experiment in physics

Some areas of science, such as certain branches of biology and psychology, are purely experimental. Facts are being gathered and knowledge increased, but there exists no body of ideas, concepts, and relationships that are collectively called theory to tie together the facts into a coherent whole, that is, to "explain" the facts. In other areas of science, for example mechanics, the equations of a well-tested theory have been so elaborately developed that such areas seem almost to be parts of pure mathematics. But every area of science, regardless of its state of mathematical development, differs in a very fundamental way from pure mathematics and from nearly every other area of human activity. Science is essentially empirical. No idea in science survives because it is esthetically pleasing, mathematically elegant, or magnificently general, although many ideas in science are all of these things. The idea must undergo the test of experiment, and not just one experiment. To stand, it must survive attack from all sides by every device that the experimenter can muster.

The essential role of experiment

Experiment is the final arbiter in science, but a science of experiment only would be a dull one. In attempting to understand nature, man has sought much more than mere empirical facts. It is the theories tying facts together that provide the challenge and reward of science. The new way of looking at nature, the unexpected relation between facts, the single equation governing a vast range of phenomena—these are the things that give science its stature and nobility.

The value of theory

Although the hard evidence of experiment can destroy a theory, no amount of experimental verification can "prove" a theory. Every theory must remain tentative, for two reasons. First, a theory is likely to be capable of making an infinite number of different predictions, but man's finite capabilities limit his ability to test the predictions. Second, no theory is unique. The possibility must always remain open that a theory will be supplanted, not because experimental evidence requires its rejection but simply because an alternative theory may be found that, although neither better nor worse experimentally, is in some way more satisfying to man. It may be conceptually simpler, possess a more economical mathematical framework, or appear in some way to be deeper, more profound, and therefore more pleasing. Human judgment of theory is as important as experiment itself in shaping the structure and progress of science. The scientist's faith in simplicity may indeed cause him to reject a complicated and cumbersome theory even if no better alternative is at hand, just because of his conviction that a simpler description of nature must exist.

The faith in simplicity

Over the last few hundred years, theory and experiment in physics have developed side by side through mutual cross-fertilization. The experimenter without ideas can discover an endless sequence of useless facts; the theorist who is unbridled by the limitations of experiment can produce a stream of fanciful ideas that have nothing to do with nature.

1.4 The relation of physics to mathematics and technology

Theory and experiment are the two essential and inseparable units of physics, but there are two additional auxiliary services nurtured by physics and nurturing physics—mathematics and technology. Technology is the outgrowth of science and provides the essential tools for further experimental research in science. In a somewhat similar way much mathematical discovery has been stimulated by physics, and mathematics is itself a tool of theoretical research and the vehicle of expression for theoretical results (Figure 1.3). Mathematics, much more than technology, has a life of its own independent of science. That mathematics can exist as a kind of scientific theory divorced from scientific fact has been realized for less than two hundred years. Nevertheless, much of mathematics can be used for the description of nature—indeed it provides the most elegant description—and part of the scientist's faith in simplicity is a faith in the possibility of expressing nature's laws in mathematical form.

Mathematics and physics enrich each other

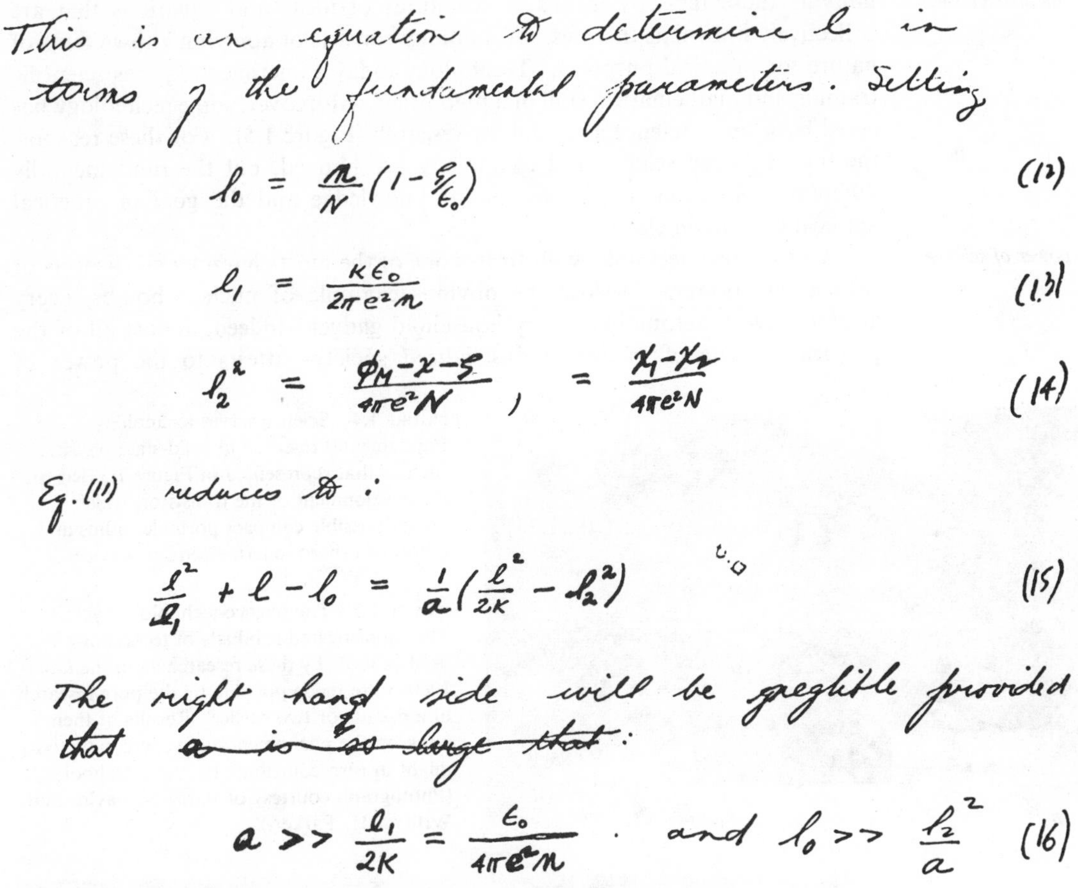

This is an equation to determine ℓ in terms of the fundamental parameters. Setting

$$\ell_0 = \frac{n}{N}\left(1 - \frac{\xi}{\epsilon_0}\right) \qquad (12)$$

$$\ell_1 = \frac{\kappa \epsilon_0}{2\pi e^2 n} \qquad (13)$$

$$\ell_2^2 = \frac{\phi_M - \chi - \xi}{4\pi e^2 N}, \quad = \frac{\chi_1 - \chi_2}{4\pi e^2 N} \qquad (14)$$

Eq. (11) reduces to:

$$\frac{\ell^2}{\ell_1} + \ell - \ell_0 = \frac{1}{a}\left(\frac{\ell^2}{2\kappa} - \ell_2^2\right) \qquad (15)$$

The right hand side will be negligble provided that ~~a is so large that~~:

$$a \gg \frac{\ell_1}{2\kappa} = \frac{\epsilon_0}{4\pi e^2 n} \quad \text{and} \quad \ell_0 \gg \frac{\ell_2^2}{a} \qquad (16)$$

FIGURE 1.3 Mathematics serves science. Notes of John Bardeen, written in 1946, on the theory of semiconductors. Before semiconductors could serve the many practical functions for which they are now known, a mathematical theory of their behavior was required. For his contributions to this theory, Bardeen received the Nobel Prize in 1956. (Notes courtesy of Professor Bardeen.)

Technology and physics enrich each other

Mathematics and the machines of technology also form a part of the bridge between pure physics and applied physics. Although the motivations for the search for new knowledge and the application of already acquired knowledge for practical purposes are entirely different, both efforts employ mathematics, quite often nearly identical mathematics, and both make use of similar mechanical devices. Through discoveries in mathematics and the development of machines, pure science and applied science have enriched each other. Not infrequently, too, the talents of individuals span and join the fields of physics, mathematics, and engineering.

Because the technological by-products of science are more readily comprehensible than fundamental science itself and also because these by-products

Science and technology must be distinguished

have a greater direct impact on our lives (Figure 1.4), technology is often confused with science. Science consists of discovering the facts of nature and unifying these facts by means of structures of ideas and equations that are collectively known as theories. Technology consists of applying known facts of nature for practical purposes. Technology today sometimes requires scientific training and mathematical skill of a high order. Moreover, some technology has acted back upon science as a tool for research (Figure 1.5). For these reasons, the lines between science and technology are blurred, but the fundamentally different motivations of the search for knowledge and the goal of practical application remain clear.

The power of science

Our modern technology illustrates one of the most characteristic aspects of science—its power. Besides the obvious example of nuclear bombs, every airplane, every automobile, every household gadget—indeed, almost all of the physical aspects of modern industrialized society—attests to the power of

FIGURE 1.4 Science serves technology. Fundamental research in solid-state physics, such as that represented in Figure 1.3, led to the development of the transistor, which makes possible compact portable radios and a host of other modern electronic devices.

FIGURE 1.5 The science-technology cycle. The sophisticated products of technology used as tools by these researchers in the late 1960s were made possible by the pure research of a decade or two earlier. Results of their work on the phenomenon of superconductivity might in turn contribute to future technology. (Photograph courtesy of Barry N. Taylor and William H. Parker.)

science. This power springs essentially from the quantitative and predictive character of science. If science merely accounted for what had been observed and experienced in the past, this power and control would not be one of its dominant features. It is the fact that simultaneously with the explanation of a previously known phenomenon comes the ability to predict new phenomena that gives science its power to modify the world around us. Occasionally, technological inventions have appeared without real scientific understanding—the steam engine is a good example—but more frequently technology has been the daughter of science. In many cases, in fact, the invention could not even have been visualized before the creation of the scientific theory upon which it rests. One can scarcely imagine the vacuum tube used in amplifiers being invented before the discovery of the electron, or the atomic bomb being developed before the theory of relativity and the discovery of the neutron.

1.5 The evolution of physics

The roots of physics, especially certain aspects of mechanics and the tools of arithmetic and geometry, go back to antiquity. But it was not until the seventeenth century that natural philosophy came alive as a major force in the development of civilization. In the year 1600 William Gilbert in England published an important work on magnetism.* In the same year, in Bohemia, Johannes Kepler was sifting data on the orbit of Mars. In Italy, Galileo was investigating motion and gravity. In the Netherlands, Simon Stevin was studying forces. In the less than four hundred years since, science, led by physics, has totally altered the face of the earth and the nature of human life.

Theories evolve gradually

The curves in Figure 1.6 show in a very approximate way the temporal development of the great theories of physics. The rising portion of a curve indicates the time of discovery leading up to the formulation of the theory. The peak of the curve locates the time of the most essential contributions to the theory. During the time indicated by the falling portion of the curve, further development, elaboration, and testing of the theory occurred. One thing this graphical presentation is meant to emphasize is that none of the major theories of physics has sprung into existence fully developed. The fact that a particular curve is shown now to have a very small value does not indicate that the theory in question is of no importance. Mechanics, for instance, is a vital, daily-used tool of the physicist and the engineer. However, it is a nearly static theory, no longer a major field of research activity. (An exception is the still-active field of hydrodynamics, which applies mechanics and thermodynamics to fluids.)

Mechanics dominated science in the seventeenth and eighteenth centuries

The high point of mechanics came in 1687 with the publication of Isaac Newton's *Principia.* Although Newton's personal contribution was enormous—comparable only to Einstein's in this century—his discoveries concerning motion and gravitation were not without antecedents. Kepler, Galileo, and others provided the foundation on which he built. After Newton, mechanics dominated physics for more than a century and contributed to the exceptional fertility of mathematical development in the eighteenth century.

* *De Magnete*, available in paperback (New York: Dover Publications, Inc., 1958).

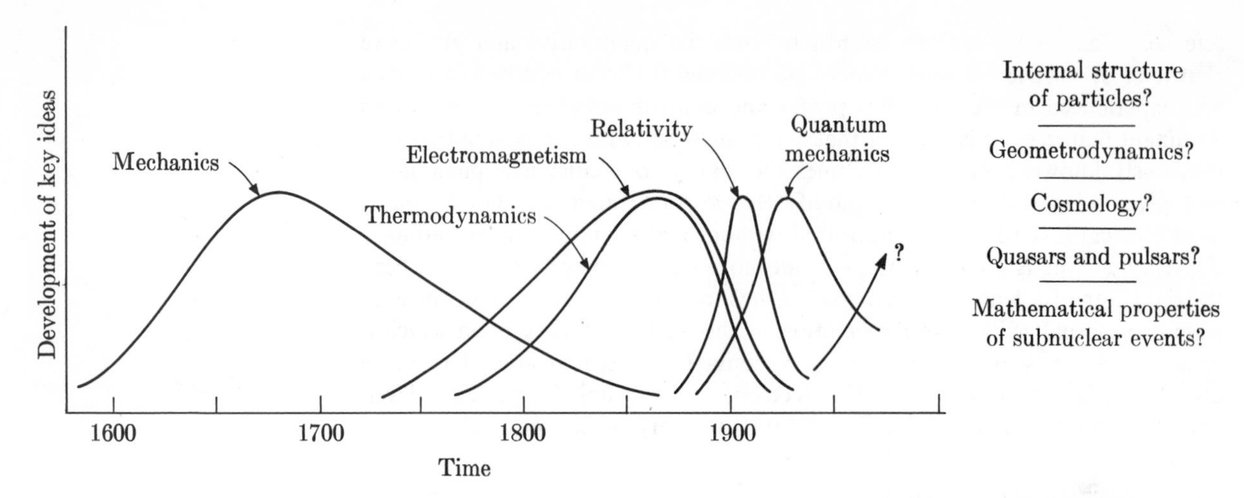

FIGURE 1.6 Schematic presentation of the development of the great theories of physics. (The curve labeled "relativity" refers to the special theory of relativity.) No one knows which current field of research might lead to a new general theory.

Thermodynamics and electromagnetism grew in the nineteenth century

Both thermodynamics and electromagnetism were brought to near-final form in the nineteenth century. Among the triumphs of thermodynamics were the general law of energy conservation and the clarification of the bulk properties of matter in terms of molecular mechanics. Maxwell's unification of electricity, magnetism, and light climaxed more than a century of discoveries in what had once been three separate fields.

Relativity and quantum mechanics belong to this century

Mechanics, thermodynamics, and electromagnetism together comprise what is called classical physics, a framework of understanding and application that at the end of the nineteenth century seemed to some scientists to be nearly complete and nearly flawless. The two great theories of this century, relativity and quantum mechanics, have revolutionized our view of nature and removed from scientific thought any such complacency about the completeness of physical understanding. Although both relativity and quantum mechanics are impressive and well-tested edifices, both have some rough edges, particularly at the points where they meet, and both are fields of active contemporary research. Most physicists believe that the next major theory of physics will emerge from the studies of elementary particles, whose domain is this meeting ground of relativity and quantum mechanics.

Contemporary research

Most current research in physics falls into one of the categories of few-body problems, astronomical problems, and many-body problems (the domains of the very small, the very large, and the very complex, respectively). Atomic physics, nuclear physics, and particle physics are concerned with the behavior and structure of single particles and the interactions among few particles. By contrast, studies of solids, fluids, plasmas, and low-temperature phenomena are studies of many-body systems. Astronomical research includes both few-body and many-body problems, as well as unique cosmological problems. In all fields of physics, the emphasis is, more often than not, on the submicroscopic constituents of the systems under study.

Summary of ideas

Physics is both a consolidated body of knowledge and an active field of human endeavor.

The consolidated parts of physics are encompassed by five general theories.

The known hierarchy of matter extends from 10^{-15} m to 10^{26} m.

Experiment and theory, the balanced faces of a science, reinforce each other.

Mathematics and technology serve and are served by physics.

Small steps are more common than giant leaps; most theories of physics have evolved gradually.

Contemporary research in physics explores frontiers of the very small, the very large, and the very complex.

QUESTIONS

Section 1.1

Q1.1 Cite evidence that structures much larger than man are less organized and less complex than man.

Q1.2 Give arguments for or against the proposition that the size of nature's most elaborate mechanisms will grow ever larger in time.

Q1.3 Why is an atom so small? (HINT: Consider why, on the atomic scale of size, a man is so large.)

Section 1.2

Q1.4 At least three basic theories of physics underlie meteorology. Name two of them and explain briefly why each is important for understanding the behavior of the atmosphere.

Q1.5 Consider any "law" with which you are familiar in another branch of science (such as the law of multiple proportions or Mendel's law). (1) What theory accounts for the law? (2) Is the theory more general or less general than the law? Is it more certain or less certain?

Q1.6 Many phenomena, scientific and nonscientific, require an "explanation." What are the principal characteristics of scientific explanation?

Q1.7 Which of the following statements constitutes a scientific explanation? Why? (1) Grass is green because it contains chlorophyll. (2) The earth rotates on its axis because living creatures need an alternation of day and night. (3) Burning is a process of oxidation.

Q1.8 In most high schools in the United States, the sequence of science courses is biology, chemistry, physics. In some high schools, the reverse sequence—physics, chemistry, biology—is offered. Give a brief argument in favor of each sequence. Which argument do you find more convincing?

Section 1.3

Q1.9 In an experiment, natural events are deliberately manipulated. Discuss what you think has probably been the significance of experiment in science, as contrasted with mere observation. In your everyday affairs, have you ever learned through experiment? How?

Q1.10 Give several aspects of your daily life in which you rely on the simplicity of nature, or at least on its orderliness and predictability.

Q1.11 Einstein has characterized a great theory as one possessed of "inner perfection" and verified by "external confirmation." Explain his meaning.

Q1.12 The physicist calls nature "simple," but by this he does not mean that it is easy to understand. Illustrate this point by example. Discuss any fact, law, or work of man that requires considerable intellectual effort to understand, but which, once understood, possesses an appealing simplicity.

Section 1.4

Q1.13 Is manned space flight primarily a technological or primarily a scientific achievement? Explain.

Q1.14 Give an argument for or against the proposition that some branches of mathematics have nothing to do with nature.

Q1.15 Give an argument for or against the proposition that some events in nature have no mathematical foundation.

Q1.16 Based on your own observation or experience in a physics laboratory, name one or two devices of technology (other than those mentioned in the text) that contribute to physics research.

Section 1.5

Q1.17 Do you think it is appropriate to characterize fundamental research in physics as creative activity? Why or why not?

Q1.18 Will physics someday come to an end, when man comprehends all the fundamental laws of nature or all that he is capable of comprehending? Why or why not?

Q1.19 Do we now know "enough" physics? Is it time for man to turn away from fundamental science and seek beneficial application of what he already knows?

EXERCISES

Numerical practice

E1.1 Carry out the following arithmetic operations: (a) $10^{13} \times 10^{-12}$; (b) $(3 \times 10^{8})/(6 \times 10^{-8})$; (c) $21{,}000 \times 3{,}000{,}000 \times 0.001$; (d) $3 \times 10^{-7} + 3 \times 10^{-8}$.

E1.2 Using the fact that light travels about 10^{16} m/year, calculate from Table 1.1 the time required for light to cross (a) a galaxy, (b) a galactic cluster, and (c) the known part of the universe.

E1.3 If a fast electron moves at 10^{8} m/sec, calculate how long it would spend traversing (a) an atomic nucleus, (b) an atom, and (c) a giant molecule.

Section 1.1

E1.4 To help you visualize some of the magnitudes in Table 1.1, calculate the following quantities approximately: (a) the time required to lay down a one-meter line of nuclei, working at one nucleus per second; (b) the number of nuclei that would be required to fill the volume of an atom; and (c) the number of earth diameters in the span of a large star.

E1.5 If an atom were enlarged from a diameter of 2×10^{-10} m to a diameter of 2 m, how large would its nucleus be? If all the atoms in a man were so enlarged, would he fit between the earth and the sun?

E1.6 Verify that the state of Kansas lies midway in the hierarchy of matter, i.e., that its dimension exceeds the radius of an elementary particle and falls short of the radius of the universe by the same factor.

E1.7 There are about 10^{59} elementary particles in a typical star, about 10^{11} stars in a galaxy, about 10^{3} galaxies in a cluster, and about 10^{9} clusters in the known part of the universe. (1) Approximately how many elementary particles comprise the known part of the universe? (2) If all of this matter were so compressed that each particle occupied only 10^{-45} m^3 (roughly the volume of a proton), what would be the approximate radius of this highly condensed universe? With what presently known structure does this dimension compare?

PROBLEM

Finding an equation to summarize data

P1.1 A law of nature summarizes in compact form a sequence of observations. Study the following observational data and try to find a single "law of rocket flight." Write the law in the form of an equation. Predict the height of each rocket after 25 sec of flight. Suggest measurements that might be expected to reveal limits to the validity of the law. (NOTE: Some of the data may contain small errors of measurement.)

Time since Launch	Height above Ground of Rocket A	Height above Ground of Rocket B	Height above Ground of Rocket C
5 sec	124 ft	0.1 mile	91 m
10 sec	500 ft	0.4 mile	365 m
15 sec	1,127 ft	0.9 mile	820 m
20 sec	2,001 ft	1.6 mile	1,455 m

2 Concepts, Units, and Dimensions

A theory of physics is a mathematical description of nature, yet it is much more than bare mathematics. A set of equations is no more a physical theory than a balance statement is the true capital of a corporation. Also essential are a set of ideas, agreements, and assumptions. These must stand beside the equations and give them life. Even then the equations are made to correspond only with experimental results; their correspondence with the deeper underlying reality in nature requires further assumptions that are tempting to make but are probably inessential to science.

Ingredients of a developed theory

A general theory of physics is made up of a set of concepts; assumptions about the mathematical representation of these concepts; mathematical relationships among the concepts; prescriptions for relating the mathematical structure to actual measurements; the accumulated evidence in favor of the assumptions and prescriptions; and a way of looking at nature provoked by the ideas and their success. Taken together all these things constitute the understanding of past facts, the predictions of new facts, and the view of nature that in concert comprise a full theory of physics. Equally important to the theory are its skeleton of equations and its flesh and blood of ideas, interpretation, and visualizable pictures.

The *concepts* of physics are of special significance because they are the links joining the mathematical, the verbal, and the experimental. This chapter is concerned with some of the important physical concepts and their dimensions and units of measurement.

2.1 Quantitative concepts

In everyday use the word "concept" usually refers to an idea that is not directly visualizable, either because it is abstract (bravery) or because it is beyond the reach of our sense (an atom). It is sometimes used in the latter sense in science, but we shall most often mean by a concept a thing or a quantity that is measurable. This may be called a quantitative concept. A quantitative concept has a name, a physical dimension, and a unit of measurement (or several possible units of measurement); it can be manipulated mathematically, and it can be assigned a numerical or other mathematical value. The simplest physical concepts are scalar quantities. Some physical concepts are vector quantities or other, more abstract, mathematical quantities. More than one number is required for their measurement. When we say that a concept *is* a scalar or vector quantity, we mean that the concept can be manipulated in equations according to the rules governing the mathematical abstraction of scalar or vector—in short, that the physical quantity behaves like the mathematical quantity. This idea will be pursued in Chapter 6.

Scalar quantities and vector quantities

Familiar scalar quantities are mass, energy, and temperature. Familiar vector quantities include force, velocity, and acceleration. We shall encounter more of both kinds. In the course of this text we shall not have occasion to study quantities of a higher level of abstraction than vectors, but it is worth knowing that such quantities (tensors, for instance) have found a place in advanced physics and have helped to simplify our description of nature.

Since physics rests ultimately on experiment, at least some of its concepts can be defined only by reference to measurement. Such concepts are called primitive concepts.* Usually mass, length, and time are included among the primitive concepts. They are defined in terms of arbitrarily chosen standards.† Temperature and electric current are also defined by reference to experiment and, accordingly, may be called primitive concepts. (The definition of current, however, makes use of no new standard.) A concept defined merely in terms of other concepts is called a derived concept. Most concepts are derived. Which concepts are primitive and which derived is entirely arbitrary, a matter of historical accident and convenient choice. Even the number of primitive concepts is arbitrary. So is the number of standards.

Primitive concepts

Derived concepts

At present, physical measurement is based on four arbitrary standards: those of length, mass, time, and temperature. It would be easy to increase or decrease this number. For example, a standard of force could be added. If it were, Newton's second law, relating force, mass, and acceleration, would take the form

Flexibility in the number of standards

$$F = Kma, \tag{2.1}$$

where K is a constant of proportionality. The existence of such a law, however, makes it unnecessary to adopt a separate standard of force. We can (and do) set $K = 1$ and define the unit of force by the equation

$$F = ma. \tag{2.2}$$

To decrease the number of arbitrary standards, we could take advantage of an experimental connection between temperature and energy. This would make it

* The phrase "basic concepts" is also used.

† See Appendix 1. For a background article, see Allen V. Astin, "Standards of Measurement," *Scientific American*, June, 1968.

possible to measure temperature in the unit of energy, which is fixed by the standards of mass, length, and time.

Distinguishing between concepts as primitive or derived need not be pursued in detail because of its arbitrariness. In fact, strictly speaking, there are not two kinds of physical concepts but three: (1) those defined by means of experimental comparison with an arbitrary standard; (2) those defined by means of experiment but without a new standard; and (3) those defined purely logically or mathematically. An example of the last type is acceleration, which is defined as the rate of change of velocity with respect to time.* All these kinds of definition are called operational definitions. A concept is operationally defined if an unambiguous procedure is given for identifying and measuring it. An operational definition may consist of a recipe in the form of an equation ($F = ma$: measure m, measure a, and multiply to get F), or it may consist of experimental procedures. To be meaningful in physics, every concept must eventually be related to measurement. Only operationally defined quantities appear in the equations of physics.

Operational definitions

It is not far from the truth to say that when the concepts of physics are understood, physics is understood. In your study of physics, it will prove a very good practice to ask yourself at every stage: Do I understand all the concepts I am working with? Understanding a concept means knowing its definition, its dimension, its unit, its typical magnitudes in physical situations, and knowing the equations into which it enters.†

Concepts are the core of physics

2.2 Units

It is a familiar fact that most physical quantities require a unit of measurement and that any particular quantity can be expressed in a variety of units. We reckon distance in inches or feet or meters or miles, time in seconds or minutes or hours, money in cents or dollars or pounds or francs, and so on. A standard set of units is any self-consistant set with one unit for each quantity. In science and engineering, three standard sets are in common use:

Three sets of standard units

1. The international system, or SI‡ (often called the mks system), based on the meter, the kilogram, the second, and the kelvin (a unit of temperature);
2. The cgs (Gaussian) system, based on the centimeter, the gram, the second, and the kelvin;
3. The British system, based on the foot, the pound, the second, and the degree Fahrenheit.

Superficially, the mks and cgs systems appear similar, for both are metric systems and their units of length and mass differ only by powers of ten. However, their electric and magnetic units differ more substantially. Some *equations* of electromagnetism also differ in these two systems. The British system has largely

Equations of electromagnetism depend on the choice of units

* $a = dv/dt$, a derivative.

† The symbols and units of many important concepts are tabulated in Appendix 2.

‡ The abbreviation is after the French, *Système International.*

disappeared from physics research, although some engineers still use it. In both physics and engineering, there has been a gradual trend toward the SI, and it is this system that we shall emphasize in this text.* Its greatest advantage, apart from the fact that it is a metric system, is that it incorporates the practical, or household, units of electricity: the volt, the ampere, the ohm, the watt, etc. To avoid confusion, we shall make no more than occasional references to other systems. Physical constants in Appendix 3 are expressed primarily in SI units. Conversion factor are given in Appendix 4. Some cgs (Gaussian) equations of electromagnetism are set forth in Appendix 5.

We choose SI units

In addition to four arbitrary standards (those of length, mass, time, and temperature), basic SI units include the ampere as the unit of electric current and the candela as the unit of luminous intensity. The unit of every physical quantity can be expressed as some combination of the six fundamental units: the meter (m), the kilogram (kg), the second (sec), the kelvin (K), the ampere (A), and the candela (cd). (The candela is mentioned here simply for the sake of completeness; we shall in fact have no further need of it.)

Many combined units are given new names. Thus the SI units of force and energy are the newton (N) and the joule (J):

SI units of force and energy

$$1 \text{ N} = 1 \text{ kg m/sec}^2, \tag{2.3}$$

$$1 \text{ J} = 1 \text{ kg m}^2\text{/sec}^2 = 1 \text{ N m}. \tag{2.4}$$

Most electrical units are named after physicists of the eighteenth and nineteenth centuries. The unit of electric charge, the A sec, is called the coulomb (C). One J/A sec is called one volt (V), and one J/sec is one watt (W). It is important to realize that all of these new names are merely shorthand for combinations of a few units. They do not imply a proliferation of new units.

Besides the widespread standard systems of units, there are special systems for special branches of science that will undoubtedly continue to flourish. The nuclear physicist will continue to find the charge of the proton a convenient unit of charge; the quantum physicist will continue to prefer a system of units in which Planck's constant is equal to unity; and the astronomer will continue to prefer a unit of length more in keeping with the scale of the universe than the centimeter or meter. We shall have occasion to use a few special units, such as the angstrom (Å) of length and the electron volt (eV) of energy.

2.3 Dimensions

Dimensionless quantities in physics are those that require no unit. More accurately stated, a dimensionless quantity is one whose numerical value does not depend on the choice of units. Conversely, a dimensional quantity is one whose numerical value can be changed by changing the set of units. Dimensionless numbers may result either from counting (the number of pages in this book is a dimensionless number) or from taking the ratio of two quantities expressed in the same unit of measurement.

* The international system was adopted by the Eleventh General Conference on Weights and Measures, which met in Paris in 1960. It has been used regularly by the U.S. National Bureau of Standards since 1964.

The geometrical idea of dimension

The idea of dimension, and the name, go back to very early thinking about geometry. Length is a concept of one dimension. It is, in geometry, a fundamental idea that cannot be expressed in terms of any others. Therefore length has the dimension of length, and nothing else can be said about it. Area, on the other hand, is a two-dimensional concept. It has the dimension of $(\text{length})^2$. Similarly, volume has the dimension of $(\text{length})^3$. This idea of the dimension of geometrical concepts is obviously related to the units of measurement of the corresponding physical quantities. Distance might be measured in m, area in m^2, and volume in m^3. Or distance in ft, area in ft^2, and volume in ft^3. Whatever the unit of length, standing behind the unit is the more abstract idea of dimension.

Generalization: the dimension of a physical quantity

Here we are making a subtle but important transition from the purely geometric idea of the dimensionality of space to the idea of the dimension of a physically measurable quantity, such as volume. Having done so, it is an easy next step to generalize the idea of dimension to nongeometric quantities. In kinematics occurs the concept of time as well as the concept of space. What is the dimension of time? In geometric terms it has no dimension. But physically it proves useful to define a dimension of time—a new and independent dimension, not expressible in terms of the dimension of length. Then kinematic quantities, such as speed and acceleration, can be assigned dimensions that are certain combinations of the new fundamental dimensions, length and time. The dimension of speed is length/time. The dimension of acceleration is $\text{length}/(\text{time})^2$. The rate of expansion of a balloon, to offer a third example, has the dimension of $(\text{length})^3/\text{time}$, or volume per unit time.

Three fundamental dimensions suffice

The adoption of one fundamental dimension, length, serves as a basis for describing the dimensions of all geometric quantities. The addition of a second dimension, time, provides a basis for assigning a dimension to every kinematic quantity. To go further, and describe the dimensionality of all the physical concepts of mechanics, a third independent dimension is necessary. Interestingly, three fundamental dimensions are sufficient. (These are not to be confused with the three dimensions of space, which are all length dimensions.) Usually, mass is chosen as the third independent dimension. Its dimension cannot be expressed as any combination of the dimensions of space and time, but once it is chosen the dimension of every other mechanical concept can be expressed as some combination of these three—length, mass, and time.

It might be supposed that as more and more physical concepts are introduced in new fields of physics, more and more fundamental dimensions must be introduced. Fortunately, this has not been necessary. Relationships among concepts have more than kept pace with the introduction of new concepts. Thanks to new insights provided by relativity and quantum mechanics, we could even reduce the number of fundamental dimensions to less than three. The fact that the international system makes use of as many as six physical dimensions is related to practical convenience, not necessity.

2.4 Dimensional consistency and units consistency

The simplest equation one can imagine has the form

$$A = B.$$

If this is to be an equation of physics and not merely a mathematical statement, it means that one physical quantity, designated by A, is equal to another physical quantity, designated by B. In particular, if A has a certain numerical value, then B has the same numerical value. To be physically meaningful in general, the equation must be true regardless of the set of units adopted for measurement. Then, if the units are changed, A and B change by the same factor and the equation remains valid. It is a common practice, one that we shall follow in this book as far as possible, to write equations of physics in such a way that their correctness does not depend on a particular choice of units. Newton's second law, for instance, written $F = ma$, is correct for mks, cgs, and British units. We therefore call it dimensionally consistent. Nevertheless, it would be possible to invent a set of units in which it is not correct—compare Equations 2.1 and 2.2. Dimensional consistency may be limited to a class of recognized standard sets of units.

Dimensional consistency: An equation is true for various sets of standard units

Many of the equations of electromagnetism are special in that their form is different for mks (SI) and cgs (Gaussian) units. Appendix 5 calls attention to some of these differences. These equations are not dimensionally consistent, so it is important to fit the units to the equations. However, they do have units consistency. With the proper set of units, the left side of each equation has the same unit as the right side.

Units consistency: Unit of the left side equals unit of the right side

■ EXAMPLE 1: Uniform motion in a straight line is described by the equation

$$s = vt, \tag{2.5}$$

i.e., distance equals speed times time. The dimensional consistency of this equation is displayed by writing another "equation":

$$[\text{length}] = \left[\frac{\text{length}}{\text{time}}\right] \times [\text{time}].$$

A dimensional "equation"

The square brackets are used to indicate that only the dimensions of the equation are being considered. Dimensional consistency alone, of course, does not prove that an equation is correct. However, dimensional inconsistency can prove that it is incorrect if it is derived entirely from dimensionally consistent equations. Checking each step of a derivation for dimensional consistency is a useful way to look for errors. ■

■ EXAMPLE 2: Since the density of water is 1 gm/cm^3, the mass of a quantity of water in grams is equal to the volume of the water in cm^3. Letting M designate mass and V designate volume, we have

$$M = V.$$

This is a correct but dimensionally inconsistent equation. It is true only so long as the unit of length is the centimeter and the unit of mass is the gram. For SI units, the same equation must be re-expressed as

$$M = 10^3 V.$$

This again is an equation that is true only for a particular choice of units. Clearly, a dimensionally consistent equation that can replace these two is

$$M = \rho V, \tag{2.6}$$

where ρ (rho) is the density of water, or mass per unit volume. The dimensional consistency is illustrated by the dimensional equation,

$$[\text{mass}] = \left[\frac{\text{mass}}{(\text{length})^3}\right] \times [(\text{length})^3].$$

In this example, the dimensionally consistent equation has the greatest flexibility. It allows not only for a change of units but also for a change of the density of water; it even allows for a change of substance. ■

UNITS ARITHMETIC

In physics problems it is possible, and almost always desirable, to carry out a kind of "units arithmetic" in parallel with numerical arithmetic. The idea will already be familiar to most readers. A few examples serve as illustrations.

■ EXAMPLE 3: A body whose mass is 2 kg experiences a net force of 5 newtons. What is its acceleration? The equation is

$$a = \frac{F}{m}.$$

Evaluation of the right side gives

$$a = 2.5 \text{ N/kg}.$$

Since $1 \text{ N} = 1 \text{ kg m/sec}^2$, this may be written

$$a = 2.5 \frac{\text{kg m/sec}^2}{\text{kg}} = 2.5 \text{ m/sec}^2.$$

In the units arithmetic, the mass units cancel. ■

■ EXAMPLE 4: Consider a simple problem of constant-speed motion with mixed units. An automobile travels at 45 ft/sec for 30 min. How many kilometers does it cover? In order to use the equation

$$s = vt,$$

one could first convert v from 45 ft/sec to 49.4 km/hr and t from 30 min to 0.5 hr, then multiply to obtain $s = 24.7$ km. An alternative method is often convenient. Substitute the given numbers in the equation, then add conversion factors as needed to make the units arithmetic work out correctly. The calculation looks as follows:

Units arithmetic with conversion factors

$$s = \frac{45 \dfrac{\text{ft}}{\text{sec}} \times 30 \text{ min} \times 60 \dfrac{\text{sec}}{\text{min}} \times 0.305 \dfrac{\text{m}}{\text{ft}}}{10^3 \dfrac{\text{m}}{\text{km}}},$$

or

$$s = 24.7 \text{ km}.$$

Conversion factors can be incorporated at liberty, since each is equivalent to unity. ■

■ EXAMPLE 5: If a particle of charge q moves with speed v perpendicular to lines of magnetic field of magnitude B, it experiences a magnetic force given by

$$F_m = qvB. \tag{2.7}$$

Does this equation have units consistency? The SI units on the right side are coulombs, m/sec, and teslas. So we ask: Is the product of units

$$\mathrm{C} \times \frac{\mathrm{m}}{\mathrm{sec}} \times \mathrm{T}$$

equal to the SI force unit, the newton? Reference to Appendix 2 shows the equivalence to basic units:

$$1\ \mathrm{C} = 1\ \mathrm{A\ sec},$$

$$1\ \mathrm{T} = 1\ \mathrm{kg/A\ sec^2}.$$

Thus the product of units is

A units "equation"

$$\mathrm{C} \times \frac{\mathrm{m}}{\mathrm{sec}} \times \mathrm{T} = \mathrm{A\ sec} \times \frac{\mathrm{m}}{\mathrm{sec}} \times \frac{\mathrm{kg}}{\mathrm{A\ sec^2}}$$

$$= \mathrm{kg\ m/sec^2} = \mathrm{N}.$$

This result verifies the units consistency of Equation 2.7 for SI units but provides no evidence regarding a more general dimensional consistency. In fact, this magnetic-force equation does *not* hold in the cgs (Gaussian) system. ■

2.5 Length

Definitions and laws of physics cannot be separated

It is not possible to set down precise definitions of all or even most of the important concepts of physics at one time and then build the theories of physics upon these, a technique that might seem appealing from a logical point of view. This is so simply because the concepts of physics cannot be divorced from its laws. Electric current can be defined with quantitative precision only in conjunction with the law of magnetic force. Inertial mass can be defined only in terms of certain mechanical measurements whose success hinges on Newton's laws of motion. Energy, the most ubiquitous concept in physics except for space and time, can be defined in its manifold forms only with the help of some knowledge of every great theory of physics. In spite of these limitations, it is important early in the study of physics to gain what the physicist would call a "feeling" for the important concepts in order to be able to think meaningfully about the natural world. In this and the next few sections, we shall define and discuss a few of the key concepts of physics in a preliminary way. Most of the definitions will undergo later refinement, but the discussion of the "meaning" of the concepts will require no alteration.

The concept of distance, or spatial separation, is probably as primitive as any idea in science other than counting.* Geometry, the first branch of mathematics that was brought to a sophisticated level, is based simply on the idea of

* Primitive or not, length has proved to be a tricky concept, as later chapters on relativity will show. According to our modern insight, the length of an object depends on whether or not it is in motion with respect to the measurer. This fact need not concern us now.

TABLE 2.1 UNITS OF LENGTH

Name	Symbol	Magnitude	Remarks
Fermi	fm	10^{-15} m	Same as femtometer
Angstrom	Å	10^{-10} m	
Micrometer	μm	10^{-6} m	Same as micron
Centimeter	cm	10^{-2} m	
Kilometer	km	10^{3} m	
Astronomical unit	A.U.	1.4960×10^{11} m	Mean radius of earth's orbit
Light-year		9.461×10^{15} m	Distance light travels in 1 year
Parsec		3.084×10^{16} m	Distance at which 1 A.U. subtends 1 sec of angle

length—or more accurately, the idea of spatial relationships. The definition of length rests on the size of some standard object. For everyday purposes, that standard object could be a yardstick or meter stick. The meter was originally defined as one ten-millionth of the distance from pole to equator. This makes it easy to remember the circumference of the earth, which is very nearly 4×10^7 m, or 40,000 km. For many years, the scientific standard of length was a metal bar kept in a temperature-controlled vault near Paris, its length—by definition—one meter. Now the standard "object" is a light wave, a particular wave of red-orange light emitted by atoms of the isotope krypton 86 (Figure 2.1). One meter is—by definition—1,650,763.73 wavelengths of this light. (This apparently odd value was chosen in order to keep the length of the new meter as close as possible to the length of the previous standard.)

The standard of length

As indicated in Table 1.1, known sizes of physical systems span a range of 10^{41}. Because of this enormous range, a number of special units of length have come to be applied to the various domains of nature. Some of these are defined in Table 2.1. Table 2.2 shows typical magnitudes of lengths and several other important quantities in the macroscopic, submicroscopic, and astronomical worlds.

2.6 Time

The idea of time is essentially an idea of periodic repetition. That time is a "visualizable" concept, a built-in part of every man's view of the world, stems from two facts: that man has a memory and that man is constantly subjected to regularly repeated stimuli. We are conscious of the alternation of seasons, the steady pattern of day and night, our own recurring hunger or fatigue. A clock amounts to nothing more than any device that counts off the number of repetitions of some recurring motion. An old-fashioned pendulum clock is an obvious example. The earth itself is a clock, because of its daily rotation and its annual trip around the sun. One second of time was defined originally as 1/86,400 of a day. Later, because the earth's orbital motion is more reliably calculable than

Definition of a clock

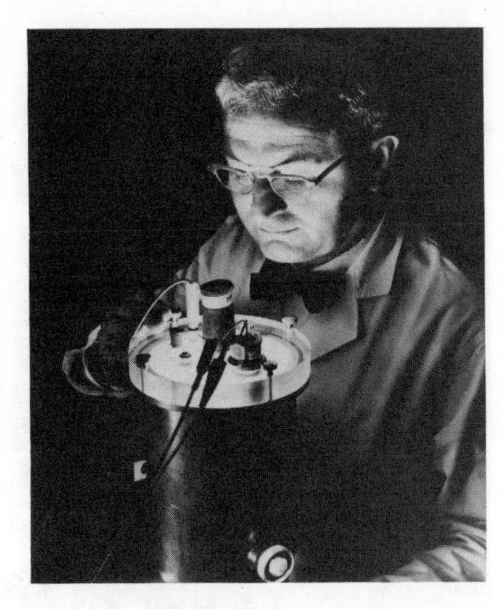

FIGURE 2.1 A krypton 86 lamp in its liquid nitrogen bath. The lamp is maintained at a temperature of $-210°C$ in order to improve the reproducibility of the standard wavelength. (Photograph courtesy of National Bureau of Standards.)

TABLE 2.2 TYPICAL MAGNITUDES OF PHYSICAL QUANTITIES

Physical Quantity	Common Unit in the Macroscopic World	Scale of the Submicroscopic World	Scale of the Astronomical World
Length	meter (about 39 in.)	Size of atom $\cong 10^{-10}$ m	Earth-sun distance = 1.5×10^{11} m
		Size of particle $\cong 10^{-15}$ m	"Radius of universe" $\cong 10^{26}$ m
Speed	meter/second (man walking)	Speed of light = 3×10^{8} m/sec	Orbital speed of earth = 3×10^{4} m/sec
			Speed of light = 3×10^{8} m/sec
Time	second (swing of a pendulum)	Natural time unit of particle about 10^{-23} sec	Period of earth = 3×10^{7} sec
		Typical lifetime of "long-lived" particle about 10^{-10} sec	"Age of universe" $\cong 3 \times 10^{17}$ sec
Mass	kilogram (mass of 1 liter of water)	Mass of electron = 9×10^{-31} kg	Mass of earth = 6×10^{24} kg
		Mass of heavy atom $\cong 4 \times 10^{-25}$ kg	Mass of sun = 2×10^{30} kg
			Mass of visible universe $\cong 10^{55}$ kg
Energy	joule (frog jumping)	Kinetic energy of air molecule about 6×10^{-21} J	Kinetic energy of earth in orbit $\cong 2.7 \times 10^{33}$ J
	kilocalorie or food calorie (4,184 J)	Proton in largest accelerator about 8×10^{-8} J	Solar energy production in 1 sec = 3.9×10^{26} J
Charge	coulomb (lights a lamp for 1 sec)	Proton charge = 1.6×10^{-19} C	Large-scale matter is neutral
Angular momentum	kg m²/sec (spinning bowling ball)	Spin of photon = $h \cong 10^{-34}$ kg m²/sec	Spin of earth = 7×10^{33} kg m²/sec
			Spin of galaxy $\cong 10^{69}$ kg m²/sec

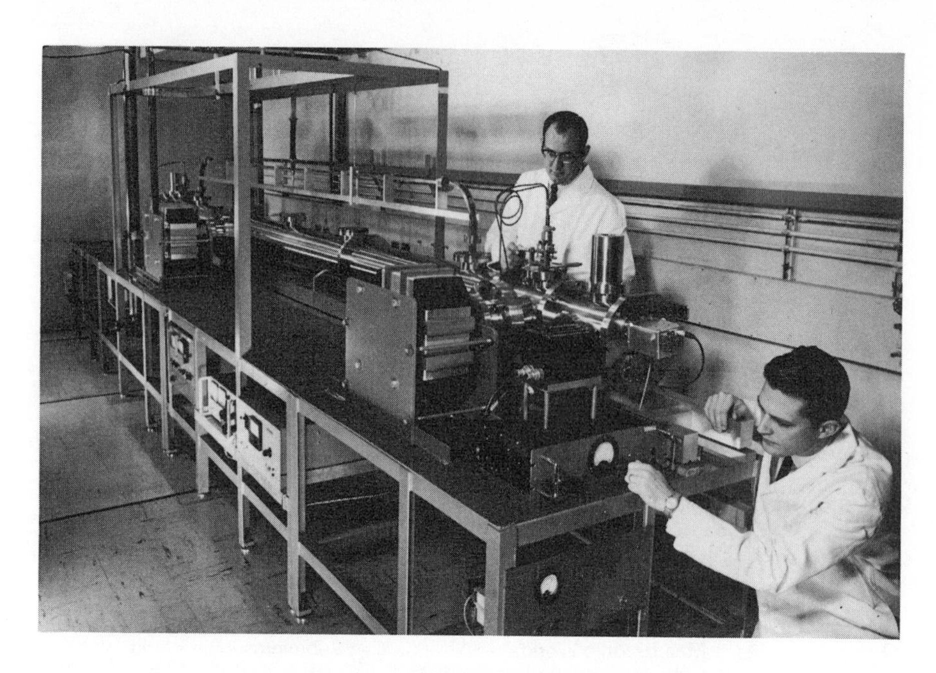

FIGURE 2.2 At the heart of this atomic clock is a beam of cesium 133 atoms. In the midsection of the tube, the atoms experience an oscillating electromagnetic field. If the oscillation is of a particular frequency, the atoms strike a target-detector at the end of the tube. If the frequency drifts from its proper value, the atoms miss the center of the target, and a feedback signal automatically corrects the frequency. With such a clock, a time standard is maintained to an accuracy of about 1 part in 10^{12} (1 sec in 30,000 years). (Photograph courtesy of National Bureau of Standards.)

its spin motion, the ephemeris second was defined as a certain fraction of the period of rotation of the earth about the sun (for the year 1900). Now the standard clock is a cesium atom (Figure 2.2). A particular frequency emitted and absorbed by cesium is chosen to be 9,192,631,770 sec^{-1} (or Hz*), which is the present definition of the second. For laboratory work and short time measurements, this is a more suitable standard than the earlier astronomical standards.

The standard of time

How can we be sure that the time intervals between successive cycles of a clock's repeating motion are the same? We cannot. The uniformity of time scale can be based only upon self-consistency (that clocks of different design continue to agree) and upon agreement with simple laws of nature (that the periodic motion is theoretically regular). Time, like every other concept, cannot escape to a status independent of the laws in which it is employed.

In dealing with time scales of the extreme domains of nature, it is essential first to get rid of preconceived notions about what constitutes a short time or a long time. In the domain of particles, a natural time unit is the time required

Natural time scales in nature

* The SI unit of frequency, the hertz (Hz), is one vibration per second, often written simply sec^{-1}.

TABLE 2.3 SOME SHORT TIME INTERVALS

Name	Symbol	Duration
Picosecond	psec	10^{-12} sec
Nanosecond	nsec	10^{-9} sec
Microsecond	μsec	10^{-6} sec
Millisecond	msec	10^{-3} sec

for light to cross the diameter of an elementary particle, about 10^{-23} sec. In terms of this unit, one-millionth of a second is an extremely long time; most particles are born, live, and die in less than 10^{-6} sec. Table 2.3 shows the names assigned to some short time intervals. The prefixes that appear in this table are part of a standard set often used as abbreviations for powers of ten. A more complete list of such prefixes is given inside the front cover. The evolution of the earth and universe is characterized by a time scale measured in billions of years (Gyr). In the thermonuclear furnace at the center of the sun, a proton has a mean lifetime of 14 Gyr (4.4×10^{17} sec) before it participates in a thermonuclear reaction.

Since the smallest object probed experimentally is about 10^{-15} m in size, it is fair to say that the shortest time interval studied is about 3×10^{-24} sec (although direct time measurements are still very far from reaching this short an interval). The longest known time is the "lifetime of the universe," that is, the apparent duration of the expansion of the universe, which is also a few times the age of the earth. This amounts to 3×10^{17} sec (10 Gyr). The ratio of these is 10^{41}, the same enormous number as the ratio of the largest and smallest distances. This is not a coincidence. The outermost reaches of the universe are moving away from us at a speed near the speed of light and the particles used to probe the submicroscopic world are moving at speeds near the speed of light. On both the cosmological and submicroscopic frontiers, the speed of light is the natural link between distance and time measurements.

The speed of light links space and time

2.7 Speed

If an object moves through distance Δs in time Δt,* its average speed $\bar{v}$ during the interval is defined by the equation

$$\bar{v} = \frac{\Delta s}{\Delta t}. \tag{2.8}$$

In the limit that Δt (and, with it, Δs) approaches zero, the ratio becomes a derivative that defines the instantaneous speed v†:

$$v = \lim_{\Delta t \to 0} \frac{\Delta s}{\Delta t} = \frac{ds}{dt}. \tag{2.9}$$

Definition of speed, a derived concept

* The symbol Δ is commonly used to mean "change of."

† The significance of this definition will be developed in Sections 5.4 and 5.7.

FIGURE 2.3 This platinum-iridium cylinder is the U.S. standard kilogram. It was compared with the international standard around 1885 and again in 1948. It defines a mass standard to an accuracy of about 1 part in 10^8. (Photograph courtesy of National Bureau of Standards.)

Thus speed is a derived concept defined mathematically in terms of length and time. In a panoramic survey of nature, speed is interesting because it has a ceiling (the speed of light). Although the speed of light, $c = 3 \times 10^8$ m/sec, may seem incredibly fast, man is not so far removed from this limit as he is from the submicroscopic and cosmological frontiers of space and time. A jet plane flies at a speed of $10^{-6}c$; the earth in its orbit moves at a speed of $10^{-4}c$. An astronaut falls short of the speed of light by a factor of 40,000. He needs an hour and a half to get once around the earth, while a photon (if it could be caused to travel in a curved path) could complete the trip in 0.13 sec.*

Atoms and molecules, executing their continual restless motion in solids, liquids, and gases on earth, move at speeds of about 10^2 to 10^4 m/sec, spanning the same range as the speed of sound does in various substances. For the still smaller elementary particles, speeds near the speed of light are common. The photon of light is itself an elementary particle and moves exactly at nature's speed limit. Because it is without mass, it achieves the limit that most other particles can only approach. There is ample evidence that no material particle or other radiation moves faster.

The speed of light as nature's speed limit

Astronauts in science fiction frequently shift into "superdrive" and cruise about the galaxy above the speed of light. Is there any chance that this will one day become reality? It is extremely unlikely. There does exist the possibility of a new kind of particle that moves *always* faster than light. For such matter, the speed of light is a lower limit, not an upper limit. Even if such particles† were to be discovered (there is presently no evidence for them), the theory of relativity predicts that the speed of light remains an impenetrable barrier from both above and below. It is easy to see why known matter cannot be accelerated beyond the speed of light. The less massive a particle is, the more easily it can be accelerated. Freight trains lumber slowly up to speed, automobiles more quickly, and protons in a cyclotron still more quickly. The massless photon requires no time at all for acceleration. It jumps instantaneously to the speed of light when it is created. If any material particle could go faster than light, then light itself, being composed of massless photons, should go faster.

2.8 Mass

The standard of mass

The international standard of mass remains a macroscopic standard. It is a platinum-iridium cylinder stored in a repository in Sèvres, France, defined to have a mass of 1 kg. Secondary standards are available in laboratories throughout the world (see Figure 2.3). Before the present standardization, the kilogram was defined to be the mass of 1 liter (10^3 cm^3, or 10^{-3} m^3) of water at 4 °C, at which temperature the density of water is maximum. It is interesting to note that our earth is a determining factor in all three standards of length, time, and

* In 1969, the average man could, for the first time, directly perceive the finite speed of light. Listening in on conversations between controllers on earth and astronauts on the moon, he could notice the 2.5-sec delay required for radio signals to reach the moon and return.

† These hypothetical particles are known as superluminal particles or tachyons. See Gerald Feinberg, "Particles That Go Faster than Light," *Scientific American,* February, 1970.

mass. The quarter-circumference of the earth was divided into 10 million meters, the day into 60 × 60 × 24 seconds. This terrestrial standard of length in turn defined the volume of water used as a mass standard.

Two kinds of mass

There are two quite distinct ways to measure mass—that is, to compare an unknown mass with a standard. First, the relative *weight* of the known and the unknown can be determined, perhaps with the help of a balance. Second, the relative *acceleration* of the known and the unknown can be determined when both experience the same net force. These two methods measure what are called gravitational mass and inertial mass. Since exceedingly accurate experiments have shown gravitational and inertial mass to be equal, they may be considered equally fundamental quantities.

Inertial mass

To consider the meaning of inertial mass, imagine two astronauts floating freely inside the cabin of their spaceship in a weightless condition (Figure 2.4). If they join hands, then push and let go, they will float apart—in a specific way. The larger man will move away more slowly than the smaller man; we attribute this to his greater mass, in particular his greater inertial mass. Inertial mass is a measure of resistance to a change of motion. If a single astronaut floating in midcabin tosses a baseball, he will recoil and drift backward slowly as the ball moves swiftly off in the opposite direction. If the ball leaves the point of separation five hundred times faster than the astronaut, it is because its mass is only one five-hundredth that of the astronaut; it has, therefore, five hundred times less resistance to being set into motion.

The recoil definition of mass

This property of mass makes it possible to establish what is called the recoil definition of inertial mass. An astronaut need only throw a standard kilogram instead of a baseball in order to measure his own mass. If he drifts from the point of separation at a speed one-ninetieth the speed of the kilogram, his mass is 90 kg. The equation is

$$\frac{m}{m_{\text{std}}} = \frac{v_{\text{std}}}{v}. \tag{2.10}$$

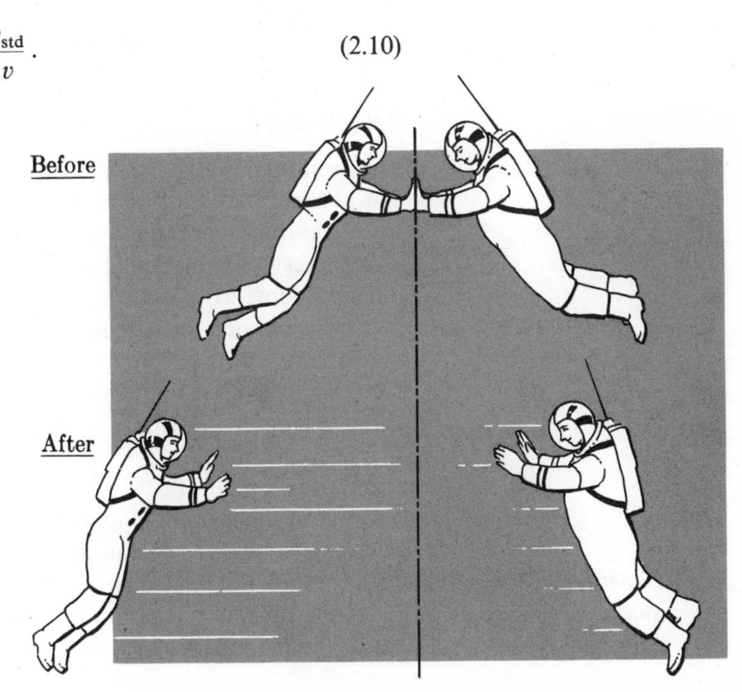

FIGURE 2.4 A way to define mass. The ratio of the masses of the astronauts is the inverse of the ratio of their recoil speeds if they push apart when initially at rest. Before: A thin astronaut and a fat astronaut start to push apart. After: They are moving apart, the thin astronaut faster than the fat astronaut.

This definition has logical appeal because neither forces nor accelerations, only velocities, need be measured. It is practical, however, only in the submicroscopic domain, and there it must be modified somewhat to take into account corrections required by the theory of relativity. The physical law supporting the value of the recoil definition is the law of momentum conservation.

Gravitational mass

In the macroscopic world, an accurate determination of gravitational mass is easier than an accurate determination of inertial mass. Thus scales and balances are the tools of mass measurement in all laboratories. In this text, we shall seldom need to distinguish between inertial and gravitational mass.

Mass of particles

The electron, the lightest particle with nonvanishing mass, has a mass of 9.1×10^{-31} kg. The proton is 1,836 times more massive; its mass is 1.67×10^{-27} kg. A single drop of water contains as many protons as there are stars in the universe. The mass of the earth is 6×10^{24} kg, and the mass of the sun is 2×10^{30} kg. It is possible very roughly to estimate the mass of the visible universe. There are about 10^{23} stars emitting light. Multiplication by 10^{32} kg, the average mass of a star, gives 10^{55} kg for the total mass of which we have any evidence. (There is some speculation that the universe may also contain a great deal of invisible mass.) In 10^{55} kg of mass there are about 10^{82} protons. The imagination of some physicists has been challenged by the facts that this enormous number, 10^{82}, is the square of another large number, 10^{41} (which measures the ratio of the size of the universe to the size of an elementary particle), and that 10^{41} is also not far from the ratio of electrical to gravitational forces between electrons. So far, however, these numerical coincidences have not shown themselves to be more than coincidences.

Mass of visible universe

A cautionary word about the British system of units is in order. In one version of that system, the pound (lb) is a unit of force, not of mass. In everyday affairs, however, it is quite common to use the pound as a unit of mass. We say that there are 454 gm in 1 lb or 2.2 lb in 1 kg. To avoid error, it is safest to carry out all calculations involving force or mass in a metric system.

2.9 Energy

The most remarkable fact about energy is its diversity. Like a clever actor who can assume many guises, energy appears in a variety of forms and can shift from one role to another. Because of this richness of form, energy appears in nearly every part of the description of nature and provides vital links among different domains of nature.

Kinetic energy

One common form of energy is energy of motion, kinetic energy, which is a measure of how much force is required to set an object into motion or to bring it to rest in a specified distance. For speeds that are much smaller than the speed of light, the formula for kinetic energy is

$$K = \tfrac{1}{2}mv^2 \qquad (v \ll c), \tag{2.11}$$

one-half of the mass m of an object multiplied by the square of its speed v. The revised formula for K, valid near the speed of light, appears in Chapter 21.

Special significance of energy: variety and conservation

The great significance of energy springs in part from the variety of its manifestations, in part from the fact that it is conserved: the total amount of energy remains always the same, the loss of one kind of energy being compensated by the gain of another kind of energy. Trace, for example, the flow of energy from sun to earth to man. When protons in the sun unite to form helium nuclei, nuclear energy is released. This energy may go first to the kinetic energy of motion of nuclei. Some of the energy is then carried away from the sun by photons, the particles that are bundles of electromagnetic energy. The energy content of the photons may be transformed, by the complicated and not yet fully understood process of photosynthesis, into stored chemical energy in plants. Either by eating plants or by eating animals that have eaten plants, man acquires this solar energy, which is then made available to power his brain and muscles and to keep him warm.

Mass energy

That mass is one of the forms of energy was first realized at the beginning of this century. Mass energy can be thought of as the "energy of being," matter possessing energy just by virtue of existing. A material particle is nothing more than a highly concentrated and localized bundle of energy. The amount of concentrated energy for a motionless particle is proportional to its mass. If the particle is moving, it has still more energy, its kinetic energy. A massless particle, such as a photon, has only energy of motion (kinetic energy) and no energy of being (mass).

Einstein's most famous equation,

$$E = mc^2, \tag{2.12}$$

provides the relation between the mass, m, of a particle and its intrinsic energy, or energy of being, E. The important statement Einstein's equation makes is that intrinsic energy is *proportional* to mass. The square of the speed of light is a constant of proportionality; it does the job of converting from the unit in which mass is expressed to the unit in which energy is expressed. Numerically,

c^2 as a conversion factor

$$c^2 = 9 \times 10^{16} \text{ m}^2/\text{sec}^2 = 9 \times 10^{16} \text{ J/kg}. \tag{2.13}$$

In the submicroscopic world of elementary particles, only kinetic energy and mass energy play significant roles. Other forms of energy, necessary to the description of nuclei, atoms, and more complex systems, are considered in later chapters.

Special units of energy

Largely because of its great variety of form, energy is measured in a great variety of units. The SI energy unit, the joule, has already been defined as 1 kg m^2/sec^2. The much smaller energy unit of the cgs system, the erg, is equal to 1 gm cm^2/sec^2. The erg and several other common energy units are related to the joule in Table 2.4. The calorie is used principally to measure heat energy: it is the energy required to raise the temperature of 1 gm of water by 1 °C. The food calorie or kilocalorie, which is 1,000 times greater, is commonly used by chemists and is also familiar to weight-watchers.* A thousand or more kilocalories are needed each day to keep the human machine running efficiently. Another unit of heat measurement is the British thermal unit, or Btu, which is defined as the energy needed to raise the temperature of 1 lb of water by 1 °F. Electric companies prefer the kilowatt hour as a unit of energy. The watt (W)

* This unit is also sometimes called the large calorie, or simply Calorie (capitalized to distinguish it from the calorie). Kilocalorie is preferred.

TABLE 2.4 SOME ENERGY UNITS

Name of Unit	Number of Joules in One Unit
Erg	10^{-7} J
Calorie	4.184 J
Food calorie or kilocalorie	4,184 J
British thermal unit (Btu)	1,054 J
Kilowatt hour	3.6×10^6 J
Electron volt (eV)	1.602×10^{-19} J
Million electron volts (MeV)	1.602×10^{-13} J
Ton	4.2×10^9 J
Megaton	4.2×10^{15} J

is the SI power unit, 1 J/sec, so the kilowatt hour is 1,000 J/sec $\times$ 3,600 sec = 3.6×10^6 J. In atomic, nuclear, and particle physics, the most commonly used energy unit is the electron volt (eV), or its multiples, the keV, the MeV, and the GeV. One electron volt is the energy acquired by an electron (or any other particle of the same charge) in being accelerated through an electrical potential difference of one volt. Chemical reactions typically involve about 1 eV per atom, whereas nuclear energies are measured in MeV. The largest present-day accelerator (in Batavia, Illinois) produces protons of energy 500 GeV. A new unit of the nuclear age is the ton (or kiloton, or megaton) used to gauge the energy of nuclear explosions. One ton is the energy equivalent of the explosion of one ton of TNT.

2.10 Charge

Electric charge is best described as being like French perfume. It is that certain something worn by particles that makes them attractive—specifically, attractive to the opposite kind of particle. Particles that do not have it are called neutral and have no influence (at least no electric influence) on other particles. Charge can lead to pairing of particles, as between the proton and electron in the hydrogen atom. More energetic particles are not held together by the electric force; it merely causes them to deviate slightly from a straight course.

This whimsical approach to charge is meant to emphasize the fact that at the deepest level we do not really understand charge very well. Why, for instance, do some particles have it and some not? Why does it come in quantum units of a certain size? What we do know is that charge may be positive or negative, that it is the seat of electric and magnetic phenomena, and that it is scalar quantity whose total magnitude remains precisely constant during processes of change.

The sign of charge, a historical accident

Which charge is called positive and which negative is entirely arbitrary and is the result of historical accident. The definition that led to electrons being considered negative and protons being considered positive probably stems from a guess made by Benjamin Franklin around the middle of the eighteenth century. His choice of nomenclature was based on the erroneous supposition that it is positive electricity that flows most readily from one object to another. We now know that it is the negative electrons that are mobile and account for the flow of electric current in metals.

Definition of the coulomb

The SI unit of charge, the coulomb (C), is defined in terms of the SI unit of electric current, the ampere (A):

$$1 \text{ C} = 1 \text{ A sec.} \tag{2.14}$$

Definition of the ampere

A current of 1 A flowing for 1 sec conducts a net charge of 1 C. The ampere is defined by a magnetic force experiment. If equal currents in two parallel wires 1 m apart cause the wires to exert on each other a force of 2×10^{-7} newtons per meter length of wire, the current in each wire is, by definition, 1 A. There is a historical reason for the particular number 2×10^{-7}, but since the definition is arbitrary, it need not concern us. Charge is defined in terms of current rather than the other way around for a practical reason. Forces between currents can be measured more accurately and more easily than can forces between stationary charged objects. Only one other unit of charge is in common use, the electrostatic unit, or esu (sometimes called the statcoulomb). This unit, which is incorporated into the cgs system, is so defined that equal charges of 1 esu located 1 cm apart exert on each other a force of 1 dyne. The esu is 3 billion times smaller than the coulomb:

$$1 \text{ C} = 3.00 \times 10^9 \text{ esu.}$$

The magnitude of the quantum unit of charge, usually designated by e, is

The quantum unit of charge

$$e = 1.602 \times 10^{-19} \text{ C} = 4.803 \times 10^{-10} \text{ esu.} \tag{2.15}$$

2.11 Angular momentum

Rotation is common in all parts of nature

Rotational motion is apparently a characteristic of most of the structures in the universe, from neutrinos to galaxies. Our earth rotates on its axis once a day and rotates once around the sun in a year. The sun itself turns on its axis once in 25 days and, along with the other stars of our galaxy, travels once around the galaxy in 230 million years. It is not known whether larger structures, such as clusters of galaxies, have an overall rotational motion, but it would be surprising if they did not.

Going down the scale, molecules rotate, and so do electrons within atoms. The remarkable fact that the electron also spins like a top about its own axis was first discovered in 1925; now we know that many particles have this property of intrinsic spin.

A particle moving with speed v in a circle of radius r about a fixed point has, by definition, an angular momentum equal to the product of its mass, its speed, and the radius of its orbit:

Definition of angular momentum for the special case of circular motion

$$L = mvr \tag{2.16}$$

(L is a commonly used symbol for angular momentum). A more general definition of angular momentum appears in Chapter 9. The SI unit of angular momentum is kg m^2/sec. This combination has been assigned no separate

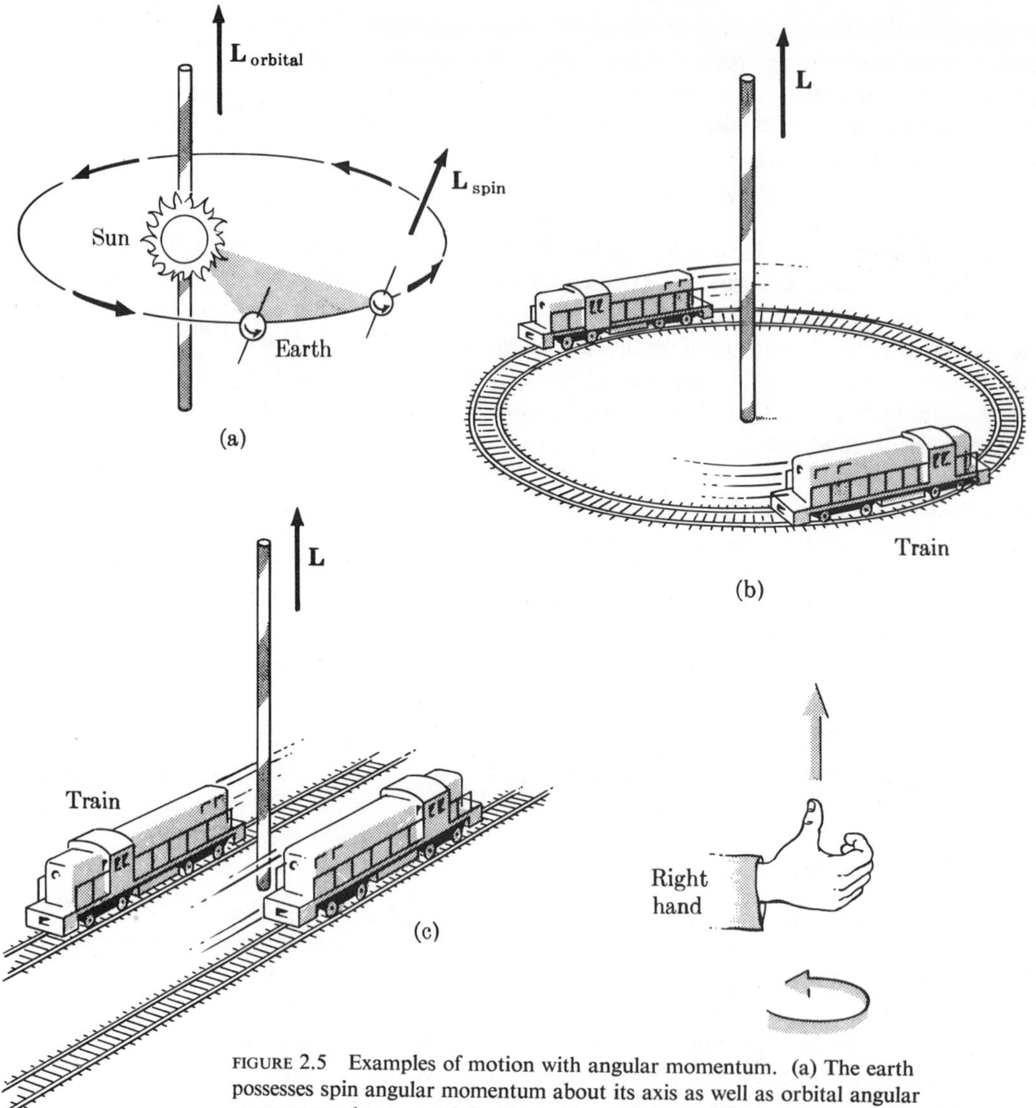

FIGURE 2.5 Examples of motion with angular momentum. (a) The earth possesses spin angular momentum about its axis as well as orbital angular momentum about an axis designated by the giant barber pole. (b) Trains on a circular track possess angular momentum about a vertical axis. (c) Even on straight tracks, angular momentum is associated with a similar relative motion of trains. When the curved fingers of the right hand designate the motion, the right thumb indicates the direction of the angular momentum.

name of its own. Angular momentum is actually a vector quantity,* its direction being assigned along the axis of rotation according to a right-hand rule (Figure 2.5). It is defined with respect to some reference point.

In qualitative terms, angular momentum can be thought of as a measure of

* More exactly, it is an *axial* vector quantity. By contrast, force, velocity, and acceleration are *polar* vector quantities. The distinction will be explained in Chapter 6.

the "intensity" of rotational motion. It is a quantity rendered greater by greater mass, or by greater speed, or by greater distance from the reference point. In analyzing motion, it has proved convenient to discriminate between two kinds of angular momentum: orbital angular momentum and spin. Orbital angular momentum arises from the motion of the center of mass of an object about some reference point. Spin is the angular momentum associated with the rotation of the object itself, or of any system, about its own center of mass. The earth possesses both kinds of angular momentum. Most elementary particles can also possess both kinds.

Two kinds of angular momentum: orbital and spin

As the electron carries nature's smallest unit of electric charge, it also carries the indivisible unit of spin, which, for historical reasons, is denoted by $\frac{1}{2}\hbar$. At the turn of this century, Max Planck discovered the existence of a constant in nature that relates the frequency and the energy of photons (this relation will be discussed in Chapter 23). We now call it Planck's constant and write it h. A dozen years later Niels Bohr discovered that this constant also has to do with the rotation of electrons about the nucleus in an atom. In their motion within the atom, the electrons have an angular momentum that is always equal to $h/2\pi$ or two times $h/2\pi$ or three times $h/2\pi$, and so on, never any value between. Since it is a nuisance to write out the extra factor of 2π whenever it occurs, the notation $\hbar$ (pronounced "h bar") has come into common use for $h/2\pi$. Finally, when Samuel Goudsmit and George Uhlenbeck discovered in 1925 that the electron spins, they found that its spin angular momentum is not $\hbar$, which had been thought to be the indivisible unit, but only $\frac{1}{2}\hbar$. The quantity $\hbar$ has been adopted as the unit of spin and angular momentum in the submicroscopic world, even though the smallest indivisible unit is only half as great. Orbital angular momentum is always an integral multiple of $\hbar$. Spin is an integral multiple of $\frac{1}{2}\hbar$.

The quantum unit of angular momentum

In the macroscopic world, spin quantization, although still valid in principle, is irrelevant because of the infinitesimal magnitude of $\hbar$ relative to everyday angular momenta:

$$\hbar = 1.055 \times 10^{-34} \text{ kg m}^2/\text{sec}.$$

A spectator turning his head at a tennis tournament might have an angular momentum of $10^{33}\hbar$. To discriminate experimentally between 10^{33} and $10^{33} + 1$ is quite out of the question. Not until man could study in detail the structure of a system as small as an atom did he have a chance to discover the quantum theory of angular momentum.

★2.12 Natural units and dimensionless physics*

The units of measurement man normally uses, even in scientific work, have been defined arbitrarily, chosen merely for convenience. They have nothing to do with "natural units" and are in no particular harmony with the basic structure of the world. Yet in this century we have learned of the existence of two natural

* At the discretion of the instructor, sections or subsections marked with a star may be considered optional. Starred material is discussed in the Preface.

c and h: two natural units

units, the speed of light c and Planck's constant h. Neither is directly a mass, a length, or a time, but they are simple combinations of these three. If they were joined by a third natural unit, they would form a basis of measurement as complete as, and much more satisfying than, the kilogram, the meter, and the second. (The thoughtful reader might propose the charge of the electron as an obvious candidate for the third natural unit. Unfortunately, it will not serve, for it is not independent of c and h—just as speed is not independent of time and distance.) Many physicists believe that a deeper understanding of submicroscopic nature awaits the discovery of another natural unit, possibly a fundamental length.

Is there a third natural unit?

It has to be recognized that every measurement is really the statement of a ratio. If you say you weigh 151 pounds, you are, in effect, saying your weight is 151 times greater than the weight of a standard object (a pint of water), which is arbitrarily chosen to be one pound. A 50-minute class is 50 times longer than the arbitrarily defined time unit, the minute. When we use natural units, the ratio is taken with respect to some physically significant unit rather than an arbitrary one. On the natural scale, a jet plane's speed of $10^{-6}c$ is very slow, and a particle speed of $0.99c$ is very fast. An angular momentum of $10{,}000\hbar$ is large, and an angular momentum of $\frac{1}{2}\hbar$ is small.

A difficult but important point to recognize here is that once the speed of light has been adopted as the unit of speed, it no longer makes any sense to ask how fast light travels. The only answer is: Light goes as fast as it goes. Since every measurement is really a comparison, there must always be at least one standard that cannot be compared with anything but itself. This leads to the idea of a "dimensionless physics." Having agreed on a standard of speed, we can say a jet plane travels at a speed of 10^{-6}, that is, one-millionth the speed of light. The 10^{-6} is a pure number to which no unit need be attached: it is the ratio of the speed of the plane to the speed of light. To make possible a dimensionless physics we need one more independent natural unit, and this has not yet emerged.

It should be added that a dimensionless physics is not so profound as it may sound, nor would it necessarily be a terminus of man's downward probing. Its lack of profundity springs from the fact that it, too, would rest on an arbitrary agreement among men about units. The hope is, however, that all scientists will be led naturally and uniquely to agree that there is only one sensible set of natural units, in contrast with the present situation, in which the fact we all agree on is that there is nothing at all special about the meter, the kilogram, and the second.

The hope for a dimensionless physics

Summary of ideas and definitions

The international system of units (SI) is based on six basic units: the meter, the kilogram, the second, the ampere, the kelvin, and the candela. These are defined in Appendix 1.

Many derived SI units have special names. These are listed in Appendix 2.

Mass, length, and time provide a sufficient set of dimensions for all physical measurement.

"Units arithmetic" provides a convenient way of checking for errors in derived equations, and it facilitates the use of conversion factors.

The definitions of concepts and the laws of physics are inextricably intertwined.

Man has explored length and time intervals spanning a factor of about 10^{41} in scale.

The speed of light is an upper-limit speed for all known matter and a lower-limit speed for hypothetical particles called tachyons.

The recoil definition of inertial mass is given by Equation 2.10.

Gravitational mass, measured by weighing, has been found by experiment to be equal to inertial mass.

For speed that is small compared with the speed of light, kinetic energy is defined by

$$K = \tfrac{1}{2}mv^2. \tag{2.11}$$

Mass energy, or intrinsic energy, is defined by

$$E = mc^2. \tag{2.12}$$

Angular momentum is a vector quantity. Its magnitude, relative to a reference point, for the special case of circular motion about the reference point, is

$$L = mvr. \tag{2.16}$$

The charge of the electron, e, is the quantum unit of charge. The spin of the electron, $\frac{1}{2}\hbar$, is half the quantum unit of angular momentum.

The speed of light c and Planck's constant h are two "natural units" that could replace arbitrary basic units such as the meter or kilogram. Physicists hope to find a third natural unit.

QUESTIONS

Section 2.1

Q2.1 (1) Give an operational definition of any concept you have encountered in another area of science, such as chemistry or biology. (2) Attempt a brief operational definition of science itself.

Q2.2 Which of the following are quantitative concepts: (a) time, (b) beauty, (c) temperature, (d) progress, (e) galaxy, (f) atomic energy, and (g) disorder?

Q2.3 A quantitative statement is not just a numerical relationship. It is any statement that is verifiable by experiment, possibly with some experimental uncertainty. Which of the following are quantitative statements? (1) $F = ma$. (2) Faulkner is a greater novelist than Hemingway. (3) Sound travels at about 330 m/sec through air. (4) Dogs can hear higher frequencies than people can. (5) Philosophers of the eighteenth century were 100 percent more effective than philosophers of the nineteenth century.

Q2.4 The thrust of a jet engine can be defined operationally. Suggest an experimental procedure to measure it.

Section 2.2

Q2.5 (1) Suggest units of speed and acceleration that are reproducible and might serve as world standards. (2) Why would such a speed-acceleration system be less satisfactory for high-precision work than the present length-time system?

Section 2.3

Q2.6 The dimension of force times distance is the same as the dimension of energy. Does this prove that force times distance *is* energy? Explain.

Section 2.4

Q2.7 The area A of a circle is expressed in terms of its radius r by $A = Kr^2$, where K is a dimensionless number. (1) Does the dimensional consistency of this equation depend on the value of K? (2) Does its correctness depend on the value of K? Explain both answers.

Section 2.5

Q2.8 Suggest an operational procedure for using a standard of length to fix the unit of time.

Section 2.6

Q2.9 Suggest an operational procedure for using a standard of time to fix the unit of length.

Q2.10 When the ammonia-beam maser oscillator was first operated, it was proclaimed the most accurate clock ever built. With it, irregularities in the rate of rotation of the earth on its axis were detected. How could the accuracy of a clock be tested if it were the world's most accurate clock?

Q2.11 How, if at all, do you think that time might be defined if there were in nature no regularly recurring phenomena whatever? Defend either one of these two positions: (a) The concept of time would be meaningless, and it would be impossible to design a clock. (b) It would still be possible to define time and design a clock.

Section 2.7

Q2.12 The "speed of sound barrier" and the "speed of light barrier" have some very important differences. Discuss two or three of their differences.

Q2.13 (1) Suggest a way to measure the distance to the moon that does not require an accurate clock. (2) Suggest a way that does require an accurate clock. Is an accurate measurement of speed necessary in either method? *Optional:* Using an outside source, report on the laser ranging technique, which can pinpoint the distance to the moon (more exactly, to one spot on the moon) with an uncertainty measured in inches.

Section 2.8

Q2.14 Mass is sometimes defined as "the quantity of matter." Is this an operational definition? Scientifically, is it a satisfactory definition?

Q2.15 Explain in simple terms why the height of a man is "only" about 10^{10} times greater than the diameter of an atom, whereas his mass is about 10^{28} times greater than the mass of a typical atom.

Q2.16 Inertial and gravitational mass are equal. Mass can be determined by weighing an object; yet in orbit, an object that is weightless retains its mass. Explain the apparent paradox.

Q2.17 Using the operational definition of gravitational mass, explain why an object has the same gravitational mass on the surface of the moon as it does on the surface of the earth.

Q2.18 Given three objects numbered 1, 2, and 3 and a standard kilogram, discuss what hypothetical experiments you might carry out to test the self-consistency of the recoil definition of mass.

Section 2.9

Q2.19 The hypothetical tachyon travels always at a speed greater than the speed of light and gains speed when it loses energy. Discuss the collision of a tachyon with a normal particle at rest.

Q2.20 Give an account of the various energy transformations that culminate in an increase of the kinetic energy of an automobile when the accelerator is depressed.

Section 2.10

Q2.21 Explain how a wire can conduct a current yet be electrically neutral.

Q2.22 Suggest a reason why the force between currents can be measured more accurately than the force between static charges.

Section 2.11

Q2.23 (1) Which is greater, the angular momentum of the moon with respect to the earth or the angular momentum of the moon with respect to the sun? (2) Which is more nearly constant? Give reasons for both answers.

Q2.24 Angular momentum and charge are examples of quantized variables. What are some variables that are not quantized (so far as we know now)?

Q2.25 If you measure a distance with a ruler or a time interval with a stop watch, explain in what sense you are determining a ratio of two distances or two times.

Section 2.12

Q2.26 For man, the earth is a special place of the utmost significance. Why, then, are physicists unwilling to attribute fundamental significance to units based on properties of the earth?

Q2.27 Is ever-increasing precision of measurement likely to be a route to fundamental new discovery in science? Or is the search for the next decimal a pedantic exercise off the mainstream of scientific advance? Defend either point of view.

EXERCISES

Section 2.2

E2.1 The following questions provide further review of exponential arithmetic: (1) The speed of light is 3×10^8 m/sec. How long does it take light to travel across an atom whose diameter is 10^{-10} m? (2) If at a certain time of year it takes light 10 min to reach Earth from Mars, how far away is Mars at that time (a) in meters? (b) in miles? (3) Express the speed of light in ft/sec, mile/sec, mile/hr, and m/μsec.

E2.2 Express (a) 100 km/hr in mile/hr, (b) 100 m in yards, (c) 880 yards in meters, and (d) 10^6 ergs in joules.

E2.3 Re-express each of the following quantities in SI units: (a) 186,000 mile/sec, (b) 1 yard, and (c) 32 ft/sec^2.

E2.4 The density of water is 8.35 lb/gallon, or 1 gm/cm^3. Useful mass conversion factors are 454 gm/lb and 2.205 lb/kg. (1) What is the density of water in kg/m^3? (2) What is the volume conversion factor linking gallons and m^3?

E2.5 Mention half a dozen different units that you commonly employ (either in everyday life or in other courses of study) that are not SI units. Find the factors that convert any two of these to their corresponding SI units.

Section 2.3

E2.6 Express the dimension of energy per unit time in terms of mass, length, and time. What is the SI unit of this quantity?

E2.7 Adopt length and speed as the two fundamental dimensions of kinematics. What, then, are the dimensions of (a) time and (b) acceleration?

E2.8 Adopt time and acceleration as the two fundamental dimensions of kinematics. What, then, are the dimensions of (a) length and (b) speed?

E2.9 Suppose that force, mass, and time are adopted as fundamental dimensions. Express length as a combination of these three.

Section 2.4

E2.10 Two objects of masses m_1 amd m_2, separated by distance r, exert on each other a gravitational force F given by

$$F = G \frac{m_1 m_2}{r^2} .$$

What is the SI unit of the gravitational constant G?

E2.11 In the nineteenth century the English physicist George Stokes studied the motion of small spheres moving slowly through liquids. He was able to show theoretically that the "drag force" of the liquid on the sphere is given by the expression

$$F_D = 6\pi\eta vr,$$

where F_D is the drag force, v is the speed of the sphere, and r is the radius of the sphere. The constant η is called the *viscosity* of the liquid. What is the dimension of η? What is its SI unit of measurement?

E2.12 Calculate and record for possible later use the following conversion factors: (a) the number of astronomical units, or A.U., in 1 light-year (the A.U. is the mean distance from earth to sun); and (b) the number of ft/sec in 1 mile/hr.

E2.13 With the help of Appendix 2, verify the units consistency of the following equation: $W = qEd$, work = charge × electric field × distance.

E2.14 If an object moves in a straight line, the net force acting on it multiplied by its speed is equal to its energy gain per unit time. Show that this is a dimensionally consistent statement. From this law derive Equation 2.4, relating the SI units of energy, force, and distance.

E2.15 A certain steak costs $4.00 per kilogram. (1) Write a dimensionally inconsistent equation to express its price in terms of its mass. (2) Replace this equation with a more general, dimensionally consistent, equation.

E2.16 The equations below involve distance (s), time (t), speed (v), mass (m), and energy (E). Which are dimensionally consistent?

(a) $t = s/v$
(b) $v^2 = E^2/m^2$
(c) $s = \frac{1}{2}vt^2$
(d) $E = mvst$
(e) $s = \frac{1}{2}vt$
(f) $E = ms^2/t^2$
(g) $t = v/s$
(h) $E^2 = m^2v^4$

E2.17 The equations below involve force (F), mass (m), length (l), time (t), and energy (E). Which are dimensionally consistent?

(a) $E = F + l$
(b) $E = Fl$
(c) $F = ml/t^2$
(d) $Ft = El$
(e) $E = F^2t^2/m$

Section 2.5

E2.18 Is 1,500 meters (sometimes known as a "metric mile") greater or less than one mile? What should be the time for a four-minute miler to cover 1,500 meters? *Optional:* Look up the current world records for the 1,500-meter run and the mile. Compare the ratio of these two times with the ratio of the distances.

Section 2.6

E2.19 Which is longer, a 50-minute lecture or a microcentury lecture?

Section 2.7

E2.20 What is the speed of an astronaut in mile/hr if he is traveling 24,500 ft/sec? (Introduce conversion factors and carry out the units arithmetic as well as the numerical arithmetic.)

E2.21 The highest speed reached by an astronaut thus far is about 7 mile/sec. (1) Express this speed in m/sec. (2) If such a speed were attained in air, what would be its Mach number (the ratio of the speed to the speed of sound)?

E2.22 At the speed of an orbital astronaut (5 mile/sec), how long would it take to reach the nearest star, Alpha Centauri, which is about 4 light-years distant? Express your answer in years.

Section 2.8

E2.23 (1) How many electrons are needed to make a mass of 1 kg? (2) Make a rough estimate of the number of electrons in the earth, remembering that for every electron in the earth, there is one proton and about one neutron.

E2.24 In continental Europe, one "pound" is half a kilogram. (1) How does this relate to the English and American pound? (2) Which is a better buy, one German pound of coffee for $1.00 or one American pound of coffee for $0.80?

Section 2.9

E2.25 (1) Calculate in joules the kinetic energy of an electron traveling at 3×10^7 m/sec (fast enough to circumnavigate the earth in 1 sec). (2) How fast would you have to walk in order to have the same kinetic energy? At this speed, how long would it take you to cover 1 m?

E2.26 A supersonic jet is flying at a speed of 600 m/sec (about Mach 1.8). What is the ratio of its intrinsic mass energy to its kinetic energy? Give an answer first in algebraic form, then in numerical form.

Section 2.10

E2.27 (1) How many electrons are needed to accumulate -1 coulomb of electric charge? (2) If 30 C of charge pass through a 15-amp fuse in $\frac{1}{2}$ sec, will the fuse "blow"?

E2.28 (1) Calculate the charge-to-mass ratio of the proton in C/kg and in esu/gm. (2) What is the total positive charge carried by all the protons in 1 gm of hydrogen? (Ignore the mass of the electrons as compared with the mass of the protons.) (3) If a 1-gm sample of hydrogen carries a net positive charge of 1 esu, what fraction of the hydrogen atoms have lost electrons?

Section 2.11

E2.29 A man holds a 15-kg child by the hands and swings the child around through the air as he turns. Estimate the angular momentum of the child relative to the center of the man. Explain the basis of your choice of input numbers.

E2.30 The kinetic energy of satellites of equal mass in circular orbits varies inversely with their distance from the center of the earth ($K \sim 1/r$). Find the dependence on radial distance of their (a) speed, (b) angular momentum, and (c) period of rotation.

Section 2.12

E2.31 In a system of units in which the speed of light is equal to 1, both time and distance can be measured in seconds. Express each of the following distances in seconds: (a) your height, (b) 1 mile, and (c) 1 light-year.

PROBLEMS

Unit chosen to simplify one equation

P2.1 Consider a set of units in which two equal unit masses that are separated by 1 m exert on each other a gravitational force of 1 N. The law of gravitational force is written as

$$F = \frac{m_1 m_2}{r^2}.$$

(1) Is this a dimensionally consistent equation? (2) If the mass unit in this system is called the heft, what is the conversion factor linking heft and kilogram? *Optional:* Evaluate the proportionality constant K in Equation 2.1 in a meter-heft-second-newton system of units.

Dimensional combination of variables

P2.2 The properties of fluid flow past an obstacle are determined in part by the Reynolds number,

$$R = \frac{av\rho}{\eta},$$

where a is the linear dimension of the object, v is the speed of the fluid, ρ is the density of the fluid, and η is its viscosity. The physical dimension of η can be obtained from the equation in Exercise 2.11. (1) Show that R is dimensionless. (2) Show that no dimensionless quantity can be constructed from any three of the four quantities a, v, ρ, and η. (3) Show that R is the *only* dimensionless combination that can be obtained from a, v, ρ, and η (except for R raised to a power or R multiplied by a number).

Dimensional analysis

P2.3 If a ball or stone is thrown with initial speed v at an angle of about 45 deg to the horizontal, its range is given approximately by

$$R = K_1 \frac{v^2}{g},$$

where g is the acceleration of gravity and K_1 is a dimensionless constant. (1) Verify that this equation is dimensionally consistent. (2) Find a similar equation for the time of flight that involves another dimensionless constant, K_2. (3) Perform crude outdoor experiments in order to find the approximate numerical values of K_1 and K_2. Describe your methods.

P2.4 The period of a pendulum might depend on the length of the pendulum (l), its mass (m), and the acceleration of gravity (g). (1) Find a combination of these variables with the dimension of time. (2) Show that it is the only such combination. (3) Make an approximate measurement of the period of a pendulum, and compare this measured time with the time calculated from your combination of variables. Interpret the fact that the two times are not the same. Does this mean that your combination of variables is without physical significance? *Optional:* Extend your measurements to pendulums of different mass and length. How do the additional results support an answer to the question in Part 3 of this problem?

P2.5 It has been found by experiment that any low-altitude earth satellite has a period of about 90 min, regardless of its mass. (1) List the variables on which the satellite period might reasonably depend, and show that the only combinations of these variables with the dimension of time are indeed mass-independent. (2) Explain why dimensional analysis alone cannot reveal whether the period of the satellite increases or decreases as its distance from the earth increases. (AUXILIARY INFORMATION: The force F on a satellite of mass m can be written $F = mgR^2/r^2$, where g is the acceleration of gravity at the earth's surface, R is the radius of the earth, and r is the distance of the satellite from the center of the earth.)

Recoil definition of mass

P2.6 A man coasting on a small flat car at $v_1 = 2.0$ m/sec throws a 2-lb stone forward at a speed $v_s = 18$ m/sec relative to the ground (see the figure). This action slows the flat car to a speed of $v_2 = 1.9$ m/sec. What was the original total mass of the system (car plus man plus stone)? (HINT: Consider the event from the viewpoint of a cyclist continuing at a steady speed v_1.)

P2.7 Astronauts on board a spacecraft drifting in empty space wish to measure the mass of their vehicle, including its occupants. They have several disposable objects of known mass on board and a radar unit that enables them to measure their speed relative to any nearby object. However, they have no way to measure their absolute speed and no way to measure acceleration. Describe carefully an operational procedure whereby they could make the desired mass measurement.

P2.8 A certain otherwise law-abiding citizen drives consistently 10 mile/hr above any speed limit, so the kinetic energy of his automobile can be written

Energy and speed

$$K = K_L + \Delta K,$$

where K_L is the kinetic energy the automobile would have if it were driven at the speed limit. (1) Obtain an algebraic formula for ΔK in terms of the speed limit v_L and the excess speed Δv. (2) Sketch a graph of ΔK vs v_L for fixed Δv. *Optional:* Is the practice of driving at 10 mile/hr above the speed limit more hazardous at low speed or high speed? Why?

P2.9 (1) Using only Planck's constant h, the speed of light c, and the mass of a proton m_p, construct quantities with the dimensions of (a) length, (b) time, and (c) energy. (2) Evaluate these three quantities in units of (a) fm, (b) sec, and (c) eV.

Combinations of fundamental constants

P2.10 In terms of mass, length, and time, Planck's constant h has the dimension ml^2/t, and the speed of light c has the dimension l/t. (1) Consider a system of measurement based on h, c, and a mass unit m_0. Show that any unit expressible as a combination of mass, length, and time units is also expressible as a combination of h, c, and m_0 units. (2) Repeat the above demonstration for a system based on h, c, and a length unit l_0. (3) Would a time unit t_0 be a suitable third natural unit to join h and c as a basis of measurement?

Natural units

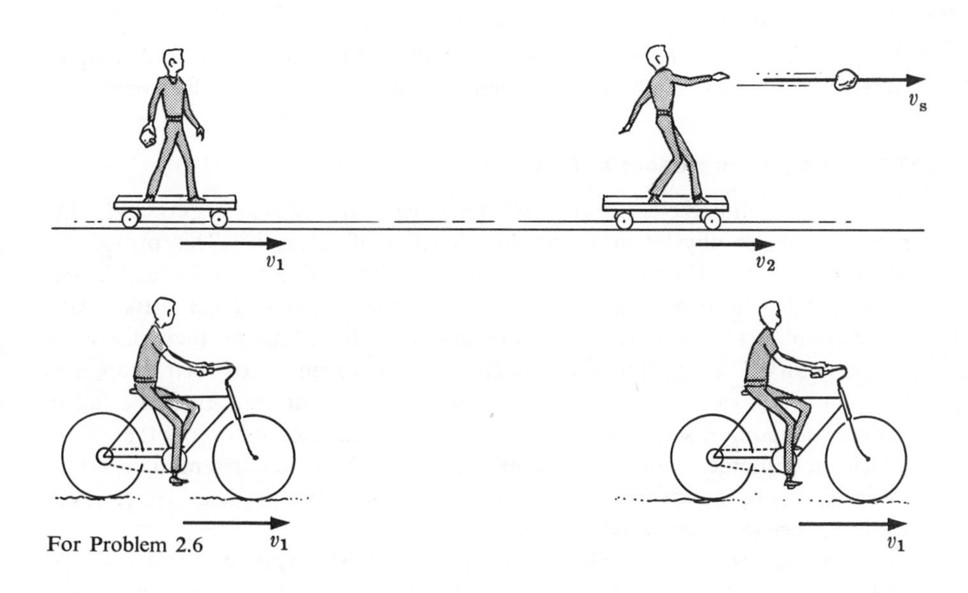

For Problem 2.6

3 Elementary Particles

The downward probing of the past few centuries in physics has in recent decades led to the world of elementary particles. Actually no scientist believes that all the particles we now know are truly elementary, but for lack of a better word, and more especially for lack of any deeper knowledge of a lower substratum of matter, they are called elementary particles. The effort to understand the particles—to somehow tie them together through deeper and simpler concepts—occupies the attention of many scientists throughout the world.

3.1 The submicroscopic frontier

Why do we examine elementary particles so early in a survey of physics? One reason is to see physics in action at a frontier of research. This provides a reference point in the present as the major theories of physics are developed, approximately in historical order, in the subsequent parts of the book. Also there is about the particles the paradoxical fact that although their discovery rested upon all the earlier developments of physics, many of their properties are simple and can be understood even by those with no background at all in science. Of course, some of the properties of the particles, especially the way in which they interact with one another, require for their understanding a rather extensive background in physics. These aspects of the particles are reserved for appropriate places in later chapters.

Despite their mystery and challenge, particles have many simple properties

The modern scientist, whether biologist, chemist, physicist, or astronomer, is decidedly microscopically oriented. He pictures the large-scale world as put together from smaller and smaller units, down to the elementary particles and someday, soon perhaps, to a still deeper unit. And he pictures events and phenomena as arising from laws that ultimately have their simplest expression in the world of the very small. If Aristotelian physics was a science of final

Particles are the structural elements of all matter

causes, modern physics is a science of microscopic causes. "Explanation" in physical science is as often as not the description of something larger in terms of something smaller. In looking first at the elementary particles, we are examining the basis of physics as well as its frontier.

Particles usefully illustrate many laws

The elementary particles are useful objects for illustrating many aspects of physics. Their very simplicity means they are better adapted for illustrating basic principles of science than are the unwieldy and more complicated objects of our everyday life. We shall see in Chapter 4 how the elementary particles illustrate the all-important conservation laws. Throughout the remainder of the book, we shall find the particles convenient for demonstrating aspects of each of the great theories of physics.

About the particles we now know a great deal, but not nearly enough. They are not bounded by a closed theory. Looking first at the particles will emphasize, as eighteenth- and nineteenth-century science could not do nearly so well, the exploratory and tentative nature of science. With the particles, the known and the unknown appear equally exciting.

To the scientist, the elementary particles provide the greatest challenge in modern physics. To the student they provide important insight into the scientist's view of the world. It is well to recall that the most familiar points of impact of physical science on man—our satellite communications, our detergents, our bombs, our household gadgets—form only a sideline off the mainstream of scientific advance. The true frontiers of science lie far more remote from everyday life, and one of these is the submicroscopic frontier inhabited by the particles.

3.2 The early particles

Analogy: particles today and atoms in 1900

Man and the familiar objects of his world are constructed from atoms and molecules. At the turn of this century, atoms were known to exist, but like today's elementary particles, the structure of the atom and the relation of one atom to another were mysteries. It was known that the atom was the smallest unit of an element such as hydrogen or oxygen or sodium or uranium; that there were some eighty-odd different kinds of atoms (today we know over a hundred); and that atoms were all of about the same size, a size such that one hundred million of them side by side would make a line less than one inch long. It was also known that groups of different kinds of atoms could join together to form tiny structures called molecules, which in turn formed the basic building blocks of the vast and wonderful variety of substances we encounter in the world.

THE ELECTRON

Discovered by studying cathode rays

In the first decade of the twentieth century, discoveries came thick and fast, leading to a giant step downward toward the subatomic world of particles. The electron, the first known particle, had been discovered in 1897 by J. J. Thomson in England. He showed that "cathode rays" that are produced by high voltage in an evacuated vessel could be deflected both electrically and magnetically. The rays were most simply interpreted as a beam of high-speed identical particles carrying negative electric charge (the same sign as the charge on an electrified rubber balloon). That the mass of each of these particles was considerably less than the mass of a single atom was verified by the ease of deflecting them. Also, that each was of a size much less than the size of an atom was shown by their power to penetrate through a gas. Thomson realized, as those who had studied cathode rays before him had not, that he was dealing with truly subatomic objects, even though he was still far from isolating an individual particle.

Links between atoms

The electron, as soon as it was identified, was strongly suspected to be a particle contained within an atom. As a common constituent of atoms, it provided the first definite link between distinct atoms. Another important link was forged in 1902 when Ernest Rutherford and Frederick Soddy suggested that a radioactive atom (one capable of spontaneously emitting powerful radiation) could transmute itself into an entirely different kind of atom. This transmutation strongly suggested that rather than independent, indivisible entities, atoms must be structures built up of some common, more elementary building blocks.

THE ATOMIC NUCLEUS

Alpha particles

Shortly thereafter, alpha particles ejected at high speed from radioactive atoms were used as the first projectiles for bombarding atoms. (These atomic "bullets," conveniently provided free by nature, are not energetic enough for modern purposes; they have since been replaced by particles pushed to greater energy in accelerators.) A result of the early bombardments was the revelation that the interior of atoms is largely empty space. By 1911 the experimenter Ernest Rutherford had discovered that the atom contained a massive, positively charged core—the nucleus—at least ten thousand times smaller than the atom as a whole, and that the remaining space was occupied by a few lightweight, negatively charged electrons. Two years later the theorist Niels Bohr provided a successful mathematical description of the motion of the electrons in the atom. Despite later modifications of detail, the essentials of this description remain as our picture of atomic structure up to the present. The electrons whirl about the nucleus in quantized orbits. The size of the atom is determined by the radius of the outermost orbits.

The Rutherford-Bohr atom

THE PROTON AND THE NEUTRON

The old view: Nuclei contain protons and electrons

The nucleus of the lightest atom, hydrogen, was christened the proton and joined the electron to bring the list of elementary particles up to two. Heavier nuclei were believed to be formed of a number of protons and electrons packed closely together. The details of this picture of nuclear structure were never clear; indeed, it was abandoned in 1932. However, it was a most attractive idea to have just two fundamental particles—the negatively charged electron and the positively charged proton—from which all of the matter in the universe was constructed. The only puzzling feature (a puzzle not yet resolved) was why the proton should be so much heavier than the electron. In any case, this idyllic two-particle situation was not to last.

Several things happened in the early 1930s to begin the disturbing increase in the number of known elementary particles that has continued until the present time. A new particle, the neutron, was discovered; it was about as massive as the proton but carried no electric charge. The neutron was quite welcome, for it was just the particle needed to join the proton to form atomic nuclei. The picture of the nucleus that was immediately adopted and is still accepted was that of a collection of protons and neutrons glued tightly together by a new strong force, simply called the nuclear force. For example, ^{235}U, the most famous isotope of uranium, has a nucleus consisting of 92 protons and 143 neutrons. The far simpler nucleus of helium (the same thing as an alpha particle*) contains two neutrons and two protons.

The new view: Nuclei contain protons and neutrons

THE POSITRON

At nearly the same time, a fourth particle announced its presence by the track it left in a cloud chamber exposed to cosmic radiation in Pasadena, California. This new particle, the positron, was as light as the electron but carried a positive rather than negative charge. Some positron tracks are shown in Figure 3.1. Like the neutron, the positron arrived at an opportune time. A few years earlier, in 1928, Paul Dirac had constructed a new theory of the electron that was brilliantly successful in accounting for the fine details of atomic structure. But Dirac's theory seemed to have a flaw in that it predicted a sister particle for the electron, alike in all respects except for the sign of its electric charge. A slot in the structure of theoretical physics was ready and waiting for the positron when Carl Anderson discovered it in 1932. (Dirac's theory also predicted a negatively charged sister for the proton, called the antiproton, but many years were to go by before it was seen. The construction of the six-billion-volt Bevatron in Berkeley, California, made possible the production of antiprotons, and they were first observed there in 1955.)

The positron is an antielectron

THE PHOTON

The advance of physical theory in the late 1920s was responsible for the "rediscovery" of an old particle, the photon. Back in 1905, the same year his first important paper on relativity was published, Albert Einstein had suggested that electromagnetic radiation is corpuscular, its energy being carried in packets or quanta. Although these energy packets, now called photons, behaved in some ways like particles, they were quite different from ordinary material particles. They carried energy but had no mass. They could be neither speeded up nor slowed down but traveled always at the same invariable (and enormous) speed. They could be born and die (that is, be emitted and absorbed), whereas ordinary particles remained in existence forever—or so it was then believed. And unlike material particles, the photon could never be isolated at a particular point except during the moment of its birth or death; otherwise it spread diffusely through space. For all of these reasons, the photon was not associated with the electron and proton as a true elementary particle.

---

* For historical reasons, a helium nucleus is called an alpha "particle" even though we now know it to be a composite of other particles.

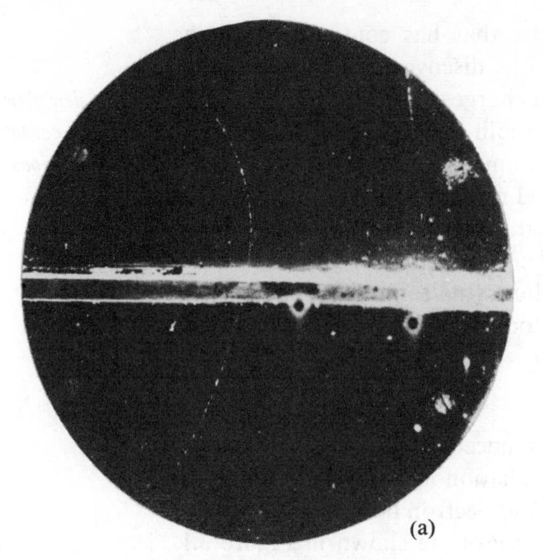

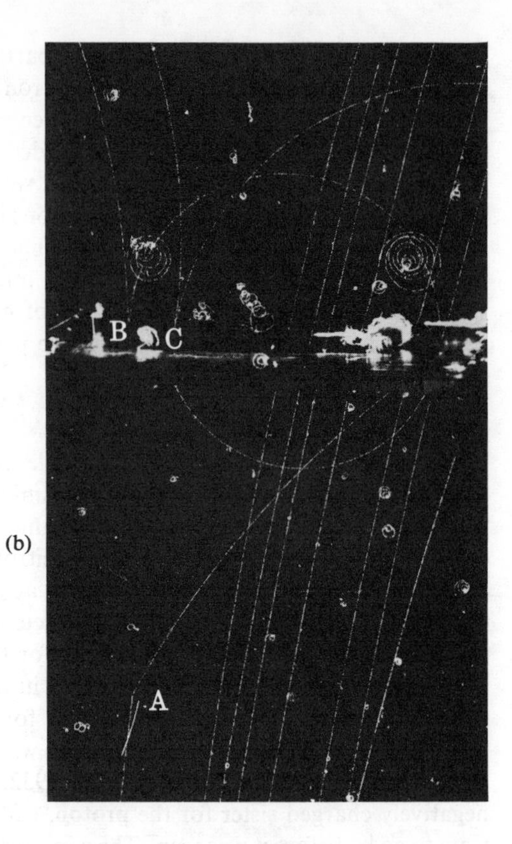

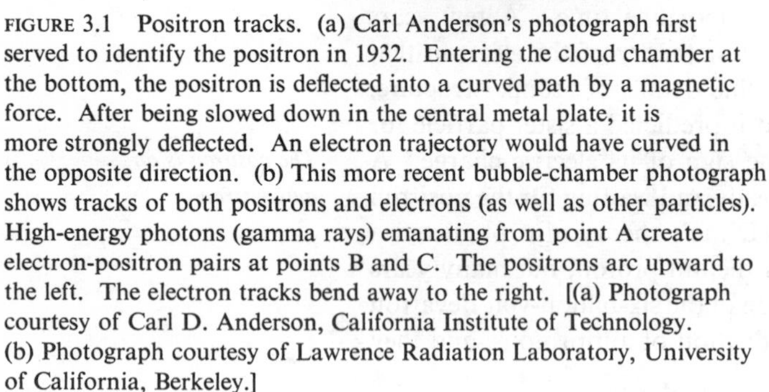

FIGURE 3.1 Positron tracks. (a) Carl Anderson's photograph first served to identify the positron in 1932. Entering the cloud chamber at the bottom, the positron is deflected into a curved path by a magnetic force. After being slowed down in the central metal plate, it is more strongly deflected. An electron trajectory would have curved in the opposite direction. (b) This more recent bubble-chamber photograph shows tracks of both positrons and electrons (as well as other particles). High-energy photons (gamma rays) emanating from point A create electron-positron pairs at points B and C. The positrons arc upward to the left. The electron tracks bend away to the right. [(a) Photograph courtesy of Carl D. Anderson, California Institute of Technology. (b) Photograph courtesy of Lawrence Radiation Laboratory, University of California, Berkeley.]

A particle without mass

The theory of quantum mechanics, discovered in 1925 and developed over the next decade, changed this view of the photon. It showed that from a fundamental point of view, the difference between a photon and a material particle was not so great as had been thought. The particle happened to have mass; the photon did not. All of their other dissimilarities could be understood simply as arising from this single difference. In particular, the quantum theory suggested that it should be possible for material particles to be created and annihilated. The photon appeared to be not so distinctive after all, and it joined the list of elementary particles.

BETA DECAY AND THE NEUTRINO

Beta particles are electrons

Very shortly afterward a theory developed by Enrico Fermi showed that man had in fact, been witnessing the creation of material particles for some time. Ever since the time of early studies of radioactivity at the beginning of this century, scientists had known that some radioactive atoms shoot out high-speed electrons, which, in this particular manifestation, were called beta particles. The radioactive transformation giving rise to these electrons was known as beta decay, but its revolutionary import was not suspected for many years.

Since atoms were known to contain electrons, it did not seem surprising that electrons should sometimes be ejected from atoms.

Even after the discovery of the atomic nucleus in 1911, when it became clear that the beta electrons must emerge from the nucleus, the significance of beta decay was not appreciated. Electrons were simply assumed to exist within the nucleus as well as in the space surrounding the nucleus. But when the discovery of the neutron in 1932, together with various theoretical difficulties, finally banished electrons from the nucleus, beta decay became perplexing. Fermi in 1934 suggested that at the instant of the radioactive transformation, an electron came suddenly into existence in the nucleus and swiftly departed, to be recorded as the beta particle. In short, beta decay, already known for many years, represented the creation of material particles. Fermi's suggestion was, of course, more than a verbal hypothesis. Couched in mathematical language within the framework of the quantum theory, it gave a satisfying explanation of beta decay. Among other things, it predicted the possibility that the beta transformation might, for some atoms, result in positrons rather than electrons; this was at once verified with artificially produced radioactive materials. Only slightly modified, Fermi's theory still gives an adequate explanation of all beta-decay phenomena.

Matter can be created

Positron decay

As so often happens with successful theories in physical science, there is an unexpected bonus. The theory does what is expected of it and then a bit more. Dirac's theory of the electron magnificently explained details of atomic structure and then surprisingly predicted the positron as well. Fermi's theory, in a similar way, accounted for beta decay and as a bonus predicted a strange new particle, the neutrino. (The first prediction of the neutrino was actually made by Wolfgang Pauli several years before Fermi's work. More accurately stated, Fermi provided a definite mathematical framework to accommodate Pauli's speculative suggestion.) According to the theory, a particle with no electric charge and little or no mass (it is now believed to be, like the photon, precisely massless) is also created at the moment of beta decay and leaves the nucleus along with the electron. This most elusive of all particles leaves no trail in a bubble chamber and travels through miles of solid material as if nothing were there. Nevertheless, theory demanded it as a new member of the elementary-particle family, and physicists would have been most upset if the neutrino had not finally been detected. The story of its observation in 1956 and the discovery of a second kind of neutrino in 1962 will be told in Chapter 27.

The neutrino: another particle without mass

By the middle 1930s the proton had been joined by the photon, the neutron, and the positron (or antielectron). The neutrino, although not directly observed, was a theoretical necessity and was added with considerable confidence to the list of particles. In addition, there was every reason to believe that antiprotons and antineutrons existed. The graviton, a particle supposed to transmit the gravitational force, had also been hypothesized, but little hope was held out then (or now) for its observation.

3.3 The pion and the muon

Before most scientists all over the world interrupted their fundamental research to devote their talents to the technology of warfare, one more particle was

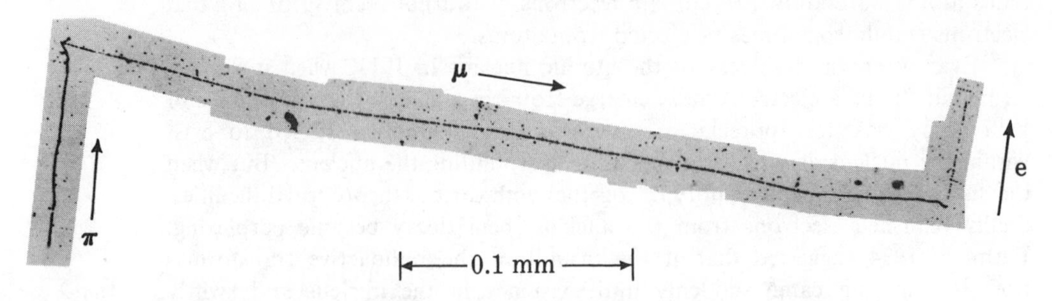

FIGURE 3.2 Early example of pion-to-muon decay. After being brought to rest in a photographic emulsion, the pion decays into a muon and an unseen neutrino. The muon, as it slows, leaves an increasingly heavy track. After being stopped, the muon also decays, as is evidenced by the track of a high-speed electron on the right. (From C. F. Powell, P. H. Fowler, and D. H. Perkins, *The Study of Elementary Particles by the Photographic Method* [New York: Pergamon Press, Inc., 1959], p. 245. Reprinted by permission of the authors and publisher.)

predicted and another was discovered. In a brilliant piece of theoretical work in Japan in 1935, Hideki Yukawa predicted the particle we now call the pi meson or pion. For this he was awarded a Nobel prize—after his prediction proved correct. Yukawa initiated a line of reasoning about the nature of forces that has become one of the key steps in the transition from our "everyday" way of looking at the world to the new way of looking at the submicroscopic world.

The idea of an exchange force

A few years after the discovery of quantum mechanics, not long before Yukawa's work, a quantum theory of electric and magnetic forces was developed. This theory, which was responsible for the new view of the photon as an elementary particle, assigned to the photon the role of carrier of the force. A proton and an electron, for example, which attract each other electrically, continually exchange photons. Each electrically charged particle is continually emitting and absorbing photons, and it is this ceaseless exchange that gives rise to the force. Yukawa gave thought to the powerful new nuclear force that acted between neutrons and protons to hold them together in nuclei. He asked the question: What if this force also arises from an exchange between the nuclear particles? The thing exchanged, he demonstrated, could be neither a photon nor any other known particle. (Yukawa's reasoning, which rests on basic ideas in the quantum theory, will be explained in Chapter 25.) The new particle should be 200 to 300 times more massive than an electron (but still 6 to 9 times lighter than a proton). Yukawa concluded his paper somewhat pessimistically: "As such a quantum with large mass and positive or negative charge has never been found by the experiment, the above theory seems to be on a wrong line. We can show, however, that in the ordinary nuclear transformation, such a quantum cannot be emitted into outer space." In fact, his insight proved flawless. Yukawa's particle was eventually discovered (in 1947), and we now recognize particle exchange as a principal mechanism of interaction among all the particles.

Analogy: photon in electric force and pion in nuclear force

Just after Yukawa's work, in 1936, a new particle *was* discovered through cloud-chamber studies of cosmic radiation. Originally called the meson (for its intermediate mass), this was the particle that is now called the mu meson or muon. At first it appeared to be a good candidate for Yukawa's nuclear exchange particle, for its mass of about 200 electron masses seemed right. However, this interpretation did not hold up. Studies over the next decade gradually made it clear that the muon's interaction with neutrons and protons is too weak to be consistent with its supposed role as a carrier of the nuclear force. Also, mounting evidence from cosmic-ray studies indicated that the cosmic-ray mesons consisted of at least two kinds. Clarification came in 1947 with the discovery of the pion by Cecil F. Powell and co-workers in Bristol, England. An early example of a pion track in a photographic emulsion is shown in Figure 3.2. Its decay product, a muon, in turn undergoes decay into a positron and unseen neutrinos. The next year, man-made pions were discovered in the products resulting from nuclear collisions in an accelerator in Berkeley, California. Today they are studied routinely by the million. There is no doubt that the pion is an exchange particle contributing to nuclear forces. The role of the muon remains unclear.*

The muon does not interact strongly

3.4 The modern particles

The collection of known particles expanded rapidly between 1947 and 1954 with the discovery of four more groups whose members are known collectively as "strange" particles. The kaons (or K particles) constitute the lightest of the new groups, each kaon being about half as heavy as a proton; each member of the other groups, the lambda (Λ), sigma (Σ), and xi (Ξ) particles, is somewhat heavier than a proton. These particles were all first seen in cosmic radiation. Studies of their properties are now in progress at the sites of high-energy accelerators in the United States, western Europe, and the Soviet Union. One of these accelerators, in Stanford, California, is pictured in Figure 3.3. Since 1954, accelerators have provided the arena of discovery for all new particles.

"Strange" particles: K, Λ, Σ, Ξ

Before the construction of these accelerators, physicists had to rely exclusively on cosmic radiation as a source of particles. The earth is continually bombarded by particles from outer space, enough to provide the physicist with particles to study but not enough to be a health hazard to the world's population because it is sheltered beneath the blanketing atmosphere. Most of the particles are protons, some of which have exceedingly great energy. When the protons strike the air, nuclear collisions occur, giving rise to a shower of various particles, including the short-lived muons, pions, and strange particles as well as photons, electrons, and positrons. Although accelerators now provide beams of particles far more copious than those available in cosmic radiation, the natural radiation is still of great interest for two reasons. First, the cosmic-ray particles striking the atmosphere are direct messengers from other parts of our galaxy, possibly from outside the galaxy as well. Second, some of the cosmic-ray particles have vastly more energy than the particles of any accelerator now operating or planned (see Figure 3.4).

Two sources of energetic particles

Accelerators provide the greatest intensity

Cosmic rays provide the greatest energy

The need for machines larger than a football field to study the tiniest

* In fact, the muon is the most perplexing of all particles. Its peculiar status will be explained in Section 27.2.

FIGURE 3.3 The Stanford Linear Accelerator. (a) Air view of the 2-mile-long accelerator, with the "switchyard" in the foreground where energetic particles are deflected into experimental research buildings on the left or the right. (b) Interior view of the accelerator. The large tube at the bottom provides support and alignment. Electrons fly through the smaller beam tube mounted atop a girder. [(a) Photograph courtesy of Roland Quintero. (b) Photograph courtesy of Stanford Linear Accelerator Center.]

(a)

(b)

things in nature is a paradox related to two remarkable connecting links discovered in this century: the connection between mass and energy and the connection between waves and particles. These subtle connections, which have so strongly altered man's view of the small-scale world, will be discussed in later chapters.

Two kinds of neutrino

Not all of the newly discovered particles have come as surprises. The electron's neutrino was identified in 1956 as the result of a carefully designed search for it (the experiment will be described in Chapter 27). A few years later (1962) the muon's neutrino, too, showed itself in response to an experiment specifically designed to find it. And in 1964 another theoretical prediction bore fruit, in the form of the discovery of the omega, a strange particle heavier than those previously known. Yet the surprises have been more numerous. During the 1960s, dozens of new supershort-lived particles were discovered. These particles come into existence and vanish again so quickly—in 10^{-20} sec or less—that they cannot even cross an atom, much less move far enough to leave a

Another strange particle: Ω

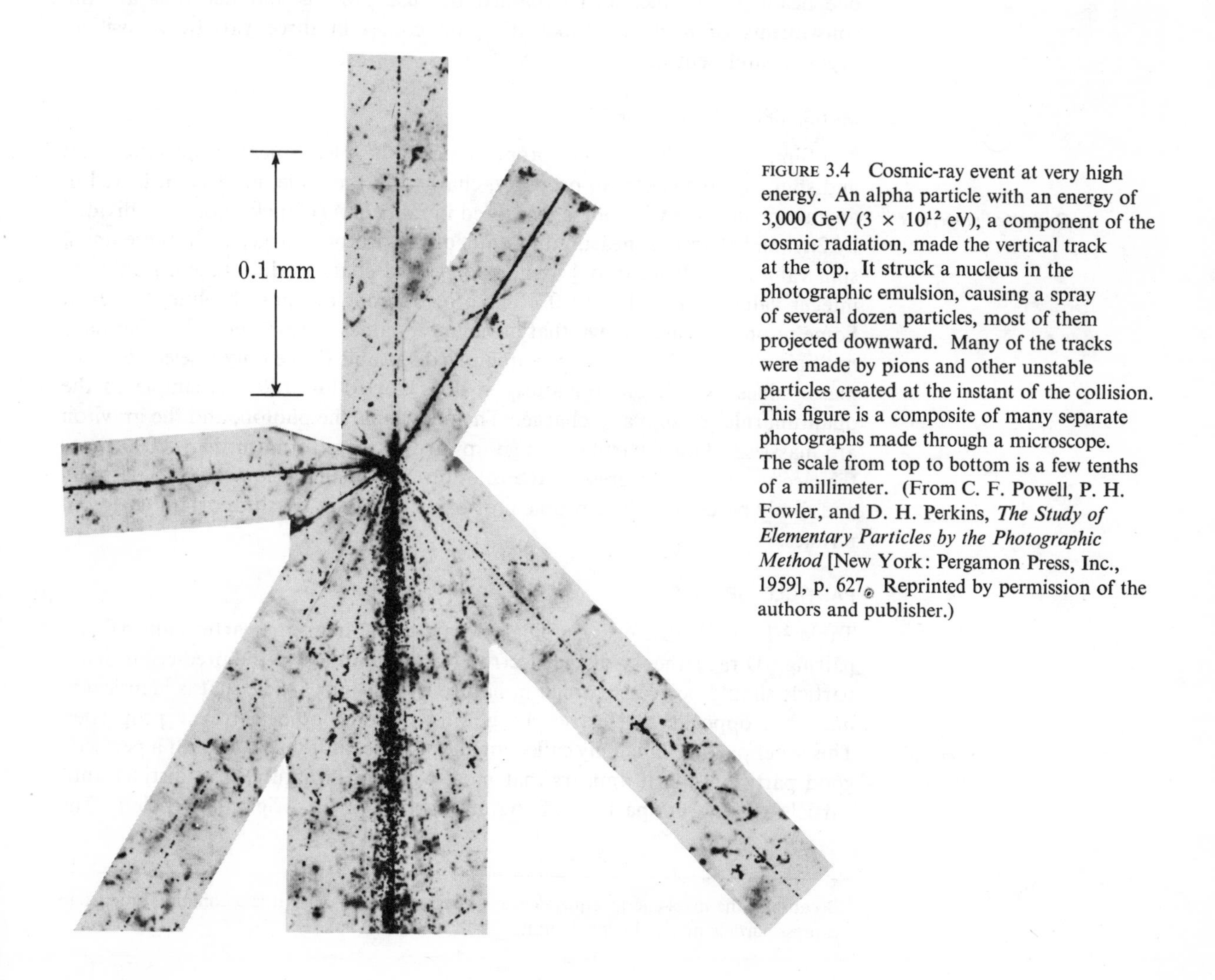

FIGURE 3.4 Cosmic-ray event at very high energy. An alpha particle with an energy of 3,000 GeV (3×10^{12} eV), a component of the cosmic radiation, made the vertical track at the top. It struck a nucleus in the photographic emulsion, causing a spray of several dozen particles, most of them projected downward. Many of the tracks were made by pions and other unstable particles created at the instant of the collision. This figure is a composite of many separate photographs made through a microscope. The scale from top to bottom is a few tenths of a millimeter. (From C. F. Powell, P. H. Fowler, and D. H. Perkins, *The Study of Elementary Particles by the Photographic Method* [New York: Pergamon Press, Inc., 1959], p. 627. Reprinted by permission of the authors and publisher.)

track or trigger a counter. They must be detected by more indirect means, through the products of their rapid decay. Ephemeral though these particles may be, there can be no doubt about their reality. They are elementary structures closely related to the particles that happen to be dignified by longer lives. Nevertheless, in the interest of avoiding complication, we shall say relatively little about the resonances, the name given to these newly discovered objects.

New supershort-lived particles are called resonances

3.5 The properties of the particles

Table 3.1 lists the names and principal identifying characteristics of all of the known particles, excluding the resonances, and of one unknown particle, the graviton (whose properties can be predicted). Overlooking the graviton, the table shows 35 distinct particles, divided into 13 kinds. Certain very closely related particles are lumped together. The electron and its antiparticle, the positron, for example, are of the same kind. The neutron and proton, along with their opposites, the antiproton and the antineutron, are also united under one heading, the nucleon (so named because protons and neutrons are the constituents of nuclei). Yukawa's pion comes in three varieties: positive, negative, and neutral.

MASS, SPIN, AND CHARGE

Quantized properties

Simpliest of the particle properties are the mechanical properties, mass and spin, and the electrical property, charge. In the table, mass is measured in an energy unit, MeV*; spin is measured in units of $\hbar$ (Planck's constant divided by 2π); and charge is measured in units of the proton charge, e. In these units, spin values vary from 0 to $\frac{3}{2}$ in steps of $\frac{1}{2}$, and charge values take on only the integer values $+1$, -1, and 0. These restrictions are not absolute, however. Some resonances are known that have larger values of spin and some that have multiple charges. Mass, too, is a quantized variable (since only a selected set of mass values exists), but obviously it does not follow rules as simple as the quantum rules of spin and charge. The neutrinos, the photon, and the graviton are massless. Other particles range upward in mass to more than 3,000 times the electron mass. At present, the rules of mass quantization are only partially and dimly perceived. The problem of particle masses is as important as any in physics.

PARTICLE GROUPS

Table 3.1 reveals various groupings of particles. First is the particle-antiparticle pairing. Dirac's theory of the electron (1928–1930) led to the prediction that a particle should be accompanied in nature by a sister particle that is identical in mass but opposite in electric charge and in some other intrinsic properties. This sister particle is usually called an antiparticle (although it is itself a perfectly good particle), and it appears that most particles in nature have distinct antiparticles. (The antiparticle of the antiparticle is the original particle.) The

Antiparticles

* Because of the mass-energy equivalence, expressed by $E = mc^2$, it is a common practice to express particle masses in energy units.

TABLE 3.1 SOME OF THE MORE IMPORTANT ELEMENTARY PARTICLES*

Family Name	Particle Name	Symbol	Mass (MeV)	Spin (unit $\hbar$)	Electric Charge (unit e)	Antiparticle	Number of Distinct Particles	Average Lifetime (sec)	Typical Mode of Decay
	Photon	γ (gamma ray)	0	1	0	Same particle	1	Infinite	
	Graviton		0	2	0	Same particle	1	Infinite	
Electron family	Electron's neutrino	ν_e	0	$\frac{1}{2}$	0	$\overline{\nu_e}$	2	Infinite	
	Electron	e^-	0.51100	$\frac{1}{2}$	−1	e^+ (positron)	2	Infinite	
Muon family	Muon's neutrino	ν_μ	0(?)	$\frac{1}{2}$	0	$\overline{\nu_\mu}$	2	Infinite	
	Muon	μ^-	105.660	$\frac{1}{2}$	−1	μ^+	2	2.20×10^{-6}	$\mu^- \to e^- + \overline{\nu_e} + \nu_\mu$
Mesons	Pion	π^+	139.58	0	+1	π^- same as the particles		2.60×10^{-8}	$\pi^+ \to \mu^+ + \nu_\mu$
		π^-	139.58	0	−1	π^+	3	2.60×10^{-8}	$\pi^- \to \mu^- + \overline{\nu_\mu}$
		π^0	134.97	0	0	π^0		0.8×10^{-16}	$\pi^0 \to \gamma + \gamma$
	Kaon	K^+	493.8	0	+1	$\overline{K^+}$ (negative)	4	1.24×10^{-8}	$K^+ \to \pi^+ + \pi^0$
		K^0	497.8	0	0	$\overline{K^0}$		0.86×10^{-10} and†	$K^0 \to \pi^+ + \pi^-$
								5.2×10^{-8}	$K^0 \to \pi^+ + e^- + \overline{\nu_e}$
	Eta	η	549	0	0	Same particle	1	2×10^{-19}	$\eta \to \gamma + \gamma$
Baryons	Nucleon	p (proton)	938.26	$\frac{1}{2}$	+1	$\overline{p}$ (negative)	4	Infinite	
		n (neutron)	939.55	$\frac{1}{2}$	0	$\overline{n}$		930	$n \to p + e^- + \overline{\nu_e}$
	Lambda	Λ^0	1,115.6	$\frac{1}{2}$	0	$\overline{\Lambda^0}$	2	2.5×10^{-10}	$\Lambda^0 \to p + \pi^-$
	Sigma	Σ^+	1,189.4	$\frac{1}{2}$	+1	$\overline{\Sigma^+}$ (negative)		0.80×10^{-10}	$\Sigma^+ \to n + \pi^+$
		Σ^-	1,197.4	$\frac{1}{2}$	−1	$\overline{\Sigma^-}$ (positive)	6	1.5×10^{-10}	$\Sigma^- \to n + \pi^-$
		Σ^0	1,192.5	$\frac{1}{2}$	0	$\overline{\Sigma^0}$		About 10^{-20}	$\Sigma^0 \to \Lambda^0 + \gamma$
	Xi	Ξ^-	1,321.3	$\frac{1}{2}$	−1	$\overline{\Xi^-}$ (positive)	4	1.7×10^{-10}	$\Xi^- \to \Lambda^0 + \pi^-$
		Ξ^0	1,315	$\frac{1}{2}$	0	$\overline{\Xi^0}$		3.0×10^{-10}	$\Xi^0 \to \Lambda^0 + \pi^0$
	Omega	Ω^-	1,673	$\frac{3}{2}$	−1	$\overline{\Omega^-}$ (positive)	2	1.3×10^{-10}	$\Omega^- \to \Xi^0 + \pi^-$
							36		

*This table includes all known particles that do not decay by means of the strong interactions.
†The K^0 meson has two different lifetimes. All other particles have only one.

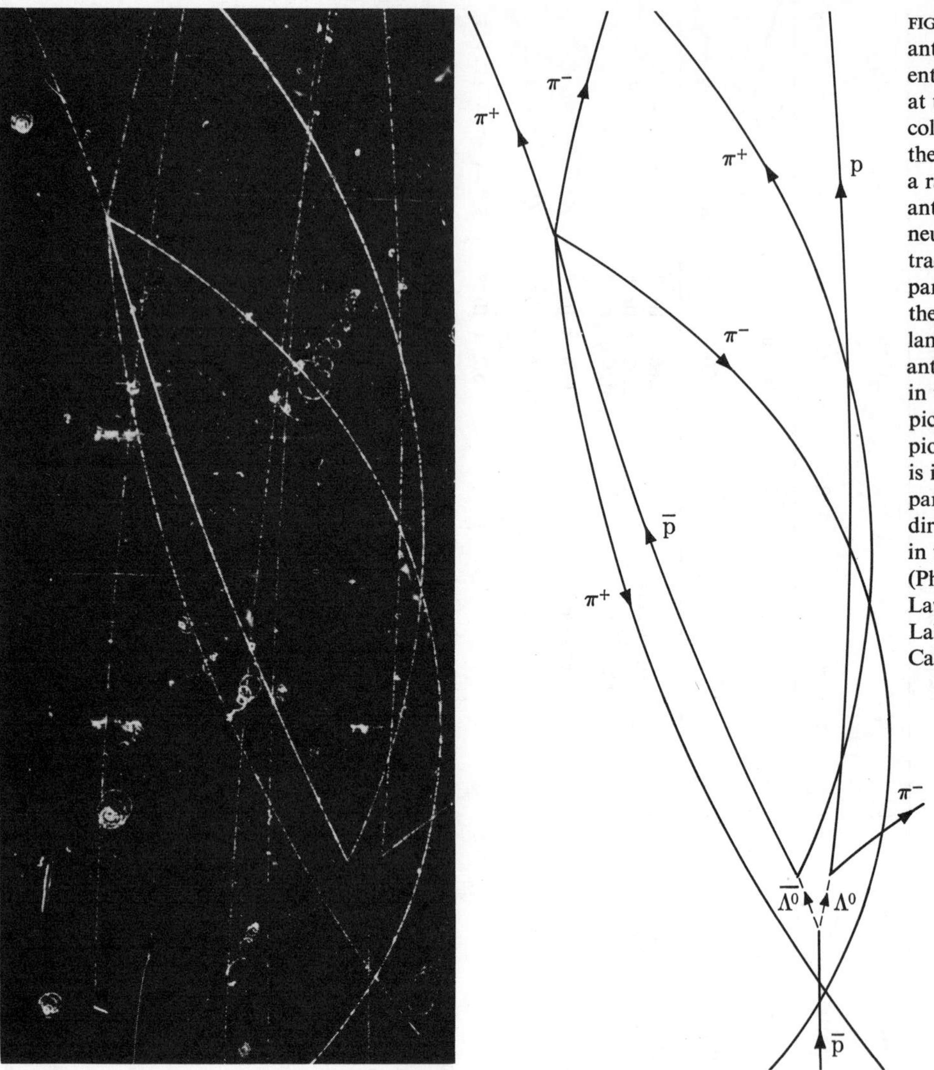

FIGURE 3.5 Particles and antiparticles. An antiproton entering the bubble chamber at the bottom vanishes in collision with a proton in the chamber, giving rise, in a rare event, to a lambda-antilambda pair. These neutral particles leave no tracks, but decay into charged particles that do. One of the products of the antilambda decay is another antiproton whose annihilation in the upper left part of the picture produces a spray of pions. Because the chamber is in a magnetic field, positive particles are deflected in one direction, negative particles in the other direction. (Photograph courtesy of Lawrence Radiation Laboratory, University of California, Berkeley.)

photon, the graviton, the neutral pion, and the eta are special; for each of them, the particle and antiparticle are identical. For all the others, particle and antiparticle differ. The antineutron, for example, is distinguishable from the neutron even though both are neutral.* The proton and antiproton are even more easily distinguishable since one is positively charged and the other negatively charged. A horizontal bar over a particle symbol is used to designate the antiparticle.

To explain the why of antiparticles is beyond the scope of this text. Dirac did not set out to predict such entities, for until 1932 there was no evidence for their existence. Rather, he sought to construct a theory that conformed both to

* Neutron and antineutron have opposite magnetic properties. They also interact very differently with protons.

quantum mechanics and to relativity. From his successful merger of these theories he was led, to his own surprise, to the idea of antiparticles.* He supposed at first that protons might be antielectrons, but by 1931 it had become theoretically clear that particle and antiparticle must have identical mass. The discovery of the positron shortly provided vindication of these ideas. There is now ample evidence for many other antiparticles. Figure 3.5 is an interesting photograph showing tracks of an antilambda and two antiprotons.

A striking example of the power of theory

Another particle grouping is referred to as charge multiplets. The three sigma particles, for instance, form a charge triplet whose members have different charge but nearly the same mass and whose other properties are either identical or very similar. It is natural, and it proves fruitful, to think of these three particles as different manifestations of the same particle. The pions form another triplet, electrically different but otherwise alike. Proton and neutron form a charge doublet. High-precision experiments have shown them to have equal properties of strong nuclear interaction despite their electrical difference. Not all particles have partners of different charge; the lambda and the omega are examples of charge singlets.

Charge multiplets

A broader grouping divides particles into families: the baryons (heavy particles), the mesons (intermediate particles), and the smaller electron and muon families. Notice in Table 3.1 that the mesons have zero spin (other mesons are known among the resonances with integer spin), whereas the members of the other families have half-odd-integer spin ($\frac{1}{2}, \frac{3}{2}, \ldots$). More important than either mass or spin, however, are conservation laws that set these families apart. Whenever one baryon disappears, another one appears in its place. Similar laws apply to the muon family and the electron family but not to the mesons, which, like photons, can be created and annihilated in arbitrary numbers. Implications of the three family-number conservation laws will be considered in Chapter 4.

Family groups

Particles of spin $\frac{1}{2}, \frac{3}{2}, \ldots$ obey conservation laws of number

Particles of spin 0, 1, 2, ... are not conserved in number

STABILITY AND INSTABILITY

The remaining information in Table 3.1 concerns particle decay (Figure 3.6). The typical fate of a particle is to decay spontaneously into two or more other particles after a mean lifetime of about 10^{-10} sec. Resonance lifetimes are much less than this. A few of the particles in Table 3.1 live longer, such as the muon, which has a mean lifetime of 2×10^{-6} sec (2 μsec), and the extraordinarily long-lived neutron, with a lifetime of about 1,000 sec.

Instability is normal

In the world of particles stability is a rarity, yet it is an essential rarity for the existence of a material world. Although stable, the massless particles are not suitable building blocks of matter because they move always at the speed of light and can be stopped only by being destroyed. This leaves only the electron and the proton, both of which are stabilized by conservation laws (see Chapter 4). In addition, by a stroke of good luck, the neutron can be stabilized when joined

Stability is rare

Stabilization of the neutron

* This surprising outcome of theoretical reasoning has caused Dirac to join Einstein in predicting that the deepest insights of the future may come not from efforts to organize and understand experiments but rather from purely mathematical efforts to simplify and generalize the structure of theory.

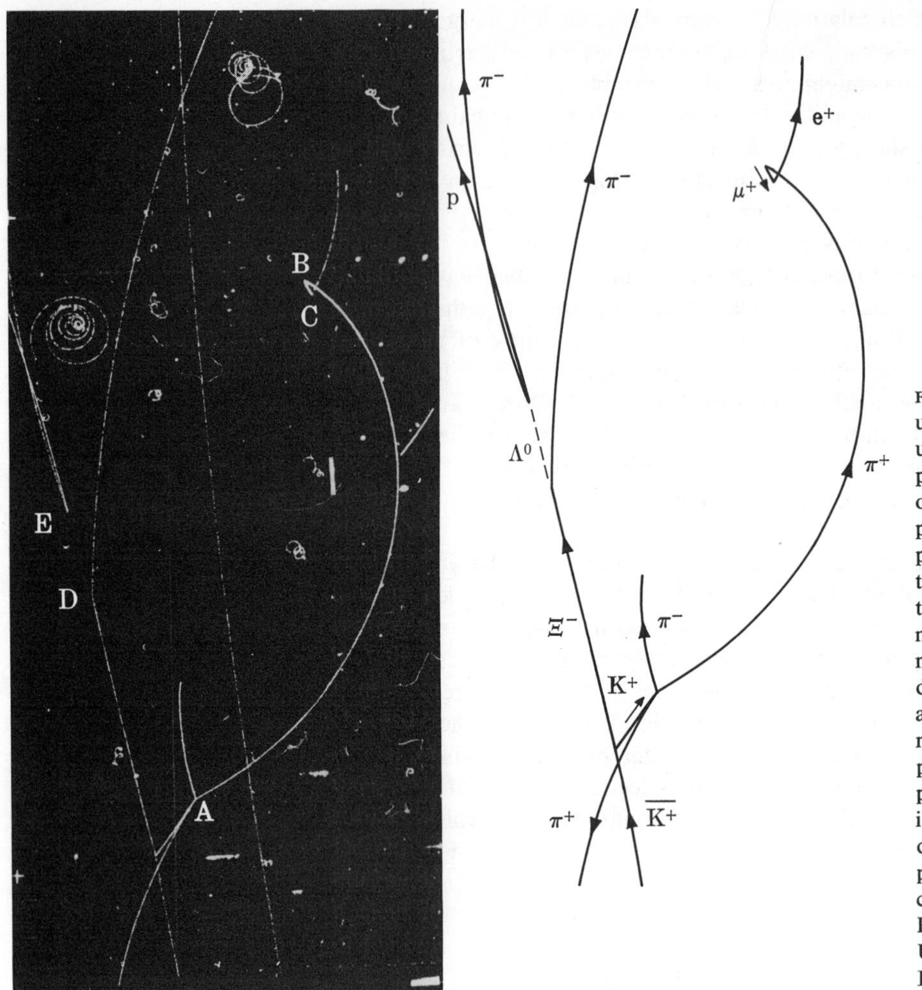

FIGURE 3.6 Decay of unstable particles. This unusual bubble-chamber photograph shows the decay of five different elementary particles. At point A, a positive kaon decays into three pions. At B, one of these pions decays into a muon and an unseen neutrino. At C, the muon decays into a positron (plus a neutrino and an anti-neutrino). At point D, a xi particle decays into a lambda particle and a pion. The invisible neutral lambda decays into a proton and a pion at point E. (Photograph courtesy of Lawrence Radiation Laboratory, University of California, Berkeley.)

with protons. The nuclear force holding these particles together in nuclei is strong enough to lower the energy of the neutron and prevent its decay. Exactly how this effect, which depends on the mass-energy equivalence, comes about will be discussed in Chapter 25. Without this stabilizing effect, the world would contain only hydrogen. The existence of every other substance in the world depends upon nuclear forces being powerful enough to stabilize the normally unstable neutron, thereby making it available as a universal building block.

The information in Table 3.1 does not exhaust the important properties of the particles. However, it is enough to identify and classify them, and it is as much as will be needed as the particles are used in illustrative examples.

★3.6 Experimental tools

The arena of modern science lies beyond direct perception

Modern science got its start with the study of objects visible to the human eye—the sun, the planets in the night sky, wooden balls rolling down inclined surfaces. The most marked characteristic of contemporary science is that it deals with domains of nature outside the range of human perception—the interior of stars, the structure of molecules, the interactions of elementary particles. Man's picture of nature must now rest on the indirect evidence supplied by instruments of his own design rather than on the direct evidence of his senses.

Actually, the fact that we cannot directly perceive the parts of nature we now study is a much less revolutionary aspect of contemporary science than it might at first appear to be. The true revolution of science was wrought in the sixteenth and seventeenth centuries when man supplemented the subjectivity of perception with the objectivity of painstaking, quantitative measurements. Our present-day dependence on instruments and machines, whose remarkable complexity and refinement reveal phenomena beyond the range of human perception, is only the natural evolutionary result of turning from qualitative observation to quantitative measurement. There has been a revolution in twentieth-century physics, but it is a revolution in our view of nature and our outlook on the form of natural laws and not really a revolution in how we study nature. The great surprise and insight afforded by science in this century is the fact that the laws of nature governing domains beyond the range of human perception violate common sense. Nature at its most fundamental behaves quite differently from nature in the macroscopic world. Extrapolation of observational techniques to new domains has been successful, but extrapolation of ideas consistent with human perception has not worked. This is the true revolution of recent scientific advance, and it will be a dominant theme in Parts 6 and 7 of this text.

The use of technical aids to observation is old

The ideas uncovered are new

Man's perceptions are limited in scope. More important for the scientist, they are severely limited in accuracy. It is by no means necessary to leave the macroscopic world in order to go beyond the range of human perception. A butcher who weighs a piece of meat on a scale is turning to a technical device to extend the range and precision of his senses. That we cannot "see" an atom is fundamentally no more serious a problem than our not being able to weigh a piece of meat by hand. In either case we must supplement human perception by mechanical devices, making us form a judgment about something in the world outside ourselves based on information supplied by these devices. Even the simple measuring tape of a housewife is a technical aid to observation. Throughout the whole history of modern science man has leaned heavily on machines, devices, and equipment whose sole purpose was to extend the range and precision of his sense perceptions. In the sixteenth century, Tycho de Brahe needed what we would now call a giant protractor to measure angles of inclination of stars and planets in the sky. Later, Galileo, in studying motion, needed an accurate technique of measuring time intervals and devised an improved method of using the water clock. With every step forward in fundamental understanding, man learned how to base new experimental equipment on his new knowledge. With every new device he could probe new domains or measure old domains more accurately, laying the experimental basis for still deeper understanding. In studying elementary particles and seeking a pattern of nature in the subatomic domain, contemporary man uses a remarkable array of technological aids to observation. These devices, the most recent and the most complicated in a long history of mechanical links between man and nature, rest on every earlier fundamental theory of nature.

Positive feedback: New knowledge provides new tools to aid the search for new knowledge

ACCELERATORS

The devices used in elementary particle research are basically of two kinds, those used to create particles and those used to detect particles. The creation of material particles requires energy, and that energy must be highly concentrated within dimensions no greater than those of an atomic nucleus. This means that the energy itself be carried by a particle. By repeated application of electric forces, an accelerator pushes charged particles to high energy. The energetic particles, concentrated in a narrow beam, strike atomic nuclei in a target; there they undergo interactions or create other particles that fly on to interact in another target or in a detector.

FIGURE 3.7 The National Accelerator at Batavia, Illinois. (a) An air view shows the main ring of the accelerator, with the smaller ring of the booster accelerator in the right foreground. (b) Magnets encase the beam tube along most of its 4-mile circumference. (Photographs courtesy of National Accelerator Laboratory.)

The earliest accelerators were called cyclotrons, Cockcroft-Walton accelerators, and Van de Graaf accelerators. The principle of operation of the cyclotron will be discussed in Chapter 16. Such accelerators are still in operation, but they do not achieve energies as great as more recent machines, called linear accelerators and synchrotrons. So far no one has succeeded in accelerating neutral particles or unstable particles—the first because charge is the only convenient "handle" for pushing a particle, the second because even the longest lived unstable particle with charge, the muon, lives but two-millionths of a second. Projectiles in accelerators have therefore been limited to electrons, protons, and atomic nuclei.

Much basic physics underlies accelerator operation

In the National Accelerator in Batavia, Illinois (Figure 3.7), the acceleration of protons to 500 GeV is an achievement resting upon the theories of mechanics and electromagnetism, the more recent theory of relativity, and a host of engineering and technical developments of this century. Electromagnetism enters in several vital ways. Bombardment of hydrogen atoms by low-energy electrons strips away the orbital electrons from the atoms, leaving the central proton alone. In a linear accelerator electric forces pull the freed protons down a straight tube until they are "injected" at modest energy (200 MeV) into the slender evacuated doughnut-shaped tube of the booster synchrotron—about 470 m around its periphery. Here a weak magnetic field causes the protons to be deflected in such a way that they follow around the curve of the beam tube without striking its walls. At intervals the protons receive impulses of electric force to accelerate them forward. At the same time the strength of the magnetic field is gradually increased, for as the protons gain energy a stronger magnetic field is required to bend their trajectories along the same curved path. After $\frac{1}{15}$ sec, the protons, having acquired an energy of 8 GeV, are guided into the tube of the main accelerator ring—5 cm $\times$ 12 cm in cross section and 6.3 km long—where the processes of magnetic deflection and electric acceleration continue. Ten seconds later, after about 500,000 trips around the ring, the protons strike a target and the cycle can begin again. At 500 GeV, the protons are traveling at 99.9998 percent of the speed of light. Since relativity is that theory describing the motion of high-speed objects, it is clear that the successful acceleration of the protons to such enormous speed while guiding them within a 5-cm channel over a distance of 3 million km would have been out of the question without an understanding of relativity.

DETECTORS

J. J. Thomson detected electrons by observing a fluorescent glow where the beam of electrons struck the glass wall of his cathode-ray tube. At about the same time (around 1900) Ernest Rutherford detected alpha particles by noticing that they electrified the air in passing through it and thereby caused an electrified object (an electroscope) to lose its charge to the air more rapidly than otherwise. Very early in this century methods were discovered to detect single particles rather than the cumulative effect of many particles. Rutherford observed through a microscope the tiny flashes of light at points on a zinc-sulphide screen where alpha particles struck. In 1908, his assistant, Hans Geiger, invented a primitive version of what we now call the Geiger counter. Within the Geiger

Single particles detected since early 1900s

The Geiger counter

counter a highly charged central wire is almost but not quite able to cause a spark to jump to or from the walls of the counter tube. The electrification of the gas within the tube caused by the passage of a single energetic charged particle through the counter is enough to trigger the spark and record the particle. Other counters have been invented since then. Especially popular detectors of more recent invention allow the experimenter not merely to detect the passage of a single particle by means of a single pulse of energy or light but to actually see (or photograph) the path of the particle as it passes through the detector. Such detectors include the cloud chamber, the spark chamber, and the bubble chamber. A cloud-chamber photograph appears in Figure 3.1, bubble-chamber photographs in Figures 3.5 and 3.6, and a spark-chamber photograph in Figure 3.9.

Cloud chambers, spark chambers, and bubble chambers show paths of particles

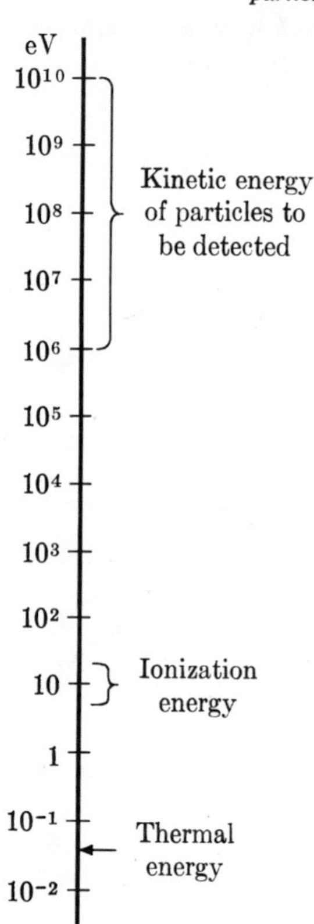

Almost every method of particle detection rests upon the fact that the particle to be detected has an energy far larger than the average energy of the atoms and molecules of the matter through which the particle is passing. The constituent atoms and molecules of ordinary matter have a ceaseless, random motion. Like a restless flock of sheep they mill about, colliding frequently with each other. The more energetic particle charges into their midst like a wolf, literally tearing some atoms apart and leaving behind a swath of destruction. (The difference is that the atoms quickly repair the damage and resume their normal behavior.) To make this comparison quantitative, we note that atoms at normal temperature move about with an energy of about 0.04 eV, whereas the particles to be detected might have energies of millions or even billions of eV. The energy required to strip an electron away from an atom is about 10 eV, several hundred times greater than the energy of random motion of the atoms. Thus in the jostling of one atom against another, not enough energy is involved in the collision to set free an electron, and the atoms remain neutral. But an energetic particle flying through the material can strip off one electron after another as it collides with successive atoms. It leaves behind a trial of freed electrons and positively charged atoms (ions). This process, called ionization, is responsible for the deleterious effects of high-energy radiation in living cells. It also makes it relatively easy to detect a high-energy charged particle.

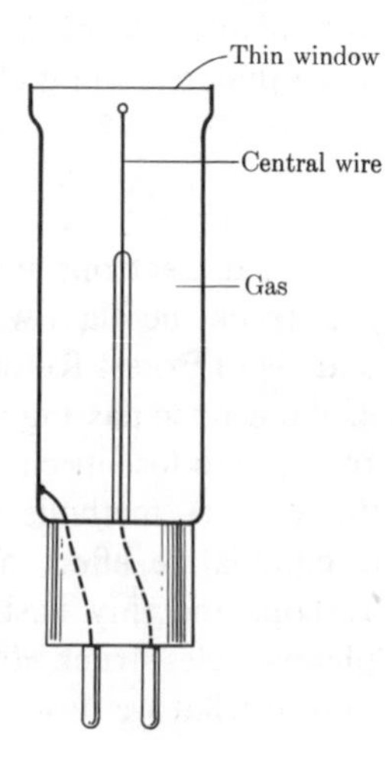

FIGURE 3.8 The essential ingredients of a small laboratory Geiger counter: a central wire maintained at high voltage relative to the wall of the tube, a low-density gas able to conduct a short burst of current, and a thin window to allow particles of relatively low energy to enter the tube.

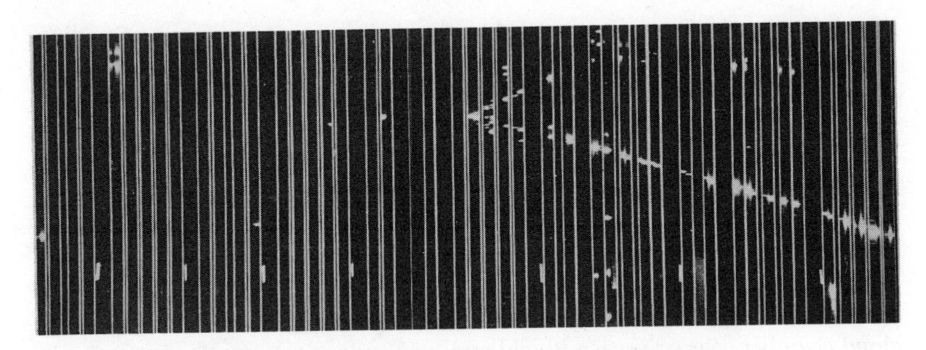

FIGURE 3.9 Spark-chamber photograph. An unseen neutrino entering the chamber from the left interacted with a neutron within an atomic nucleus in one of the plates to produce a muon and a proton. The longer line of sparks traces the path of the muon; the shorter path is that of the proton. [Photograph courtesy of European Organization for Nuclear Research (CERN).]

Operation of the Geiger counter

In a Geiger counter (Figure 3.8), the ions and electrons left in the trail of the energetic particle are set into motion by the electric field acting between the central wire and the outer wall of the counter. They gain enough energy to do some more ionizing themselves, and a cascade of electric current rapidly builds up; that is, a spark jumps the gap. A similar principle operates in the more modern spark chamber. The wire and the wall of the Geiger counter are replaced by two parallel metal plates, which are also charged in such a way that a spark is almost but not quite able to jump the gap between them. Then many more plates are added parallel to the first pair. They are charged alternately positive and negative to make possible the jumping of a spark between any plate and either of its neighbors. In research applications, the plates are charged for only a few millionths of a second just after the passage of one or more particles through the chamber. A sufficiently energetic particle can penetrate the entire stack of parallel plates. Each time it crosses the gas-filled gap between two plates, it leaves a trail of ions that serves as a localized trigger of sparking. Looking in from the side (Figure 3.9), one sees a series of sparks that provide a visual record of the path followed by the particle.*

The spark chamber also uses high voltage

Ions are centers of condensation in a cloud chamber

The process of ionization is also crucial to the operation of the cloud chamber and the bubble chamber. In the cloud chamber, ions left in the wake of the fast charged particles form centers of condensation for vapor in the gas-filled chamber. Tiny liquid droplets form in a line along the track of the particle and are photographed before they have a chance to drift away. (Ordinary water droplets in the air are also frequently centered about ions that are formed by random cosmic rays or by ultraviolet light from the sun.) The particle trails seen in a bubble chamber are gas bubbles in a liquid rather than

* In another kind of spark chamber, called a wire chamber, a computer instead of a camera locates the sparks. The chamber contains arrays of charged wires. When a spark jumps between any pair of wires, the resulting pulse of current sends a signal to a computer. From many such signals, a computer program determines the paths of charged particles through the chamber.

Ions are centers of boiling in a bubble chamber

liquid droplets in a gas. In the bubble chamber, ions act as centers of boiling. With timing even more critical than in a cloud chamber, the thin line of boiling liquid along the trail of the particle is photographed before the boiling spreads throughout the liquid. It should be added that both the cloud chamber and the bubble chamber must be prepared to be in a specific critical condition just before the photograph is taken. The vapor in the cloud chamber must be supersaturated. It is more than ready to condense into liquid and will readily do so on any convenient center of condensation, such as an ion. In a somewhat similar way, the bubble-chamber liquid must be superheated; that is, it must be more than ready to boil. Supersaturation is achieved by sudden cooling of the cloud chamber. Superheating is achieved by a sudden release of pressure in the bubble chamber. In Figure 3.10 one of the first bubble chambers is compared with a more recent installation.

The detection of neutral particles

In passing through matter, a gamma ray ionizes only very weakly. It can be detected by a Geiger counter and by some other counters, but it leaves no visible track in a bubble chamber or spark chamber. The paths of other neutral particles also remain invisible in these chambers, although their presence can be inferred indirectly (such inference is possible in the photographs of Figures 3.5 and 3.6). To be detected, a neutron must first interact with a nucleus. This interaction can produce gamma rays or secondary protons, which eventually cause detectable ionization. Only the neutrino now offers a serious challenge to physicists wanting to record particles.

3.7 The significance of the particles

In view of the extraordinarily short lifetime of most of the particles, one may be tempted to ask why they are considered so important. A proton is obviously important; all matter, animate and inanimate, contains protons. But why a lambda particle, which is not a constituent of anything we know? In a sufficiently violent nuclear collision brought about with the help of an accelerator, a lambda particle may be created. It travels a few centimeters in less than one-billionth of a second, then decays into a nucleon and a pion. The pion, after a slightly longer time, decays into a muon and a neutrino. The muon soon decays into an electron, a neutrino, and an antineutrino. In a total time of about a millionth of a second, and all within a few feet of the point of the initial collision, a sequence of transient particles has been born and has died, with no net effect beyond the addition to the universe of a few more neutrinos.

There are two reasons why physicists believe that the short-lived unstable particles are fully as important and interesting as the few stable particles that compose our world. In the first place, the unstable particles may have a vitally important effect on the properties of the stable particles. To give the most important example, the force that holds nuclei together (and therefore makes possible the existence of all atoms heavier than hydrogen) arises from the exchange of unstable pions between the nuclear particles. The second, perhaps deeper, reason is that it appears to be entirely a matter of "chance" which particles are stable and which are unstable. The muon and the electron, for example, appear to be nearly identical in all respects except that the muon happens to be heavier than the electron. It can therefore release its extra mass

Unstable particles contribute crucially to the forces between stable particles

Stability and instability seem to be matters of "chance"

energy and decay spontaneously into an electron (and neutrinos). The muon lives two-millionths of a second; the electron apparently lives forever. Yet to the physicist this difference is less striking than the many points of similarity between muon and electron. It seems very unlikely that the "true" nature of the electron will ever be understood unless the closely related muon is understood, too.

All the elementary particles seem to belong to one big family. None of them is independent of all the others. The "normal" thing is for a particle to undergo decay and transmute itself into other lighter particles. For reasons that are not yet fully understood, there are two "abnormal" particles, the proton and the electron, which are prohibited from decaying. According to this larger view of the particles, there are certain rules of nature, described in Chapter 4, that prevent the decay of these two particles. Because of this chance, the construction of a material world is possible.

Of course, since there is only one universe and one set of natural laws, it makes little sense to say that a particular state of affairs in the world exists by chance. But this view of the multiplicity of particles continues the process, begun by Copernicus, of making man feel more and more humble when facing the design of nature. We and our world exist by the grace of certain conservation laws that stabilize a few particles and permit an orderly structure to be built upon the normal chaos of the submicroscopic world.

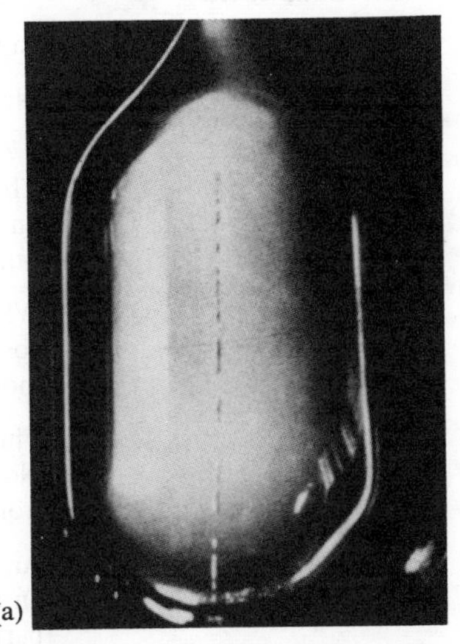

(a)

(b)

FIGURE 3.10 Bubble chambers. (a) An early model (1953) of the bubble chamber invented by Donald Glaser. It is a little more than 1 in. long and less than half an inch across. (b) A modern installation, the 80-in. liquid hydrogen bubble chamber at Brookhaven. [(a) Photograph courtesy of Donald Glaser, University of California, Berkeley. (b) Photograph courtesy of Brookhaven National Laboratory.)]

Summary of ideas

The submicroscopic domain, inhabited by elementary particles, is probably the most fundamental frontier of physics.

With some simple properties, particles can usefully illustrate many laws of physics; with some mysterious properties, particles put us in touch with the unknown.

The electron, the first subatomic particle to be discovered, showed itself in cathode rays.

The proton is the simplest atomic nucleus.

Nuclei are composed of protons and neutrons.

The positron is an antielectron.

The photon, a quantum of radiant energy, is an elementary particle without mass.

In beta decay, electrons (or positrons) are created, together with antineutrinos (or neutrinos).

The pion contributes to the exchange force between nucleons.

The muon is a mysterious heavy partner to the electron.

Some of the "strange" particles have long lifetimes of about 10^{-10} sec.

Most of the newly discovered particles, known as resonances, have short lifetimes of less than 10^{-20} sec.

Mass, spin, and charge are quantized particle properties.

Various family groupings help to unify the study of particles.

Accelerators and detectors are the primary tools of particle research.

Links among particles suggest that transitory particles are fully as important to physical theory as the few stable particles.

QUESTIONS

Section 3.1

Q3.1 In what ways is an elementary particle "simpler" than a golf ball?

Q3.2 Name several objects within the realm of your own experience that may be "explained" by describing their composition in terms of smaller units.

Q3.3 The physicist explains the structure of an object in terms of its parts. Cite another way in which structure may be "explained," and illustrate this other way with an example.

Q3.4 What are "final causes" (sometimes called teleology)? Why are final causes consistent with Aristotle's organic view of nature but inconsistent with the modern scientific view of nature?

Q3.5 Find and examine a list of Nobel Prize winners in physics. How many Nobel Prizes have been awarded for discoveries on the submicroscopic frontier of physics? What fraction is this of the total number of prizes?

Section 3.2

Q3.6 An alpha particle is the same thing as a nucleus of a helium atom (^{4}He). Is an alpha particle an elementary particle? Is a beta particle an elementary particle? Explain.

Q3.7 What aspect of Rutherford's experiments with alpha particles suggested that an atom possesses a small positively charged core, or nucleus? (Consult another text if necessary, or see Section 26.2 and Figure 26.3.)

Q3.8 How did Anderson know that the track left in his cloud chamber [(Figure 3.1(a)] was made by a positron going from bottom to top and not an electron going from top to bottom?

Q3.9 In the early days of quantum theory, what we now call photons were called "quanta" and were distinguished from material particles or "corpuscles." Why is such a distinction no longer made?

Q3.10 Do you think it would be possible to make an "atom" out of a proton and a negative muon? out of a proton and a negative pion? out of a positive electron and a negative electron? out of a proton and a positron?

Section 3.3

Q3.11 About the still hypothetical pion, Yukawa wrote: "We can show, however, that in the ordinary nuclear transformation, such a quantum cannot be emitted into outer space." The "ordinary nuclear transformation" to which he referred is one in which the nucleus loses no more than a few MeV of energy. (1) Use energy considerations to explain the meaning of Yukawa's statement. (2) Now we know that a nucleus *can* be caused to emit a pion. How does this come about?

Q3.12 Why is a process in which one pion strikes another pion nearly impossible to achieve in the laboratory? Can you think of a way in which physicists might learn something about the interaction of one pion with another?

Q3.13 Explain exactly what it means to say "the mass of the electron is 0.511 MeV." Is this really the *mass* of the electron?

Q3.14 Our part of the universe—perhaps all of the universe—contains many more particles than antiparticles (for example, many more electrons than positrons). What might the world be like if particles and antiparticles existed in about equal numbers?

Section 3.5

Q3.15 Some past advances in physics have come mainly from the evidence of accumulated experimental data. Some have come more from a philosophical-mathematical consideration of the way nature "ought" to behave. If you had to place a bet on the way in which deeper insight into the inner structure of elementary particles will be gained, which of these two possibilities would you choose? Why? (Your answer to this question must of course be a guess, based on very little information. However, it will be of interest to your instructor for what it reveals about your outlook on science.)

Q3.16 Mention several ways in which you use technological aids to observation to increase either the accuracy or the scope of your sense perceptions.

Section 3.6

Q3.17 Give any example you may have learned about of a device whose invention rested on basic science and which itself became a tool of basic science. (Do not repeat examples given in the text.)

Q3.18 Why are accelerators of high energy needed for elementary particle research?

Q3.19 Explain why an elementary particle can leave a visible trail or cause a scintillation flash visible to the unaided eye even though the particle itself is far below the limit of visibility of even the best microscope.

Q3.20 It is stated in the text that a typical ionization energy is about 10 eV. However, in passing through matter, a fast charged particle loses about 30 eV per ionization event. Suggest one or more reasons for the difference between these two numbers.

Q3.21 Comment in a short paragraph on this statement: "It is not reasonable to believe that all the particles we know are truly elementary."

Section 3.7

Q3.22 It was stated in this chapter that the neutron, by a "stroke of good luck," is stabilized within the nucleus and that it is a matter of "chance" which particles are stable. Such seemingly nonscientific phrases reflect a general point of view about nature that goes hand in hand with an actual scientific theory or a particular level of scientific understanding. Give one or more other examples—from your own experience, if you wish—of such a partnership between quantitative understanding and a qualitative point of view.

EXERCISES

Section 3.2

E3.1 (1) When ^{60}Co, whose nucleus contains 27 protons and 33 neutrons, undergoes beta decay (an electron and an antineutrino are emitted), what nucleus is formed? (2) When ^{238}U (92 protons and 146 neutrons) undergoes alpha decay, what nucleus is left behind? (A periodic table appears in Chapter 24.)

Section 3.3

E3.2 According to the theory of exchange forces, the approximate range of the force is equal to $\hbar/mc$, where m is the mass of the particle exchanged. (1) Verify that $\hbar/mc$ (known as the reduced Compton wavelength) has the dimension of length. (2) Evaluate $\hbar/mc$ for the pion. (3) Why should the pion be the most important exchange particle for the nuclear force? (HINT: Does the muon interact strongly with nuclei? Does the electron?)

Section 3.4

E3.3 A resonance particle is created within an atomic nucleus. If it moves at close to the speed of light during its lifetime of about 10^{-20} sec, describe its approximate location at the time it decays.

Section 3.5

E3.4 (1) A charged particle that moves 10 micrometers (microns) in a photographic emulsion leaves a clearly discernible track. If the particle traveled at close to the speed of light (3×10^8 m/sec), how long must it live in order to move 10 micrometers? Compare this time with the average lifetimes of charged particles in Table 3.1. (2) Moving at 10^7 m/sec, how far does a neutral sigma progress in its average lifetime? Is this distance greater or less than the diameter of a single atom?

E3.5 (1) Moving at 10^8 m/sec, which of the particles in Table 3.1 move less than 10 μm before decaying? (2) at the same speed, which particle moves an average distance greater than the diameter of the earth before decaying?

E3.6 Study Table 3.1 in order to answer the following questions: (1) What are the final end products of a Ξ^0 particle after the successive decays have run their course? (2) As one moves from nucleon to omega through successively heavier baryons, what happens to the average electric charge of each particle type?

E3.7 (1) Examine with care the five pion tracks in Figure 3.6. (a) What rule relates the sign of the charge to the direction of curvature? (b) What is the correlation between intensity of track and speed of particle? (A slower particle is more strongly deflected.) (2) A small spiral track in the upper left part of the picture was made by an electron. Was this electron positive or negative? How can you deduce its direction of motion?

Section 3.6

E3.8 Without *any* measuring devices (clocks, rulers, scales, etc.), about how accurately can you estimate (a) times of a few seconds; (b) times of a few hours; (c) distances of a few inches; (d) distances of several yards; and (e) masses of a few pounds? (SUGGESTION: Make some such estimates, then check your accuracy with measuring devices. Report the results.)

E3.9 If an alpha particle passing through a cloud chamber loses 30 eV of energy for each molecule that it ionizes, how many ions does it leave in its wake for each MeV of energy lost?

PROBLEMS

Nuclear structure

P3.1 According to theory, if the neutron and the proton had equal mass and if only the strong nuclear force acted within nuclei, (a) all stable nuclei would have $N \cong Z$ (neutron number approximately equal to proton number), and (b) the periodic table of the elements would not end because stable nuclei of arbitrarily high atomic number could exist. With this as background, account for the following facts: (1) For light nuclei (e.g., ${}^{16}_{8}O$), $N \cong Z$. (2) For heavy nuclei (e.g., ${}^{238}_{92}U$), $N > Z$. (3) The nucleus ${}^{3}_{2}He$, with $Z = 2$ and $N = 1$, is the only stable nucleus with $Z > N$. (4) The periodic table does end; no atoms with $Z > 92$ are found in nature.

Particle masses

P3.2 It has been argued that the masses of elementary particles lie closer to integral multiples of a certain mass unit m_0 than can be explained by chance. The mass m_0 is in the range $65m_e < m_0 < 70m_e$, where m_e is the mass of an electron. Find the "best" value of m_0 in this range, and explain your criterion for choosing it. Compare the particle masses in Table 3.1 with integral multiples of m_0. Do you consider the agreement better than chance? You may give either a somewhat intuitive argument or a mathematical one.

Mean free path

P3.3 Take the diameter of a nucleus to be 5 fm (5×10^{-15} m) and the spacing between nuclei in a solid to be 2 Å (2×10^{-10} m). What is the approximate average distance that a neutron can move through solid matter before it strikes a nucleus? Would you expect charged particles to penetrate a greater or lesser distance in solid matter? Why?

Development of accelerators

P3.4 The characteristics of circular particle accelerators as a function of time are shown in the accompanying table. Assuming the trend continues, describe the projected accelerator scheduled for the year 2000. (SUGGESTION: Plot the data on semilog graph paper.) The cost of recent accelerators has been about 0.1 ¢ per eV (or $1 million per GeV). At this rate, what would be the price of the projected accelerator in 2000? Comment on the reasonableness of the forecast.

Accelerator	Year of Completion	Energy	Diameter
Lawrence/Livingston cyclotron	1932	1.2 MeV	11 in.
Columbia University cyclotron	1938	14 MeV	36 in.
Berkeley synchrocyclotron	1946	350 MeV	184 in.
Berkeley Bevatron	1954	6 GeV	100 ft
Brookhaven AGS	1960	33 GeV	840 ft
National Accelerator	1972	500 GeV	6,560 ft (1.24 mile)

4 Conservation Laws

Over the course of the past few centuries, conservation laws have risen from a minor supporting role in physics to a starring role. What we now understand about the interactions and transformations of particles comes in large part through certain conservation laws that govern elementary-particle behavior. In the large-scale world as well, conservation laws are the source of our deepest insights into the simple regularity of nature. And as often as not, conservation laws have proved valuable tools for solving practical problems.

4.1 Absolute conservation laws

A conservation law specifies constancy during change

A conservation law is a statement of constancy in nature—in particular, constancy during change. If for an isolated system a quantity can be defined that remains precisely constant, regardless of what changes may take place within the system, the quantity is said to be absolutely conserved. Partially conserved quantities, which remain constant only for certain kinds of changes, have also become interesting in recent years, but in this chapter we shall restrict our attention to seven quantities believed to obey absolute conservation laws. These are:

1. Energy (including mass)
2. Momentum
3. Angular momentum, including spin
4. Charge
5. Electron-family number
6. Muon-family number
7. Baryon-family number.

Conservation of the first four quantities was known in classical physics. Conservation laws for the last three have been postulated and tested only in recent decades. Even the classical conservation laws receive their most stringent tests in the submicroscopic world.

Two kinds of quantities are conserved, properties of motion and intrinsic properties

There are in the list two different kinds of quantities, which can be called (a) properties of motion and (b) intrinsic properties of particles; however, the two are not completely separated. Angular momentum includes both orbital angular momentum and intrinsic spin. Energy includes both kinetic energy and mass, an intrinsic property. Momentum, however, is exclusively a property of motion, and the other four are exclusively intrinsic properties.

The interactions and transformations of the elementary particles admirably illustrate the conservation laws, and we shall focus attention on the particles for illustrative purposes in this chapter. Applications to other systems will occur later. The particles provide the best possible testing ground for conservation laws because we expect any conservation law satisfied by small numbers of particles to be satisfied for all larger collections of particles, including the macroscopic objects of our everyday world. Justification for extrapolating the submicroscopic conservation laws on into the cosmological domain is uncertain since gravity, whose effects in the particle world appear to be entirely negligible, becomes of dominant importance in the astronomical realm.

Various intrinsic properties of the particles were discussed in Chapter 3; we shall examine first the conservation laws that have to do with the intrinsic properties.

4.2 Charge conservation

Charge before = charge after

The law of charge conservation requires that the total charge remain unchanged during every process of interaction or transformation. For any event involving particles, then, the total charge before the event must add up to the same value as the total charge after it. In the decay of a lambda into a neutron and a pion.

$$\Lambda^0 \rightarrow \mathrm{n} + \pi^0,$$

the charge is zero both before and after. In the positive pion decay,

$$\pi^+ \rightarrow \mu^+ + \nu_\mu ,$$

the products are a positive muon and a neutral neutrino. A possible high-energy nuclear collision might proceed as follows:

$$\mathrm{p} + \mathrm{p} \rightarrow \mathrm{n} + \Lambda^0 + \mathrm{K}^+ + \pi^+.$$

Neither positively charged proton survives the collision, but the net charge +2 appears on the particles created.

Conservation may be related to quantization

Note that the law of charge conservation provides a partial explanation for the fact that particle charges come in only one size. If the charge on a pion were 0.73 electron charge, it would be quite difficult to balance the books in transformation processes and maintain charge conservation. Actually, according to the present theory of elementary processes, the charge is conserved not only from "before" to "after" but also at every intermediate stage of the process. One can visualize a single charge as an indivisible unit that, like a baton in a relay race, can be handed off from one particle to another but never dropped or divided.*

Stability of the electron

Perhaps the most salutary effect of the law of charge conservation in human affairs is the stabilization of the electron. The electron is the lightest charged particle, and for this reason alone it cannot decay. The only lighter particles, the photon and neutrinos (and graviton), are neutral; a decay of the electron would therefore necessarily violate the law of charge conservation. The stability of the electron is one of the simplest yet most stringent tests of the law of charge conservation. If the law were almost, but not quite, valid, the electron would have a finite lifetime. An experiment reported in 1965 places the electron lifetime beyond 10^{21} years.† Actually, we have no right to say what might happen in 10^{21} years, a time vastly longer than what we call the lifetime of the universe. What this number means experimentally is that in a collection of 10^{21} electrons, no more than one per year undergoes decay. Charge conservation must be regarded as at least a very good approximation to an absolute law.

4.3 Family-number conservation laws

Unlike the other four laws, which were already known in the macroscopic world, the three laws of family-number conservation were discovered through studies of particle transformation. We can best explain their meaning through examples.

BARYON FAMILY

Recall that the proton and all heavier particles in Table 3.1 are called baryons; that is, they belong to the baryon family. In the decay of the unstable Λ particle,

Baryon conservation

$$\Lambda^0 \rightarrow p + \pi^-,$$

one baryon, the Λ, disappears, but another, the proton, appears. Similarly, in the decay of the Σ^0,

$$\Sigma^0 \rightarrow \Lambda^0 + \gamma,$$

the number of baryons is conserved. Note that in one of these examples, a pion is created; in the other, a photon. Pions and photons belong to none of the special family groups and can come and go in any number. In a typical proton-proton collision the number of baryons (two) remains unchanged, as in the following example:

$$p + p \rightarrow p + \Sigma^+ + K^0.$$

---

* Hypothetical particles called quarks, carrying fractional charges of $\frac{1}{3}e$ and $\frac{2}{3}e$, have been postulated. As this text is being written, extensive experimental searches for quarks are underway, and some preliminary reports indicate that they may have been seen. If their existence is established, they will upset neither the law of charge conservation nor the fact of charge quantization. They will merely reveal a quantum unit of charge three times smaller than what we now call the smallest unit.

† For a technical report on this experiment, see M. K. Moe and F. Reines, *Physical Review* **140**, B992 (1965).

These and numerous other examples have made it appear that the number of baryons remains forever constant—in every single event, and therefore, of course, in any larger structure.

Each of the Ω, Ξ, Σ, and Λ particles and the neutron undergoes spontaneous decay into a lighter baryon. But the lightest baryon, the proton, has nowhere to go. The law of baryon conservation stabilizes the proton and makes possible the structure of nuclei and atoms and, therefore, of our world. From the particle physicist's point of view, this is a truly miraculous phenomenon, for the proton stands perched at a mass nearly 2,000 times the electron mass, having an intrinsic energy of about 1 GeV, while beneath it lie several unstable particles. Only the law of baryon conservation holds this enormous energy locked within the proton and makes it a suitable building block for the universe. The proton appears to be absolutely stable. If it is unstable it has, according to experiment,* a half life greater than 2×10^{28} years. To express this result in terms of a probability rather than in terms of such an unimaginably long time: The chance that a proton decays in one year is less than 0.5×10^{-28} (no definite example of proton decay has been seen).

Stability of the proton

The statement that for every baryon that disappears another baryon appears is incomplete because it leaves antibaryons out of the reckoning. A typical antiproton-production event at the Berkeley Bevatron might go as follows:

$$\mathrm{p} + \mathrm{p} \rightarrow \mathrm{p} + \mathrm{p} + \mathrm{p} + \bar{\mathrm{p}}.$$

(The bar over the letter designates the antiparticle. Since the antiproton has negative charge, the total charge of +2 is conserved.) It appears that we have transformed two baryons into four. Similarly, in the antiproton annihilation event,

$$\mathrm{p} + \bar{\mathrm{p}} \rightarrow \pi^+ + \pi^- + \pi^0,$$

two baryons have apparently vanished. The obvious way to patch up the law of baryon conservation is to assign to the antiparticles baryon number −1 and to the particles baryon number +1. Then the law reads: In every event the total number of baryons *minus* the total number of antibaryons is conserved:

Antibaryons and negative baryon number

$$\text{total baryon number} = N_{\mathrm{B}} - N_{\bar{\mathrm{B}}} = \textit{constant}. \tag{4.1}$$

It may seem that with so many arbitrary definitions—which particles should be called baryons, which should not, and which should have negative baryon numbers—it is no wonder that a conservation law can be constructed. Actually, it is not easy to find an absolute conservation law. To find any quantity absolutely conserved in nature is so important that it easily justifies a few arbitrary definitions. The arbitrariness at this stage of history only reflects our lack of any deep understanding of the reason for baryon conservation, but it does not detract from the obvious significance of baryon conservation as a law of nature. Moreover, the mathematics of the quantum theory shows that the use of a negative baryon number for antiparticles is perfectly natural; in fact, it is demanded

Pragmatic justification of negative baryon number

Theoretical justification

* H. S. Gurr, W. R. Kropp, F. Reines, and B. Meyer, "Experimental Test of Baryon Conservation," *Physical Review* **158**, 1321 (1967).

by the theory. This is because the description of the appearance of an antiparticle is "equivalent" (in a mathematical sense we cannot here delve into*) to the description of the disappearance of a particle; conversely, antiparticle annihilation is "equivalent" to particle creation.

ELECTRON FAMILY AND MUON FAMILY

Two more family-number conservation laws

The electron family contains only the electron and its neutrino, the muon family only the muon and its neutrino. For each of these small groups, there is a conservation of family members exactly like the conservation of baryons. The antiparticles must be considered negative members of the families, the particles positive members. These light-particle conservation laws are not nearly as well tested as the other absolute conservation laws because of the difficulties of studying neutrinos, but there are no known exceptions to them.

Neutron decay illustrates several conservation laws

The beta decay of the neutron,

$$\mathrm{n} \rightarrow \mathrm{p} + \mathrm{e}^- + \overline{\nu}_\mathrm{e},$$

nicely illustrates the conservation laws we have discussed. Initially, the single neutron has charge zero, baryon number 1, and electron-family number zero. The oppositely charged proton and electron preserve zero charge; the single proton preserves the baryon number; and the electron with its antineutrino $\overline{\nu}_\mathrm{e}$ together preserve zero electron-family number. In the pion decay processes,

Pion decay

$$\pi^+ \rightarrow \mu^+ + \nu_\mu \quad \text{and} \quad \pi^- \rightarrow \mu^- + \overline{\nu}_\mu,$$

muon-family conservation demands that a neutrino accompany the μ^+ antimuon and that an antineutrino accompany the μ^- muon. The muon, in turn, decays into three particles, for example,

Muon decay

$$\mu^- \rightarrow \mathrm{e}^- + \nu_\mu + \overline{\nu}_\mathrm{e},$$

which conserves the members of the muon family and electron family.

Chaos and order in the particle world

A strong hint emerging from studies of elementary particles is that the only inhibition imposed upon the chaotic flux of events in the world of the very small is that imposed by the conservation laws. Everything that *can* happen without violating a conservation law *does* happen.† Until 1962, there was a notable exception to this rule; its resolution has beautifully strengthened the idea that conservation laws play a central role in the submicroscopic world. The decay of a muon into an electron and a photon,

$$\mu^- \rightarrow \mathrm{e}^- + \gamma,$$

The μ-e-γ puzzle

has never been seen, a circumstance that had come to be known as the μ-e-γ puzzle. Before the discovery of the muon's neutrino it was believed that electron, muon, and one neutrino formed a single family (called the lepton family) with a

---

* In Chapter 27 antiparticles will be described as particles moving backward in time.

† Some physicists borrow a phrase from T. H. White's *The Once and Future King* (New York: G. P. Putnam's Sons, 1939) and express it more strongly: Everything not forbidden is compulsory. Such statements must be regarded as ways of looking at nature, not as laws of nature.

single family-conservation law. Had this been the case, no conservation law would have prohibited the decay of muon into electron and photon, since the lost muon was replaced with an electron and since charge and all other quantities were also conserved. According to the classical view of physical law, the absence of this process ought to have caused no concern. There was, after all, no known law *requiring* that it occur. There was only the double negative: No conservation law was known to prohibit the decay.

However, the view of the fundamental role of conservation laws in nature as the only inhibition on physical processes had become so ingrained in the thinking of physicists that the absence of this particular decay mode of the muon was regarded as a significant mystery. It was largely this mystery that stimulated the search for a second neutrino that belonged exclusively to the muon. The discovery of the muon's neutrino established as a near certainty that the electron and muon belong to two different small families that are separately conserved. The electron and muon being governed by two separate laws of conservation, the prohibition of the μ-e-γ decay became immediately explicable, and the faith that what can happen does happen was further bolstered.

Discovery of the muon's neutrino resolved the puzzle

We turn now to the conservation laws that involve properties of motion (the first three in the list on page 70).

4.4 Energy conservation

The discovery in the 1840s that heat is a form of energy led quickly to the recognition of the general significance of the law of energy conservation. Christened the first law of thermodynamics, it became a pillar of the developing theory of thermodynamics. Since then other forms of energy have been discovered, all interconvertible, like the currencies of the world, and all contributing to the constant energy sum. Throughout chemistry, biology, and engineering, as well as all branches of physics, the law of energy conservation is both a basic law and a practical tool.

Energy: Its many forms give it widespread relevance

Despite this wide range of applications of energy conservation, we shall restrict discussion at this stage to the world of particles, in which only two kinds of energy are relevant, kinetic energy and mass energy. For two particles, for example,

Only two forms of energy for particles

$$E = K_1 + m_1c^2 + K_2 + m_2c^2. \tag{4.2}$$

In transformation processes, kinetic energies and mass energies change, but the total energy is conserved:

$$E_{\text{before}} = E_{\text{after}}. \tag{4.3}$$

For the spontaneous decay of unstable particles, a simple consequence of energy conservation is the requirement that the total mass of the products must be less than the mass of the parent. For each of the following decay processes, the masses on the right add up to less than the mass on the left:

The "downhill" rule of particle decay: Parent mass exceeds product masses

$$K^+ \to \pi^+ + \pi^+ + \pi^-,$$

$$\Xi^- \to \Lambda^0 + \pi^-,$$

$$\Lambda^0 \to \text{n} + \pi^0.$$

These are allowed processes, which are "downhill" in mass; "uphill" decays are forbidden. An unstable particle at rest has only its energy of being, no energy of motion. The difference between this parent mass and the mass of the product particles is transformed into kinetic energy, which the product particles carry away.

Even in motion, a parent particle cannot escape the downhill rule

One might suppose that if the parent particle is moving when it decays, it has some kinetic energy of its own that might be transformed into mass. The conservation of momentum prohibits this. The extra energy of motion is called unavailable energy. To take an extreme case, if all of the energy of the parent particle, kinetic energy plus mass energy, were to be transformed into mass energy of the product particles, these product particles would be motionless (zero kinetic energy). Momentum conservation would surely be violated then because the product particles would have no momentum to match the momentum of the parent particle. It turns out that momentum and energy conservation taken together forbid uphill decays into heavier particles no matter how fast the initial particle might be moving. If two particles collide, on the other hand, some—but not all—of their energy of motion is available to create mass. It is in this way that the various unstable particles are manufactured in the laboratory.

4.5 Momentum conservation

Momentum is purely a property of motion: that is, if there is no motion, there is no momentum. In classical mechanics, momentum is defined as the product of mass and velocity:

Momentum: a vector quantity

$$\mathbf{p} = m\mathbf{v} \tag{4.4}$$

($\mathbf{p}$ is the usual symbol for momentum). In relativistic mechanics (where even massless particles have momentum), its definition is slightly more complicated.* Momentum is simpler than energy in that it has no multiplicity of form, yet in one respect it is more complicated than energy because it is a vector quantity and not a scalar quantity. The addition of vectors, an essential aspect of the law of momentum conservation, will be reviewed in Chapter 6. For present illustrative purposes, we need only take advantage of the fact that the sum of two or more vectors may have a wide range of values, depending on the relative orientation of the vectors. In particular, the sum of two vectors equal in magnitude and opposite in direction is zero.

Consider the law of momentum conservation applied to the decay of a kaon into muon and neutrino,

$$K^+ \rightarrow \mu^+ + \nu_\mu .$$

Two-body decays: Products emerge with equal and opposite momenta in rest frame of parent

Before decay, suppose the kaon is at rest [Figure 4.1(a)]. After decay, momentum conservation requires that muon and neutrino fly off with equal magnitudes of momenta *and* that the momenta be oppositely directed [Figure 4.1(b)]. Only in this way can the vector sum of the two final momenta be equal to the original

* See Equations 21.26 and 21.30.

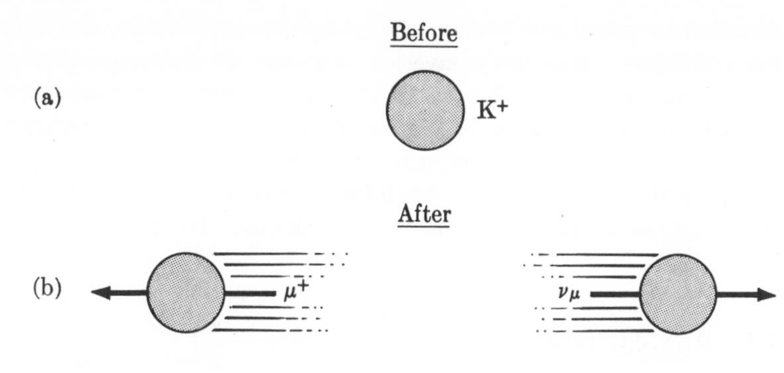

FIGURE 4.1 Momentum conservation in kaon decay. The total momentum is zero both before and after the decay.

momentum, namely, zero. This type of decay, called a two-body decay, is rather common and is always characterized by particles emerging in exactly opposite directions.

Three-body decays: no simple rules of direction

In a three-body decay, the emerging particles have more freedom. Look ahead to Figure 8.14 to see the decay of a kaon at rest into three pions, the tracks pointing in three different directions. The sum of the three final momenta is zero, but there is no fixed rule about their relative orientation. Momentum conservation on a grander scale is shown in Figure 4.2, in which eight particles emerge from a single event.

One-body decays: prohibited by combined laws of energy and momentum conservation

One vital prohibition of the law of momentum conservation is that against one-body decays. Consider the possibility of

$$K^+ \rightarrow \pi^+,$$

the transformation of kaon to pion. It satisfies the laws of charge and family-number conservation. It is consistent with energy conservation because it is

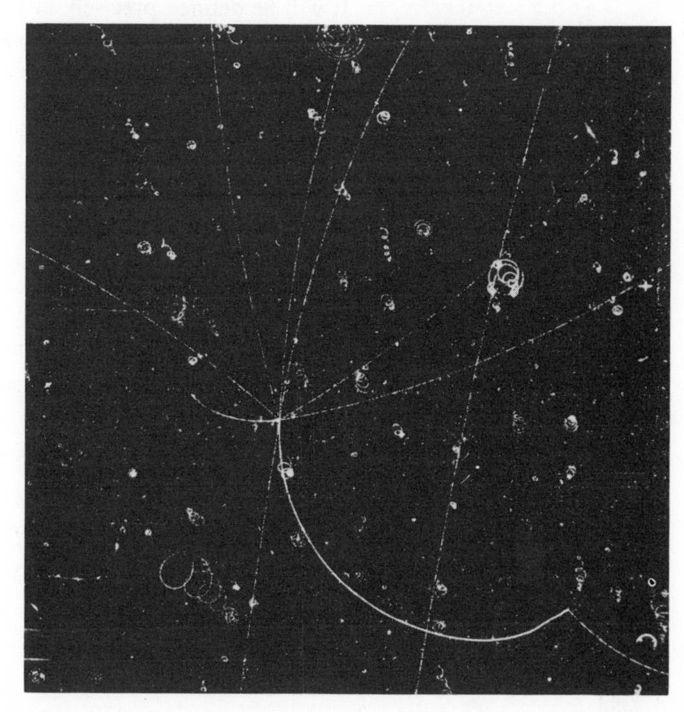

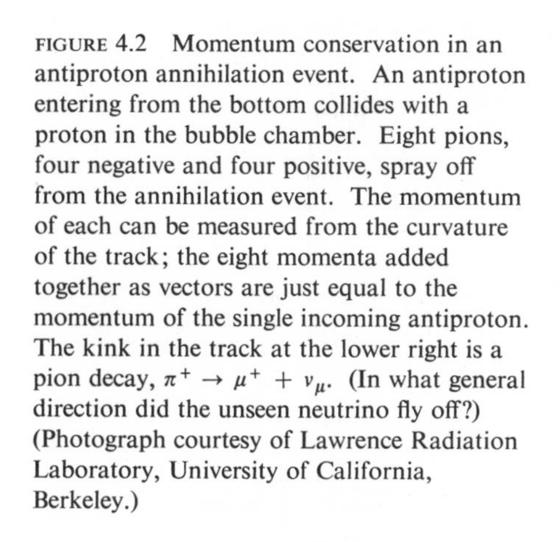
FIGURE 4.2 Momentum conservation in an antiproton annihilation event. An antiproton entering from the bottom collides with a proton in the bubble chamber. Eight pions, four negative and four positive, spray off from the annihilation event. The momentum of each can be measured from the curvature of the track; the eight momenta added together as vectors are just equal to the momentum of the single incoming antiproton. The kink in the track at the lower right is a pion decay, $\pi^+ \rightarrow \mu^+ + \nu_\mu$. (In what general direction did the unseen neutrino fly off?) (Photograph courtesy of Lawrence Radiation Laboratory, University of California, Berkeley.)

downhill in mass, and it also satisfies spin conservation. But the kaon-pion mass difference must get converted to energy of motion, so if the kaon were at rest, the pion would fly away. In whatever direction it moves it has some momentum, and it therefore violates momentum conservation since the kaon had none. On the other hand, if we enforce the law of momentum conservation and keep the pion at rest, we shall have violated energy conservation because in this case the extra energy arising from the mass difference will be unaccounted for.

4.6 Angular-momentum conservation

Kepler's law of areas is equivalent to the conservation of angular momentum

Early in the seventeenth century Johannes Kepler published three laws of planetary motion. The second of these, called the law of areas, is a conservation law. According to this law, an imaginary straight line drawn from the earth to the sun sweeps out area in space at a constant rate. During a single day this line sweeps across a thin triangular region with its apex at the sun and its base along the earth's orbit. The area of this triangle is the same for every day of the year. So when the earth is closer to the sun, it must move faster in order to define a triangle with the same area. It speeds up just enough, in fact, to maintain a constant value of its angular momentum, and the law of areas can now be derived as a simple consequence of the law of conservation of angular momentum. Actually the concept of angular momentum did not enter physics until more than a hundred years after Kepler's discoveries, and only with new mathematical formulations of mechanics in the nineteenth century did it seem to be one of the few fundamental concepts of mechanics. Like momentum, angular momentum is purely a property of motion and is a vector quantity. It will be defined precisely in Chapter 9.

Spin angular momentum in particle decay

For elementary particles, as for the earth, both spin and orbital angular momentum are important. A simple example involving spin only is the decay of the spinless pion into muon and neutrino, each with spin $\frac{1}{2}$ (in units of $\hbar$). In Figure 4.3 we use artistic license and represent the particles as little spheres with arrows indicating their direction of spin. Muon and neutrino spin oppositely in order to preserve the total zero angular momentum.

Coupling of spin and orbital angular momentum

Another two-body decay, that of the Λ, illustrates the coupling of spin and orbital motion. The Λ, supposed initially at rest [Figure 4.4(a)], has spin $\frac{1}{2}$.

Before (no spin)

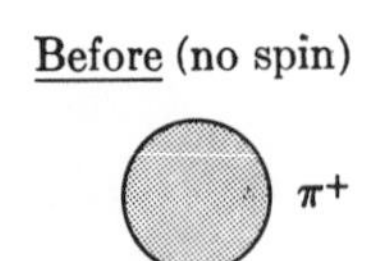

After (cancelling spin)

FIGURE 4.3 Angular-momentum conservation in pion decay. The total angular momentum is zero before and after the decay.

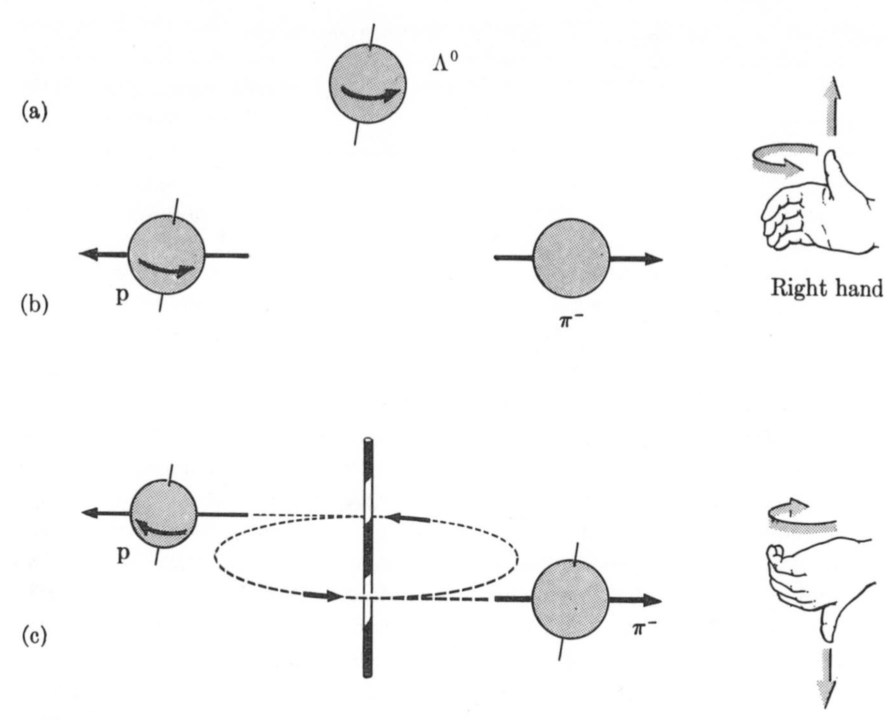

FIGURE 4.4 Angular-momentum conservation in lambda decay. Use of the right-hand rule shows that the particle spin is up in diagrams (a) and (b) and down in diagram (c); the orbital angular momentum is up in diagram (c).

One of its possible decay modes is

$$\Lambda^0 \to p + \pi^- .$$

This may proceed in two ways. The proton and pion may move apart with no orbital angular momentum, the proton spin being directed upward to match the initial Λ spin [Figure 4.4(b)]; or the proton spin may be flipped so it points downward while proton and pion separate with one unit of orbital angular momentum that is directed upward [Figure 4.4(c)]. In the first case,

$$\text{original spin } \tfrac{1}{2} \text{ (up)} \to \text{final spin } \tfrac{1}{2} \text{ (up)}.$$

In the second case,

$$\text{original spin } \tfrac{1}{2} \text{ (up)} \to \text{final spin } \tfrac{1}{2} \text{ (down)} + \text{orbital angular momentum 1 (up)}.$$

A special quantum rule for summing spin magnitudes

Because of the vector nature of angular momentum, the simple before-and-after arithmetic that works for charge or baryon number cannot be applied to particle spins. However, there is one general rule: The sum of the initial spin magnitudes and the sum of the final spin magnitudes must be either both integral or both half-odd-integral. This comes about because of quantum laws governing vector orientations (Chapter 24). The rule means, for instance, that a particle of spin $\frac{1}{2}$ could not decay into two particles each of spin $\frac{1}{2}$, although a particle of spin 1 might decay into two particles each of spin 1.

★4.7 Conservation laws and symmetry principles

The aspect of conservation laws that makes them appear to the theorist and the philosopher to be the most beautiful and profound statements of natural law is their connection with principles of symmetry in nature. Otherwise stated, energy, momentum, and angular momentum are all conserved because space and time are isotropic (the same in every direction) and homogeneous (the same at every place). This is a breathtaking statement when one reflects upon it, for it says that three of the seven absolute conservation laws arise solely because empty space has no distinguishing characteristics and is everywhere equally empty and equally undistinguished. (Because of the relativistic link between space and time, we actually mean spacetime.) It seems, in the truest sense, that we are getting something from nothing.

Isotropy and homogeneity are the basic symmetries of space and time

Yet there can be no doubt about the connection between the properties of empty space and the fundamental conservation laws. This connection raises philosophical questions that we will mention but not pursue at any length. On the one hand, it may be interpreted to mean that conservation laws, being based on the most elementary and intuitive ideas, are the most profound statements of natural law. On the other hand one may argue, as Bertrand Russell has done, that the connection only demonstrates the hollowness of conservation laws ("truisms," according to Russell), energy, momentum, and angular momentum all being defined in just such a way that they must be conserved.* Now, in fact, it is not inconsistent to hold both views at once. If the aim of science is the self-consistent description of natural phenomena based upon the simplest set of basic assumptions, what could be more satisfying than to have basic assumptions so completely elementary and self-evident (e.g., the uniformity of spacetime) that the laws derived from them can be called truisms? Since the scientist is generally inclined to regard as most profound that which is most simple and most general, he is not above calling a truism profound. Speaking more pragmatically, we must recognize the discovery of *anything* that is absolutely conserved as something of an achievement, regardless of the arbitrariness of definition involved. Looking at those conservation laws whose basis we do not understand (the three family-number conservation laws) also brings home the fact that it is easier to call a conservation law a truism after it is understood than before. It seems quite likely that we shall have gained a deeper understanding of nature and of natural laws before the conservation of baryon number appears to anyone to be a self-evident truth.

Are conservation laws truisms or profound statements?

They may be both

Before trying to clarify through simple examples the connection between conservation laws and the uniformity of space, we shall consider the question "What is symmetry?" In the most general terms, symmetry means that in spite of some particular change in one thing, something else remains unchanged (see Figure 4.5). A symmetrical face is one whose appearance would remain the same if its two sides were interchanged. If a square figure is rotated through 90 deg, its appearance is not changed. Among plane figures, the circle is the most symmetrical because if it is rotated about its center through any angle

Spatial symmetry: Some geometric changes leave appearance unchanged

* Bertrand Russell, *The ABC of Relativity* (New York: New American Library, 1959).

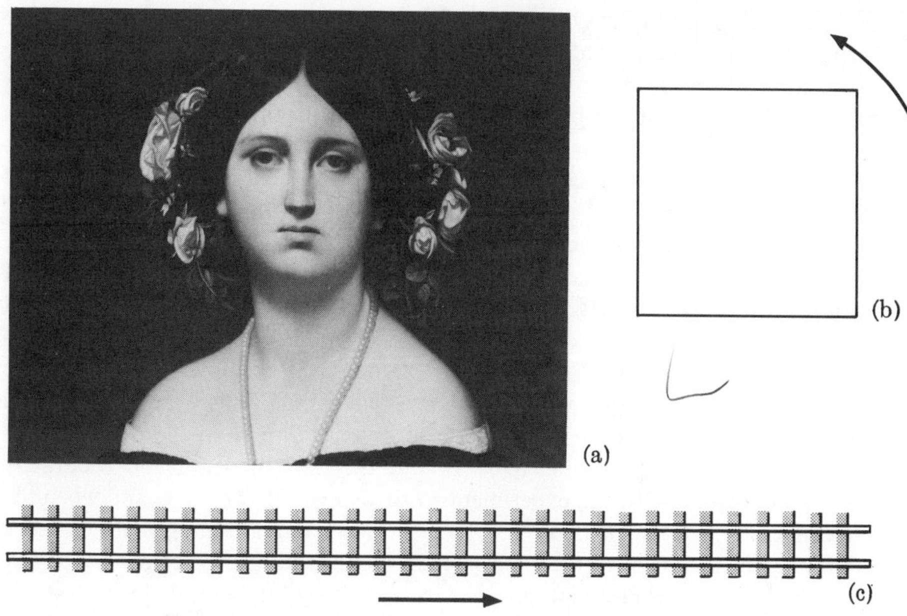

FIGURE 4.5 Three kinds of symmetry. (a) Reflection symmetry. A symmetrical face is unchanged if left and right are interchanged (reflection through a vertical plane). (b) Rotation symmetry. The appearance of a square is unchanged if it is rotated through 90 deg or any integral multiple of 90 deg. (c) Translation symmetry. A straight railroad line, hypothetically infinite in extent, is unchanged if it is shifted by any integral number of ties. [(a) *Madame Moitessier*, Jean-Auguste-Dominique Ingres, National Gallery of Art, Washington, D.C., Samuel H. Kress Collection.]

whatever, it remains indistinguishable from the original circle—or, in the language of modern physics, its form remains invariant.

We are accustomed to think of symmetry in spatial terms. The symmetry of the circle, the square, and the face are associated with rotations or inversions in space. Symmetry in time is an obvious extension of spatial symmetry; the fact that nature's laws appear to remain unchanged as time passes is a fundamental symmetry of nature. However, there are subtler symmetries, and it is reasonable to guess that the understanding of baryon conservation, for example, will come through the discovery of new symmetries that are not directly connected with space and time.

Symmetry in physics: Laws of nature are unaffected by certain changes

In the symmetry of interest to the physicist, the unchanging thing—the invariant element—is the form of natural laws. The thing changed may be orientation in space, position in space or time, or some more abstract change (not necessarily realizable in practice), such as the interchange of two particles. The inversion of space and the reversal of the direction of flow of time are other examples of changes not realizable in practice but nonetheless of interest for the symmetries of natural law. These latter two phenomena will be discussed in Chapter 27.

If scientists in Chicago, Serpukhov, and Geneva perform the same experiment and get the same answer (within experimental uncertainty), they are

Homogeneity of spacetime: Laws are unaffected by change of place or time

demonstrating one of the symmetries of nature, the homogeneity of space. If the experiment is later repeated with the same result, no one is surprised because we have come to accept the homogeneity of time. The laws of nature are the same, so far as we know, at all points in space and for all times. This invariance is important and is related to the laws of conservation of energy and momentum, but ordinary experience conditions us to expect such invariance so that it seems at first to be trivial or self-evident. It might seem hard to visualize any science at all if natural law changed from place to place and time to time; in fact, though, quantitative science would be possible even without the homogeneity of spacetime. Imagine yourself, for example, on a merry-go-round that speeded up and slowed down according to a regular schedule. If you carried out experiments to deduce the laws of mechanics and had no way of knowing that you were on a rotating system, you would conclude that falling balls were governed by laws that varied with time and with position (distance from the central axis), but you would be quite able to work out the laws in detail and accurately predict the results of future experiments provided you knew where and when the experiment was to be carried out. Thanks to the actual homogeneity of space and time, the results of future experiments can, in fact, be predicted without any knowledge of the where or when.

Physics would be possible without this symmetry

Symmetry implies invariance

We have been discussing the first link in the connected chain: symmetry → invariance → conservation. The symmetry of space and time, or possibly some subtler symmetry of nature, implies the invariance of physical laws under certain changes associated with the symmetry. In the simplest case, for example, the symmetry of space, which we call its homogeneity, implies the invariance of experimental results when the apparatus is moved from one place to another. This invariance, in turn, implies the existence of a certain conservation law. The link between invariance and conservation is what we will now attempt to illuminate through two examples, although an adequate discussion of this important topic requires methods that belong to more advanced courses in mechanics.

★4.8 The uniformity of space

A hypothetical isolated system: binary stars in intergalactic space

Imagine a pair of stars that interact gravitationally and orbit about one another.* Suppose that this system is located deep in intergalactic space, so far removed from outside influences that it can be regarded as truly isolated. Let us pretend that we can observe these stars with any degree of precision we choose. If we look closely enough, what do we see? The two stars trace out fairly complicated paths in space, combinations of rotational and translational motion, rather like the two ends of a tennis racket sent spinning through the air [Figure 4.6(a)]. However, if we back off until the whole binary system appears as a single point of light, its motion is seen to be very simple: it moves in a straight line with constant velocity [Figure 4.6(b)]. The system as a whole possesses a simplicity that its individual parts lack. Is this a significant fact? It is, for it is related to

* Such double stars, or binaries, are rather common in our galaxy.

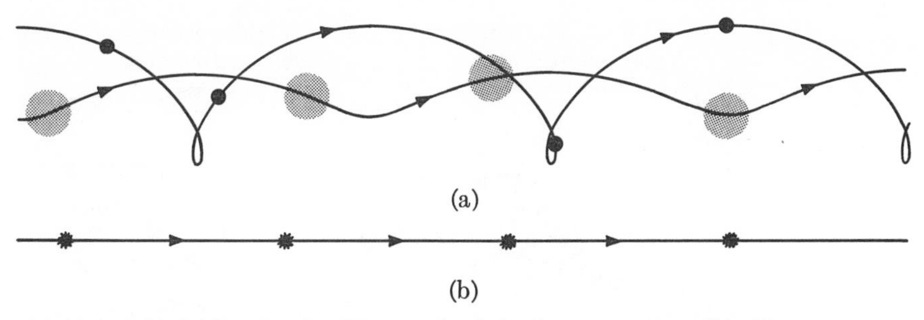

FIGURE 4.6 (a) A binary star system, examined closely, executes complicated motion. (b) Seen from a great distance, the system appears as a point moving with constant velocity. In both diagrams, the stars are pictured at equal time intervals.

Newton's laws of mechanics, to the law of momentum conservation, and to the homogeneity of space.

The simple motion of the whole system (actually, of its center of mass) is hardly surprising because we have postulated that the system is free of outside influences. Nevertheless, if we focus attention on the motion of the individual stars, the simple motion of their center of mass is not obvious. Each star is accelerated and each exerts a force on the other. Why do these forces not combine to accelerate the whole system? Or, in an example closer to home, why do the countless billions of interatomic forces acting within a book on a table cancel each other so precisely that the book never tends to move spontaneously?

Why does complicated internal dynamics not influence bulk motion?

The student who has already studied some physics knows that the answers to these questions are provided by Newton's third law: All forces in nature occur in equal and opposite matched pairs. The force of star A on star B is exactly equal and opposite to the force of star B on star A. The vector sum of the two forces is zero and no additional forces act from outside, so the system as a whole is unaccelerated. From another classical law, Newton's second law, it is possible to derive the fact that the total momentum of a system is constant if the total force acting on the system is zero. For the binary star system, we can write

One answer: Newton's third law

Internal forces sum to zero

$$\mathbf{p}_1 + \mathbf{p}_2 = \textbf{constant}, \qquad \text{if} \qquad \mathbf{F}_{\text{total}} = 0. \tag{4.5}$$

From Newton's second law, zero force implies constant momentum

The individual momenta, $\mathbf{p}_1$ and $\mathbf{p}_2$, are continually changing, but their vector sum is constant. Finally, one can prove that the center of mass of a system moves with constant velocity if the total momentum of the system is constant (this proof appears in Chapter 8).

Constant momentum means constant velocity of center of mass

In brief, this is the classical approach to understanding the motion of an isolated system. By "discovering" and applying two laws, Newton's second and third laws, we derive the law of momentum conservation. This conservation law in turn explains the constant velocity of the center of mass (see Figure 4.7).

The classical approach starts with Newton's laws

A modern approach to the same problem starts in quite a different way, by first seeking a principle that explains why the system as a whole does *not* accelerate. Recall the chain of key ideas referred to above: symmetry → invariance → conservation. Let us consider these three ideas for the binary star system.

A modern approach starts with symmetry

Newton's third law. Total internal force vanishes.

↓

Total momentum is constant for an isolated system.

↓

The center of mass of the system moves with constant velocity.

FIGURE 4.7 Classical chain of reasoning (top to bottom) regarding the motion of an isolated system. Newton's third law is basic. The constant velocity of the center of mass is derived.

Newton's third law. Total internal force must vanish.

↑

CONSERVATION: The total momentum is constant.

↑

The center of mass moves with constant velocity.

↑

INVARIANCE: No aspect of the motion of the system depends on the location of its center of mass.

↑

SYMMETRY: Space remote from matter is homogeneous.

FIGURE 4.8 A modern chain of reasoning (bottom to top) regarding the motion of an isolated system. The homogeneity of space is basic. Newton's third law is derived.

Symmetry: the homogeneity of space

The symmetry of interest is the homogeneity of space. We assume that in the depths of intergalactic space* where we have placed our stars, all points in space are exactly equivalent and indistinguishable.

Invariance: Laws of motion are independent of the location of the center of mass

This symmetry leads to an invariance principle: No law governing the motion of the system depends on the location of its center of mass. (The center of mass of a system is the average position of all of its mass.) This statement gives physical meaning to the idea of homogeneity. Translating the system from one part of empty space to another cannot change the nature of its motion.

Conservation: Center-of-mass velocity and total momentum are constant

What, then, is conserved? If the center of mass has certain velocity at one point, it must have the same velocity at a different point; otherwise, there would be a place-dependence in the laws of motion. So the center-of-mass velocity is constant. This implies in turn that momentum is conserved. Finally, to ensure momentum conservation, we must assume that internal forces cancel each other (Newton's third law). The final point in this chain of reasoning (summarized in Figure 4.8) is the same as the starting point in what we called the classical chain of reasoning (Figure 4.7).

The modern approach ends with Newton's laws

Choosing between these two ways of reasoning about an isolated system is more than a matter of taste. The modern argument proceeds directly from the symmetry principle to the conservation law without invoking Newton's laws. If in Figure 4.8 we go only as far as the conservation law, we have an argument that holds for the theories of relativity and quantum mechanics. Even where Newton's laws fail, momentum conservation remains valid and can

The symmetry argument is deeper and more general

* For experiments in which gravity is unimportant, as in the domain of electricity or atomic phenomena, space at the surface of the earth may be considered homogeneous.

be linked to the homogeneity of space.

Angular-momentum conservation rests on the isotropy of space

Through similar examples it is possible to relate the law of conservation of angular momentum to the isotropy of space. A compass needle that is held pointing east and then released will swing toward the north because of the action of the earth's magnetic field upon it. But if the same compass needle is taken to the depths of empty space, far removed from all external influences, and set to point in some direction, it will remain pointing in that direction. A swing in one direction or the other would imply a nonuniformity* of space. If the uniformity of space is adopted as a fundamental symmetry principle, it can be concluded that the total angular momentum of all the atomic constituents of the needle must be a constant. Otherwise, the internal motions within the needle could set the whole needle into spontaneous rotation, and its motion would violate the symmetry principle.

Energy conservation rests on the homogeneity of time

In a way that is admittedly not easy to see, energy conservation is related to the homogeneity of time. Thus all three conservation laws—of energy, momentum, and angular momentum—are "understood" in terms of the symmetry of spacetime.

* Strictly considered, momentum conservation rests on the *homogeneity* of space (uniformity of place), and angular-momentum conservation rests on the *isotropy* of space (uniformity of direction). The distinction is not important for our purposes, and it is satisfactory to think of space simply as everywhere the same, homogeneity and isotropy being summarized by the word uniformity.

Summary of ideas

A conservation law is a law of constancy during change. By itself it does not dictate a specific sequence of events. It places a general constraint on a system.

Charge conservation is known in the macroscopic world and has been accurately tested in the submicroscopic world. It accounts for the stability of the electron.

Baryon number is defined as $N_B - N_{\bar{B}}$, the number of baryons minus the number of antibaryons.

Baryon conservation accounts for the stability of the proton.

Electron-family-number conservation and muon-family-number conservation are believed valid but have not been tested to high precision.

Energy conservation is important in all branches of physics and in most other sciences.

Energy conservation dictates the downhill rule of particle decay and limits the production of particles at accelerators.

Momentum is a vector quantity. Its conservation prohibits one-body decays and makes two-body decays have simple properties.

Angular momentum conservation involves spin angular momentum and orbital angular momentum.

Symmetries in physics are related to invariance of the laws of nature.

Momentum conservation is based on the homogeneity of space (uniformity of place).

Angular momentum conservation is based on the isotropy of space (uniformity of direction).

Energy conservation is based on the homogeneity of time.

These conservation laws can be derived classically from Newton's laws, but they are more general and remain valid in non-Newtonian modern physics.

QUESTIONS

Section 4.1

Q4.1 The conservation of matter is an important principle in chemistry, ecology, and other sciences. It states that the number of atoms of a given kind is unchanged in any chemical or biological process. Why is this principle not included as one of the absolute conservation laws of this chapter?

Q4.2 Give an example of a quantity that is the same at different locations within some system but changes with time. Such a quantity is not called a conserved quantity.

Q4.3 Formulate a law of conservation of value (money, goods, and services) as applied to the finances of an individual. Is it an absolute conservation law?

Section 4.2

Q4.4 If you walk across a rug on a dry day, your body acquires an electric charge. Why does this *change* of charge not violate the principle of charge conservation?

Section 4.3

Q4.5 It has been postulated that the total baryon number of the universe is zero. Explain what this means and what the implications are for the nature of some other galaxies.

Q4.6 Suppose that the neutron were slightly less massive than the proton instead of slightly more massive. What would be the implications for the structure of matter?

Q4.7 Why is the *non*conservation of photons so vitally important in our world?

Q4.8 The idea that what *can* happen *does* happen sounds at first like a very unscientific doctrine. Explain how a physical world governed by simple laws can be described in this way. (HINT: The idea of probability should enter the discussion.)

Section 4.4

Q4.9 For a system that is *not* isolated, energy need not be conserved. Give a simple example of such nonconservation.

Q4.10 Is it possible in principle to convert a 1-kg piece of matter entirely into energy? Is it possible in practice to do so? Explain.

Q4.11 In some physics laboratories, gross violations of mass conservation are seen every day. In chemistry laboratories, measuring instruments of the highest available precision can reveal no deviations from a law of mass conservation. Explain.

Q4.12 In the operation of an oven, heat is conserved. The heat supplied by the heating element is equal to the sum of the heat gained within the oven and the heat transferred to the surroundings. Is the conservation of heat an absolute conservation law? Give a reason for your answer.

Section 4.5

Q4.13 An unstable particle is moving through the laboratory. Explain why at least one of the products of its decay must also be in motion.

Q4.14 A shell fired vertically upward explodes and breaks into two equal fragments just at the moment it reaches its greatest height (at which moment it is at rest). In how many directions may its fragments fly apart? Explain why the two fragments must start apart with exactly equal kinetic energy.

Q4.15 When a ball is thrown into the air, the magnitude of its momentum decreases and then increases. Why do these changes not violate the principle of momentum conservation?

Q4.16 An astronaut floating in space fires a gun with a rifled barrel, which gives the bullet a spin. Discuss the motion of the astronaut after the bullet is fired, assuming that he was motionless before. — Section 4.6

Q4.17 Mention a few simple ways in which you rely daily on the homogeneity of space and of time. — Section 4.7

Q4.18 Suppose that the universe is not uniform but instead has a definite center and boundaries. Explain why this property of the universe at large might stimulate physicists to search the submicroscopic world for small violations of the law of momentum conservation. — Section 4.8

Q4.19 The next-to-the-last paragraph in this chapter explains the connection between the isotropy of space and the conservation of angular momentum. Write a similar paragraph of about the same length explaining as briefly as possible the connection between the homogeneity of space and the conservation of momentum. Your paragraph might begin as follows, for example: "A rock that is held above the ground and then released will be accelerated downward because of the action of the earth's gravitational field upon it. But if the same rock is taken to the depths of empty space. . . ."

Q4.20 Suggest simple experiments you might carry out inside a closed room to find out whether or not the room is rotating. In a rotating room, what normally conserved quantity would seem not to be conserved?

Q4.21 When measured by observers in different frames of reference, some quantities are found to be the *same* even though they are *not* conserved quantities. Name one such quantity. Give a reason for your choice. — *Moving observers*

Q4.22 When measured by observers in different frames of reference, some quantities are found to be *different* even though they *are* conserved quantities. Name one such quantity. Give a reason for your choice.

EXERCISES

E4.1 Suppose that the mean life of the electron is 10^{25} years. How many electrons are needed in a sample in order for the decay rate of the whole sample to be 1 electron per sec? About how much mass of material would this be? (Assume that for each electron in the sample, there is one proton and one neutron.) — Section 4.2

E4.2 An antineutron strikes a proton. Write several possible outcomes of this collision. Explain why each is possible. — Section 4.3

E4.3 Write a reaction formula that expresses the decay of μ^+. Show that it satisfies the relevant conservation laws.

E4.4 If E is the total energy in a closed system, the law of energy conservation can be written algebraically as $E = constant$. Write the law in an alternative form that makes use of a derivative. — Section 4.4

E4.5 A certain proton-proton collision proceeds as follows:

$$\mathrm{p} + \mathrm{p} \rightarrow \mathrm{p} + \mathrm{p} + \text{pions}.$$

If the available energy for making new mass is 600 MeV, write four possible results of the collision that conserve energy and charge.

E4.6 (1) What is the dimension of momentum? (2) What is its SI unit? (3) Show that the product of force and time has the same dimension as momentum. — Section 4.5

E4.7 A proton moving at 10 percent of the speed of light strikes a 1-gm piece of matter and is brought to rest in the matter. At what speed does the bit of matter recoil as a result of the blow delivered by the proton?

Section 4.6

E4.8 Two automobiles approaching an intersection from opposite directions both turn right, as shown in the figure. What is the direction of their relative angular momentum (a) before the turn, (b) during the turn, and (c) after the turn? Justify each part of your answer.

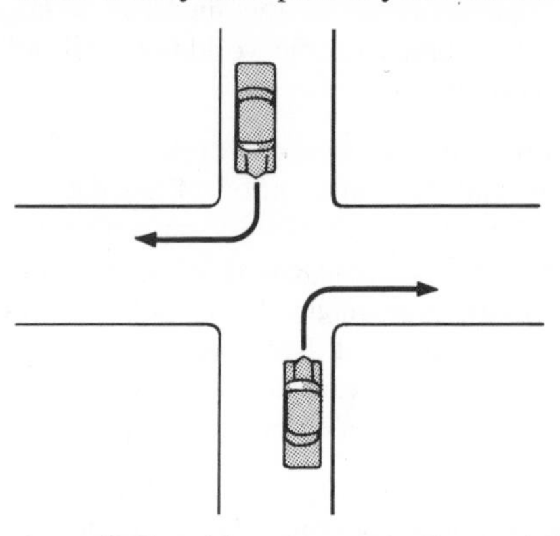

E4.9 At perihelion (closest to sun), the earth is 91.4 million miles from the sun. At aphelion (farthest from sun), the earth-sun distance is 94.5 million miles. (1) According to the law of areas, is the earth's orbital speed greater at perihelion or at aphelion? (2) What is the percentage difference of the earth's orbital speed at these two points?

E4.10 Make a drawing analogous to Figure 4.3 showing "before" and "after" diagrams for neutron decay, a three-body decay. Pay special attention to momentum conservation and angular momentum conservation.

Allowed and forbidden processes

E4.11 Table 3.1 shows that the Σ^+ particle can decay into a neutron and a positive pion. Give one other way in which it should be able to decay. Justify your answer by pointing out how each relevant conservation law is satisfied.

E4.12 Two of the following reaction processes do occur and two do not. Select the two allowed reactions and explain how the relevant conservation laws are satisfied. (Assume that sufficient energy is available in the collision to provide for any necessary mass increase.)

(a) $\mathrm{n} + \mathrm{p} \to \Lambda^0 + \mathrm{n} + \mathrm{K}^+ + \pi^0$
(b) $\mathrm{p} + \mathrm{p} \to \mathrm{p} + \Lambda^0 + \Sigma^+$
(c) $\mathrm{e}^+ + \mathrm{e}^- \to \mu^+ + \pi^-$
(d) $\mathrm{e}^+ + \mathrm{e}^- \to \gamma + \gamma$

E4.13 Each of the following particle decays, except one, violates some conservation law. Which one does not violate any laws and which law or laws is violated by each of the others?

(a) $\mu^+ \to \mathrm{e}^+ + \mathrm{e}^-$
(b) $\pi^- \to \mathrm{K}^- + \nu_\mu$
(c) $\mathrm{K}^- \to \mu^- + \overline{\nu_\mu}$
(d) $\mathrm{n} \to \pi^+ + \pi^-$
(e) $\mathrm{e}^- \to \nu_e + \gamma$
(f) $\Lambda^0 \to \mathrm{p} + \mathrm{e}^-$

E4.14 Gives at least two conservation laws violated by each of the following forbidden decay processes:

(a) $\mathrm{e}^- \to \nu_\mu + \gamma$
(b) $\mathrm{p} \to \mathrm{e}^- + \pi^+ + \pi^+$
(c) $\mu^- \to \pi^- + \gamma$
(d) $\pi^+ \to \mu^+ + \mathrm{e}^-$.

PROBLEMS

P4.1 *Allowed processes*

Write down several possible results of a proton-proton collision from which four particles emerge. (Assume that enough energy is at hand to create the new particles.) For each, explain why you consider the result possible.

P4.2 *Forbidden processes*

Pick any four of the fundamental conservation laws; for each, name one decay or transformation process that it forbids. For at least two of the four, pick a process forbidden *only* by this one conservation law.

P4.3 *π^0 decay*

A neutral pion (π^0) at rest has a mass energy of 135 MeV. It decays into two photons. (1) Give (a) the energy of each photon in MeV and in joules, (b) the momentum of each photon in kg m/sec (for a photon, $E = pc$), and (c) the relative direction of the photons. (2) What conservation laws are relevant in the decay? (3) If the pion had decayed in flight instead of at rest, how would the results have differed qualitatively?

P4.4 *Three-body decay*

(1) Explain explicitly how the beta decay of the neutron conforms to the conservation laws of energy, charge, electron-family number, and baryon number. (2) If a neutron at rest decays, the initial paths of the three product particles must be in a plane. What conservation law imposes this condition? Explain how it does so.

P4.5 *Prohibition of one-body decay*

In a hypothetical decay process, a particle of mass m_1 moving with speed v_1 transforms into a single product particle of mass m_2 moving with speed v_2. Using the laws of energy and momentum conservation and the requirement that neither v_1 nor v_2 can exceed the speed of light c, prove that the decay is forbidden. (Use the nonrelativistic formulas, $K = \frac{1}{2}mv^2$ and $p = mv$, and include the mass energy mc^2 in the energy balance.)

P4.6 *Prohibition of uphill decay*

In a hypothetical decay process, a particle of mass m is transformed into two other particles, each of the same mass m. If the initial particle is at rest, energy conservation evidently prohibits this decay. Prove that energy and momentum conservation, together with the requirement that particle speeds cannot exceed the speed of light c, prohibit the decay even if the initial particle is moving. Assume, for simplicity, that the initial and final particles move along the same straight line. (Use nonrelativistic formulas $K = \frac{1}{2}mv^2$ and $p = mv$, and include the mass energy mc^2 in the energy balance.) *Optional problems:* Extend the proof to unequal masses of the final particles and/or to arbitrary directions of the final particles.

P4.7 *Another kind of conservation*

Some conservation laws do not require that a quantity have a fixed value but do require it to have some other fixed property, such as oddness or evenness. Consider the guests at a reception who mix and from time to time shake hands. At any particular instant, a certain number of them will have shaken hands an odd number of times and the rest will have shaken hands either not at all or else an even number of times. Fix attention on the number who have shaken hands an odd number of times. (1) Explain why this number need not be constant. (2) What property does this number have that is conserved? Explain the basis of your answer.

PART TWO

Mathematics

5 Mathematics in Science

Throughout most of human history, mathematics and physics (or natural philosophy) have been inseparably joined, each discipline enriching the other. Today we find them still closely linked but with a relationship considerably altered from what it was two hundred or even one hundred years ago. Mathematics has grown into a more clearly distinct discipline, and its dependence on physical ideas has diminished. Physics without mathematics, on the other hand, remains inconceivable; in fact, more and more of mathematics is finding a place in physics. The student of physics or engineering must learn a considerable amount of mathematics and how to apply it. He must also learn to distinguish clearly between mathematics and physics. Unless the nature of mathematics as a separate human enterprise is understood, its role in science cannot be fully appreciated.

In this chapter and the next, we will be concerned mainly with practical aspects of mathematics—tools for the study of nature. We begin, however, in Sections 5.1 and 5.2 with a more general discussion of the nature of mathematics and its relationship to science.

★5.1 Two kinds of mathematical truth

If a beginning student of mathematics is asked why two plus two equals four, he is likely to respond that it is simply a self-evident truth. If given more time to reflect on the question, he would perhaps recognize that he accepts the statement as true for two different reasons. The first is a purely logical or mathematical reason; the numerical concepts "two" and "four" and the operational concepts "plus" and "equal" have all been defined in such a way that the statement is true. We may say that two plus two equals four by definition. Actually it is not true directly by definition but by certain definitions plus rules of logical deduction from these definitions. This is the truth of the mathematician. We might call it the "inner truth" of the mathematical statement.

The inner truth, a truth of logic

There is quite another kind of truth in the statement "two plus two equals four" that we might call the "outer truth." This is the more convincing to the nonmathematically trained person, and it is the reason he calls the truth of the statement self-evident. We all know from our own experience that if we have two objects and add to them two more objects, we will have four objects. "Two plus two equals four" is not just a logical statement about a relationship among mathematical symbols; it is a statement about the "real" world, about nature, and is therefore a scientific truth. In human terms the inner truth of mathematics is a purely intellectual truth, a truth of agreement among men. The outer truth of mathematics is a truth about the physical world, the world of man's sense perceptions.

The outer truth, a truth of physical confirmation

It is too commonly believed that a mathematical fact, if it is true at all, possesses both inner truth and outer truth. This belief is hardly surprising, for all of the elementary mathematics usually taught in schools does indeed possess both kinds of truth. Yet from the point of view of pure mathematics, the outer truth is a bonus, by no means necessary. The mathematician requires of his facts only that they are logically consistent with other facts and are deducible from his basic definitions and assumptions (axioms). If by chance they happen to conform to something known in the physical world and thereby become useful tools of the scientist or the practical man, it makes the facts neither more nor less acceptable to the mathematician. From the scientist's point of view, it should be regarded as the miracle of nature that mathematics—and often very simple mathematics—can be used successfully to describe natural phenomena in quantitative detail.

To the mathematician, inner truth is enough

Inner truth and outer truth have not always been distinguished in mathematics. Their clear separation came very late indeed in the history of mathematics, in the period from the mid-nineteenth century to the early twentieth century. Perhaps the most important single discovery ever made in mathematics was the discovery that mathematics is a logical structure independent of the physical world. The "discovery," made neither by one man nor at one time, came over a period of decades while "nonphysical" mathematics was gradually being accepted as a significant part of mathematics and while "physical" mathematics was being placed on a solid logical basis independent of experiment. For many centuries, nonphysical mathematics had been knocking at the door waiting to be let in. In some ancient civilizations, negative numbers were known but were regarded as being outside the scope of mathematics because negative length and quantity seemed to have no physical meanings. (Negative numbers were, in fact, a bit suspect as late as the seventeenth century.) The Greeks discovered "irrational" lengths but rejected irrational numbers because they involved the nonphysical concept of infinity. Imaginary numbers, first studied carefully in the sixteenth century, had to wait until the nineteenth century to become completely acceptable members of the arithmetic family.

Mathematics without outer truth, a recent discovery

The particular developments of the nineteenth century that led to the explosive expansion of mathematics into nonphysical domains were the study of non-Euclidean geometry and the study of spaces of more than three dimensions. The geometry of Euclid is the branch of mathematics whose outer truth is the

Non-Euclidean geometry, the first nonphysical mathematics

spatial relationships of our three-dimensional world. It is at one and the same time a branch of mathematics (a logically self-consistent structure of ideas) and a branch of physics (a quantitative description of distances and shapes in our physical world). Mathematicians first broke free of the requirement of outer truth when they courageously altered one of the basic axioms of Euclidean geometry and built a new self-consistent theoretical structure upon the new axioms. Then followed geometries of more than three dimensions and algebras that violated the laws of arithmetic. Finally the old familiar branches of "physical" mathematics were recast in a purely logical form, as sets of basic axioms and relationships derived from those axioms, so it no longer mattered whether or not they conformed to observations in the physical world.

Gauss found macroscopic space to be Euclidean

In the process of surveying a sizeable part of northern Germany in the 1820s, Karl Friedrich Gauss found that within his uncertainty of measurement of about 1 sec of arc, the interior angles of large triangles (many kilometers on a side) summed to 180 deg. Apart from its remarkable accuracy for the time, this result is of special interest because Gauss was himself one of the originators of non-Euclidean geometry. From remarks in his correspondence, we can be confident that he recognized he was putting Euclidean geometry to an experimental test. Yet being a man who shunned controversy, Gauss chose not to comment in print on this aspect of his surveying work.

A definition of mathematics

Numbers and vectors, examples of mathematical "objects"

From the modern point of view, mathematics can be defined roughly in this way: It is the study of logical relationships among mathematical "objects" based upon a few basic assumptions about the properties of these objects. The "objects" of mathematical study are abstract concepts whose basic nature is determined exclusively by the axioms that specify their behavior. Vectors are examples of such mathematical objects. So are the ordinary numbers of mathematics—integers, positive numbers, negative numbers, real numbers, rational numbers, imaginary numbers. Most numbers do not seem formidably abstract to us only because we happen to know that the rules governing the behavior of numbers are the same as the rules governing combinations of real physical objects in the external world. One of the axioms of the arithmetic of numbers, for example, is called the commutative law of addition. It states:

An axiom of arithmetic

$$a + b = b + a;$$

the result of adding two numbers is the same regardless of the order of the addition. For millennia this was regarded not as an assumption about the behavior of abstract concepts but as a self-evident truth about perfectly concrete things called numbers. You know that if you deposit ten dollars in your checking account today and deposit twenty dollars tomorrow, your balance will increase by exactly the same amount as if you had deposited first twenty and then ten. It is "obviously" so. The great discovery of the nineteenth century was the realization that the commutative law could be regarded as an entirely arbitrary assumption. Being arbitrary, it could be discarded or changed at will, and a whole new arithmetic could be developed upon the basis of new axioms. In what is called noncommutative algebra, $a + b$ is *not* the same as $b + a$. In this nonphysical branch of mathematics, the symbols a and b represent objects that are no longer visualizable, since we know of no physical objects obeying a noncommutative law of addition. From a purely mathematical point of view, however, the noncommuting objects are no more abstract or unreal than the commuting objects. It just happens that the commuting objects, which we call numbers, bear an exact relationship to things we perceive in the physical world, and the noncommuting objects do not.* This fact affords an important insight into the nature of the physical world, but it in no way affects the structure of mathematics.

Noncommuting objects are as real to the mathematician as commuting objects

A big scientific surprise of the twentieth century has been that some "nonphysical" branches of mathematics turned out to be physical after all. We shall return to this point in the next section.

★5.2 Mathematics and nature

Mathematics as a game, a deep analogy

Pure mathematics, as an independent structure of thought, can best be looked at as a game because it is precisely that. Its axioms define the game, which is played according to rules of logic. The mathematician's symbols are the pieces to be moved about. The wealth of relationships that can be derived from the basic axioms correspond to the rich variety of situations that can arise in a game with comparatively simple rules. In some particularly simple area of mathematics, such as the study of a group of only three objects, all possible consequences of the axioms can be derived, just as every possibility in tic-tac-toe can be spelled out. In more complex fields of mathematics, the seemingly limitless wealth of consequences that flow from the axioms continue to challenge the minds of generations of mathematicians, just as the inexhaustible possibilities in bridge and chess have fascinated generations of players. To pursue this deep-going analogy a bit further, consider the game of chess. To make it correspond to a branch of mathematics, we must regard the board as the writing tablet of the mathematician and the chessmen as the symbols representing the abstract objects of study. One set of chessmen may be of ivory, another of wood, another of plastic, but every knight, regardless of his size or shape or material, represents the same essence. There is but one single abstract object called "knight." The piece on the board is no more that basic concept than is a pencil stroke on a piece of paper the abstract object of mathematical study. The statements of the way each chessman can move are the axioms of the game. A minor change in a single axiom produces a drastic change in the game, just as a single alteration in the axioms of arithmetic produces a whole new kind of mathematics. Finally, there is the question of what constitutes complete knowledge of a concept or of a chessman. When the way a knight moves, the way he takes other pieces, and the way he can be taken are specified, that constitutes a complete definition of the abstract concept of a knight. There is nothing else to know. Questions such as how a knight would move on a triangular board are meaningless because they are outside the game of chess. There is still much that can be learned about the usefulness of a knight and his relation to other chessmen, but nothing more that can be said to define what he is. In the same way, a mathematical concept is completely defined in terms of a few

* Some noncommutative algebra has found a place in modern physics. It is nonphysical so far as the description of the everyday world is concerned.

axioms. Its very abstraction rests on the fact that so few of its properties are specified. The beginning student of mathematics might say, "Yes, I understand how vectors behave, but what *is* a vector?" The only possible answer is that a vector is an idea defined by a few rules of its behavior. A scientist, more interested in making the idea visualizable and useful than in sticking to logical rigor, might answer, "Well, force, for example, is a vector." What he means by this is that there exists a physical concept called "force" that behaves like the mathematical concept called "vector." The mathematical concept has been found to correspond to a physical concept and thereby to have acquired outer truth as well as inner truth. His answer is exactly the equivalent of saying about a particular carved piece of ivory, "This is a knight." More precisely, the piece is a physical object that behaves like the abstract concept of knight. The essence of knight is an idea quite independent of whether the piece is black or white, large or small, ivory or plastic. In the same way, the mathematical ideas of number, vector, length, or angle are abstractions quite independent of any correspondence to things in the "real" world.

Mathematical vectors and physical vectors

Mathematics can be regarded, then, as a creation of the human mind independent of science and independent of nature. As such, it is not merely *like* a game; it *is* a game. Its rules are arbitrary, and its only criterion of truth is the inner truth of self-consistency. In terms of the nature of truth, mathematics and science can be clearly separated. Mathematics, no matter how numerous its applications to the real world, finds its ultimate criterion of validity only in the harmony of inner truth. Science, no matter how abstract its concepts or how theoretical its reasoning, is ultimately justified only by the outer truth of experimental confirmation. In human terms, the dichotomy can be expressed this way: Science rests basically upon man's awareness of an orderly world outside himself; mathematics rests basically upon man's awareness of an orderly world within himself.

Inner truth justifies mathematics

Outer truth justifies science

To the scientist, of course, mathematics is more than a game. Science without mathematics is unthinkable, for it is mathematics that gives science its quantitative character and its predictive power. But taking the modern point of view about the nature of mathematics, it must be regarded as a miraculous chance that mathematics has found useful application in the description of the physical world.

The practical utility of mathematics: necessity or lucky chance?

Or is it a chance? Of course, it is no surprise that arithmetic and algebra and Euclidean geometry have something to do with the real world. They were invented and developed for practical ends, in order to describe reality. Application came first, abstraction later. But the question goes deeper than that. Basically it is this: Is man capable of conceiving the nonphysical? Since man is himself a part of the physical world, does it make any sense to distinguish between his inner world and his outer world? Must any mathematics the human mind is capable of inventing have some connection with the physical world, whether or not that connection has yet been discovered? In other words, is the popular "misconception" about the outer truth of mathematics perhaps not a misconception after all? Whatever the future holds, the fact is that there now exist branches of mathematics with no known outer truth; that part of mathematics remains, so far, an intellectual exercise.

At the same time, much of the "nonphysical" mathematics of the nineteenth century has become the physical mathematics of the twentieth century. Our three-dimensional space has been generalized to a four-dimensional spacetime. The general theory of relativity has shown that space is not really Euclidean, although it seems so in any small region. The mathematical formulation of the quantum theory has required the use of imaginary numbers, of quantities that do not commute with each other, and of vectors in a space of infinitely many dimensions. Impressed by the great expansion in the mathematical basis of physics, Paul Dirac wrote in 1931

Physics uses ever-broader ranges of mathematics

> Non-Euclidean geometry and noncommutative algebra, which were at one time considered to be purely fictions of the mind and pastimes for logical thinkers, have now been found to be very necessary for the description of general facts of the physical world. It seems likely that this process of increasing abstraction will continue in the future and that advance in physics is to be associated with a continual modification and generalization of the axioms at the base of the mathematics rather than with a logical development of any one mathematical scheme on a fixed foundation.*

Dirac does not say that *all* mathematics will eventually turn out to be useful in the scientific description of nature, but he expresses his belief that ever wider ranges of mathematics will prove to be physical mathematics, that is, to have an outer as well as an inner truth. Even if it were true that any self-consistent mathematical scheme that man can invent necessarily bears some correspondence to the physical world (a proposition that seems unlikely to this author), it seems impossible that man could prove this to himself. In the race for truth, the mathematicians would always be a lap ahead of the scientists. At all times there would exist some branches of mathematics that had as yet found no physical application and would remain, at least for a time, "purely fictions of the mind and pastimes for logical thinkers."

5.3 Coordinate systems and frames of reference

The arena of physics is spacetime. For the present, however, we can set time aside for separate consideration and limit our attention to three-dimensional space. Most of physics involves statements about location in space. Normally a point in space is identified by means of coordinates. To locate a point on a line, one coordinate suffices. Two coordinates are required to fix a point in a plane, and three are needed to fix a point in a three-dimensional space. A coordinate system consists of a specified point, the origin; specified directions, or axes; and instructions that define how points in space shall be labeled relative to the origin and the axes. There is no end to the number of coordinate systems that can be defined, and many special systems are convenient for special problems. Relatively

Coordinate system defined

---

* *Proceedings of the Royal Society (London)*, **A133,** 60 (1931). Paul Dirac, an outstanding theoretical physicist in England, was himself responsible for the introduction into physics of some previously nonphysical mathematics. In 1928 he discovered that the electron must be described not only by its position in ordinary space and time but also by its position in a peculiar abstract "spinor space." This new mathematics led to an improved theory of the electron and to the prediction of the positron. For this work, Dirac received the Nobel Prize in physics in 1933.

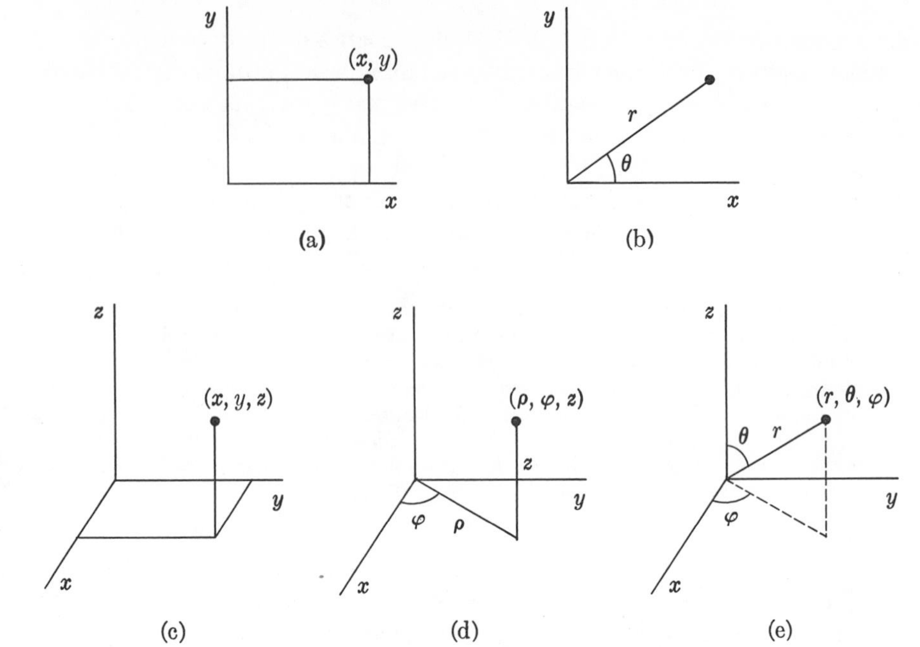

FIGURE 5.1 Common coordinate systems. (a) Cartesian coordinates in two dimensions. (b) Polar coordinates in two dimensions. (c) Cartesian coordinates in three dimensions. (d) Cylindrical coordinates. (e) Spherical coordinates.

few systems are used widely, however; among the most common are the Cartesian, the polar, the cylindrical, and the spherical (Figure 5.1). We shall find that Cartesian and polar coordinates suffice for almost all the applications in this book.

In two dimensions the Cartesian coordinates are conventionally called x, y and the polar coordinates r, θ. In Figure 5.1 we follow the common practice of using the same symbols x, y to label the axes and also to specify a particular point. These two sets of plane coordinates are related by the equations

Relation of polar to Cartesian coordinates

$$x = r \cos \theta, \tag{5.1}$$

$$y = r \sin \theta. \tag{5.2}$$

(Formulas of trigonometry are reviewed in Appendix 6.) The third coordinate, z, generalizes the Cartesian system from two to three dimensions and also extends the two-dimensional polar coordinates to the three-dimensional cylindrical coordinates (with a change of symbols from r, θ to ρ, φ). Spherical coordinates include one distance, r, and two angles, θ and φ. We shall not pursue the details of this coordinate system.

Definition of a right-handed system

The Cartesian system shown in Figure 5.1(c) is what is called a right-handed system. If the fingers of the right hand point in the direction of the x axis and are then curved to point in the direction of the y axis, the right thumb indicates the direction of the z axis. Alternatively, think of swinging the x axis toward the y axis, with the curved fingers of the right hand showing the direction of rotation. Again the thumb points along the z axis.* Note that this definition

* The z axis is also the direction in which a right-handed screw, if twisted in this way, would advance.

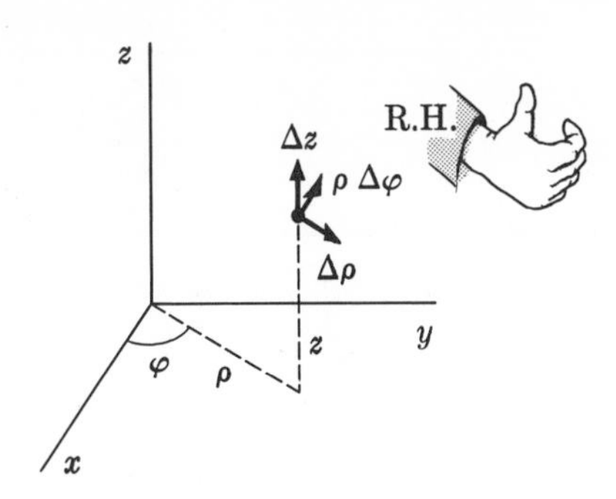

FIGURE 5.2 A cylindrical coordinate system is right-handed if the coordinates are ordered ρ, φ, z. Since the displacements $\Delta\rho$, $\rho\,\Delta\varphi$, and Δz are mutually perpendicular, it is also an *orthogonal* system.

of handedness depends on ordering the coordinates: x first, y second, z third. As a matter of convention, right-handed coordinate systems are used almost exclusively in physics. The cylindrical coordinate system shown in Figure 5.1 is also right-handed if the ordering is chosen to be ρ, φ, z. Figure 5.2 shows why this is so. Consider what happens to a point at ρ, φ, z if ρ alone is increased by $\Delta\rho$, if φ alone is increased by $\Delta\varphi$, or if z alone is increased by Δz. The three separate changes of the point are indicated by arrows in the figure. These three directions of increase form a tiny right-handed set of axes at the point in question. Because these three directions are mutually perpendicular, the cylindrical system is what is called an orthogonal system, just as the Cartesian system is.

Orthogonal systems

Note that the non-Cartesian coordinates are still defined with respect to a set of axes, which we continue to call the x, y, and z axes even though the coordinates used to specify the locations of points are no longer x, y, and z. In general an origin and a set of axes define what is called a frame of reference. A frame of reference is a more general idea than a coordinate system, for within one frame many different coordinate systems are possible. Two different frames of reference may retain a common origin, differing only in the orientation of their axes [Figure 5.3(a)]; or their origins may be separated by a fixed distance [Figure 5.3(b)]; or one frame may be moving relative to the other [Figure 5.3(c)]; or they may differ in any combination of these ways. In physical applications, frames of reference are most often chosen to be attached to some material thing. We speak of an earth-fixed frame, or a sun-fixed frame, or the frame of reference of a moving observer. Clearly, whenever the coordinates of a particle are specified, it is important to know with respect to what frame of reference the coordinates are measured.

Frame of reference defined

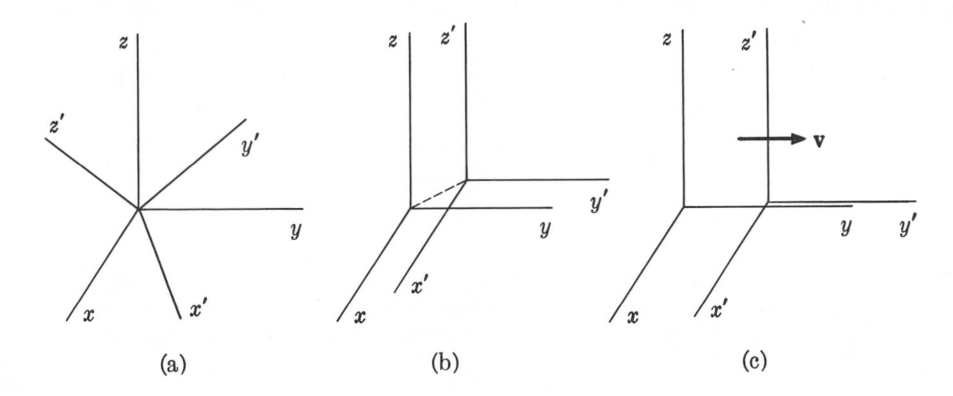

FIGURE 5.3 Axes xyz and axes $x'y'z'$ define two frames of reference. They may differ by (a) rotation, (b) translation, (c) relative velocity, or in any combination of these ways.

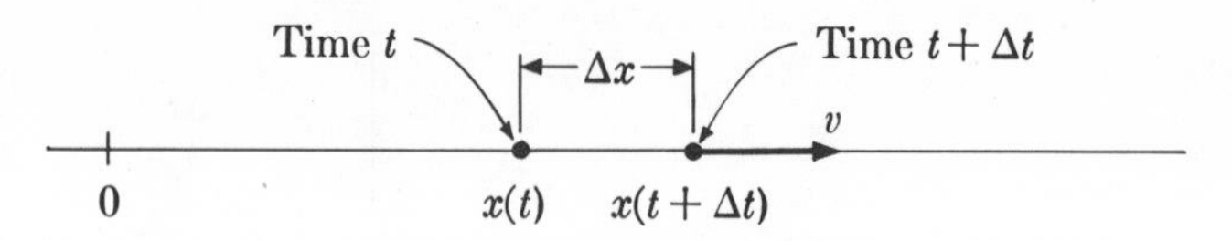

FIGURE 5.4 A particle moves along the x axis. Between times t and $t + \Delta t$, its displacement is Δx. The quotient $\Delta x/\Delta t$ is its average x component of velocity in this interval.

Not only the coordinates of a particle but also other physical quantities may depend upon the frame of reference. A question such as "What is the force in the x direction?" can be answered only if the direction of the x axis is known. The velocity of an object may also depend on the frame of reference. The object may be stationary in one frame of reference and moving with respect to another frame.

5.4 Speed; the derivative*

Consider a particle moving along a straight line, which we may define to be the x axis (Figure 5.4). Its position is a function of time; we designate it by $x(t)$. As indicated in the figure, the particle at time t is at $x(t)$; at time $t + \Delta t$, it is at $x(t + \Delta t)$. In the time interval Δt, it has experienced a displacement Δx given by

$$\Delta x = x(t + \Delta t) - x(t). \tag{5.3}$$

Its average x component of velocity during this interval is, by definition, the displacement divided by the time interval,

Average component of velocity

$$\overline{v_x} = \frac{\Delta x}{\Delta t} = \frac{x(t + \Delta t) - x(t)}{\Delta t}. \tag{5.4}$$

Its average *speed*, a positive quantity, is the magnitude of $\overline{v_x}$. (Compare this with Equation 2.8, in which Δs is assumed to be positive.)

If the particle moves in a given direction at constant speed v_0, its displacement is proportional to the time interval:

$$\Delta x = \pm v_0 \, \Delta t.$$

(The positive sign is for motion to the right in Figure 5.4; the negative sign is for motion to the left.) Then Equation 5.4 yields

$$\overline{v_x} = \frac{\pm v_0 \, \Delta t}{\Delta t} = \pm v_0 ,$$

a constant, independent of the choice of times t and $t + \Delta t$. If, however, the particle moves with variable speed, its average x component of velocity does

* Readers already familiar with elementary calculus may skip this section or scan it quickly for review. Students just beginning a course in calculus may want to read this section and Sections 5.8 and 5.9 on integration now and return to them for later reference as the mathematics and physics courses progress.

TABLE 5.1 NUMERICAL EXAMPLE OF LIMITING PROCESS

Hypothetical measured times and positions and calculated average speeds.* At $t = 1$ sec, $x = 10$ m.

$t + \Delta t$ (sec)	Δt (sec)	$x(t + \Delta t)$ (m)	Δx (m)	$\overline{v_x} = \frac{\Delta x}{\Delta t}$ (m/sec)
2	1	32.502	22.5020	22.5020
1.5	0.5	19.68825	9.68825	19.3765
1.1	0.1	11.68761	1.68761	16.8761
1.01	10^{-2}	10.163135	0.163135	16.3135
1.001	10^{-3}	10.0162573	0.0162573	16.2573
1.0001	10^{-4}	10.00162516	0.00162516	16.2516
1.00001	10^{-5}	10.000162511	0.000162511	16.2511
1.000001	10^{-6}	10.0000162510	0.0000162510	16.2510

* The numbers in the table are based on the arbitrarily chosen formula $x = 3.749t + 6.251t^2$.

depend on the times t and $t + \Delta t$. For such motion, a quantity of special interest is the component of instantaneous velocity at time t. It is defined by a limiting process:

$$v_x = \lim_{\Delta t \to 0} \frac{\Delta x}{\Delta t} = \lim_{\Delta t \to 0} \frac{x(t + \Delta t) - x(t)}{\Delta t}. \tag{5.5}$$

Component of instantaneous velocity

Since Δx approaches zero as Δt approaches zero, this is the limit of the ratio of two quantities as both approach zero. If this limit exists, it is called the *derivative* of x with respect to t; it is written dx/dt, and we have

$$\frac{dx}{dt} = \lim_{\Delta t \to 0} \frac{x(t + \Delta t) - x(t)}{\Delta t}. \tag{5.6}$$

It is a derivative

This basic definition of a derivative can be applied to other functions as well.* Note that the symbol dx/dt does *not* mean dx divided by dt. The notation is chosen to be suggestive of the limiting process by which the derivative is defined. In Table 5.1, a numerical example illustrates this limiting process. As the increments Δx and Δt separately approach zero, their ratio approaches a specific finite value.

Only "smooth" functions have derivatives

The derivative defined by Equation 5.6 has a well-defined value if x changes "smoothly." Physically, this means that the particle suffers no sudden changes of position or velocity. If the particle were by some miracle transported from one place to another in no time, its speed need not have a unique value at the moment of transport. If it were dealt an idealized hammer blow of zero time duration, the derivative—or the speed—would not be defined at the instant of the hammer blow. The speed would, however, be well-defined at other times,

* For example, if y is a function of x, then

$$\frac{dy}{dx} = \lim_{\Delta x \to 0} \frac{y(x + \Delta x) - y(x)}{\Delta x}.$$

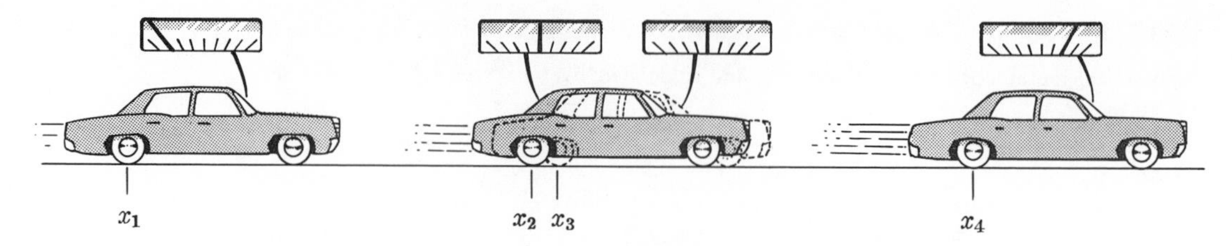

FIGURE 5.5 An accelerating automobile. Its speedometer provides a measure of its instantaneous speed, dx/dt. For a sufficiently small displacement, as from x_2 to x_3, its average speed $[(x_3 - x_2)/(t_3 - t_2)]$ is very nearly the same as its instantaneous speed at x_2 (or at x_3).

just before and just after the blow. An example of "smooth" motion appears in Figure 5.5.

The derivative is a rate

Whenever the derivative exists, it can be interpreted as the rate at which one quantity changes as another quantity is varied. In particular, speed is the rate at which position changes as time varies—or, as it is usually expressed, the rate of change of position with respect to time. The derivative also has important significance as the slope of a line or curve in a graph. This aspect of the derivative will be considered in Section 5.6.

■ EXAMPLE 1: A particle moving along the x axis is described by the equation

$$x = x_0 + v_0 t. \tag{5.7}$$

Let us show explicitly, from the definition of the derivative, that its x component of velocity is v_0. We have

$$\frac{\Delta x}{\Delta t} = \frac{[x_0 + v_0 \cdot (t + \Delta t)] - [x_0 + v_0 t]}{\Delta t} = \frac{v_0\, \Delta t}{\Delta t}.$$

This quotient, for all Δt, no matter how small, is v_0. The derivative is therefore

$$\frac{dx}{dt} = \lim_{\Delta t \to 0} \frac{v_0\, \Delta t}{\Delta t} = v_0.$$

■

■ EXAMPLE 2: A falling stone obeys the equation

$$z(t) = h - \tfrac{1}{2} g t^2, \tag{5.8}$$

where z is its height above the ground and where g and h are constants. What is its speed as a function of time? From Equation 5.5, with x replaced by z, we get

$$\begin{aligned} v_z &= \lim_{\Delta t \to 0} \frac{z(t + \Delta t) - z(t)}{\Delta t} \\ &= \lim_{\Delta t \to 0} \frac{[h - \frac{1}{2} g \cdot (t + \Delta t)^2] - [h - \frac{1}{2} g t^2]}{\Delta t} \\ &= \lim_{\Delta t \to 0} \frac{-gt\, \Delta t - \frac{1}{2} g \cdot (\Delta t)^2}{\Delta t}. \end{aligned} \tag{5.9}$$

The ratio $\Delta t/\Delta t$ is equal to 1 and remains 1 in the limit $\Delta t \to 0$. The quotient,

$(\Delta t)^2/\Delta t$, is equal to Δt and approaches zero in the limit. Therefore, only the first term on the right side of Equation 5.9 survives in the limit. The z component of velocity is

$$v_z = \frac{dz}{dt} = -gt.$$

The speed is $|v_z| = gt$. ■

SOME RULES OF DIFFERENTIATION

An important function is a variable raised to a constant power:

$$f(x) = x^n.$$

The derivatives of x^1 and x^2 are, in essence, derived in the preceding Examples 1 and 2. Their derivatives (note that x now replaces t) are

$$\frac{dx}{dx} = 1 \tag{5.10}$$

and

Some simple derivatives

$$\frac{d(x^2)}{dx} = 2x. \tag{5.11}$$

As another example, consider $f(x) = x^3$. The determination of its derivative proceeds as follows:

Example of a limiting process

$$\begin{aligned}\frac{df}{dx} &= \lim_{\Delta x \to 0} \frac{f(x + \Delta x) - f(x)}{\Delta x} \\ &= \lim_{\Delta x \to 0} \frac{(x + \Delta x)^3 - x^3}{\Delta x} \\ &= \lim_{\Delta x \to 0} \frac{3x^2\,\Delta x + 3x(\Delta x)^2 + (\Delta x)^3}{\Delta x},\end{aligned}$$

the last expression resulting from the binomial expansion of $(x + \Delta x)^3$. In the limit, the terms proportional to $(\Delta x)^2/\Delta x$ and $(\Delta x)^3/\Delta x$ vanish, leaving only $3x^2\,\Delta x/\Delta x$, whose limit is $3x^2$. Therefore,

$$\frac{d(x^3)}{dx} = \frac{d}{dx}(x^3) = 3x^2. \tag{5.12}$$

To the right of the first equal sign is shown another common notation for the derivative. It is left as an exercise to prove, using the definition of the derivative and the binomial theorem, that for any constant n,

$$\frac{d}{dx}(x^n) = nx^{n-1}. \tag{5.13}$$

Derivative of x^n

This and some other common derivatives are listed in Table 5.2. The trigonometric, exponential, and logarithmic functions that appear in the table will be discussed in later sections.

TABLE 5.2 SOME COMMON DERIVATIVES

$f(x)$	$\frac{df}{dx}$
$a = constant$	0
x	1
x^2	$2x$
$1/x$	$-1/x^2$
$1/x^2$	$-2/x^3$
$\sqrt{x}$	$1/(2\sqrt{x})$
x^n	nx^{n-1}
$\sin x$	$\cos x$
$\sin ax$	$a \cos ax$
$\cos x$	$-\sin x$
$\cos ax$	$-a \sin ax$
$\tan x$	$\sec^2 x = 1/\cos^2 x$
$\tan ax$	$a \sec^2 ax = a/\cos^2 ax$
e^x	e^x
e^{-x}	$-e^{-x}$
e^{ax}	ae^{ax}
$\ln x$	$1/x$
$\ln ax$	$1/x$

The process of finding the derivative of a function is called differentiating the function. *Differentiation* obeys several simple rules that are worth committing to memory. *The derivative of a constant times a function is the constant times the derivative of the function; the derivative of the sum of two functions is the sum of their derivatives.* Together, these two rules provide what is called the *linearity property* of differentiation:

The linearity property of differentiation

$$\frac{d}{dx}[af(x) + bg(x)] = a\frac{df}{dx} + b\frac{dg}{dx}. \tag{5.14}$$

Here a and b are constants. Differentiating a function is an example of "operating" on a function. Because of the property expressed by Equation 5.14, differentiation is called a linear operation.

One of the most important rules of differentiation expresses the derivative of the product of two functions $f(x)$ and $g(x)$. It is

Derivative of a product

$$\frac{d}{dx}(fg) = f\frac{dg}{dx} + \frac{df}{dx}g. \tag{5.15}$$

To derive this result, consider the quotient

$$\frac{\Delta(fg)}{\Delta x} = \frac{f(x + \Delta x)g(x + \Delta x) - f(x)g(x)}{\Delta x}.$$

This expression may be rewritten as

$$\frac{f(x + \Delta x)[g(x + \Delta x) - g(x)] + [f(x + \Delta x) - f(x)]g(x)}{\Delta x}.$$

The limit defining the derivative is then

$$\frac{d}{dx}(fg) = \lim_{\Delta x \to 0} \left[f(x + \Delta x) \left(\frac{g(x + \Delta x) - g(x)}{\Delta x} \right) + \left(\frac{f(x + \Delta x) - f(x)}{\Delta x} \right) g(x) \right].$$

In the limit, the quantities in large parentheses become the derivatives dg/dx and df/dx, and their multipliers become $f(x)$ and $g(x)$ respectively. Equation 5.15 is the result. A closely related formula, whose proof is left as a problem, gives the derivative of the quotient of two functions:

$$\frac{d}{dx}\left(\frac{f}{g}\right) = \frac{\frac{df}{dx} g - f \frac{dg}{dx}}{g^2}. \qquad (5.16)$$

Derivative of a quotient

One other rule of differentiation is often useful in physical applications. Suppose that f is a is a function of u, which in turn is a function of t. The derivative df/dt can be written as the product of two derivatives:

$$\frac{df}{dt} = \frac{df}{du}\frac{du}{dt}. \qquad (5.17)$$

The chain rule

This is the so-called chain rule of differential calculus.

■ EXAMPLE 3: The kinetic energy of a body of mass m is $K = \frac{1}{2}mv^2$. Suppose that its speed v is proportional to time: $v = at$. What is the time rate of change of its kinetic energy? Application of Equation 5.17 gives

$$\frac{dK}{dt} = \frac{dK}{dv}\frac{dv}{dt},$$

which can be written

$$\frac{dK}{dt} = mv \times a.$$

Since $v = at$, this is the same as

$$\frac{dK}{dt} = ma^2t.$$

How else might this result have been obtained? ■

The derivative of a derivative is called a second derivative. It is written

$$\frac{d}{dx}\left(\frac{df}{dx}\right) \quad \text{or} \quad \frac{d^2f}{dx^2}.$$

The second derivative

Third and higher derivatives may also be defined. Successive differentiation involves no new ideas. It is straightforward once first derivatives have been mastered. The second time derivative of the kinetic energy in Example 3, for instance, is $d^2K/dt^2 = ma^2$.

■ EXAMPLE 4: What is the second derivative of $f(x) = ax^n$? Since the first derivative of this function is $df/dx = anx^{n-1}$, its second derivative is

$$\frac{d^2f}{dx^2} = \frac{d}{dx}(anx^{n-1}) = an(n - 1)x^{n-2}.$$ ■

5.5 Angles and angular speed

Two units for angles

1 radian $= \frac{180}{\pi}$ deg

There are two common units of measurement for angles: degrees (deg) and radians. A complete revolution contains 360 deg, or 2π radians. The radian, equal to $(180/\pi)$ deg, or approximately 57.3 deg, is defined as the angle that subtends a circular arc equal in length to the radius of a circle [Figure 5.6(a)]. As illustrated in Figure 5.6(b), any angle is expressed in radians by

$$\theta = \frac{s}{r}, \tag{5.18}$$

Radian measure is preferred

where r is the radius and s is the length of circular arc. The radian is the preferred unit for scientific work, and we shall generally adhere to it in this book. Whenever the unit of angle is unstated, the radian may be assumed.

If an angle is changing with time, we may define an angular speed. The usual unit for angular speed is radian/sec, and a frequently used symbol for it is ω (omega). Consider a particle moving around the circumference of a circle of radius r (Figure 5.7). If, in a certain time, it sweeps out an angle θ, it covers in the same time a distance s, which, according to Equation 5.18, is given by

Distance and angle simply related for circular motion

$$s = r\theta. \tag{5.19}$$

Suppose that in a time interval Δt, the change of s is Δs and that the change of θ is $\Delta\theta$. These increments are related by

$$\Delta s = r\,\Delta\theta.$$

Division of both sides by Δt gives

$$\frac{\Delta s}{\Delta t} = r\frac{\Delta\theta}{\Delta t}. \tag{5.20}$$

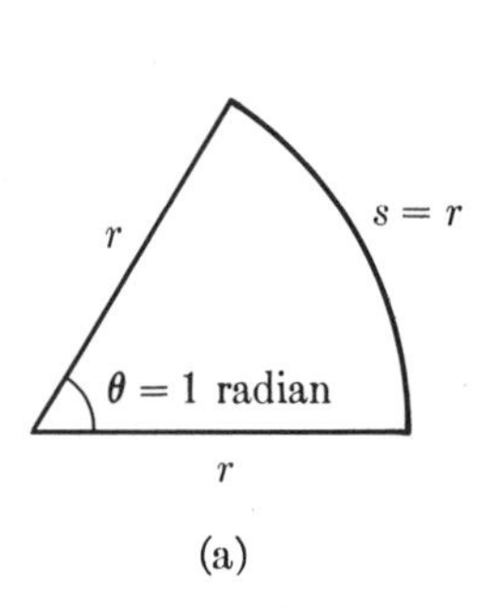

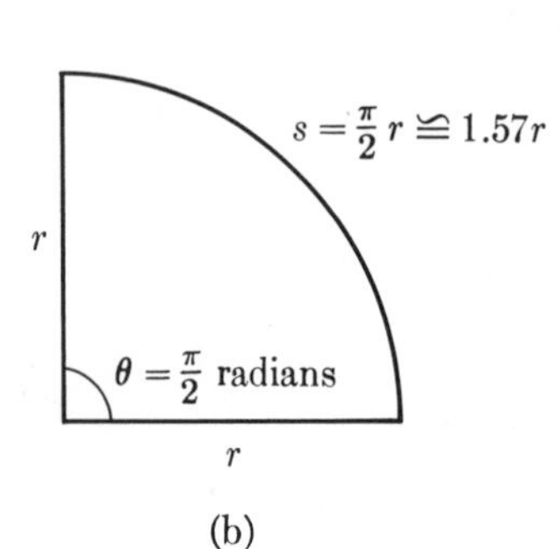

FIGURE 5.6 Definition of the radian. (a) If the length of a circular arc is equal to the radius, the angular opening is 1 radian, or about 57.3 deg. (b) For any other angle, the ratio of the length of a circular arc to the radius is equal to the angle measured in radians: $\theta = s/r$.

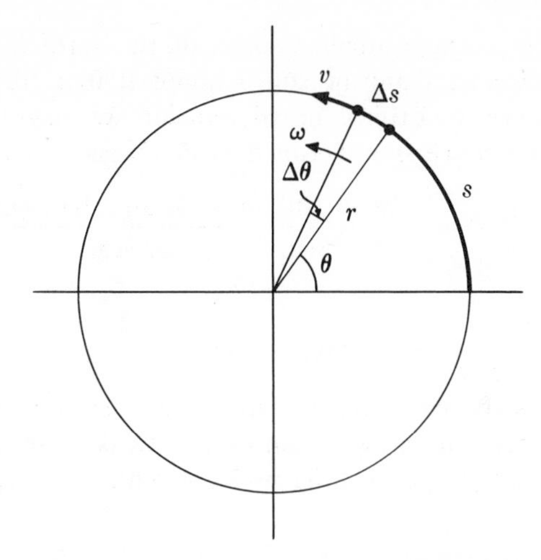

FIGURE 5.7 A particle moving in a circle. Its angular speed is $\omega = d\theta/dt = v/r$.

In the limit that Δt approaches zero, the left side becomes the instantaneous speed v. Angular speed is defined in a similar way:

Definition of angular speed

$$\omega = \lim_{\Delta t \to 0} \frac{\Delta\theta}{\Delta t} = \frac{d\theta}{dt}. \tag{5.21}$$

Therefore, in the limit of small Δt, Equation 5.20 becomes

$$v = r\omega \qquad \text{(constant } r\text{)}. \tag{5.22}$$

This simple and useful formula connecting v and ω may be derived more easily by direct application of calculus to Equation 5.19. Since r is constant, differentiation of both sides of Equation 5.19 with respect to time gives

$$\frac{ds}{dt} = r\frac{d\theta}{dt} \qquad \text{(constant } r\text{)}. \tag{5.23}$$

Equations 5.22 and 5.23 are, of course, equivalent.

If we specialize further to circular motion executed with constant speed, distance and angle both increase in proportion to time. Then, measuring from $t = 0$,

$$s = vt, \tag{5.24}$$

$$\theta = \omega t. \tag{5.25}$$

$\begin{pmatrix}\text{uniform}\\ \text{circular}\\ \text{motion}\end{pmatrix}$

■ EXAMPLE 1: A shaft rotates at n revolutions per second. What is its angular speed? Introduce the conversion factor 2π radian/revolution and write

$$\omega = n\,\frac{\text{revolution}}{\text{sec}} \times 2\pi\,\frac{\text{radian}}{\text{revolution}},$$

which gives

$$\omega = 2\pi n \text{ radian/sec}. \tag{5.26}$$

■

■ EXAMPLE 2: The average angular speed of the earth in its orbit is 6.28 radian/year. Its distance from the sun is about 1.50×10^{11} m. What is its speed in orbit? Since the orbit is nearly circular, we may use Equation 5.22. With an appropriate conversion factor, the calculation looks as follows:

$$v = r\omega = \frac{1.50 \times 10^{11}\ \text{m} \times 6.28\ \text{radian/year}}{3.16 \times 10^{7}\ \text{sec/year}},$$

hence

$$v = 2.98 \times 10^{4}\ \text{m/sec}. \tag{5.27}$$

This speed can also be written 29.8 km/sec, or 18.5 mile/sec. (Actually the orbital speed of the earth varies because the orbit is not exactly circular. Its maximum and minimum speeds differ by 3 percent.) ■

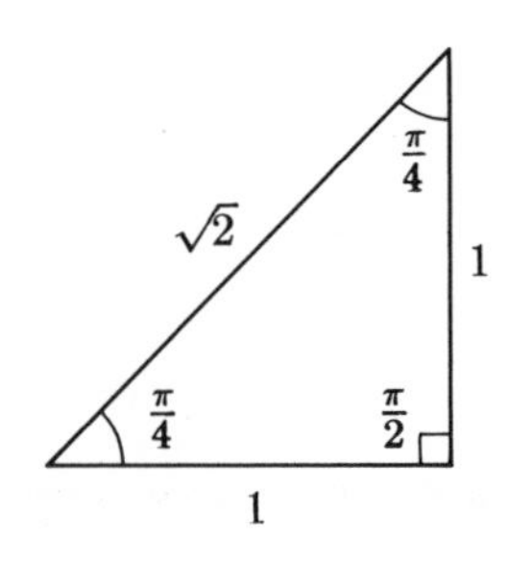

TRIGONOMETRIC FUNCTIONS*

The trigonometric functions of an angle are normally defined with the help of a right triangle. For angles greater than 90 deg, a slight generalization of the definitions is required. In Figure 5.8, the hypotenuse of a right triangle serves as the radial line in a circle drawn about the origin. If the angle θ is the angle of this radial line measured counter-clockwise from the x axis, the most-used trigonometric functions are

Equations valid for any angle

$$\sin\theta = \frac{y}{r}, \tag{5.28}$$

$$\cos\theta = \frac{x}{r}, \tag{5.29}$$

$$\tan\theta = \frac{y}{x}. \tag{5.30}$$

The signs implied by these definitions are summarized in Table 5.3. Compare Equations 5.28 and 5.29 with Equations 5.1 and 5.2. These equivalent pairs of equations serve both to define the sine and the cosine and to relate plane Cartesian coordinates to polar coordinates. In the latter role they are called transformation equations.

The cosine is important in the geometrical process of projection. Consider a line segment of length L inclined at an angle θ away from the x axis (Figure 5.9). The projection of L onto the x axis is defined as the line segment L_x, which is formed by dropping lines perpendicular to the axis from both ends of the segment L. From the geometry of the right triangle ABC in Figure 5.9, it is clear that

A projection equation

$$\cos\theta = \frac{L_x}{L} \quad \text{or} \quad L_x = L\cos\theta. \tag{5.31}$$

Similar reasoning applies to the projection L_y. Since the angle φ between

* See also Appendices 6 and 7.

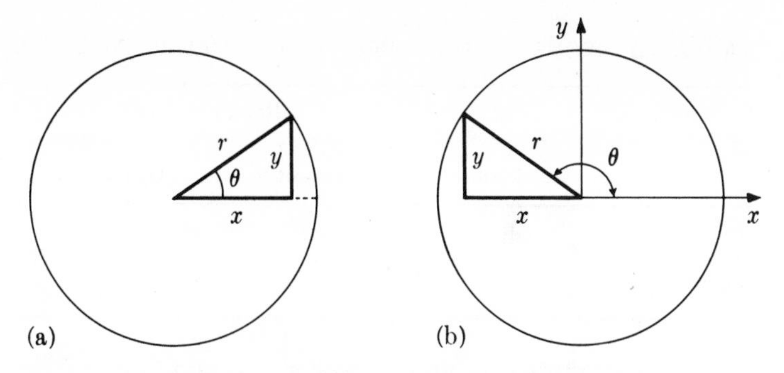

FIGURE 5.8 Diagrams for general definitions of trigonometric functions.

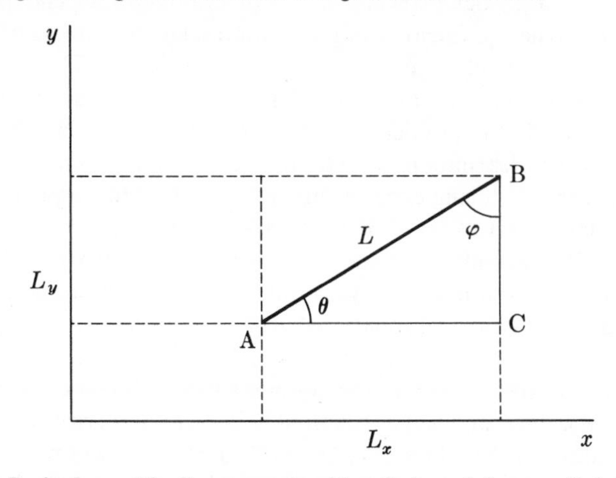

FIGURE 5.9 Projections. The line segments of length L_x and L_y are called the projections onto the x and y axes of the line segment AB, of length L.

the segment L and the y axis is equal to $\frac{1}{2}\pi - \theta$, we have

$$L_y = L \cos \varphi = L \sin \theta. \tag{5.32}$$

Although $\sin \theta$ appears here, it is well to remember that every projection is defined by the *cosine* of the angle between the line segment and the axis in question. The projections may be thought of as the shadows of the line segment on the axes. If the line segment represents a vector, its projections are called the components of the vector. Components are considered in Section 6.5.

Vector components are projections

5.6 Functions, tables, and graphs

A key idea in mathematical analysis and in physics is the idea of dependence. One quantity depends on another if the variation of one of them is accompanied by a variation of the other. Mathematicians speak of the independent variable and the dependent variable. In physics it is better to think of *interdependent* quantities. The idea that one variable quantity is independent and another dependent is, like all ideas of cause and effect in science, often useful as a way of describing events but not strictly necessary. Logically, only the *relationship* of different physical quantities is meaningful.

TABLE 5.3 SIGNS OF PRINCIPAL TRIGONOMETRIC FUNCTIONS

Function	Range of Angle			
	0–$\pi/2$ Quadrant I	$\pi/2$–π Quadrant II	π–$3\pi/2$ Quadrant III	$3\pi/2$–2π Quadrant IV
sin	+	+	−	−
cos	+	−	−	+
tan	+	−	+	−

Functions

If one quantity depends on another, it is said to be a function of the other. The average stock market price is a function of time. The speed of an accelerating automobile is a function of its position. The brightness of a spot on a television tube is a function of the speed of the electrons. Functional relationships permeate every part of science, for most of science is concerned with connections among different physical concepts. Much of the experimental work in science is devoted to learning in quantitative detail exactly how one quantity depends on another. Reducing experimental relationships to simple mathematical functions is a large part of the theoretical work in science.

Tables, graphs, and equations can express functions

The dependence of one quantity on another can be given quantitative expression in three different ways: in tabular presentation, in graphical presentation, and in mathematical equations. Although an equation provides the most concise expression of a functional relationship, tables and graphs are often useful as well. A table of data is appropriate when the precise mathematical relationship is not known or when it is desired to present numerical values to a high degree of accuracy. Like tables, graphs can be used to present relationships, such as those discovered experimentally, whose exact mathematical form is not known. Even if the equation *is* known, a graph can do much to bring it alive and make its meaning more evident. We shall frequently employ graphs in this book for just that purpose—to clarify an equation by making it pictorial.

LINEAR FUNCTION

The simplest functional relationship is a direct proportionality. The distance covered by an automobile moving at constant speed, for example, is proportional to the elapsed time. Data for a particular example of such motion could be presented numerically, as in Table 5.4. The distance is a function of the time. (Alternatively, we could say that the time is a function of the distance, but this is an uncommon way to look at motion. Why is it more natural to regard time as the independent variable and distance as the dependent variable?) The exact mathematical relationship between time and distance in this example will not be immediately obvious while examining the table. This is one of the disadvantages of tabular presentation. Although the numerical values can be precisely specified, they do not at once convey a clear picture of how the variables are related. A graph does this job much better.

Tables: accuracy, even though overall form of relationship may be unclear

In Figure 5.10 the same data are presented graphically. The variable chosen to be regarded as independent—in this case the time—is plotted horizontally; the variable chosen to be regarded as dependent is plotted vertically. Each pair of numbers in the table gives a single point on the graph.

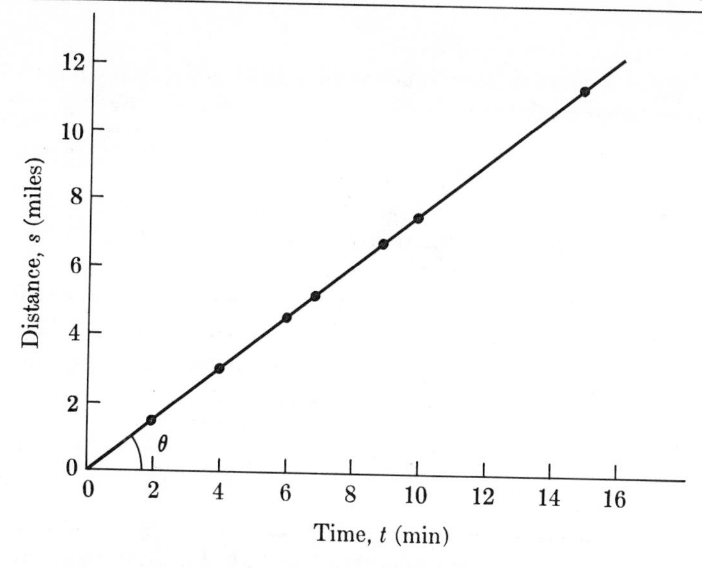

FIGURE 5.10 Graph of distance vs time for the data of Table 5.4.

Graphs: clarity of overall view

It is immediately obvious that the points may all be joined by a single straight line. It is not necessarily true that every point on that line reveals exactly where the automobile was at a particular time. We were supplied with data only for certain selected points. However, the fact that those points lie on a line is very suggestive. It is at least a reasonable supposition that the straight line also represents the location of the automobile at intermediate points. Apparently the motion was carried out with constant speed.

The equation that fits these tabular and graphical data is

$$s = 0.75t,$$

where s represents the distance in miles and t represents the time in minutes. This equation, lacking dimensional consistency and therefore tied to a particular set of units, is better replaced by

Equations: concise and general

$$s = v_0 t, \tag{5.33}$$

where v_0 stands for a constant whose value in this example is

$$v_0 = 0.75 \text{ mile/min} = 45 \text{ mile/hr}.$$

TABLE 5.4 TIME AND DISTANCE DATA FOR MOTION AT CONSTANT SPEED

Elapsed Time (min)	Distance (miles)
0	0
2	1.50
4	3.00
6	4.50
7	5.25
9	6.75
10	7.50
15	11.25

An equation of this type is called a *linear equation* because its graph is a straight line. A more general linear equation is

A linear equation

$$s = v_0 t + s_0, \tag{5.34}$$

where s_0 is another constant, whose physical meaning in this example is the initial distance from some arbitrarily chosen origin. Thus a direct proportionality is a special case of a linear relationship.

Definition of slope of line

The parameter v_0 appearing in Equation 5.33 is also equal to the *slope* of the straight line in Figure 5.10. The slope of a line in a graph is defined as the tangent of the angle that the line makes with the horizontal axis. This angle is designated by θ in Figure 5.10. For the special example being considered,

$$\tan \theta = \frac{s}{t} = v_0 . \tag{5.35}$$

Slope is dimensional

It is important to notice that we are generalizing the idea of the tangent from its purely geometric meaning, for here the quantity called $\tan \theta$ is a dimensional quantity, length divided by time. We shall always measure slopes as a vertical increment divided by a horizontal increment on a graph, each increment being measured in the appropriate unit for the quantity in question. With this understanding, the slope of a line is independent of the scales chosen to prepare the graph.

GRAPHICAL SIGNIFICANCE OF THE DERIVATIVE

Another consequence of this definition of slope is that the slope has a simple physical meaning. It is the rate of change of the quantity being plotted vertically with respect to the quantity being plotted horizontally, that is, a derivative. If $s = v_0 t$ (Equation 5.33), the derivative of s with respect to t is

Slope is a derivative

$$\frac{ds}{dt} = v_0 . \tag{5.36}$$

This equation states that the rate of change of position with respect to time is v_0, the same as the slope of the line in the graph. When the quantity plotted horizontally is time, rate has its common meaning—rate of change with respect to time. More generally, however, rate may be defined with respect to quantities other than time.

QUADRATIC FUNCTION

A physicist is studying the thermal properties of a certain substance. He theorizes that if a spherical piece of the material is uniformly chilled and is then returned to a room-temperature environment, its internal temperature after a time will be given by

A quadratic function

$$T = T_0 + Ar^2. \tag{5.37}$$

Here T is the temperature, r is the radial distance from the center of the sphere, and T_0 and A are constants. To render this equation pictorial, he makes a graph of T vs r (Figure 5.11). Because r^2 appears on the right side of Equation

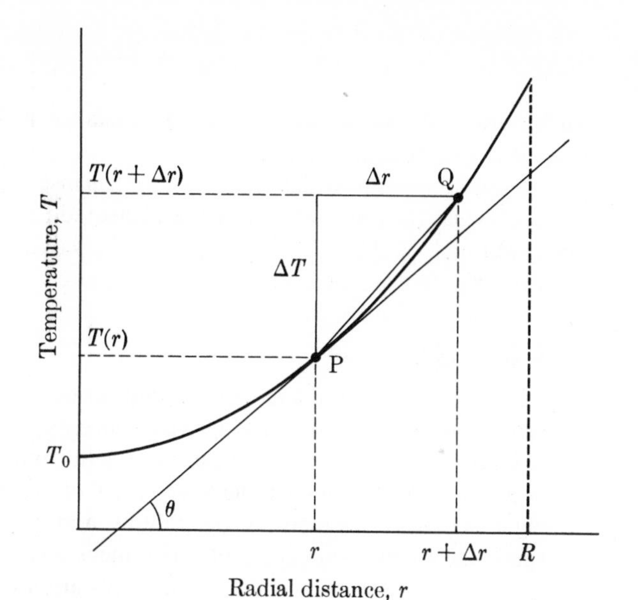

FIGURE 5.11 Hypothetical temperature distribution within a sphere of radius R. At the point P, the slope of the curve is $dT/dr = \tan\theta$.

5.37, T is called a *quadratic function* of r.* It could equally well be called a parabolic function because the graph of T vs r is a parabola.

Its graph is a parabola

The slope of a curve is defined by a limiting process. On the curve in Figure 5.11 select two points P and Q, and draw a straight line through them. The slope of this line is $\Delta T/\Delta r$. The limit of this slope as Q approaches P is defined to be the slope of the curve at point P:

$$\text{slope at point P} = \lim_{\Delta r \to 0} \frac{\Delta T}{\Delta r} = \lim_{\Delta r \to 0} \frac{T(r + \Delta r) - T(r)}{\Delta r}. \tag{5.38}$$

This is evidently a derivative (see Equation 5.6):

$$\text{slope at point P} = \frac{dT}{dr}. \tag{5.39}$$

Definition of slope of curve

As Figure 5.11 makes clear, the slope defined in this way is the same as the slope of a straight line drawn tangent to the curve at point P

$$\text{slope at point P} = \tan\theta. \tag{5.40}$$

The figure also clarifies the "smoothness" criterion mentioned in Section 5.4. In order that the derivative of a function be uniquely defined at a particular point, it is necessary that the graph of the function have a slope that does not change discontinuously at that point.

In this example, $\tan\theta$ has the dimension temperature/distance (recall the discussion below Equation 5.35). It could be evaluated graphically. However, since Equation 5.37 provides an explicit expression for $T(r)$, it is easier to obtain the slope by differentiation:

* In general, y is a quadratic function of x if $y = a + bx + cx^2$, with a, b, and c constant.

A gradient: derivative with respect to distance

$$\frac{dT}{dr} = 2Ar. \tag{5.41}$$

Such a rate of change with respect to distance is usually called a *gradient*. Here it is a temperature gradient.

Let us suppose now that our physicist has inserted probes in a sphere of the material he is studying and that he has measured its internal temperature as it is being warmed. Table 5.5 shows one set of his data. It is left as a problem to discover whether or not Equation 5.37 adequately represents these data.

5.7 One-dimensional kinematics

Kinematics uses length and time

Kinematics can be defined as the description of motion using only two basic concepts, length and time. Its derived concepts are velocity and accleration. Galileo, early in the seventeenth century, was the first to give a clear definition of acceleration. He discovered the kinematics of freely falling objects, of balls rolling down inclined planes, of pendulums, and of projectiles. In this section, we specialize to the kinematics of one-dimensional motion. This provides an introduction to mechanics and a useful application of mathematics to nature.

For convenience, we suppose, as in Section 5.4, that a particle moves along the x axis. If between times t_1 and t_2 it moves from x_1 to x_2, its average velocity* during this time is defined to be

Average velocity

$$\bar{v} = \frac{\Delta x}{\Delta t} = \frac{x_2 - x_1}{t_2 - t_1}. \tag{5.42}$$

Except for notational changes, this is the same as Equation 5.4. The instantaneous velocity of the particle (Equations 5.5 and 5.6) is

Instantaneous velocity

$$v = \lim_{\Delta t \to 0} \frac{\Delta x}{\Delta t} = \frac{dx}{dt}. \tag{5.43}$$

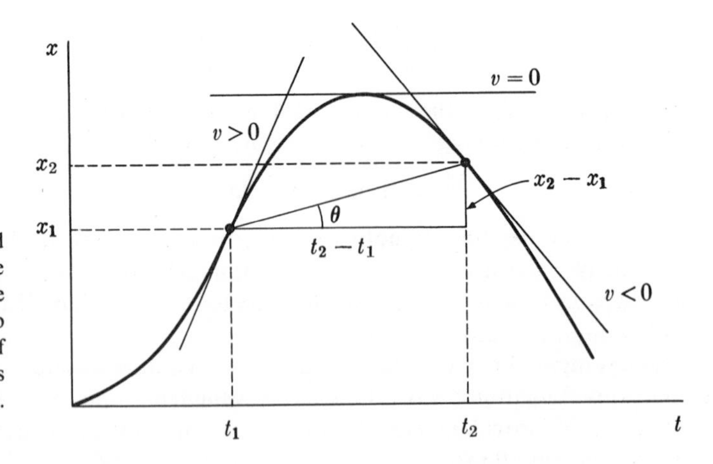

FIGURE 5.12 Average velocity and instantaneous velocity. The slope of the line drawn between the two dots gives the average velocity in the time interval t_1 to t_2: $\bar{v} = (x_2 - x_1)/(t_2 - t_1)$. The slopes of the tangent lines give the instantaneous velocities at these points.

* To simplify phraseology in this section, we use "velocity" to mean x component of velocity and "acceleration" to mean x component of acceleration. Also, for notational convenience we drop the subscript x from v and a.

TABLE 5.5 TEMPERATURE MEASUREMENTS IN A HYPOTHETICAL EXPERIMENT

Distance, r, from Center of Sphere (cm)	Temperature, T (K)
0	251.0
1	252.0
2	254.6
3	259.0
4	265.3
5	273.7
6	283.4
7	295.0

These operations are indicated graphically in Figure 5.12. In a similar way, acceleration is defined as rate of change of velocity with time. In some finite interval, the average acceleration is given by

Definition of average acceleration

$$\bar{a} = \frac{\Delta v}{\Delta t} = \frac{v_2 - v_1}{t_2 - t_1}; \tag{5.44}$$

the instantaneous acceleration is defined by

Instantaneous acceleration

$$a = \lim_{\Delta t \to 0} \frac{\Delta v}{\Delta t} = \frac{dv}{dt}. \tag{5.45}$$

Acceleration may also be written as a second derivative:

$$a = \frac{d}{dt}\left(\frac{dx}{dt}\right) = \frac{d^2x}{dt^2}. \tag{5.46}$$

Nature is simple: Mechanics needs only first and second derivatives

It is a remarkable feature of the theory of mechanics that all motion can be described without reference to derivatives higher than the second.

Figure 5.13 shows graphs of x vs t, v vs t, and a vs t for an example of motion that follows no simple mathematical law. Such graphs are helpful in visualizing motion. Note the characteristic features that relate the three graphs. The slope of the position curve is equal to the height of the velocity curve. The slope of the velocity curve is equal to the height of the acceleration curve. When the position curve reaches a maximum or minimum, the velocity is zero. When the velocity reaches a maximum or minimum, the acceleration is zero. Without the intermediary of the velocity curve, there is also a simple connection between the position curve and the acceleration. When the acceleration is negative (directed to the left), the curvature of the position curve is negative—like an upside-down cup. For positive acceleration, the position curve has positive curvature. At a point of zero acceleration, the position curve has an inflection point (zero curvature).

MOTION WITH CONSTANT VELOCITY

For constant-velocity motion, the special case of greater mathematical simplcity occurs when the particle is located at the origin, $x = 0$, at the initial time,

$t = 0$. More generally, the particle might be somewhere else, at x_0 when $t = 0$. The equations for the special and general cases are:

Special General

$$x = v_0 t, \quad (5.47) \qquad x = x_0 + v_0 t. \quad (5.48)$$

Both

Equations of constant velocity

$$v = \frac{dx}{dt} = v_0, \quad (5.49)$$

$$a = \frac{dv}{dt} = 0. \quad (5.50)$$

Since $v = v_0$, the symbols v and v_0 may be used interchangeably. For this simplest example of motion, the graph of position vs time is a straight line of slope v (Figure 5.10), and the graph of velocity vs time is a straight horizontal line (zero slope).

MOTION WITH CONSTANT ACCELERATION

For constant acceleration, we distinguish again between a special case and the general case. The special case is characterized by $x = 0$ and $v = 0$ when

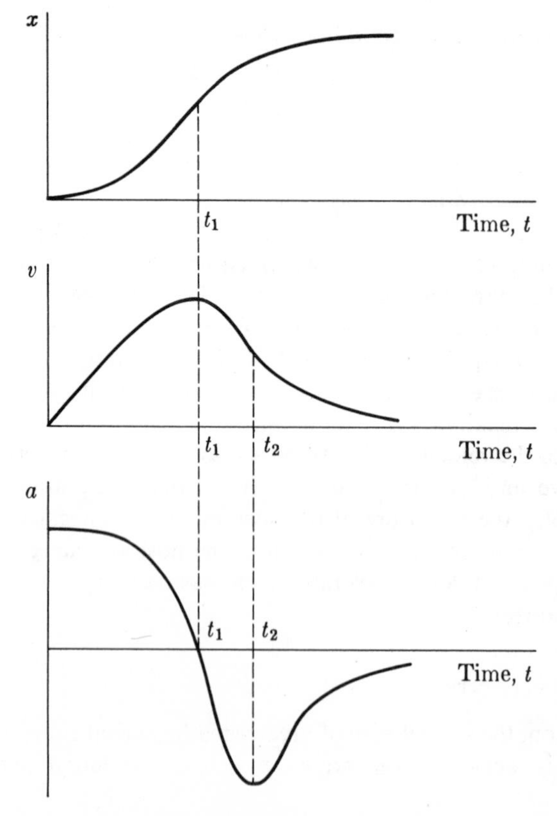

FIGURE 5.13 Graphs of position, velocity, and acceleration for an example of motion in one dimension. Each of the quantities plotted vertically is actually the component of a vector quantity. Vertical dashed lines are drawn at t_1, the time of maximum positive velocity and zero acceleration, and at t_2, the time of maximum rate of decrease of velocity (peak negative acceleration).

$t = 0$, motion starting initially from rest at the origin. For the general case, neither the initial position x_0 nor the initial velocity v_0 need be zero. *All* the important equations governing uniformly accelerated motion can be written down in a small space. We shall list them first and then discuss them. They are:

Equations of constant acceleration

Special		General	
$x = \frac{1}{2}a_0t^2,$	(5.51)	$x = x_0 + v_0t + \frac{1}{2}a_0t^2,$	(5.55)
$v = \frac{dx}{dt} = a_0t,$	(5.52)	$v = \frac{dx}{dt} = v_0 + a_0t,$	(5.56)
$a = \frac{dv}{dt} = a_0,$	(5.53)	$a = \frac{dv}{dt} = a_0,$	(5.57)
$v^2 = 2a_0x,$	(5.54)	$v^2 - v_0^2 = 2a_0(x - x_0).$	(5.58)

For notational convenience, the constant a_0 may be replaced by a in any of the above equations. Differentiation of the first equation in each column yields the second equation in the column. Differentiation of the second yields the third, confirming the constancy of the acceleration. The fourth equations are useful relationships between v and x, obtained by eliminating the time between the first two equations in each column. It is instructive to compare Equations 5.52 and 5.54. The first states that the velocity increases in proportion to time; the second states that the *square* of the velocity increases in proportion to distance, or equivalently, that the velocity increases in proportion to the square root of the distance.*

An important example, free fall from rest

■ EXAMPLE 1: A simple example of motion with constant acceleration is free fall from rest. Time and distance data for such motion are given in Table 5.6, with x measured downward from the starting point. Friction is ignored. A graph of x vs t constructed from these numbers is a parabola passing through the origin (Figure 5.14). Its equation is

$$x = \tfrac{1}{2}gt^2, \tag{5.59}$$

with

$$a = g = 9.8\ \text{m/sec}^2.$$

The straight line in Figure 5.14 has a slope equal to the average velocity in the interval 0 to t. This is

$$\bar{v} = \frac{x}{t} = \tfrac{1}{2}gt.$$

---

* In *Dialogues Concerning Two New Sciences*, Galileo confesses that when he first began his study of uniformly accelerated motion, he believed that velocity increases in proportion to the distance traversed, an error that he soon corrected.

TABLE 5.6 TIME AND DISTANCE DATA FOR MOTION WITH CONSTANT ACCELERATION

Time, t (sec)	Distance, x (m)
0	0
1	4.9
2	19.6
3	44.1
4	78.4
6	176.4
10	490.0

Interestingly, for free fall from rest, the average velocity up to a given time is exactly half of the instantaneous velocity at that time,

$$v = gt. \tag{5.60}$$

A graph of v vs t is shown in Figure 5.15. What is its slope? ■

■ EXAMPLE 2: The takeoff roll of a jet is executed with approximately constant acceleration, $a = 1.5$ m/sec², and the duration of the roll is 40 sec. What length of runway is required? What is the takeoff speed? The special equations, 5.51–5.54, suffice. Since acceleration and time are given, the takeoff speed may be calculated directly from Equation 5.52. It is

$$v = at = 1.5 \text{ m/sec}^2 \times 40 \text{ sec} = 60 \text{ m/sec}.$$

(This is 134 mile/hr.) Either Equation 5.51 or Equation 5.54 may be used to

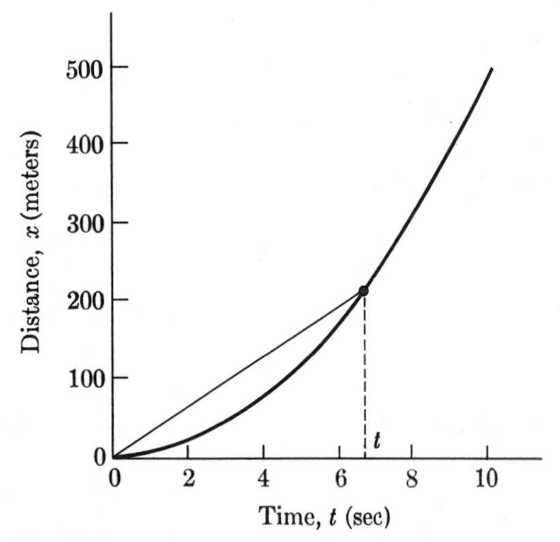

FIGURE 5.14 Graph of distance vs time for free fall without friction. The curve is a parabola. The slope of the straight line is the average velocity in the interval 0 to t. It proves to be half the instantaneous velocity at t.

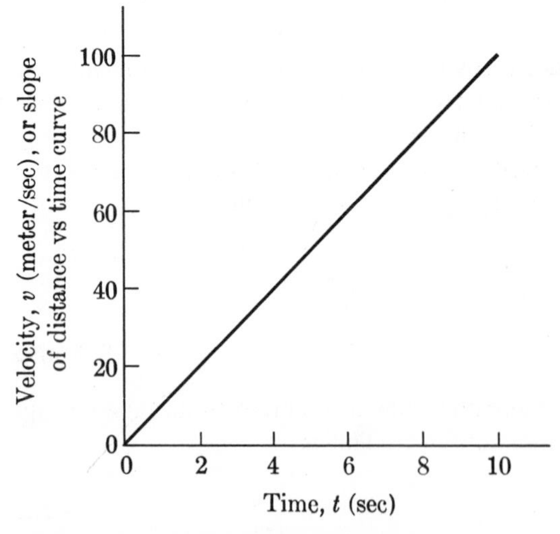

FIGURE 5.15 Graph of velocity vs time for free fall without friction. The velocity, plotted vertically, is the same as the slope of the curve in Figure 5.14.

obtain the takeoff distance. Let us use both. From Equation 5.51, we get

$$x = \tfrac{1}{2}at^2 = \tfrac{1}{2}(1.5 \text{ m/sec}^2)(1{,}600 \text{ sec}^2) = 1{,}200 \text{ m}.$$

From Equation 5.54, we get

$$x = \frac{v^2}{2a} = \frac{3{,}600 \text{ m}^2/\text{sec}^2}{2 \times 1.5 \text{ m/sec}^2} = 1{,}200 \text{ m} = 1.2 \text{ km}.$$

An interesting implication of Equation 5.54 is that when half the takeoff distance has been used up, the speed has reached 70.7 percent of the takeoff speed. If a jet pilot finds that he has reached only 60 percent of his takeoff speed when half the runway is used up, should he continue the takeoff or attempt to stop? ■

Example with x_0 and v_0 not zero

■ EXAMPLE 3: A baseball is thrown vertically upward, leaving the hand of the thrower 2.0 m above the ground with a speed of 15 m/sec. What is the maximum height above the ground reached by the ball? Before the equations of uniform acceleration can be applied, an origin of the x coordinate and an origin of time must be chosen, and the values of the three constants x_0, v_0, and a must be known. The time origin may conveniently be chosen as the instant when the ball leaves the thrower's hand; the x origin may be chosen to be at ground level, with positive x being measured upward from the ground (opposite to the x direction in Example 1). With these choices, the constants are

$$x_0 = 2.0 \text{ m},$$

$$v_0 = 15 \text{ m/sec},$$

$$a = -9.8 \text{ m/sec}^2.$$

The first, x_0, is the initial height above ground, the value of x when $t = 0$. The second, v_0, is the initial velocity, positive because it is directed upward. The acceleration caused by gravity is a constant vector of magnitude 9.8 m/sec^2 directed vertically downward. In this example its x component, or upward component, a, is therefore negative. These numbers may be substituted in Equations 5.55–5.58 to provide equations describing the subsequent history of the baseball (until it collides with the ground). To answer the particular question posed above, we must utilize the fact that the velocity of the ball is zero at the moment it reaches its maximum height. At that moment the velocity equation, $v = v_0 + at$, becomes

$$0 = 15 - 9.8t.$$

Its solution, $t = 1.53$ sec, is the time when the ball is at its peak height. The general position equation, $x = x_0 + v_0t + \tfrac{1}{2}at^2$, is, in this example,

$$x = 2.0 + 15t - 4.9t^2.$$

At the time $t = 1.53$ sec, $x = 13.5$ m. This is the maximum height of the ball. ■

As this example makes clear, equations are tools that do not provide automatic right answers. It is necessary to choose a frame of reference (in this case the origins of x and t), to find the correct constants, to pick the right equation, to apply the right criterion for the solution (in this case the fact that $v = 0$ at the maximum height), to work with a consistent set of units, and to interpret the solution correctly. Diagrams are frequently helpful—either pictorial diagrams of the problem in question or graphs of the variables of greatest interest. Both kinds of diagrams for the example of the thrown baseball are shown in Figure 5.16.

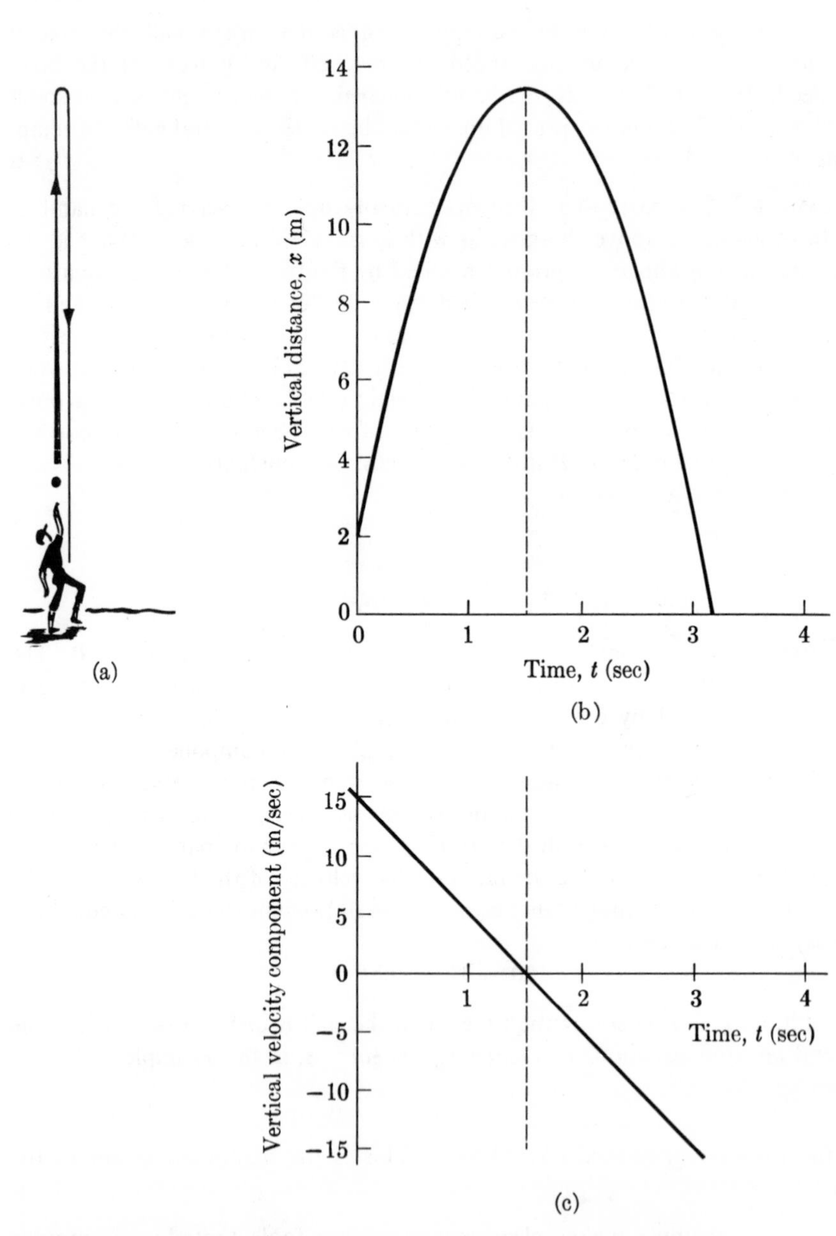

FIGURE 5.16 Example of vertically thrown baseball. (a) Pictorial diagram. (b) Graph of position vs time. The curve is a parabola. (c) Graph of vertical component of velocity vs time.

5.8 The indefinite integral

Integration: the inverse of differentiation

The operation that is the inverse of differentiation is called *integration*. The derivative of a function $I(x)$ is itself a function, which we shall call $f(x)$:

$$f(x) = \frac{dI}{dx}. \tag{5.61}$$

Integration consists in finding the function $I(x)$ whose derivative is equal to a given function $f(x)$. For the moment we are concerned with functions $f(x)$ that have a simple mathematical form. In the next section we examine the graphical significance of integration; this will open up the possibility of coping also with functions $f(x)$ that are mathematically complicated or are defined only by means of tabular data.

■ EXAMPLE 1: A derivative function is $f(x) = 0$. What function $I(x)$ is related to this f by Equation 5.61? A knowledge of simple derivatives (see Table 5.2) is sufficient to find the answer. The function whose derivative is zero is a constant:

$$I(x) = C,$$

where C is *any* fixed number. ■

■ EXAMPLE 2: The function $f(x) = ax$ (where a is a constant) is the derivative of what function $I(x)$? Examination of Table 5.2 suggests the proportionality $I \sim x^2$. Trial and error shows the correct coefficient of x^2 to be $\frac{1}{2}a$, so an answer is

$$I(x) = \tfrac{1}{2}ax^2.$$

However, this is not the most general answer. Since any constant C added to I leaves its derivative unchanged, a more general (in fact the *most* general) answer is

$$I(x) = \tfrac{1}{2}ax^2 + C.$$

To check an indefinite integral, differentiate it

The correctness of this answer can be verified by differentiating it:

$$\frac{dI}{dx} = ax = f(x).$$ ■

In the examples above, I is called the *indefinite integral* of f. It is "indefinite" because it is arbitrary to within an additive constant (see Figure 5.17). The notation for the indefinite integral is

Notation for the indefinite integral

$$I(x) = \int f(x)\,dx. \tag{5.62}$$

In this expression, $f(x)$ is called the *integrand*. It is the derivative of I (Equation 5.61), and I is the indefinite integral* of f. (The function I is also known sometimes as the *antiderivative* of f.) The symbols on the right side of Equation 5.62 together represent a single entity. They do not mean, for example, that $f(x)$ is multiplied by dx. The reason for this choice of notation for the integral will

* The modifier "indefinite" is often omitted.

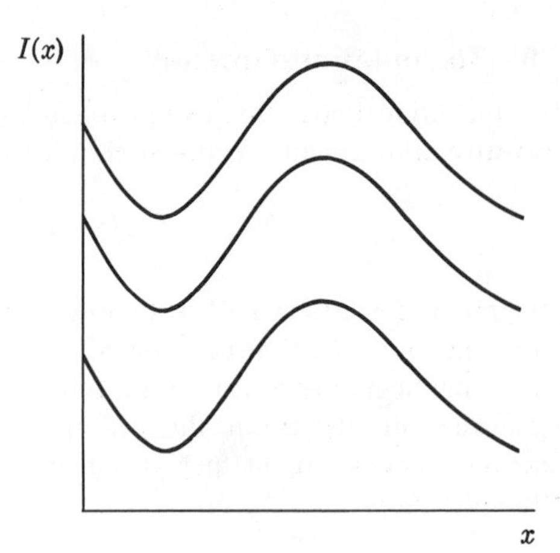

FIGURE 5.17 Functions with identical derivatives may differ from one another by a constant.

become clear in the next section.

The results of Examples 1 and 2 may be expressed by the equations

Indefinite integrals contain an arbitrary constant

$$\int 0 \, dx = C,$$

$$\int ax \, dx = \tfrac{1}{2}ax^2 + C.$$

These and other common integrals are listed in Table 5.7. Note that moving "backward" in this table (right column to left column) is equivalent to differentiating. In Table 5.2, moving from right column to left column is equivalent to integrating.

Integration, like differentiation, is a linear operation. This property of linearity can be expressed by the equation

The linearity property of integration

$$\int [af(x) + bg(x)] \, dx = a \int f(x) \, dx + b \int g(x) \, dx, \tag{5.63}$$

in which a and b are constants. There are, however, no simple rules analogous to Equations 5.15 and 5.16 for integrating products or quotients. Integration is, in general, a more complex operation than differentiation.*

■ EXAMPLE 3: In physical applications, the constant of integration is usually fixed with the help of initial conditions, or "boundary conditions." Consider the motion of a particle along the x axis with constant acceleration. Its velocity is given by

$$v = a_0 t,$$

and at time $t = 0$ it is located at $x = x_0$. What is its position at some later time t? Since

$$v = \frac{dx}{dt}, \tag{5.64}$$

with v being given and x being the unknown, this is a problem in integration.

* In numerical work, however, such as in computer calculations, integration is generally more accurate than differentiation.

TABLE 5.7 SOME INDEFINITE INTEGRALS

$f(x)$	$I(x) = \int f(x)\, dx$
0	C
$a = \text{constant}$	$ax + C$
x	$\frac{1}{2}x^2 + C$
x^2	$\frac{1}{3}x^3 + C$
$1/x$	$\ln x + C$
$1/x^2$	$-(1/x) + C$
$\sqrt{x}$	$\frac{2}{3}x^{3/2} + C$
$1/\sqrt{x}$	$2\sqrt{x} + C$
$x^n \ (n \neq -1)$	$\dfrac{x^{n+1}}{n+1} + C$
$\sin x$	$-\cos x + C$
$\sin ax$	$-\dfrac{1}{a}\cos ax + C$
$\cos x$	$\sin x + C$
$\cos ax$	$\dfrac{1}{a}\sin ax + C$
$\tan x$	$-\ln \lvert\cos x\rvert + C$
$\tan ax$	$-\dfrac{1}{a}\ln \lvert\cos ax\rvert + C$
e^x	$e^x + C$
e^{ax}	$\dfrac{1}{a}e^{ax} + C$
e^{-x}	$-e^{-x} + C$
e^{-ax}	$-\dfrac{1}{a}e^{-ax} + C$
$\ln x$	$x \ln x - x + C$
$\ln ax$	$x \ln ax - x + C$

Equations 5.64 and 5.61 are to be compared. The correspondence is:

Equation 5.64	Equation 5.61
v	f
x	I
t	x.

The equivalent of Equation 5.62 is then

$$x(t) = \int v(t)\, dt. \tag{5.65}$$

In this example,

$$x(t) = \int a_0 t \, dt = \tfrac{1}{2}a_0 t^2 + C. \tag{5.66}$$

Now the initial condition may be introduced. In Equation 5.66, set $t = 0$ and $x = x_0$ to obtain

An initial condition fixes the integration constant

$$x_0 = 0 + C.$$

The constant C is determined to be x_0, so for this example of motion, Equation 5.66 becomes

$$x = x_0 + \tfrac{1}{2}a_0 t^2.$$

The result conforms to Equation 5.55, with $v_0 = 0$. Note, incidentally, that Equation 5.65 could also be written

A notational reminder: The indefinite integral is an antiderivative

$$x(t) = \int \frac{dx}{dt} \, dt.$$

The simplicity of this form means that the detailed comparison of notation in Equations 5.64 and 5.61 is not really necessary. ■

★INTEGRATION AND DIFFERENTIAL EQUATIONS

Equations 5.55–5.58 summarize the features of motion with constant acceleration. In considering such motion, it is somewhat more logical to step up from Equation 5.57 to Equation 5.56 and from Equation 5.56 to Equation 5.55 by successive integration, rather than step down by differentiation. In general, for motion in one dimension (with acceleration not necessarily constant), velocity is obtained from acceleration by the integral formula

$$v = \int \frac{dv}{dt} \, dt = \int a \, dt, \tag{5.67}$$

Integral formulas of one-dimensional kinematics

and position is obtained from velocity by the analogous formula

$$x = \int \frac{dx}{dt} \, dt = \int v \, dt. \tag{5.68}$$

In the absence of specific knowledge about initial conditions, these are indefinite integrals, each introducing an arbitrary constant of integration. For the special case of constant acceleration, the integrations may be performed readily. Then Equation 5.67 yields

$$v = a \int dt = a(t + C) = at + v_0 . \tag{5.69}$$

Here C is the constant of integration, and the combination aC is rewritten as another constant, v_0. This result shows that Equation 5.56 is the most general implication of Equation 5.57. If, in Equation 5.68, we replace v by $at + v_0$, we obtain

$$x = \int (at + v_0) \, dt = \tfrac{1}{2}at^2 + v_0 t + x_0 , \tag{5.70}$$

where x_0 is the new constant of integration. This result agrees with Equation 5.55.

Another way to look at the route from acceleration to velocity to position is with the help of the idea of differential equations. An equation such as

$$\frac{dv}{dt} = a_0 \tag{5.71}$$

A simple differential equation

is a differential equation. This means simply that it is an equation containing derivatives as well as possibly algebraic expressions. This particular equation can be looked upon as a statement about the function $v(t)$. Were $v(t)$ not already known, the equation would pose the following question: What function $v(t)$ has a first derivative equal to the constant a_0? Finding the function (or functions) $v(t)$ is called solving the differential equation. Here we have a very simple example and know the answer to be obtainable directly by integration. It is

Solution of the differential equation

$$v(t) = v_0 + a_0 t.$$

In general, however, solving a differential equation may require more complicated techniques. Our purpose here is not to go into these techniques but only to call attention to the fact that some of the important equations of physics are differential equations.

Equation 5.71 can be rewritten

$$\frac{d^2x}{dt^2} = a_0 , \tag{5.72}$$

which is called a second-order differential equation. Its general solution is given by Equation 5.70. It is characteristic of the solution of a second-order differential equation to contain two constants of integration.

5.9 The definite integral

Differentiation, as we have learned in earlier sections, has a simple graphical interpretation. It is equivalent to finding the slope of a curve. Integration also has a simple graphical interpretation. It is related to finding the area under a curve. In this section we will first *define* the definite integral as the area under a curve in a graph; then we will explore properties of the definite integral so defined; finally we will demonstrate the simple and important connection between the definite and the indefinite integrals.

DEFINITION OF THE DEFINITE INTEGRAL

Definition of area under a curve

If a function $f(x)$ is presented graphically in the form $f(x)$ vs x, the "area under the curve between the limits a and b" means the area bounded by the curve of $f(x)$, the x axis, and the two lines $x = a$ and $x = b$ (Figure 5.18). The area under the graph of a positive function is defined to be positive. The area "under" (actually above) the graph of a negative function is defined to be negative. As shown in Figure 5.18(c), positive and negative areas add algebraically and may cancel. The total area between definite limits of x is called a *definite integral.* The notation for the definite integral is

Area = definite integral

$$D = \int_a^b f(x)\, dx. \tag{5.73}$$

Notation for definite integral

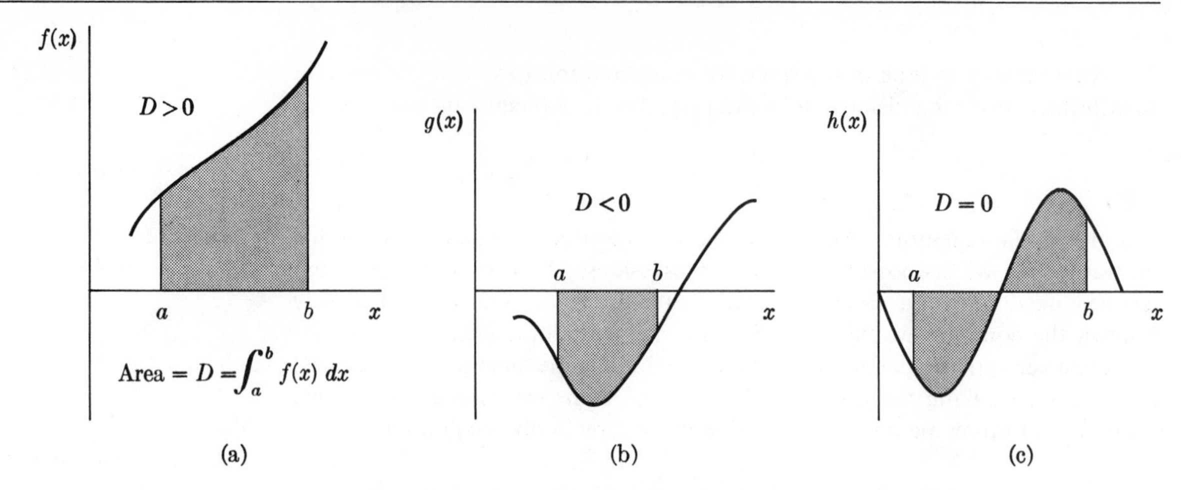

FIGURE 5.18 Definition of the definite integral D. It is equal to the area bounded by the x axis, a plotted curve, and arbitrarily chosen limits of x such as $x = a$ and $x = b$. Area above the x axis is defined to be positive; area below the x axis is negative. Shown here are examples with (a) $D > 0$, (b) $D < 0$, and (c) $D = 0$.

This area is also called the integral of $f(x)$ from a to b. Despite the obvious similarity of this notation to the notation for the indefinite integral, we have *no reason*, based on what has been presented so far, to see a connection between the two kinds of integrals. They *are* closely related, of course, but that relationship will not be established until the end of this section. For the moment, let us regard the notational similarity as a coincidence and proceed to examine properties of definite integrals—or, equivalently, of the areas under curves.

SOME PROPERTIES OF DEFINITE INTEGRALS

Generalized area may have any unit

It is important to notice that a definite integral between fixed limits is a fixed quantity, not a function. It has a specific numerical value and generally has a unit, which need not be a unit of area such as m^2 or cm^2. Just as the idea of *slope* is generalized from its purely geometric meaning and acquires a unit determined by the *quotient* of the vertically plotted quantity and the horizontally plotted quantity, the idea of *area* is also generalized and acquires a unit determined by the *product* of the vertically and horizontally plotted quantities. If, for instance, $f(x)$ were expressed in force units, newtons, and x were expressed in distance units, meters, the area under the curve of $f(x)$ would be expressed in units of force times distance, N m. The notation $\int_a^b f(x)\,dx$ makes this easy to remember.

■ EXAMPLE 1: A function $f(x)$ is equal to a constant C. What is its definite integral from $x = a$ to $x = b$? As shown in Figure 5.19, the definite integral

$$D = \int_a^b f(x)\,dx = \int_a^b C\,dx$$

To find some areas, simple geometry suffices

is equal to the area of a rectangle. Its base is the distance $b - a$; its height is equal to C. The area is therefore

$$D = (b - a)C.$$

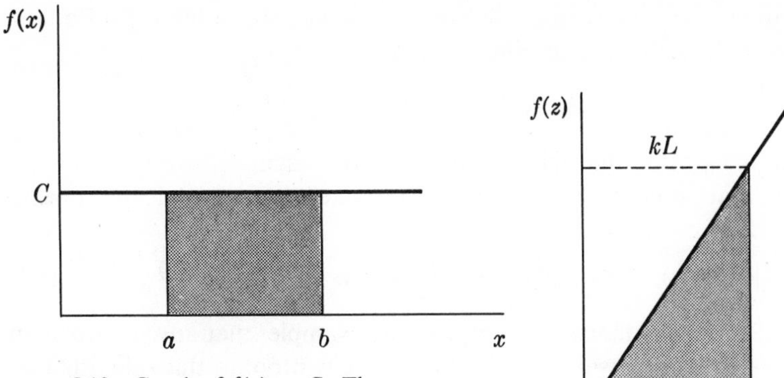

FIGURE 5.19 Graph of $f(x) = C$. The definite integral from a to b is easily calculable as the area of a rectangle.

FIGURE 5.20 Graph of $f(z) = kz$. The definite integral from $z = 0$ to $z = L$ is equal to the area of the shaded triangle.

The unit of D is evidently the unit of x multiplied by the unit of f. ■

■ EXAMPLE 2: A certain force function is given by $f(z) = kz$, where z is distance measured in meters and where k is a constant expressed in N/m. Evaluate the definite integral

$$D = \int_0^L f(z)\,dz = \int_0^L kz\,dz.$$

The appropriate graph (Figure 5.20) shows that this definite integral is equal to the area of a right triangle of base L and height kL. This area is

$$D = \tfrac{1}{2} \times L \times kL = \tfrac{1}{2}kL^2.$$

The unit of D is N m^2. ■

As shown in Figure 5.21, definite integrals possess a simple additive property expressed by

An additive property

$$\int_a^b f(x)\,dx + \int_b^c f(x)\,dx = \int_a^c f(x)\,dx. \tag{5.74}$$

In the figure, $a < b < c$. However, this additive property holds for any relative magnitudes of a, b, and c provided we define

Definition of "right-to-left" integration

$$\int_b^a f(x)\,dx = -\int_a^b f(x)\,dx. \tag{5.75}$$

This states that a definite integral carried out in the direction of decreasing x is (by definition) equal in magnitude and opposite in sign to the integral carried

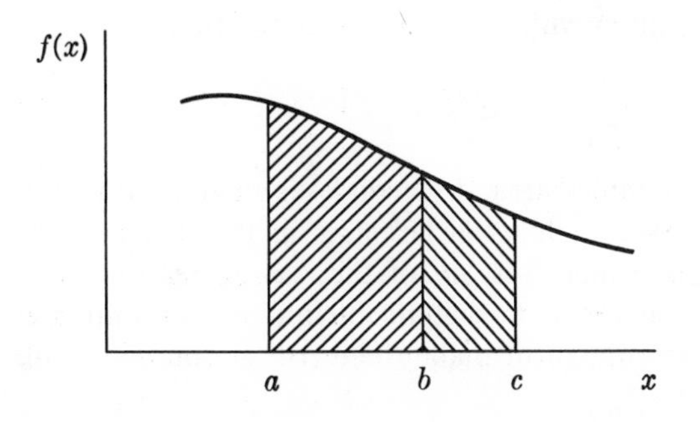

FIGURE 5.21 Graphical demonstration of Equation 5.74. The integral from a to c is the sum of integrals from a to b and b to c.

out in the direction of increasing x between the same pair of limits. In Equation 5.74, for example, set $c = a$ to obtain

$$\int_a^b f(x)\,dx + \int_b^a f(x)\,dx = \int_a^a f(x)\,dx.$$

The definite integral on the right, from a to a (no distance), is zero.

The definite integral shares with the indefinite integral the linearity property:

The linearity property again

$$\int_a^b [\alpha f(x) + \beta g(x)]\,dx = \alpha \int_a^b f(x)\,dx + \beta \int_a^b g(x)\,dx, \tag{5.76}$$

where α and β are constants. This means, for example, that any multiplicative constant can be taken outside an integral sign. The proof of this rule is left as a problem.

THE DEFINITE INTEGRAL AS THE LIMIT OF A SUM

In the two examples above, the definite integrals could be evaluated readily using simple geometric formulas for the areas of a rectangle and a triangle. But what if the function $f(x)$ is more complicated, perhaps having a graph like that in Figure 5.18? Or what if $f(x)$ has no known mathematical form at all, being only a curve drawn through experimentally determined points? Simple geometric formulas for area will not suffice. A general procedure that does work, for all but very "pathological" functions, is illustrated in Figure 5.22. Divide the interval from $x = a$ to $x = b$ into n subintervals of width Δx_1, $\Delta x_2, \ldots, \Delta x_n$. (These subintervals may be of equal width, but they need not be.) The left edges of the successive subintervals are at $x_1, x_2, x_3, \ldots, x_n$. From the figure, it is obvious that

$$x_1 = a,$$
$$x_2 = x_1 + \Delta x_1,$$
$$x_3 = x_2 + \Delta x_2,$$
$$\vdots$$

Now form a set of rectangles, the first of base Δx_1 and height $f(x_1)$, the second of base Δx_2 and height $f(x_2), \ldots,$ the last of base Δx_n and height $f(x_n)$. The area of the first rectangle is $f(x_1)\,\Delta x_1$; the total area of all the rectangles is

$$A_n = f(x_1)\,\Delta x_1 + f(x_2)\,\Delta x_2 + \cdots + f(x_n)\,\Delta x_n.$$

The subscript n designates the number of rectangles used to calculate the area. This sum is more conveniently written with the summation notation:

Area of a set of rectangles

$$A_n = \sum_{i=1}^{n} f(x_i)\,\Delta x_i. \tag{5.77}$$

It approximates the area under the curve

It is clear in Figure 5.22 that this summed area of rectangles is an approximation to the true area under the curve and that the larger is n, the better is the approximation (provided that as n increases, *all* of the Δx_i decrease). In a practical situation with an empirical curve, the definite integral may be evaluated in this way, n being chosen large enough to make the error acceptably small.

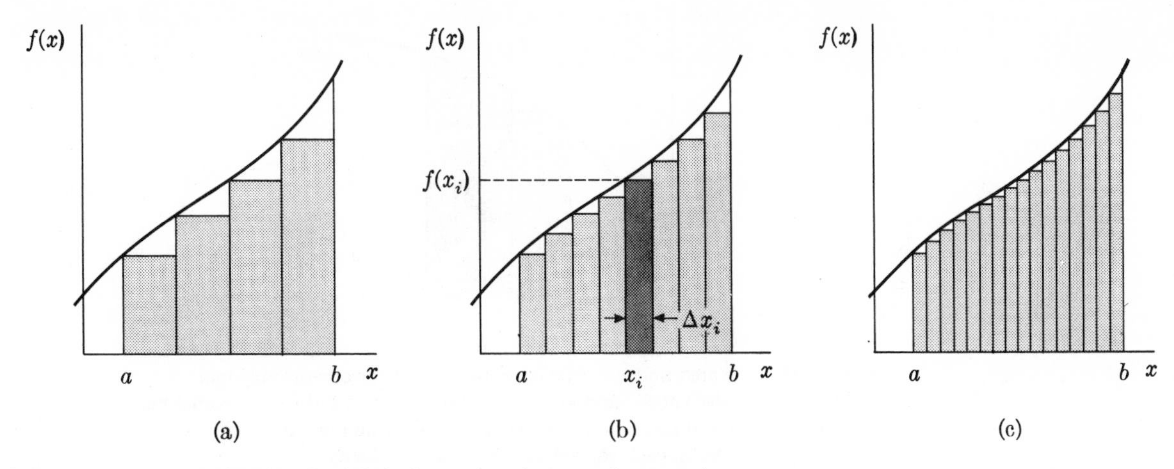

FIGURE 5.22 Approximating the definite integral as the sum of rectangular areas for (a) $n = 4$, (b) $n = 8$, and (c) $n = 16$. As n approaches infinity, the total area of the rectangles approaches the true area under the curve.

Integration on a computer is usually performed in much this way, except that the rectangles may be replaced by figures whose tops are slanting or curved to follow the graph of the function more closely.

It is an important theorem of calculus that for a wide class of functions, the true area under the curve is the limit of the area A_n as n approaches infinity and as all Δx_i approach zero:

$$D = \int_a^b f(x)\,dx = \lim_{n\to\infty} \sum_{i=1}^{n} f(x_i)\,\Delta x_i\,. \tag{5.78}$$

Definite integral as the limit of a sum

(It is to be understood that the limit also includes the condition $\Delta x_i \to 0$ for all i; this is most easily ensured by setting all of the Δx_i equal to the same magnitude, $(b - a)/n$.) Equation 5.78 now makes clear the reason for the notation used for the integral. Symbolically, the integral sign $\int$ (derived from the letter S for sum) replaces the Greek sigma, Σ, and dx replaces Δx_i. The limits of integration, a and b, replace the limits of the sum, 1 and n.

■ EXAMPLE 3: A particle moves along the x axis with a velocity $v(t)$ that is a known function of time. What is the displacement of the particle from time t_a to time t_b? We can use physical reasoning to show that the displacement can be written as a definite integral.

In this discussion, take note of the fact that the independent variable, called x in the preceding general discussion, is now the time t. The correspondence is:

General discussion	This example
x	t
$f(x)$	$v(t)$.

In this example, the coordinate x is another dependent variable, which can be written $x(t)$. However, we seek only the particular displacement $x_b - x_a$.

At any time t_i, the particle has velocity $v(t_i)$. For a sufficiently short ensuing

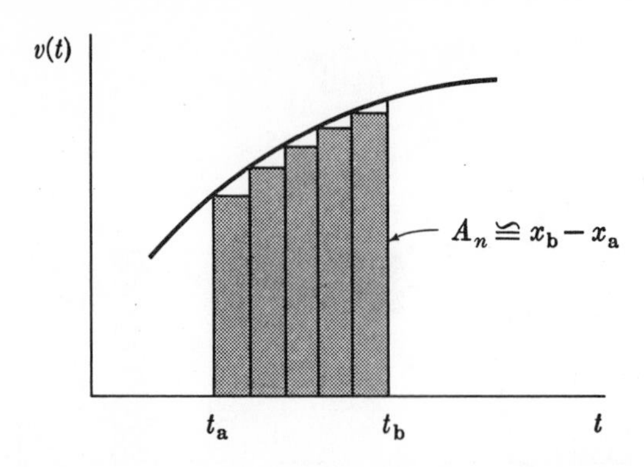

FIGURE 5.23 The area under a velocity-vs-time curve for one-dimensional motion is the displacement during the time interval considered. Approximating the area by a series of rectangles is equivalent to dividing the motion into a succession of segments, each carried out at constant velocity.

time Δt_i, this velocity will change very little (see Figure 5.5). The displacement of the particle during this short interval can be calculated approximately as if the particle were moving with constant velocity:

Constant-velocity approximation in a brief interval

$$\Delta x_i \cong v(t_i)\, \Delta t_i \, .$$

(The symbol $\cong$ means "approximately equals.") A similar formula gives the displacement during the next interval, and so on for other intervals. If the total time period, $t_b - t_a$, is divided into n intervals, $\Delta t_1, \Delta t_2, \ldots, \Delta t_n$, an approximate formula for the total displacement is

$$x_b - x_a = \sum_{i=1}^{n} \Delta x_i \cong \sum_{i=1}^{n} v(t_i)\, \Delta t_i \, . \tag{5.79}$$

This formula is represented graphically in Figure 5.23. It evidently gives the rectangular approximation to the definite interval. Since the approximation of constant-velocity motion becomes increasingly accurate as the duration of a time interval decreases, Equation 5.79 becomes exact in the limit $n \to \infty$ and $\Delta t_i \to 0$. It is

$$x_b - x_a = \lim_{n \to \infty} \sum_{i=1}^{n} v(t_i)\, \Delta t_i \tag{5.80}$$

($\Delta t_i \to 0$ is also to be understood). According to Equation 5.78, this is a definite integral:

Displacement in a fixed time interval

$$x_b - x_a = \int_{t_a}^{t_b} v(t)\, dt. \tag{5.81}$$

Since $v = dx/dt$, this result can also be stated

$$x_b - x_a = \int_{t_a}^{t_b} \frac{dx}{dt}\, dt. \tag{5.82}$$

Here we see a strong hint of the connection between the definite and indefinite integrals (compare Equation 5.68). ■

■ EXAMPLE 4: A particle moves with constant acceleration a_0 along the x axis, its velocity being given by

$$v(t) = v_0 + a_0 t. \tag{5.83}$$

What is its displacement from $t = 0$ to $t = T$? The algebraic Equation 5.55 provides the answer very simply. It is

$$x_T - x_0 = v_0 T + \tfrac{1}{2} a_0 T^2. \tag{5.84}$$

Let us obtain the answer in a different way, using Equation 5.81 and the idea of the definite integral. Substituting the right side of Equation 5.83 for $v(t)$ in Equation 5.81 gives

$$x_T - x_0 = \int_0^T (v_0 + a_0 t)\, dt$$

The integrand, $v_0 + a_0 t$, is plotted as a function of t in Figure 5.24. The total shaded area is the required definite integral. It is the sum of the area of a rectangle,

$$A_1 = v_0 T,$$

and the area of a triangle,

$$A_2 = \tfrac{1}{2} \times T \times a_0 T = \tfrac{1}{2} a_0 T^2.$$

The displacement is

$$x_T - x_0 = A_1 + A_2 = v_0 T + \tfrac{1}{2} a_0 T^2,$$

which is the same as Equation 5.84. ■

RELATIONSHIP OF THE DEFINITE INTEGRAL TO THE INDEFINITE INTEGRAL

A definite integral may be converted from a fixed quantity to a variable quantity by allowing its upper limit to vary. We may write a definite integral as a function of x, for example, in the following way:

Definite integral as a function of its upper limit

$$D(x) = \int_{x_0}^{x} f(x')\, dx'. \tag{5.85}$$

Let us examine how D depends on x. As indicated in Figure 5.25, if x is increased by the small increment Δx, the increase of the definite integral is given approximately by

$$\Delta D = D(x + \Delta x) - D(x) \cong f(x)\, \Delta x.$$

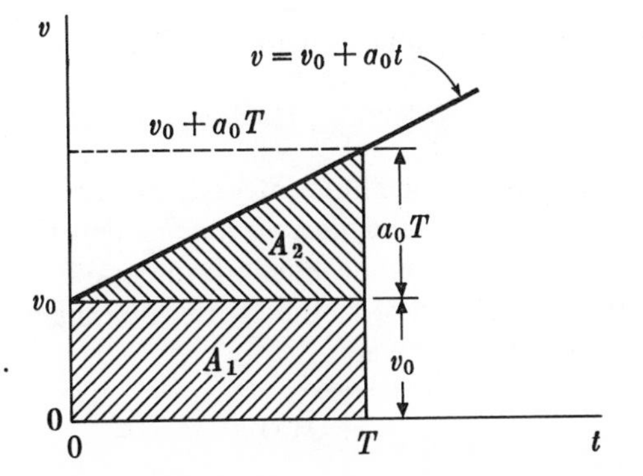

FIGURE 5.24 Graphical method of evaluating a definite integral. The area under the curve (which here is a straight line) is $A_1 + A_2$.

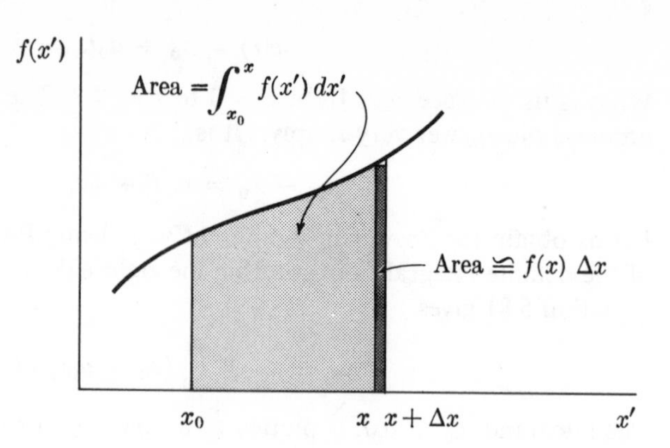

FIGURE 5.25 As the upper limit of a definite integral changes from x to $x + \Delta x$, the value of the definite integral changes approximately by $f(x)\,\Delta x$.

This can be written

$$f(x) \cong \frac{D(x + \Delta x) - D(x)}{\Delta x}. \tag{5.86}$$

In the limit that Δx approaches zero, this approximate equality becomes exact because in that limit the rectangular approximation for ΔD (Figure 5.25) becomes exact. We therefore obtain

Its derivative is the integrand

$$f(x) = \lim_{\Delta x \to 0} \frac{D(x + \Delta x) - D(x)}{\Delta x} = \frac{dD}{dx}. \tag{5.87}$$

Let us write this result next to Equation 5.61:

Same property for definite and indefinite integrals

$$f(x) = \frac{dD}{dx}, \qquad f(x) = \frac{dI}{dx}.$$

The first states that the derivative of a *definite* integral with respect to a variable upper limit is equal to the integrand $f(x)$. The second states that the derivative of the *indefinite* integral is equal to the same integrand $f(x)$. Since D and I have the same derivative, they can differ only by a constant, and we can write

$$D(x) = \int_{x_0}^{x} f(x')\,dx' = I(x) + C, \tag{5.88}$$

where C is a constant. To find C, set $x = x_0$. This means that the range of integration, from x_0 to x_0, is zero, so the definite integral is zero; hence

$$0 = I(x_0) + C.$$

Therefore $C = -I(x_0)$, and Equation 5.88 can be rewritten

The fundamental theorem relates the definite and indefinite integrals

$$D(x) = \int_{x_0}^{x} f(x')\,dx' = I(x) - I(x_0). \tag{5.89}$$

Recall that $I(x)$ is the indefinite integral, or antiderivative, of $f(x)$. This important result is known as the *fundamental theorem of integral calculus*. Whenever the function $f(x)$ is such that its indefinite integral $I(x)$ is known, the fundamental theorem provides the easiest method for evaluating a definite integral. Another notation for Equation 5.89 is

An alternative notation

$$\int_a^b f(x)\,dx = I(x)\Big|_a^b. \tag{5.90}$$

The notation to the right of the equal sign is shorthand for $I(b) - I(a)$.

■ EXAMPLE 5: Let us carry out Example 4, using the fundamental theorem. For the motion depicted in Figure 5.24, the displacement is

$$x_T - x_0 = \int_0^T (v_0 + a_0 t)\, dt.$$

Instead of evaluating the area geometrically, we can write

$$x_T - x_0 = v_0 t + \tfrac{1}{2} a_0 t^2 \Big|_0^T.$$

Since the terms on the right vanish at the lower limit, $t = 0$, the result is

$$x_T - x_0 = v_0 T + \tfrac{1}{2} a_0 T^2.$$

■

■ EXAMPLE 6: A parabola is described by $y = x^2$. What is the area under the parabola from $x = 2$ to $x = 3$ (see Figure 5.26)? Applying Equation 5.90, we have

$$D = \int_2^3 x^2\, dx = \frac{1}{3} x^3 \Big|_2^3 = \frac{27}{3} - \frac{8}{3}.$$

The answer is

$$D = \frac{19}{3}.$$

■

■ EXAMPLE 7: Show that the change of velocity and the acceleration of a body moving in one dimension are related by

$$\Delta v = \int_{t_1}^{t_2} a(t)\, dt. \tag{5.91}$$

Since $a = dv/dt$, then v is the indefinite integral of a (Equation 5.67). Therefore,

$$\int_{t_1}^{t_2} a\, dt = v(t)\Big|_{t_1}^{t_2} = v(t_2) - v(t_1).$$

The difference, $v(t_2) - v(t_1)$, is Δv, the change of velocity from t_1 to t_2. This verifies the correctness of Equation 5.91. ■

■ EXAMPLE 8: The acceleration of a hydraulic lift, starting from rest, is given for the first few seconds by

$$a(t) = \beta t^2,$$

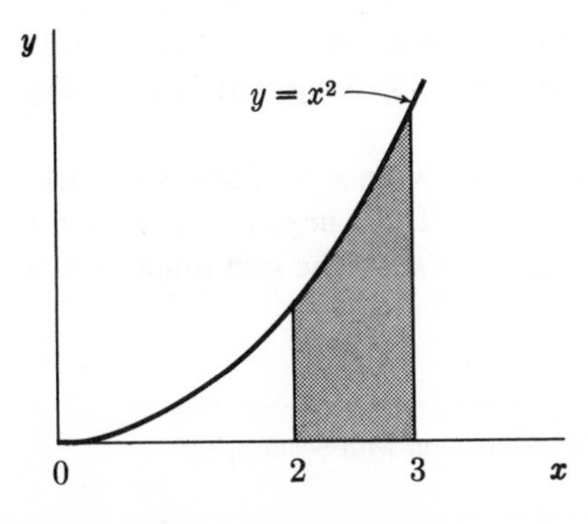

FIGURE 5.26 To find the area under a parabolic curve is a simple problem in integration.

where β is a constant equal to 10^{-2} m/sec^4. What is the displacement of the lift after 2 sec? From Equation 5.91, its velocity at time t (the same as its change of velocity) is

$$v(t) = \int_0^t \beta t^2 \, dt$$

$$= \tfrac{1}{3}\beta t^3 \Big|_0^t = \tfrac{1}{3}\beta t^3.$$

Note here that we have used the same symbol t for the variable in the integrand and for the upper limit of the integration. This is a common practice that need not cause confusion if one is aware of the double usage. The displacement in time t (see Equation 5.81) is

$$\Delta x = \int_0^t v(t) \, dt = \int_0^t \tfrac{1}{3}\beta t^3 \, dt$$

$$= \tfrac{1}{12}\beta t^4 \Big|_0^t = \tfrac{1}{12}\beta t^4.$$

Substitution of $t = 2$ sec gives

$$x = 0.0133 \text{ m} = 1.33 \text{ cm}.$$

■

5.10 Sine and cosine functions*

Besides algebraic functions, several other functions are remarkably ubiquitous in physics and deserve special attention. We consider here the sine and cosine functions, and in Section 5.11 we will examine the exponential and logarithmic functions.

A general sine function need have nothing to do with angles

The quantity $\sin \theta$ is a function of the angle θ, which is called the argument of the function. There is, however, no reason why the physical significance of the argument need be an angle. If we write the function

$$\sin x,$$

its argument x could be any dimensionless quantity, not necessarily an angle. The expression is nevertheless given definite meaning with the help of an angle. If, for example, x has the numerical value 2, then $\sin x$ means the sine of the angle 2 radians. Note that the generalization from angles to other quantities rests on an understanding that the radian shall be the standard unit of angular measurement. The idea of a trigonometric function that has nothing to do with angles and triangles is less mysterious than it might seem at first. It simply means that the mathematical essence has been distilled out of the original physical (or, in this case, geometrical) application. It is not very different from using the quadratic equation $s = \frac{1}{2}gt^2$ to describe something that has nothing to do with free fall.

The sine function is periodic

A sine function is plotted in Figure 5.27(a). Between $x = 0$ and π radians, the sine is positive, with a peak value of 1 at $x = \pi/2$. Its negative peak of -1 is reached at $x = 3\pi/2$. Beyond 2π, the pattern repeats. The sine function is a periodic function, with period 2π.

---

* The material in this section will be needed first in conjunction with Section 6.10.

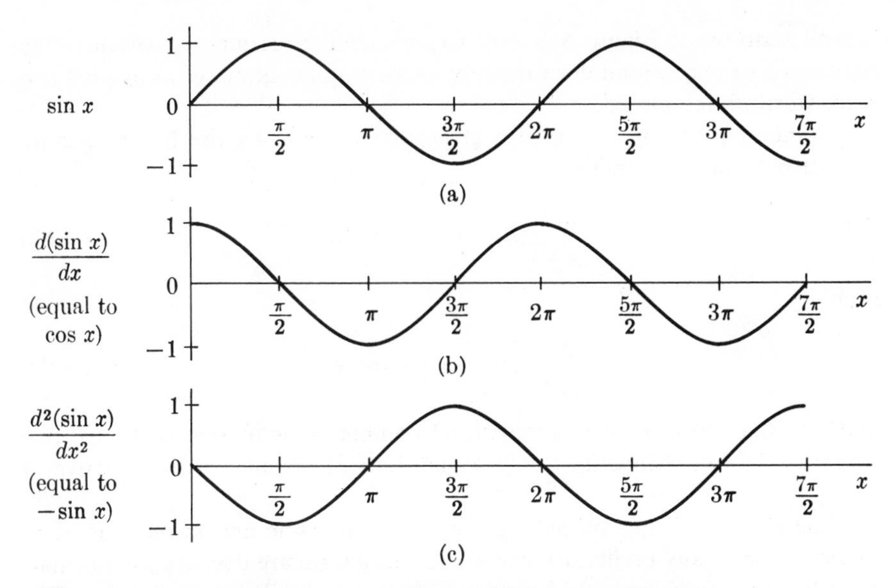

FIGURE 5.27 (a) Graph of $\sin x$ vs x. The horizontal scale is expressed in radians. (b) The derivative of the sine function is equal to the cosine function. Its graph is also the same as a sine function displaced horizontally to the left through $\frac{1}{2}\pi$. (c) The second derivative of the sine function is a negative sine function. Its graph is the same as a sine function displaced horizontally through π.

Relationships of sine and cosine functions

Inspection of Figure 5.27(a) makes clear that the derivative of the sine function (or the slope of its graph) must also be a periodic function with period 2π. This derivative is in fact a cosine function:

$$\frac{d}{dx}(\sin x) = \cos x. \tag{5.92}$$

The function $\cos x$ is plotted in Figure 5.27(b). Note that its graph is identical to the graph of $\sin x$ except for a displacement to the left through $\frac{1}{2}\pi$. This property is expressed by the formula

$$\cos x = \sin\left(\frac{\pi}{2} + x\right). \tag{5.93}$$

As illustrated in Figure 5.27(c), the derivative of the cosine function is the negative of the sine function:

$$\frac{d}{dx}(\cos x) = -\sin x. \tag{5.94}$$

Again differentiation is equivalent to shifting the curve leftward through $\frac{1}{2}\pi$:

$$-\sin x = \cos\left(\frac{\pi}{2} + x\right). \tag{5.95}$$

The double shift from the sine curve to its negative is expressed by

$$-\sin x = \sin(\pi + x). \tag{5.96}$$

Careful scrutiny of Figure 5.27 will pay dividends; learning to visualize the graphs of first and second derivatives by inspecting the graph of a function is a talent worth developing.

Combining the two derivatives given above, we can write for the second derivative of the sine function

$$\frac{d^2}{dx^2}(\sin x) = -\sin x. \tag{5.97}$$

Simple and special properties of second derivatives

Similarly,

$$\frac{d^2}{dx^2}(\cos x) = -\cos x. \tag{5.98}$$

Only the sine and cosine functions (and combinations of the two) have this property. No other function has a second derivative equal to the negative of itself.

The reason for the special significance of the sine and cosine curves in physics is that many oscillatory phenomena in nature are described mathematically by these functions. It is sufficient here to consider one of them. The curve of Figure 5.28 shows the displacement of a pendulum bob from its central position as a function of time. The equation for this curve is

Equation of oscillatory motion

$$s = A \sin \omega t, \tag{5.99}$$

whose two variables are s, the displacement, and t, the time. The other two quantities, A and ω, are constants. The quantity A, with the dimension of length, is the amplitude of the oscillation. It is equal to the maximum value achieved by s at the limit of the swing. The constant ω, with the dimension of inverse time (so that the product ωt is dimensionless), is related to the period T. A full oscillation of the sine function is completed when the product ωt is equal to 2π. But at that moment, one period has elapsed, so $t = T$. Therefore, $\omega T = 2\pi$, or, solving for ω,

Angular frequency

$$\omega = \frac{2\pi}{T}. \tag{5.100}$$

The quantity ω is called the angular frequency. The true frequency (oscillations per unit time), for which the usual symbol is ν (nu), is the inverse of the period T:

Frequency

$$\nu = \frac{1}{T}. \tag{5.101}$$

Therefore, the angular frequency and the frequency are related by

$$\omega = 2\pi\nu. \tag{5.102}$$

One reason for using ω in Equation 5.99 is simply for compactness of notation. In terms of frequency or period, the same equation is written

$$s = A \sin(2\pi\nu t) = A \sin\left(\frac{2\pi t}{T}\right). \tag{5.103}$$

In this example, the derivative of s with respect to t is the velocity com-

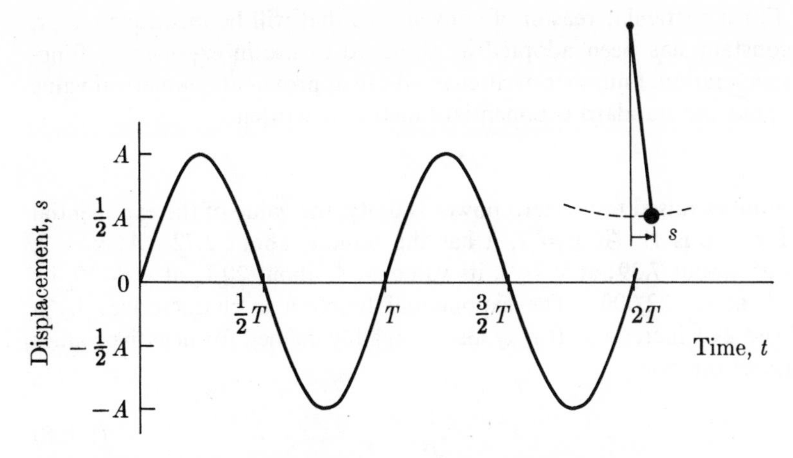

FIGURE 5.28 Graph of displacement vs time for a pendulum swinging through a small angle. This is a sine curve, as expressed by Equation 5.99.

ponent of the pendulum bob. Differentiating Equation 5.99, we get

$$v = \frac{ds}{dt} = A\omega \cos \omega t. \tag{5.104}$$

A second differentiation gives the acceleration of the bob:

$$a = \frac{dv}{dt} = \frac{d^2s}{dt^2} = -A\omega^2 \sin \omega t, \tag{5.105}$$

which is proportional to the negative of the displacement s.

Approximations to the sine and cosine functions at small argument are often useful. These are

Useful approximations

$$\sin x \cong x \qquad \text{if} \qquad x \ll 1, \tag{5.106}$$

$$\cos x \cong 1 - \tfrac{1}{2}x^2 \qquad \text{if} \qquad x \ll 1. \tag{5.107}$$

(The symbol $\ll$ means "much less than.") Near $x = 0$, the cosine is parabolic and the sine is linear. Thus the curve in Figure 5.28 starts out approximately as a straight line following the equation

$$s \cong A\omega t.$$

What does this imply about the speed of the pendulum bob as it passes through the central point of its swing?

5.11 Exponential and logarithmic functions*

An exponential function is one in which the variable appears in an exponent. Whereas t^2 is called an algebraic function, 2^t is called an exponential function. Any other constant raised to a variable power would also be an exponential

* The material in this section will be needed first in conjunction with Section 8.10.

function. For a particular reason of convenience that will be mentioned later, a certain constant has been adopted as standard to use in exponential functions. It is an irrational number, written e, whose approximate numerical value is 2.7183. Thus the standard exponential function is written

The special constant e

$$e^x.$$

Since any number raised to the zero power is unity, the value of the exponential function at $x = 0$ is 1. At $x = 1$, it has the value e, about 2.72. At $x = 2$, its value is e^2, about 7.39; at $x = 3$, its value is e^3, about 20.1; at $x = 10$, its value is e^{10}, about 22,000. The exponential function is characterized by a very rapid rise as x increases. It may also be readily defined for negative values of x because of the rule

$$e^{-x} = \frac{1}{e^x}. \tag{5.108}$$

Thus at $x = -1$, the exponential function has the value $e^{-1} = 1/2.72 = 0.368$. At $x = -10$, its value is $e^{-10} = 4.54 \times 10^{-5}$. Obviously, the exponential function falls very rapidly toward zero as the value of x becomes more negative. A graph of the exponential function e^x is shown in Figure 5.29. For comparison, the reflected function e^{-x} is also plotted in the graph.

Since the exponential function ascends ever more rapidly as x increases, the numerical value of the slope grows as the numerical value of the function grows. In fact, the exponential function has an important property that it shares with no other function: Its slope is directly proportional to its value. If one point on an exponential curve has twice the magnitude of another point, the slopes at these two points also differ by a factor of two. This is true of any exponential function, such as 2^x or 3^{5t}. The particular exponential function e^x has an even more special property: Its slope is exactly equal to its value. Expressing this property in terms of a derivative, we have

Derivative of e^x

$$\frac{d}{dx}(e^x) = e^x. \tag{5.109}$$

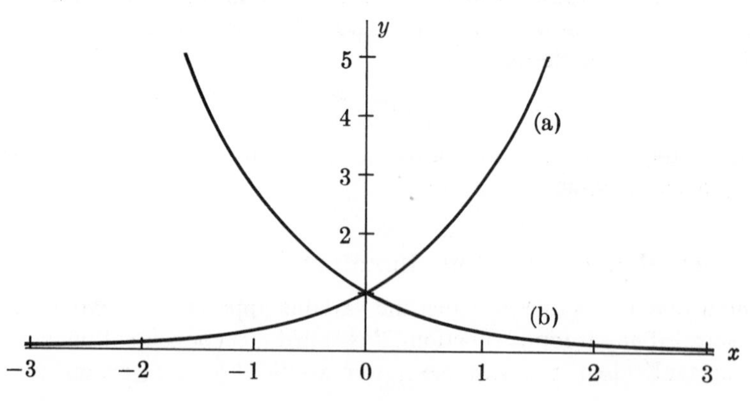

FIGURE 5.29 Graphs of exponential functions. (a) Rising exponential, $y = e^x$. (b) Falling exponential, $y = e^{-x}$.

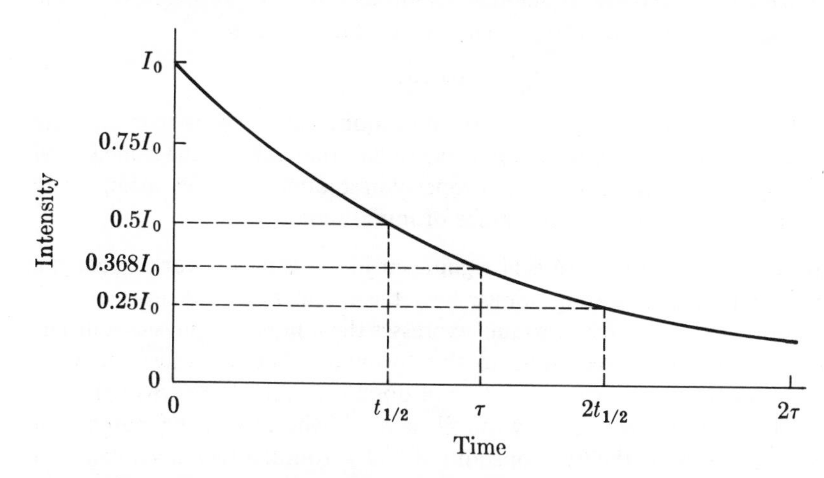

FIGURE 5.30 Exponential decay of intensity for sample of radioactive material. The mean life is τ; the half life is $t_{1/2}$.

Thus at $x = 0$ in Figure 5.29, both the value of the function and the numerical magnitude of its slope are equal to 1. For the descending exponential, the derivative is given by

$$\frac{d}{dx}(e^{-x}) = -e^{-x}. \tag{5.110}$$

It is because of these exceptionally simple properties that the constant e has been adopted as the standard constant for exponential functions.

Example of decreasing exponenential: radioactive decay

The exponential function is encountered often in physics, as well as in chemistry, biology, and the social sciences. An important example that involves the decreasing exponential is the decay of a sample of radioactive material. The intensity of the radioactivity of a given sample is plotted as a function of time in Figure 5.30. The curve in this figure is a graph of the function

$$I = I_0 e^{-t/\tau}, \tag{5.111}$$

whose variables are the intensity I and the time t. The constant I_0 is the initial intensity. The constant τ (tau), with the dimension of time, is called the mean time of the decay. When the time t reaches the value τ, then $I = I_0 e^{-1} = 0.368I_0$. Therefore, τ is the time required for the intensity to drop to 36.8 percent of its initial value.* Another characteristic time associated with an exponential

* The quantity τ is actually the mean time or average time of the decay in the following sense. Initially, the radioactive sample contains a large number of radioactive nuclei. Each of these undergoes decay after some time interval that is unpredictable and perhaps different for every nucleus. If the elapsed time until decay were measured for each nucleus separately and if all of these different times were then averaged, their average would be the mean time τ.

decay curve is the half life, $t_{1/2}$. This constant is defined as the time required for the intensity to decrease to one-half its initial value. It is obviously a time somewhat less than the mean time. The exact relationship is

Half life

$$t_{1/2} = 0.693\tau. \tag{5.112}$$

An important mathematical property of the exponential curve is that the time required for the curve to fall from any value to half that value is the same for all parts of the curve. Stating the same property in another way, the mean decay time is the same regardless of the choice of initial time.

Example of increasing exponential: physics research activity

■ EXAMPLE: In 1910, there were 11 significant journals in the world reporting the results of physics research. Since then the number of such journals has doubled every 16 years. What formula expresses the number of physics journals as a function of time? According to this formula, what was the number of physics journals in 1970? The given fact of doubling every 16 years implies an exponential function. We could write $N = Ae^{t/\tau}$, where N is the number of journals, and determine the two constants A and τ from the two given pieces of data. It is slightly more convenient to write

$$N = N_0 e^{(t-1910)/\tau}, \tag{5.113}$$

where t is the time in years A.D. Then, for $t = 1910$, $N = N_0$, so

$$N_0 = 11 \text{ journals.}$$

The parameter τ is the e-folding time (the time for N to increase by a factor 2.72). This is longer than the doubling time, t_2. The relation between τ and t_2, closely analogous to Equation 5.112, is

Doubling time

$$t_2 = 0.693\tau. \tag{5.114}$$

Since $t_2 = 16$ years, we have

$$\tau = 23.1 \text{ years.}$$

Finally, we can see what the formula has to say about 1970. Then

$$\frac{t - 1910}{\tau} = \frac{60}{23.1} = 2.60;$$

hence

$$N = 11e^{2.60} = (11)(13.5),$$

so

$$N \cong 148 \text{ journals.}$$

This number is in rather close agreement with fact for 1970. As this is being written, there is no sign of any deviation from this exponential behavior of research publication in physics. However, it is clear that this exponential, like a number of others in human affairs, cannot persist very long into the future. ■

Appendix 6 gives a series expansion for e^x that is useful for $|x| < 1$. For small $|x|$,

A useful approximation

$$e^x \cong 1 + x \qquad (|x| \ll 1). \tag{5.115}$$

A numerical table of the exponential function is found in Appendix 8.

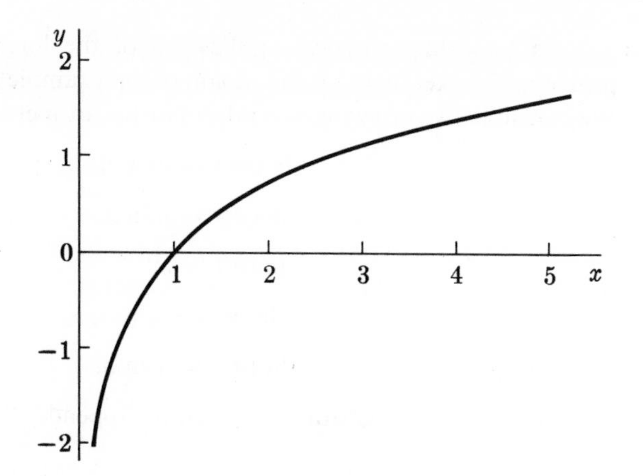

FIGURE 5.31 Graph of the logarithmic function, $y = \ln x$.

LOGARITHMS

Logarithms are defined with respect to an arbitrarily chosen constant, called the base. If, for example, the base is chosen to be 10, the equation

Any base is possible

$$y = \log_{10} x \tag{5.116}$$

means that y is the power to which 10 must be raised to give x, so

$$x = 10^y. \tag{5.117}$$

In scientific work, the standard base is not 10; it is the same constant e used to define the exponential function. A logarithm to the base e is written $\log_e x$, or, more commonly, $\ln x$. Such logarithms are called natural logarithms or Napierian logarithms.* The equation

Natural logarithms are used in most scientific work

$$y = \ln x \tag{5.118}$$

means that y is the power to which e must be raised to give x:

$$x = e^y. \tag{5.119}$$

A graph of $\ln x$ is shown in Figure 5.31. It approaches $-\infty$ at $x = 0$; it approaches $+\infty$ as $x \to +\infty$; and it is equal to zero at $x = 1$. Near $x = 1$, a useful approximation to the logarithm is

$$\ln x \cong x - 1 \quad \text{if} \quad |x - 1| \ll 1. \tag{5.120}$$

Approximation for x near 1

The derivative of the logarithmic function is given by

$$\frac{d}{dx}(\ln x) = \frac{1}{x}. \tag{5.121}$$

Derivative of ln x

The simplicity of the expressions on the right sides of Equations 5.120 and 5.121, containing no extra constants, are among the reasons why e is the most convenient base of logarithms.

---

* It is not uncommon to see natural logarithms denoted by $\log x$. In this book we shall use only the notation $\ln x$.

An important physical application of the logarithmic function is to the problem of rocket flight. A discussion of this example is postponed to Chapter 8. We conclude by reviewing several important properties of logarithms:

Properties of logarithms

$$\ln(xy) = \ln x + \ln y; \tag{5.122}$$

$$\ln(x^n) = n \ln x; \tag{5.123}$$

$$\ln(e^x) = x; \tag{5.124}$$

$$\ln(e) = 1; \tag{5.125}$$

$$\ln(a^x) = x \ln a. \tag{5.126}$$

A table of natural logarithms is given in Appendix 9.

★5.12 Probability, experimental error, and uncertainty

Probability is a broad subject in both mathematics and physics. Here we wish to discuss only one aspect of probability, that associated with experimental error. Perfect straight lines and perfect parabolas may exist in physical theory, but they do not exist in physical measurement because perfect precision of measurement is impossible. Where there is measurement there is error, and where there is error the concept of probability plays a role.

Probability is concerned with ignorance

Basically, probability is concerned with ignorance. Probability can be discussed quantitatively because exact statements can be made even about ignorance. Although we do not know in advance the result of the toss of a coin, we do know that the probability of heads is 0.5. Insurance companies, ignorant of the fate of an individual policyholder, can nevertheless set premiums intelligently. A gambling casino, ignorant of the outcome of any game, can adjust odds to assure a profit. In an experimental measurement, the thing we are ignorant of is the magnitude of the error. By chance a particular measurement could be precisely correct, or it might have a significant error. In order to make a meaningful comparison between experiment and theory, the physicist must be able to assess the probable magnitude of the error. To every experimentally determined quantity A is assigned an uncertainty ΔA, and the magnitude of the quantity, with its uncertainty, is written

Notation for uncertainty

$$A \pm \Delta A.$$

Suppose a college catalogue states that the area of a single student's room is 200 ft^2, and a student decides to check this number. He measures the length of the room and its breadth, then multiplies these two numbers together and obtains, perhaps, 194 ft^2. Has he demonstrated that the college catalogue is wrong? Not necessarily. It depends on the uncertainty of his measurement.

The magnitude of uncertainty is often only an educated guess

Determination of the uncertainty is usually not easy, and it may amount to little more than an educated guess. He might guess, for instance, that his length and breadth measurements are accurate to within about 2 percent. This would imply that his area determination should be accurate to within 4 percent. He would then write down for his experimentally determined area

$$(194 \pm 8)\ \text{ft}^2,$$

indicating his opinion that the true area probably lies somewhere between 186 ft^2 and 202 ft^2. Therefore his measurement would be consistent with "theory," as set forth in the catalogue.

A somewhat better, but still not foolproof, way to determine uncertainty is by repeated measurement. If several measurements of the area of the room agree to within 8 ft^2, the student can be more confident that the correct value lies within 8 ft^2 of his first measurement. However, repeated measurement reveals only random error. Another kind of error, called systematic error, could result from a defective ruler or a consistently faulty measuring technique. These sources of error would act always in the same direction rather than randomly in both directions. Having assessed all sources of error, the physicist tries to assign an uncertainty defined as follows: It should bracket a range of values within which the correct value has a 66 percent chance to lie and outside of which the correct value has a 34 percent chance to lie. Thus if it had been possible to define the uncertainty with this much certitude for the room area measurement, the expression (194 ± 8) ft^2 would mean that with a probability of 0.66, the true area has a value between 186 ft^2 and 202 ft^2; but with a probability of 0.34, it might be less than 186 ft^2 or greater than 202 ft^2.

Two kinds of error: random and systematic

For only one kind of error, called statistical error, is it actually possible to determine accurately the uncertainty of a measurement. Statistical error results from the measurement of purely random events. It might be determined, for example, that in a particular radioactive sample exactly 100 nuclei decay in a 1-min interval. In one sense, this measurement is completely free of error, for we may suppose that the number 100 was determined with complete precision. In another sense, however, it does have "error." for it may differ from the average number of decays in other identical samples. The quantity of real physical importance is the average number of decays per minute in a large number of identical samples. It is this average rather than the actual number of decays for a particular sample that can be meaningfully compared with theory. Thus the idea of error enters into the search for an answer to the question: By how much is the average number of decays for all such samples likely to differ from the actual number measured for this sample? With this understanding of what is meant by the "error" of a precise measurement, a very simple answer to the question can be given.* The uncertainty associated with the measurement of n random events is $\sqrt{n}$. If a particular radioactive sample undergoes n decays in a given time interval, the average number for all such samples can be said to be equal to

Statistical error is associated with the number of random events

$$n \pm \sqrt{n}; \tag{5.127}$$

Statistical uncertainty follows the "square-root-of-n rule"

that is, the correct average has a 66 percent chance to lie within the range $n - \sqrt{n}$ to $n + \sqrt{n}$. If one student in a laboratory measures 100 decays in a minute, he predicts that the average number measured by all of the students working with identical samples will be 100 ± 10. There is a good chance (probability two-thirds) that the true average lies between 90 and 110. This

* The answer, although simple in form, is not at all self-evident. It rests on the theory of random processes.

also means that about two-thirds of the students should record values within this range. It is important to remember that this "square-root-of-n rule" applies *only* to one kind of variable, the number of random events.

No other kind of uncertainty is so simple

The discussion to this point has concerned experimental errors associated with single measurements. Experimental error also plays a very important role in the determination of functional relationships. A curve cannot sensibly be drawn precisely through each experimentally determined point in a graph because the experimental errors in the measurements are reflected in an uncertainty of location of the points in the graph. Suppose that we wish to determine the functional relationship implied by the following set of time and distance data for a moving particle.

Time (sec)	Distance (cm)
1	280
2	450
3	1,000
4	1,300
5	1,450
6	1,840

Suppose further, for the sake of illustration, that the time measurements are so precise that their errors can be ignored and that the distance measurements are rather crude, with the uncertainty of each estimated to be ± 70 cm. Figure 5.32 shows both the wrong way and the right way to handle these data graphically. In the wrong way shown in Figure 5.32(a), the measured values are simply joined by straight-line segments. This is wrong because (1) the experimental errors mean that the plotted points need not represent the actual position of the particle and (2) the data provide no evidence about exactly how the motion

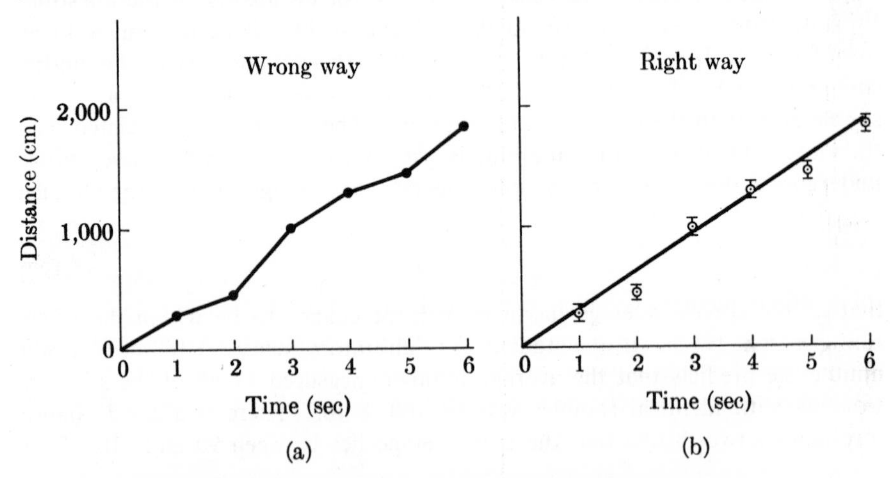

FIGURE 5.32 (a) The wrong way and (b) the right way to present experimental data graphically. In this example, the experimental uncertainties are sufficiently large that it is not justified to assume that the graph is anything other than a straight line.

occurred between the times of measurement. In the right way shown in Figure 5.32(b), the errors are incorporated into the graph by attaching to each datum point an error bar extending 70 cm above and 70 cm below the point. These bars at once provide a picture of where the particle most probably was at each time. Next a smooth line or curve is drawn among the points with no more details of shape and curvature than is justified by the magnitude of the uncertainties. In this example, a straight line adequately fits the data, and it would not be justifiable to seek a "better" fit with a curve. This, of course, does not mean that the straight line necessarily represents the actual motion accurately. We can say only that within the accuracy of the measurements, it is consistent to make the simplest assumption, namely, that the motion occurred with uniform speed such that the distance vs time graph is a straight line. If the uncertainty of ± 70 cm indeed spans a range of values containing the correct distance value with 66 percent probability, the line drawn among the points should cut through approximately two-thirds of the error bars and miss one-third of them.* The fact that the straight line in Figure 5.32(b) cuts four error bars and misses two tends to support the assumption that the straight line provides an adequate representation of the data. The slope of the straight line then provides a value for the speed of the particle. In Figure 5.32(b), the slope of the line is 300 cm/sec. It is left as a problem to estimate the uncertainty of this speed determination.

The error bar is used in graphs

Curves should show no more detail than the error bars can justify

---

We end this chapter with some remarks about practical calculations and some words of advice. Power and insight in science come only through a proper balance between qualitative understanding and mathematical description. Obviously, mathematics provides the power of science. It also provides an important part of the insight into the true nature of science, for much of the beauty of science springs from the remarkable mathematical simplicity of some of its fundamental laws. On the other hand, the mathematics of physical laws is itself only the skeleton of physics, sterile without the flesh and blood of ideas, models, and measurements. An important problem for the student is to separate the difficulties of conceptual understanding from the difficulties of practical calculation. The ideas of physics not bolstered by calculation provide at best a hollow understanding of the subject. Equally lacking in substance is a foundation of calculation by blindly applied rules. In this text we seek to strike a balance between the conceptual and the calculational.

Mathematics is an essential and beautiful part of physics

Mathematics alone is not enough

The advice is this: Think about calculations with the same degree of intensity applied to thinking about ideas expressed verbally. Profit from numbers. Every practicing scientist learns to think about mathematics pictorially and to think about pictures (graphs) mathematically. For the student, learning to rapidly bridge the gaps among tables and graphs and equations is a talent worth cultivating. A single word or phrase can evoke a host of references to the past, tapping a lifetime of experience in the use of words. Even for mathematically

It is usually important to think about mathematical calculation

---

* Perhaps from a sense of caution, many physicists tend to overestimate uncertainties. In physics research journals it is common to see graphs in which curves cut all—or nearly all—error bars.

adept students, there is no comparable background of mathematical experience to enrich the meaning of an equation or a numerical result. The enlargement of this background should be deliberately forced. A number should stimulate comparison; an equation should evoke an image; a calculation should invite tests of its "reasonableness." To learn that a slow neutron moves at 2,000 m/sec is to learn very little if that speed is not mentally placed in a framework of familiar speeds. How does it compare with the speed of an airplane, astronaut, or photon? An equation such as $s = vt$ is a cold fact if not warmed by the image of a moving object or at least of a straight-line graph. A calculation is a risky undertaking if not made with a wide-awake mind asking if it makes sense. It would be easy to calculate (mistakenly) that 1 mile is equal to 0.6 km. Does it make sense? No. The traveler might recall that in Europe his automobile covers more kilometers in an hour than it covers miles here at home. The sports enthusiast might recall that the 1,000-meter (1 km) run is shorter than a mile. The laboratory student should remember that a meter stick is about 1 yard long, hence that 1,000 m add up to about 3,000 ft, less than the 5,280 ft in a mile.

It is sometimes important to "turn the mathematical crank"

Despite the great importance of thinking about equations and their interpretations, there is also a time to turn the mathematical crank. An important part of the power of science stems from the fact that deduction and derivation can be carried out according to a set of mathematical rules without the necessity of careful thinking in physical terms about the meaning of each step. In practice, the student is more likely to err on the side of too little thinking about equations than too much. Nevertheless, worth reflecting upon is Alfred North Whitehead's emphatic statement, quoted in Question 5.35, on the merits of turning the crank.

Summary of ideas and definitions

Mathematics has an inner truth of logic that justifies it as a separate intellectual discipline.

Mathematics has an outer truth of physical confirmation that justifies its use in science.

The applicability of mathematics to science can be decided only by experiment.

The common coordinate systems—Cartesian, cylindrical, and spherical—are right-handed and orthogonal.

A derivative is defined by a limiting process (Equation 5.6).

The derivative is interpretable as a rate of change and as the slope of a curve.

A derivative with respect to distance is called a gradient.

By definition, one revolution, or 360 deg, equals 2π radians.

Angular speed is defined by $\omega = d\theta/dt$. Its unit is radian/sec.

The projection of a line segment L on an axis is $L_x = L \cos \theta$ if the line segment makes angle θ with the axis.

In a graph, the meaning of $\tan \theta$ is generalized; it is equal to a slope, df/dx, and may have any unit.

Kinematics is the description of motion using only the concepts of length and time.

For one-dimensional motion, average velocity is defined by $\bar{v} = \Delta x/\Delta t = (x_2 - x_1)/(t_2 - t_1)$ and instantaneous velocity by $v = dx/dt$; average acceleration is defined by $\bar{a} = \Delta v/\Delta t = (v_2 - v_1)/(t_2 - t_1)$ and instantaneous acceleration by $a = dv/dt = d^2x/dt^2$.

Motion in one dimension with constant speed is described by the linear equation $x = x_0 + v_0t$. A graph of x vs t is a straight line.

Motion in one dimension with constant acceleration is described by the quadratic equation $x = x_0 + v_0t + \frac{1}{2}a_0t^2$. A graph of x vs t is a parabola.

Notation for the indefinite integral is $I(x) = \int f(x)\,dx$. Notation for the definite integral is $D = \int_a^b f(x)\,dx$.

The indefinite integral, or antiderivative, of a function $f(x)$ has a derivative equal to $f(x)$: $dI/dx = f(x)$.

The indefinite integral contains an arbitrary additive constant.

To go from acceleration to velocity, or from velocity to position, requires integration (Equations 5.67 and 5.68).

The definite integral is the area under a curve between two limits; such a generalized area may have any unit of measurement.

The definite integral may be approximated by a sum of rectangular areas and is exactly equal to a limit of this sum (Figure 5.22 and Equation 5.78).

As a function of its upper limit, the definite integral satisfies the equation $dD/dx = f(x)$ (Equation 5.87).

According to the fundamental theorem of integral calculus, the definite integral may be expressed in terms of the indefinite integral by

$$D = \int_a^b f(x)\,dx = I(x)\Big|_a^b = I(b) - I(a). \qquad (5.90)$$

Differentiation and integration are linear operations (Equations 5.14, 5.63, and 5.76).

The functions $\sin x$ and $\cos x$ are periodic with period 2π. These functions have the property $d^2f/dx^2 = -f$.

Angular frequency, frequency, and period are related by $\omega = 2\pi\nu = 2\pi/T$.

The functions e^x and e^{-x} change by fixed fractions for given Δx. These functions have the property $d^2f/dx^2 = +f$.

The logarithm to base e, called the natural logarithm, is most common in scientific work; the notation for it is $\ln x$.

Two kinds of error are distinguished: random error and systematic error.

The uncertainty of a measurement is often no more than an educated guess.

If a quantity is written $A \pm \Delta A$, the uncertainty ΔA is usually defined so that the true value has a probability of about $\frac{2}{3}$ to lie between $A + \Delta A$ and $A - \Delta A$.

For the measurement of n random events, the "error" is defined as the difference between n and the average number $\bar{n}$ for many such measurements. The uncertainty of n is $\Delta n = \sqrt{n}$. This is a special result, not applicable to other kinds of uncertainty.

The student of physics must learn how to interpret mathematics and appreciate its physical content. He must also know when to turn the mathematical crank.

QUESTIONS

Section 5.1

Q5.1 Briefly discuss this proposition (pro or con): "Science without mathematics is unthinkable."

Q5.2 In 1817, K. F. Gauss wrote: "One would have to rank geometry not with arithmetic, which stands a priori, but approximately with mechanics." Interpret this statement in terms of the ideas of the inner and outer truth of mathematics.

Q5.3 Certain physical "operations" are not commutative: that is, operation A followed by operation B does not give the same result as B followed by A. Give an example of noncommuting operations.

Q5.4 Give an example of two physical operations A and B that are commutative (i.e., the net result is the same for A followed by B as for B followed by A).

Section 5.2

Q5.5 Some humanists display their ignorance of mathematics with a certain pride. Is this a justifiable attitude for someone whose concern is with human character and human relations? Do you think such an attitude reflects a misconception about the true nature of mathematics?

Q5.6 Name one discipline other than geometry that can be considered a branch of both mathematics and physics.

Q5.7 Do you support the idea that every consistent logical structure that man can build must correspond to something in nature? Why or why not?

Q5.8 If an automobile traveling at 30 mile/hr adds 30 mile/hr to its speed, it moves at 60 mile/hr. Does this statement possess the inner truth of pure mathematics, the outer truth of physical verification, or both? Think about this with care, and explain your answer.

Section 5.3

Q5.9 Latitude and longitude are a set of orthogonal coordinates on the surface of the earth. Explain why.

Q5.10 Describe qualitatively the motion of the moon in (a) a moon-fixed frame of reference, (b) an earth-fixed frame of reference, and (c) a sun-fixed frame of reference.

Section 5.4

Q5.11 An operation such as

$$\lim_{\Delta x \to 0} \frac{\Delta f}{\Delta x}$$

is a purely mathematical operation with no true counterpart in physical measurement since only finite intervals can be measured. Why is it nevertheless both useful and legitimate to use derivatives in describing nature?

Q5.12 When we use derivatives in physics, what assumption are we making about nature? In particular, when we use dx/dt for speed, what are we assuming about the properties of space and time?

Q5.13 According to one form of Zeno's paradox, a hare can never overtake a tortoise because in order to reach the tortoise, the hare must first cover half the remaining distance between them; yet no matter how many times the hare covers half the remaining distance, there still remains some distance to go. Criticize this statement of the paradox.

Section 5.5

Q5.14 (1) Can Equation 5.19 ($s = r\theta$) be valid for motion that is not executed in a circle? (2) Can it be true for motion that is not carried out at constant speed?

Q5.15 Answer Question 5.14 for Equation 5.24 ($s = vt$).

Q5.16 Answer Question 5.14 for Equation 5.25 ($\theta = \omega t$).

Section 5.6

Q5.17 The population of the world increases with time. What is the independent variable? What is the dependent variable? Is it possible to interchange the role of these two variables?

Q5.18 On a graph, the surface area of a reservoir is plotted as a function of time. What would be a suitable unit of measurement for expressing the slope of this curve?

Q5.19 Name a variable other than temperature whose gradient is of physical importance. Name a variable other than position or displacement whose time rate of change is of physical importance.

Q5.20 Give a physical example in which a derivative with respect to something other than time or distance is important.

Section 5.7

Q5.21 Name some branch or subbranch of science, other than kinematics, that uses only one or two basic concepts in its theoretical structure.

Q5.22 Describe in words the motion represented by the graphs in Figure 5.13.

Q5.23 Two balls are projected upward from ground level, one with twice the initial speed of the other. (1) What is the ratio of their maximum heights? (2) What is the ratio of the lengths of time they spend in the air?

Q5.24 An automobile is moving forward with speed ds/dt. (1) If d^3s/dt^3 is negative, what can you conclude about the acceleration of the automobile? (2) Is its speed increasing or decreasing, or is the given information insufficient to know?

Section 5.9

Q5.25 Whenever a very large number of small contributions are to be summed, integration is likely to be the tool that is needed. Give an example from science or engineering that illustrates this point.

Q5.26 If a man's annual salary is designated S and his age in years is designated t, what is the meaning of the integral $\int_a^b S\,dt$? What is its unit of measurement?

Q5.27 Explain why numerical integration of tabular data is generally more accurate than numerical differentiation of tabular data.

Section 5.10

Q5.28 What is the physical significance of the slope of the curve in Figure 5.28?

Q5.29 Where might sine and/or cosine functions be encountered in the description of nature, apart from pendulum motion? Give at least one, and preferably two, examples.

Section 5.11

Q5.30 If intensity in Figure 5.30 is measured in number of radioactive disintegrations per second, what is the significance of the area under the curve in that figure?

Q5.31 The present doubling time of the world population is about 30 years. Roughly how many human beings are expected to be living on earth in the year 2000? Do you expect the present rate of exponential growth to continue past the year 2000? Why or why not?

Q5.32 About how long do you think Equation 5.113, describing physics research publication in the world, might be valid in the future, with the given parameters N_0 and τ? Give reasons for your answer.

Section 5.12

Q5.33 Approximately what percentage uncertainty would you assign to (a) a typical bathroom scale, (b) a typical automobile speedometer, and (c) a typical measurement with a yardstick? Give a reason for each answer.

Q5.34 What is suspicious about the representation of experimental uncertainty in the accompanying graph?

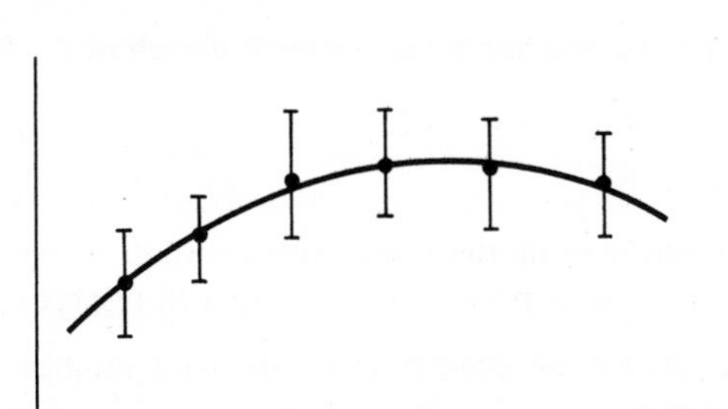

Q5.35 Alfred North Whitehead wrote: "It is a profoundly erroneous truism, repeated by all copy books and by eminent people when they are making speeches, that we should cultivate the habit of thinking of what we are doing. The precise opposite is the case. Civilization advances by extending the number of important operations which we can perform without thinking about them." Illustrate Whitehead's point with some examples of your own "thoughtless" use of mathematics and machines.

EXERCISES

Section 5.2

E5.1 (1) Construct the complete multiplication table for the four mathematical objects $+1$, -1, $\sqrt{-1}$, and $-\sqrt{-1}$. (2) Evaluate $\frac{1}{2}(1 + \sqrt{-1})^2$. For simplicity, use the symbol i for $\sqrt{-1}$.

E5.2 The three complex numbers,

$$a = -\frac{1}{2} + \frac{\sqrt{3}}{2}i, \qquad b = -\frac{1}{2} - \frac{\sqrt{3}}{2}i, \qquad c = 1$$

(where $i = \sqrt{-1}$), have the property that any combination of products of a, b, and c is equal to a, b, or c. (1) Demonstrate this fact to your own satisfaction by a succession of trial products. (2) Give an example of a set of three physical operations with a similar property—namely, that any number of these operations carried out in any order produces the same result as a single one of the operations.

Section 5.3

E5.3 (1) Sketch two different perspective views of a right-handed Cartesian coordinate system. (2) Sketch one view of a left-handed Cartesian system.

E5.4 A circle of radius a is centered at the origin. (1) What is the equation of the circle in Cartesian coordinates? (2) What is its equation in polar coordinates?

E5.5 (1) Show that polar coordinates are expressed in terms of Cartesian coordinates by

$$r = \sqrt{x^2 + y^2},$$

$$\theta = \arctan\left(\frac{y}{x}\right).$$

These equations, like Equations 5.1 and 5.2, are called transformation equations. (2) Discuss the ambiguity inherent in each of these equations. How is the ambiguity resolved?

Section 5.4

E5.6 Two given functions are $f(x) = 3x$ and $g(x) = 4x^3$. Find dy/dx (a) if $y = fg$, (b) if $y = f/g$, and (c) if $y = f^2$.

E5.7 Evaluate the first and second derivatives of each of the following functions:

(1) $f(x) = x^2 + 2ax + a^2$, (3) $h(z) = e^{-z^2}$,

(2) $g(y) = \sqrt{ay}$, (4) $x(t) = A \sin \omega t + B \cos \omega t$.

E5.8 A particle moves along the x axis with constant acceleration. What is the third derivative of its position coordinate with respect to time, d^3x/dt^3?

E5.9 A particle has kinetic energy $K = \frac{1}{2}mv^2$ and momentum $p = mv$. Show that $dK/dp = v$.

E5.10 If x is a function of t, then t can be considered a function of x. (1) Using the basic definition of a derivative (Equation 5.6), prove that $(dt/dx) = 1/(dx/dt)$. (2) Illustrate this result, using the function $x(t) = \frac{1}{2}at^2$, in which a is constant.

E5.11 A particle moves in the xy plane with velocity components $v_x = dx/dt$ and $v_y = dy/dt$. (1) Using the basic definition of the derivative (Equation 5.6), prove that

$$\frac{dy}{dx} = \frac{(dy/dt)}{(dx/dt)}.$$

(2) What is the physical meaning of dy/dx?

E5.12 Using the basic definition of a derivative (Equation 5.6) and the binomial theorem (Appendix 6), prove that

$$\frac{d}{dx}(x^n) = nx^{n-1}. \tag{5.13}$$

E5.13 Prove that

$$\frac{d}{dx}[af(x)] = a\frac{df}{dx},$$

where a is a constant.

E5.14 Prove that

$$\frac{d}{dx}[f(x) + g(x)] = \frac{df}{dx} + \frac{dg}{dx}.$$

E5.15 What is the conversion factor linking radian/sec and rpm (revolutions per minute)? (SUGGESTION: Save the result by adding it to Appendix 4.)

Section 5.5

E5.16 (1) Arrange the following in order of increasing angular speed: (a) the minute hand of a clock, (b) the hour hand of a clock, (c) the earth in its orbit around the sun, (d) a wheel 2 ft in diameter on a car traveling 60 mile/hr, (e) an astronaut circling the earth in 90 min, and (f) the crankshaft of an engine turning at 600 rpm. (2) Give the angular speed ω in radian/sec for any three of these.

E5.17 A helicopter rotor blade that measures 4 m from shaft to tip turns at 400 rpm. How does its tip speed compare with the speed of sound in air?

E5.18 What are the following trigonometric quantities: (a) sin (360 deg), (b) tan (π radians), (c) cos (135 deg), (d) sin ($\pi/2$ radians), and (e) cos (3π radians)?

E5.19 (1) What are the projections L_x and L_y onto the x and y axes of a line segment L of length 4 units that makes an angle of 30 deg with the x axis and 60 deg with the y axis? Include a sketch with your answer. (2) What algebraic relationship is satisfied by L_x, L_y, and L?

E5.20 Prove the trigonometric identities

(a) $\tan\theta = \dfrac{\sin\theta}{\cos\theta}$, (b) $\sin^2\theta + \cos^2\theta = 1$.

E5.21 Prove that

$$\frac{1}{\cos^2\theta} - \tan^2\theta = 1.$$

Section 5.6

E5.22 Sketch on a graph straight lines whose slopes are 0, 0.5, 1, 2, and −1.

E5.23 Sketch a curve whose slope varies continuously from zero to infinity.

E5.24 The volume of a cube is expressed as a function of the length of one of its sides by $V = l^3$. (1) Sketch a graph of V vs l. (2) What is the derivative dV/dl? What is its physical meaning? What is its dimension?

E5.25 The average speed $\bar{v}$ of a molecule in a gas is proportional to the square root of the absolute temperature T. (1) Write an equation for $\bar{v}$ as a function of T (introduce a constant if necessary). (2) Sketch a graph of $\bar{v}$ vs T. (3) Sketch a graph of $d\bar{v}/dT$ vs T.

E5.26 The motion of a particle along the x axis is represented approximately by the graph shown here. (1) What is the x component of velocity of the particle at $t = 2$ sec? (2) What is dx/dt at $t = 5$ sec? (3) Discuss the motion near $t = 4$ sec. What happened to the particle at this time? Can a true discontinuity of slope be realized in nature? Explain.

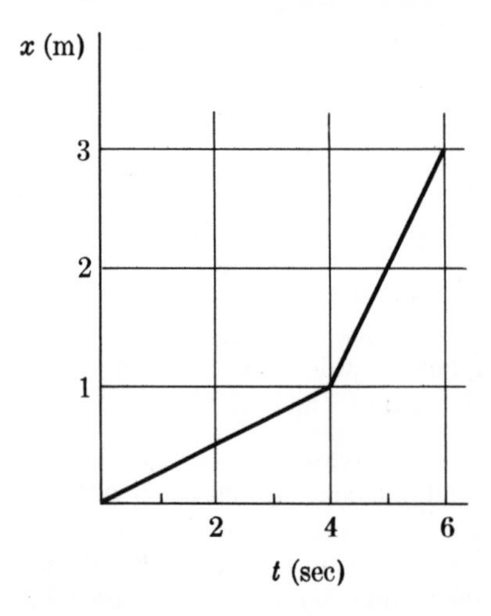

E5.27 (1) Sketch a graph of the temperature gradient, dT/dr, vs r, using Equation 5.41 with $A > 0$. (2) What is the dimension of A?

E5.28 Suppose that the coefficient A in Equation 5.37 is negative. (1) Sketch a graph of T vs r. (2) What physical situation might this represent? (3) Evaluate the gradient, dT/dr, and sketch a graph of dT/dr vs r.

Section 5.7

E5.29 The motion of a particle moving with constant speed along the x axis is described by Equation 5.48. At $t = 1$ sec, $x = 3.5$ m; at $t = 2$ sec, $x = 5.0$ m. (1) What is the particle's speed v_0? (2) What is the magnitude of the constant x_0? (3) Where is the particle at $t = 10$ sec?

E5.30 Sketch graphs of v vs x for the special and general cases of motion with constant acceleration (Equations 5.54 and 5.58). Take note of the slopes, dv/dx, at $x = 0$ for the two cases.

E5.31 For the general case of motion with constant acceleration (Equations 5.55–5.58), express dv/dx (a) as a function of v, (b) as a function of x, and (c) as a function of t.

E5.32 Starting at $x = 0$, a bug crawls in the direction of positive x at a speed of

3 cm/sec for 2 sec, then stops for 2 sec, then crawls again in the same direction at 2 cm/sec for 2 sec, then reverses course and scampers back to its starting point at 5 cm/sec. (1) Draw a graph of x vs time t for the bug. (2) Draw a graph of v_x vs t. (3) According to this description, what is the maximum acceleration of which the bug is capable?

E5.33 The speed of a runner in the 100-yard dash varies as shown in the figure. (1) Sketch a graph showing his acceleration as a function of time. (2) Estimate roughly his average speed during the race.

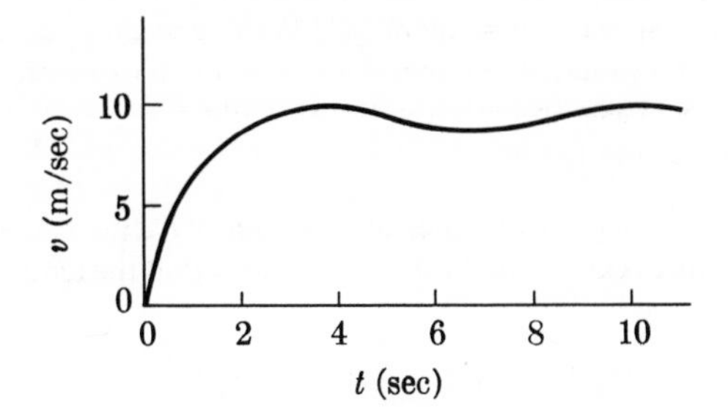

E5.34 A china cup drops from a height of 5 cm onto a wooden table. (1) With what speed does it strike the table? (2) If the table yields by 10^{-5} m as it brings the cup to rest, what is the magnitude of the acceleration of the cup as it is stopped (assuming a uniform acceleration)? (3) How long does it take the table to stop the cup? (For calculational convenience, use $g = 10\ \text{m/sec}^2$.)

E5.35 The component of gravitational acceleration along a certain inclined air track is 3 m/sec^2. A cart is started up the track from $s = 0$ at $t = 0$ with initial speed $v_0 = 2$ m/sec. (1) Write equations for its position s and speed v as a function of time. (2) How far up the track does the cart move before starting back? (3) When does the cart return to its starting point?

E5.36 Leaning over the edge of a cliff, a man throws a rock vertically with initial speed v_0. It strikes the ground with speed v_f a distance h below the point from which it was thrown. Assume that its motion is executed with constant downward acceleration g. (1) For a given initial speed v_0, prove that v_f is independent of the initial direction of the throw (up or down). (2) Write an expression for its final kinetic energy ($K_f = \frac{1}{2}mv_f^2$) as a function of its initial kinetic energy ($K_0 = \frac{1}{2}mv_0^2$), and sketch a graph of K_f vs K_0.

E5.37 An elevator starts from $y = 0$ at $t = 0$. It is programmed to follow the kinematic formula $y = Et^3$ for the first 3 sec, where E is a constant. (1) Give formulas for its speed and acceleration during this time. (2) Suggest a reasonable value for E, and, using this value, tabulate y, v, and a at $t = 0$, 1 sec, 2 sec, and 3 sec.

E5.38 The motion of a car along a straight road is described approximately by the equation

$$x = \frac{\frac{1}{2}at_0^2t^2}{t_0^2 + t^2},$$

where a and t_0 are constants and t is time. (1) Sketch a graph of x vs t. (2) Give the dimensions of a and t_0 and discuss their physical meaning.

Section 5.8

E5.39 Evaluate the following indefinite integrals:

$$(1)\ \int (4x^3 + 5x^4)\,dx, \qquad (3)\ \int (A \sin \omega t + B \cos \omega t)\,dt,$$

$$(2)\ \int \sqrt{ay}\,dy, \qquad (4)\ \int (1 + \ln k\theta)\,d\theta.$$

E5.40 A particle moving in one dimension is known to have zero acceleration. Show that the most general expression for its position as a function of time is $x = x_0 + v_0 t$.

E5.41 The definition of angular speed is $\omega = d\theta/dt$. (1) Write θ as an integral over time. (2) If ω is constant, evaluate the integral and obtain an expression for θ that includes Equation 5.25 as a special case. (3) If $d^2\theta/dt^2 = \alpha$ (a constant), what is the general expression for θ?

E5.42 The temperature gradient in a piece of material is given by $dT/dx = A + Bx^2$; at $x = x_0$, the temperature is known to be $T = T_0$. Show that the temperature must be given by

$$T = T_0 + A(x - x_0) + \tfrac{1}{3}B(x^3 - x_0^3).$$

Section 5.9

E5.43 Evaluate the following definite integrals:

$$(1)\ \int_1^2 (5 + 4x^3)\,dx, \qquad (3)\ \int_a^{-a} (bz + cz^3)\,dz,$$

$$(2)\ \int_{-1}^{1} u^2\,du, \qquad (4)\ \int_\pi^{2\pi} \cos\theta\,d\theta.$$

E5.44 In Figure 5.13: (1) What is the physical meaning of the area under the a vs t curve from $t = 0$ to $t = t_1$? (2) What must be the area under the a vs t curve from $t = 0$ to $t = \infty$? (3) What is the physical meaning of the area under the v vs t curve from $t = 0$ to $t = \infty$?

E5.45 A trench of fixed width w and variable depth d extends over a total distance L. Express the volume of the trench as an integral. Is it a definite or an indefinite integral?

Section 5.10

E5.46 The angular frequency ω of a pendulum is $\omega = 10\ \text{sec}^{-1}$. (1) What is the frequency ν of the pendulum? (2) What is its period T? (3) If the amplitude of the pendulum is 0.03 m, what is the maximum speed of its swing?

E5.47 Differentiate Equations 5.106 and 5.107 (small-angle approximations for $\sin x$ and $\cos x$), and show that the results are consistent with Equations 5.92 and 5.94.

E5.48 The derivative of $\sin x$ is, by definition,

$$\frac{d}{dx}(\sin x) = \lim_{\Delta x \to 0} \frac{\sin(x + \Delta x) - \sin x}{\Delta x}.$$

Prove that the limit on the right is $\cos x$. To begin the proof, use the trigonometric expansion for $\sin(A + B)$.

E5.49 An equation of a vibrating string is

$$y = A \sin\left(\frac{2\pi}{\lambda} x\right) \sin \omega t,$$

where y is its transverse displacement, x is distance measured along the string, and t is time. (1) Sketch graphs of y vs x for several times t. (2) What are the physical meanings of λ and ω? (3) What are the nodal points, or stationary points, on the string?

E5.50 The function $f(x) = \sin(1/x)$ has unusual properties near $x = 0$. (1) Give an expression for df/dx. (2) Sketch a graph of $f(x)$ vs x, and point out its features near $x = 0$.

Section 5.11

E5.51 When a comic strip character needed about 10^6 dollars, he was advised to get a dollar and then double his money n times. How big is n? Express this n-fold doubling process by means of a simple equation.

E5.52 For the decreasing exponential function, $e^{-t/\tau}$, show that $t_{1/2} = \tau \ln 2$; and for the increasing exponential function, $e^{t/\tau}$, that $t_2 = \tau \ln 2$. (The quantities $t_{1/2}$ and t_2 are the values of t at which the functions are equal to 0.5 and 2 respectively.)

E5.53 Suppose that the number of neutrons in an exploding atomic bomb is given by $N = N_0 e^{t/\tau}$, with $N_0 = 1$ and $\tau = 10^{-8}$ sec. At this rate of growth, how long would it take N to grow to 6×10^{23} (one mole of neutrons)?

E5.54 At what date, according to Equation 5.113, was there only one important physics journal in the world? (At that date, and earlier, the number of journals is too crude a measure of physics research activity. Journals then were multidisciplinary. The number of journals in *all* fields of science follows an exponential law back to about 1750.)

E5.55 Using the data given in Problem 3.4, obtain approximate formulas, of exponential form, for (a) the energy of circular accelerators as a function of time and (b) the diameter of such accelerators as a function of time. What are the approximate doubling times of these two variables?

E5.56 On graph paper plot the function

$$y(x) = e^x + e^{-x}$$

from $x = -3$ to $x = +3$. This function defines a *catenary*, the curve taken by a hanging chain supported at its two ends.

E5.57 (1) For the catenary defined in Exercise 5.56, show that $d^2y/dx^2 = y$. (2) What function $z(x)$ has the property $d^2z/dx^2 = -z$?

E5.58 Derive Equation 5.109 for the derivative of e^x. Use the fact that $e^{x+\Delta x} = e^x e^{\Delta x}$, and write $e^{\Delta x}$ in approximate form with the help of Equation 5.115.

E5.59 Derive Equation 5.121 for the derivative of $\ln x$. You will need to use one or more of the properties listed in Equations 5.122–5.126 and also the approximate expression given by Equation 5.120.

E5.60 During some period of time, the position s of a car is expressed as a function of time by $s = a \ln(t/t_0)$, where a and t_0 are constants. (1) What is the dimension of a? (2) Can the equation be valid at $t = 0$? (3) What is the speed of the car at $t = t_0$? (4) Does the car ever come to rest?

Section 5.12

E5.61 If $\sqrt{n}$ is the uncertainty associated with the statistical error in the measurement of n random events (Equation 5.127), what is the *fractional* uncertainty? By what fraction must the number of counts n be increased in order to halve the fractional uncertainty?

E5.62 A student makes 7 measurements of the wavelength of the yellow light of sodium and gets results shown in the table. Measurements of much higher precision show the correct value to be 5,892 Å. (1) What is the approximate magnitude of the random error in these measurements? (2) Is there any evidence for a systematic error? (See also Problem 5.21.)

Trial	Wavelength (Å)
1	5,889
2	5,874
3	5,881
4	5,895
5	5,892
6	5,881
7	5,883

PROBLEMS

Axiomatic arithmetic

P5.1 The following are axioms of positive integers:

1. (integer) + 1 = (next integer),
2. $a \cdot 1 = a$,
3. $a + b = b + a$,
4. $ab = ba$,
5. $a(b + c) = ab + ac$.

Prove that $2a = a + a$. State carefully which axiom is used at each step of the proof.

P5.2 To the axioms of positive integers given in Problem 5.1, we may add:

6. $a + (b + c) = (a + b) + c$,
7. $a(bc) = (ab)c$.

(1) Working directly from the axioms, prove (a) $(ab)(c + d) = a(bc) + d(ba)$, and prove (b) $(a + b)(a + b) = a^2 + 2ab + b^2$. (2) Why is the physicist not normally concerned with proofs of this kind?

Observers in different frames of reference

P5.3 An unstable particle is moving through the laboratory. Consider its observation from two frames of reference, one at rest with respect to the laboratory, the other at rest with respect to the particle. Based on this consideration, give a carefully reasoned argument that explains why the particle must obey the "downhill rule" of decay—the masses of the product particles sum to less than the mass of the parent.

Spherical coordinates

P5.4 (1) Find three equations analogous to Equations 5.1 and 5.2 that express the Cartesian coordinates x, y, and z in terms of the spherical coordinates r, θ, and φ. (2) Show that r, θ, and φ (in that order) are a right-handed, orthogonal coordinate system.

Derivative of a quotient

P5.5 Prove that

$$\frac{d}{dx}\left[\frac{f(x)}{g(x)}\right] = \frac{\dfrac{df}{dx}g - f\dfrac{dg}{dx}}{g^2}. \tag{5.16}$$

Discontinuity of slope

P5.6 The graph accompanying Exercise 5.26 shows a discontinuity of slope. Using Equation 5.6 and graphical illustration, explain carefully why the slope of the curve is not well defined at $t = 4$ sec but is well defined at any other point.

P5.7 The accompanying table shows the mass of a particular child as a function of his height over the years. (1) Graph these data, plotting height horizontally and mass vertically. Join the plotted points by a smooth curve. (2) Which part of the curve has maximum slope? What is the approximate magnitude of this slope? (3) Extend your curve beyond the range of the data in order to estimate the mass of the unborn child when his height was 35 cm and to predict his later mass if he grows to a height of 175 cm.

Graphical representation of tabular data

Height (cm)	Mass (kg)
50	3.5
75	11
100	20
125	31
150	45

P5.8 (1) Plot distance vs time for the data in the accompanying table. (2) Describe the motion in words. (3) What is the speed of the motion near $t = 8$ sec? (4) Can s be written as a simple mathematical function of t?

t (sec)	s (m)
0	0
1	2.5
2	10.0
3	22.5
4	40
5	60
6	80
7	100
8	120

P5.9 A physicist wishes to test whether or not the data of Table 5.5 conform to the expression $T = T_0 + Ar^2$. He believes the experimental uncertainty in T to be about 0.1 K and the experimental uncertainty in r to be negligible. (1) Do these data conform adequately to the theoretical formula? (2) If they do, what are the values of T_0 and A? If they do not, is there any other reasonably simple formula that fits the data better?

Equation to represent tabular data

P5.10 An object starting from rest at point A slides down a nearly frictionless track. Its speed is measured at point B 1 m from A and at point C 2 m from A. The results of four trial slides are shown in the following table. (1) Estimate the approximate percentage error of the random measurements. (2) Find the ratio of speeds v_C/v_B for each trial. Is it consistent to assume that the ratio is a constant within experimental error? (3) Hypothesize an equation of the form $(v_C/v_B) = (x_C/x_B)^n$ relating the speed ratio v_C/v_B to the distance ratio x_C/x_B (which in this instance is 2). If you believe in the simplicity of the laws of nature, the exponent n should be a "simple" number. What value of n fits the experimental results?

Trial	Speed at B (cm/sec)	Speed at C (cm/sec)
1	24.8	35.0
2	24.2	34.5
3	25.0	34.8
4	24.6	35.2

The nature of free fall

P5.11 In Galileo's *Dialogues Concerning Two New Sciences*, Sagredo suggests that the free fall of a body under the action of gravity "is such that its speed increases in proportion to the space traversed; so that, for example, the speed acquired by a body in falling four cubits would be double that acquired in falling two cubits and this latter speed would be double that acquired in the first cubit." Express Sagredo's hypothesis in mathematical terms, and prove that it is inconsistent not only with accurate measurements associated with free fall but also with the crudest observations of the properties of free fall. (If you wish to see how Salviati, the "voice of Galileo," deals with Sagredo's hypothesis, see the Dialogue of the Third Day, under the heading, "Naturally Accelerated Motion.")

One-dimensional kinematics

P5.12 (1) For one-dimensional motion, prove that the slope of a graph of $\frac{1}{2}v^2$ plotted as a function of x is equal to the acceleration. (2) For what other graph is slope equal to acceleration?

P5.13 A particle moving along the x axis with constant acceleration passes the origin ($x = 0$) at $t = 0$ with velocity $v_0 = 10$ m/sec. Later, at $t = 4$ sec, it passes the origin again. (1) What is its acceleration? (2) What is its velocity the second time it passes the origin? (3) Does it pass the origin a third time? If so, when? (4) What is the greatest value of x reached by the particle?

P5.14 The boy shown in Figure 5.16 wants to throw one ball to a height of 6 m above the ground, and another to a height of 12 m above the ground. Both leave his hand at a height of 2 m. (1) What must be the ratio of the initial speeds of the two balls? (2) What must be the ratio of their initial kinetic energies?

P5.15 An automobile is moving initially at a speed of 15 m/sec. As it is being braked to a stop, its x component of acceleration is given by (see the accompanying graph)

$$a_x = \frac{a_0(t - t_0)}{t_0} \qquad (0 < t \le t_0),$$

where $a_0 = 3$ m/sec^2 and $t_0 = 10$ sec. (1) Show that its speed reaches zero at the same moment its acceleration reaches zero. (2) How far does the automobile move during the 10 sec it is being braked? (3) Write formulas for $v_x(t)$ and $x(t)$ that include appropriate boundary conditions and are valid for $0 < t \le t_0$. (4) Assuming that the automobile moves at a steady speed of v_0 for $t < 0$ and that it remains at rest for $t > t_0$, sketch graphs of v_x vs t and x vs t that include times before and after the braking period.

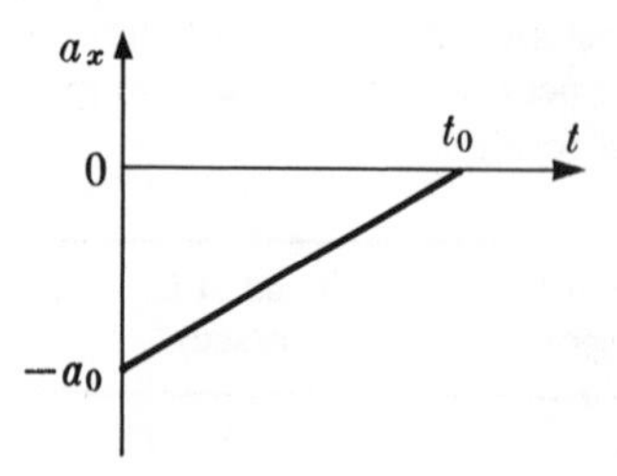

P5.16 The speed of a racing car during a time trial is given by

$$v = A(1 - e^{-t/\tau}),$$

where $A = 150$ mile/hr and $\tau = 6$ sec. (1) Obtain expressions for the position of the car and for its acceleration as a function of time. (2) Sketch graphs of x vs t, v vs t, and a vs t. (3) Obtain approximate expressions for x, v, and a, that are valid for $t \ll \tau$.

Average velocity

P5.17 For one-dimensional motion, the average velocity in the time interval 0 to T is

$$\bar{v} = \frac{1}{T}\int_0^T v(t)\,dt.$$

(1) Evaluate this expression for (a) constant-velocity motion and (b) uniformly accelerated motion. (2) Explain the graphical meaning of the time average.

Motion in two dimensions

P5.18 The x and y coordinates of a pendulum bob free to swing in two dimensions are given by

$$x(t) = a \cos \omega t,$$

$$y(t) = b \sin \omega t,$$

where a and b are unequal positive constants. (1) Sketch the path of the bob in the xy plane. Show that it is a closed orbit. (2) Find the velocity components, $v_x = dx/dt$ and $v_y = dy/dt$, and the speed, $v = \sqrt{v_x^2 + v_y^2}$. Where is the speed maximum? Where is it minimum? (3) Obtain an expression for dy/dx, the slope of the trajectory in the xy plane. At what points is dy/dx zero? At what points is it infinite? Is this infinite slope physically unrealistic?

Linearity of the definite integral

P5.19 The definite integral may be defined as either a geometrical area or the limit of a sum. Use either definition or a combination of both to prove the following:

$$(1)\quad \int_a^b [F(x) + G(x)]\,dx = \int_a^b F(x)\,dx + \int_a^b G(x)\,dx,$$

$$(2)\quad \int_a^b cf(x)\,dx = c\int_a^b f(x)\,dx,$$

where c is a constant. (3) Assemble these two proofs in order to establish the linearity property of the definite integral (Equation 5.76).

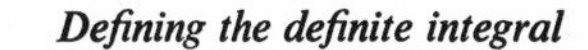
Defining the definite integral

P5.20 In the figure, the lightly shaded rectangle, whose area is $\Delta A = f(x)\,\Delta x$, represents the approximate area under the curve from x to $x + \Delta x$. It differs from the true area by an amount δA (darkly shaded area). Approximate $f(x)$ as a linear function in the region of interest. Obtain an expression for δA, and show that the fractional error in the rectangular approximation, $\delta A/\Delta A$, approaches zero as ΔA approaches zero. *Optional:* Justify the linear approximation for $f(x)$.

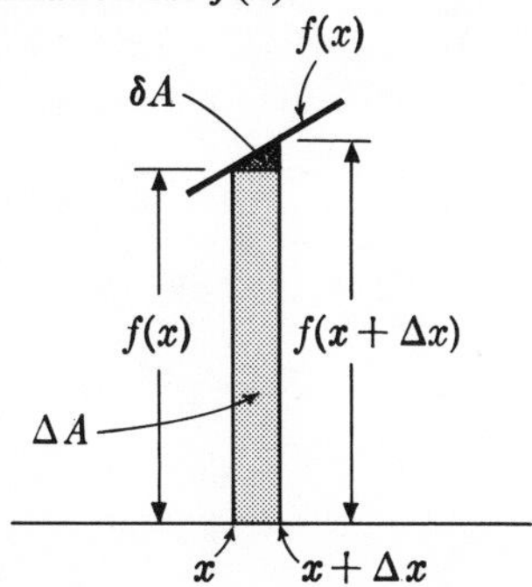

rms deviation

P5.21 (1) For the data of Exercise 5.62, find the average, or mean, value of the wavelength λ, and call it $\bar{\lambda}$. (2) Find the root-mean-square deviation of the data from the mean, defined by

$$\Delta\lambda = \sqrt{\frac{1}{n}\sum_{i=1}^{n}(\lambda_i - \bar{\lambda})^2},$$

where n is the number of measurements. If there were no systematic error, $\Delta\lambda$ would provide a good measure of the uncertainty of each measurement. (3) Prove that $(\Delta\lambda)^2 = \overline{\lambda^2} - \bar{\lambda}^2$, where $\overline{\lambda^2}$ is defined by

$$\overline{\lambda^2} = \frac{1}{n}\sum_{i=1}^{n}\lambda_i^2.$$

Optional: Consult another reference to learn about standard deviation, and obtain a value for the standard deviation of the *average* of the seven measurements in Exercise 5.62.

Uncertainty of slope

P5.22 Estimate the uncertainty of the slope of the straight line graph in Figure 5.32(b). To do this use any method you like, but explain how you reach your conclusion.

6 Vectors

A vector is a mathematical object with both numerical and geometrical properties. The physicist has been able to make good use of it in his description of nature. In this chapter we will be concerned with vector algebra and some vector calculus. The vector properties discussed here will find application throughout the rest of the text.

6.1 Scalars and numerics: geometrical arithmetic

Numeric: a quantity represented by a single number

Scalar: a numeric that is independent of the coordinate system

A numeric is quantity with magnitude only, a quantity that is represented by a single number. A scalar is a special kind of numeric, one whose numerical value is independent of the choice of a frame of reference or a coordinate system. Longitude on the earth's surface is an example of a numeric. It is specified by a single number, but that number is determined by the choice of an origin, the zero of longitude passing through Greenwich, England. The mass of an object, on the other hand, is independent of the choice of a frame of reference. Mass is a scalar quantity. (Its magnitude may, of course, depend on the choice of units.) Whether or not a physical quantity is a scalar quantity can be decided only by experiment.

The rules for manipulating numerics are the familiar rules of arithmetic. In school one learns of the correspondence between numbers and distances measured along a line. Because of this correspondence, the operations of arithmetic can be represented geometrically by rules of displacement along a straight line. Here we briefly review this geometrical arithmetic.

The real axis, a labeled line

Figure 6.1 shows a line with equally spaced points labeled by the positive and negative integers and zero. To every real number there corresponds a point on this line, and to every point on the line there corresponds a real number; this is a one-to-one correspondence. Every point on the line is said to be "labeled" with a number. This labeled line may be called the real axis.

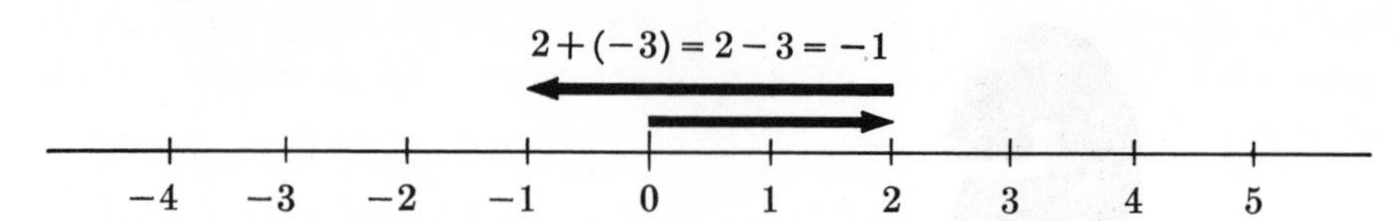

FIGURE 6.1 The real axis and geometrical arithmetic. There is a one-to-one correspondence between points on the line and real numbers. Displacements represent addition and subtraction.

Geometrical arithmetic

Addition and subtraction of numbers is equivalent to marching up and down the real axis according to the following rules: To add a positive number or subtract a negative number, execute a displacement to the right of a distance corresponding to the number in question; to subtract a positive number or add a negative number, execute a displacement of appropriate distance to the left. The result of a series of additions and subtractions is represented by the net displacement (positive or negative) from the starting point to the end point. Beginning at the origin, for example, moving 2 units to the right and then 3 units to the left (ending 1 unit to the left of the origin) represents the operation $2 - 3 = -1$, or $2 + (-3) = -1$. The operation $1.5 + 1.5 - (-3)$ is represented by successive displacements of 1.5, 1.5, and 3 to the right, ending 6 units to the right of the origin.

These geometrical rules can be used to illustrate in a simple way the following basic properties of numbers:

$$a = -(-a), \tag{6.1}$$

Commutative law

$$a + b = b + a, \tag{6.2}$$

Associative law

$$a + (b + c) = (a + b) + c. \tag{6.3}$$

The second is the commutative property of addition; the third is the associative property of addition.

Multiplication is equivalent to stretching and shrinking

There is also an instructive way to look at multiplication geometrically—in terms of stretching and shrinking. To perform the operation $3 \times 4 = 12$, imagine a string stretched from the origin to the point labeled 4. Then stretch the string to triple its length so that it extends from the origin to the point labeled 12. To carry out the multiplication $0.35 \times 10 = 3.5$, imagine a string first extended from the origin 10 units to the right, then contracted to 35 percent of its original length so that it reaches only 3.5 units from the origin. Thus multiplication by a number larger than 1 corresponds to stretching, and multiplication by a number smaller than 1 corresponds to shrinking.

Another formal property of numbers, the distributive property, has to do with a combination of addition and multiplication. Algebraically, the distributive law may be written

Distributive law

$$c(a + b) = ca + cb. \tag{6.4}$$

Our everyday experience with numbers tells us that the distributive law is a "self-evident truth." To the mathematician, it is instead an arbitrary assumption, one of the defining axioms of numerical arithmetic. Because addition and multiplication of numbers can be represented geometrically, the distributive law can be illustrated geometrically. Adding together the length of two strings, for example, then stretching the total length by a certain factor, produces the same result as first stretching each string by the given factor and then adding their new lengths together.

Numbers are so familiar and seemingly so concrete that it is easy to forget that they exist as abstract concepts. One reason for discussing geometrical arithmetic is to provide a useful analogy to vector arithmetic. Another reason is to provide a reminder of the essential difference between an abstract concept and its geometrical representation, a real difference despite their one-to-one correspondence. Actually, physicists think of numbers in three different ways. Although they (including this author) may sometimes mix these different ways verbally, it is important that they be clearly separated mentally. First is the abstract concept, a mathematical object whose reality and existence are defined by the axioms of arithmetic. Second is the geometrical representation, a point on a line or a distance from the origin to the point. The geometrical representation makes the abstract concept pictorial, and is useful because rules for combining lengths along a straight line can be made to correspond exactly to the rules for combining abstract numbers. Third is the physical number, a measure of the size or quantity of some measurable concept.

Three ways to think of numbers

1. *An abstract concept*

2. *A geometrical representation*

3. *A physical measure*

Usually, when a number is assigned to a physical concept, it means more than that the concept is quantitative (measurable). It also means that the quantity in question obeys the laws of arithmetic. If we say that a piece of meat weighs 4 lb, it is a useful statement only because it is an experimental fact that weights combine like numbers. A 4-lb piece of meat and a 3-lb piece of meat together weigh 7 lb. If this were not so—and there is no self-evident reason that it must be so—then numbers might not usefully be attached to weight. The physical concept, weight, might then have to be associated with some mathematical concept more abstract than numbers. This has indeed happened with some of the concepts of modern physics, which have to be represented by vectors or other more complicated mathematical objects. Nevertheless, a great many physical concepts are measured in terms of numbers simply because, experimentally, numbers have been found to suffice. The assignment of a number to a physical concept is provisional in exactly the same sense that a law of nature is provisional; it is accepted so long as it is successful and free from contradiction.

Which physical quantities are numerics? Only experiment can decide

6.2 Vectors: addition and subtraction

A vector is defined mathematically in terms of its transformation properties—the way it changes when the frame of reference is changed. We shall expand on this idea in Section 6.5. It is convenient first to survey vector arithmetic by taking advantage of its equivalence to certain rules of geometrical combination. A vector is a quantity with both magnitude and direction, and it can be represented by a directed line segment: an arrow whose length, in any convenient unit, is the magnitude of the vector, and whose direction is the direction of the vector (Figure 6.2). The space in which the vector direction is defined may be a fictional mathematical space and may have any number of dimensions. We are concerned here only with vectors in three-dimensional Euclidean space. Such

A vector is represented by an arrow

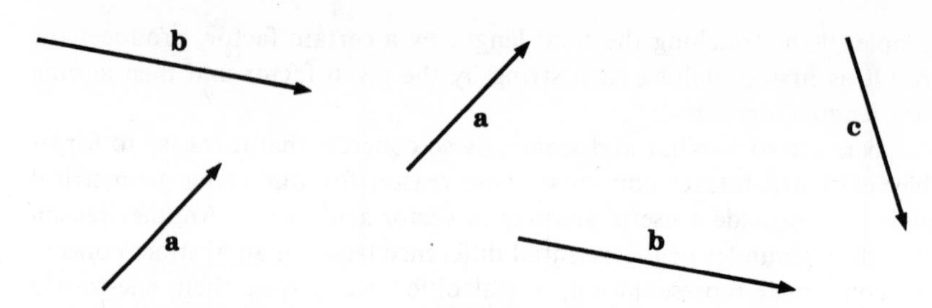

FIGURE 6.2 Vectors are represented by arrows. The position of the arrow is unimportant. Only the magnitude and the direction matter. The two arrows marked **a** represent the same vector. The two arrows marked **b** are also equivalent.

vectors are called three-vectors.* (In many physical applications, only two of the three dimensions are relevant.)

Three ways to think of vectors

1. Abstract object

2. Geometrical representation

3. Physical quantity

With vectors, as with numbers, it is important to distinguish three separate ideas: the abstract mathematical object, the concrete geometrical representation, and the physical quantity. With numbers, the geometrical representation is interesting but hardly necessary. With vectors, the geometrical representation is almost a necessity. Only with the help of arrow diagrams can we think easily about vectors and vector arithmetic. A vector quantity is a physical quantity that behaves like a vector. This means that in the equations governing its physical behavior the quantity must be treated mathematically as a vector. Velocity, force, momentum, and angular momentum are examples of vector quantities. Such quantities are themselves sometimes called vectors. However, it is preferable to reserve the word "vector," like the word "number," to refer to the mathematical entity. The physical concept (force, for instance) that "behaves like" a vector we call a vector quantity.

Arrow diagrams represent vector addition

An arrow represents a vector in exactly the same sense that a displacement along the real axis represents a number. Between the formal rules of vector arithmetic and the geometrical rules of arrow diagrams there is a one-to-one correspondence. The addition of two vectors is represented by arrow diagrams of the kind shown in Figure 6.3. With its length and direction fixed, the representative arrow of the first vector is placed arbitrarily. The second representative arrow is placed with its tail at the head of the first arrow. Then an

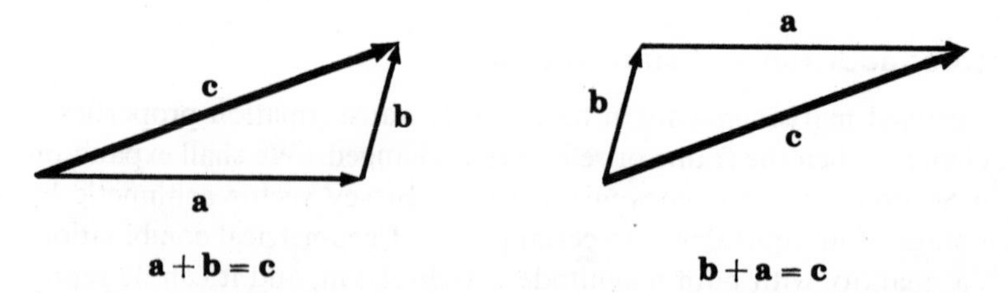

FIGURE 6.3 Vector addition. The diagrams show the commutative law of vector addition: $\mathbf{a} + \mathbf{b} = \mathbf{b} + \mathbf{a}$.

* Four-vectors will be introduced in Section 21.5.

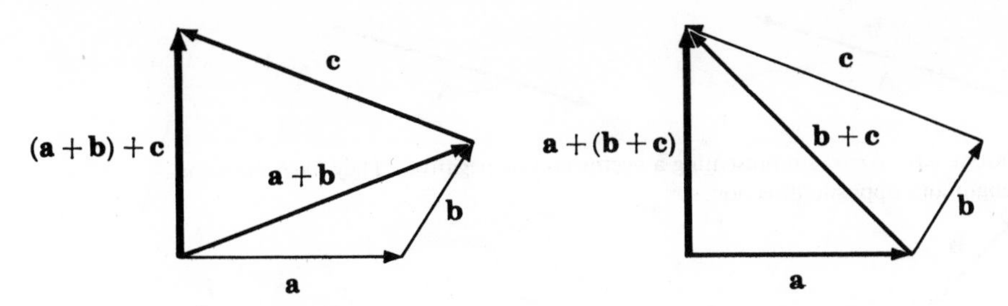

FIGURE 6.4 The associative property of vector addition, illustrated geometrically: $(\mathbf{a} + \mathbf{b}) + \mathbf{c} = \mathbf{a} + (\mathbf{b} + \mathbf{c})$.

arrow drawn from the tail of the first arrow to the head of the second represents the sum. Algebraically, the vector sum is written

$$\mathbf{a} + \mathbf{b} = \mathbf{c}. \tag{6.5}$$

Boldface type is conventionally used to indicate vectors. In handwritten work, vectors are usually designated by arrows ($\vec{a}$) or by wavy underlining ($\underset{\sim}{a}$), which is the printer's symbol for boldface. It is essential to have a special notation for vectors, since numerical equations and vector equations have entirely different meanings. We shall consistently use boldface type for vector quantities and italic type for numerical quantities. If a particular vector is represented by **F**, its magnitude is represented by the same symbol in italic type, *F*, or by the boldface symbol between vertical bars, $|\mathbf{F}|$. The latter notation serves as a reminder that the magnitude of a vector is always a positive quantity.

As indicated in Figure 6.3, vector addition is commutative:

Commutative law

$$\mathbf{a} + \mathbf{b} = \mathbf{b} + \mathbf{a}. \tag{6.6}$$

The associative property of addition is also valid for vectors:

Associative law

$$(\mathbf{a} + \mathbf{b}) + \mathbf{c} = \mathbf{a} + (\mathbf{b} + \mathbf{c}). \tag{6.7}$$

This property is demonstrated geometrically in Figure 6.4. Because of the associative property, parentheses may be omitted in vector addition, and the arrow diagram may be extended to the addition of any number of vectors. Figure 6.5 represents the addition of four forces, which produces a total force $\mathbf{F}_T$.

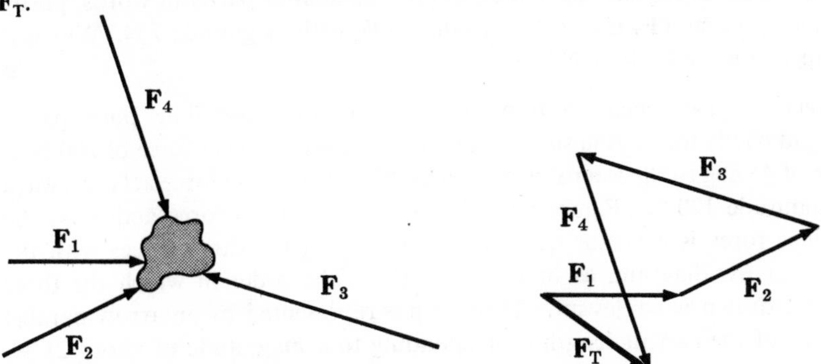

FIGURE 6.5 An object experiences four forces. The total force is the vector sum, $\mathbf{F}_T = \mathbf{F}_1 + \mathbf{F}_2 + \mathbf{F}_3 + \mathbf{F}_4$.

FIGURE 6.6 Arrows representing a vector and its negative. They have the same length and opposite direction.

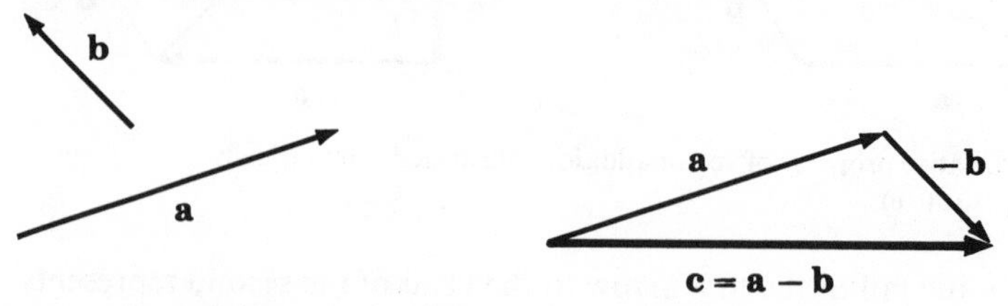

FIGURE 6.7 Subtraction of vectors: $\mathbf{a} - \mathbf{b}$ is defined to be $\mathbf{a} + (-\mathbf{b})$. Note that the *difference* of two vectors may have a magnitude greater than the magnitude of either vector.

The negative of a vector is defined as a vector of the same magnitude and opposite direction (Figure 6.6). This definition makes clear that the negative of the negative of a vector is the original vector:

$$\mathbf{A} = -(-\mathbf{A}). \tag{6.8}$$

Other properties shared with numbers

Subtraction of a vector is defined as addition of the negative vector. Vector subtraction, which can be written

$$\mathbf{c} = \mathbf{a} - \mathbf{b} = \mathbf{a} + (-\mathbf{b}), \tag{6.9}$$

is illustrated in Figure 6.7.

■ EXAMPLE 1: Force $\mathbf{F}_1$ has magnitude 3 N (newtons), and force $\mathbf{F}_2$ has magnitude 4 N. What is the magnitude of the sum $\mathbf{F}_T = \mathbf{F}_1 + \mathbf{F}_2$? As phrased, the question has no definite answer. The sum $\mathbf{F}_T$ may have any magnitude from 1 N to 7 N, depending on the relative orientation of $\mathbf{F}_1$ and $\mathbf{F}_2$. It is a common error to suppose that if $\mathbf{F}_T = \mathbf{F}_1 + \mathbf{F}_2$, then the magnitudes also add: that is, $F_T = F_1 + F_2$. In general, this is *not* true. ■

$|\mathbf{F}_1 + \mathbf{F}_2|$ need not equal $F_1 + F_2$

■ EXAMPLE 2: The two forces referred to in Example 1 act in the same direction. What is the sum $\mathbf{F}_T$? We can give the answer partly in words, partly in symbols. The sum $\mathbf{F}_T$ is a vector parallel to $\mathbf{F}_1$ with magnitude 7 N. Why is it *not* correct to write $\mathbf{F}_T = 7$ N? ■

■ EXAMPLE 3: Two men on opposite sides of an automobile push partly inward and partly forward in such a way that each is exerting a force of 100 N at an angle of 45 deg to the long axis of the car, and a frictional force acts rearward with magnitude 120 N [Figure 6.8(a)]. What is the total force acting on the car? Since force is a vector quantity, we may sum the three forces with the aid of an arrow diagram, as in Figure 6.8(b). The order in which the three arrows are drawn is irrelevant. Their sum is represented by an arrow parallel to the axis of the car, its length corresponding to a magnitude of about 21 N. This answer could be obtained graphically if the drawing were done with great care. But it is better to use trigonometry, as illustrated in Figure 6.9. As the figure makes clear,

Addition of three vectors: trigonometry is helpful

$$|\mathbf{F}_1 + \mathbf{F}_2| = 2F_1 \cos (45 \text{ deg}) = 141.4 \text{ N}.$$

To the vector $\mathbf{F}_1 + \mathbf{F}_2$ is added the oppositely directed vector $\mathbf{F}_f$, of magnitude 120 N. The sum is a forward-pointing vector of magnitude 21.4 N. ■

A problem of relative motion

■ EXAMPLE 4: Consider this problem of relative motion. A passenger walks at 5 ft/sec inside a train that is moving forward at a speed of 12 ft/sec. What is the velocity of the passenger with respect to the ground? The general answer is given by the vector equation

$$\mathbf{v} = \mathbf{v}_p + \mathbf{v}_t, \tag{6.10}$$

where $\mathbf{v}_p$ is the velocity of the passenger with respect to the train and $\mathbf{v}_t$ is the velocity of the train with respect to the ground. If the passenger walks forward [Figure 6.10(a)], $\mathbf{v}$ has magnitude 17 ft/sec and is directed forward. If he walks to the rear [Figure 6.10(b)], $\mathbf{v}$ has magnitude 7 ft/sec and is directed forward. If he walks transverse to the train's motion [Figure 6.10(c)], the arrows representing $\mathbf{v}_p$ and $\mathbf{v}_t$ form the sides of a right triangle whose hypotenuse is the arrow representing $\mathbf{v}$. Therefore,

$$v = \sqrt{v_p^{\,2} + v_t^{\,2}} = 13 \text{ ft/sec}.$$

The vector $\mathbf{v}$ makes an angle of approximately 23 deg with the vector $\mathbf{v}_t$. ■

■ EXAMPLE 5: From radio-navigation measurements, a pilot determines that he is moving due north with respect to the earth at a speed of 120 mile/hr. Call this velocity $\mathbf{v}$. It is the sum of the airplane's velocity with respect to the air $\mathbf{v}_a$, and the velocity of the air with respect to the ground (the wind velocity), $\mathbf{v}_w$:

$$\mathbf{v} = \mathbf{v}_a + \mathbf{v}_w. \tag{6.11}$$

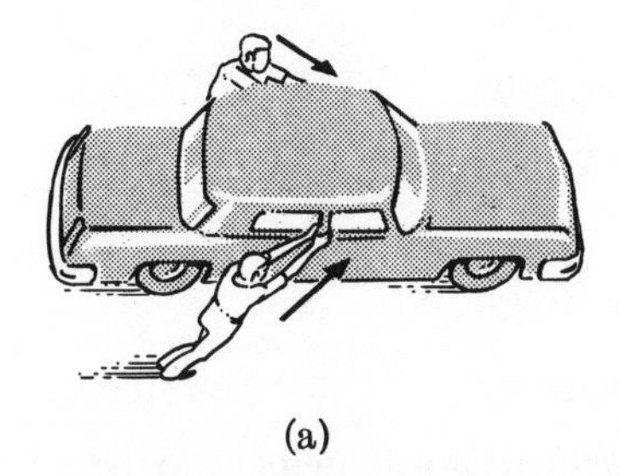

(a)

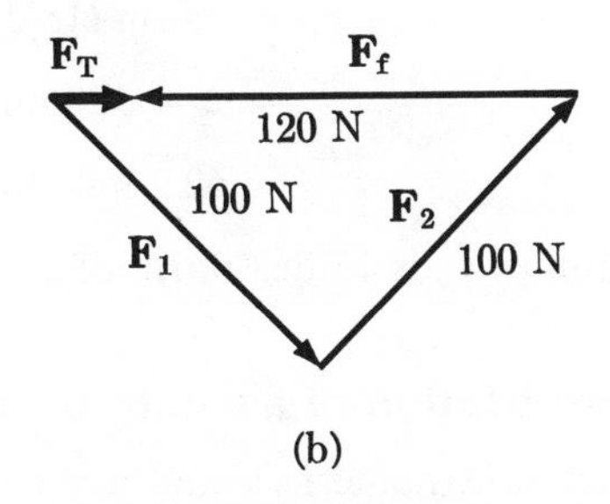

(b)

FIGURE 6.8 Two men exert forces $\mathbf{F}_1$ and $\mathbf{F}_2$ on the car, and the frictional force is $\mathbf{F}_f$. The total force on the car is $\mathbf{F}_T = \mathbf{F}_1 + \mathbf{F}_2 + \mathbf{F}_f$.

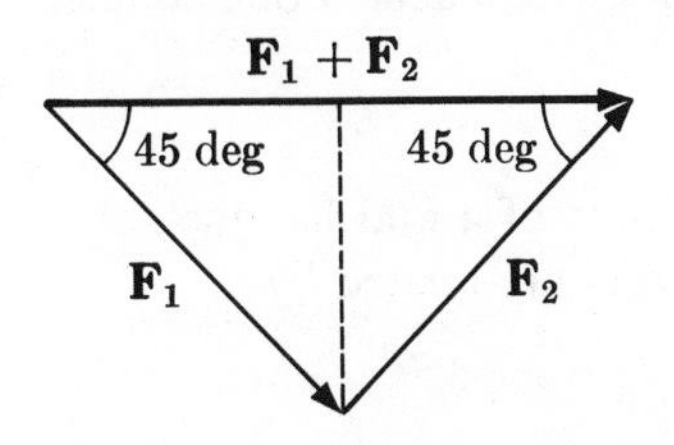

$\mathbf{F}_1 + \mathbf{F}_2$

$\mathbf{F}_T$ $\mathbf{F}_f$

FIGURE 6.9 Use of trigonometry to sum the force vectors in Example 3. Forces $\mathbf{F}_1$ and $\mathbf{F}_2$ are added first. Their sum is then easily added to $\mathbf{F}_f$.

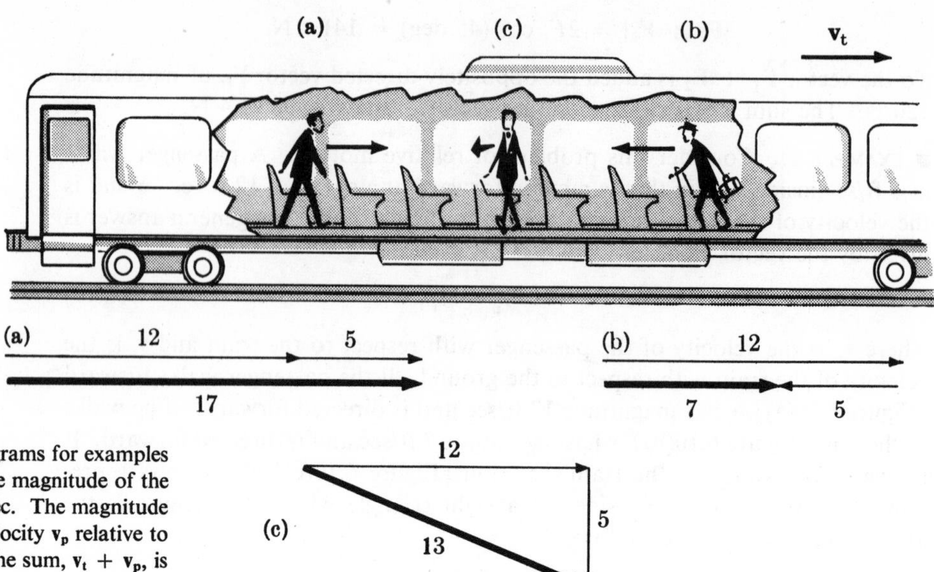

FIGURE 6.10 Vector diagrams for examples of relative motion. The magnitude of the train's velocity $\mathbf{v}_t$ is 12 ft/sec. The magnitude of the passenger's velocity $\mathbf{v}_p$ relative to the train is 5 ft/sec. The sum, $\mathbf{v}_t + \mathbf{v}_p$, is the passenger's velocity relative to the ground.

From his compass and airspeed indicator he determines that $\mathbf{v}_a$ has magnitude 120 mile/hr and is directed 30 deg to the west of north [Figure 6.11(a)]. What is the wind velocity? This is a problem in vector subtraction:

$$\mathbf{v}_w = \mathbf{v} - \mathbf{v}_a . \tag{6.12}$$

The wind triangle

The arrow diagram is shown in Figure 6.11(b). We take advantage of the fact that it forms an isosceles triangle and write

$$\begin{aligned} v_w &= 2v \sin(15 \text{ deg}) \\ &= 2 \times 120 \text{ mile/hr} \times 0.259 \\ &= 62.2 \text{ mile/hr.} \end{aligned}$$

The direction of $\mathbf{v}_w$ is 15 deg north of east. ■

6.3 Multiplication of a vector by a numeric

The product of a numeric α and a vector $\mathbf{V}$ (written $\alpha\mathbf{V}$) is defined to be a vector of magnitude $|\alpha|\, V$, parallel to $\mathbf{V}$ if α is positive and antiparallel to $\mathbf{V}$ if α is negative. This operation can be viewed geometrically as a stretching or shrinking of a vector without changing the line along which it acts. For example, the vector equation

$$\mathbf{b} = 3\mathbf{a}$$

means that the vector $\mathbf{b}$ has three times the magnitude of $\mathbf{a}$ and has the same direction as $\mathbf{a}$ (Figure 6.12). If another pair of vectors are related by

$$\mathbf{c} = 0.25\mathbf{d},$$

the vector $\mathbf{d}$, if shrunk to one quarter of its magnitude without change of direction, yields the vector $\mathbf{c}$. The equation

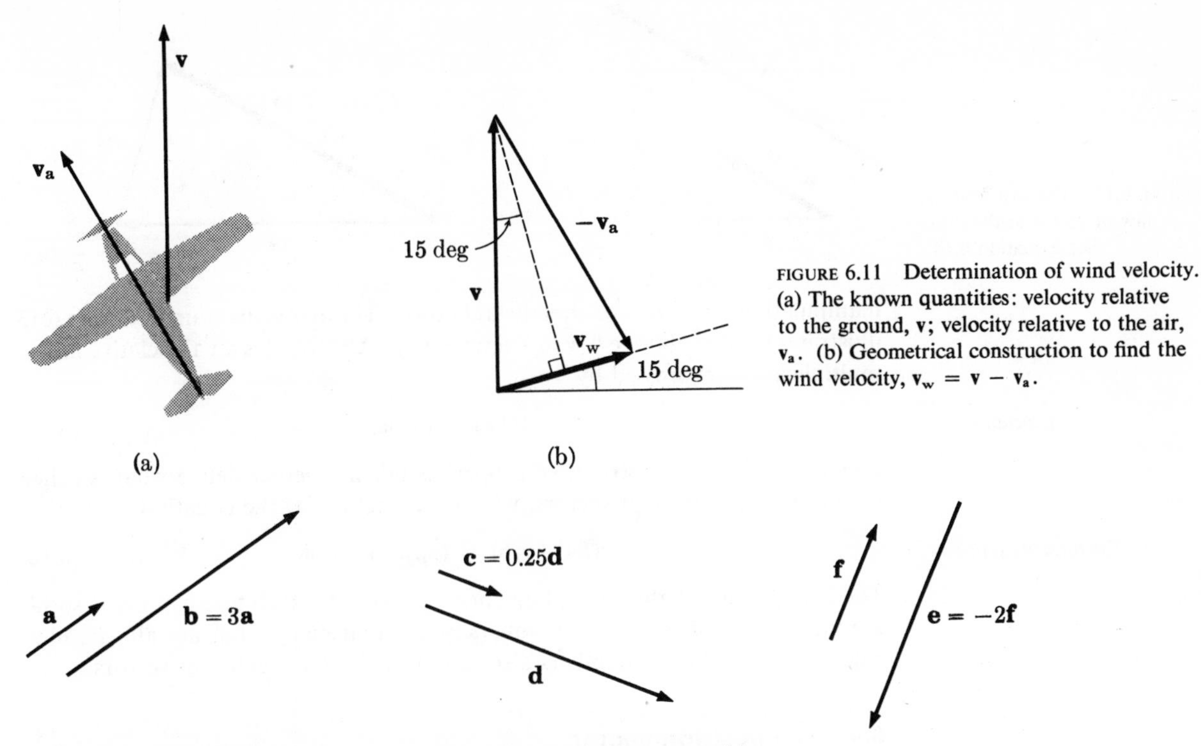

FIGURE 6.11 Determination of wind velocity. (a) The known quantities: velocity relative to the ground, **v**; velocity relative to the air, $\mathbf{v}_a$. (b) Geometrical construction to find the wind velocity, $\mathbf{v}_w = \mathbf{v} - \mathbf{v}_a$.

FIGURE 6.12 Multiplication of a vector by a numeric is equivalent to stretching or shrinking the vector. The new vector may be oppositely directed, if the numeric is negative; and it may have a different unit of measurement, if the numeric is not dimensionless.

$$\mathbf{e} = (-2)\mathbf{f}$$

has the same meaning as the equation

$$\mathbf{e} = 2(-\mathbf{f}).$$

The vector **e** has twice the magnitude of **f** and is oppositely directed. A simple physical example of the multiplication of a vector by a numeric is afforded by the definition of momentum:

$$\mathbf{p} = m\mathbf{v}. \tag{6.13}$$

The velocity vector **v** multiplied by the scalar mass m gives the momentum vector **p**. Note that the numeric appearing in such multiplication need not be dimensionless; the dimensions of **p** and **v** are different.

The operations of addition and multiplication by a numeric may be combined to express the distributive law of vector arithmetic:

Distributive law

$$n(\mathbf{a} + \mathbf{b}) = n\mathbf{a} + n\mathbf{b}. \tag{6.14}$$

Summing two vectors and then multiplying the sum by a numeric gives the same result as separately multiplying the two vectors by the numeric before adding them. The distributive law may be regarded as a fundamental axiom of vector arithmetic, but its truth follows from the geometrical rules of vector

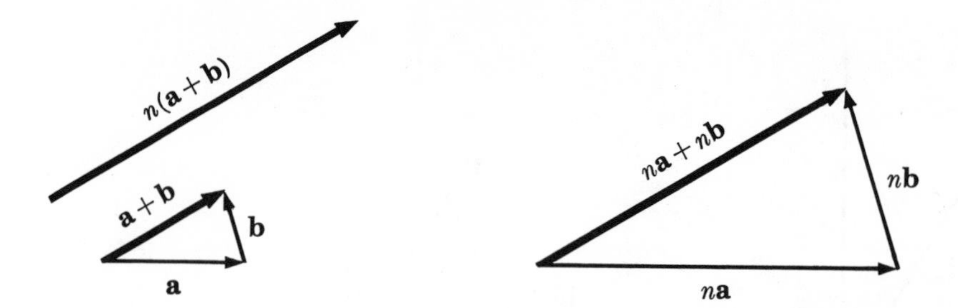

FIGURE 6.13 The distributive law of vector arithmetic. See Equation 6.14.

manipulation that we have already laid down. The arrow diagram in Figure 6.13 illustrates the distributive law in a simple way. Also valid is an associative law of multiplication, expressed by

Associative law

$$\alpha(\beta\mathbf{a}) = (\alpha\beta)\mathbf{a}. \tag{6.15}$$

The associative and distributive properties taken together define what is called the linearity property of vectors, which is expressed by the equation

The linearity property

$$\alpha(\beta\mathbf{a} + \gamma\mathbf{b}) = (\alpha\beta)\mathbf{a} + (\alpha\gamma)\mathbf{b}. \tag{6.16}$$

The fact that all of the general algebraic properties of vectors expressed so far are identical to properties of numbers means that many—but not all—familiar rules of the algebra of numbers also hold true for the algebra of vectors.

6.4 The position vector

Special features of the position vector

Among physical vector quantities, one of the most important is position. The location in space of a particle (or for that matter of any point, whether occupied by a particle or not) is a vector quantity. Its magnitude is the distance between the point and an arbitrarily chosen origin, and its direction is the direction of the point from the origin. The position vector is special in several ways. First, unlike most vector quantities, its magnitude as well as its direction has geometrical significance. Second, also unlike most vector quantities, its definition depends on the choice of an origin. Third, as we shall discuss in the next section, it is a standard vector whose properties help to define other vectors. A common notation for the position vector is **r**.

The position vector, like most vectors, is defined at a point

Most vectors, including the position vector, are defined at a single point in space. This fact is sometimes clouded by the fact that the representative arrow in a diagram must span a finite distance. Consider, for example, three forces whose sum is zero, acting at a point. Although the vectors are defined only at that point, the arrow diagram representing their sum (Figure 6.14) is necessarily spread out. Or consider an electron moving along some path in space (Figure 6.15). Its velocity is a vector quantity and may be represented at any instant by an arrow. This arrow, whatever its length or placement, refers to a single point on the path.

It is conventional, as indicated in Figure 6.15, to draw the representative arrow of a position vector from the origin to the point in question. This is often convenient, since it ensures getting the right direction and relative length for the arrow, but it is important to remember that the arrow applies only to a single point and that its actual placement in the diagram is not important. The

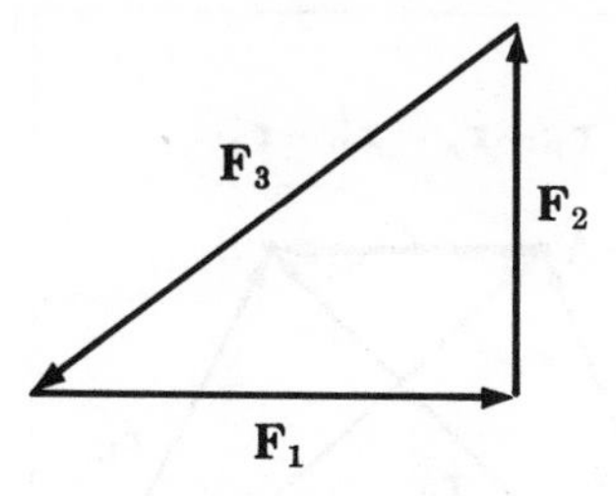

FIGURE 6.14 Balanced forces on a particle. Despite the spatial extent of the diagram, the forces all act at a single point.

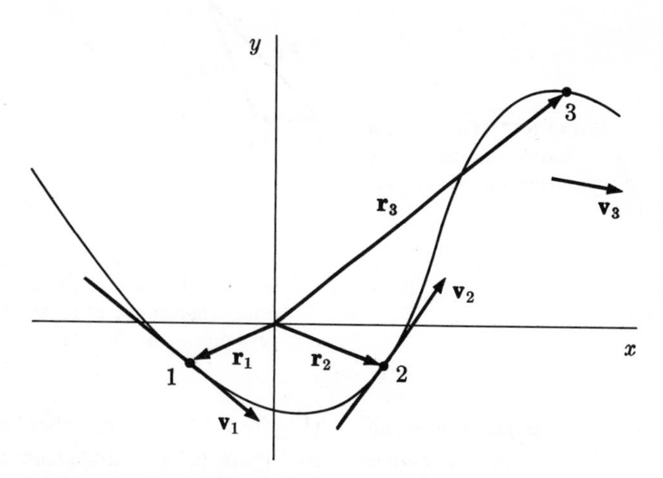

FIGURE 6.15 An electron track in space and representations of position vectors and velocity vectors at three points along the track.

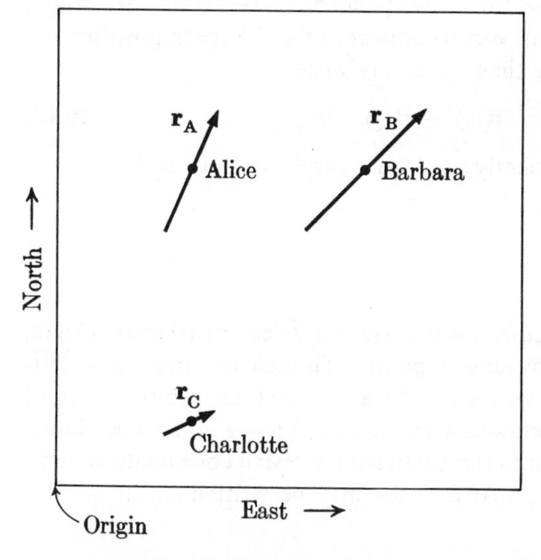

FIGURE 6.16 Position vectors. The three arrows represent the positions of Alice, Barbara, and Charlotte in a frame of reference whose origin is the southwest corner of the room.

position vector of a point is quite distinct from a line joining the origin to the point. Another example of position vectors is illustrated in Figure 6.16. Alice, Barbara, and Charlotte stand at points A, B, and C in a square room whose walls are aligned north-south and east-west. Choosing the southwest corner as the origin, we call the three position vectors $\mathbf{r}_A$, $\mathbf{r}_B$, and $\mathbf{r}_C$ and show their representative arrows in the figure, this time with the arrows chosen deliberately shorter than the distance from the origin to points A, B, and C.

DISPLACEMENT VECTOR

The difference of two position vectors is called a displacement vector. It is the position vector of one point with respect to the other point rather than the origin. Thus $\mathbf{r}_B - \mathbf{r}_A$ is the position vector of Barbara with respect to Alice

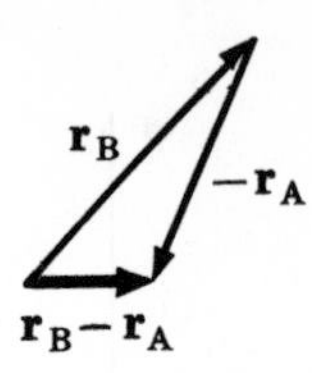

FIGURE 6.17 Geometrical construction to find the displacement vector $\mathbf{r}_B - \mathbf{r}_A$.

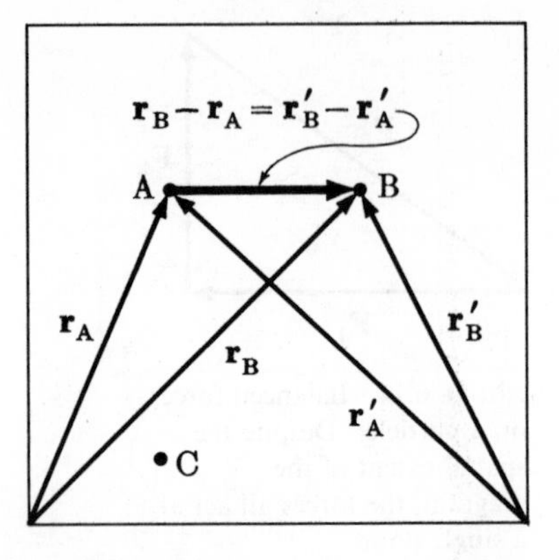

FIGURE 6.18 For a change of origin, position vectors change—$\mathbf{r}_A$ to $\mathbf{r}'_A$ and $\mathbf{r}_B$ to $\mathbf{r}'_B$ in this example—but displacement vectors do not change: $\mathbf{r}_B - \mathbf{r}_A = \mathbf{r}'_B - \mathbf{r}'_A$.

Displacement vector independent of origin

(Figure 6.17). It may also be called the displacement from A to B. As shown in Figure 6.18, a displacement vector is independent of the choice of origin. Frequently in physics, displacement vectors appear as the change in position of a particle from one *time* to another time. We may write

$$\Delta\mathbf{r} = \mathbf{r}(t_2) - \mathbf{r}(t_1). \tag{6.17}$$

The notation **s** or $\Delta\mathbf{s}$ is also frequently used for a displacement vector.

6.5 Components

Coordinates: not necessary but convenient for studying vectors

In order to define the position vector, it is necessary to select an arbitrary origin, which may be located at any convenient point. Though not necessary, it is often very convenient in studying vectors to also select an arbitrary set of reference directions—that is, to choose a frame of reference and a coordinate system. We shall restrict attention to right-handed Cartesian coordinate systems characterized by x, y, and z axes. Any vector may be written as the sum of vectors parallel to the axes:

$$\mathbf{a} = \mathbf{a}_x + \mathbf{a}_y + \mathbf{a}_z\,. \tag{6.18}$$

Unit vectors

This process, called the resolution of a vector into components, is illustrated for a vector in a plane in Figure 6.19. A further decomposition is also convenient, in which each of the two or three component vectors is written as the product of a numeric and a vector of unit length. These unit vectors, **i**, **j**, and **k**, which are parallel to the x, y, and z axes respectively, are shown in Figure 6.20. In terms of the unit vectors, a vector **a** may be written

Resolution of a vector into components

$$\mathbf{a} = a_x\mathbf{i} + a_y\mathbf{j} + a_z\mathbf{k}. \tag{6.19}$$

The numerics a_x, a_y, and a_z may be positive or negative (or zero). It is these numerics that are usually called the components of the vector. Geometrically, a component is equal in magnitude to the length of the projection of a vector onto an axis (Figure 6.21). This projection is expressible as the magnitude of the vector multiplied by the cosine of the angle between the vector and the axis; for instance,

$$a_x = a\cos\theta_x\,. \tag{6.20}$$

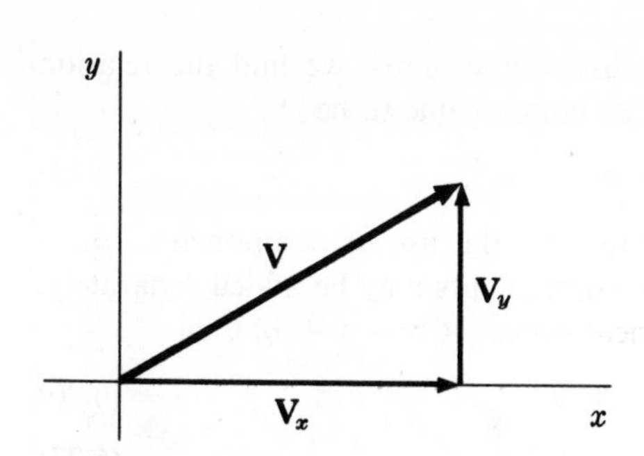

FIGURE 6.19 Components. A vector **V** is expressed as the sum of a vector $\mathbf{V}_x$ parallel to the x axis and a vector $\mathbf{V}_y$ parallel to the y axis.

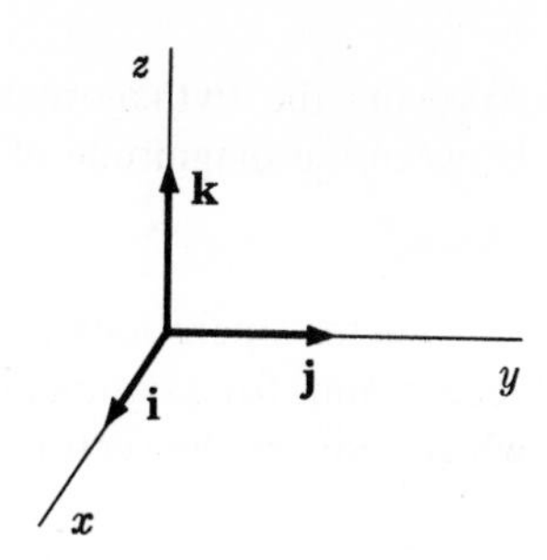

FIGURE 6.20 The unit vectors, **i**, **j**, and **k**.

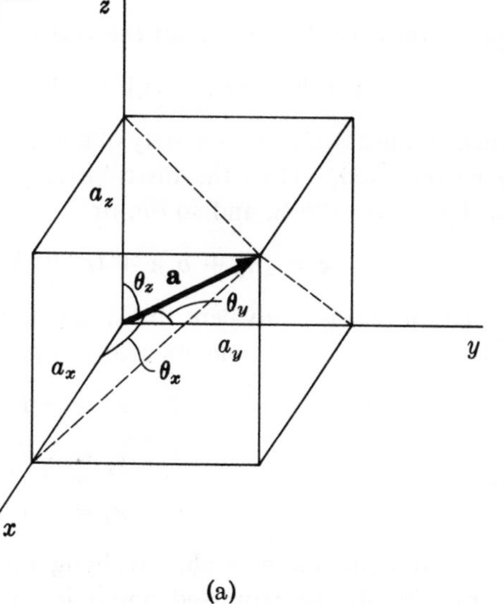

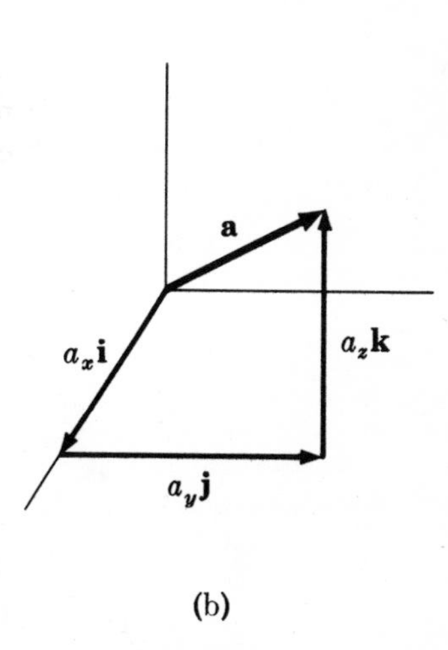

FIGURE 6.21 (a) The components of a vector are the magnitudes of its projections onto the x, y, and z axes. (b) A vector can be written as the sum of three perpendicular vectors: $\mathbf{a} = a_x\mathbf{i} + a_y\mathbf{j} + a_z\mathbf{k}$.

FIGURE 6.22 Components in two dimensions: $B_x = B\cos\theta$ and $B_y = B\cos\varphi = B\sin\theta$.

In two dimensions (Figure 6.22), we may write

$$B_x = B\cos\theta, \tag{6.21}$$

$$B_y = B\cos\varphi, \tag{6.22}$$

if θ is the angle between **B** and the x axis and if φ is the angle between **B** and the y axis. Alternatively, this pair of equations may be written

Equations for components

$$B_x = B\cos\theta, \tag{6.23}$$

$$B_y = B\sin\theta. \tag{6.24}$$

Applying the Pythagorean theorem in three dimensions, we find the relation between the magnitude of a vector and its components to be

$$a^2 = a_x^2 + a_y^2 + a_z^2. \tag{6.25}$$

Vector operations are often facilitated by the use of components. In a vector sum, for example, corresponding components may be added separately, which reduces the vector sum to numerical sums. If $\mathbf{c} = \mathbf{a} + \mathbf{b}$, then

One vector equation gives three component equations

$$c_x = a_x + b_x, \tag{6.26}$$

$$c_y = a_y + b_y, \tag{6.27}$$

$$c_z = a_z + b_z. \tag{6.28}$$

To prove these results, write out the vector sum explicitly:

$$\mathbf{a} + \mathbf{b} = (a_x\mathbf{i} + a_y\mathbf{j} + a_z\mathbf{k}) + (b_x\mathbf{i} + b_y\mathbf{j} + b_z\mathbf{k}). \tag{6.29}$$

Because of the associative property of addition, the parentheses in Equation 6.29 may be removed. Then the distributive property makes it possible to write $a_x\mathbf{i} + b_x\mathbf{i} = (a_x + b_x)\mathbf{i}$, and so on, or

$$\mathbf{c} = (a_x + b_x)\mathbf{i} + (a_y + b_y)\mathbf{j} + (a_z + b_z)\mathbf{k}, \tag{6.30}$$

an equation that is equivalent to Equations 6.26–6.28.

If $\mathbf{e} = -\mathbf{f}$, all components change sign:

$$e_x = -f_x, \tag{6.31}$$

$$e_y = -f_y, \tag{6.32}$$

$$e_z = -f_z. \tag{6.33}$$

The vector equation $\mathbf{g} = \alpha\mathbf{h}$, involving the multiplication of a vector by a numeric, can also be expressed simply in terms of components:

$$g_x = \alpha h_x, \tag{6.34}$$

$$g_y = \alpha h_y, \tag{6.35}$$

$$g_z = \alpha h_z. \tag{6.36}$$

Although seemingly obvious, these equations do require proof, which is left as a problem.

The components of the position vector of a point are simply the Cartesian coordinates of the point:

Components of position vector

$$\mathbf{r} = x\mathbf{i} + y\mathbf{j} + z\mathbf{k}. \tag{6.37}$$

A displacement vector, $\mathbf{s} = \mathbf{r}_2 - \mathbf{r}_1$, can then be expressed in component form by the equations

Components of displacement vector

$$s_x = x_2 - x_1, \tag{6.38}$$

$$s_y = y_2 - y_1, \tag{6.39}$$

$$s_z = z_2 - z_1. \tag{6.40}$$

These displacement components are independent of the choice of origin provided the reference directions are held fixed.

■ EXAMPLE 1: In Figure 6.16, Alice is located 10.9 ft from the origin, and her position vector makes an angle of 66.5 deg with the eastward direction. What are the components of her position vector? Choose the x axis to point east and the y axis to point north. For notational convenience, write $(r_A)_x = x_A$ and $(r_A)_y = y_A$. Then

$$x_A = r_A \cos(66.5 \text{ deg}) = 10.9 \text{ ft} \times 0.399 = 4.35 \text{ ft},$$

$$y_A = r_A \cos(23.5 \text{ deg}) = 10.9 \text{ ft} \times 0.917 = 10.0 \text{ ft}.$$

The second result could also be obtained from $y_A = r_A \sin(66.5 \text{ deg})$. ■

■ EXAMPLE 2: The components of Barbara's position vector are $x_B = y_B = 10.0$ ft. What is Barbara's distance from the origin, and what angle does her position vector make with respect to the x axis (directed eastward)? From Equation 6.25, the magnitude of her position vector is given by

$$r_B = \sqrt{x_B^2 + y_B^2} = \sqrt{200 \text{ ft}^2} = 14.14 \text{ ft}.$$

From Equation 6.20, the cosine of the angle between this vector and the x axis is given by

$$\cos\theta = \frac{x_B}{r_B} = \frac{10.0}{14.14} = 0.707,$$

from which it follows that

$$\theta = 45 \text{ deg} = \frac{\pi}{4}.$$

■

■ EXAMPLE 3: What is the displacement vector from Alice to Barbara? This question is answered graphically in Figure 6.17. It is answered more simply and accurately in terms of components. Define $\mathbf{s} = \mathbf{r}_B - \mathbf{r}_A$. Then

$$s_x = x_B - x_A = 10.0 \text{ ft} - 4.35 \text{ ft} = 5.65 \text{ ft},$$

$$s_y = y_B - y_A = 10.0 \text{ ft} - 10.0 \text{ ft} = 0.$$

■

■ EXAMPLE 4: An electron flying toward a stationary proton moves along the x axis with momentum $p_1 = 3.0 \times 10^{-24}$ kg m/sec (Figure 6.23). The electron emerges from the collision with momentum $\mathbf{p}_3$ directed along the y axis. Measurement shows that the magnitude of its momentum is unchanged:* $p_3 = p_1$. What is the final momentum $\mathbf{p}_4$ of the proton? Mathematically, we may phrase the problem in this way:

$$\mathbf{p}_1 = p\mathbf{i}, \qquad \mathbf{p}_2 = 0,$$

$$\mathbf{p}_3 = p\mathbf{j}, \qquad p_4 = ?,$$

where $p = 3.0 \times 10^{-24}$ kg m/sec. The answer is obtained by using the vector equation of momentum conservation,

$$\mathbf{p}_1 + \mathbf{p}_2 = \mathbf{p}_3 + \mathbf{p}_4 \,. \tag{6.41}$$

Momentum conservation: a vector law

* In reality, $p_3 < p_1$, but the difference is less than 0.1 percent.

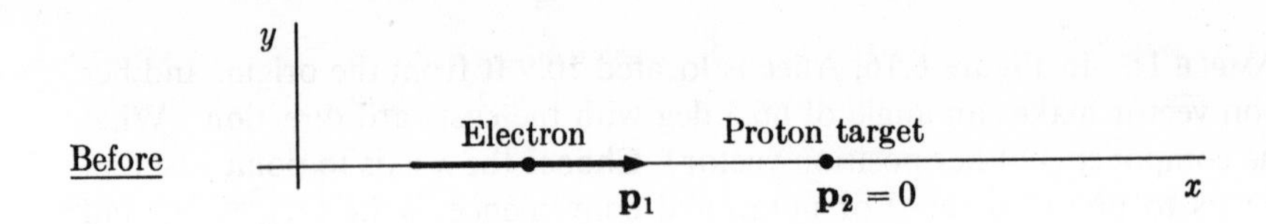

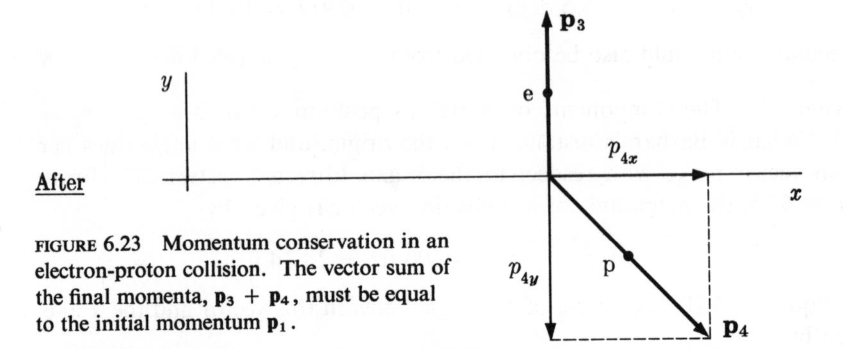

FIGURE 6.23 Momentum conservation in an electron-proton collision. The vector sum of the final momenta, $\mathbf{p}_3 + \mathbf{p}_4$, must be equal to the initial momentum $\mathbf{p}_1$.

Since the total momentum vector on the left is equal to the total momentum vector on the right, their components may also be equated:

$$p_{1x} + p_{2x} = p_{3x} + p_{4x}, \tag{6.42}$$

$$p_{1y} + p_{2y} = p_{3y} + p_{4y}. \tag{6.43}$$

A vector equation is always equivalent to two or three component equations. Rewriting Equation 6.42 gives

$$p + 0 = 0 + p_{4x},$$

and from Equation 6.43 we get

$$0 + 0 = p + p_{4y}.$$

These two equations are readily solved to give

$$p_{4x} = p = 3.0 \times 10^{-24} \text{ kg m/sec},$$

$$p_{4y} = -p = -3.0 \times 10^{-24} \text{ kg m/sec}.$$

It follows that $\mathbf{p}_4$ is directed at 45 deg to the x axis and has magnitude

$$p_4 = \sqrt{p^2 + p^2} = 4.24 \times 10^{-24} \text{ kg m/sec}.$$

Momentum-conservation analysis of this kind is applied to elementary-particle tracks in bubble chambers and helps to establish the existence of unseen neutral particles. ■

★TRANSFORMATION OF COMPONENTS

One of the advantages of a vector equation such as $\mathbf{a} = \mathbf{b}$ is that it refers to no particular coordinate system and is valid in any coordinate system. As an abstract entity, the vector $\mathbf{a}$ is independent of coordinates. To give it concrete expression, however, we must define its direction relative to chosen axes or give its components. The components do depend on the choice of coordinates. Consider in particular the effect of a rotation of plane Cartesian coordinates on the components of a position vector. Figure 6.24(a) shows a position vector $\mathbf{r}_1$ and its components x_1 and y_1 referred to a particular coordinate system xy. With respect to a different coordinate system $x'y'$ [Figure 6.24(b)], the same vector $\mathbf{r}_1$ has different components, x_1' and y_1'. Thus we can write two expressions for the same vector:

$$\mathbf{r}_1 = x_1\mathbf{i} + y_1\mathbf{j}, \tag{6.44}$$

$$\mathbf{r}_1 = x_1'\mathbf{i}' + y_1'\mathbf{j}'. \tag{6.45}$$

It is left as a problem to prove that if the coordinate system $x'y'$ is rotated through an angle θ with respect to the coordinate system xy, the two sets of components are related by

Transformation equations

$$x_1' = x_1 \cos\theta + y_1 \sin\theta, \tag{6.46}$$

$$y_1' = -x_1 \sin\theta + y_1 \cos\theta. \tag{6.47}$$

Together these equations comprise what is called a transformation (from components x_1, y_1 to components x_1', y_1'). Formally, a vector is defined in terms of its transformation properties. The position vector is taken to be the standard representative vector. Any other quantity with components that transform in the same way is, by definition, also a vector. Shown in Figure 6.25, for instance, is a force vector $\mathbf{F}$. Its components in the first frame of reference are F_x, F_y, and in the second frame of reference are F_x', F_y'. Because $\mathbf{F}$ is a vector, these components are related by equations exactly like Equations 6.46 and 6.47:

The position vector is the prototype vector

$$F_x' = F_x \cos\theta + F_y \sin\theta, \tag{6.48}$$

$$F_y' = -F_x \sin\theta + F_y \cos\theta. \tag{6.49}$$

Geometrically, a vector quantity is an entity with a magnitude and a direction in space, neither of which properties is changed by a change of coordinates.

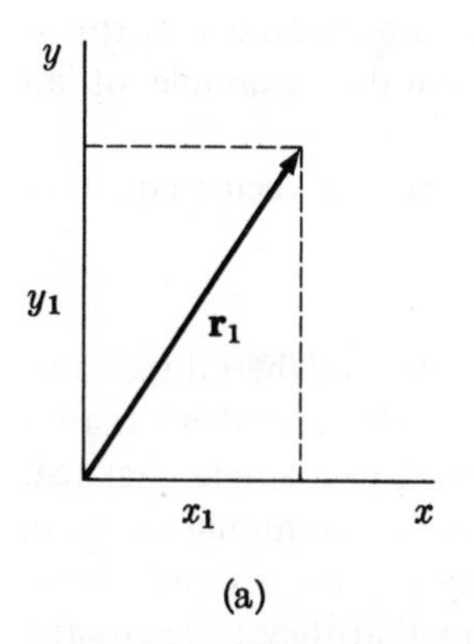

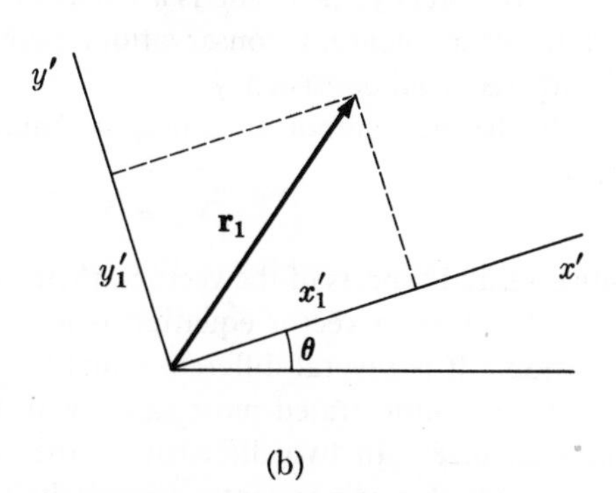

FIGURE 6.24 The components of a vector depend on the choice of coordinate system. The vector $\mathbf{r}_1$ has components x_1 and y_1 with respect to one coordinate system and components x_1' and y_1' with respect to another.

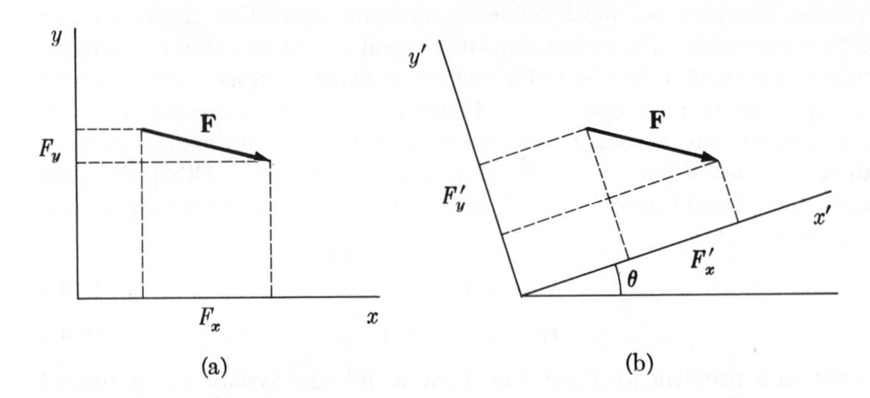

FIGURE 6.25 Transformation of components. The components F'_x and F'_y are related to F_x and F_y by Equations 6.48 and 6.49. These are identical in form to Equations 6.46 and 6.47, which transform the components of a position vector. A vector may be *defined* as an entity whose components transform in this way. The representative arrow of a vector is unaffected by a rotation of the coordinate axes.

6.6 Vectorial consistency

Terms added and equated must be of the same kind

Besides consistency of units and dimensions, equations of physics must have vectorial consistency to be generally useful. Technically, this means that both sides of an equation, or terms added together in an equation, must have the same transformation properties if the frame of reference is changed. Put more simply, vectors can be added only to vectors and equated only to vectors. Scalars can be added only to scalars and equated only to scalars. Newton's second law is an example of an equation with vectorial consistency:

$$\mathbf{F} = m\mathbf{a}.$$

Each side is a vector quantity. Similarly, the formula for kinetic energy,

$$K = \tfrac{1}{2}mv^2,$$

has vectorial consistency; each side is a scalar quantity. Equation 6.41, expressing the law of momentum conservation, provides another example of an equation with vectorial consistency.

Probably the most important thing to remember about a vector equation of the form

$$\mathbf{V}_1 = \mathbf{V}_2$$

A vector equation is two or three equations in one

is that *both* essential aspects of the vectors, their magnitudes *and* their directions, are equal. Therefore, a vector equation is a more powerful statement than a scalar equation. It is several different equalities wrapped in a single package. This fact can be demonstrated most easily with the help of components, as in Equations 6.26–6.28. In two dimensions, the equality of two vectors means that they have equal x components; separately it means that they have equal y components. In three dimensions, a vector equation is equivalent to three different equations:

$$V_{1x} = V_{2x}, \tag{6.50}$$

$$V_{1y} = V_{2y}, \tag{6.51}$$

$$V_{1z} = V_{2z}. \tag{6.52}$$

These equations have a special kind of vectorial consistency. The component V_{1x} is a numeric, neither a scalar nor a vector. However, if the frame of reference is rotated, V_{1x} and V_{2x} change by the same factor and remain equal.

6.7 The scalar product

Although numbers are multiplied in only one way, there are several ways to define vector multiplication. Two products are of special importance in physics: the scalar product, considered in this section, and the vector product, considered in Section 6.8. Geometrically, the scalar product, written $\mathbf{a} \cdot \mathbf{b}$, is defined as the product of the magnitudes of the two vectors and the cosine of the angle between them (Figure 6.26):

$$\mathbf{a} \cdot \mathbf{b} = ab \cos \theta. \tag{6.53}$$

A scalar quantity

Since none of the three factors on the right depends on the choice of coordinates, this product is a scalar quantity; its magnitude is unchanged by a change of coordinates. The definition makes clear that this product is commutative:

$$\mathbf{a} \cdot \mathbf{b} = \mathbf{b} \cdot \mathbf{a}. \tag{6.54}$$

Commutative property

The scalar product of a vector with itself is the square of the magnitude of the vector:

$$\mathbf{a} \cdot \mathbf{a} = a^2. \tag{6.55}$$

The scalar product of perpendicular vectors is zero. The unit vectors, for example, have the following simple scalar products:

$$\mathbf{i} \cdot \mathbf{i} = \mathbf{j} \cdot \mathbf{j} = \mathbf{k} \cdot \mathbf{k} = 1; \tag{6.56}$$

$$\mathbf{i} \cdot \mathbf{j} = \mathbf{j} \cdot \mathbf{k} = \mathbf{k} \cdot \mathbf{i} = 0. \tag{6.57}$$

Note in Equation 6.53 and Figure 6.26 that the combination $a \cos \theta$ is the component of the vector $\mathbf{a}$ along the direction of $\mathbf{b}$. If we call this component $a_{\parallel}$ (meaning the component of $\mathbf{a}$ parallel to $\mathbf{b}$), we may write

$$\mathbf{a} \cdot \mathbf{b} = a_{\parallel} b. \tag{6.58}$$

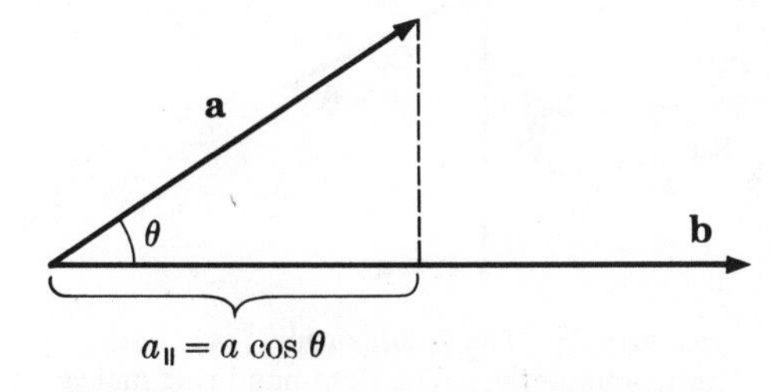

FIGURE 6.26 The scalar product is defined by $\mathbf{a} \cdot \mathbf{b} = ab \cos \theta$. It is also equal to the component of $\mathbf{a}$ parallel to $\mathbf{b}$ ($a_{\parallel} = a \cos \theta$) multiplied by b.

Similarly, by defining $b_{\|} = b \cos\theta$, the same product can be written

$$\mathbf{a}\cdot\mathbf{b} = ab_{\|}. \tag{6.59}$$

The scalar product possesses the linearity property expressed by

The linearity property

$$\mathbf{c}\cdot(\alpha\mathbf{a} + \beta\mathbf{b}) = \alpha\mathbf{c}\cdot\mathbf{a} + \beta\mathbf{c}\cdot\mathbf{b}. \tag{6.60}$$

We can take advantage of this property to express the scalar product in terms of components. First we write

$$\mathbf{a}\cdot\mathbf{b} = (a_x\mathbf{i} + a_y\mathbf{j} + a_z\mathbf{k})\cdot(b_x\mathbf{i} + b_y\mathbf{j} + b_z\mathbf{k}).$$

Because of the linearity property, this product can be expanded according to the usual rules of algebra. Terms containing $\mathbf{i}\cdot\mathbf{j}$, $\mathbf{i}\cdot\mathbf{k}$, and $\mathbf{j}\cdot\mathbf{k}$ vanish (Equation 6.57). The remaining terms contain $\mathbf{i}\cdot\mathbf{i}$, $\mathbf{j}\cdot\mathbf{j}$, and $\mathbf{k}\cdot\mathbf{k}$, all of which are equal to 1 (Equation 6.56). As a result, the product simplifies to

Expression in terms of components

$$\mathbf{a}\cdot\mathbf{b} = a_xb_x + a_yb_y + a_zb_z. \tag{6.61}$$

This is often a useful form of the scalar product.

6.8 The vector product

From two vectors another vector may be formed that is perpendicular to the plane containing the first two. This vector product, or cross product, is written

$$\mathbf{c} = \mathbf{a}\times\mathbf{b}. \tag{6.62}$$

The direction of **c**, perpendicular to both **a** and **b**, is defined by a right-hand rule (Figure 6.27). If the fingers of the right hand point first along **a** and are then curved to point along **b** (via the lesser angle between **a** and **b**), the right thumb indicates the direction of **c**. The magnitude of **c** is given by

$$c = ab\sin\theta, \tag{6.63}$$

where θ is the lesser angle between **a** and **b**. Note that the vector product of

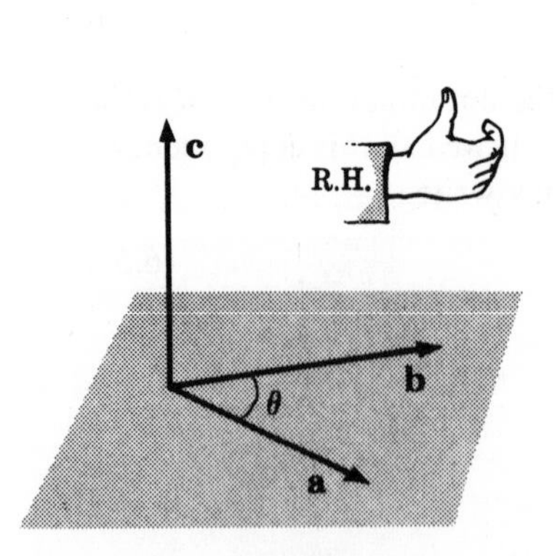

FIGURE 6.27 The vector product, $\mathbf{c} = \mathbf{a}\times\mathbf{b}$. The vector **c** is perpendicular to the plane containing **a** and **b**; its direction is defined by a right-hand rule. The magnitude of **c** is $ab\sin\theta$.

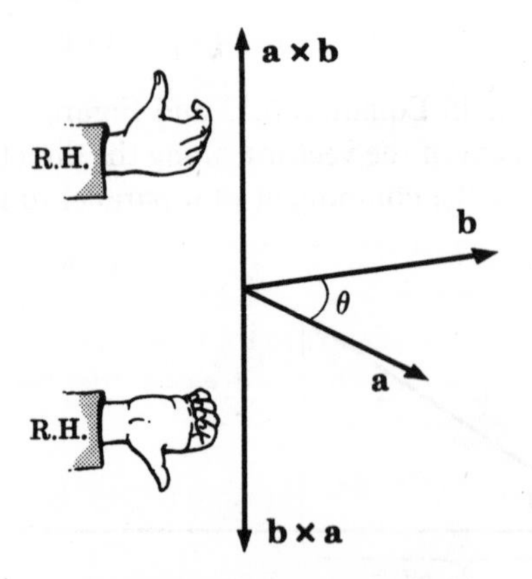

FIGURE 6.28 The vector product is anticommutative. The right-hand rule makes clear that $\mathbf{a}\times\mathbf{b}$ and $\mathbf{b}\times\mathbf{a}$ are oppositely directed.

parallel vectors is zero. Since θ lies between 0 and π, $\sin\theta \geq 0$, and Formula 6.63 correctly gives a positive (or zero) magnitude for the vector **c**. That the quantity **c** so defined is indeed a vector we state without proof. Actually, it differs from **a** and **b** in one subtle but important way; the difference will be explained at the end of this section.

The algebraic definition of the vector product, in terms of components, is given by

Definition in terms of components

$$c_x = a_y b_z - a_z b_y, \tag{6.64}$$

$$c_y = a_z b_x - a_x b_z, \tag{6.65}$$

$$c_z = a_x b_y - a_y b_x. \tag{6.66}$$

The equivalence of Equations 6.63 and 6.66 for vectors **a** and **b** lying in the *xy* plane is left as a problem. A useful aid to memory is the component definition expressed as a symbolic determinant:

Vector product can be expressed as a determinant

$$\mathbf{c} = \mathbf{a} \times \mathbf{b} = \begin{vmatrix} \mathbf{i} & \mathbf{j} & \mathbf{k} \\ a_x & a_y & a_z \\ b_x & b_y & b_z \end{vmatrix}. \tag{6.67}$$

Evaluation of this determinant, using the unit vectors as co-factors, shows it to be equivalent to Equations 6.64–6.66.

The vector product possesses a linearity property:

A linearity property

$$\mathbf{c} \times (\alpha\mathbf{a} + \beta\mathbf{b}) = \alpha\mathbf{c} \times \mathbf{a} + \beta\mathbf{c} \times \mathbf{b}. \tag{6.68}$$

However, it is not commutative. Rather

Anticommutative property

$$\mathbf{a} \times \mathbf{b} = -\mathbf{b} \times \mathbf{a}; \tag{6.69}$$

the vector product is said to be anticommutative. This property follows from the algebraic definitions, Equations 6.64–6.66 or Equation 6.67. It also follows readily from the geometrical definition, as indicated in Figure 6.28. It is therefore important that the order of factors be preserved in vector products. Since other noncommutative operations occur in advanced physics, it is wise to form the general habit of preserving the order of factors in vector expressions even when it is not necessary (as in Equation 6.60).

The two kinds of products just defined may be pyramided to give more complicated products, such as $\mathbf{a} \cdot (\mathbf{b} \times \mathbf{c})$ and $\mathbf{a} \times (\mathbf{b} \times \mathbf{c})$. We mention the latter to display an example of the failure of the associative property of multiplication. The expressions $\mathbf{a} \times (\mathbf{b} \times \mathbf{c})$ and $(\mathbf{a} \times \mathbf{b}) \times \mathbf{c}$ are not equal. For example, $\mathbf{i} \times (\mathbf{i} \times \mathbf{j}) = -\mathbf{j}$, whereas $(\mathbf{i} \times \mathbf{i}) \times \mathbf{j} = 0$. The so-called mixed product, $\mathbf{a} \cdot (\mathbf{b} \times \mathbf{c})$, has a simple meaning if **a**, **b**, and **c** are all displacement vectors. The scalar quantity $\mathbf{s}_1 \cdot (\mathbf{s}_2 \times \mathbf{s}_3)$ is equal to the volume of a parallelepiped whose three edges are defined by $\mathbf{s}_1$, $\mathbf{s}_2$, and $\mathbf{s}_3$ emanating from a point (Figure 6.29).

Triple product $\mathbf{a} \times (\mathbf{b} \times \mathbf{c})$

Mixed product $\mathbf{a} \cdot (\mathbf{b} \times \mathbf{c})$

■ EXAMPLE 1: The position vector of a particle is

$$\mathbf{r} = 6\mathbf{i}\ \text{m}, \tag{6.70}$$

and its velocity vector (Figure 6.30) is

$$\mathbf{v} = (3\mathbf{i} + 5\mathbf{j})\ \text{m/sec}. \tag{6.71}$$

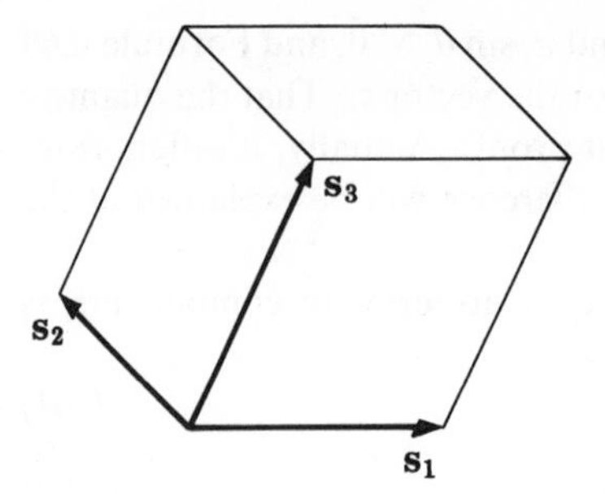

FIGURE 6.29 The "mixed product" of three displacement vectors, $\mathbf{s}_1 \cdot (\mathbf{s}_2 \times \mathbf{s}_3)$, is equal in magnitude to the volume of a parallelpiped whose edges are defined by $\mathbf{s}_1$, $\mathbf{s}_2$, and $\mathbf{s}_3$. (If $\mathbf{s}_1$, $\mathbf{s}_2$, and $\mathbf{s}_3$ are in "right-handed" order, as in this diagram, the mixed product is equal to the volume. If they are in left-handed order, it is equal to the negative of the volume.)

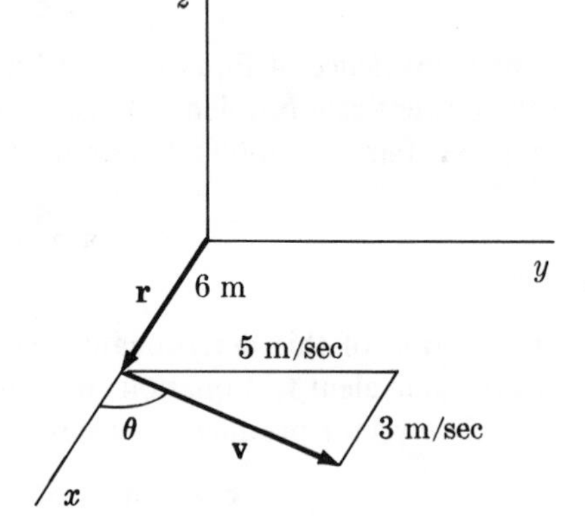

FIGURE 6.30 Position vector and velocity vector of a particle.

What are all of the following products?

(a) $\mathbf{r} \cdot \mathbf{r}$ (e) $\mathbf{r} \cdot \mathbf{v}$
(b) $\mathbf{v} \cdot \mathbf{v}$ (f) $\mathbf{r} \times \mathbf{v}$
(c) $\mathbf{r} \times \mathbf{r}$ (g) $\mathbf{r} \cdot (\mathbf{r} \times \mathbf{v})$
(d) $\mathbf{v} \times \mathbf{v}$ (h) $\mathbf{r} \times (\mathbf{r} \times \mathbf{v})$

(a) and (b). The scalar product of a vector with itself is the square of the vector's magnitude. Therefore $\mathbf{r} \cdot \mathbf{r} = r^2 = 36\ \text{m}^2$, and $\mathbf{v} \cdot \mathbf{v} = v^2 = v_x^2 + v_y^2 = (9 + 25)\ \text{m}^2/\text{sec}^2 = 34\ \text{m}^2/\text{sec}^2$.

(c) and (d). Since the angle between a vector and itself is zero, $\sin\theta = 0$, and $\mathbf{r} \times \mathbf{r} = \mathbf{v} \times \mathbf{v} = 0$.

(e) and (f). The definitions in terms of components are easiest to apply. (Verification that the geometrical definitions yield the same results is left as an exercise.) The scalar product is $\mathbf{r} \cdot \mathbf{v} = xv_x + yv_y = 18\ \text{m}^2/\text{sec}$. Since $z = v_z = 0$, Equations 6.64 and 6.65 show that the x and y components of $\mathbf{r} \times \mathbf{v}$ vanish. The z component is $xv_y - yv_x = 30\ \text{m}^2/\text{sec}$. Therefore we can write

$$\mathbf{r} \times \mathbf{v} = 30\mathbf{k}\ \text{m}^2/\text{sec}.$$

The direction, parallel to the z axis, confirms that $\mathbf{r} \times \mathbf{v}$ is perpendicular to the plane containing $\mathbf{r}$ and $\mathbf{v}$, and the direction along the positive z axis confirms the requirement of the right-hand rule.

(g). Since $\mathbf{r}$ and $\mathbf{r} \times \mathbf{v}$ are perpendicular, $\mathbf{r} \cdot (\mathbf{r} \times \mathbf{v}) = 0$.

(h). Application of the right-hand rule shows that $\mathbf{r} \times (\mathbf{r} \times \mathbf{v})$ must be directed along the negative y axis. Since $\mathbf{r}$ and $\mathbf{r} \times \mathbf{v}$ are perpendicular, $\sin\theta = 1$, and the magnitude of this triple product is $|\mathbf{r}||\mathbf{r} \times \mathbf{v}| = 180$ m^3/sec. Written as a vector equation, the answer is

$$\mathbf{r} \times (\mathbf{r} \times \mathbf{v}) = -180\mathbf{j}\ \text{m}^3/\text{sec}. \qquad ■$$

★POLAR VECTORS AND AXIAL VECTORS

The position vector and all other vectors with identical transformation properties are called polar vectors. Velocity, acceleration, force, and momentum are examples of polar vector quantities. The vector product of two polar vectors is called an axial vector. Angular momentum is an example of an axial vector quantity. An axial vector differs from a polar vector in just one respect: If the axes of a Cartesian coordinate system are reversed, x to $-x$, y to $-y$, and z to $-z$, all components of a polar vector change sign, whereas all components of an axial vector remain unchanged. (If $\mathbf{c} = \mathbf{a} \times \mathbf{b}$, a reversal of both $\mathbf{a}$ and $\mathbf{b}$ leaves $\mathbf{c}$ unchanged.) Such a reversal of coordinates is called an inversion of the coordinate system. It differs from any rotation or combination of rotations because it changes a right-handed system into a left-handed system. Polar vectors and axial vectors have identical properties with respect to rotation and opposite properties with respect to inversion.*

Polar vectors and axial vectors are distinguished by their inversion properties

Inversion changes the "handedness" of a coordinate system

An axial vector may itself enter into a vector product. The vector product of an axial vector and a polar vector is a polar vector. What is the vector product of two axial vectors?

Like the scalar product of two polar vectors, the scalar product of two axial vectors is a true scalar, a numeric that is unchanged by any transformation of the coordinate system, including an inversion. However, the scalar product of a polar vector and an axial vector is something slightly different. It is called a pseudoscalar. It is a numeric that is unchanged by rotations of the coordinates but whose sign is changed by an inversion of the coordinates.

A pseudoscalar defined

One important consequence of the distinction between polar vectors and axial vectors, and between scalars and pseudoscalars, is an extension of the idea of vectorial consistency. Throughout classical physics, if one side of an equation is a polar vector, the other side is also a polar vector. Similarly, axial vectors are equated only to axial vectors, scalars only to scalars, and pseudoscalars only to pseudoscalars.

Vectorial consistency is generalized

■ EXAMPLE 2: Characterize the vector or scalar nature of each of the eight quantities listed in the example on page 182. We may begin from the fact that both $\mathbf{r}$ and $\mathbf{v}$ are polar vectors.

(a) and (b). The products $\mathbf{r} \cdot \mathbf{r}$ and $\mathbf{v} \cdot \mathbf{v}$ are true scalars. Under an inversion both factors change sign, so the product does not change.

(c) and (d). The products $\mathbf{r} \times \mathbf{r}$ and $\mathbf{v} \times \mathbf{v}$ would be axial vectors if they were not zero.

* A note on nomenclature: Polar vectors are sometimes called vectors, and axial vectors are sometimes called pseudovectors. In this text, the word "vector" alone means either a polar or an axial vector. For many purposes it is not important to distinguish between them.

(e) and (f). For reasons already cited, $\mathbf{r} \cdot \mathbf{v}$ is a scalar, and $\mathbf{r} \times \mathbf{v}$ is an axial vector.

(g). Since $\mathbf{r} \cdot (\mathbf{r} \times \mathbf{v})$ is the scalar product of a polar vector and an axial vector, it must be a pseudoscalar.

(h). The triple product $\mathbf{r} \times (\mathbf{r} \times \mathbf{v})$ is the cross product of a polar vector and an axial vector; it is therefore a polar vector. ■

6.9 Vectors changing in time; the derivative of a vector

Much of physics is concerned with motion and change. Therefore, rate of change is a crucial idea for both scalars and vectors. First, recall how we deal with scalars that change. Let $m(t)$ be the mass of an accelerating rocket, a scalar function. The change of mass from time t_1 to time t_2 is written

$$\Delta m = m(t_2) - m(t_1). \tag{6.72}$$

The average rate of change of m during the time interval $t_2 - t_1$ is

$$\frac{\Delta m}{\Delta t} = \frac{m(t_2) - m(t_1)}{t_2 - t_1}. \tag{6.73}$$

The rate of change at time t_1 is the derivative

Rate of change of a scalar quantity

$$\frac{dm}{dt} = \lim_{\Delta t \to 0} \frac{m(t_1 + \Delta t) - m(t_1)}{\Delta t}. \tag{6.74}$$

Exactly analogous formulas apply to vector quantities. For example, if $\mathbf{r}(t)$ designates the position of the rocket, a function of time, the change of position from time t_1 to time t_2 is given by Equation 6.17, a definition that involves only the subtraction of one vector from another (Figure 6.31). The average rate of change of position during the time interval $t_2 - t_1$ is

$$\bar{\mathbf{v}} = \frac{\Delta \mathbf{r}}{\Delta t} = \frac{\mathbf{r}(t_2) - \mathbf{r}(t_1)}{\Delta t}. \tag{6.75}$$

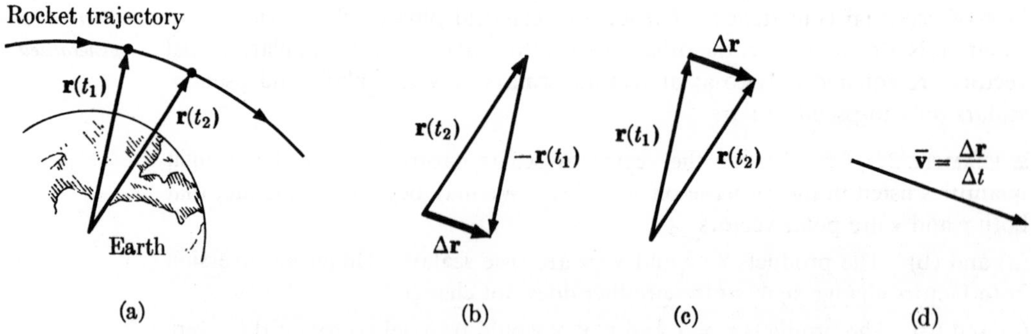

FIGURE 6.31 (a) A rocket has position vectors $\mathbf{r}(t_1)$ and $\mathbf{r}(t_2)$ at times t_1 and t_2. (b) A geometrical construction defines its displacement: $\Delta\mathbf{r} = \mathbf{r}(t_2) - \mathbf{r}(t_1)$. (c) A similar arrow diagram shows the later position as the sum of the earlier position and the displacement: $\mathbf{r}(t_2) = \mathbf{r}(t_1) + \Delta\mathbf{r}$. (d) Multiplication of $\Delta\mathbf{r}$ by $(1/\Delta t)$ gives the average velocity in the interval, $\bar{\mathbf{v}}$. In the limit that Δt and $\Delta\mathbf{r}$ approach zero, this becomes the instantaneous velocity, $d\mathbf{r}/dt$.

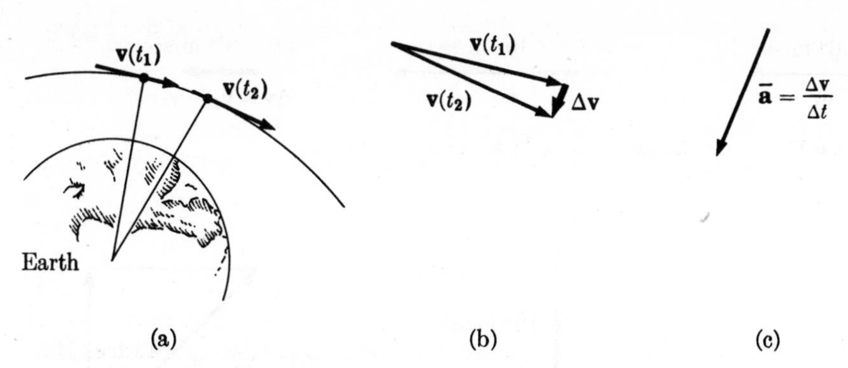

FIGURE 6.32 (a) A rocket has velocities $\mathbf{v}(t_1)$ and $\mathbf{v}(t_2)$ at times t_1 and t_2. (b) Its change of velocity is defined by $\Delta\mathbf{v} = \mathbf{v}(t_2) - \mathbf{v}(t_1)$. The arrow diagram represents the vector addition, $\mathbf{v}(t_2) = \mathbf{v}(t_1) + \Delta\mathbf{v}$. (c) Multiplication of $\Delta\mathbf{v}$ by $(1/\Delta t)$ gives the average acceleration in the interval, $\bar{\mathbf{a}}$. In the limit that Δt and $\Delta\mathbf{v}$ approach zero, this becomes the instantaneous acceleration, $\mathbf{a} = d\mathbf{v}/dt = d^2\mathbf{r}/dt^2$.

The derivative of the position vector is defined as the limit of this quotient as Δt approaches zero:

Rate of change of a vector quantity

$$\frac{d\mathbf{r}}{dt} = \lim_{\Delta t \to 0} \frac{\mathbf{r}(t_1 + \Delta t) - \mathbf{r}(t_1)}{\Delta t}. \tag{6.76}$$

The derivatives of other vectors are defined in a similar way. Note in Equations 6.75 and 6.76 that division by Δt is the same as multiplication by the numeric $(1/\Delta t)$, an operation that has been defined.

The derivative of the position vector defines velocity:

Definition of velocity

$$\mathbf{v} = \frac{d\mathbf{r}}{dt}. \tag{6.77}$$

The derivative of the velocity vector in turn defines acceleration (Figure 6.32):

Definition of acceleration

$$\mathbf{a} = \frac{d\mathbf{v}}{dt} = \frac{d^2\mathbf{r}}{dt^2}. \tag{6.78}$$

These are fundamental definitions of vector kinematics. Note their close similarity in appearance to the definitions for one-dimensional motion, which appeared in Section 5.7.

Since a vector is characterized by magnitude and direction, a change in either one of these properties (or in both) constitutes a change of the vector. The simplest change, analogous to the change of a numeric, is a change in magnitude only. An electron in a linear accelerator might increase its speed from 1.0×10^8 m/sec to 1.5×10^8 m/sec without change of direction. The change in its velocity is a vector in the same direction as the velocity, with magnitude 0.5×10^8 m/sec (Figure 6.33). On the other hand, a proton in a circular accelerator might change its velocity from 1.0×10^8 m/sec eastbound to 1.0×10^8 m/sec northbound (Figure 6.34)—a change of direction without change of magnitude. In that case the change of velocity, $\Delta\mathbf{v}$, has magnitude 1.41×10^8 m/sec and a direction northwestbound. This is deduced graphically

Example: change of magnitude only

Example: change of direction only

1×10^8 m/sec $\quad$ 1.5×10^8 m/sec $\quad$ 0.5×10^8 m/sec

$\mathbf{v}_1$ $\quad$ $\mathbf{v}_2$ $\quad$ $\Delta\mathbf{v} = \mathbf{v}_2 - \mathbf{v}_1$

FIGURE 6.33 Vector changing in magnitude but not in direction.

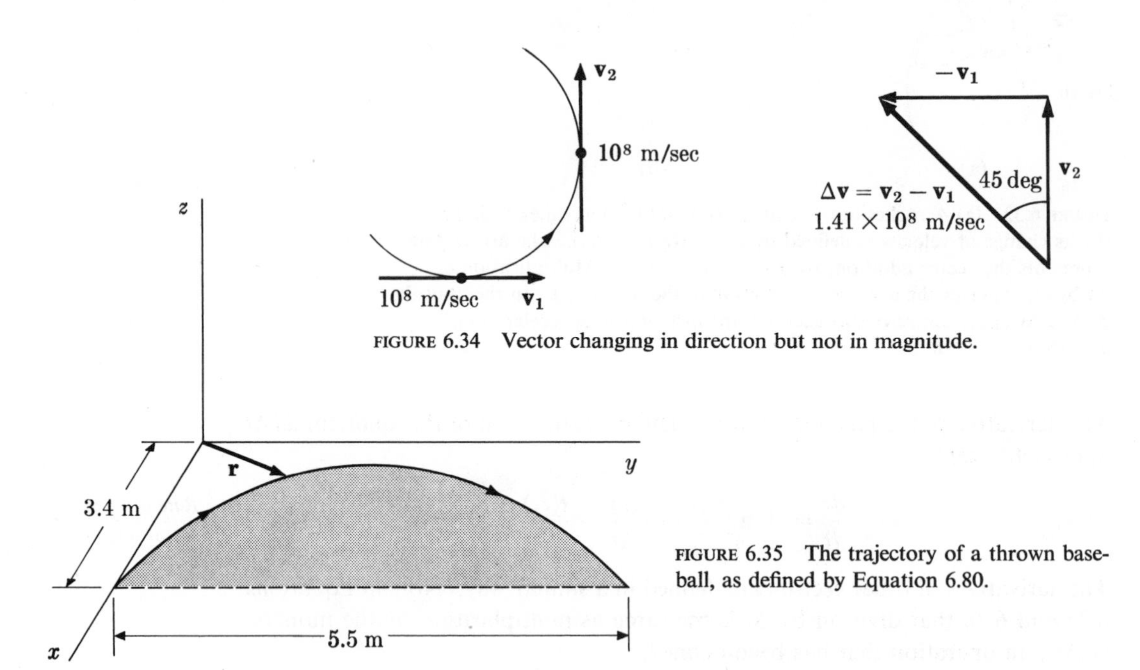

FIGURE 6.34 Vector changing in direction but not in magnitude.

FIGURE 6.35 The trajectory of a thrown baseball, as defined by Equation 6.80.

in Figure 6.34. To obtain the result algebraically, let the x axis point eastward and the y axis point northward. Then $\mathbf{v}_1 = 1.0 \times 10^8\mathbf{i}$ m/sec, $\mathbf{v}_2 = 1.0 \times 10^8\mathbf{j}$ m/sec, and

$$\Delta\mathbf{v} = \mathbf{v}_2 - \mathbf{v}_1 = (1.0 \times 10^8\mathbf{j} - 1.0 \times 10^8\mathbf{i}) \text{ m/sec}.$$

In the laws of nature, changes of direction are as important as changes of magnitude. Note in the example shown in Figure 6.34 that even when the magnitude of a vector undergoes no change, the vector itself, through change of direction, may experience a change that has a large magnitude.

The derivative of a numerical quantity with respect to time has a simple graphical meaning. It is the slope of the curve produced by graphing the quantity as a function of time. The derivative of a vector quantity has no such simple graphical interpretation. However, the components of a vector may be separately differentiated, and each behaves as a numerical quantity. If a vector $\mathbf{V}$ is expressed in terms of its components,

$$\mathbf{V} = V_x\mathbf{i} + V_y\mathbf{j} + V_z\mathbf{k},$$

the rate of change of V may be written

Derivative of a vector in terms of components

$$\frac{d\mathbf{V}}{dt} = \frac{dV_x}{dt}\mathbf{i} + \frac{dV_y}{dt}\mathbf{j} + \frac{dV_z}{dt}\mathbf{k}. \tag{6.79}$$

■ EXAMPLE: In a coordinate system with x and y axes horizontal and z axis directed vertically upward, the position vector of a thrown baseball (Figure 6.35) is given by

$$\mathbf{r} = 3.4\mathbf{i} + 5.5t\mathbf{j} + (4.9t - 4.9t^2)\mathbf{k}, \tag{6.80}$$

where $\mathbf{r}$ is expressed in meters and t in seconds. What are the velocity and the acceleration of the ball? Straightforward differentiation of the components provides the answers:

$$\mathbf{v} = \frac{d\mathbf{r}}{dt} = 5.5\mathbf{j} + (4.9 - 9.8t)\mathbf{k}, \tag{6.81}$$

$$\mathbf{a} = \frac{d\mathbf{v}}{dt} = -9.8\mathbf{k}, \tag{6.82}$$

where the unit of $\mathbf{v}$ is m/sec and the unit of $\mathbf{a}$ is m/sec^2. The acceleration is constant and directed vertically downward. ■

6.10 Uniform motion in a circle

A physical example of considerable practical importance that nicely demonstrates the power of vector kinematics is the uniform motion of a particle in a circle. Circular, or nearly circular, motion is encountered at many places in nature—a stone on the end of a rope, a proton in an accelerator, the earth in its orbit around the sun. Here we specialize to uniform circular motion and examine the three vector quantities: position, velocity, and acceleration.

Consider a particle moving with constant speed v counterclockwise around a circle of radius r (Figure 6.36). For convenience, we choose the origin to lie at the center of the circle and let the arrow representing $\mathbf{r}$ lie along a radial line from the origin to a point on the circle. Clearly $\mathbf{r}$ has constant magnitude. It is rotating at a constant rate with angular speed ω given by

$$\omega = \frac{v}{r} \tag{6.83}$$

(see Equation 5.22). For the position vector we write

$$\begin{aligned}\mathbf{r} &= x\mathbf{i} + y\mathbf{j}\\ &= r\cos\theta\,\mathbf{i} + r\sin\theta\,\mathbf{j},\end{aligned} \tag{6.84}$$

where θ is the polar angle shown in Figure 6.36 (we have used Equations 5.1 and 5.2). If we select the zero of time such that $\theta = 0$ when $t = 0$, we have, from Equation 5.25,

$$\theta = \omega t. \tag{6.85}$$

Then the position vector, as a function of time, can be written

$$\mathbf{r} = r\cos\omega t\,\mathbf{i} + r\sin\omega t\,\mathbf{j}. \tag{6.86}$$

The position vector

This important formula describes a vector of fixed magnitude rotating at a uniform rate.

Velocity is the time derivative of position. Differentiation of Equation 6.86

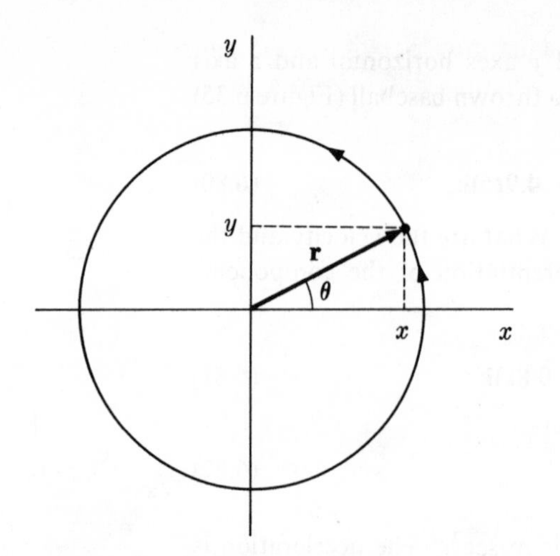

FIGURE 6.36 Representation of the position vector for a particle moving uniformly in a circle. With a suitable choice of the zero of time, the angle θ between **r** and the x axis can be written $\theta = \omega t$. The speed is given by $v = r\omega$.

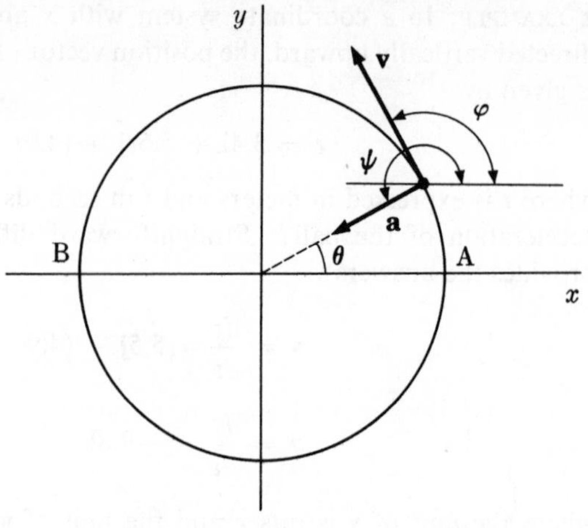

FIGURE 6.37 Velocity and acceleration in uniform circular motion. The angle φ between **v** and the x axis is given by $\varphi = \frac{1}{2}\pi + \theta = \frac{1}{2}\pi + \omega t$. The angle ψ between **a** and the x axis is given by $\psi = \pi + \theta = \pi + \omega t$.

(recalling that r and ω are constants) yields

Differentiate to get the velocity vector

$$\mathbf{v} = \frac{d\mathbf{r}}{dt} = -r\omega \sin \omega t\, \mathbf{i} + r\omega \cos \omega t\, \mathbf{j}. \tag{6.87}$$

Since we are dealing with uniform circular motion, we know that **v** must be a vector of constant magnitude, tangent to the circle and turning at a constant rate. Let us verify these properties explicitly. To check the magnitude of **v**, we can form the scalar product $\mathbf{v} \cdot \mathbf{v}$:

$$\begin{aligned} \mathbf{v} \cdot \mathbf{v} &= r^2\omega^2 \sin^2 \omega t + r^2\omega^2 \cos^2 \omega t \\ &= r^2\omega^2 (\sin^2 \omega t + \cos^2 \omega t) = r^2\omega^2. \end{aligned} \tag{6.88}$$

Properties of the velocity vector

Thus the magnitude of **v** is the constant $r\omega$, which is in agreement with Equation 6.83. Next, form the scalar product $\mathbf{v} \cdot \mathbf{r}$:

$$\mathbf{v} \cdot \mathbf{r} = -r^2\omega \sin \omega t \cos \omega t + r^2\omega \sin \omega t \cos \omega t = 0. \tag{6.89}$$

Since this product vanishes, **v** must be perpendicular to **r** and therefore tangent to the circle. The angle φ that **v** makes with the x axis can be determined from the projection formulas

$$v_x = v \cos \varphi, \tag{6.90}$$

$$v_y = v \sin \varphi. \tag{6.91}$$

Consider the first of these two. It can be rewritten

$$-r\omega \sin \omega t = v \cos \varphi.$$

Since $v = r\omega$, this is equivalent to

$$\cos \varphi = -\sin \omega t. \tag{6.92}$$

Two physically distinct angles φ satisfy this equation: $\varphi = \frac{1}{2}\pi + \omega t$ and $\varphi = -\frac{1}{2}\pi - \omega t$ (other angles that satisfy the equation differ from one of these by 2π and are not physically distinct). It is left as a problem to prove, using both Equations 6.90 and 6.91, that the correct solution is

$$\varphi = \tfrac{1}{2}\pi + \omega t. \tag{6.93}$$

This means that **v** is directed counterclockwise around the circle (Figure 6.37) and that it turns at the constant rate ω since

$$\frac{d\varphi}{dt} = \omega. \tag{6.94}$$

v rotates at ω radian/sec

To obtain the acceleration, we differentiate Equation 6.87 with respect to time:

$$\mathbf{a} = \frac{d\mathbf{v}}{dt} = -r\omega^2 \cos \omega t\, \mathbf{i} - r\omega^2 \sin \omega t\, \mathbf{j}. \tag{6.95}$$

Differentiate again to get the acceleration vector

Comparison of this result with Equation 6.86 for **r** leads at once to a very simple and interesting expression for **a**:

$$\mathbf{a} = -\omega^2\mathbf{r}. \tag{6.96}$$

The acceleration has constant magnitude

$$a = \omega^2 r \tag{6.97}$$

and is directed toward the center of the circle, antiparallel to **r**. Since $\omega = v/r$, we may also write

Two expressions for centripetal acceleration

$$a = \frac{v^2}{r}. \tag{6.98}$$

An acceleration directed toward a fixed center, as in this example, is called a centripetal acceleration.

■ EXAMPLE 1: What is the centripetal acceleration of the earth (approximating its motion as circular)? Substituting data from Appendix 3 in Equation 6.98, we find

$$a = \frac{v^2}{r} = \frac{(2.98 \times 10^4 \text{ m/sec})^2}{1.496 \times 10^{11} \text{ m}}$$

$$= 5.94 \times 10^{-3} \text{ m/sec}^2, \tag{6.99}$$

a quantity that is small compared with the acceleration of free fall at the earth's surface. The ratio is

$$\frac{a}{g} = \frac{5.94 \times 10^{-3} \text{ m/sec}^2}{9.80 \text{ m/sec}^2} = 6.06 \times 10^{-4}.$$

■

■ EXAMPLE 2: In uniform circular motion, what are the average speed and the average velocity during half a revolution? Consider in particular the interval $\theta = 0$ to π, or $t = 0$ to π/ω (recall that the full period is $T = 2\pi/\omega$). Since the speed is constant, the average speed is the same as the instantaneous speed, v.

To get the average velocity, we make use of Equation 6.75. The change of position is

$$\Delta\mathbf{r} = \mathbf{r}(\pi/\omega) - \mathbf{r}(0)$$
$$= -r\mathbf{i} - r\mathbf{i} = -2r\mathbf{i}.$$

The change of time is $\Delta t = \pi/\omega$. The quotient, which is the average velocity, is

$$\bar{\mathbf{v}} = \frac{\Delta\mathbf{r}}{\Delta t} = -\frac{2}{\pi}\,\omega r\mathbf{i}.$$

Since $v = \omega r$, this can be written more simply as

$$\bar{\mathbf{v}} = -\frac{2}{\pi}\,v\mathbf{i}.$$

This average velocity is directed across the circle (from A to B in Figure 6.37) and has magnitude $(2/\pi)v$, or about $0.637v$. Note that it would be quite incorrect to write $\bar{\mathbf{v}} = \frac{1}{2}[\mathbf{v}(\pi/\omega) + \mathbf{v}(0)]$. Why? ■

6.11 Vector calculus

In this section we summarize, without proof, several important formulas of vector calculus. The proofs, which follow from the basic definition of a derivative, are in fact not difficult. Two of them are among the problems. For illustrative purposes, we write derivatives with respect to time t, but the formulas are valid for differentiation with respect to any other numerical variables. Differentiation is a linear operation, as expressed by

$$\frac{d}{dt}(\alpha\mathbf{a} + \beta\mathbf{b}) = \alpha\frac{d\mathbf{a}}{dt} + \beta\frac{d\mathbf{b}}{dt}, \tag{6.100}$$

where α and β are constants. The derivative of the product of a numerical quantity and a vector quantity is given by

$$\frac{d}{dt}(f\mathbf{V}) = \frac{df}{dt}\mathbf{V} + f\frac{d\mathbf{V}}{dt}, \tag{6.101}$$

which is exactly analogous to a corresponding formula for the derivative of the product of two numerics. From Equations 6.100 and 6.101 follows Equation 6.79, which expresses the derivative of a vector in terms of components. Both scalar and vector products have simple rules of differentiation. They are

$$\frac{d}{dt}(\mathbf{a}\cdot\mathbf{b}) = \frac{d\mathbf{a}}{dt}\cdot\mathbf{b} + \mathbf{a}\cdot\frac{d\mathbf{b}}{dt}, \tag{6.102}$$

and

$$\frac{d}{dt}(\mathbf{a}\times\mathbf{b}) = \frac{d\mathbf{a}}{dt}\times\mathbf{b} + \mathbf{a}\times\frac{d\mathbf{b}}{dt}. \tag{6.103}$$

In the latter formula the order of factors is important. These are but a sampling of many equations of vector calculus. A few others of importance in this book are introduced at later points when physical applications require them. Appendix 6 includes a summary of such formulas.

Summary of ideas and definitions

A numeric is a quantity with magnitude only, represented by a number (for example, the component of a vector).

A scalar is a numeric that is independent of the coordinate system (for example, mass).

A vector is a quantity with magnitude and direction, whose components transform from one coordinate system to another in the same manner as the components of the position vector.

The position vector is the prototype vector. Its properties define the nature of a vector.

Numbers and vectors may be conceived as (1) abstract concepts, (2) geometrical representations, and (3) physical quantities.

Vectors may be represented by arrows, and vector arithmetic by arrow diagrams.

A vector is written **V**. Its magnitude is written $|\mathbf{V}|$, or V. Its components are written V_x, V_y, V_z.

A component of a vector is given by $V_x = V \cos\theta$, where θ is the angle between **V** and the x axis.

Scalars can be equated only to scalars and added only to scalars. Vectors can be equated only to vectors and added only to vectors.

The scalar product is defined geometrically by

$$\mathbf{a}\cdot\mathbf{b} = ab\cos\theta$$

and algebraically, in terms of components, by

$$\mathbf{a}\cdot\mathbf{b} = a_x b_x + a_y b_y + a_z b_z .$$

The vector product is defined geometrically by the right-hand rule and the equation

$$|\mathbf{a}\times\mathbf{b}| = ab\sin\theta$$

and algebraically, in terms of components, by

$$c_x = a_y b_z - a_z b_y ,$$
$$c_y = a_z b_x - a_x b_z ,$$
$$c_z = a_x b_y - a_y b_x .$$

Vectors addition satisfies
a commutative law, $\mathbf{a} + \mathbf{b} = \mathbf{b} + \mathbf{a}$, and
an associative law, $(\mathbf{a} + \mathbf{b}) + \mathbf{c} = \mathbf{a} + (\mathbf{b} + \mathbf{c})$.

The multiplication of a vector by a numeric satisfies
a distributive law, $n(\mathbf{a} + \mathbf{b}) = n\mathbf{a} + n\mathbf{b}$,
an associative law, $\alpha(\beta\mathbf{a}) = (\alpha\beta)\mathbf{a}$, and
a linearity property, $\alpha(\beta\mathbf{a} + \gamma\mathbf{b}) = (\alpha\beta)\mathbf{a} + (\alpha\gamma)\mathbf{b}$.

The scalar product satisfies
a commutative law, $\mathbf{a}\cdot\mathbf{b} = \mathbf{b}\cdot\mathbf{a}$, and
a linearity property, $\mathbf{c}\cdot(\alpha\mathbf{a} + \beta\mathbf{b}) = \alpha\mathbf{c}\cdot\mathbf{a} + \beta\mathbf{c}\cdot\mathbf{b}$.

The vector product satisfies
an anticommutative property, $\mathbf{a}\times\mathbf{b} = -\mathbf{b}\times\mathbf{a}$, and
a linearity property,

$$\mathbf{c}\times(\alpha\mathbf{a} + \beta\mathbf{b}) = \alpha\mathbf{c}\times\mathbf{a} + \beta\mathbf{c}\times\mathbf{b}.$$

A polar vector is a vector that behaves exactly like the position vector.

An axial vector is a vector that behaves like the position vector for rotation of the coordinate system but behaves oppositely for an inversion of the coordinate system.

A pseudoscalar is a numeric that is unchanged by rotations of a coordinate system but whose sign is changed by an inversion of the coordinate system.

The derivative of a vector is defined as the limit of a quotient, in exact analogy to the definition of the derivative of a numeric.

Velocity is defined by $\mathbf{v} = d\mathbf{r}/dt$.

Acceleration is defined by $\mathbf{a} = d\mathbf{v}/dt = d^2\mathbf{r}/dt^2$.

The derivative is a linear operator, with the property

$$\frac{d}{dt}\mathbf{V} = \frac{dV_x}{dt}\mathbf{i} + \frac{dV_y}{dt}\mathbf{j} + \frac{dV_z}{dt}\mathbf{k}.$$

The position vector,

$$\mathbf{r} = r\cos\omega t\,\mathbf{i} + r\sin\omega t\,\mathbf{j},$$

describes uniform motion in a circle.

For uniform motion in a circle, the centripetal acceleration has magnitude given by

$$a = \omega^2 r = \frac{v^2}{r}.$$

QUESTIONS

Section 6.1

Q6.1 Which of the following are numerical quantities, which are vector quantities, and which are neither: (a) velocity, (b) money, (c) momentum, (d) mass, (e) shape, (f) color, (g) force, and (h) age?

Q6.2 Which of the following quantities are scalar quantities and which are numerical quantities but not scalar: (a) the x component of a position vector, (b) mass, (c) speed, (d) temperature, and (e) latitude?

Q6.3 Is there any reason why the line representing real numbers (Figure 6.1) must be a straight line? Explain.

Q6.4 Name a set of physical *operations* (or experimental procedures) that obey an associative property analogous to the associative property of numbers (Equation 6.3). Name a set of operations that do not obey an associative property.

Section 6.2

Q6.5 Two vectors sum to zero. How must they be related?

Q6.6 Three vectors sum to zero: $\mathbf{A} + \mathbf{B} + \mathbf{C} = 0$. If $\mathbf{A}$ and $\mathbf{B}$ are of equal magnitude ($A = B$), what is the maximum value of C? What is the minimum value of C?

Q6.7 Define a zero vector (sometimes called a null vector). Give an argument to show that the zero vector is distinct from the number zero.

Q6.8 Can the sum of two vectors be a scalar? Can it be a numeric?

Q6.9 If $\mathbf{c} = \mathbf{a} - \mathbf{b}$, which of the following relations among magnitudes can be achieved by suitable choices of $\mathbf{a}$ and $\mathbf{b}$: (a) $c = a - b$, (b) $c = a + b$, (c) $c > a$ and $c > b$, (d) $c < a$ and $c < b$, and (e) $c < a - b$?

Q6.10 A photographic emulsion shows the decay of an unstable particle at rest. Two charged-particle tracks emerge from the decay point, the angle between them being 130 deg. Was a neutral particle also produced in the decay? Why? Is it possible that *two* neutral particles were produced?

Q6.11 Check a dictionary for meanings of the word "vector" other than the meaning used in this chapter. Is there any connection between a biological vector and a physical vector?

Section 6.3

Q6.12 Is the result of multiplying a zero vector by a finite number different from the result of multiplying a finite vector by zero? Why or why not?

Q6.13 If a particle had negative mass, how would its velocity and momentum be related?

Q6.14 (1) Can a vector be divided by a numeric? (2) Can a vector be added to a numeric?

Section 6.4

Q6.15 A ball thrown from point O (chosen as an origin) follows a parabolic arc. (1) Does its velocity vector change in magnitude only, in direction only, or in both magnitude and direction? (2) Answer the same question for its position vector.

Q6.16 An object moves with constant speed v. Explain why the magnitude of its displacement in a certain time divided by the time might have any value from 0 to v but cannot be less than 0 or greater than v.

Q6.17 As a kind of shorthand, we often say that "displacement is a vector," or we refer to "the displacement vector." Technically, it is more accurate to say that "displacement is a quantity that behaves like a vector." Explain the difference in these modes of description. (NOTE: The former phraseology is permissible provided we understand that it is shorthand for the latter phraseology.)

Section 6.5

Q6.18 For what vector is a *component* the same as a *coordinate*?

Q6.19 (1) When a coordinate system is rotated, what aspects of a vector change? What aspects do not change? (2) What aspects of a scalar are changed by a rotation of coordinates? What aspects are not changed?

Q6.20 Give an example of a numeric whose value depends on the choice of coordinates.

Section 6.6

Q6.21 Explain in what sense a vector equation contains more information than a scalar equation.

Q6.22 Kinetic energy is a scalar quantity. Is it possible that some other form of energy might be a vector quantity? Why or why not?

Q6.23 Why is vectorial consistency a requirement for the equations of physics? Give an example of an equation that lacks consistency, and explain why it cannot be correct.

Section 6.8

Q6.24 Torque is a quantity defined by a vector product: $\mathbf{T} = \mathbf{r} \times \mathbf{F}$, where $\mathbf{r}$ is a position vector and $\mathbf{F}$ is a force. Is torque a polar vector quantity or an axial vector quantity?

Q6.25 Is the vector product of two axial vectors a polar vector, an axial vector, or neither? Give a reason for your answer.

Q6.26 Is the scalar product of force and displacement a true scalar or a pseudoscalar?

Section 6.9

Q6.27 In an automobile that is initially at rest, you shift into reverse, back up a short distance, then start forward, and round a turn to the left. Give the direction of the acceleration during each part of this motion.

Section 6.10

Q6.28 List half a dozen examples of motion that are well approximated as uniform motion in a circle (or the arc of a circle).

Q6.29 For the uniform motion of a particle around a circle, what is the direction of the rate of change of the acceleration vector?

Q6.30 Doubling the speed of motion around a circle increases the centripetal acceleration fourfold. Explain in words why the *rate of change* of velocity increases by a factor of 4 when the *magnitude* of velocity increases by a factor of 2.

Q6.31 Doubling the radius of circular motion at fixed speed halves the centripetal acceleration. Explain in words why the rate of change of velocity *decreases* by a factor of 2 when the radius *increases* by a factor of 2.

EXERCISES

Section 6.1

E6.1 Sketch diagrams of "geometrical arithmetic" for these operations: (a) $3 + 7$, (b) $3 - 7$, (c) $3 + 0.5 - 5$, and (d) $-4 - 3 - 2 - 1$.

Section 6.2

E6.2 Three forces whose sum is zero act on a particle. One is directed eastward and has magnitude 3 N. One is directed northward and has magnitude 5.2 N. What are the direction and magnitude of the third force?

E6.3 (1) Each of two vectors has unit magnitude. Find the magnitude of their sum (a) when they are directed parallel, (b) when they are directed oppositely, and (c) when one is directed at a right angle to the other. (2) What must their relative direction be in order that their sum also have unit magnitude? With each of your four answers, sketch a vector arrow diagram.

E6.4 You are driving north on a city street at a steady speed of 10 m/sec. You slow down, round a 90-deg corner to the right, and accelerate back to a steady speed of 10 m/sec. What is the net change in your velocity vector?

E6.5 A pilot whose plane flies at 100 mile/hr wants to fly north on a day when a wind of 50 mile/hr is blowing from the west. (1) In what direction must he point his plane? (2) What is his speed over the ground? Draw a vector diagram showing the velocity with respect to the ground as the sum of the velocity with respect to the air plus the velocity of the air.

E6.6 An airplane with an airspeed of 300 mile/hr is maintaining a northeast heading (that is, the nose of the plane is pointing directly northeast). The wind is from the north at 100 mile/hr. Graphically, find the vector giving the airplane's velocity over the ground. What is its approximate ground speed, as determined by measurement with a ruler?

E6.7 A railroad train is traveling east at 10 m/sec. A boy on the train throws a baseball straight out the window toward the south at a speed of 20 m/sec relative to the train. A man on the ground catches the ball. From what direction and at what speed does he see it come?

E6.8 The sum of the magnitudes of two vectors cannot be equated to the magnitude of the sum. In the circle below,

$$|\mathbf{F}_1 + \mathbf{F}_2| \quad \bigcirc \quad |\mathbf{F}_1| + |\mathbf{F}_2|,$$

insert the appropriate mathematical symbol ($>$, $\geq$, $<$, or $\leq$), and explain the reason for your choice.

Section 6.3

E6.9 Pick any three numbers, α, β, and γ, and any two vectors, **a** and **b**. With the help of appropriate arrow diagrams, show that these quantities satisfy the linearity property expressed by Equation 6.16.

E6.10 Consider two vectors, **a** and **b**. Using the laws of vector algebra, write the expression $6(\mathbf{a} + \mathbf{b}) - (10\mathbf{a} - \mathbf{b}) + 3(\mathbf{b} - 2\mathbf{a})$ in terms of the vector $\mathbf{b} - \mathbf{a}$ alone. Demonstrate graphically the correctness of your result if **a** is a vector of unit length pointing east and if **b** is a vector of unit length pointing north.

Section 6.4

E6.11 Starting from an origin, a man walks 4 m north, then 12 m east, then 9 m south. Draw a reasonably accurate diagram of his trip. (1) What is his final position vector? (2) What is his net displacement vector?

E6.12 Choose as the origin of the room containing Alice, Barbara, and Charlotte the northwest corner instead of the southwest corner, and redraw Figure 6.16 accordingly. In a careful sketch, show approximately that the displacement vector, $\mathbf{r}_B - \mathbf{r}_A$, remains the same as shown in Figure 6.17.

E6.13 In the accompanying figure, point A has position vector $\mathbf{r}_1$ in one frame of reference and position vector $\mathbf{r}_1'$ in another frame of reference. Point B has position vectors $\mathbf{r}_2$ and $\mathbf{r}_2'$ in these two frames. (1) Relate $\mathbf{r}_1$ to $\mathbf{r}_1'$ and $\mathbf{r}_2$ to $\mathbf{r}_2'$. (2) Show algebraically that $\mathbf{r}_2 - \mathbf{r}_1 = \mathbf{r}_2' - \mathbf{r}_1'$, a fact that is geometrically obvious in this figure or in Figure 6.18. This proves that a displacement vector is independent of the choice of origin.

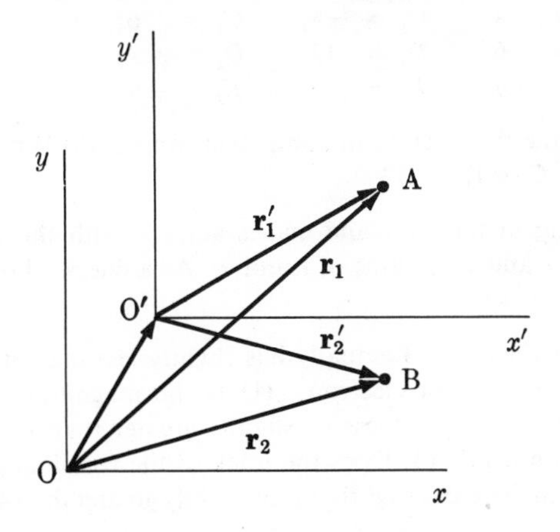

E6.14 The minute hand of a clock measures 10 cm from its axis to its tip. What is the displacement vector of its tip (a) from a quarter after the hour until half past, (b) in the next half hour, and (c) in the next hour?

E6.15 Consider the pair of vectors shown in the figure. (1) Give the components of **A** and of **B**. (2) Find the sum **A** + **B**. (3) Find the difference **A** − **B**.

Section 6.5

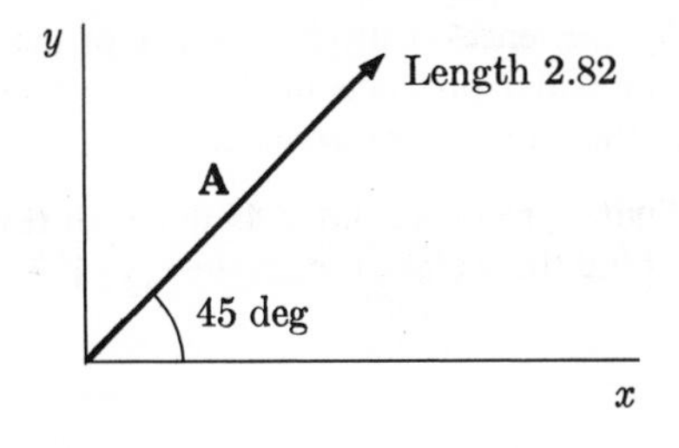

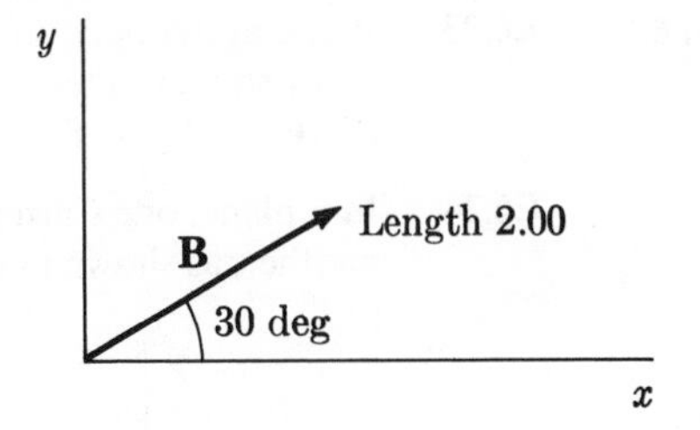

E6.16 Consider the pair of vectors **A** and **B** in the preceding exercise. (1) Sketch vector sum diagrams for $\mathbf{C} = 2\mathbf{A} + 3\mathbf{B}$ and for $\mathbf{D} = 0.5\mathbf{A} + 4\mathbf{B}$. (These need not be precise.) (2) Use components to find the vectors **C** and **D**. (3) Is it possible to multiply the vectors **A** and **B** by scalars α and β in such a way that the sum $\alpha\mathbf{A} + \beta\mathbf{B}$ is equal to zero? If so, what are the values of α and β?

E6.17 Three forces act on an object. Each force has a magnitude of 10 N. The directions in which the forces act are to the east, to the north, and to the northeast. Choose the x axis in the eastward direction and y axis in the northward direction. (1) Write the x and y components of *each* force. (2) Write the x and y components of the *sum* of the three forces. (3) What is the magnitude of the sum, and what is its direction?

E6.18 Solve Exercise 6.6 using components; specify the east (x) and north (y) components of the velocity over the ground.

E6.19 (1) Which of the following vectors are equal? Which have equal magnitude? Which have the same direction? Which are oppositely directed?

$$\begin{array}{llll} \mathbf{A}: & A_x = 2, & A_y = 4, & A_z = -3; \\ \mathbf{B}: & B_x = 4, & B_y = 3, & B_z = 2; \\ \mathbf{C}: & C_x = 4, & C_y = -8, & C_z = 6; \\ \mathbf{D}: & D_x = 6, & D_y = 12, & D_z = -9; \\ \mathbf{E}: & E_x = -8, & E_y = -6, & E_z = -4. \end{array}$$

(2) Express the sum of the five vectors in component form. (3) Which has greater magnitude, $\mathbf{E} + \mathbf{C}$ or $\mathbf{D} - \mathbf{B}$?

E6.20 A position vector $\mathbf{r}$ lying in the xy plane makes angle α with the x axis. Express its components x and y in terms of r and α. As a check, show that your formulas satisfy $x^2 + y^2 = r^2$.

E6.21 The recoiling electron in Figure 6.23 actually has slightly less magnitude of momentum than does the incident electron. (1) Is the magnitude of the proton's momentum actually slightly less or slightly greater than the value calculated below Equation 6.43? (2) Does the track of the recoiling proton actually make an angle with the x axis slightly less or slightly greater than 45 deg?

E6.22 If a plane Cartesian coordinate system is rotated through angle θ, components of a position vector transform in accordance with Equations 6.46 and 6.47. From these equations, prove that

$$x_1'^2 + y_1'^2 = x_1^2 + y_1^2.$$

This is a simple example of an invariance principle. The magnitude of the vector is invariant under the rotation.

Section 6.7

E6.23 If the automobile in Figure 6.8 experiences a displacement $\mathbf{s}$ of magnitude 15 m straight ahead, what are the scalar products (a) $\mathbf{F}_1 \cdot \mathbf{s}$, (b) $\mathbf{F}_2 \cdot \mathbf{s}$, and (c) $(\mathbf{F}_1 + \mathbf{F}_2 + \mathbf{F}_f) \cdot \mathbf{s}$? Include the unit of measurement.

E6.24 In a plane, one Cartesian coordinate system is rotated 45 deg with respect to another, as shown in the figure. Find the scalar products $\mathbf{i} \cdot \mathbf{i}'$, $\mathbf{i} \cdot \mathbf{j}'$, $\mathbf{j} \cdot \mathbf{i}'$, and $\mathbf{j} \cdot \mathbf{j}'$.

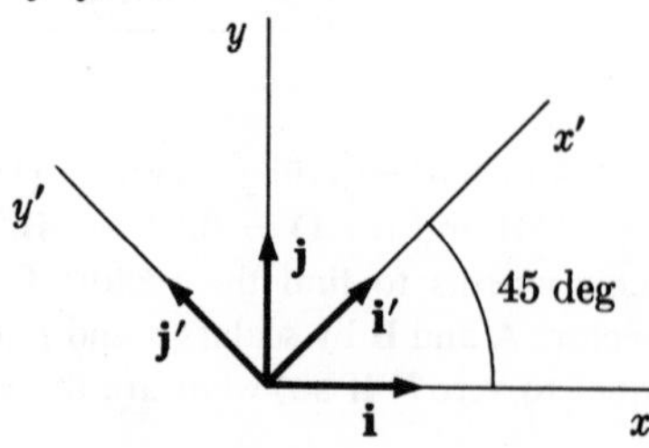

E6.25 Use both the geometrical and the algebraic definitions of the scalar product to show that the magnitude of a vector $\mathbf{a}$ can be written $|\mathbf{a}| = \sqrt{\mathbf{a} \cdot \mathbf{a}}$.

E6.26 Three vectors sum to zero, as shown in the figure. Find the scalar products (a) $\mathbf{A} \cdot \mathbf{B}$, (b) $\mathbf{B} \cdot \mathbf{C}$, and (c) $\mathbf{C} \cdot \mathbf{A}$.

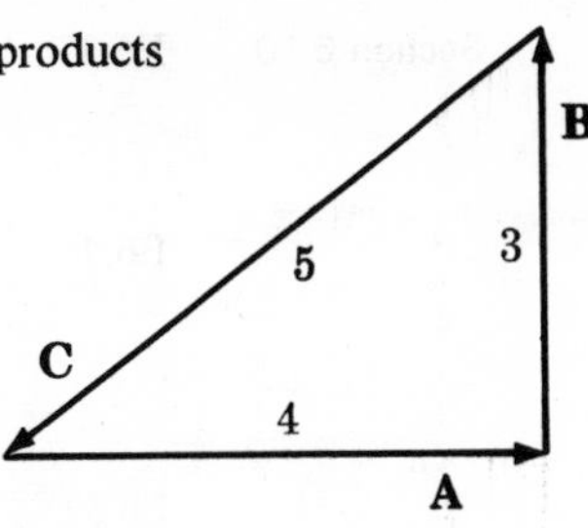

Section 6.8

E6.27 For the vectors shown in the preceding exercise, find the vector products (a) $\mathbf{A} \times \mathbf{B}$, (b) $\mathbf{B} \times \mathbf{C}$, (c) $\mathbf{C} \times \mathbf{A}$, and (d) $\mathbf{B} \times \mathbf{A}$. (Take the plane of the figure to be the xy plane.)

E6.28 For the vectors of Exercise 6.26, find (a) $\mathbf{A} \cdot (\mathbf{B} \times \mathbf{C})$, (b) $\mathbf{A} \times (\mathbf{B} \times \mathbf{C})$, and (c) $(\mathbf{A} \times \mathbf{B}) \times \mathbf{C}$.

E6.29 Find the scalar and vector products of the two vectors shown in Exercise 6.15.

E6.30 (1) Find the following vector products of unit vectors: (a) $\mathbf{i} \times \mathbf{i}$, (b) $\mathbf{j} \times \mathbf{j}$, (c) $\mathbf{i} \times \mathbf{j}$, (d) $\mathbf{i} \times \mathbf{k}$, and (e) $\mathbf{k} \times \mathbf{i}$. (2) Explain how a vector product of unit vectors could be used to define a right-handed Cartesian coordinate system.

E6.31 Find the following products of unit vectors: (a) $\mathbf{i} \times (\mathbf{j} \times \mathbf{k})$, (b) $\mathbf{k} \times (\mathbf{k} \times \mathbf{j})$, (c) $(\mathbf{k} \times \mathbf{k}) \times \mathbf{j}$, (d) $\mathbf{i} \cdot (\mathbf{j} \times \mathbf{k})$, and (e) $\mathbf{i} \cdot (\mathbf{i} \times \mathbf{j})$.

E6.32 Equations 6.70 and 6.71 give a pair of arbitrarily chosen vectors $\mathbf{r}$ and $\mathbf{v}$. (1) Find the magnitudes r and v. (2) Find the angle between $\mathbf{r}$ and $\mathbf{v}$. (3) Work from the geometrical definitions of the scalar and vector products to find $\mathbf{r} \cdot \mathbf{v}$ and $\mathbf{r} \times \mathbf{v}$ (answers, obtained by using components, are given on page 182).

Section 6.9

E6.33 The position vector of a thrown baseball is given by Equation 6.80. (1) Where is the ball at $t = 0$? (2) At what time ($t > 0$) does the ball reach the xy plane? (3) At what time does the ball reach its maximum height? Where is it at this time?

E6.34 A man standing in the xy plane at the fixed point x_1y_1 throws a ball vertically upward (see the figure). (1) Write an expression for the position vector $\mathbf{r}$ of the ball as a function of time. (2) Differentiate this expression to obtain vector formulas for the velocity $\mathbf{v}$ and acceleration $\mathbf{a}$.

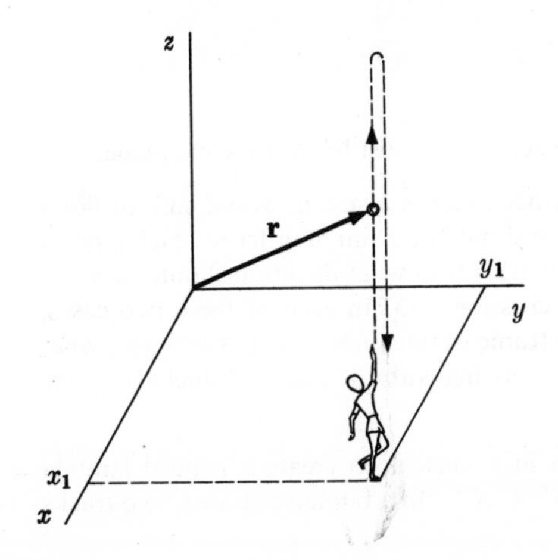

Section 6.10

E6.35 (1) From the equation $\mathbf{r} = r\cos\theta\,\mathbf{i} + r\sin\theta\,\mathbf{j}$ (Equation 6.84), verify that $\mathbf{r}\cdot\mathbf{r} = r^2$. (2) For what kind of motion other than uniform motion in a circle might this expression for the position vector $\mathbf{r}$ be useful?

E6.36 A particle moves in a circle of radius 3 m at a constant angular speed of 2 radian/sec. (1) Write an expression for the position vector of the particle, $\mathbf{r}(t)$, assuming the center of its circular track is at the origin. (2) Write an expression for its velocity vector.

E6.37 A racing car executes a turn of radius 200 m at a speed of 40 m/sec (about 90 mile/hr). What is its inward acceleration? Is this reasonable? Why?

E6.38 An artificial earth satellite moves in a circle of circumference 8×10^7 m with a period of 1.44×10^4 sec. (1) What is the magnitude of its acceleration? How does this compare with the acceleration of gravity at the earth's surface? (2) In order to help yourself visualize the characteristics of this satellite, calculate its altitude above the earth's surface in miles and its period in minutes.

E6.39 An airplane is flying in a horizontal circle. Its speed is 100 m/sec (about 220 mile/hr). If its acceleration toward the center of the circle is equal to g (9.8 m/sec^2), what is the radius of the circle? Express the answer in meters and in miles. *Optional:* What is (or should be) the angle of bank of the airplane?

E6.40 Use Equations 6.87 and 6.95 to show that for uniform circular motion, $\mathbf{a}\cdot\mathbf{v} = 0$. Does this prove that $\mathbf{a}$ is directed toward the center of the circle if $\mathbf{v}$ is known to be directed tangent to the circle?

E6.41 For uniform circular motion, show that the magnitudes of acceleration and velocity are related by $a = \omega v$.

E6.42 A particle moves in a circle of constant radius r with constant angular acceleration such that the angle θ (Figure 6.36) is given by $\theta = \frac{1}{2}\alpha t^2$. Write vector expressions for its position vector $\mathbf{r}$ and its velocity $\mathbf{v}$. From the expression for $\mathbf{v}$, verify that the speed of the particle is $v = r\alpha t$.

Section 6.11

E6.43 A rocket has both variable mass, $m(t)$, and variable velocity, $\mathbf{v}(t)$. (1) What is the time rate of change of its momentum, $\mathbf{p} = m\mathbf{v}$? (2) During a certain period of time, the momentum of the rocket is constant. What statement can you make about the magnitude and direction of the rocket's acceleration during this time?

PROBLEMS

$\mathbf{a} + \mathbf{b} + \mathbf{c} = 0$

P6.1 Prove that if three vectors sum to zero, they must lie in the same plane.

Relative motion

P6.2 Swimming at 1 mile/hr, a man wishes to cross a stream whose current flows at 0.5 mile/hr. (1) In what direction should he swim in order to reach a point directly across from his starting point? (2) In what direction should he swim in order to minimize his time of crossing? (3) In each of these two cases, what is his speed in an earth-fixed frame of reference? (4) In each case, what is his speed in a frame of reference moving with the current? Include vector diagrams with your answers.

Momentum conservation

P6.3 A positive pion strikes a neutron in a nucleus to create a neutral lambda and a positive kaon: $\pi^+ + \mathrm{n} \rightarrow \Lambda^0 + \mathrm{K}^+$. In a bubble chamber, two tracks are seen, as in the figure. The magnitude of the K^+ momentum is determined to be half the magnitude of the π^+ momentum. Along what direction must the unseen Λ^0 have flown?

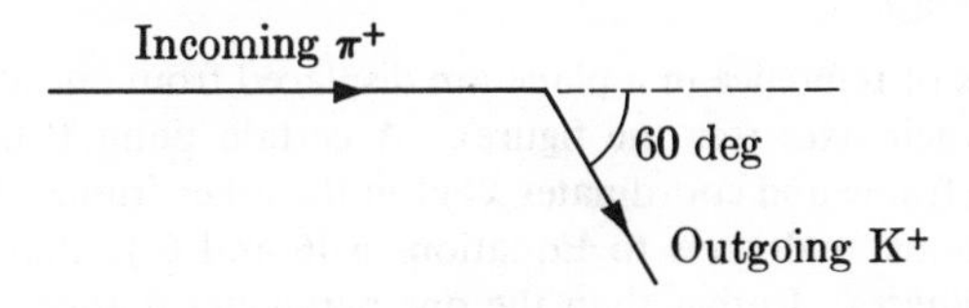

P6.4 An eastbound proton strikes a proton at rest. After the collision one proton flies off to the northeast, the other to the southeast, the path of each one at 45 deg to the direction of motion of the original projectile proton. Use the vector character of momentum and the law of momentum conservation to prove that the magnitudes of the two final momenta must be equal. (A diagram should be helpful.)

P6.5 A long-range cannon is mounted on a railway flatcar (see the figure). The shells have a mass of 300 kg and a muzzle velocity of 10^3 m/sec. The total mass of flatcar plus cannon is 4×10^4 kg. (1) If the cannon is fired horizontally along the track, what is the recoil speed of the flatcar? (2) If the barrel is elevated 30 deg above the horizontal, what is the recoil speed of the flatcar after the cannon is fired?

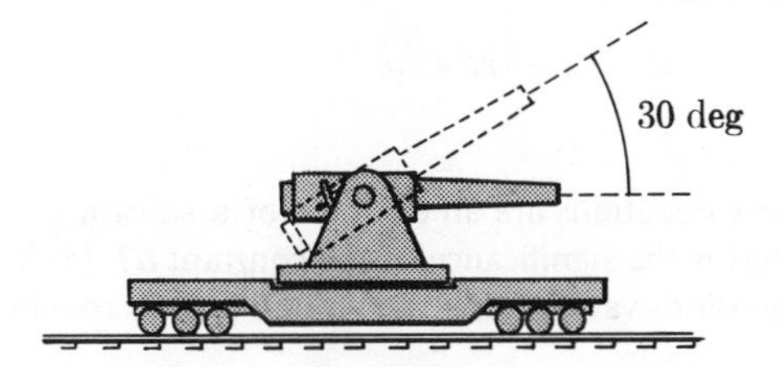

P6.6 Two balls of putty collide in midair and stick together (see the figure). At the moment of their collision, ball 1, of mass $m_1 = 1$ kg, is moving horizontally with speed $v_1 = 3$ m/sec; ball 2, of mass $m_2 = 0.5$ kg, is moving vertically with speed $v_2 = 8$ m/sec. What is the velocity (magnitude and direction) of the mass of putty just after the collision?

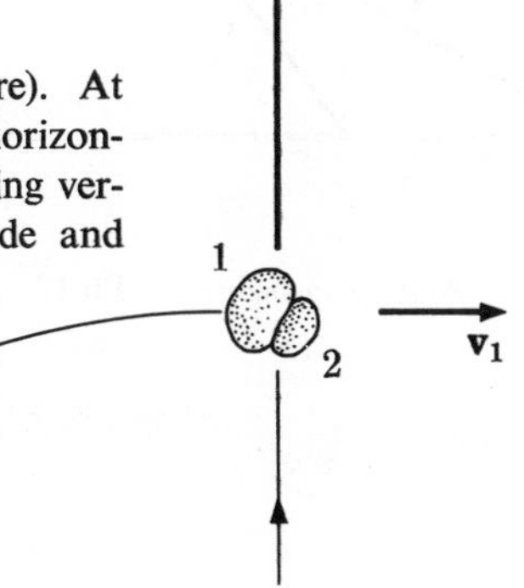

P6.7 Carefully prove Formulas 6.34–6.36 for the multiplication of components by a numeric. As axioms, use basic vector properties, such as the linearity property, and state which such properties are used in your proof.

Multiplication of a vector by a numeric

P6.8 Figure 6.24 shows a vector $\mathbf{r}_1$ referred to two sets of axes. Prove that the components are related by Equations 6.46 and 6.47. (Note that the angle θ in these equations and in Figure 6.24 is not the same as the angle θ in Equations 6.23 and 6.24.) (SUGGESTION: Give the angle between $\mathbf{r}_1$ and the x axis a name and apply trigonometric reasoning to the diagrams in Figure 6.24.)

Rotation of axes

P6.9 Working from the diagrams of Figure 6.25, prove that vector components transform according to Equations 6.48 and 6.49. What basic *physical* assumption underlies your proof. (SUGGESTION: Solve Problem 6.8 before attempting this problem.)

Translation of axes

P6.10 Two Cartesian frames of reference in a plane are displaced from one another without rotation of their axes (see the figure). A certain point P has coordinates x_1y_1 in one frame and coordinates $x_1'y_1'$ in the other frame. Derive transformation equations analogous to Equations 6.46 and 6.47 that relate the two sets of coordinates. Rather than the one parameter θ appearing in Equations 6.46 and 6.47, two parameters must appear in the new equations. What are they?

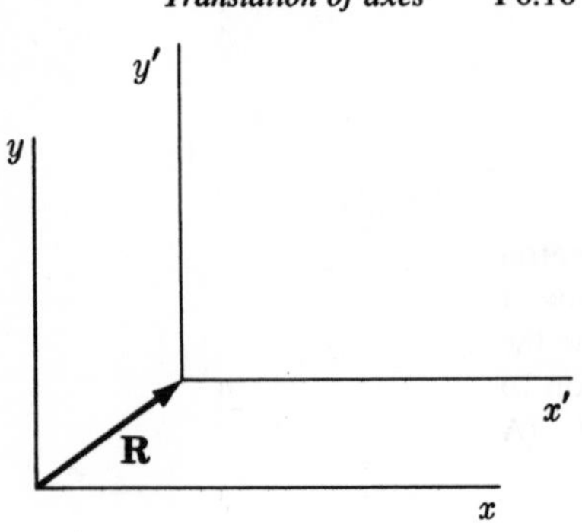

Galilean transformation

P6.11 In a frame of reference S, with axes x and y, the position vector of a point P is $\mathbf{r} = x\mathbf{i} + y\mathbf{j}$. In a frame of reference S′, with axes x' and y', the position vector of the same point is $\mathbf{r}' = x'\mathbf{i}' + y'\mathbf{j}'$. The axes are aligned as shown, and frame S′ is moving relative to frame S in the positive x direction with speed v. (1) Explain why $\mathbf{i} = \mathbf{i}'$ and $\mathbf{j} = \mathbf{j}'$. (2) Show that the coordinates of P in the two frames are related by

$$x' = x - vt + a,$$

$$y' = y,$$

where t is the time. These equations are an example of a so-called Galilean transformation. (3) What is the significance of the constant a? (4) Why are these transformation equations valid even if the point P is itself moving in an arbitrary way?

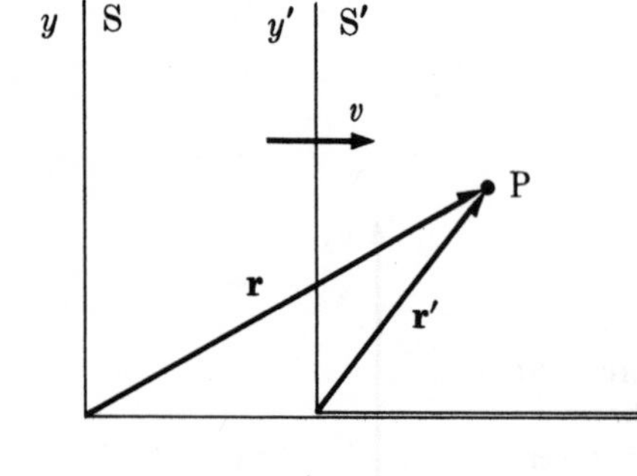

Polar components

P6.12 For use with polar coordinates, unit vectors $\mathbf{i}_r$ and $\mathbf{i}_\theta$ may be defined (see the figure). Prove that these unit vectors are related to the Cartesian unit vectors by the equations

$$\mathbf{i} = \mathbf{i}_r \cos\theta - \mathbf{i}_\theta \sin\theta,$$

$$\mathbf{j} = \mathbf{i}_r \sin\theta + \mathbf{i}_\theta \cos\theta.$$

(SUGGESTION: Consider the arrow diagram of the vector addition $\mathbf{a} + \mathbf{b} = \mathbf{i}$, where $\mathbf{a}$ is a vector parallel to $\mathbf{i}_r$ and where $\mathbf{b}$ is a vector antiparallel to $\mathbf{i}_\theta$. What must be the magnitudes of $\mathbf{a}$ and $\mathbf{b}$?)

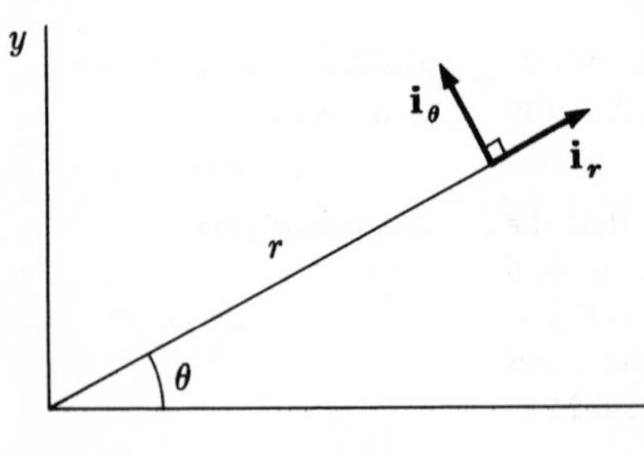

P6.13 In Cartesian and in polar coordinates, the same vector is expressed by the two equations

$$\mathbf{V} = V_x\mathbf{i} + V_y\mathbf{j},$$

$$\mathbf{V} = V_r\mathbf{i}_r + V_\theta\mathbf{i}_\theta.$$

Use the results of the preceding problem in order to show that

$$V_r = V_x \cos\theta + V_y \sin\theta,$$

$$V_\theta = -V_x \sin\theta + V_y \cos\theta.$$

Scalar product

P6.14 Working from the definition of the scalar product, Equation 6.53, prove that $\mathbf{c}\cdot(\alpha\mathbf{a}) = \alpha(\mathbf{c}\cdot\mathbf{a})$. What other facts about vectors are needed for the proof?

P6.15 With the help of Equations 6.48 and 6.49, demonstrate that $a_xb_x + a_yb_y$, the scalar product of vectors **a** and **b** in a plane, is equal to $a_x'b_x' + a_y'b_y'$, where the primed components are referred to a coordinate system rotated through angle θ with respect to an unprimed coordinate system. What is the significance of this equality?

Vector product

P6.16 Using the linearity property of the vector product (Equation 6.68) and the properties of the unit vectors, derive the components of $\mathbf{a} \times \mathbf{b}$ (Equations 6.64–6.66) from the expression

$$(a_x\mathbf{i} + a_y\mathbf{j} + a_z\mathbf{k}) \times (b_x\mathbf{i} + b_y\mathbf{j} + b_z\mathbf{k}).$$

P6.17 Consider two vectors **a** and **b** in the xy plane, as shown, with components

$$a_x = a\cos\theta, \qquad b_x = b\cos\varphi,$$

$$a_y = a\sin\theta, \qquad b_y = b\sin\varphi.$$

Using trigonometry, prove that

$$a_xb_y - a_yb_x = -ab\sin(\theta - \varphi),$$

an equality implied by Equations 6.63 and 6.66 and the right-hand rule.

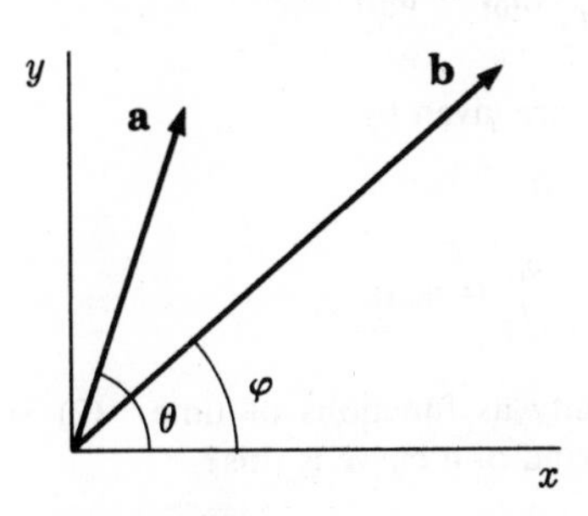

P6.18 Prove that $|\mathbf{s}_1 \times \mathbf{s}_2|$ is the area of a parallelogram with sides defined by the displacement vectors $\mathbf{s}_1$ and $\mathbf{s}_2$.

P6.19 Two vectors **a** and **b** lie in a plane. In a coordinate system defined by axes xy, they have components a_xa_y and b_xb_y. In a rotated coordinate system defined by axes $x'y'$ (Figure 6.24), they have components $a_x'a_y'$ and $b_x'b_y'$. (1) With the help of Equations 6.48 and 6.49, show that

$$a_xb_y - a_yb_x = a_x'b_y' - a_y'b_x'.$$

(2) What is the significance of this equality?

Vector kinematics

P6.20 A ball moving with constant velocity **v** crosses the x axis at $x = x_1$ at time $t = 0$ (see the figure). Later it crosses the y axis at $y = y_1$. (1) Write an expression for its position vector **r** as a function of time. (2) Differentiate this expression to verify that its velocity is constant. (3) Suggest a coordinate system $x'y'$ in which the description of this motion would be simpler, and write formulas for **r** and **v** in this system.

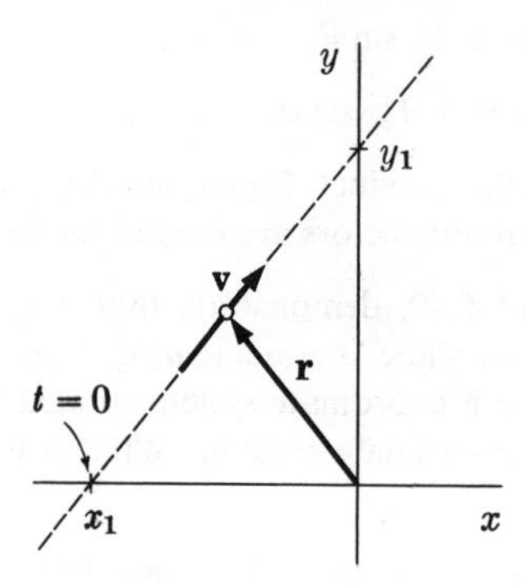

P6.21 The position vector of an object moving in the xy plane is

$$\mathbf{r} = v_0 t\mathbf{i} + a_0 e^{bt}\mathbf{j}.$$

(1) Find its velocity vector and its speed as a function of time. (2) Find its acceleration as a function of time and as a function of its y coordinate. (3) Sketch its trajectory in the xy plane.

P6.22 The acceleration of a particle moving in the xy plane is

$$\mathbf{a} = \frac{1}{\sqrt{2}}(a_0\mathbf{i} - a_0\mathbf{j}).$$

Its position and velocity, at $t = 0$, are given by

$$\mathbf{r}(0) = 0,$$

$$\mathbf{v}(0) = \frac{v_0}{\sqrt{2}}(\mathbf{i} + \mathbf{j}).$$

(1) Express its position and velocity as functions of time. (2) Sketch its trajectory in the xy plane. What kind of a curve is this?

Uniform motion in a circle

P6.23 For uniform circular motion, $\mathbf{v} \cdot \mathbf{r} = 0$ (Equation 6.89), which proves that the velocity **v** is tangent to the circle. Use equations from Section 6.10, especially Equations 6.90 and 6.91, to prove that **v** is directed counterclockwise around the circle.

P6.24 Consider a geometrical approach to the analysis of uniform circular motion. (1) For the interval shown in the diagram, with $\Delta\theta = \omega\,\Delta t = v\,\Delta t/r$, show that the magnitude of the average velocity is given by

$$|\bar{\mathbf{v}}| = \left|\frac{\Delta\mathbf{r}}{\Delta t}\right| = \frac{2v\sin(\Delta\theta/2)}{\Delta\theta}$$

and that $|\bar{\mathbf{v}}| = v$ in the limit of small $\Delta\theta$. (2) Make a similar geometrical construction to calculate $|\bar{\mathbf{a}}| = |\Delta\mathbf{v}/\Delta t|$ for the same interval, and show that $|\mathbf{a}| = v^2/r$ in the limit of small $\Delta\theta$. Comment on any advantages or disadvantages you see in this method of analysis compared with the method of Section 6.10, which uses components.

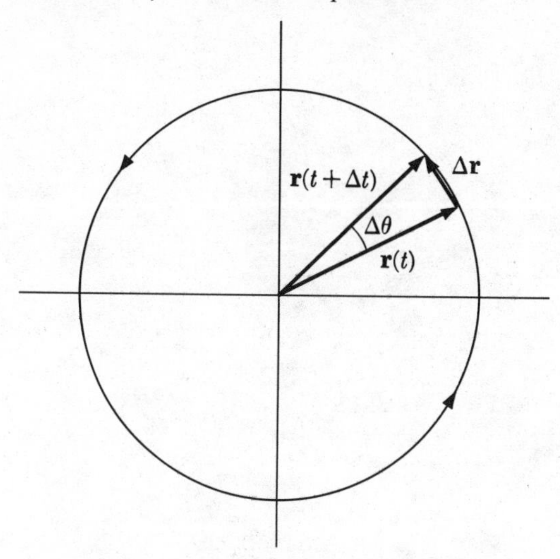

P6.25 The position vector of a moving particle is given by $\mathbf{r} = a \cos \omega t\, \mathbf{i} + b \sin \omega t\, \mathbf{j}$, where a and b are unequal constants. (1) Find expressions for the velocity and acceleration of the particle. (2) What is the direction of the acceleration? Show that its magnitude can be written as a simple function of r. (3) Find an orbit equation that contains x and y but not t.

Noncircular orbit

P6.26 Prove the rule of vector calculus

Vector calculus

$$\frac{d}{dt}(f\mathbf{V}) = \frac{df}{dt}\mathbf{V} + f\frac{d\mathbf{V}}{dt}.$$

Work from the basic definition of a vector derivative, Equation 6.76. (SUGGESTION: Refer to an analogous derivation given on on page 104–105.)

P6.27 Prove the rule of vector calculus

$$\frac{d}{dt}(\mathbf{a} \cdot \mathbf{b}) = \frac{d\mathbf{a}}{dt} \cdot \mathbf{b} + \mathbf{a} \cdot \frac{d\mathbf{b}}{dt}.$$

(SUGGESTION: Complete the proof of Problem 6.26 before attempting this.)

P6.28 (1) Working from the definition of an integral as the limit of a sum, show that

$$\int_a^b [f(t)\mathbf{i} + g(t)\mathbf{j}]\, dt = \left[\int_a^b f(t)\, dt\right] \mathbf{i} + \left[\int_a^b g(t)\, dt\right] \mathbf{j}.$$

(2) What are the components of the vector $\mathbf{D}$, defined by

$$\mathbf{D} = \int_a^b \mathbf{F}(x)\, dx?$$

P6.29 Discuss the connection between complex numbers and vectors in a plane. What complex number operations, if any, correspond to the vector operations of (a) addition? (b) subtraction? (c) the scalar product? (d) the vector product?

Complex numbers and vectors

PART THREE

Mechanics

7 Force and Motion: Newton's First and Second Laws

In the foregoing chapters, many of the concepts and ideas of mechanics and some of its laws—the conservation laws—have been presented. It is the task of this part of the book to sharpen the definition of some mechanical concepts, to assemble the concepts and laws into the coherent structure called the theory of mechanics, and to apply this theory and its laws to the description and explanation of various kinds of motion.

7.1 Motion without force; Newton's first law

It is a common observation that to cause a body to move horizontally on earth requires that the body be acted upon in some way—pushed, pulled, or mechanically propelled. When whatever is causing the motion is removed or turned off, the body comes to rest. This fact led Aristotle and scholars after him to suppose that horizontal motion requires continual propulsion. Aristotle believed, for example, that air acts on a flying arrow to propel it forward. Galileo, early in the seventeenth century, was the first to see the error in this thinking. He realized that if frictional forces could be removed, an object could slide indefinitely over a smooth surface without slowing down. Newton generalized this idea to the universe at large, and wrote as his first law: "*Every body continues in its state of rest, or of uniform motion in a straight line, unless it is compelled to change that state by forces impressed upon it.*" Expressed mathematically, Newton's first law is

Newton's first law

$$\text{if} \quad \mathbf{F} = 0 \quad \text{then} \quad \mathbf{a} = 0. \tag{7.1}$$

Zero force implies zero acceleration (or, equivalently, constant velocity). This law was of great historical importance, for it showed the necessity of finding a mechanical explanation for the motion of the moon and planets. Since these bodies experience acceleration, forces must act upon them.

Newton's first law can be applied to objects on earth. If two or more forces acting on a body sum to zero, the body moves with constant velocity. Conversely, if a body is observed to move without acceleration, the net force acting on it must be zero. However, the law is of special significance when applied to an *isolated* body. Imagine a meteorite in empty space far removed from all outside influences [Figure 7.1(a)]. It moves with constant velocity. In effect Newton's first law *defines* what kind of motion is "natural" or undisturbed motion by specifying how a body moves if it is free of all outside influences. In this way it delimits what kind of motion requires further explanation. The constant horizontal component of velocity of an arrow (when friction is negligible) corresponds to free motion and requires no explanation in terms of force. The acceleration of the moon toward the earth shows that the moon is not free. A force is needed to explain its motion.

It defines "natural" motion for isolated bodies

Another reason for applying Newton's first law to an isolated body is that the law then takes on significance without the necessity of defining force. We can define the *absence* of force by the fact of isolation. A modern statement of Newton's first law is: *An isolated body free of external influences moves with constant velocity.* In this form, the law makes no reference to force, and in fact it is valid even in domains where Newton's second and third laws (concerned with force) do not apply. An elementary particle, for example, does not follow the laws of classical mechanics when it interacts. Yet when it is free of interaction, it moves with constant velocity, in accordance with Newton's first law.* As far as we know, Newton's first law for isolated bodies is universally valid.†

Newton's first law without reference to force

It is valid even outside the domain of classical mechanics

Since common experience with motion conditions us to think more in

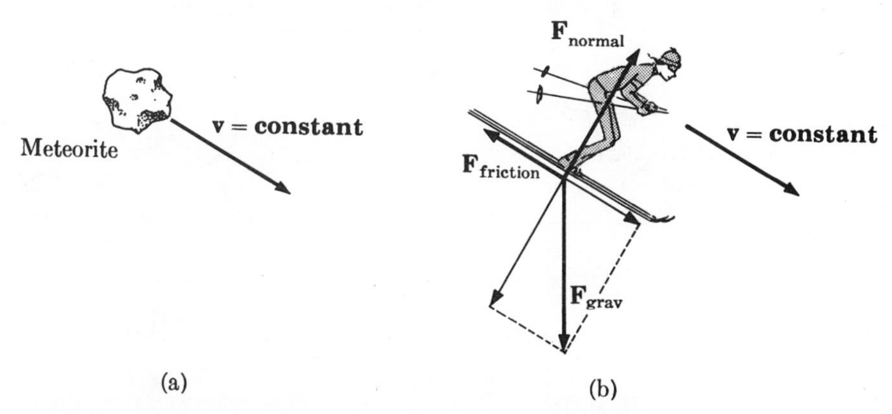

FIGURE 7.1 Motion without force. (a) An isolated body, free of all outside influences, moves with constant velocity. (b) A system that is not isolated also moves with constant velocity if the sum of the forces acting on it is zero.

---

* Near the earth an elementary particle cannot be truly "free" because it is accelerated by gravity. However, its speed is usually so enormous that the change in its velocity produced by gravity is negligible (see Exercise 7.45).

† This is related to the homogeneity of space. See Section 4.8.

Aristotelian than in Newtonian terms, application of Newton's first law to everyday situations can sometimes be confusing, despite the great simplicity of the law. We can ask, for example, whether any force acts on a skier going downhill at constant velocity. The natural tendency is to answer yes. He is pulled down by gravity. In fact, no *net* force acts on the skier. The uphill frictional force must precisely cancel the downhill component of gravitational force, and forces normal to the slope must also cancel [Figure 7.1(b)].

INERTIAL FRAMES OF REFERENCE

Two important points about Newton's first law remain to be discussed. One is its generalization to isolated *systems* whose internal parts may execute complicated motions. This will be dealt with in Section 8.5. The other is its

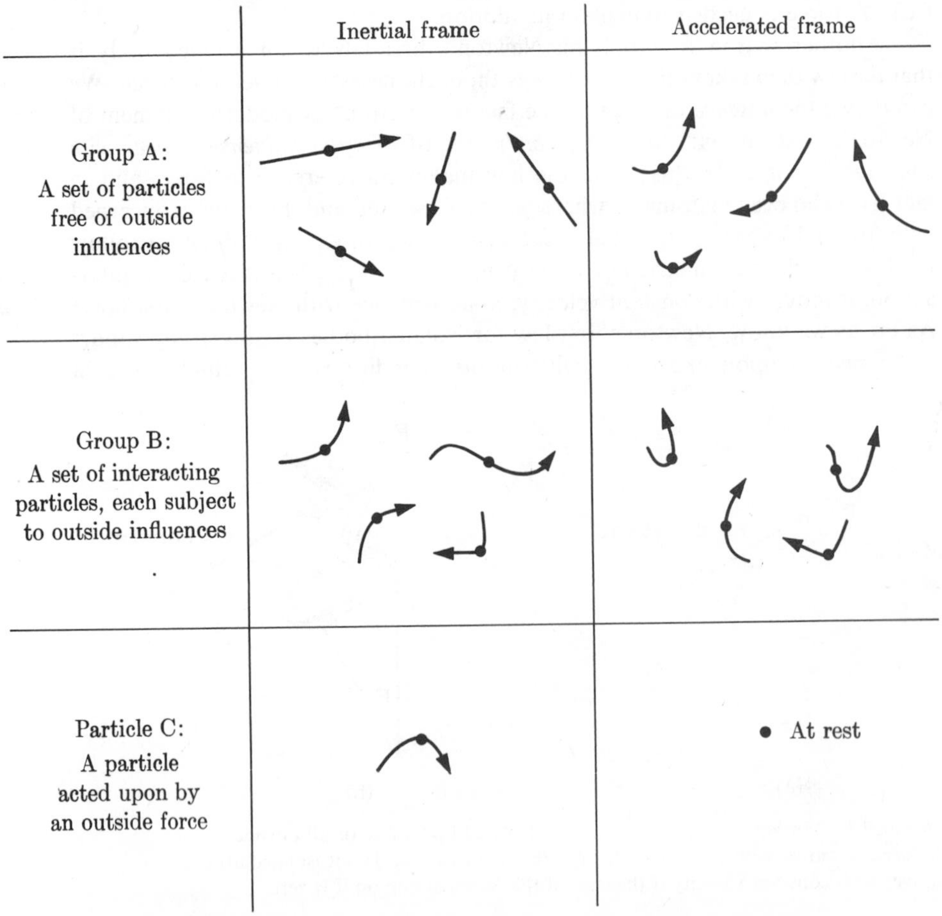

	Inertial frame	Accelerated frame
Group A: A set of particles free of outside influences		
Group B: A set of interacting particles, each subject to outside influences		
Particle C: A particle acted upon by an outside force		• At rest

FIGURE 7.2 The motion of particles in two frames of reference. In an inertial frame (left), all objects free of outside influences move with constant velocity. In an accelerated frame (right), the motion of a particular object, such as particle C, might be greatly simplified, but nearly all other motions appear accelerated, including the motion of free particles (Group A).

role in defining inertial frames of reference. The acceleration of an object depends on the frame of reference of the observer. A motionless object seen by an accelerated observer seems to be accelerated. An accelerated object is, relative to an observer accompanying it, at rest. These facts mean that isolated bodies are unaccelerated only for certain observers. The frames of reference of these observers are called inertial frames of reference. Only in inertial frames are Newton's first and second laws valid. The "universal validity" of Newton's first law referred to earlier must therefore be understood to mean validity for all bodies, all scales of size, and all velocities, but restricted to certain frames of reference.

Inertial frame defined

Newton's first law in fact plays a dual role in physics. It provides at one and the same time a definition of an inertial frame of reference and a significant statement about nature. This dual role can be appreciated by stating the law in this way: In a frame of reference in which one free object is unaccelerated, all other free objects are also unaccelerated. The first half of the statement defines an inertial frame; the second half expresses the law of inertial (undisturbed) motion. Inertial frames of reference are those in which isolated systems move in the simplest possible way (Figure 7.2). Further discussion of frames of reference is found in Section 7.10.

Newton's first law defines inertial frames

It is also a meaningful law in inertial frames

7.2 The concept of force

In nontechnical usage, it is permissible to speak of the force of an author's words, the force of a man's personality, or the force of troops stationed abroad. We may read that an atomic explosion occurs with enormous force (energy is meant) or that a flywheel spins with great force (angular momentum is meant), two examples of incorrect technical usage. Force has a single, specific—and rather simple—meaning in physics, just as it must if it is to be a useful quantitative concept. The physicist takes one of the most common everyday meanings of force, a push or a pull, and sharpens it into a definition precise enough to form a basis of measurement.

The idea of force: a push or pull

Since, according to Newton's first law, zero acceleration implies zero force, it follows that a nonzero acceleration must demonstrate the existence of a nonzero force. Acceleration can therefore be used to define force. The SI unit of force, the newton (N), is defined in this way. It is the force required to impart an acceleration of 1 m/sec^2 to a mass of 1 kg.* It is expressed in terms of basic units (repeating Equation 2.3) by

One way to define force: accelerate a standard mass

$$1 \text{ N} = 1 \text{ kg m/sec}^2. \tag{7.2}$$

The unit of force

This turns out to be a convenient macroscopic unit of force. It is equal to the weight of 102 gm at the surface of the earth. In lifting a 2-kg book, you exert a force of about 20 N. A 150-lb man weighs about 670 N.

Although we at once adopt this *unit* of force, we prefer to define the *concept* of force, at least provisionally, by reference to a static measurement.

* That this is a *useful*, as well as a possible, definition depends on the existence of Newton's second law, which will be discussed in the next section.

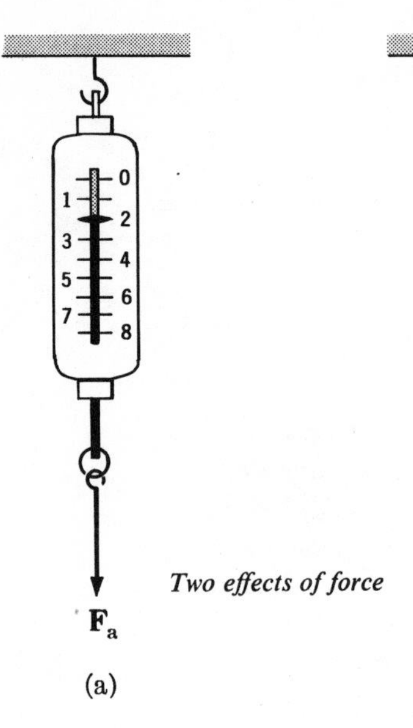

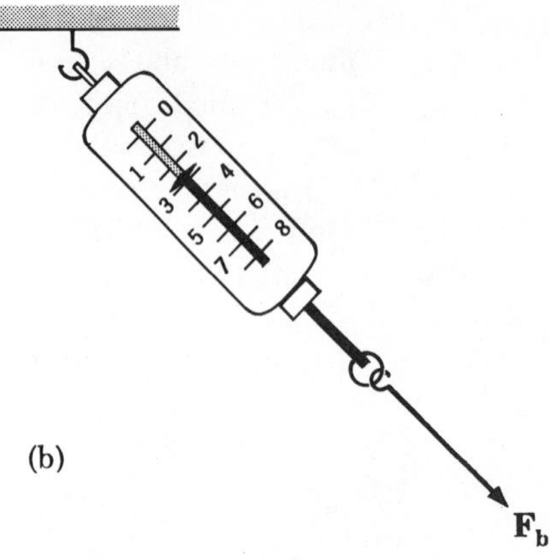

FIGURE 7.3 Operational definition of force. A spring obeying Hooke's law is one for which equal increments of force add equal increments of spring extension. The two diagrams show that the spring can, in principle, define and measure force as a vector quantity, with direction as well as magnitude.

Two effects of force

Besides producing acceleration, forces can change the shape or size of an object. A toy balloon subjected to a pair of forces between two hands has its shape altered. A wooden plank supported at both ends is pulled downward by the force of gravity and is bent into a bow. The shape of a soft pillow is changed according to the forces acting on it. A tree is bent from the force of the wind. A spring is elongated when pulled and compressed when pushed. This general kind of effect of force we may call deformation. No object is so rigid that it does not undergo at least slight deformation when acted upon by external forces.

Another way to define force: the extension of a standard spring

The deforming effect of force can be used to define force. Imagine a well-designed standard spring whose elongation when pulled can provide a measure of force (Figure 7.3). One end of the spring is held fixed in position (this requires some force, but it is not the force of interest). The force to be measured is applied to the other end of the spring. We could decide to call the force required to extend the spring by 1 cm the unit force. Two such forces, each of 1 unit, acting in the same direction, provide a force of 2 units. If equal and parallel forces together add to 1 unit, each of them is equal to 0.5 unit. By such combinations any unknown force can in principle be related to the unit force and the chosen spring extension. If the spring happens to have the happy property that its extension is proportional to the applied force, it is a practical and useful force-measuring device, not just a definer of force in principle. Then, for instance, an extension of 2.5 cm would measure a force of 2.5 units, an extension of 0.78 cm would measure a force of 0.78 unit, and so on. Such springs, which can in fact be constructed to a high degree of accuracy, are said to obey Hooke's law, which states that deformation is proportional to force.

Ideal springs obey Hooke's law

Since force is a vector quantity, its definition must include this fact. This means first that force has direction as well as magnitude; both properties can be included in the spring definition (Figure 7.3). But the vector nature of force means more. It means that two or more forces must combine physically in exactly the same way that vectors combine mathematically. As illustrated in Figure 7.4, the standard spring may easily be used to test the laws of combination of several forces in order to verify their vector character. It can be verified, for instance, that two equal-magnitude forces directed at right angles to each

Force is a vector quantity

other exert a total force directed halfway between the two combining forces with a magnitude 41 percent greater than the magnitude of each combining force. And it can be verified that two forces equal in magnitude and opposite in direction exert no net force on the spring.

■ EXAMPLE 1: A block of weight 14 N sits motionless on a plank inclined at 30 deg to the horizontal (Figure 7.5). What frictional force acts on the block? Since the block is unaccelerated, it experiences no net force. We may therefore write

$$\mathbf{W} + \mathbf{F} + \mathbf{N} = 0, \tag{7.3}$$

Zero force if $\mathbf{v} = 0$

where **W** is the weight, **F** is the frictional force directed parallel to the plank, and **N** is the force exerted by the plank perpendicular to its surface. Since $\mathbf{F} + \mathbf{N} = -\mathbf{W}$, the vertical dashed arrow in Figure 7.5 has length W, and

$$F = W \sin (30 \text{ deg}).$$

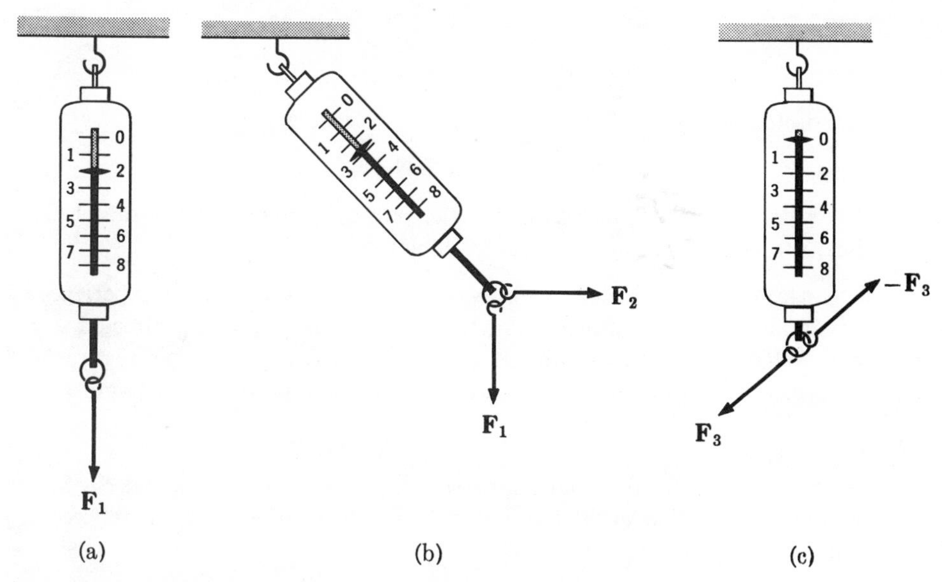

FIGURE 7.4 Direct test of the vector nature of force. (a) A force $\mathbf{F}_1$ acting alone has magnitude 2 on this scale. (b) Two equal-magnitude forces, $\mathbf{F}_1$ and $\mathbf{F}_2$, acting at right angles, produce a force directed halfway between them with magnitude 2.8. (c) Two equal-magnitude forces directed oppositely add to zero.

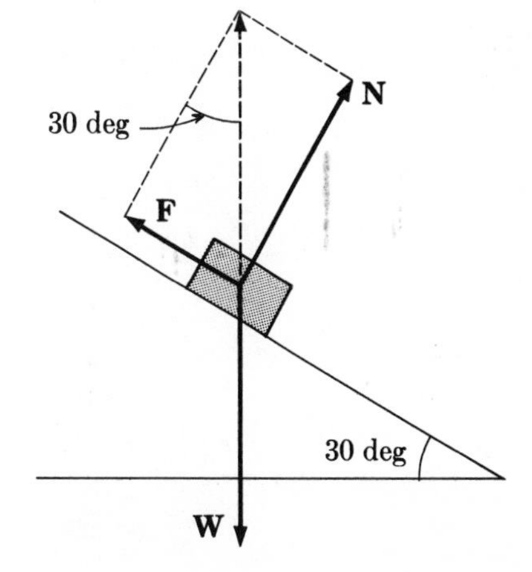

FIGURE 7.5 A block at rest on an inclined plank. The total force acting on it is zero.

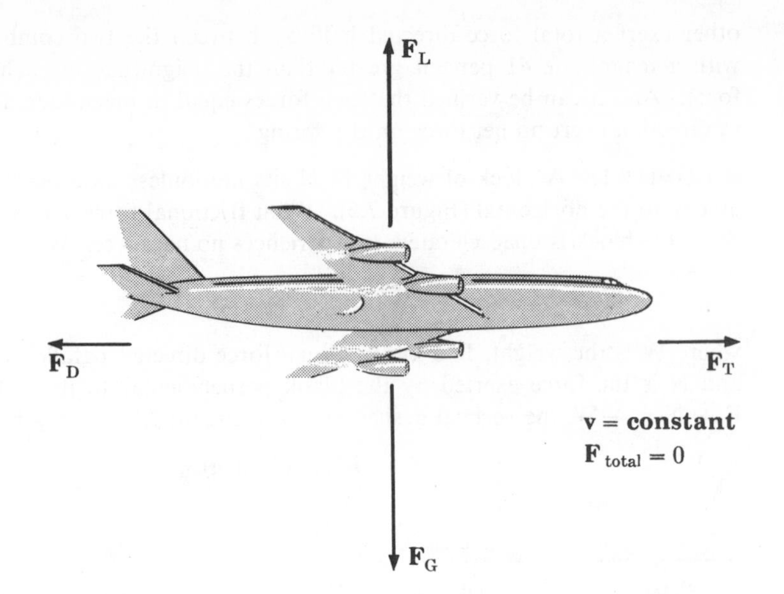

FIGURE 7.6 An airplane moves with constant velocity. The vanishing of the total force acting on it requires that x and y components vanish separately.

Substituting $W = 14$ N and sin (30 deg) $= 0.5$, we get

$$F = 7 \text{ N}.$$

If the plank is tipped to a greater angle, yet not enough to cause the block to slide, will $\mathbf{F}$ change in magnitude? in direction? ■

■ EXAMPLE 2: An airplane is cruising at constant velocity. It experiences a gravitational force $\mathbf{F}_G$ directed downward and a drag force $\mathbf{F}_D$ directed to the rear. What are the forces of lift $\mathbf{F}_L$ and thrust $\mathbf{F}_T$ acting on it (see Figure 7.6)? Again, from Newton's first law, we know that the total force must vanish:

Zero force if $\mathbf{v}$ = **constant**

$$\mathbf{F}_G + \mathbf{F}_D + \mathbf{F}_L + \mathbf{F}_T = 0. \tag{7.4}$$

Because this is a vector equation, *two* unknown magnitudes, F_L and F_T, can be found from it. If we let the unit vector $\mathbf{i}$ point forward and let the unit vector $\mathbf{j}$ point upward, the equation reads

$$(F_T - F_D)\mathbf{i} + (F_L - F_G)\mathbf{j} = 0.$$

If a vector vanishes, both of its components vanish. Therefore

$$F_T = F_D,$$
$$F_L = F_G.$$ ■

7.3 Newton's second law; inertial mass

The acceleration of a body must be a function of the net force applied to it, and could in principle depend on velocity or other variables as well. In fact, experiments show the relation between acceleration and force to be the simplest one imaginable, a direct proportionality:

Nature is simple: Acceleration is proportional to force

$$\mathbf{a} \sim \mathbf{F}. \tag{7.5}$$

Note that this is a vector proportionality. Acceleration always has the direction of the net force, regardless of the direction of the velocity. In uniform circular motion, for example, velocity is directed tangentially around the circle; acceleration, and therefore force, are directed inward toward the center of the circle. To reinforce a point made earlier, we further remark that this proportionality implies no causal connection between force and acceleration. It is usually rather natural (and not harmful) to think of force as the cause and acceleration as the result. Logically, it is as valid to say that acceleration causes force, and sometimes it is in fact easier to think in this way. We can say that the forward acceleration of an automobile "causes" us to feel an extra force from the back of the seat. In any case, as a law of nature, Formula 7.5 states only that force and acceleration always go together in a certain way.

Properties of accelerated motion can conveniently be studied in the laboratory, using apparatus such as that pictured in Figure 7.7. A car rides on a horizontal track with very little friction. On the car can be placed weights of various sizes. The car and its weights are drawn along the track with a known force, and the acceleration is measured. Since there is no vertical component of acceleration, there is no net vertical component of force, and only the horizontally applied force need be considered. The first result of such an experiment is that a constant force produces constant acceleration. The second result is that if the car is loaded in a particular way, its acceleration is proportional to the force applied. These results verify the proportionality of Formula 7.5.

Since both **F** and **a** can be measured, the constant of proportionality linking them can be determined. For any particular loading of the car, the motion is described by an equation:

Newton's second law

$$\mathbf{F} = m\mathbf{a}. \tag{7.6}$$

This is *Newton's second law.** Its implications for motion in general will be discussed in later sections. Here we are concerned with its role in fixing the basic concepts of mechanics.

It defines inertial mass

The constant of proportionality in Equation 7.6 is called the inertial mass (or simply the mass) of the object under study, in this case the car plus its load.

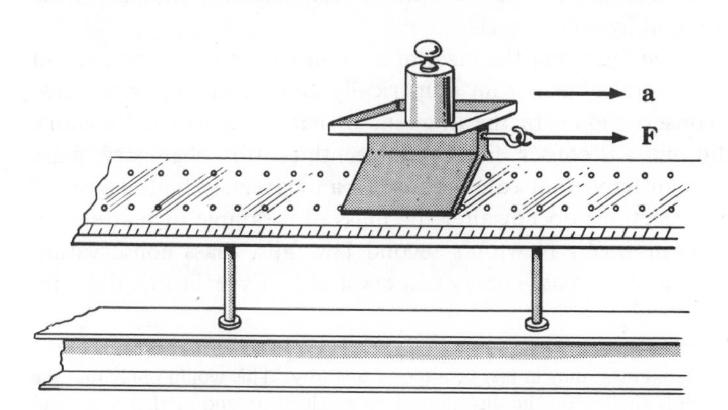

FIGURE 7.7 Experimental arrangement to verify Newton's second law and to measure inertial mass. A car and its load, supported on a nearly frictionless air track, responds with constant acceleration **a** to a constant force **F**.

* Newton's own version of his second law, although equivalent in content to Equation 7.6, was phrased differently. We use a modernized version suited to our needs in this chapter.

Evidently, from Equation 7.6, the dimension of mass must be force × (time)^2/length. In SI units, the corresponding relation among units is given by Equation 7.2.

If Equation 7.6 is to have vectorial consistency, we expect mass to be a scalar quantity. However, the scalar nature of mass requires experimemtal confirmation.* First, experimentally determined masses are independent of the direction of motion and independent of coordinate systems used to define motion. Second, masses combine in accordance with the laws of numerical arithmetic. Suppose that it is ascertained by acceleration experiments that the mass of the unloaded car is 0.2 kg, the mass of the car with load A is 0.5 kg, and the mass of the car with load B is 0.7 kg. Using C to designate the car, we have

An experimental fact: Mass is a scalar quantity

$$m(\mathrm{C}) = 0.2\ \mathrm{kg},$$

$$m(\mathrm{C} + \mathrm{A}) = 0.5\ \mathrm{kg},$$

$$m(\mathrm{C} + \mathrm{B}) = 0.7\ \mathrm{kg}.$$

If masses combine physically in the same way that numbers combine arithmetically, relations such as

$$m(\mathrm{C} + \mathrm{A}) = m(\mathrm{C}) + m(\mathrm{A}) \tag{7.7}$$

should be true, so we can deduce the separate masses of the loads:

$$m(\mathrm{A}) = 0.3\ \mathrm{kg},$$

$$m(\mathrm{B}) = 0.5\ \mathrm{kg}.$$

To test this assumption that masses add numerically, we fill the car with loads A and B together, then measure the acceleration produced by a known force. The mass so determined should be

$$m(\mathrm{A} + \mathrm{B} + \mathrm{C}) = m(\mathrm{A}) + m(\mathrm{B}) + m(\mathrm{C}) = 1.0\ \mathrm{kg}.$$

Many such experiments could be performed to verify the fact that the inertial mass of a group of objects taken together is numerically equal to the sum of the inertial masses of the individual objects.

Lurking behind the facts that the mass of a single object is a constant and that the masses of several objects sum numerically is another important law, the law of mass conservation. In the domains of nature in which Newton's second law is valid, mass is conserved. In nineteenth-century chemistry, mass conservation played an important role because to an extremely high degree of precision, chemical reactions do not alter the mass of a sample of matter. In subatomic domains, in which Newton's second law fails, mass conservation also fails (and is replaced by mass-energy conservation). Even in this domain,

The conservation of mass is a law of classical physics

* It has been speculated that mass may in fact be a tensor quantity. This would imply that the mass of a body depends slightly on the direction of its acceleration and/or that force and acceleration are not precisely parallel. If mass were discovered to have such properties—there is at present no evidence for them—the discovery would be interpreted in terms of an anisotropy of the universe (see the remarks in the last paragraph of this chapter and the discussion of Mach's principle in Section 20.11).

however, a particle is still characterized by a scalar quantity—its rest mass.

To review: Force has been defined in terms of the static extension of a pulled spring; mass has been defined in terms of the acceleration of an object subjected to a known force. In this formulation, force is a primitive concept and mass is a derived concept. Although both concepts are operationally defined, the unit of force is an arbitrary unit referred to an arbitrary standard spring, whereas the unit of mass is defined as a certain combination of the units of force, length, and time. However, because of the experimental connection between force and mass set forth mathematically by Equation 7.6, it is easy to see that mass could have been chosen as the primitive concept and force as the derived concept. A standard kilogram does in fact define the mass unit for modern scientific work.

In practice, because of the equivalence of inertial and gravitational mass, unknown masses are compared with the standard mass by measurements of relative weight. However, direct determinations of mass by acceleration experiments are possible in principle. Most simply, imagine a standard kilogram and an unknown mass subjected to the same force (whose magnitude need not be known). The ratio of the masses may be defined to be the inverse ratio of their accelerations,

Relative acceleration provides one way to measure mass

$$\frac{m}{m_s} = \frac{a_s}{a}, \tag{7.8}$$

where a is the acceleration of the unknown mass m and where a_s is the acceleration of the known standard mass m_s. For such a measurement, no force unit is required, although evidently Newton's second law is playing a role.

The choice of basic concepts is arbitrary

So far as the logical framework of mechanics is concerned, it is not important whether mass is regarded as a primitive or a derived concept. The only important things are that the theory provides a self-consistent set of equations, that the basic concepts are chosen in such a way that these equations are as simple as possible, and that there is a clearly understood connection between physical measurements and mathematical symbols. When these things are achieved, the theory provides a *simple description* of some part of nature and the *power of prediction.* Nothing else can be asked of it. Modern science neither tries nor succeeds to answer questions such as: What are "really" the most basic concepts? or Why does mass behave in the way that it does? The interweaving of definitions and laws in the structure of modern science, although occasionally confusing, is a necessary aspect of a science whose mathematics is securely tied to nature by the bonds of operational definitions and experimental confirmation.

7.4 Applications of Newton's second law

Since Equation 7.6 is a vector equation, it may be re-expressed as three numerical equations:

Component form of Newton's second law

$$F_x = ma_x, \tag{7.9}$$

$$F_y = ma_y, \tag{7.10}$$

$$F_z = ma_z, \tag{7.11}$$

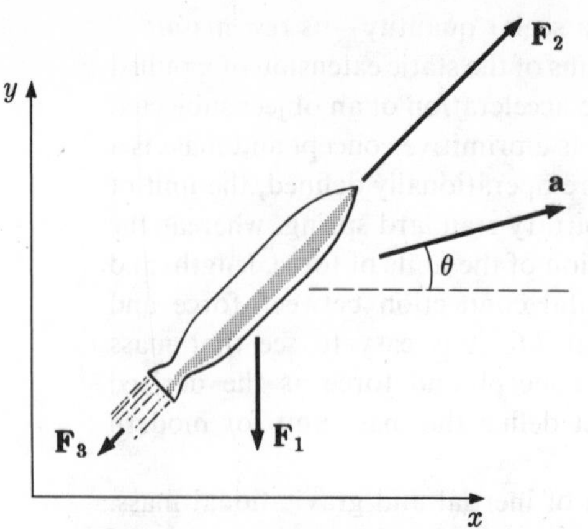

FIGURE 7.8 A rocket experiences acceleration **a** in response to forces of gravity ($\mathbf{F}_1$), thrust ($\mathbf{F}_2$), and frictional air drag ($\mathbf{F}_3$).

where x, y, and z refer to three mutually perpendicular Cartesian axes. If the motion is known to occur along a single straight line, the direction of the line, whatever it may be, can be chosen to be the x direction. Then only one numerical equation is required. If the motion occurs in a plane, two of the three axes can be chosen to lie in that plane.

In this section we will first work out two numerical examples involving motion in a plane, then consider two kinds of motion in one dimension. Other examples of motion in one and two dimensions will be studied in later sections.

■ EXAMPLE 1: A rocket of mass 1.5×10^5 kg is moving through the air with its velocity directed 45 deg above the horizontal (Figure 7.8). At a particular instant, it is subjected to three forces: its weight, $\mathbf{F}_1$, acting vertically downward and of magnitude 1.47×10^6 N; its thrust, $\mathbf{F}_2$, directed along its flight path and of magnitude 3×10^6 N; and frictional air drag, $\mathbf{F}_3$, directed opposite to its velocity and of magnitude 0.2×10^6 N. What is the acceleration of the rocket? Choosing x and y axes as in Figure 7.8, we may write

$$\mathbf{F}_1 = -1.47 \times 10^6\mathbf{j},$$

$$\mathbf{F}_2 = 2.12 \times 10^6\mathbf{i} + 2.12 \times 10^6\mathbf{j},$$

$$\mathbf{F}_3 = -0.14 \times 10^6\mathbf{i} - 0.14 \times 10^6\mathbf{j},$$

where **i** and **j** are unit vectors along the x and y axes. The total force, $\mathbf{F}_T = \mathbf{F}_1 + \mathbf{F}_2 + \mathbf{F}_3$, is

$$\mathbf{F}_T = 1.98 \times 10^6\mathbf{i} + 0.51 \times 10^6\mathbf{j}, \quad (7.12)$$

measured in newtons. Division by the mass of the rocket yields an expression for its acceleration:

Acceleration calculated from force

$$\mathbf{a} = \frac{\mathbf{F}_T}{m} = (13.2\mathbf{i} + 3.4\mathbf{j})\ \text{m/sec}^2. \quad (7.13)$$

This is a vector of magnitude 13.6 m/sec² (about $1.4g$), directed 14.4 deg above the horizontal ($\tan\theta = 3.4/13.2 = 0.258$). ■

■ EXAMPLE 2: Newton's second law need not always be used to answer the question, Given the force, what is the motion? It can be equally well used to answer the question, Given the motion, what is the force? Newton himself used it in this way. He made use of the knowledge about lunar and planetary motion supplied by Kepler's laws in order to deduce the law of gravitational force. Here is a more homely example. An automobile with a speed of 10 m/sec rounds a curve whose radius of curvature is 30 m. What is the force experienced by a 75-kg passenger? The study of the kinematics of circular motion in Section 6.10 revealed that the magnitude of the acceleration is given by

$$a = \frac{v^2}{r},$$

where v is the speed and r is the radius of the circle. In this example, the acceleration of the automobile is therefore

$$a = \frac{(10 \text{ m/sec})^2}{30\ m} = 3.33 \text{ m/sec}^2.$$

The passenger, following the same path as the automobile, has the same acceleration. The force acting on him, according to Newton's second law, must therefore be

Force calculated from acceleration

$$F = ma = 75 \text{ kg} \times 3.33 \text{ m/sec}^2 = 250 \text{ N}. \tag{7.14}$$

Force and acceleration are directed along the same line, toward the center of the circular path. This *net* force is horizontal. Since the passenger experiences no vertical component of acceleration, the vertical forces acting on him (the downward force of gravity and the upward force of the seat) must sum to zero. ■

MOTION WITH CONSTANT FORCE

Suppose a particle moving in a straight line experiences a constant force **F**. Choose the x axis to lie along the direction of the force. Then $a_y = a_z = 0$, and

Constant force means constant acceleration

$$a_x = \frac{F}{m} = constant. \tag{7.15}$$

The kinematics of constantly accelerated motion in one dimension was studied in Section 5.7. Equations 5.55 and 5.56 provide what is called the "solution for the motion," that is, answers to the questions, Where is the particle and how fast is it moving at all times? The same force can produce different motions, depending on initial conditions. Suppose, for example, that a constant force of 6 N acts in the positive x direction on an object of mass 2 kg. Its acceleration will be $a_x = 3$ m/sec^2. If it starts from rest ($v_0 = 0$) at the origin ($x_0 = 0$) at the initial time ($t = 0$), its later position is given by the equation

$$x = \tfrac{1}{2}a_x t^2,$$

which in mks units is

$$x = 1.5t^2.$$

The object moves away to the right with ever increasing speed (Figure 7.9). If, however, it had been located at some other point x_0 at the initial time and had possessed some initial speed v_0 at this time, its subsequent history would be given by the more complicated equation

$$x = x_0 + v_0 t + \tfrac{1}{2}a_x t^2.$$

Two examples of such motion are also illustrated in Figure 7.9, one with $x_0 = 5$ m and $v_0 = 9$ m/sec, one with $x_0 = 15$ m and $v_0 = -9$ m/sec. For positive initial velocity, the object accelerates away to larger values of x, maintaining a higher speed than when it started from rest. For negative initial velocity, the object begins to move to the left (to smaller values of x) but is slowed by the rightward acceleration. After 3 sec it is brought to rest, its direction of motion is reversed, and it accelerates away to larger values of x. The essential points illustrated by this example are these: The one-dimensional motion of an object subject to a constant force proceeds with constant acceleration. Precise details of the motion depend on initial conditions—where the particle was located and how fast it was moving initially. However, for a specific acceleration, all possible motions follow a parabolic graph of position as a function of time; indeed the parabolas are, except for horizontal and vertical shifts on the graph, all the same. Motion begun from rest or in the direction of the acceleration will continue in the same direction. An object moving initially in the direction opposite to the acceleration will be stopped and turned around after some time.

FORCE VARYING WITH TIME

Since $\mathbf{a} = d\mathbf{v}/dt$ and $\mathbf{v} = d\mathbf{r}/dt$, Newton's second law may also be written in either of the two following forms:

Newton's second law can be expressed as a differential equation

$$m\frac{d\mathbf{v}}{dt} = \mathbf{F}, \tag{7.16}$$

or

$$m\frac{d^2\mathbf{r}}{dt^2} = \mathbf{F}. \tag{7.17}$$

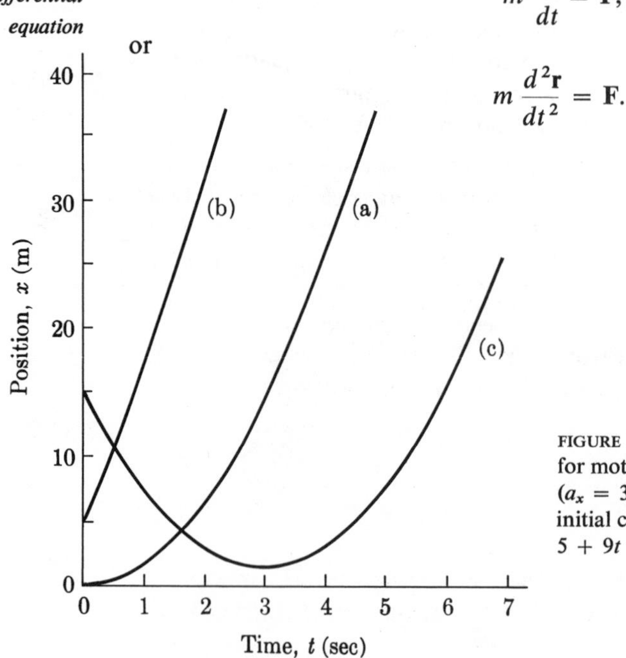

FIGURE 7.9 Three graphs of position vs time for motion with the same constant acceleration ($a_x = 3$ m/sec^2), but with three different initial conditions: (a) $x = 1.5t^2$; (b) $x = 5 + 9t + 1.5t^2$; (c) $x = 15 - 9t + 1.5t^2$.

These are differential equations. In order to draw attention to this differential-equation aspect of Newton's second law, we consider a hypothetical force directed along the x axis and increasing in proportion to the time:

A soluble special case: F_x proportional to time

$$F_x = \alpha t,$$

where α is a constant of proportionality. The equation $F_x = ma_x$ may be written

$$m\,\frac{dv_x}{dt} = \alpha t. \tag{7.18}$$

This differential equation is asking, in effect, What function $v_x(t)$ has a first derivative equal to $(\alpha/m)t$? Clearly the answer is a quadratic function. Including an arbitrary constant, v_0, it is

$$v_x = \frac{1}{2}\left(\frac{\alpha}{m}\right)t^2 + v_0. \tag{7.19}$$

Differentiation of Equation 7.19 yields Equation 7.18. This solution for the velocity is in turn itself a differential equation:

$$\frac{dx}{dt} = \frac{1}{2}\left(\frac{\alpha}{m}\right)t^2 + v_0.$$

The function x whose derivative is the right side of Equation 7.19 may be found by writing

$$x = \int\left(\frac{dx}{dt}\right)dt = \int\left[\frac{1}{2}\left(\frac{\alpha}{m}\right)t^2 + v_0\right]dt.$$

The result is

$$x = \frac{1}{6}\left(\frac{\alpha}{m}\right)t^3 + v_0t + x_0. \tag{7.20}$$

Each of the two steps—from Equation 7.18 to 7.19 and from Equation 7.19 to 7.20—is an integration, and in each step a constant of integration is introduced. Equation 7.20 is called a solution of the second-order differential equation

$$\frac{d^2x}{dt^2} = \left(\frac{\alpha}{m}\right)t.$$

Characteristically, the solution contains two arbitrary constants. Whenever the force $\mathbf{F}$ is known, Newton's second law may be looked upon as a differential equation (or a set of three differential equations) for the unknown quantities x, y, and z. Only very few force functions lead to solutions as simple as the one considered here.

7.5 The harmonic oscillator

A law of force that occurs rather often for one-dimensional motion in nature is this:

The force law of a harmonic oscillator

$$F_x = -kx. \tag{7.21}$$

It says that the force is directed toward the origin (its component is negative for positive x and positive for negative x) and that the magnitude of the force, which is zero at the origin, increases in proportion to the distance away from the origin. The quantity k is called the force constant. As applied to a spring, this force law is called *Hooke's law*: Force is proportional to displacement.* It is a good approximation for many springs and explains why in principle a spring can be used to define force as well as measure it. The same force law is also approximately valid for a pendulum or a swing. In general, any system described by this force law is called a harmonic oscillator.

For springs, it is called Hooke's law

Figure 7.10 shows an idealized version of a harmonic oscillator—a cart of mass m resting on a frictionless horizontal surface and connected to one end of a spring whose other end is held fixed.† The mass of the spring is assumed negligible in comparison with the mass of the cart. As indicated in Figure 7.10, the position coordinate x is defined as the displacement of the cart away from its equilibrium position. Applied to the cart, Newton's second law may be written

$$ma_x = -kx, \tag{7.22}$$

or, equivalently,

The law of motion

$$a_x = \frac{d^2x}{dt^2} = -\left(\frac{k}{m}\right)x. \tag{7.23}$$

Since the ratio k/m is constant, the interesting thing this equation states is that the acceleration is proportional to the negative of the position coordinate. Again we are dealing with a second-order differential equation. There are formal ways to proceed to find its solution, but we shall instead use a trial-and-error method, simply asking, Are we familiar with any function whose second derivative is proportional to the negative of the function (as required by Equation 7.23)? As discussed in Section 5.10, the sine and cosine functions have this property (indeed they are the only such functions). We may therefore postulate, as a solution for the position coordinate,

A special case of harmonic motion

$$x = A \sin \omega t, \tag{7.24}$$

or

$$x = A \sin\left(\frac{2\pi t}{T}\right) \tag{7.25}$$

(compare Equations 5.99 and 5.103). The constants are

$$A = \text{amplitude} \qquad [\text{dimension, length}], \tag{7.26}$$

$$\omega = \text{angular frequency} \qquad [\text{dimension, (time)}^{-1}], \tag{7.27}$$

$$T = \frac{2\pi}{\omega} = \text{period} \qquad [\text{dimension, time}]. \tag{7.28}$$

By differentiating Equation 7.24, for the x component of velocity we find

$$v_x = \frac{dx}{dt} = \omega A \cos \omega t. \tag{7.29}$$

* The force in question is the force required to displace one end of a spring if the other end is held fixed.

† Because of the sag of an actual spring, this setup is not practical in the laboratory. However, see Exercise 7.25.

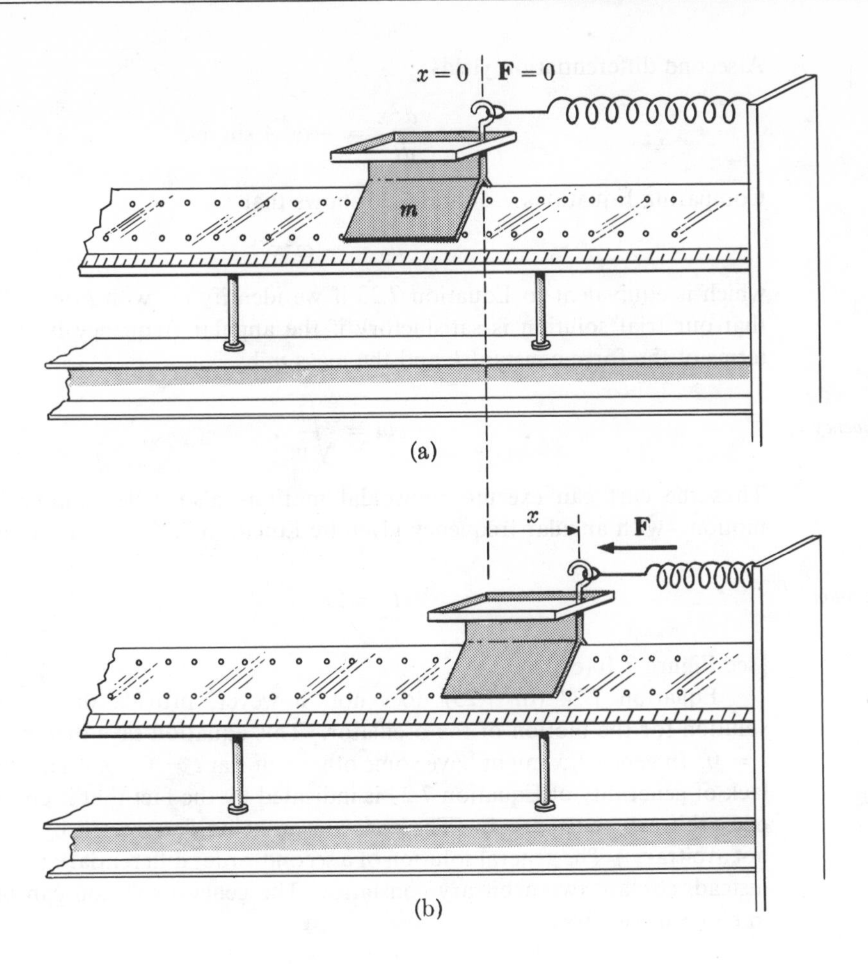

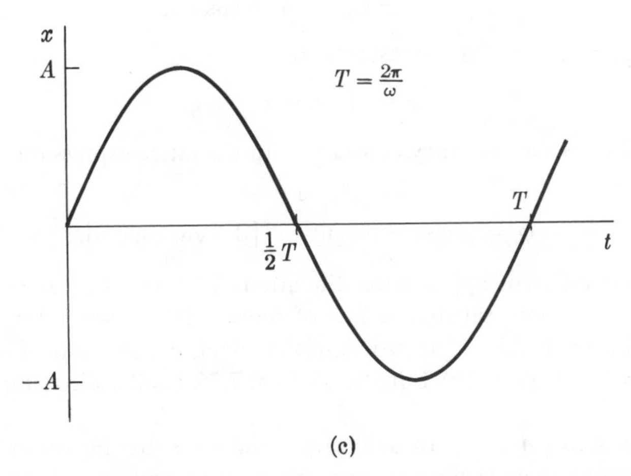

(c)

FIGURE 7.10 An idealized harmonic oscillator. (a) At the equilibrium position of the spring, the cart experiences no force. (b) For displacement x away from equilibrium, the force acts opposite to the displacement, and has magnitude proportional to the displacement ($F_x = -kx$). (c) A graph of x vs t for the oscillator (Equation 7.24).

A second differentiation yields

$$a_x = \frac{d^2x}{dt^2} = -\omega^2 A \sin \omega t. \tag{7.30}$$

Comparing Equations 7.24 and 7.30 shows that

$$a_x = -\omega^2 x, \tag{7.31}$$

which is equivalent to Equation 7.23 if we identify ω^2 with k/m. This means that our trial solution is satisfactory if the angular frequency is expressed in terms of the force constant k and the mass m by

Angular frequency

$$\omega = \sqrt{\frac{k}{m}}. \tag{7.32}$$

Thus the cart can execute sinusoidal motion—also called simple harmonic motion—with angular frequency given by Equation 7.32, or period given by

Period of the motion

$$T = 2\pi \sqrt{\frac{m}{k}} \tag{7.33}$$

[see Figure 7.10(c)].

Equation 7.24 (or 7.25) does not, however, provide the most general solution for the motion of the oscillator. This equation says that $x = 0$ when $t = 0$. In general, x might have some other value at $t = 0$. Mathematically, the lack of generality of Equation 7.24 is indicated by the fact that it contains only one arbitrary constant, A. (The constant ω, being determined by k and m, is not arbitrary.) The general solution of a second-order differential equation must, instead, contain two arbitrary constants. The general solution can be written in either of two ways:

$$x = a \sin \omega t + b \cos \omega t, \tag{7.34}$$

General case of harmonic motion

where a and b are arbitrary constants; or

$$x = A \sin (\omega t + \varphi), \tag{7.35}$$

with A and φ as the arbitrary constants. In the latter expression, A is still the amplitude, and

$$\varphi = \text{phase constant} \qquad [\text{dimensionless}]. \tag{7.36}$$

Despite their different appearance, Equations 7.34 and 7.35 are equivalent to each other, and both satisfy the law of motion (Equation 7.23), with ω still given by Equations 7.32. The proofs of these facts are left to the exercises. The general motion, described by Equation 7.34 or 7.35, is still called simple harmonic motion.

Initial conditions determine amplitude and phase

The constants A and φ are arbitrary in the sense that Equation 7.35 satisfies the law of motion regardless of the values of A and φ. However, for any particular initial conditions, A and φ must be chosen to match these conditions. Suppose at $t = 0$ that $x = x_0$ and that $v = dx/dt = v_0$. At $t = 0$, Equation 7.35 then reads

$$x_0 = A \sin \varphi, \tag{7.37}$$

and its first derivative with respect to time reads

$$v_0 = \omega A \cos \varphi. \tag{7.38}$$

Equations 7.37 and 7.38 provide two equations for the two unknowns, A and φ. Taking the ratio of the two equations eliminates A and gives

$$\frac{x_0}{v_0} = \frac{1}{\omega}\frac{\sin \varphi}{\cos \varphi} = \frac{1}{\omega}\tan \varphi.$$

This equation has two solutions for φ between 0 and 2π. Both can be written

$$\varphi = \arctan\left(\frac{\omega x_0}{v_0}\right). \tag{7.39}$$

Solution for the phase

To find A, square both sides of Equation 7.37; then

$$x_0^2 = A^2 \sin^2 \varphi. \tag{7.40}$$

Then do the same to Equation 7.38, first dividing both sides by ω:

$$\left(\frac{v_0}{\omega}\right)^2 = A^2 \cos^2 \varphi. \tag{7.41}$$

Summing Equations 7.40 and 7.41 gives

$$x_0^2 + \left(\frac{v_0}{\omega}\right)^2 = A^2(\sin^2 \varphi + \cos^2 \varphi) = A^2.$$

Because of a trigonometric identity, φ disappears. The solution for A is therefore

$$A = \sqrt{x_0^2 + \left(\frac{v_0}{\omega}\right)^2}. \tag{7.42}$$

Solution for the amplitude

It is conventional to choose the positive root. Then only one of the two solutions for φ is appropriate, and Equation 7.37 can be used to decide which one.

The harmonic oscillator has the simple and important property that its period is independent of its amplitude. In Formula 7.33, T depends only on m and k, not on A (or on φ or x_0 or v_0). A larger value of k (a stiffer spring) produces a smaller period. A larger value of m (greater inertia) produces a larger period.

An important property: Period is independent of amplitude

Oscillatory phenomena are common in nature—a vibrating tuning fork, a child's rocking horse, waves on water, sound transmitted through the air, an entire building swaying in the wind—the number of examples are almost unlimited; and for nearly all of them, the simple law of force expressed by Equation 7.21 is a good approximation to the true force acting toward the central equilibrium point of the motion. Therefore, a vast variety of swaying, swinging, vibrating, or oscillatory motions are rather accurately described by sine functions.

Harmonic motion is common throughout nature

Not all sine functions are identical, of course. They may be differently displaced in time (different φ); they may have large magnitude or small (different A); and they may have the long stately period of a swaying suspension bridge or the frenetic million-times-per-second vibration of a small crystal (different ω or T). They do, however, have an underlying mathematical identity. It is remarkable that a single simple mathematical function describes an enormous range of different types and sizes and speeds of oscillatory motion. The reason, according to Newton's second law, is that forces proportional to displacement and opposite in direction to the displacement are common in all parts of nature.

7.6 Motion near the earth

The gravitational force experienced by a body is called its weight. The force acts whether or not the body is in motion, but it is most conveniently measured when the body is at rest. Every scale that records weight is a force-measuring device. The units painted on the dial of a scale are often mass units, not force units. Nevertheless, the deflection of the needle of the scale is a measure of the gravitational force experienced by the object being weighed. Or, in a pan balance, known weights are adjusted until the gravitational forces on the two pans are equal. Since the measurement of weight is a static measurement, it provides important information about a fundamental force of nature independent of the laws of motion.

Properties of the gravitational force

Studies of the gravitational force reveal the following facts: (1) Near the surface of the earth, the weight of an object is, to a good approximation, independent of its position. (2) The gravitational force on a body is independent of its state of motion. (3) Most important, weight is proportional to mass. This proportionality, apparently universal for all substances of all sizes, has been tested experimentally for aluminum and gold to an accuracy of three parts in one hundred billion.

Because of the observed proportionality between weight and mass (inertial mass), there is often confusion between these concepts. Recall that inertial mass, defined as the constant of proportionality between force and acceleration in Newton's second law, is a measure of the resistance to a change in motion of an object subjected to a force—*any* force. Weight, on the other hand, is specifically a measure of the intensity of response of an object to a *gravitational* force. There is no obvious reason at all for a connection between these two concepts. Their direct proportionality, discovered by Newton, remained a mysterious coincidence for over two centuries. How Einstein's general theory of relativity made use of this proportionality and removed its mystery will be discussed in Chapter 22.

The experimental proportionality between gravitational force and inertial mass can be written

It is proportional to mass

$$F_{\text{grav}} = mg, \tag{7.43}$$

where g is a constant of proportionality. If the z axis of a Cartesian coordinate system (with its unit vector $\mathbf{k}$) is chosen to point vertically upward, the vector equation is

$$\mathbf{F}_{\text{grav}} = -mg\mathbf{k}. \tag{7.44}$$

Although g varies slightly from place to place on the earth's surface, at any one place it is precisely constant, independent of the size or nature of the matter being weighed. Comparing this law of force with Newton's second law shows at once that g must have the same dimension as acceleration. Its numerical value in the international system is

The acceleration of gravity

$$g = 9.80 \text{ m/sec}^2. \tag{7.45}$$

If the gravitational force is the *only* force acting on an object, its acceleration is related to that force by Newton's second law,

$$\mathbf{F}_{\text{grav}} = m\mathbf{a}. \tag{7.46}$$

Since $\mathbf{F}_{\text{grav}} = -mg\mathbf{k}$, it follows at once that

$$\mathbf{a} = -g\mathbf{k}, \tag{7.47}$$

a constant acceleration, directed downward and of magnitude 9.8 m/sec^2. Accordingly, g is usually called "the acceleration of gravity." It is important to remember, however, that if gravity is only one of several forces acting on an object, the acceleration of the object need bear no simple relation to the constant g.

The equations governing the uniformly accelerated motion of free fall have been discussed before and are set forth in Equations 5.51–5.58. No motion that takes place in the air is truly "free" fall because air resistance always acts to exert some force in a direction opposite to the motion. For sufficiently low velocity, or for a very dense, aerodynamically shaped object such as a bomb, this retarding force may be so small compared with the gravitational force that it can be ignored. Then the motion, if it is in one dimension, is successfully described by Equations 5.51–5.54 or 5.55–5.58 as uniformly accelerated motion (with $a = g$). In an opposite limit, when the retarding force of air resistance becomes so great as to be equal in magnitude (but opposite in direction) to the gravitational force, terminal speed is achieved; the acceleration is zero.

Constant acceleration when $\mathbf{F} = \mathbf{F}_{\text{grav}}$

Terminal speed when $\mathbf{F} = 0$

■ EXAMPLE 1: A skydiver of mass $m = 80$ kg, falling vertically downward, experiences a force of air resistance given approximately by $F_{\text{D}} = \gamma v^2$, with $\gamma = 0.25\ \text{N sec}^2/\text{m}^2$. For about how long a time and through how great a distance does the skydiver fall before the retarding force builds up to 5 percent of his weight? What is his terminal speed? To answer the first question, set $F_{\text{D}} = 0.05mg$, which gives

$$v = \sqrt{\frac{0.05mg}{\gamma}}\,.$$

Numerically, this is

$$v = \sqrt{\frac{(0.05)(80\ \text{kg})(9.8\ \text{m/sec}^2)}{0.25\ \text{N sec}^2/\text{m}^2}}$$

$$= 12.5\ \text{m/sec}.$$

Since the retarding force is still relatively small at this speed, the equations of constant acceleration are approximately valid. For the time to reach this speed, Equation 5.52 gives

$$t \cong \frac{v}{a} = \frac{12.5\ \text{m/sec}}{9.8\ \text{m/sec}^2} \cong 1.3\ \text{sec}.$$

From Equation 5.54, the distance covered is given as

$$x \cong \frac{v^2}{2a} = \frac{(12.5)^2}{2 \times 9.8} \cong 8\ \text{m}.$$

To answer the second question, note that the net downward force on the skydiver is

$$F = mg - \gamma v^2.$$

His terminal speed v_T, determined by the condition $F = 0$, is

$$v_T = \sqrt{\frac{mg}{\gamma}} = 56 \text{ m/sec.}$$

This is about 125 mile/hr. ■

Omit

★THE SIMPLE PENDULUM*

According to legend, Galileo, around 1582, observed swinging chandeliers in a cathedral and noticed that they oscillated with a fixed period, independent of amplitude. From this observation, he was supposedly led to investigate simple harmonic motion. He later suggested that the pendulum could serve as the heart of a clock mechanism. The pendulum clock did finally come into general use around 1660 and was widely used until very recent times. The pendulum is useful in a clock because, to a good approximation, it executes simple harmonic motion, with period independent of amplitude.

Consider an idealized pendulum (Figure 7.11). It is a rod of length l and negligible mass, pivoted at one end and supporting a bob of mass m at the other end.† Two forces act on the bob: $\mathbf{F}_{\text{grav}}$, of magnitude mg, directed vertically downward; and $\mathbf{F}_{\text{rod}}$, the force exerted by the rod, directed along the rod. The total force on the bob is

$$\mathbf{F} = \mathbf{F}_{\text{grav}} + \mathbf{F}_{\text{rod}}. \tag{7.48}$$

Tilted axes are a convenient choice

It is useful to focus attention on an instant during the motion when the rod makes angle θ with the vertical and to choose reference directions defined by a pair of axes so oriented that at this instant, the rod is aligned along axis 2 and the direction of motion of the bob is along axis 1 (see Figure 7.11). Referred to these axes, Newton's second law is expressed by the two equations

$$ma_1 = F_1, \tag{7.49}$$

$$ma_2 = F_2, \tag{7.50}$$

Only the components F_1 and a_1 need to be considered

where the subscripts designate the components of **a** and **F** along the reference directions. The reason for choosing the tilted axes is that it is then sufficient to consider only the first of these two equations. Although F_2 and a_2 need not be zero (see Problem 7.15), we can find the motion of the pendulum without using Equation 7.50.

---

* The problem of the simple pendulum is more easily treated with the help of the energy concept. This will be done in Section 10.11. The treatment here, which uses the force concept, is therefore optional.

† Such an ideal pendulum, with a weightless rod and a point mass, swinging in a plane, is called a "simple pendulum" and is distinguished from a "physical pendulum" or a "compound pendulum," one with distributed mass.

What, specifically, are F_1 and a_1? Since $\mathbf{F}_{rod}$ has no component along axis 1, only $\mathbf{F}_{grav}$ contributes to F_1. As shown in Figure 7.11, this force component is

$$F_1 = -mg \sin \theta.$$

A convenient coordinate to locate the position of the pendulum bob is s, the displacement of the bob away from its equilibrium position, measured along the arc of its motion. From the basic definition of an angle measured in radians, θ is simply related to s:

$$\theta = \frac{s}{l}. \tag{7.51}$$

Therefore the force component F_1 can be written

$$F_1 = -mg \sin \left(\frac{s}{l}\right). \tag{7.52}$$

The component of acceleration a_1 can also be expressed in terms of the displacement s. It is*

$$a_1 = \frac{dv_1}{dt} = \frac{d}{dt}\left(\frac{ds}{dt}\right) = \frac{d^2s}{dt^2}. \tag{7.53}$$

Substituting these expressions for F_1 and a_1 in Equation 7.49 leads to the following equation of motion of the pendulum:

The equation of motion of the pendulum

$$m\frac{d^2s}{dt^2} = -mg \sin \left(\frac{s}{l}\right). \tag{7.54}$$

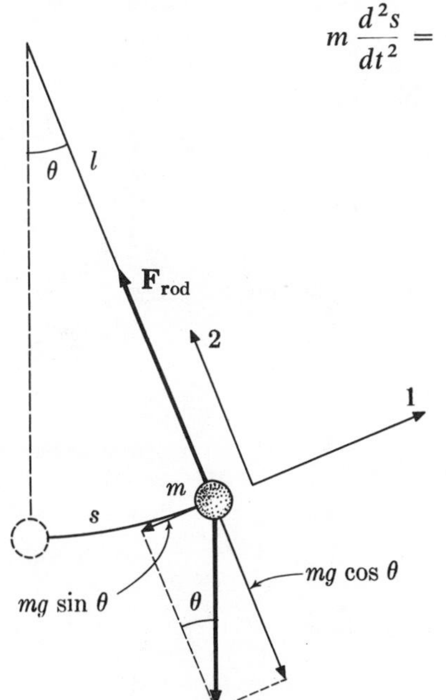

FIGURE 7.11 A simple pendulum. Tilted axes provide convenient reference directions. At the instant shown, when the displacement of the bob along its arc is s, the velocity of the bob is parallel to axis 1. It may have nonzero components of acceleration along both axes.

* Although this formula for a_1 may appear "obvious," it does require proof. The proof is left as a problem.

This is a differential equation for the function $s(t)$. Its exact solution involves what are called elliptic functions. Here we shall consider its approximate form for small displacements ($\theta \ll 1$, or $s \ll l$). According to Equation 5.106,

$$\sin\left(\frac{s}{l}\right) \cong \frac{s}{l} \qquad (s \ll l).$$

With the help of this approximation, Equation 7.54 can be written

The equation of motion for small displacement ($s \ll l$)

$$m\frac{d^2s}{dt^2} = -\frac{mg}{l}s, \tag{7.55}$$

which is valid for small angular displacement. This is the simple harmonic oscillator equation, equivalent to Equation 7.23, with the force constant k given by

$$k = \frac{mg}{l}. \tag{7.56}$$

With this substitution for k, any of the equations of Section 7.5 may be applied to the pendulum. Its motion is described by the equivalent of Equation 7.35,

Small oscillations are harmonic

$$s = A\sin(\omega t + \varphi). \tag{7.57}$$

From Equations 7.32 and 7.33, its angular frequency is given by

$$\omega = \sqrt{\frac{g}{l}} \tag{7.58}$$

and its period by

The period is independent of both mass and amplitude

$$T = 2\pi\sqrt{\frac{l}{g}}. \tag{7.59}$$

Note that the mass of the bob is absent from these expressions. This is characteristic of motion controlled by gravity.

■ EXAMPLE 2: What is the length of a simple pendulum whose period is 2 sec? From Equation 7.59, we get

$$l = g\left(\frac{T}{2\pi}\right)^2.$$

Numerically, this formula gives

$$l = 9.80\ \text{m/sec}^2 \times (2\ \text{sec}/2\pi)^2$$
$$= 0.993\ \text{m} = 99.3\ \text{cm}.$$

A pendulum of this length is called a "seconds pendulum." ■

The pendulum can be used to measure g

A pendulum of fixed length carried from one place to another on the earth will show a slight change in its period because of a change in g. The pendulum provides an excellent way to measure g—and especially to measure changes in g from one place to another.

■ EXAMPLE 3: A pendulum swings to a maximum deflection of 20 deg. At this angle, what is the fractional error in the right side of Equation 7.55? To get

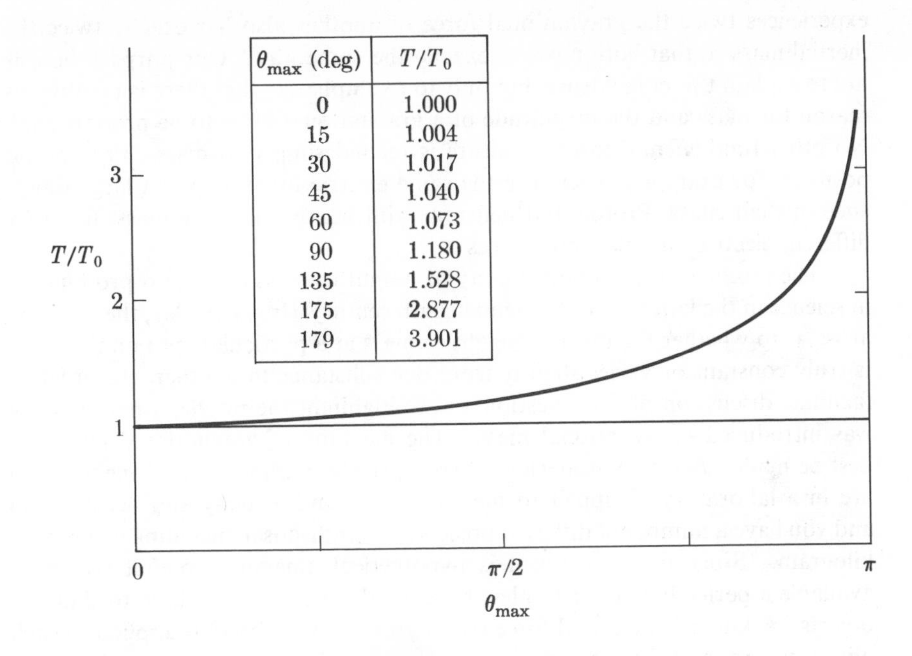

θ_{max} (deg)	T/T_0
0	1.000
15	1.004
30	1.017
45	1.040
60	1.073
90	1.180
135	1.528
175	2.877
179	3.901

FIGURE 7.12 The dependence of the period of a pendulum on its maximum angle of swing. The quantity T_0 is $2\pi\sqrt{l/g}$, the approximate period for small amplitude. The inset table shows that T differs from T_0 by only 1.7 percent for a swing of 30 deg, and by 4 percent for a swing of 45 deg. As θ approaches π, T approaches infinity.

Equation 7.55 from the accurate Equation 7.54, we replaced $\sin\theta$ by θ on the right side. The fractional error at any angle θ is therefore

$$\frac{\theta - \sin\theta}{\sin\theta}. \tag{7.60}$$

With the help of a table of sines, this ratio can be evaluated at $\theta = 20$ deg $=$ 0.349 radian. It is

$$\frac{0.349 - 0.342}{0.349} = 0.02,$$

or 2 percent. For another way to evaluate this ratio, see Exercise 7.35. ■

For larger amplitude, the period increases

As the preceding discussion and derivation make clear, the simple pendulum is a harmonic oscillator only for small angles of vibration. For large amplitude, its period is amplitude-dependent. Figure 7.12 shows how the period depends on the maximum angle of swing.

Suggestion: With a set of keys or other weight hung from a string, verify the proportionality $T \sim \sqrt{l}$ (Equation 7.59) for small amplitude, check the length of a seconds pendulum, and investigate the dependence of period on amplitude for larger amplitude.

7.7 Gravitational mass

The fact that objects near the earth on which the frictional force is negligible all fall with the same constant acceleration is so familiar that its significance is easily overlooked. It ought to be regarded as truly remarkable. Consider a number of different weights. On each the pull of gravity is different. Yet all fall with precisely the same acceleration. How does it come about that a body that experiences twice the gravitational force of another also has exactly twice the inertial mass so that both move in exactly the same way? Our purpose here is not to explain this coincidence, but only to re-emphasize that there is no obvious reason for mass and the magnitude of a gravitational force to be proportional. No other fundamental force in nature is related simply to mass. Proton and positron, for example, experience the same electric force despite a large difference in their mass. Proton and neutron, with nearly the same mass, feel very different electric and magnetic forces.

Force proportional to mass, a unique feature of gravity

The puzzle of the proportionality of weight to mass came into prominence in science in the latter part of the nineteenth century. In particular, the question arose as to whether the ratio of weight to mass at a particular spot on the earth is truly constant or varies slightly from one substance to another. In order to facilitate discussion of this question and to highlight the puzzle, a new concept was introduced—gravitational mass. The meaning of gravitational mass can best be made clear by considering a hypothetical experiment. Imagine that you are in a laboratory equipped to measure force and velocity and acceleration and you have a number of different objects at your disposal, including a standard kilogram. Since the experiment is hypothetical, imagine too that you have available a perfectly flat, frictionless horizontal surface upon which to slide the objects. A known horizontal force (*not* a gravitational force) is applied to each object in turn, and the acceleration is measured. By virtue of Newton's second law, the force and acceleration provide a measurement of the inertial mass m_I (the subscript I is introduced to emphasize that it is inertial mass being measured):

Operational definition of inertial mass, using motion

$$m_I = \frac{F}{a} . \tag{7.61}$$

Having defined and measured the inertial mass of each object by means of experiments independent of gravity, you turn attention to the nature of the gravitational force acting on each of your test objects. Here a fundamental postulate enters. Its truth will be tested by further experiments. The postulate is this: The magnitude of gravitational force experienced by a particular object depends on only one property of that object, a scalar quantity. This is obviously a powerful postulate about the simplicity of nature. It says that a single number is enough to reveal how much gravitational force a body will experience, regardless of its shape or color or chemical composition and regardless of whether it contains liquid, solid, or gas. To this numerical quantity the name gravitational mass is given. It is to be regarded as an intrinsic property of a body, completely analogous to electric charge. Whereas charge measures the electric interaction strength of a body, gravitational mass measures the gravitational interaction strength. Indeed gravitational charge might be a better name for the quantity.

A significant analogy: gravitational mass and electric charge

According to the postulate, the gravitational force experienced by any one of the test objects should be proportional to its gravitational mass, m_G. The mathematical expression of the postulate is

$$F_{\text{grav}} = g m_G , \tag{7.62}$$

where g is a constant of proportionality (at a given location). The new concept, gravitational mass, needs a unit of measurement and a standard for this unit. As a matter of convenience, you may decide to measure gravitational mass in the same unit as inertial mass, the kilogram, and to adopt as its standard the same one-kilogram object. Having done this, you proceed to weigh each of your test objects: that is, you determine by a static measurement the gravitational force on each. The weight of the standard kilogram determines the constant g through the equation

$$g = \frac{F_{\text{grav}}}{1\ \text{kg}}.$$

The weight of any other object determines its gravitational mass through the equation

Operational definition of gravitational mass, using static force measurement

$$m_G = \frac{F_{\text{grav}}}{g}. \tag{7.63}$$

To review: Using an experiment *with* motion but *without* gravity, you have measured the inertial masses of a number of objects. Using an experiment *without* motion but *with* gravity, you have measured the gravitational masses of the same set of objects. Despite the confusing facts that both quantities are measured in the same unit and referred to the same standard, it is clear that their operational definitions are entirely distinct. Therefore, if you discover the equality of inertial and gravitational mass for every different object,

$$m_I = m_G, \tag{7.64}$$

you have discovered a very significant fact about nature, by no means a mere matter of definition.

The success of Newton's law of gravitation in describing planetary motion implies that gravitational and inertial mass are equal. The constant acceleration of all freely falling objects implies the same thing. However, the verification of their equality to an extremely high degree of precision—now to within three parts in 10^{11}—rests on more specialized experiments, first carried out by Roland von Eötvös in the 1880s and refined by Robert Dicke of Princeton University in the early 1960s. Dicke sought to discover whether or not different materials experience identical gravitational acceleration toward the sun. His sophisticated experiment, although complex in execution, is simple in principle.* Two weights made of different material are suspended from a long fiber in such a way that if one of the weights were more strongly accelerated toward the sun than another, the fiber would twist (Figure 7.13). Because of the rotation of the earth, the sun's pull would twist the fiber (if at all) first in one direction, then in the other direction. Such a daily alternation of the direction of twist would be easier to detect than a twist in only one direction, and it could be attributed with reasonable certainty to the action of the sun rather than to unknown perturbing influences. Except for inessential complications produced by the rotation of the earth and the pull of the moon, the suspended weights are, in effect, engaged in "free fall" toward the sun (along with the rest of the apparatus, the experimenter, and the earth itself). The acceleration of each toward the sun is

* For a good account of experiments to test the equality of gravitational and inertial mass, see Robert Dicke, "The Eötvös Experiment," *Scientific American*, December, 1961.

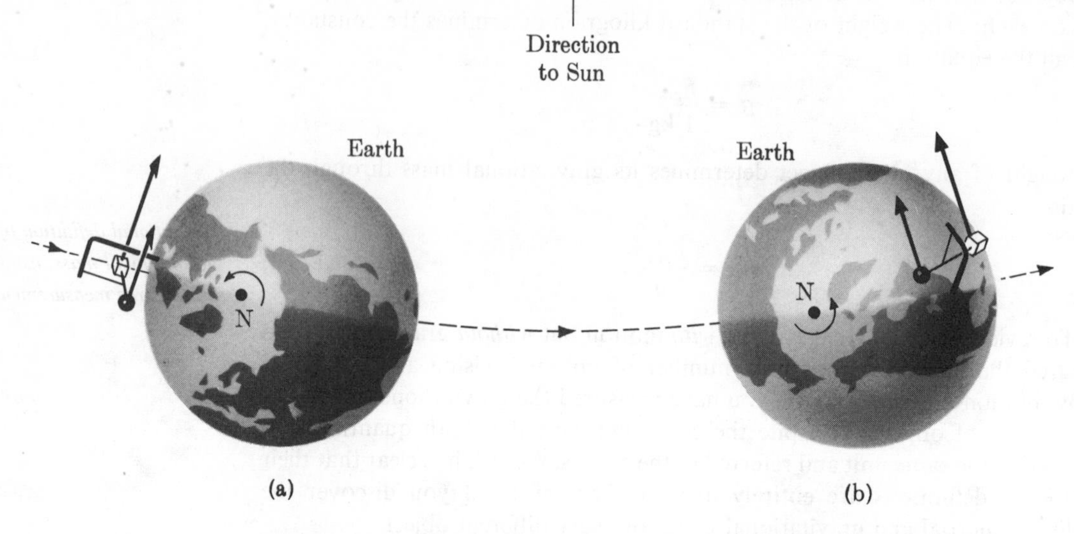

FIGURE 7.13 The Dicke experiment sought to discover a difference in the gravitational acceleration toward the sun of two objects made of different material, here shown for convenience as a cube and a sphere. Both are suspended from a long fiber. If the two objects are not equally accelerated toward the sun, the fiber will be twisted, one direction at sunset (a), the opposite direction at sunrise (b). The diagrams show a hypothetical greater acceleration of the cube than of the sphere. In fact, no differences were detected.

proportional to the ratio of its gravitational mass to its inertial mass:

Equal gravitational acceleration for two bodies implies $\left(\frac{m_G}{m_I}\right)_1 = \left(\frac{m_G}{m_I}\right)_2$

$$a \sim \frac{m_G}{m_I}.$$

To a remarkable degree of precision, Dicke verified the equality of accelerations and therefore the equality of the ratios of gravitational mass to inertial mass for gold, aluminum, and other substances.

Throughout most of the rest of this text, we shall accept the equivalence of inertial and gravitational mass as an established fact of nature and, accordingly, refer simply to "mass," designating either or both kinds of mass by the same symbol m.

7.8 Motion in two dimensions

Many of the most important examples of motion in nature take place in two dimensions: the collision and interaction of a pair of elementary particles, the deflection of an electron by a nucleus, the orbit of a satellite relative to the earth, the orbit of the earth about the sun. For each of these kinds of motion, the trajectories are traced out in an imaginary plane in space because there are no forces at work able to deflect the particles or objects out of the plane of their motion. The general principles governing motion in three dimensions are no different from the principles for two-dimensional motion. However, they are more simply illustrated for two-dimensional motion, which is all that we shall consider. It is mainly the vector character of Newton's second law that accounts for interesting features of multidimensional motion not found in one-dimensional motion.

UNIFORM CIRCULAR MOTION

The kinematics of uniform circular motion was considered in Section 6.10. There we discovered that the acceleration of an object moving in a circle at constant speed is directed toward the center of the circle and has a magnitude given by

$$a = \frac{v^2}{r} . \tag{7.65}$$

Newton's second law, $\mathbf{F} = m\mathbf{a}$, reveals at once what force is required to produce such motion. Since the magnitude of the acceleration is a constant, the magnitude of the force is also a constant, given by

$$F = \frac{mv^2}{r} . \tag{7.66}$$

Centripetal force

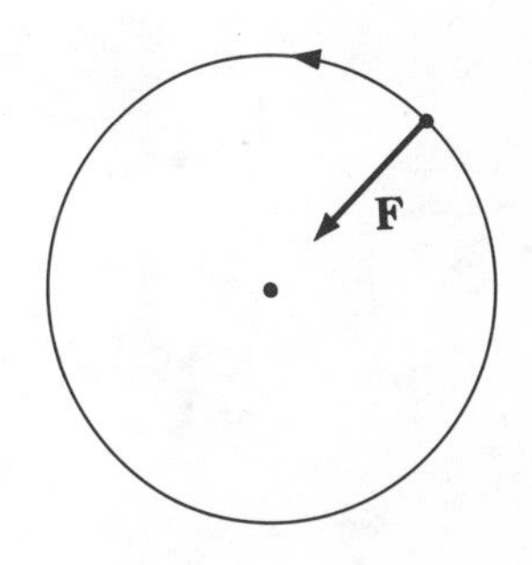

Moreover, since force and acceleration are vector quantities, both with the same direction, the force must be directed toward the center of the circle. Such a force, directed always toward a fixed center, is called a central force. It is also often known by Newton's original name for it, a centripetal (center-seeking) force.* For a stone swung on the end of a rope, the centripetal force is transmitted by the rope. Without the help of a material intermediary, gravitation and electricity act across empty space to produce central forces. The moon is held in an almost circular orbit by a force directed toward the center of the earth. The electron in a hydrogen atom experiences a centripetal force toward the central proton in the atom.

■ EXAMPLE 1: In a road race, a sports car must make a turn with a radius of curvature of 50 m on a flat road surface. If the greatest force that the road can exert on the tires before they skid is 40 percent of the weight of the car, what is the maximum speed at which the car can execute the turn without skidding? We set the force of the road on the tires, $0.4mg$, equal to the centripetal force, and write

$$0.4mg = \frac{mv^2}{r} .$$

The speed v is then given by

$$v = \sqrt{0.4gr}.$$

---

* Motion in a central field of force need not be circular.

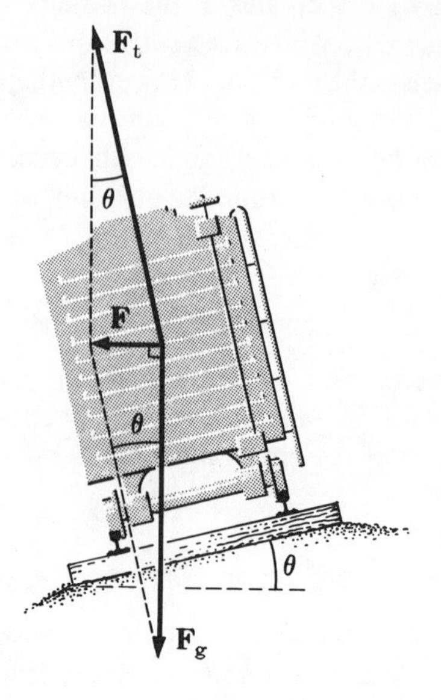

FIGURE 7.14 As a freight car rounds a curve, the gravitational force on it, $\mathbf{F}_g$, is directed vertically downward and has known magnitude. The total force **F** is determined by the acceleration: It is horizontal and has magnitude mv^2/r. The force exerted by the tracks, $\mathbf{F}_t$, is equal to $\mathbf{F} - \mathbf{F}_g$ and must have the direction shown, regardless of the bank. The bank is correct if $\mathbf{F}_t$ acts perpendicular to the roadbed.

Numerically, it is

$$v = \sqrt{0.4 \times 9.8 \text{ m/sec}^2 \times 50 \text{ m}}$$
$$= 14.0 \text{ m/sec}.$$

This is about 31 mile/hr. ■

■ EXAMPLE 2: A section of railroad track lies along the arc of a circle of radius 500 m. Trains round this curve at an average speed of 80 km/hr (about 50 mile/hr). At what angle should the track be banked? The total force on a train car is

$$\mathbf{F} = \mathbf{F}_g + \mathbf{F}_t, \tag{7.67}$$

A reminder: Total force acts in the direction of the acceleration

where $\mathbf{F}_g$ is the gravitational force and where $\mathbf{F}_t$ is the force exerted by the tracks on the car (see Figure 7.14). Since the car is executing a circular arc in a horizontal plane, the total force **F** acting on it must be directed horizontally and have magnitude

$$F = \frac{mv^2}{r}.$$

The principle of banking: $\mathbf{F}_t$ has no sidewise component on the vehicle

The force $\mathbf{F}_g$ is of course directed downward, with magnitude mg. The force $\mathbf{F}_t$ is therefore equal to $\mathbf{F} - \mathbf{F}_g$, whatever is needed to satisfy Equation 7.67. If the track is properly banked, $\mathbf{F}_t$ will act perpendicular to the roadbed, as shown in Figure 7.14. The geometry of the figure makes clear that in this case, magnitudes are related by

$$F_t = \frac{F_g}{\cos\theta}$$

and

$$F = F_g \tan\theta, \tag{7.68}$$

where θ is the angle of bank. Putting $F = mv^2/r$ and $F_g = mg$, we get

$$\tan\theta = \frac{v^2}{rg}. \tag{7.69}$$

The correct angle of bank

For this example, including an appropriate conversion factor,

$$\tan\theta = \frac{(80 \times 10^3 \text{ m/hr})^2}{500 \text{ m} \times 9.80 \text{ m/sec}^2 \times (3600 \text{ sec/hr})^2}$$

$$= 0.101.$$

The angle is $\theta = 0.101$ radian $= 5.77$ deg. ■

■ EXAMPLE 3: What are the orbital speed and the period of a low-altitude earth satellite? If the orbit is circular, we can equate the centripetal force to the gravitational force:

$$\frac{mv^2}{r} = mg. \tag{7.70}$$

The mass of the satellite is therefore irrelevant. Its inward acceleration is the same regardless of its mass. This result is for a very good reason the same as that discovered by Galileo for freely falling objects. A satellite *is* a freely falling object, in fact more precisely so than a thrown baseball or a falling stone since air resistance for a satellite is negligible, or very nearly so. From Equation 7.70, the satellite speed is given by

$$v = \sqrt{rg}. \tag{7.71}$$

Although it is unrealistic, a hypothetical satellite skimming the surface of the earth is an interesting limit for calculational purposes. Then r is the radius of the earth, 6.37×10^6 m, and g is 9.80 m/sec^2. In this limit,

$$v = \sqrt{6.37 \times 10^6 \text{ m} \times 9.80 \text{ m/sec}^2}$$

$$= 7.90 \times 10^3 \text{ m/sec}, \tag{7.72}$$

or about 24 times the speed of sound in air, which is 3.3×10^2 m/sec. The period of rotation is the circumference divided by the speed,

$$T = \frac{2\pi r}{v}. \tag{7.73}$$

Alternatively, with the help of Equation 7.71, this can be written

$$T = 2\pi\sqrt{\frac{r}{g}}. \tag{7.74}$$

Period of a hypothetical satellite at zero altitude

Substituting the above values for r and g in this equation yields

$$T = 5{,}066 \text{ sec} = 84 \text{ min}. \tag{7.75}$$

It is left as a problem to show that a satellite at an altitude of 100 miles has a speed only 100 m/sec less and a period only 4 min greater than these calculated figures. An hour and a half is a good round number for the period of a satellite within a few hundred miles of the earth's surface. ■

■ EXAMPLE 4: A communications satellite, at an altitude of 22,240 miles above the equator, seems to hover over a fixed point on the earth. It completes one circular orbit in 23 hr, 56 min.* By what factor is the force of gravity decreased in going up from the surface of the earth to this altitude? To find out, we need only calculate the acceleration of the satellite and compare it with g, the acceleration of gravity at the earth's surface. The satellite's acceleration is

$$a = \frac{v^2}{r} = \omega^2 r = \left(\frac{2\pi}{T}\right)^2 r. \tag{7.76}$$

The last form is most useful here. The distance of the satellite from the center of the earth is

$$r = 26{,}240 \text{ miles} = 4.216 \times 10^7 \text{ m};$$

its period is

$$T = 8.616 \times 10^4 \text{ sec}.$$

These numbers in Equation 7.76 give

$a = g/43.7$ when $r = 6.61R$

$$a = 0.2242 \text{ m/sec}^2, \tag{7.77}$$

about 44 times less than g at the earth's surface. It was a similar calculation for the moon's orbit that led Newton to the inverse-square law of gravitational attraction ($F \sim 1/r^2$). ■

PROJECTILE MOTION

Consider a pebble projected off the edge of a horizontal table, as shown in Figure 7.15. If we call the horizontal direct the x direction and the vertically upward direction the z direction, the relevant equations of motion are

$$F_x = ma_x, \tag{7.78}$$

$$F_z = ma_z. \tag{7.79}$$

After the pebble leaves the edge of the table, $F_x = 0$ and $F_z = -mg$ if air resistance is negligible. Its path is therefore constrained by the equations

Free motion in the x direction

$$0 = m\frac{d^2x}{dt^2}, \tag{7.80}$$

Uniform acceleration in the z direction

$$-mg = m\frac{d^2z}{dt^2}. \tag{7.81}$$

They may be treated separately. The first states that there is no horizontal acceleration. This means, of course, that the horizontal component of velocity, v_x, is a constant. If the origin of the coordinate system is located at the point

* This time duration is called a sidereal day. It is the time required for the earth to complete one revolution in an inertial frame of reference (relative to the stars). A solar day is slightly longer. An extra 4 min is required for a point on the earth to return to the same position relative to the sun. Figure 7.13 helps to explain why this is so.

at which the pebble leaves the table (Figure 7.15) and if it leaves the table at $t = 0$, the subsequent increase of the x coordinate is given simply by

$$x = v_x t. \tag{7.82}$$

At $t = 0$, we have $z = 0$ and $dz/dt = 0$ (zero vertical component of velocity). Therefore, the z coordinate is described by the simple special case of Equation 5.51,

$$z = -\tfrac{1}{2}gt^2, \tag{7.83}$$

which is negative since the acceleration is in the negative z direction. Thus the horizontal component of the motion is the same as if there were no fall, and the vertical part of the motion is the same as if the fall were straight down.

Equations 7.82 and 7.83 may be combined to yield an equation for the trajectory of the pebble. For this purpose, the time must be eliminated. According to Equation 7.82, $t = x/v_x$; substituting x/v_x for t in Equation 7.83 gives the trajectory equation,

$$z = -\left(\frac{g}{2v_x^{\,2}}\right) x^2. \tag{7.84}$$

z(x), the trajectory function

This is the equation of a parabola. The combination in parentheses is a constant; therefore, z is proportional to x^2.

The analysis for projectile motion in general is the same in principle—constant horizontal component of velocity combined with constant vertical acceleration—but mathematically slightly more complex. The general x-coordinate equation is

$$x = x_0 + v_{x0}t. \tag{7.85}$$

The general z-coordinate equation is

$$z = z_0 + v_{z0}t - \tfrac{1}{2}gt^2. \tag{7.86}$$

Differentiation twice with respect to time shows at once that these expressions for x and z satisfy Newton's second law, given by Equations 7.80 and 7.81.

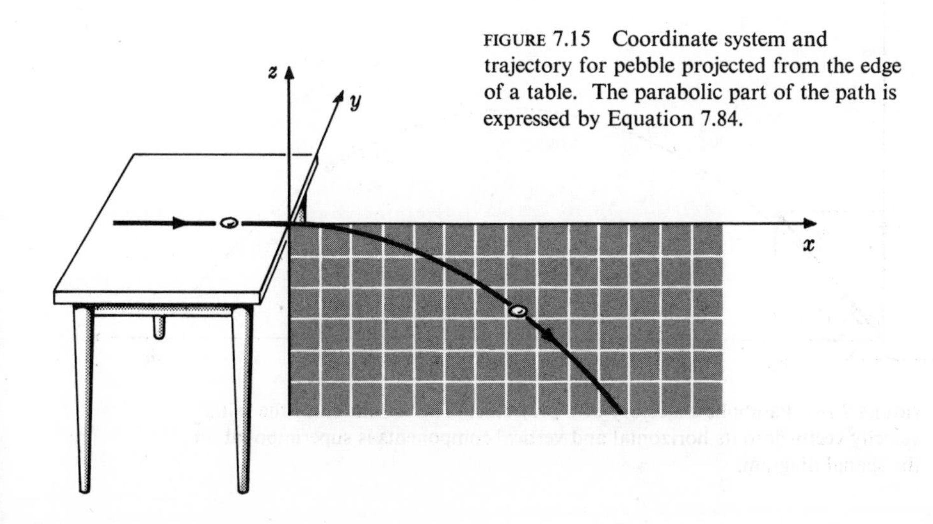

FIGURE 7.15 Coordinate system and trajectory for pebble projected from the edge of a table. The parabolic part of the path is expressed by Equation 7.84.

■ EXAMPLE 5: A projectile is fired from the origin with initial velocity components v_{x0} and v_{z0} (Figure 7.16). What is the range of the projectile? We can set $t = 0$ at the moment of firing. Then $x_0 = z_0 = 0$, and the position coordinates as functions of time are

$$x(t) = v_{x0}t, \tag{7.87}$$

$$z(t) = v_{z0}t - \tfrac{1}{2}gt^2. \tag{7.88}$$

At the impact point, $x = R$ and $z = 0$. Therefore, to get the flight time, set $z = 0$ in Equation 7.88. This gives two solutions for t: $t = 0$, which is the initial time, and

$$t = \frac{2v_{z0}}{g}, \tag{7.89}$$

which is the desired flight time. Substituting this time in Equation 7.87 gives, for the range,

Range of a projectile

$$R = \frac{2v_{x0}v_{z0}}{g}. \tag{7.90}$$

■

■ EXAMPLE 6: If there were no air friction, what would be the maximum range of a projectile fired at Mach 1 (the speed of sound)? At what angle should it be fired to achieve this range? These two questions are best answered in reverse order. As Figure 7.16 makes clear,

$$v_{x0} = v_0 \cos \theta,$$

$$v_{z0} = v_0 \sin \theta.$$

Therefore, the expression for the range, Equation 7.90, can be rewritten as

$$R = \frac{2v_0^2}{g} \cos \theta \sin \theta. \tag{7.91}$$

The mathematics can be simplified by using the trigonometric formula

$$\sin 2\theta = 2 \sin \theta \cos \theta.$$

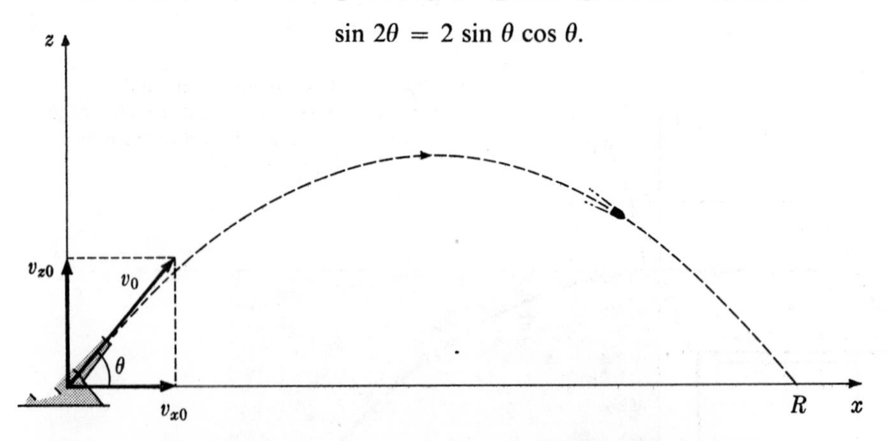

FIGURE 7.16 Parabolic trajectory of a projectile. The resolution of the initial velocity vector into its horizontal and vertical components is superimposed on the spatial diagram.

Then the range is given by

$$R = \left(\frac{v_0^2}{g}\right) \sin 2\theta. \tag{7.92}$$

Another expression for the range

In this example, v_0 and g are constant. The range is therefore maximum when $\sin 2\theta$ is maximum. Since $\sin x$ peaks at $x = \pi/2$, the condition of maximum range is $2\theta = \pi/2$, or

$$\theta = \frac{\pi}{4} = 45 \text{ deg}. \tag{7.93}$$

Alternatively, we could apply the extremal condition, $dR/d\theta = 0$, or

$$\left(\frac{2v_0^2}{g}\right) \cos 2\theta = 0.$$

Within the physical range of interest ($\theta = 0$ to $\pi/2$), this equation has one solution, at $\theta = \pi/4$.

Substituting $\pi/4$ for θ in Equation 7.92 gives, for the maximum range,

$$R_{\text{max}} = \frac{v_0^2}{g}. \tag{7.94}$$

For a projectile fired at Mach 1, $v_0 = 330$ m/sec. Then

$$R_{\text{max}} = \frac{(3.3 \times 10^2)^2 \text{ m}^2/\text{sec}^2}{9.8 \text{ m/sec}^2}$$

$$= 1.11 \times 10^4 \text{ m} = 11.1 \text{ km}, \tag{7.95}$$

or about 7 miles. This is, of course, a considerable overestimate of the *actual* range of such a projectile because we have ignored air friction. ■

★7.9 The motion of charged particles in electric and magnetic fields

The laws of electric and magnetic force will be introduced and discussed in detail in Chapters 15 and 16. However, at this point, we may take these laws as given facts and use them with Newton's second law to study the motion of charged particles.

In an electric field **E** (a vector quantity), a particle of charge q experiences a force given by

$$\mathbf{F}_e = q\mathbf{E}. \tag{7.96}$$

The law of electric force

(With **E** expressed in volt/m and q in coulombs, $\mathbf{F}_e$ is given in newtons.) Newton's second law for the particle is therefore

$$m\mathbf{a} = q\mathbf{E}. \tag{7.97}$$

For the special case of a constant electric field (Figure 7.17), the particle experiences constant acceleration, parallel to the electric field and of magnitude $|q|E/m$. Mathematically, its motion is the same as the motion of a projectile near the earth's surface.

Constant **E** *means constant* **a**

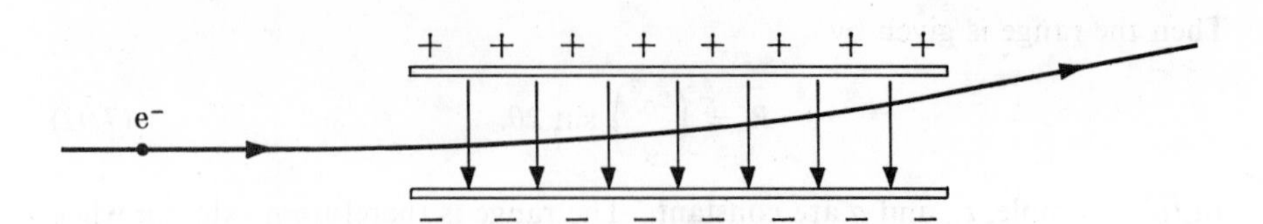

FIGURE 7.17 In a region of constant electric field, such as exists between closely spaced parallel charged plates, a charged particle experiences constant acceleration and follows a parabolic trajectory. In this illustration, the particle is negatively charged. The vertical arrows symbolize the electric field E.

■ EXAMPLE 1: In a television tube, an electron is accelerated in an electric field of magnitude 2,000 V/cm (2×10^5 V/m). What is its acceleration? What speed does it acquire in a distance of 1 cm, starting from rest? Using the electron's charge-to-mass ratio from Appendix 3, we have, for the acceleration,

$$\begin{aligned} a &= \left(\frac{e}{m}\right) E \\ &= (1.76 \times 10^{11}\ \mathrm{C/kg})(2 \times 10^5\ \mathrm{V/m}) \\ &= 3.52 \times 10^{16}\ \mathrm{m/sec^2}. \end{aligned}$$

This is, of course, enormous compared with any accelerations known for macroscopic objects. To get the speed of the electron in its "free fall" of 1 cm, we use Equation 5.54:

$$\begin{aligned} v &= \sqrt{2ax} \\ &= \sqrt{2 \times 3.52 \times 10^{16}\ \mathrm{m/sec^2} \times 10^{-2}\ \mathrm{m}} \\ &= 2.65 \times 10^7\ \mathrm{m/sec}, \end{aligned}$$

about 9 percent of the speed of light. ■

The law of magnetic force

The law of magnetic force involves a vector product. It is

$$\mathbf{F}_{\mathrm{m}} = q\mathbf{v} \times \mathbf{B}, \tag{7.98}$$

where q is the charge of the particle, $\mathbf{v}$ is its velocity, and $\mathbf{B}$ is the magnetic field. (With q in coulombs, $\mathbf{v}$ in m/sec, and $\mathbf{B}$ in teslas (T), $\mathbf{F}_{\mathrm{m}}$ is in newtons.) The law of motion for a charged particle in a magnetic field is therefore

$$m\mathbf{a} = q\mathbf{v} \times \mathbf{B}, \tag{7.99}$$

which can also be written

Acceleration in a magnetic field

$$\frac{d\mathbf{v}}{dt} = \frac{q}{m}\mathbf{v} \times \mathbf{B}. \tag{7.100}$$

Although this equation looks complicated (and, in some applications, may indeed be complicated), it has one very simple and interesting implication: the *speed* of a charged particle in any magnetic field remains constant. The proof of this fact makes use of some simple vector operations. First, form the scalar product of both sides of Equation 7.100 with the velocity vector $\mathbf{v}$:

$$\mathbf{v} \cdot \frac{d\mathbf{v}}{dt} = \frac{q}{m} \mathbf{v} \cdot (\mathbf{v} \times \mathbf{B}).$$

Since $\mathbf{v} \times \mathbf{B}$ is a vector perpendicular to $\mathbf{v}$, the scalar product on the right side is zero (regardless of the magnitude or direction of $\mathbf{B}$). Therefore,

$$\mathbf{v} \cdot \frac{d\mathbf{v}}{dt} = 0. \qquad (7.101)$$

Now, making use of Equation 6.102 (and the fact that the scalar product is commutative), we note that

$$\frac{d}{dt}(\mathbf{v} \cdot \mathbf{v}) = \frac{d\mathbf{v}}{dt} \cdot \mathbf{v} + \mathbf{v} \cdot \frac{d\mathbf{v}}{dt}$$

$$= 2\mathbf{v} \cdot \frac{d\mathbf{v}}{dt}. \qquad (7.102)$$

Comparison of Equations 7.101 and 7.102 shows that the time derivative of $\mathbf{v} \cdot \mathbf{v}$ must vanish. However, $\mathbf{v} \cdot \mathbf{v}$ is the same as v^2. Therefore,

A simple result: A particle in a magnetic field has constant speed

$$\frac{d}{dt}(v^2) = 0. \qquad (7.103)$$

This means that the magnitude of $\mathbf{v}$ is not changing in time—that is, $v = constant$. Looking back to Equation 7.98, we can see the "reason" for this result. Since the magnetic force on a moving charged particle is always perpendicular to the velocity of the particle, the force *turns* the velocity vector without changing its magnitude. In order to change the speed of a particle, a force component must act parallel (or antiparallel) to the velocity vector.

MOTION IN A CONSTANT MAGNETIC FIELD

According to Equation 7.98 (and with the help of Equations 6.64–6.66), the components of the magnetic force on a particle are

$$F_{mx} = q(v_yB_z - v_zB_y), \qquad (7.104)$$

$$F_{my} = q(v_zB_x - v_xB_z), \qquad (7.105)$$

$$F_{mz} = q(v_xB_y - v_yB_x). \qquad (7.106)$$

If the magnetic field is constant and if the z axis is chosen to lie along $\mathbf{B}$, a great simplification results. Then $B_z = B$, a constant; and $B_x = B_y = 0$. The components of the equation of motion, $d\mathbf{v}/dt = \mathbf{F}_m/m$, become

Equations of motion in a constant magnetic field

$$\frac{dv_x}{dt} = \left(\frac{qB}{m}\right) v_y, \qquad (7.107)$$

$$\frac{dv_y}{dt} = -\left(\frac{qB}{m}\right) v_x, \qquad (7.108)$$

$$\frac{dv_z}{dt} = 0. \qquad (7.109)$$

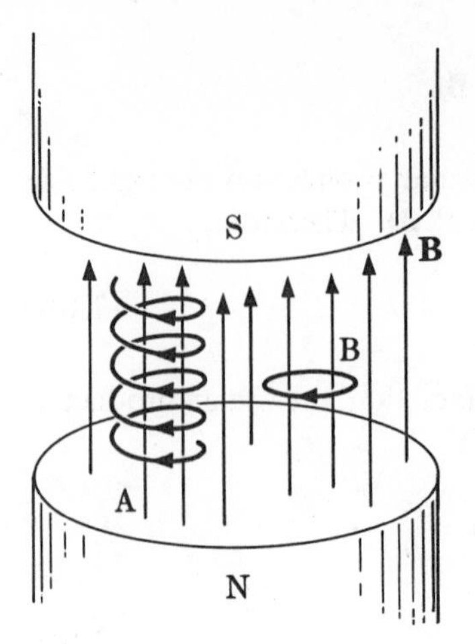

FIGURE 7.18 In a region of constant magnetic field, such as exists between the pole faces of a large magnet, a charged particle follows a helical path (A). If its velocity component along the field is initially zero, it remains zero and the trajectory is a circle (B). In this illustration, the particle is positively charged. The vertical arrows symbolize the magnetic field **B**.

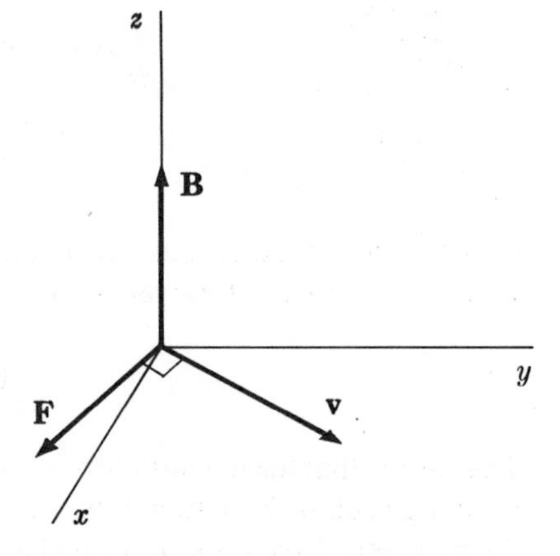

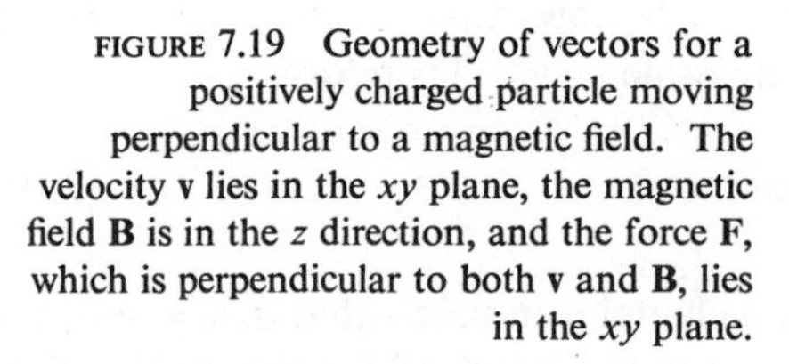

FIGURE 7.19 Geometry of vectors for a positively charged particle moving perpendicular to a magnetic field. The velocity **v** lies in the *xy* plane, the magnetic field **B** is in the *z* direction, and the force **F**, which is perpendicular to both **v** and **B**, lies in the *xy* plane.

It is an interesting and challenging problem (see Problem 7.27) to solve these equations and prove that the particle orbit is a helix—that is, circular motion in the *xy* plane superimposed on uniform-velocity motion in the *z* direction. Such an orbit is shown in Figure 7.18. Here we specialize to motion in a plane, and proceed in a somewhat different way.

From Equation 7.109, it follows at once that

Motion parallel to the field is free motion

$$v_z = \text{constant}. \tag{7.110}$$

Therefore, if v_z is initially zero, it remains zero, and the motion is confined to the *xy* plane. From Equation 7.98, the magnitude of the force is $F_m = qvB \sin\theta$, but since **v** lies in the *xy* plane, it is perpendicular to **B** (Figure 7.19); therefore $\sin\theta = 1$, and

$\mathbf{F}_m$ has constant magnitude

$$F_m = qvB. \tag{7.111}$$

Since the speed v is constant, this force magnitude is also constant. This means that the acceleration is constant in magnitude. What we have, then, is a particle of constant speed experiencing constant acceleration, with the acceleration vector perpendicular to the velocity vector. This is exactly the condition of uniform circular motion. It is left as an exercise to show that the equations of uniform circular motion (see Equation 6.86),

$\mathbf{F}_m$ is perpendicular to $\mathbf{v}$

Motion perpendicular to the field is circular motion

$$x = r\cos\omega t \tag{7.112}$$

and

$$y = r\sin\omega t, \tag{7.113}$$

do satisfy Equations 7.107 and 7.108. Accepting this fact, we can find the radius of curvature from the condition

$$F_m = \frac{mv^2}{r}.$$

Replacing F_m on the left side by qvB (from Equation 7.111), we get

$$r = \frac{mv}{qB}. \qquad (7.114)$$

The radius of curvature if motion confined to a plane

The angular frequency of the particle is $\omega = v/r$, or

$$\omega = \left(\frac{q}{m}\right) B. \qquad (7.115)$$

The cyclotron frequency, measured in radian/sec

This is called the *cyclotron frequency* (note that it is actually an angular frequency). Interestingly, it depends only on the charge-to-mass ratio of the particle and on the magnetic-field strength, not on the particle speed or its radius of curvature.

■ EXAMPLE 2: What is the cyclotron frequency of a proton in the Van Allen radiation belt at a place where the magnetic field is $B = 10^{-5}$ T? What is its radius of curvature if its energy is 1 MeV? With the charge-to-mass ratio of the proton taken from Appendix 3, the cyclotron frequency is

$$= 9.58 \times 10^7 \text{ C/kg} \times 10^{-5} \text{ T}$$
$$= 958 \text{ radian/sec}. \qquad (7.116)$$

This corresponds to about 150 revolutions/sec. The speed of the proton (from $K = \frac{1}{2}mv^2$) is

$$v = \sqrt{\frac{2K}{m}}$$
$$= \sqrt{\frac{2 \times 10^6 \text{ eV} \times 1.60 \times 10^{-19} \text{ J/eV}}{1.67 \times 10^{-27} \text{ kg}}}$$
$$= 1.38 \times 10^7 \text{ m/sec}.$$

The radius of curvature of its trajectory (if it moves in a plane perpendicular to **B**) can be obtained from Equation 7.114, or from $r = v/\omega$. Using the latter expression, we get

$$r = \frac{1.38 \times 10^7 \text{ m/sec}}{0.958 \times 10^3 \text{ sec}^{-1}}$$
$$= 1.44 \times 10^4 \text{ m} = 14.4 \text{ km}, \qquad (7.117)$$

or about 9 miles. In a typical cyclotron, the magnetic-field strength is about 1 T instead of 10^{-5} T. Thus a 1-MeV proton in a cyclotron has an angular frequency about 10^5 times greater than that calculated in Equation 7.116 and a radius of curvature 10^5 times smaller than the value of Equation 7.117. ■

For the general case of helical motion in a constant magnetic field, the preceding theory requires only slight modification. Then the particle velocity can be written

$$\mathbf{v} = \mathbf{v}_{\parallel} + \mathbf{v}_{\perp}, \tag{7.118}$$

where $\mathbf{v}_{\parallel}$ is parallel to $\mathbf{B}$ and where $\mathbf{v}_{\perp}$ is in the plane perpendicular to $\mathbf{B}$. The part $\mathbf{v}_{\parallel}$ remains constant. The part $\mathbf{v}_{\perp}$ rotates uniformly at the cyclotron angular frequency. Thus Equation 7.115 is unchanged. In Equation 7.114, v is replaced by $v_{\perp}$ so that the radius of curvature is

The radius of curvature in general

$$r = \frac{mv_{\perp}}{qB}. \tag{7.119}$$

7.10 Frames of reference

To be meaningful, every law of nature must be defined with respect to some frame of reference or, more often, to some set of frames of reference. In classical mechanics, the frame of reference is purely a *spatial* frame of reference; time is assumed to be independent of the frame of reference. The frame may be moving and thereby itself depend on time, but the time coordinate is supposed to be uninfluenced by the motion. To make this point clear, imagine yourself an observer in an airplane enroute to San Francisco. In your frame of reference, San Francisco is moving toward you—its position vector is changing with time. To an observer in New York, San Francisco is motionless a fixed distance away. The concept of time is important in distinguishing the two frames of reference. However, the *measurement* of time is presumed to be the same for you and the ground-based observer. Careful comparison would reveal no detectable difference in the two scales of time. In this century we have learned that for two frames of reference with sufficiently high relative speed, there is in fact a perceptible difference in time scales as well as a difference in position measurements. This circumstance, explained by the special theory of relativity, has led to the idea of a four-dimensional frame of reference, making both space and time measurements depend on the state of motion of the observer. The new insight is of practical importance, however, only at speeds near the speed of light. Even in the frame of reference of an orbital astronaut, the measurement of time is so little different from on earth that the difference may be ignored. In the domain of nature where classical mechanics is valid, time may be regarded as absolute, and only spatial measurement is dependent on the frame of reference.

In classical physics, time is absolute

Newton, and Galileo before him, realized that the laws of motion are the same in all inertial frames of reference. This does *not* mean that all aspects of the description of motion are the same in different reference frames. To see the effect of the frame of reference on the description of motion, we can study a particular example. Observer A [Figure 7.20(a)] drops a ball from rest at the origin of his coordinate system. If friction is negligible, the ball falls with constant acceleration g, and he describes its motion by the equation

Motion in frame A

$$x_{\mathrm{A}} = \tfrac{1}{2}gt^2. \tag{7.120}$$

Three other observers examine the same motion. Observer B [Figure 7.20(b)]

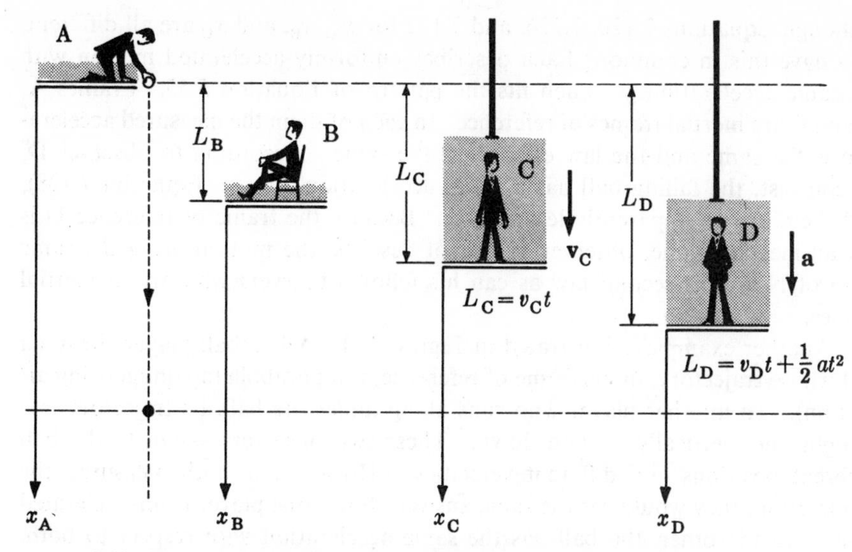

FIGURE 7.20 Frames of reference. The motion of a falling ball is studied by four observers. Observers A, B, and C are in inertial frames of reference. Observer D is in an accelerated frame of reference.

uses a coordinate system whose origin is a fixed distance L_B below the origin of reference frame A. Observer C, in an elevator descending with constant speed v_C [Figure 7.20(c)], describes the motion relative to his origin, whose distance L_C from the starting point of the motion increases uniformly with time:

$$L_C = v_C t. \tag{7.121}$$

Finally, observer D [Figure 7.20(d)], riding a downward accelerated elevator, measures the position of the falling ball relative to the floor level of his elevator. This elevator floor, with downward acceleration a, increases its distance L_D from the origin of frame A according to the equation

$$L_D = v_D t + \tfrac{1}{2}at^2. \tag{7.122}$$

(For simplicity, we assume that both elevator floors passed the origin of reference frame A at $t = 0$.) Now we can write expressions for x_B, x_C, and x_D, the position coordinates of the falling ball in the other three frames of reference. Each is equal to the total distance of fall, x_A, less the distance to the new origin (L_B, L_C, or L_D):

$$x_B = x_A - L_B, \tag{7.123}$$

$$x_C = x_A - L_C, \tag{7.124}$$

$$x_D = x_A - L_D. \tag{7.125}$$

Substitutions from Equations 7.120–7.122 yield

$$x_B = -L_B + \tfrac{1}{2}gt^2, \tag{7.126}$$

Motion in other frames

$$x_C = -v_C t + \tfrac{1}{2}gt^2, \tag{7.127}$$

$$x_D = -v_D t + \tfrac{1}{2}(g - a)t^2. \tag{7.128}$$

A, B, and C are inertial frames

Although Equations 7.120, 7.126, and 7.127 for x_A, x_B, and x_C are all different, they have this in common: Each describes uniformly accelerated motion with the same acceleration g. Each fits the pattern of Equation 5.55. Frames A, B, and C are inertial frames of reference. In each of them the measured acceleration is the same and the law of force is the same. According to observer D, by contrast, the falling ball has a lesser acceleration, $g - a$ (Equation 7.128), and therefore an apparently lesser force. Because the frame of reference D is not an inertial frame, observer D cannot describe the motion using the same form of Newton's second law as can his fellow observers who are in inertial frames.

Another example is illustrated in Figure 7.21. A baseball player throws a ball whose trajectory, in his frame of reference, is a parabola (again ignoring air friction). To another player who runs along under the ball, its trajectory is a straight line, vertically up and down. These two observers assign to the ball different positions and different velocities. However, if both measured the acceleration, they would get the same answer. Since one player is unaccelerated relative to the other, the ball has the same acceleration with respect to both. Newton's second law too is valid for both.

Galilean relativity: Laws of mechanics are the same in inertial frames

The principle that the laws of mechanics are the same in all inertial frames of reference is called the principle of Galilean relativity. It is actually a principle of invariance, a statement that the laws governing motion are the *same* in different inertial frames of reference even though the description of a particular example of motion might be different in the different reference frames. The significance of Galilean relativity can perhaps best be emphasized by stating the principle in this way: If a particular example of motion is possible (that is, consistent with the laws of motion) in one inertial frame of reference, it is also possible in every other inertial frame of reference. This way of looking at Galilean relativity can be illustrated with the help of Figure 7.21. The motion of the ball as observed by player B is straight-line vertical motion. Although this is *not* the motion observed by player A, it is a possible motion that player A *could* produce in his own frame of reference if he wished to—simply by throwing the ball straight up. Similarly, the parabolic motion observed by player A is a motion that could be observed in player B's moving frame of reference if the ball were thrown differently. The class of all possible motions for observer A is precisely the same as the class of all possible motions for observer B. Therefore, they agree about the laws of motion.

An observational way to define the equivalence of inertial frames

An important consequence of the principle of Galilean relativity is that it is impossible to discover the existence of a truly stationary frame of references, a frame at rest with respect to a hypothetical ether filling all of space. Since all physical phenomena (at least mechanical phenomena) follow the same laws in a frame moving uniformly with respect to this supposed ether as in a frame at rest with respect to the ether, there is no mechanical experiment that can distinguish the two. Galileo himself used the principle to support the Copernican view of the solar system. He argued that the laws of motion would be no different on a moving earth than on a stationary earth, so one might as well adopt the simpler Copernican view of a moving earth in place of the far more complicated Ptolemaic view of a solar system built around a stationary earth. Before Galileo, almost all scientists and philosophers from Ptolemy onward had argued that the

Mechanical experiments cannot reveal a preferred rest frame

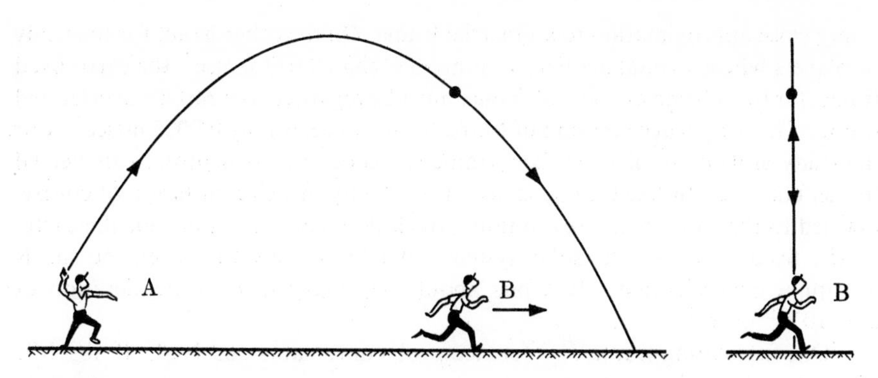

FIGURE 7.21 Description of motion in two inertial frames of reference. Observer A, at rest with respect to the ground, and observer B, running beneath the thrown baseball, assign to the ball different trajectories and different velocities, but they agree about its acceleration and about the law of force governing its motion.

straight vertical fall of a dropped stone demonstrated that the earth must be at rest. According to Galileo's principle of relativity, it proved no such thing.*

It cannot be said that we really know of the existence of any ideal inertial frame of reference since it is impossible to completely isolate an object from outside influences and find in what frame it is unaccelerated. However, the ideal can be almost attained. For certain purposes, it is useful to redefine an inertial frame as one in which an object is unaccelerated if it experiences only a gravitational force (rather than no force). Such a frame is provided by a freely falling laboratory, a spaceship coasting in orbit. Although in some ways it is the most fundamental inertial frame, the orbiting laboratory is special in two ways. First, it casts gravity in a favored role; objects in it respond only to forces *other* than gravity. Second, it is "local"; it acts as an inertial frame only over a region of space small enough so that the gravitational field can be regarded as constant. For motion in the small, the orbiting laboratory is a good inertial frame.

A practical guide for approximating an inertial frame

For describing motion on a larger scale, the earth itself is sometimes an adequate approximation to an inertial frame. A sun-fixed frame is better, and a frame in which the galaxy as a whole is unaccelerated is still better. A workable rule is this: If the acceleration of a particular frame of reference with respect to an ideal inertial frame is much smaller than the accelerations being studied, that frame of reference is a suitable approximation to an inertial frame. The surface of our earth has an average acceleration with respect to a sun-fixed frame of about 0.027 m/sec^2, less than 1 percent of the gravitational acceleration g. For any particular application, it must be decided whether this magnitude, 0.027 m/sec^2, is sufficiently small. If it is, an earth-fixed frame adequately approximates an inertial frame. Elementary particles experience accelerations enormously larger than 0.027 m/sec^2. For their study, the surface of the earth defines

* Einstein's principle of relativity is more general. It asserts that inertial frames of reference are equivalent for *all* the laws of nature, not just for mechanics; see Chapter 21.

a very close approximation to an inertial frame. On the other hand, for the study of planets whose orbital accelerations are less than 0.027 m/sec^2, the earth-fixed frame, far from being an inertial frame, must be regarded as a rapidly accelerated frame. The orbital acceleration of Mars, for instance, is only 0.0025 m/sec^2. For the study of its motion, a sun-fixed frame is a good approximation to an inertial frame, but an earth-fixed frame is not. It is exactly this circumstance, of course, that led to the Copernican revolution, which designated the sun, not the earth, as the fixed center of the solar system. We know now that even the sun is not quite unaccelerated. It orbits about the galaxy with an acceleration of 2×10^{-10} m/sec^2.

Why does absolute acceleration have meaning?

The distribution of matter and energy in the universe must play a role

One fascinating frame-of-reference problem remains unsolved. If absolute position and absolute velocity are without meaning in empty space, why does absolute acceleration seem to have a meaning? Another way to phrase the question is this: Why are inertial frames of reference a special preferred set? Think of two observers in empty space, one in an inertial frame of reference, one in an accelerated frame. In the inertial frame, an isolated object moves with constant velocity. In the accelerated frame, it does not; rather, it seems to be responding to forces. In fact, the laws of motion are very different in the two frames of reference; they are much simpler in the inertial frame. However, the inertial frame can be said to be accelerated *relative* to the accelerated frame, just as a stationary observer is accelerated relative to a falling object. Each frame is accelerated in exactly the same way relative to the other (equally but opposite). Why, then, is the inertial frame a privileged frame of reference? No satisfactory answer to this question has been found. It seems very likely that the distribution of matter spread throughout the universe must in some way be the material framework with respect to which absolute acceleration (but not absolute velocity) has a meaning. From this point of view, an inertial frame is one unaccelerated with respect to the material framework of the universe.

Consider our own rotating earth. In a frame of reference fixed with respect to the earth, the stars in the sky execute circular and therefore accelerated motion. This was the ancient view of the universe. In a frame fixed with respect to the sun, on the other hand, it is the earth's surface that executes circular accelerated motion; the acceleration of the stars is much less than in the earth-fixed frame. Evidently, nature prefers a frame of reference in which the stars have little or no average acceleration. Despite the impossibility of defining absolute motion or absolute rest, there does seem to be a material framework in the form of distant matter. It has been speculated that in an otherwise empty universe, a material object would possess no mass—that a body's resistance to acceleration, the property we call its mass, is a resistance to its being moved out of one of nature's preferred inertial frames of reference. Perhaps, continues the speculation, even the existence of measurable intervals of space and time depends upon the existence of matter and energy in the universe. According to this fascinating point of view, the properties and behavior of the elementary particles—nature's tiniest bits of matter—are influenced in an important way by by the bulk matter of distant galaxies. No problem in science is more intriguing than the search for links between the submicroscopic and cosmological domains of the physical world.

Summary of ideas and definitions

Newton's first law: A body on which no force acts (or an isolated body free of outside influences) moves with constant velocity.

Newton's second law: $\mathbf{F} = m\mathbf{a}$, the total force acting on a body is the product of its mass and its acceleration.

Newton's first and second laws are valid in inertial frames of reference.

An inertial frame is one in which any isolated body is unaccelerated.

The concept of force can be defined by means of a standard spring or by accelerating a standard mass.

The SI unit of force is the newton (N), equal to 1 kg m/sec^2. It is the force needed to give to a mass of 1 kg an acceleration of 1 m/sec^2.

The mass appearing in Newton's second law is inertial mass.

Gravitational mass, analogous to electric charge, measures the strength of gravitational force that a body experiences or exerts.

Experimentally, gravitational mass and inertial mass are equal.

Mass is a scalar quantity. In classical physics, mass is conserved.

Newton's second law can be expressed as a differential equation, either $m d\mathbf{v}/dt = \mathbf{F}$ or $m d^2\mathbf{r}/dt^2 = \mathbf{F}$. It is also called the law of motion.

An object experiencing the force law $F_x = -kx$ is called a harmonic oscillator. Applied to a spring, this force law is called Hooke's law. The constant k is the force constant.

A harmonic oscillator executes simple harmonic motion, described by $x = A \sin(\omega t + \varphi)$. Its period, independent of amplitude, is $T = 2\pi\sqrt{m/k}$.

Near the earth, the force of gravity is directed vertically downward and has magnitude $F_{\text{grav}} = mg$. The acceleration of gravity g varies slightly from place to place and is approximately 9.8 m/sec^2.

A pendulum, for small amplitude of vibration, is approximately a harmonic oscillator. Its period, independent of mass, is $T = 2\pi\sqrt{l/g}$, where l is its length.

Centripetal force, in uniform circular motion, is given by $F = mv^2/r$.

For a vehicle rounding a curve, the correct angle of bank is $\theta = \arctan(v^2/rg)$.

The period of a low-altitude satellite is $T = 2\pi\sqrt{r/g}$, about an hour and a half.

In the absence of friction, a projectile follows a parabolic path. It achieves maximum range at a firing angle of $\pi/4$.

In a constant electric field, a charged particle experiences constant acceleration.

In any magnetic field, a charged particle maintains constant speed.

In a constant magnetic field, a charged particle executes circular or helical motion with angular frequency $\omega = (q/m)B$, where q is the charge of the particle, m is its mass, and B is the magnetic field strength.

The laws of mechanics are the same in all inertial frames of reference. This is called Galilean relativity.

The class of all possible motions is the same in all inertial frames.

Absolute acceleration has meaning. Absolute velocity does not.

QUESTIONS

Section 7.1

Q7.1 The vector sum of all forces acting on a body is zero. What can you conclude about the magnitude of its velocity? What can you conclude about the direction of its velocity?

Q7.2 "To keep a vehicle moving at constant speed on a level surface requires a steady force; when the force no longer acts, the vehicle slows and comes to rest." Criticize this statement. Why is Newton's first law a more useful statement about motion than this one?

Q7.3 An object at rest in an inertial frame of reference is known to be experiencing two forces, $\mathbf{F}_1$ and $\mathbf{F}_2$, which are not parallel. What is the least number of other forces that must also be acting on the body? What is the greatest number of other forces that may be acting on it?

Q7.4 (1) Is a parachutist falling at constant velocity an isolated body, that is, one free of outside influences? What tests could you perform to find this out (the tests need not be practical)? (2) Is a photon in intergalactic space an isolated particle? How could you check on this (again, in principle)?

Q7.5 A physicist works in a laboratory located at the South Pole. The laboratory, which has no windows, is fully equipped with sensitive measuring instruments. (1) Explain how the physicist might verify that the laboratory is not at rest in an inertial frame of reference. (2) Could the physicist construct an inertial frame within his laboratory? If so, how?

Q7.6 Within an orbiting spacecraft objects float freely, unaccelerated with respect to the craft. Does the spacecraft define an inertial frame of reference? (This is a subtle question, which will be taken up again in Chapter 22. If properly justified, either a yes answer or a no answer may be considered correct.)

Section 7.2

Q7.7 Suggest deformations other than that of a spring that could be used, in principle, to define force.

Q7.8 Consider a certain spring used to define force. It responds to a force F with an extension Δx. If another force causes an extension $2\,\Delta x$, how could you decide whether or not the second force is $2F$, if this spring is the only available force-measuring device? In short, how could you test whether the spring used to *define* force obeys Hooke's law?

Q7.9 Describe the forces acting on a car (a) moving at constant speed on a level highway, (b) moving at constant speed up a hill, and (c) skidding to the side at nearly constant speed on an icy pavement. Include vector diagrams with your answers.

Q7.10 The plank in Figure 7.5 is tipped to a greater angle with the horizontal, but not enough to cause the block to slide. (1) The force $\mathbf{F}$, parallel to the plank, changes in direction. Does it change in magnitude? (2) The force $\mathbf{N}$, perpendicular to the plank, changes in direction. Does it change in magnitude? (3) Does $\mathbf{F} + \mathbf{N}$ change (a) in direction? (b) in magnitude?

Q7.11 A solid surface can exert a frictional force on a stationary object. Can a motionless liquid exert a frictional force on a stationary object? Can a motionless gas do so?

Q7.12 Just before opening his parachute, a sport parachutist is moving straight down with a constant speed of 60 m/sec. Shortly after opening his parachute, he is moving straight down with a constant speed of 4 m/sec. Discuss all the forces acting on him at both times. From the earlier to the later time does the magnitude of the net force increase, decrease, or stay the same?

Section 7.3

Q7.13 The velocity of an automobile is represented by **v** and its acceleration by **a**. Describe conditions for which (a) **a** is parallel to **v**, (b) **a** is antiparallel to **v**, (c) **a** is perpendicular to **v**, (d) **a** is zero and **v** is not zero, and (e) **a** is not zero and **v** is zero. In each of these cases, what is the direction of the net force acting on the automobile?

Q7.14 Shortly after jumping from an airplane, a skydiver has a downward acceleration of 9 m/sec^2. Later, his downward acceleration is 1 m/sec^2. Has the net force acting on him changed? If so, by what factor?

Q7.15 State the law of mass conservation and explain why it is important in chemistry.

Q7.16 According to Stokes's law (stated in Exercise 2.11), a sphere moving slowly through a liquid experiences a force proportional to its speed. According to Newton's second law, the force on the sphere is proportional to its acceleration, not its speed. Reconcile these two statements.

Q7.17 In order to reduce his arithmetic labor, an enterprising engineer assigns to each of the objects he works with a "zeal," defined in such a way that the force applied to an object multiplied by its zeal gives its acceleration. What is the relation of zeal to mass? If two objects are fastened together, is the zeal of the combination equal to the sum of the zeals of the two objects?

Q7.18 (1) Describe carefully how force may be defined either as a primitive concept or as a derived concept. (2) Why is there arbitrariness in whether a concept is primitive or derived?

Q7.19 Describe carefully how mass may be defined either as a primitive concept or as a derived concept.

Section 7.4

Q7.20 A mass vibrating at the end of a spring experiences a net force that varies with time. Nevertheless, the force exerted by the spring is usually called a "time-independent" force. Explain why the force may be characterized in this way.

Q7.21 Explain why walking on a concrete floor is more tiring than walking on a wooden floor, which in turn is more tiring than walking on a carpeted floor. (2) Why does this trend not continue? Why is walking in high grass or snow, for example, not easier than walking on a carpet?

Section 7.5

Q7.22 Observe any common vibratory motion (other than the motion of a pendulum or of a mass on a spring) and decide whether it is simple harmonic. If it is, its period will be independent of its amplitude. Report on your observations.

Q7.23 Name several changes you could make in a harmonic oscillator that would double the maximum speed of the vibrating mass.

Section 7.6

Q7.24 How can you infer that the weight of an object is the same when it is accelerating as when it is stationary or moving uniformly?

Q7.25 An astronaut on the moon has an object of known mass and various measuring devices. How can he determine the acceleration of gravity at the moon's surface in a *static* experiment (one that does not involve any observations of falling bodies)?

Q7.26 One way to appreciate the meaning of a familiar law is to imagine what the world would be like if a different law held. What would be some of the interesting consequences for everyday living if weight on earth were proportional to mass squared?

Q7.27 Estimate the terminal speed in air of some very light object, such as a bit of paper or an inflated balloon, and of a light but somewhat denser object, such as a ping-pong ball or a deflated balloon.

Q7.28 What everyday observations show that the force of frictional air drag increases as speed increases?

Q7.29 The acceleration of gravity at the earth's surface is an important physical quantity. Why is it considered less "fundamental" than quantities such as the speed of light and the charge of an electron?

Q7.30 An eccentric astronaut carries a pendulum-driven clock to the moon. Discuss its timekeeping characteristics (a) on the way to the moon and (b) on the surface of the moon ($g_{\text{moon}} = \frac{1}{6}g_{\text{earth}}$). Would the astronaut run any risk if he used this clock to time his exploration of the moon's surface?

Section 7.7

Q7.31 If the ratio m_I/m_G differed significantly from one substance to another, what would be some of the immediately obvious consequences? Would such an effect be relevant to the design of vehicles?

Q7.32 On a hypothetical planet where inertial and gravitational masses are not equal, two stones of equal inertial mass and unequal gravitational mass are dropped together from rest. (1) Are both uniformly accelerated? (2) Do they have equal weight? (3) Do they have equal acceleration?

Q7.33 Answer Question 7.32 for stones of equal gravitational mass and unequal inertial mass.

Q7.34 Briefly outline a set of experiments that might lead to the result $m_I = Km_G$, where K is a constant not equal to 1. Why is the result $m_I = Km_G$ physically equivalent to the result $m_I = m_G$ (Equation 7.64)?

Section 7.8

Q7.35 If the rate of rotation of the earth changed, would the weight of a parcel measured by a spring balance change? Would its weight measured by a pan balance change?

Q7.36 How long would a pendulum have to be in order to have a period of 84 min, equal to the period of a low-altitude earth satellite? (SUGGESTION: Look at Equations 7.59 and 7.74.)

Q7.37 When the curvature of the earth is taken into account, would you expect the range of an intercontinental ballistic missile to be greater or less than the range given by Equation 7.90? Explain the reason for your answer.

Q7.38 For maximum range, including the effect of air resistance, should a projectile be launched at an angle to the horizontal greater or less than 45 deg? Give a reason for your answer.

Section 7.9

Q7.39 Name a device in which electric acceleration of charged particles is employed. Name one in which magnetic acceleration of charged particles is employed.

Q7.40 Protons from space bombard the earth from all directions. Considering the effect of the earth's magnetic field on these particles, discuss any differences in the behavior of protons headed toward the earth from above a geomagnetic pole and from above the equator.

Section 7.10

Q7.41 A man (a) sits in a stationary car, then (b) accelerates to speed v_0 along a straight road, then (c) drives at constant speed along the straight road, and then (d) turns to the left while holding his speed constant. For each period explain why his car does or does not define an inertial frame of reference. (Ignore the rotation of the earth.)

Q7.42 Suppose that Newton's second law read $\mathbf{F} = m\mathbf{v}$ instead of $\mathbf{F} = m\mathbf{a}$. Would there be a principle of Galilean relativity in mechanics? Explain.

Q7.43 Newton's first law can be stated in this way: If $\mathbf{F} = 0$, then $d^n\mathbf{r}/dt^n = 0$, where $n = 2$. Discuss some properties of a world in which $n = 1$. Would absolute velocity have meaning in such a world? Would infinite acceleration be possible in such a world?

Q7.44 With reference to the preceding question, discuss some properties of a world in which $n = 3$. In such a world, could a frame of reference accelerated with respect to an inertial frame be an inertial frame?

Q7.45 A physics student traveling by air at night occupies his time by swinging a small weight at the end of a string. His fellow passenger considers this activity rather purposeless until the student turns and says, "Something must be wrong; we have just completed a turn of about 180 degrees and are heading back toward our starting point." Explain how the student's "inertial guidance system" works.

EXERCISES

Section 7.1

E7.1 An automobile moves along a straight road with constant velocity. It experiences a downward force of gravity of 15,000 N and a rearward force of wind resistance of 1,000 N. What is the sum of other forces that act on the car? What is the source of these other forces?

Section 7.2

E7.2 Estimate roughly in newtons the force required to (a) raise an automobile on a hydraulic lift, (b) push a thumbtack into a pine plank, (c) slice butter, and (d) push a baby carriage up a 10-deg slope.

E7.3 (1) What is the total force, $\mathbf{F} + \mathbf{N}$, exerted by the plank on the stationary block in Figure 7.5 if $W = 7$ N? (2) Answer the same question if the block is not stationary but is sliding down the plank with a constant speed of 0.5 m/sec.

E7.4 Mrs. W. attempted to hang an object weighing 50 N from the center of a clothesline 20 m long. The line sagged until its center was 0.5 m below the level of its ends and then broke. She wrote an irate letter to the manufacturer because the line was guaranteed to hold 250 N. Write a letter in response, explaining by how much the line outperformed its guarantee.

Section 7.3

E7.5 A jet plane starts its takeoff roll with an acceleration of 2 m/sec^2. Each of its two jet engines delivers a thrust of 10^5 N. What is the approximate mass of the plane?

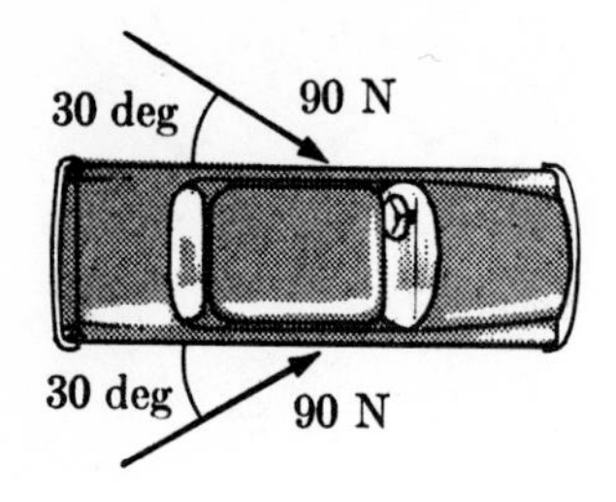

E7.6 The mass of a stalled car is 1,300 kg. Two men push on the car with the forces that are shown in the diagram. If their effort results in an acceleration of the car of 0.03 m/sec^2, what frictional force is acting on the car?

E7.7 When acted on separately by the same force F, three objects respond with accelerations of 2.0 m/sec^2, 3.0 m/sec^2, and 4.0 m/sec^2. (1) If the two lighter objects are placed on one pan of a balance and the heaviest of the three objects is placed on the other pan, which pan sinks? (2) If the three objects are fastened together and acted on by force F, what is their acceleration?

E7.8 A cart in a laboratory is "weighed" in the following way: Starting from rest, it is pushed with a steady force of 8 N. After 3 sec, its speed is measured to be 4 m/sec. (1) If friction is negligible, what is the mass of the cart? (2) If friction is not negligible, is the true mass greater or less than this calculated value?

Section 7.4

E7.9 A catapult applies a constant force to accelerate an 8,000–kg airplane to a speed of 60 m/sec in 1.5 sec. What is the force?

E7.10 A skier moves for a time at constant speed, then accelerates to higher speed, then decelerates back to the original speed, which he again maintains for a time. Sketch a graph of the net force acting on the skier as a function of time.

E7.11 A three-link chain, each link of mass 10 kg, hangs from a hook. The hook is drawn upward with a constant acceleration of 2 m/sec^2. What is the net force acting on the center link?

E7.12 The block and plank of Figure 7.5 are replaced by a small car and frictionless air track also inclined at 30 deg to the horizontal. The vertical gravitational force on the car is 6 N. (1) What is the mass of the car? (2) How much time passes as the car slides from rest a distance of 2 m along the track? (3) What is its speed at the end of this slide?

E7.13 The rocket shown in Figure 7.8 later moves outside the atmosphere, with its axis still inclined at 45 deg to the vertical and with the same thrust $\mathbf{F}_2$. The gravitational force then has magnitude $F_1 = 1.2 \times 10^6$ N and the frictional force $\mathbf{F}_3$ is zero. What is the angle between the axis of the rocket and its acceleration?

E7.14 In earlier days, horses pulled barges down canals in the manner shown in the figure. Suppose that at a particular instant a horse is exerting a force $\mathbf{F}$ of magnitude 8,000 N at an angle of 20 deg to the direction of motion of the barge, and that the 10,000-kg barge is accelerating along the canal at 0.1 m/sec^2. What are the x and y components of the force exerted by the water on the barge?

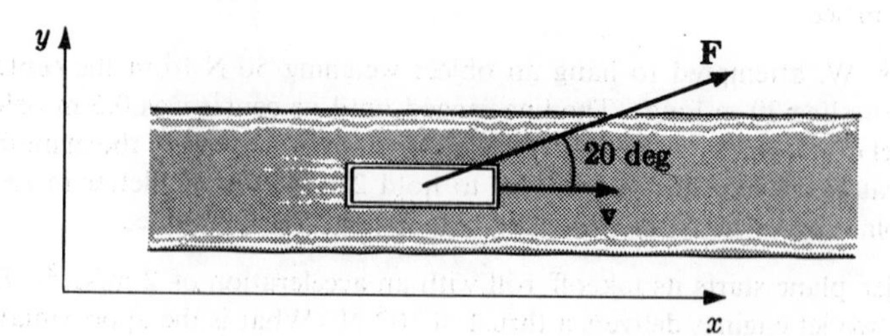

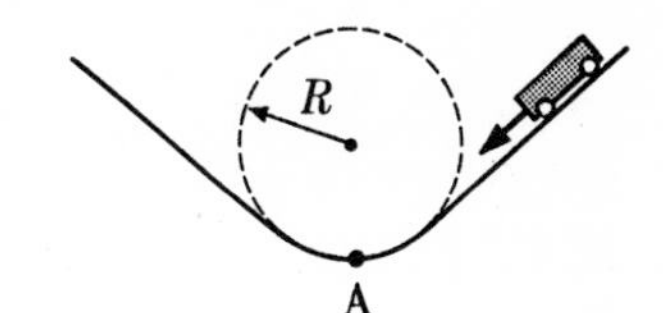

E7.15 The track of a roller coaster has a portion that is part of a vertical circle of radius $R = 25$ m (see the figure). What force does the track exert on a 500-kg car moving at 20 m/sec past point A, the lowest point of the track?

E7.16 A man of mass m rides a Ferris wheel of radius r that turns at an angular speed ω. In terms of m, r, and ω, express the total force on the man when he is at the top of the wheel and when he is at the bottom. If the two expressions are the same, explain why they are the same. If they are different, explain why they are different.

E7.17 An object of mass m moves along the x axis. At $t = 0$, it is at $x = 0$ and its x component of velocity is $v_x = v_0$. It experiences a constant force of magnitude F directed toward negative x ($F_x = -F$). (1) For $v_0 > 0$, find expressions for (a) its x coordinate as a function of time, (b) its maximum x coordinate, and (c) the time when it is again at $x = 0$. (2) Repeat for $v_0 < 0$. (3) Sketch graphs of x vs t for $v_0 > 0$ and $v_0 < 0$.

E7.18 A particle starting from rest at the origin is acted upon by the force $F_x = Ae^{-bt}$, where A and b are constants. Find an expression for the speed of the particle as a function of time t.

E7.19 A particle moving along the x axis is acted upon by the force $F_x = A - \alpha t$, where A and α are constants. Find a general expression for the position of the particle as a function of time t.

E7.20 In a stationary freight elevator a spring balance supporting a bunch of bananas reads 10 kg. Shortly after the elevator starts to move, the spring balance reads 12 kg. (1) Did the elevator start to move upward or downward? (2) What is its acceleration?

E7.21 An unknown mass m oscillates with a period of 2 sec at the end of a spring that obeys Hooke's law. An experimenter finds that if he increases the mass by 1 kg, the period becomes 4 sec. What is the initial mass m?

E7.22 A rubber rope hanging from the top of a tower extends halfway down the tower. A circus performer starts down the rope; when he reaches its end, he is on the ground. If the performer weighs 750 N and his rope obeys Hooke's law, with a spring constant $k = 60$ N/m, how tall is the tower?

E7.23 For the harmonic oscillator, verify that the dimension of $\sqrt{m/k}$ is time, as required by Equation 7.33.

E7.24 What is the speed of a harmonic oscillator when its displacement is half of its amplitude? (2) At what displacement is its speed half of its maximum speed?

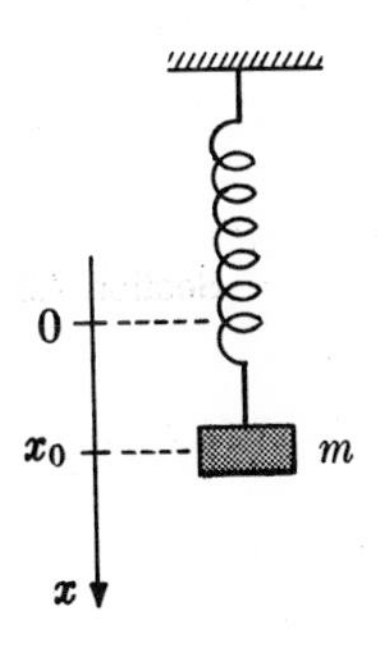

E7.25 As shown in the figure, a weight of mass m is supported by a spring. The force exerted on the weight by the spring is $F_x = -kx$. (1) Show that the equilibrium position of the weight is $x_0 = mg/k$. (2) Show that the total force on the weight is $F_x' = -kx'$, where $x' = x - x_0$. (3) What is the nature of the motion of the weight if it is displaced from its equilibrium position and released?

E7.26 For the harmonic oscillator, show that the two ways of writing $x(t)$, given by Equations 7.34 and 7.35, are equivalent. Express the constants a and b in terms of the constants A and φ.

E7.27 The initial position and velocity of a harmonic oscillator are $x_0 = -B$,

$v_0 = 0$. (1) What are the constants A and φ in Equation 7.35? (2) What are the constants a and b in Equation 7.34? (3) Answer the same pair of questions if the initial conditions are $x_0 = 0$, $v_0 = C$.

E7.28 A harmonic oscillator has angular frequency $\omega = 200\ \text{sec}^{-1}$. (1) If its initial conditions are $x_0 = 0.002$ m, $v_0 = 4$ m/sec, what are its amplitude A and phase constant φ? (2) If its amplitude A is 0.002 m and its phase constant φ is $\pi/3$, what are its initial conditions x_0 and v_0?

E7.29 With a simple change of variable, Equation 7.35 can be written $x = A \sin \omega t'$. (1) Express t' in terms of t. (2) Sketch a graph of x vs t for an arbitrary phase constant φ. How does the change of time variable show itself in the graph?

E7.30 Show that both of the expressions for $x(t)$, Equations 7.34 and 7.35, satisfy the equation of motion of the harmonic oscillator (Equation 7.23) if the angular frequency is given by $\omega = \sqrt{k/m}$.

Section 7.6

E7.31 To prepare for a metric future, calculate your height in meters, your mass in kilograms, and your weight in newtons.

E7.32 Obtain a formula for the maximum speed of a pendulum. Evaluate it numerically for typical parameters of a child's swing.

E7.33 Obtain a formula for the speed of a pendulum as a function of its displacement.

E7.34 A pendulum has a period of 1 sec at the North Pole. What is its period at the equator? (Accurate values of g appear in Appendix 3.) *Optional:* Develop an approximate formula that gives the fractional change of period, $\Delta T/T$, for a small fractional change of the acceleration of gravity, $\Delta g/g$.

E7.35 Using a series expansion for $\sin \theta$, show that the fractional error in the right side of Equation 7.55, which is the restoring force on a simple pendulum, is approximately

$$\frac{1}{6}\,\theta^2, \qquad \text{or} \qquad \frac{1}{6}\left(\frac{s}{l}\right)^2.$$

For $\theta = 20$ deg, verify that this result agrees with the result found in Example 3 of Section 7.6. *Optional:* Extend the approximate expression to another term, showing that

$$\frac{\theta - \sin\theta}{\sin\theta} \cong \frac{1}{6}\,\theta^2 + \frac{7}{360}\,\theta^4 \qquad (\theta \ll 1).$$

Section 7.7

E7.36 (1) An object subjected to gravitational forces, from whatever sources, experiences a force proportional to its own gravitational mass m_G. Show that if it experiences no other forces, its acceleration is proportional to the ratio of its gravitational mass to its inertial mass ($a \sim m_G/m_I$). (2) Use this fact to explain why a pair of satellites can "fly formation," remaining close together without the use of power.

Section 7.8

E7.37 A satellite is in a circular orbit at an altitude of 100 miles. Show that its speed is 7,800 m/sec and its period is 88 min. (For comparative figures, see Equations 7.72 and 7.75. For an algebraic approach, see Problem 7.18.)

E7.38 For uniform motion in a circle, centripetal acceleration is usually written $a = v^2/r$ or $a = \omega^2 r$. Express a also as a function of (a) ω and v and (b) r and the period T. (SUGGESTION: Keep these two formulas in a convenient place. They are likely to prove useful.)

E7.39 If the earth increased its rate of rotation (without changing its shape) until residents at the equator became "weightless," what would be the length of a day?

E7.40 What is the acceleration of the moon toward the earth? How does this acceleration compare with g? (For necessary data, see Appendix 3D.)

E7.41 What is the acceleration of the earth toward the sun? How does this acceleration compare with g? How does it compare with the acceleration of the earth's surface produced by the earth's daily rotation (about 0.034 m/sec^2 at the equator)? (For necessary data, see Appendix 3D.)

E7.42 A spacecraft circles the moon at low altitude. (1) What is its speed? (2) What is its period? (NOTE: The acceleration of gravity at the surface of the moon is approximately $\frac{1}{6}g$.)

E7.43 A straight section of model railroad track is joined to a section that is the arc of a circle of radius 0.4 m. The model locomotive, whose mass is 1.5 kg, moves at a constant speed of 1.2 m/sec. (1) What is the total force on the locomotive on the straight section? Why? (2) What is the total force on the locomotive on the curved section? What is the source of the force? (Because the forces on the straight section and the curved section are not equal, actual railroads do not join circuluar arcs directly to straight tracks. Instead, "railroad curves" of gradually changing radius of curvature are used.)

E7.44 A man swings a mass of 1 kg in a horizontal circle of radius 2 m (see the figure). (1) Assuming that he can exert a maximum inward force of 200 N, what is the maximum speed he can impart to the circling mass? (2) If he shortens the cord attached to the mass and continues to exert maximum inward force, does the speed of the mass increase or decrease? Does his angular speed increase or decrease? (Ignore the minor effect of gravity and assume that the man pulls horizontally.)

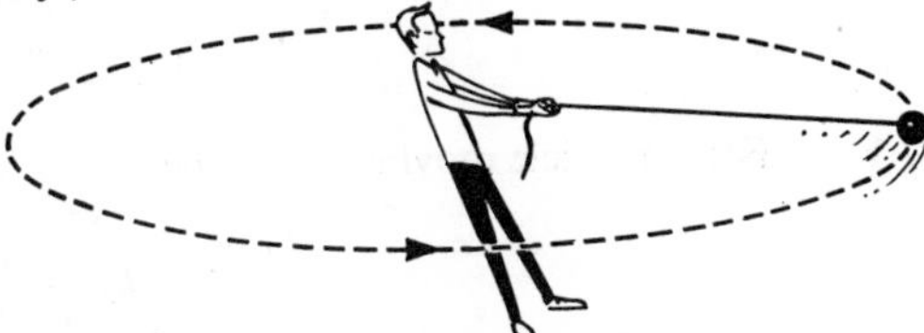

E7.45 An electron is accelerated horizontally in a television tube to a speed of 10^7 m/sec. It then flies freely through 0.2 m until it strikes the face of the tube. Initially it was aimed directly at a central point on the tube face. How much below this point does it actually strike because of the downward force of gravity? Is this a significant effect that tends to blur television pictures?

E7.46 A marble rolling at 80 cm/sec rolls off the edge of a table and hits the floor at a distance of 40 cm from the table. How high is the table?

E7.47 With what speed must a ball be thrown vertically upward if it is to reach a height of 25 m? What is the maximum horizontal distance that a ball can be thrown with this same initial speed?

E7.48 (1) For the projectile motion illustrated in Figure 7.16 and described by Equations 7.87 and 7.88, write algebraic equations for the horizontal and vertical components of velocity, v_x and v_z. (2) Derive an expression for the maximum height of the trajectory reached at its midpoint.

E7.49 (1) Find the trajectory equation for the projectile shown in Figure 7.16. This is an equation containing x and z but not t. (2) Eliminate t from Equations 7.85 and 7.86 in order to express z as a function of x for the general case of parabolic trajectory motion.

Section 7.9

E7.50 (1) Compare the magnitudes of acceleration of an electron and a proton in the same electric field. (2) After having been accelerated from rest through the same distance in the same direction by electric fields of the same magnitude, an electron and a proton enter a region containing a magnetic field directed perpendicular to their velocities. Do the electron and proton bend in the same direction or opposite directions? What is the ratio of the radii of their circular paths in the region of the magnetic field?

E7.51 (1) Show that the equations of uniform circular motion,

$$x = r \cos \omega t,$$

$$y = r \sin \omega t,$$

where r and ω are constants, satisfy Equations 7.107 and 7.108, which govern the motion of a charged particle in a constant magnetic field. (2) Give an expression for ω. (3) Show that the speed v is constant. (4) How is the radius r related to the speed v?

Section 7.10

E7.52 In one frame of reference, the position vector of a particle is $\mathbf{r}$. In another frame of reference, the position vector of the same particle is $\mathbf{r}' = \mathbf{r} + \mathbf{r}_0 + \mathbf{v}_0 t$, where $\mathbf{r}_0$ and $\mathbf{v}_0$ are constant vectors. (1) How does the velocity of the particle in one frame of reference differ from its velocity in the other frame? (2) How does its acceleration in one frame differ from its acceleration in the other frame? (3) Describe in words the relationship between these two frames of reference.

PROBLEMS

Force and kinetic energy

P7.1 For a particle moving along the x axis, prove that

$$F_x = \frac{dK}{dx},$$

where F_x is the x component of force acting on the particle and K is its kinetic energy.

Averages over space and time

P7.2 The average of a function F over time may be defined by

$$(\bar{F})_t = \frac{1}{T}\int_0^T F\,dt;$$

its average over position may be defined by

$$(\bar{F})_x = \frac{1}{X}\int_0^X F\,dx.$$

(1) Show that the two averages are the same if $F = $ *constant:*

$$(\bar{F})_t = (\bar{F})_x = F.$$

(2) Show by example that the two averages may differ if F is not constant. In particular, let F be the force function $F = \alpha t$ acting in the positive x direction

on a particle that is at rest at the origin at $t = 0$. Find $x(t)$ for this particle and then show that as the particle moves from $x = 0$ to $x = X$ in time T, the two averages are

$$(\bar{F})_t = \tfrac{1}{2}\alpha T \quad \text{and} \quad (\bar{F})_x = \tfrac{3}{4}\alpha T.$$

P7.3 The motion of a particle on the x axis is described by the equation

Given $x(t)$, find $F(x)$

$$x = Ae^{-Bt},$$

where A and B are constants and t is time. (1) What is the dimension of B? (2) What is the physical meaning of A? (3) Sketch graphs of position x vs t, velocity v vs t, and acceleration a vs t. (4) Write the force F as a function of x.

P7.4 A particle moving along the x axis experiences the force $F_x = F_0 \sin \omega t$, where F_0 and ω are constants. (1) Obtain a general expression for $x(t)$, the position of the particle as a function of time. (2) If $v(0) = 0$ and $x(0) = 0$, what is $x(t)$? Why does the particle move increasingly far from the origin even though the time average of the force F_x is zero over a period of its oscillation? (3) What initial velocity $v(0)$ is required in order that $x(t)$ be periodic?

Given $F(t)$, find $x(t)$

P7.5 Spring 1, whose equilibrium length is l_1, and spring 2, whose equilibrium length is l_2, may be connected to a cart on an air track as shown in the figure. When spring 1 acts alone, the cart oscillates with frequency ν_1; when spring 2 acts alone, the cart oscillates with frequency ν_2. Then both springs are connected to the cart. (1) The distance between the points of support of the two springs is adjusted to match their equilibrium lengths: $L = l_1 + l_2$. Show that the cart oscillates with frequency $\nu = \sqrt{\nu_1^2 + \nu_2^2}$. (2) The distance L is changed so that when the cart is in equilibrium neither spring is in equilibrium. Show that the oscillation frequency of the cart is still $\nu = \sqrt{\nu_1^2 + \nu_2^2}$.

Oscillator with two springs

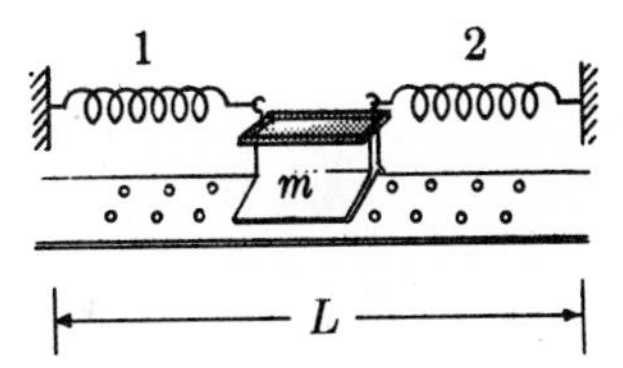

P7.6 Extend the theory given by Equations 7.37–7.42 and show how to choose the phase constant φ of the harmonic oscillator for any initial conditions x_0 and v_0. In particular, show that φ lies in the second quadrant if $x_0 > 0$ and $v_0 < 0$ (with $A > 0$). (SUGGESTION: Sketch a graph of arc tan u vs u, where $u = \omega x_0/v_0$.) *Optional:* If the negative root is chosen for A in Equation 7.42, show that φ can be chosen in such a way that Equation 7.35 for $x(t)$ is unchanged.

Amplitude and phase of harmonic oscillator

P7.7 A shallow trough has a parabolic cross section given by $y = \beta x^2$. In it a block of mass m slides without friction, experiencing the force of gravity **W**, directed vertically downward, and the force of the trough **F**, directed perpendicular to the trough (see the figure). (1) For small oscillations, any of the following approximations are valid: $\theta \ll 1$, $\cos\theta \cong 1$, and $\sin\theta \cong \tan\theta$ (θ is defined in the figure). Explain why the approximation $|d^2y/dt^2| \ll mg$ is also valid. (2) For small oscillations, show that the x component of the total force on the block is given approximately by $F_x = -kx$. Give an expression for k. (3) Give a formula for the period of oscillation T. (4) Briefly discuss the dependence of T on β, on g, and on m.

Another harmonic oscillator

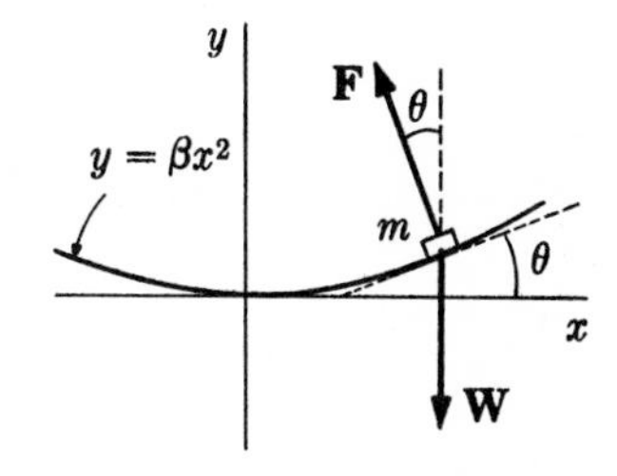

P7.8 A stone is dropped from the roof of a building with zero initial speed. It takes 0.10 sec to pass by a window whose height is 1.03 m. How far below the roof is the top of the window?

Falling bodies

P7.9 After dropping a stone from the top of a building, a man waits 1.5 sec, then hurls a baseball downward with an initial speed of 20 m/sec. (1) If the stone and the ball reach the ground together, how high is the building? (2) For a given initial speed of the ball, show that there is a maximum waiting time, after which the ball cannot catch the stone, no matter how high the building. For $v_0 = 20$ m/sec, what is the maximum waiting time?

Chain sliding from a table

P7.10 A chain of length l, part of which hangs over the edge of a table, is initially at rest (see the figure). When the chain is released, it slides off the table with negligible friction. (1) What is its acceleration as a function of time? (2) Show that the time required for the chain to leave the table is

$$T = \sqrt{\frac{l}{g}} \ln\left(\frac{l + \sqrt{l^2 - x_0^2}}{x_0}\right),$$

where x_0 is the length of chain that initially hangs from the edge of the table. Discuss this expression in the limits $x_0 \to 0$ and $x_0 \to l$. (3) A 1-m chain slides from a table in 1 sec. What is x_0?

Tension in a vertical rope

P7.11 The tension in a rope is the force that one part of the rope exerts on an adjacent part—for instance, the force F_t exerted by the unshaded portion of rope on the shaded portion of rope in the figure. As a function of y, find the tension F_t for (a) a massless rope supporting a stationary load of mass m_0, (b) a rope of fixed mass per unit length τ supporting the same stationary load, and (c) a rope of fixed mass per unit length τ accelerating the load of mass m_0 upward with acceleration a.

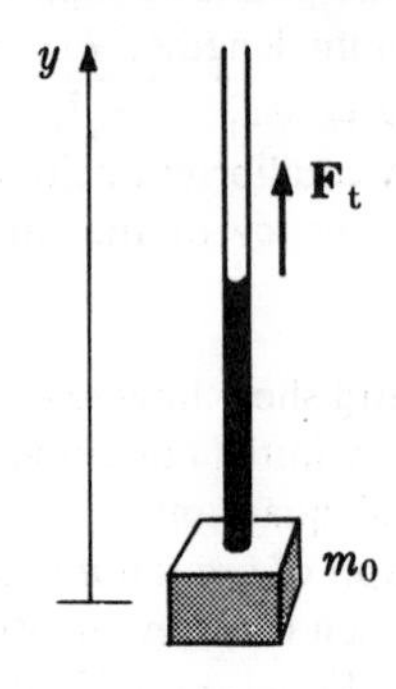

Optimal design of vertical cable

P7.12 The greatest tension that a certain cable can safely withstand is $F_t = \alpha A$, where A is its cross-sectional area and α is a constant. (1) Discuss the design of an optimal cable (one using the least material) to support a load of mass M. (2) Show that the cross-sectional area of this cable can be written

$$A(y) = \frac{Mg}{\alpha} e^{(\rho g/\alpha)y},$$

where ρ is the density of the cable material and y is distance measured upward from the load. (3) For what length of cable is the weight of the cable equal to the weight of the load it is supporting?

Safety-rope system

P7.13 Design a safety-rope system for a 100-kg man working where he may slip and fall. He must have freedom of movement over a distance of 10 m while he is working and must experience acceleration no greater than $8g$ if he falls. Give attention to practical as well as theoretical aspects of your system.

Arbitrary motion in a circle

P7.14 A particle moves with nonconstant speed around a circle of radius l. Its position vector is given by $\mathbf{r} = l\cos\theta\,\mathbf{i} + l\sin\theta\,\mathbf{j}$, where $\theta = s/l$ (see the figure). (1) Show that the x and y components of acceleration of the particle are given by

$$a_x = -l\left[\frac{d^2\theta}{dt^2}\sin\theta + \left(\frac{d\theta}{dt}\right)^2\cos\theta\right],$$

$$a_y = l\left[\frac{d^2\theta}{dt^2}\cos\theta - \left(\frac{d\theta}{dt}\right)^2\sin\theta\right].$$

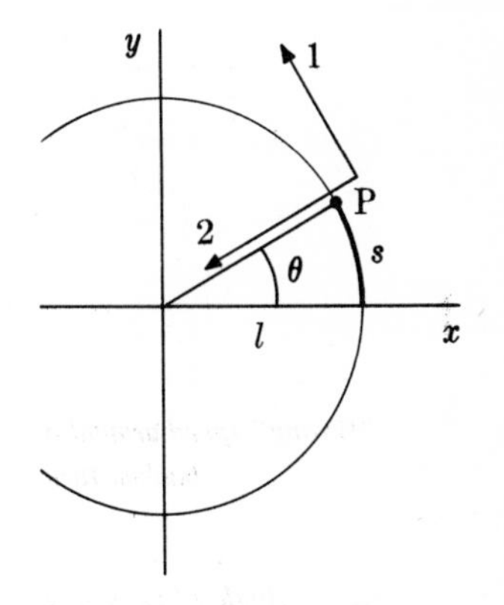

(2) When the particle is at point P, show that its components of acceleration along axes 1 and 2 are related to a_x and a_y by

$$a_1 = -a_x\sin\theta + a_y\cos\theta,$$

$$a_2 = -a_x\cos\theta - a_y\sin\theta.$$

(3) Using these results, show that

$$a_1 = l\left(\frac{d^2\theta}{dt^2}\right) = \frac{d^2s}{dt^2},$$

$$a_2 = l\left(\frac{d\theta}{dt}\right)^2 = \frac{\left(\frac{ds}{dt}\right)^2}{l}.$$

The expression for a_1 confirms Equation 7.53. The expression for a_2 shows that centripetal acceleration in circular motion is v^2/r, whether or not v is constant.

Simple pendulum

P7.15 For a simple pendulum, the bob experiences a component of acceleration toward the pivot point (except at the end points of the swing). (1) Why? (2) Show that F_{rod}, the magnitude of the force exerted by the rod on the bob, can be written

$$F_{\text{rod}} \cong mg\left(1 + \frac{A^2}{l^2} - \frac{3}{2}\frac{s^2}{l^2}\right),$$

where s is the displacement of the bob and A is its amplitude (refer to Figure 7.11 and Equation 7.57). (HINT: The last result stated in Problem 7.14 may be useful.)

Centripetal acceleration

P7.16 A jet trainer executes a loop in a vertical circle of radius 800 m. Going around the loop, the speed of the airplane varies in such a way that at the bottom of the loop the pilot experiences 4 "g's" (that is, his seat exerts an upward force on him 4 times greater than that during normal level flight), and at the top of the loop he is "weightless" (his seat exerts no force on him). (1) Find the centripetal acceleration of the jet at the top and bottom of the loop. (2) Find the speed of the jet at the top and bottom of the loop.

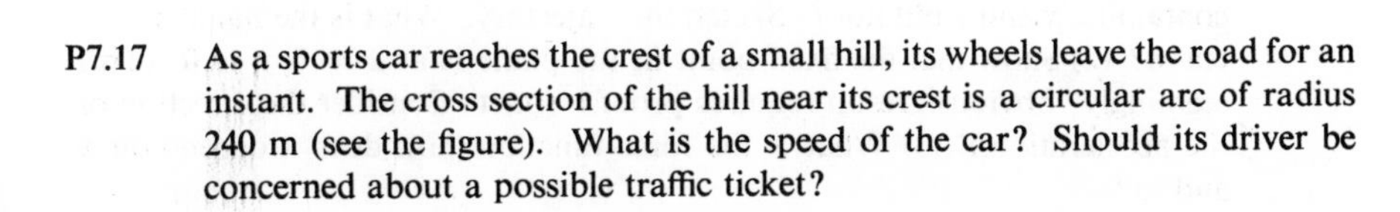

P7.17 As a sports car reaches the crest of a small hill, its wheels leave the road for an instant. The cross section of the hill near its crest is a circular arc of radius 240 m (see the figure). What is the speed of the car? Should its driver be concerned about a possible traffic ticket?

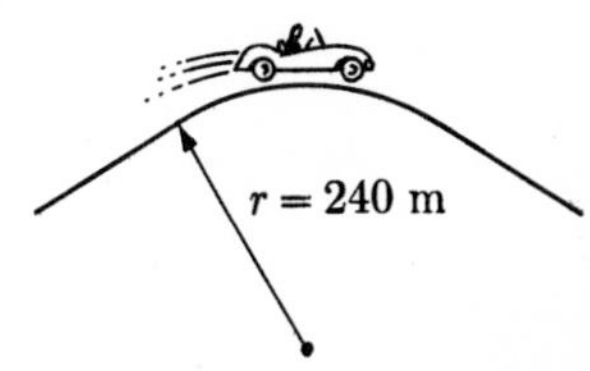

Low-altitude satellites

P7.18 (1) Taking into account the fact that the acceleration of gravity varies inversely with the square of the radial distance r, show that the *period* of a satellite in a circular orbit (Equation 7.74) can be written $T = T_0(r/R)^{3/2}$, where R is the radius of the earth and T_0 is the period of a hypothetical satellite skimming the earth's surface (see Equation 7.74). (2) For altitudes h that are small compared with R, show that

$$\frac{T}{T_0} \cong 1 + \frac{3}{2}\frac{h}{R}.$$

(3) Starting with Equation 7.71, develop similar formulas for the *speed* of a satellite in a circular orbit.

"Wrong" speed around a banked turn

P7.19 What is the net force acting on a railroad car if it moves on the track described in Example 2 in Section 7.8 (see also Figure 7.14) at a speed of 125 km/hr?

Angle of bank and "g forces"

P7.20 As an airplane turns in a horizontal circle, a passenger experiences a net acceleration $\mathbf{a}_c = \mathbf{g} + \mathbf{a}$, where g is the acceleration of gravity and $\mathbf{a}$ is the acceleration provided by the airplane (see the figure). (1) Show that the correct angle of bank of the airplane is given by

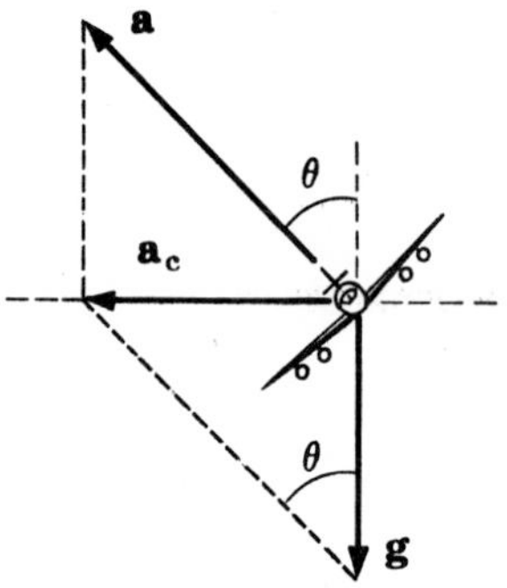

$$\theta = \text{arc tan}\left(\frac{\omega v}{g}\right),$$

where v is the speed of the airplane and ω is its angular speed in its circular arc. (2) The "g force" on the passenger may be defined as the force, measured in units of his weight, exerted on him by his seat. Show that the "g force" is equal to a/g, and may be written

$$\frac{a}{g} = \sec\theta = \sqrt{1 + \left(\frac{\omega v}{g}\right)^2}.$$

(3) A commercial jet airplane, with $v = 250$ m/sec, executes a "standard-rate" turn, for which $\omega = \pi/60$ radian/sec. Find (a) the angle of bank, (b) the time to complete a full circle, (c) the diameter of the circle, and (d) the "g force" on a passenger.

Orbit in a plane

P7.21 One end of an elastic band is attached to a fixed point in a horizontal plane. The other end is attached to a weight that can slide without friction on the plane. The motion of the weight is described by the equations

$$x = 5A\cos\left(\frac{2\pi t}{T}\right),$$

$$y = 2A\sin\left(\frac{2\pi t}{T}\right),$$

where A and T are constants and t is time. (1) Obtain the trajectory equation containing x and y but not t. Sketch the trajectory. What is the name of this curve? (2) Show that the fixed point in the plane is at $x = 0, y = 0$. (This seemingly "obvious" result requires proof.) (HINT: Consider the direction of the acceleration.) (3) What is the force function (how does it depend on x and y)?

Spherical pendulum

P7.22 A pendulum is free to swing in all directions. If the weight at its lower end traces out a horizontal circle, prove that its angular frequency is

$$\omega = \sqrt{\frac{g}{l \cos \theta}},$$

where l and θ are defined in the figure. (NOTE: This result does not require a small-angle approximation.) *Optional:* Prove that an object sliding without friction inside a sphere is a system that is physically equivalent to a spherical pendulum, provided the pendulum bob is supported by a flexible cord, not a rigid rod.

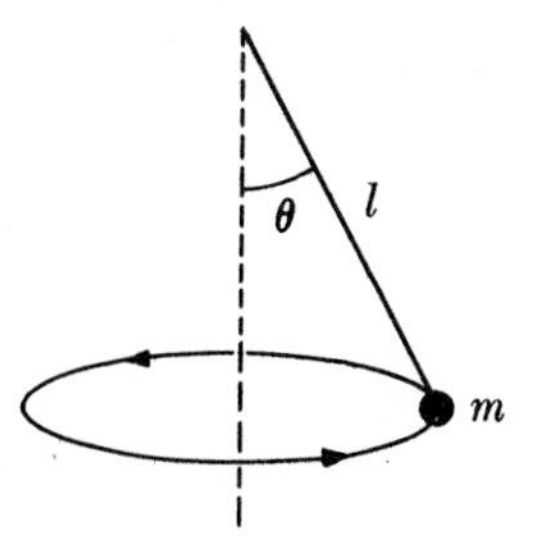

Projectile motion

P7.23 Prove that a projectile reaches its highest point when it has covered exactly one half of its horizontal range.

P7.24 A metal ball is launched toward a suspended can at the same moment that the can is released and starts to fall (see the figure). Prove that the ball will hit the can in midair (ignoring friction) regardless of the initial speed of the ball.

P7.25 (1) Show that a projectile of given range R and given initial speed v_0 may be launched with either of two initial angles if $R < R_{max}$. Call these angles θ_1 and θ_2. (2) Prove that the ratio of flight times for these two trajectories is

$$\frac{t_1}{t_2} = \tan \theta_1 .$$

Record home run

P7.26 In the 1971 All-Star Game at Detroit, Reggie Jackson of the Oakland Athletics hit what may have been a record home run. According to one newspaper account, Jackson's "herculean clout . . . left the field at the 415-foot mark and at a point 132 feet high. It was estimated to have traveled 600 feet." Show that if the ball left Jackson's bat at an angle of 45 deg to the horizontal and passed the specified point, it would have traveled slightly more than 608 ft (in the absence of air resistance). Use $g = 32.2$ ft/sec^2 if it is needed. *Optional:* (1) Prove that the theoretical range of the ball is given by

$$R = \frac{L^2}{L - H},$$

where L is the horizontal distance to the point where the ball left the field and H is the ball's height at that point. (Note the independence of g. Does this mean that Jackson could hit a ball no farther on the moon?) (2) Assume the uncertainties in L and H to be 2 ft. What is the uncertainty in R?

Particle in a constant magnetic field

P7.27 The equations of motion of a charged particle in a constant magnetic field are given by Equations 7.107–7.109. Solve these equations to find $x(t)$, $y(t)$, and $z(t)$, and show that the path of the particle is a helix. (For notational convenience, set $qB/m = \gamma$. It may be fruitful to differentiate one or more of the equations. Pay special attention to constants that appear in the solution. Are they all independent?)

Cathode-ray tube

P7.28 In a cathode-ray tube, electrons of mass m, charge $q = -e$, and initial speed v_0 pass between parallel plates of length l. Between the plates a constant electric field $\mathbf{E}$ is directed downward (see the figure); elsewhere the electric field is zero. At a distance L beyond the plates, the electrons strike a fluorescent screen. Derive an expression for the net deflection D, which is defined in the figure. (Gravity may be ignored.)

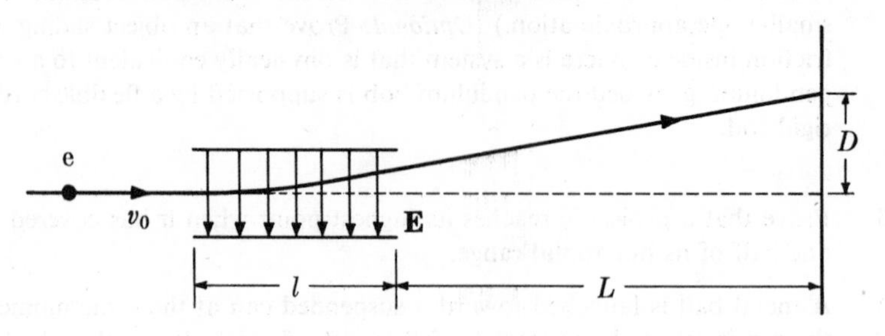

Mass spectrometer

P7.29 In one form of mass spectrometer, positively charged ions enter a region of uniform electric field $\mathbf{E}$, where they are accelerated through distance d, then enter a region of zero electric field and uniform magnetic field $\mathbf{B}$, where they follow a semicircular path to a detecting plane S (see the figure). Suppose that the ions entering the device are of two kinds, with masses m and $m + \Delta m$, all with the same charge q. If the speed of the ions is negligible when they enter the region of the electric field, what will be the separation Δx between the ion beams at the detecting plane?

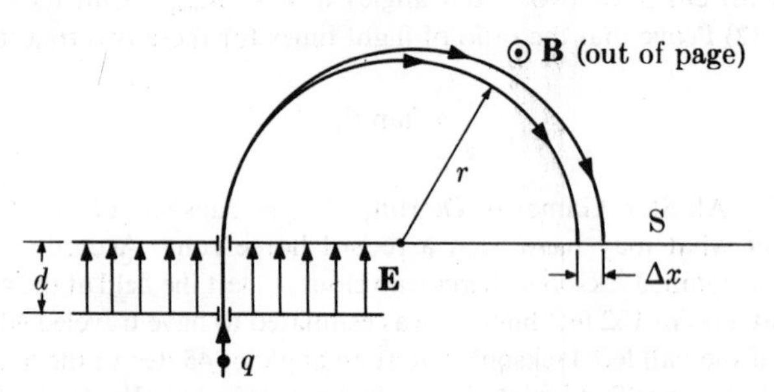

Galilean transformation

P7.30 In Figure 7.21, let player A be at the origin of a Cartesian coordinate system xy, and let player B be at the origin of a system $x'y'$ that moves in the x direction at speed v (see the figure). A point P has coordinates xy in one frame and coordinates $x'y'$ in the other frame. (1) Show that

$$x' = x - vt,$$

$$y' = y,$$

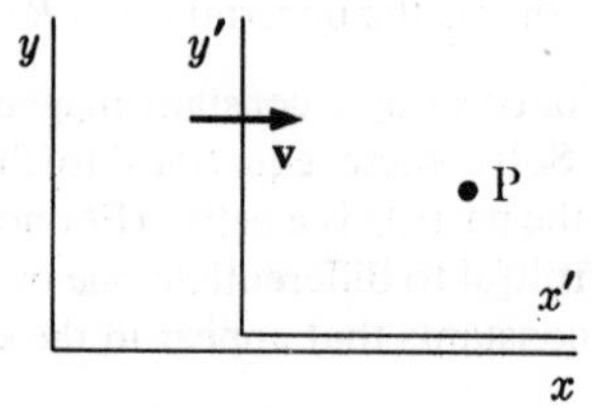

if the origins coincide at $t = 0$. These two equations, together with a statement of equal time measurements, $t' = t$, constitute a Galilean transformation. (2) Explain why the following equations also constitute a Galilean transformation:

$$x' = x + x_0 - vt,$$

$$y' = y + y_0,$$

$$t' = t + t_0.$$

Describe the relationship of the two frames of reference.

Description of motion in different frames of reference

P7.31 At $t = 0$, observer A in Figure 7.20(a) throws a ball downward from the origin of his coordinate system with initial speed $v_0 = 3$ m/sec. At this same moment, the origin of observer C's coordinate system [Figure 7.20(c)] is passing A's origin with a constant downward velocity of 5 m/sec. (1) Write the ball's equation of motion from C's point of view (x_C vs t). (2) At what time does the ball overtake and pass the origin of C's frame of reference? (3) Where is C relative to A at this time? (4) How should A throw a second ball in order that his description of its motion be identical to C's description of the motion of the first ball?

P7.32 In a building, let the origin of a vertical z axis be at the first-floor level and the origin of a vertical z' axis be at the floor level of an elevator. The elevator is ascending with constant speed v_0. At the moment when the two origins coincide, a fixture breaks loose from the ceiling of the elevator and falls. At that moment, the fixture is at $z = z' = h$. (1) Write expressions for $z(t)$ and $z'(t)$ that describe the fall of the fixture. (2) When and where does the fixture land in the building-fixed frame of reference? (3) When and where does it land in the elevator-fixed frame of reference?

P7.33 Two hockey players skate on parallel paths 3 m apart, Kurt at 10 m/sec and Lucas at 20 m/sec. Jason watches from the grandstand. Just as Lucas passes Kurt, Kurt shoots the puck across the ice to him. Lucas receives it 1.5 sec later. Describe the motion of the puck in the frames of reference of Jason, Kurt, and Lucas. What are the speeds v_J, v_K, and v_L of the puck in these three frames of reference?

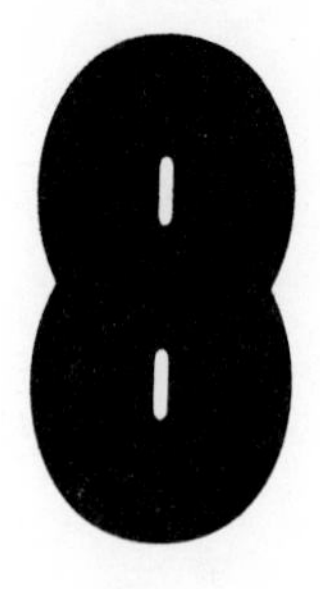

Momentum and the Motion of Systems: Newton's Third Law

Quantitative concepts form the core of a theory

Chapter 7 was built around four central concepts of mechanics—force, mass, length, and time. The number of important mechanical concepts is not large. Counting such derived concepts as velocity and acceleration, probably less than a dozen suffice for a complete foundation of mechanics. It is not far from the truth to say that when these concepts are understood, the science of mechanics is understood. But understanding a concept involves more than knowing simply a definition or an equation. To "understand" a concept thoroughly, one must know its operational definition, its dimension and unit, its relationship through laws and equations to other important concepts, its role in various parts of the physical world, and its typical magnitudes. In addition, one must acquire something of an intuitive "feeling" for the concept, a kind of immediate recognition and appreciation that can come only after the concept has been looked at from enough different angles and has been seen at work in enough different situations to make it seem a familiar friend. A word of caution: This worthwhile kind of scientific familiarity founded on quantitative study is very different from the illusory familiarity bred by nontechnical everyday usage of such words as force, energy, and acceleration.

Momentum is a key concept of mechanics

In a competition for "most fundamental concept" of mechanics, space and time would probably share first prize. Close behind would be the three key concepts: momentum, energy, and angular momentum. Each of these acquired importance in mechanics because of its appearance in a conservation law. Each proved useful in solving problems of motion. Each has weathered the twentieth-century revolution in physics and has emerged as a key concept in the new theories of relativity and quantum mechanics. Each has found a more solid foundation in physics through its relationship to a principle of symmetry in nature. Although we are concerned in this part of the book only with the classical mechanics of

Newton, it is worth knowing which parts of this theory remain valid on the contemporary frontiers of physics (the conservation of momentum, for example), and which parts have been found to have limited scope (the conservation of mass, for example).

In earlier chapters the conservation laws of momentum, energy, and angular momentum have been discussed. In this chapter we seek a deeper insight into the concept of momentum and related aspects of mechanics. Angular momentum is the theme of Chapter 9, and energy is the subject of Chapter 10.

8.1 Momentum and Newton's second law; impulse

As stated in Chapter 4, the momentum of a particle is defined, in classical mechanics, as the product of the mass and the velocity of the particle:

Classical definition of momentum

$$\mathbf{p} = m\mathbf{v}. \tag{8.1}$$

The usual notation for momentum is **p**. It is a vector quantity whose direction is the same as the direction of the velocity. Its physical dimension is mass × length/time; its SI unit is kg m/sec.

At speeds near the speed of light, the definition of Equation 8.1 has been found to be "wrong"—that is, not useful. The relativistic definition of momentum is given by Equation 21.26. It is important too to remember that Equation 8.1 is completely inappropriate for massless particles, which do possess momentum in spite of having no mass. In this chapter, however, we shall be concerned with realms of nature where Equation 8.1 adequately defines the momentum of a particle.

In earlier sections we have emphasized several times that momentum is a significant concept worth defining and studying because, under certain circumstances (in isolated systems), it is conserved. There is one other reason to define momentum. It is the property of motion most simply and directly related to force. The rate at which momentum changes is equal to the force applied. Differentiate Equation 8.1 to obtain (since m is constant)

$$\frac{d\mathbf{p}}{dt} = m\frac{d\mathbf{v}}{dt}. \tag{8.2}$$

The right side is mass times acceleration, which, according to Newton's second law, is equal to the force acting on the particle. Therefore Equation 8.2 may be replaced by

Newton's second law in terms of momentum

$$\mathbf{F} = \frac{d\mathbf{p}}{dt}. \tag{8.3}$$

This is another form of Newton's second law: *force equals rate of change of momentum.*

IMPULSE

If a constant force **F** acts on a particle, the momentum changes at a constant rate. After a time interval Δt, the change of momentum $\Delta\mathbf{p}$ will be given by

Change of momentum if **F = constant**

$$\Delta\mathbf{p} = \mathbf{F}\,\Delta t. \tag{8.4}$$

The combination $\mathbf{F}\,\Delta t$ on the right is called the "impulse." The change of momentum is equal to the impulse. More generally, impulse, for which we use the symbol **I**, is defined by means of an integral. We can write

$$\mathbf{p} = \int \left(\frac{d\mathbf{p}}{dt}\right) dt.$$

According to Equation 8.3, $d\mathbf{p}/dt$ can be replaced by **F** so that

$$\mathbf{p} = \int \mathbf{F}\, dt. \tag{8.5}$$

The integral on the right is an indefinite integral. In applications it is preferable to choose specific limits of time, t_1 and t_2, and make this a definite integral. Then

$$\mathbf{p}_2 - \mathbf{p}_1 = \int_{t_1}^{t_2} \mathbf{F}\, dt, \tag{8.6}$$

where $\mathbf{p}_2 - \mathbf{p}_1$ is the change of momentum in the time interval $t_2 - t_1$. The impulse **I** delivered to the particle in this interval is defined by

Definition of impulse

$$\mathbf{I} = \int_{t_1}^{t_2} \mathbf{F}\, dt. \tag{8.7}$$

If we abbreviate $\mathbf{p}_2 - \mathbf{p}_1$ by $\Delta\mathbf{p}$, Equation 8.6 becomes

Newton's second law in terms of impulse

$$\Delta\mathbf{p} = \mathbf{I}. \tag{8.8}$$

This is closest to the form in which Newton actually stated his second law.* He worked more often with impulse and momentum change than with force and acceleration. Note that for an infinitesimal time interval, denoted by dt, the impulse is $d\mathbf{I} = \mathbf{F}\, dt$ so that

$$d\mathbf{p} = \mathbf{F}\, dt. \tag{8.9}$$

This equality of infinitesimals, valid for any force, is analogous to the equality of finite quantities in Equation 8.4, valid for a constant force.

Impulse is closely related to average force. In general, the time average of a scalar quantity Q is defined by

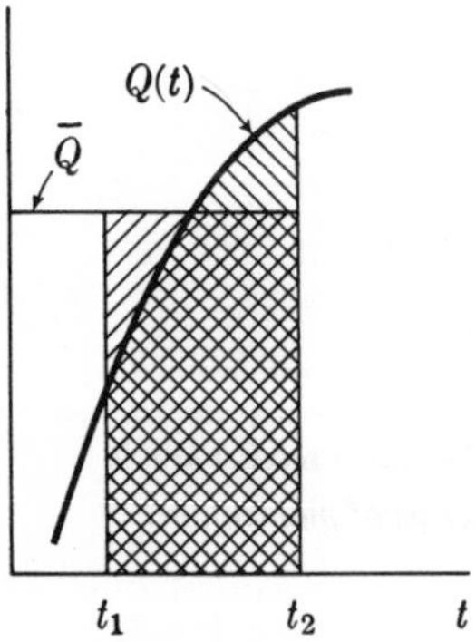

$$\bar{Q} = \frac{1}{t_2 - t_1} \int_{t_1}^{t_2} Q\, dt. \tag{8.10}$$

This means that the constant $\bar{Q}$ and the variable Q have the same time integrals from t_1 to t_2:

$$\int_{t_1}^{t_2} \bar{Q}\, dt = \bar{Q} \int_{t_1}^{t_2} dt = \bar{Q} \times (t_2 - t_1) = \int_{t_1}^{t_2} Q\, dt.$$

In a similar way, the time average of a vector quantity **V** is defined by

* Newton's statement is "The change of motion is proportional to the motive force impressed; and is made in the direction of the right line in which that force is impressed." This is a translation of Newton's original Latin. Since he uses "motion" for momentum and "force" for impulse, we must translate further into modern terminology and render his statement in this way: The change of momentum is proportional to the applied impulse; and its direction is parallel to the impulse.

$$\overline{\mathbf{F}} = \frac{1}{t_2 - t_1}\int_{t_1}^{t_2} \mathbf{F}\, dt. \tag{8.11}$$

(When not otherwise specified, "average" usually means time average. However, it is possible and sometimes useful to define averages with respect to variables other than time.) Comparison of Equations 8.11 and 8.7 shows that impulse can be simply expressed in terms of average force:

Another expression for impulse

$$\mathbf{I} = \overline{\mathbf{F}}\, \Delta t. \tag{8.12}$$

Here Δt stands for $t_2 - t_1$. This relation makes it possible to re-express Equation 8.8 in the form

$$\Delta \mathbf{p} = \overline{\mathbf{F}}\, \Delta t. \tag{8.13}$$

Compare Equation 8.13 with Equations 8.4 (constant force) and 8.9 (any force).

■ EXAMPLE 1: A particle of mass m moves along the x axis with speed v_0. Then a constant force of magnitude **F** acts in the direction of motion of the particle for a time Δt. What impulse is delivered to the particle? What is the final velocity of the particle? The magnitude of the impulse is $F\,\Delta t$. In vector form, the impulse is

$$\mathbf{I} = F\,\Delta t\,\mathbf{i},$$

where **i** is a unit vector in the x direction. The vector equation giving the change of momentum is

$$\mathbf{p}_2 - \mathbf{p}_1 = \mathbf{I}.$$

Its x component is, in this example,

$$mv_f - mv_0 = F\,\Delta t, \tag{8.14}$$

where v_f is the final speed of the particle. Solution of Equation 8.14 for v_f yields

$$v_f = v_0 + \frac{F}{m}\Delta t.$$

The direction of $\mathbf{v}_f$ is along the x axis. The same result could, of course, have been obtained using the equations of uniformly accelerated motion in the interval Δt. ■

■ EXAMPLE 2: A ball moving eastward with a momentum of 5 kg m/sec is struck a blow from the side that imparts to the ball a northward impulse of 10 N sec. What is the final momentum of the ball? If the blow lasted 1 msec (10^{-3} sec), what average force acted on the ball during this time? Let the x axis point to the east and the y axis point to the north (Figure 8.1). According to Equation 8.8, we may write

$$\mathbf{p}_2 - \mathbf{p}_1 = \mathbf{I},$$

or, equivalently,

$$\mathbf{p}_2 = \mathbf{p}_1 + \mathbf{I}. \tag{8.15}$$

Since $\mathbf{p}_1$ and **I** are given, it is a problem in vector addition to find $\mathbf{p}_2$. Note that the unit N sec is the same as the unit kg m/sec. The vector equation is

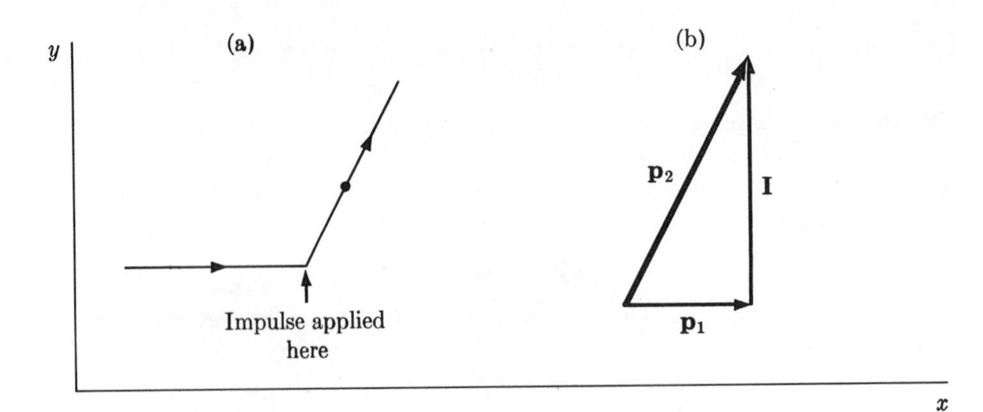

FIGURE 8.1 Particle deflected by an impulsive force. (a) Path of the particle. (b) Arrow diagram to obtain the final momentum $\mathbf{p}_2$ as the sum of the initial momentum and the impulse (Equation 8.15).

$$\mathbf{p}_2 = (5\mathbf{i} + 10\mathbf{j}) \text{ kg m/sec}.$$

Thus the final momentum $\mathbf{p}_2$ has magnitude equal to $\sqrt{5^2 + 10^2}$ kg m/sec = 11.2 kg m/sec. It is directed at an angle θ to the x axis defined by $\tan\theta = 2$ (see Figure 8.1). This angle is 63.4 deg. The second question is answered with the help of Equation 8.12. In terms of magnitudes,

$$\bar{F} = \frac{I}{\Delta t}.$$

Since $I = 10$ N sec and since $\Delta t = 10^{-3}$ sec, then $\bar{F} = 10^4$ N. The concept of impulse is especially useful in problems of this kind, in which a large force acts for a short time. ■

For a single particle on which no forces act, the conservation of momentum is equivalent to the conservation (or constancy) of velocity (Newton's first law). Only for combinations of two or more interacting particles—that is, for systems—does the conservation of momentum take on special interest and significance. Before pursuing conservation in general (Section 8.7), we must study the motion of systems and Newton's third law (Sections 8.2–8.6).

8.2 Center of mass

A system is any collection of two or more things, usually interacting with one another, such as the solar system or a collection of molecules making up a solid object. It is both fruitful and easy to study systems in quite general terms. Consider a set of interacting particles—any number of particles. They are labeled 1, 2, 3, and so on (Figure 8.2). Particle 1 has mass m_1, position vector $\mathbf{r}_1$, velocity $\mathbf{v}_1$, etc. The variables of other particles are similarly labeled. The total mass of the system we call M:

$$M = m_1 + m_2 + m_3 + \cdots. \tag{8.16}$$

Besides M, another important property of the system as a whole is its *center of*

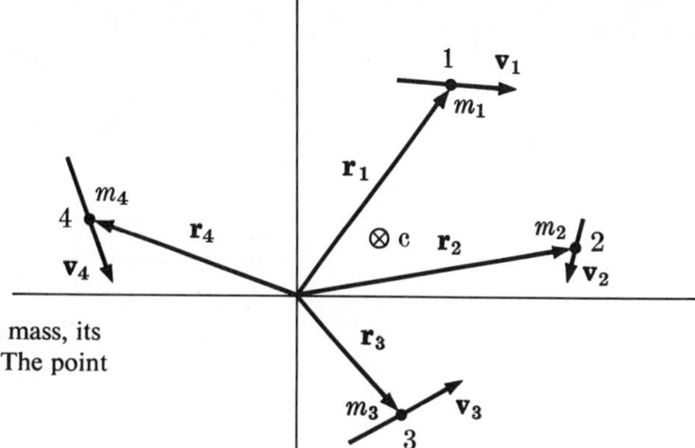

FIGURE 8.2 A system of particles. Each is characterized by its mass, its position vector, and its velocity, as well as by other variables. The point labeled c is the center of mass of the system.

mass, a point whose location is defined by

$$\mathbf{r}_c = \frac{m_1\mathbf{r}_1 + m_2\mathbf{r}_2 + m_3\mathbf{r}_3 + \cdots}{M}. \tag{8.17}$$

Definition of the center of mass

The center-of-mass vector is a *weighted average* of the individual position vectors. (The weighting factors are m_1/M, m_2/M, m_3/M, etc. Their sum is 1.) The center of mass can therefore be described as the average position of all of the mass in the system. In general, it need not be located at any material point of the system (see Figure 8.3).

It is a weighted average position

■ EXAMPLE 1: Where is the center of mass of a system consisting of just two particles? If the particles are equally massive, Equation 8.17 reduces to an average in the usual sense:

$$\mathbf{r}_c = \tfrac{1}{2}\mathbf{r}_1 + \tfrac{1}{2}\mathbf{r}_2.$$

This locates a point in space halfway between the two particles [Figure 8.3(a)]. For particles of unequal mass, the position vector of the more massive particle is weighted more heavily in the sum. Thus, if particle 1 is a proton ($m_1 = 1.67 \times 10^{-27}$ kg) and particle 2 is a deuteron ($m_2 = 3.34 \times 10^{-27}$ kg), the center-of-mass vector is

$$\mathbf{r}_c = \tfrac{1}{3}\mathbf{r}_1 + \tfrac{2}{3}\mathbf{r}_2.$$

Specifically, suppose that the proton is located at the origin of a coordinate system and that the deuteron is located on the x axis a distance 6 fm (6×10^{-15} m) away. Then $\mathbf{r}_1 = 0$, $\mathbf{r}_2 = 6\mathbf{i}$ fm, and

$$\mathbf{r}_c = 4\mathbf{i} \text{ fm},$$

a point 4 fm from the proton and 2 fm from the deuteron [Figure 8.3(b)]. ■

Centers of mass can often be accurately guessed without recourse to Equation 8.17. The center of mass of a uniform rod, for example, is at the center of the rod [Figure 8.3(c)]; that of a rectangular slab is at the center of the slab. For composite systems, a useful result, which we state without proof, is the following: If the centers of mass of each of several parts of a system are known, the center-of-mass vector of the whole system can be calculated as if the mass of each part is located at the center of mass of that part. For the

Centers of mass of symmetrical objects

Theorem: Subsystems can be treated as point particles

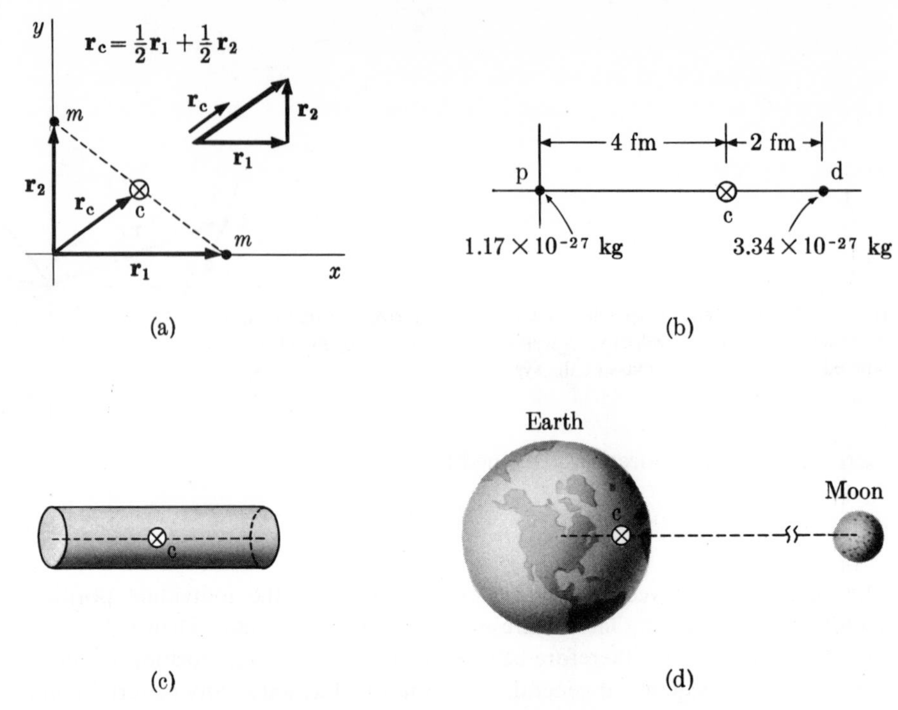

FIGURE 8.3 Centers of mass. (a) The center of mass of a pair of equally massive particles is a point in space halfway between the particles. To choose a coordinate axis along the line joining the particle would be easier, but it is not necessary. (b) The center of mass of a proton and a deuteron is located two-thirds of the way from proton to deuteron. (c) The center of mass of a uniform cylinder is located at its geometric center. (d) The center of mass of the earth-moon system is a point about 2,900 miles from the center of the earth.

earth-moon system, for instance, the center of mass may be found by first placing the mass of the earth at its center and the mass of the moon at its center, then calculating as though it were a two-particle system. The result shows the center of mass of the combined system to lie 2,900 miles from the center of the earth [Figure 8.3(d)].

As the particles comprising a system move, their center of mass moves. Even though it may be a point in space not attached to any matter, the center of mass is a well-defined point with a well-defined velocity. Differentiation of Equation 8.17 gives a formula for the center-of-mass velocity:

Velocity of center of mass

$$\mathbf{v}_c = \frac{d\mathbf{r}_c}{dt} = \frac{m_1\mathbf{v}_1 + m_2\mathbf{v}_2 + m_3\mathbf{v}_3 + \cdots}{M}. \tag{8.18}$$

The center-of-mass velocity is evidently a weighted average of the individual particle velocities. Similarly, the weighted average acceleration is the same as the acceleration of the center of mass:

Acceleration of center of mass

$$\mathbf{a}_c = \frac{d\mathbf{v}_c}{dt} = \frac{m_1\mathbf{a}_1 + m_2\mathbf{a}_2 + m_3\mathbf{a}_3 + \cdots}{M}. \tag{8.19}$$

■ EXAMPLE 2: A rigid body undergoes translational motion without rotation so that each part of it has the same acceleration **a** (Figure 8.4). What is the acceleration of the center of mass? It is almost obvious that the answer must be $\mathbf{a}_c = \mathbf{a}$. To prove this, subdivide the body into a multitude of "particles," and use the fact that $\mathbf{a}_1 = \mathbf{a}_2 = \mathbf{a}_3 = \cdots = \mathbf{a}$. Then Equation 8.19 becomes

$$\mathbf{a}_c = \frac{\mathbf{a}(m_1 + m_2 + m_3 + \cdots)}{M} = \mathbf{a}.$$

■

■ EXAMPLE 3: A billiard ball of mass m rolls with speed v toward another billiard ball, also of mass m (Figure 8.5). The position vectors of the two balls are

$$\mathbf{r}_1 = vt\mathbf{i} + y_0\mathbf{j},$$

$$\mathbf{r}_2 = x_0\mathbf{i} + y_0\mathbf{j}.$$

Find the position, velocity, and acceleration of their center of mass. Because the masses are equal, $\mathbf{r}_c = \frac{1}{2}\mathbf{r}_1 + \frac{1}{2}\mathbf{r}_2$, or

$$\mathbf{r}_c = (\tfrac{1}{2}x_0 + \tfrac{1}{2}vt)\mathbf{i} + y_0\mathbf{j}. \tag{8.20}$$

This is a moving point halfway between the two balls. Instead of taking the weighted average velocity, we may differentiate Equation 8.20 directly to obtain the center-of-mass velocity. It is

$$\mathbf{v}_c = \frac{d\mathbf{r}_c}{dt} = \tfrac{1}{2}v\mathbf{i},$$

half the velocity of the moving ball. Since $\mathbf{v}_c$ is constant, $\mathbf{a}_c = 0$. ■

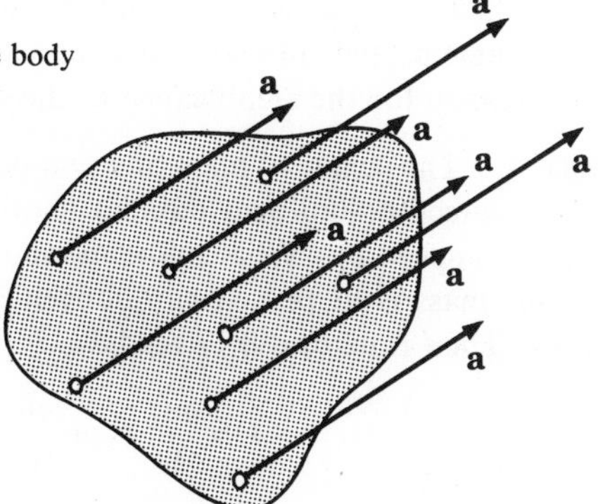

FIGURE 8.4 A rigid body undergoing translational motion. Every part of the body experiences the same acceleration **a**.

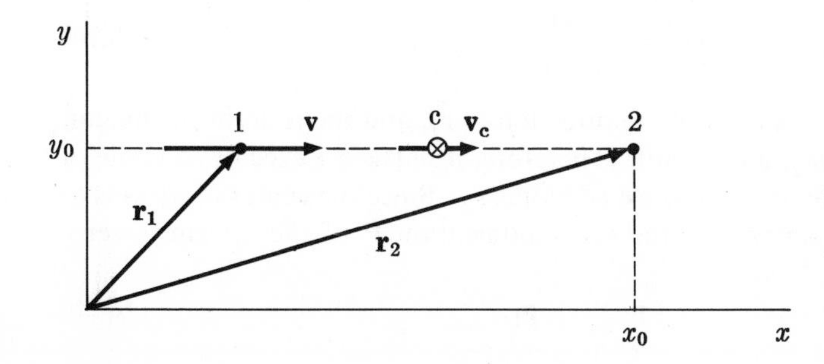

FIGURE 8.5 Billiard ball 1 moves with velocity **v** toward stationary billiard ball 2. The velocity of the center of mass is $\mathbf{v}_c = \frac{1}{2}\mathbf{v}$.

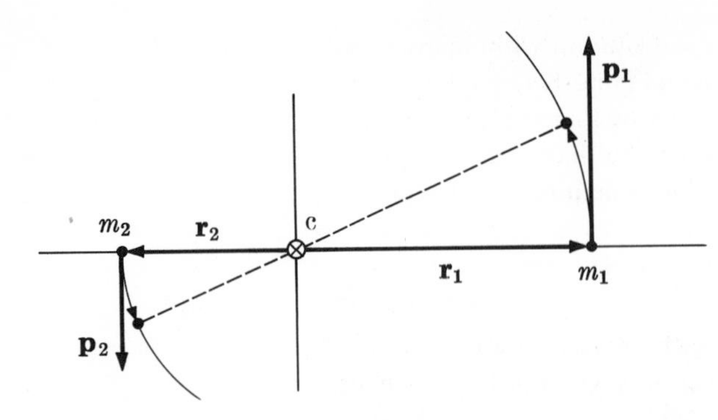

FIGURE 8.6 Two particles executing uniform circular motion about their fixed center of mass.

RELATIONSHIP OF MOMENTUM TO CENTER OF MASS

Returning to the consideration of an arbitrary number of particles, for the total momentum of a system we can write

$$\mathbf{P} = m_1\mathbf{v}_1 + m_2\mathbf{v}_2 + m_3\mathbf{v}_3 + \cdots. \tag{8.21}$$

Comparison of Equations 8.18 and 8.21 shows at once that the total momentum is simply related to the center-of-mass velocity:

An important expression for total momentum

$$\mathbf{P} = M\mathbf{v}_c. \tag{8.22}$$

From this follows also a simple expression for the time rate of change of total momentum:

$$\frac{d\mathbf{P}}{dt} = M\frac{d\mathbf{v}_c}{dt} = M\mathbf{a}_c. \tag{8.23}$$

These equations bear an obvious resemblance to the corresponding equations for a single particle: $\mathbf{p} = m\mathbf{v}$, and $d\mathbf{p}/dt = m\mathbf{a}$. One way to express the meaning of Equation 8.22 in words is to say that the momentum of a system is the same as that of a single particle of mass M located at the center of mass of the system. This is one reason for the significance of the center of mass.

The center of mass moves like a point particle of mass M

A two-body problem

■ EXAMPLE 4: The center of mass of a two-particle system is at rest. Particle 1, of mass m_1, executes uniform circular motion about the center of mass at distance r_1 and speed v_1. What is the motion of particle 2, of mass m_2? Since the center of mass is at rest, it can conveniently be placed at the origin (see Figure 8.6). Then Equation 8.17 reads

$$0 = \frac{m_1}{M}\mathbf{r}_1 + \frac{m_2}{M}\mathbf{r}_2.$$

Hence

$$\mathbf{r}_2 = -\left(\frac{m_1}{m_2}\right)\mathbf{r}_1. \tag{8.24}$$

The position vector $\mathbf{r}_2$ is opposite in direction to $\mathbf{r}_1$, and the ratio of the magnitudes is $r_2/r_1 = m_1/m_2$, a constant. Therefore, if particle 1 executes a circle of radius r_1, particle 2 executes a circle of radius r_2. Since the center of mass is at rest, Equation 8.22 states that the total momentum $\mathbf{P}$ of the system is zero. This means that

$$\mathbf{p}_2 = -\mathbf{p}_1. \tag{8.25}$$

The two momenta are equal in magnitude and opposite in direction. Therefore, $m_1v_1 = m_2v_2$, or

$$v_2 = \left(\frac{m_1}{m_2}\right) v_1. \tag{8.26}$$

If, for instance, particle 1 is an electron and particle 2 is a proton, with $m_1/m_2 = 1/1{,}836$ (and if we pretend that a classical description of their motion is permissible), the proton's orbital radius is 1,836 times smaller than the electron's, and its speed is 1,836 times less than the speed of the electron. Note, however, that their momenta are equal. ■

8.3 Newton's second law for systems

How to apply Newton's second law to a nonrigid system—the solar system, for instance—is not at once clear. The total force acting on the system and the total mass of the system are clear ideas. But what is "the" acceleration of a system whose individual parts are moving in different directions with different accelerations? As we shall see, the center-of-mass concept provides a way to answer this question.

As in the last section, let us consider a general system of any number of particles. These particles have momenta $\mathbf{p}_1, \mathbf{p}_2, \mathbf{p}_3, \ldots$, and experience forces $\mathbf{F}_1, \mathbf{F}_2, \mathbf{F}_3, \ldots$ (Figure 8.7). These forces may come from within the system or from without. The total force acting on all parts of the system is

$$\mathbf{F} = \mathbf{F}_1 + \mathbf{F}_2 + \mathbf{F}_3 + \cdots. \tag{8.27}$$

Newton's second law may be applied to each particle separately, giving the equations of motion

Each particle obeys Newton's second law

$$\mathbf{F}_1 = m_1\mathbf{a}_1,$$
$$\mathbf{F}_2 = m_2\mathbf{a}_2,$$
$$\vdots$$

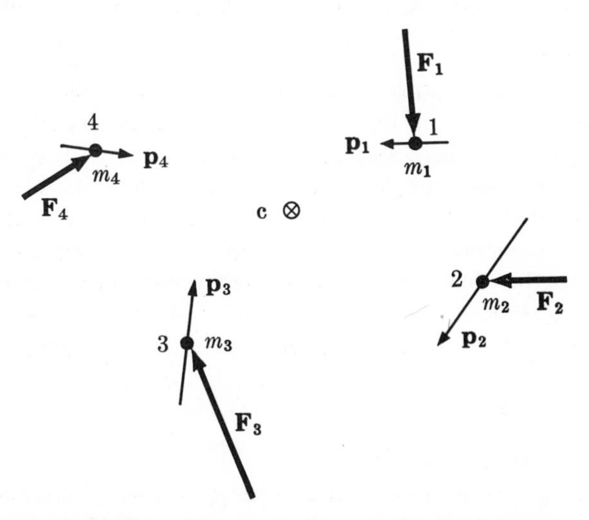

FIGURE 8.7 A system of particles with momenta $\mathbf{p}_i$ experiencing forces $\mathbf{F}_i$.

If the sum of the left sides of these equations is set equal to the sum of the right sides, the result is

$$\mathbf{F} = m_1\mathbf{a}_1 + m_2\mathbf{a}_2 + m_3\mathbf{a}_3 + \cdots. \tag{8.28}$$

According to Equation 8.19, the sum on the right is the total mass of the system times the acceleration of the center of mass. Therefore, Equation 8.28 can be rewritten very compactly:

Newton's second law for the total system

$$\mathbf{F} = M\mathbf{a}_c. \tag{8.29}$$

This we may call Newton's second law for the system as a whole. It says that the total force acting on all parts of the system is equal to the total mass times the acceleration of the center of mass. To express this interesting and beautifully simple result another way: The center of mass moves as if all of the mass of the system were concentrated there and as if all of the forces were applied there (even if there is, in fact, neither mass nor force there). The center of mass is obviously a very special point. It might be called the perfect representative point, the only point whose motion accurately reflects the average motion of the whole system. Figure 8.8 shows a wrench sliding across a smooth horizontal surface. Its center of mass, indicated by a black cross, executes straight-line motion at constant velocity, just as a single particle would, while the motion of every other part of the wrench is more complicated. This implies, as an experimental fact (with the help of Equation 8.29), that $\mathbf{F} = 0$. The sum of all the forces acting on all the constituent atoms in the wrench is zero. Such simplicity must be significant. This significance will be explored in the next two sections.

Why does the total force on the wrench vanish?

Because of the connection between momentum and center of mass brought out in the last section (see, in particular, Equation 8.23), Equation 8.29 may also be written

Another form of Newton's second law for the total system

$$\mathbf{F} = \frac{d\mathbf{P}}{dt}, \tag{8.30}$$

another form of Newton's second law for the system as a whole.

■ EXAMPLE: In the two-particle system considered in Example 4 of the preceding section and illustrated in Figure 8.6, what forces act on the particles? What is the total force? Either Equation 8.29 or Equation 8.30 implies that the total force is $\mathbf{F} = 0$ since the center of mass was said to be at rest, which means $\mathbf{a}_c = 0$ or $d\mathbf{P}/dt = 0$. The vanishing of the total force can also be verified by considering the individual forces. The nature of the forces need not be known (that is, whether they are electrical, gravitational, or something else) since the known accelerations determine the forces. Particle 1 moves uniformly in a circle. It therefore experiences a centripetal force of magnitude (from Equation 7.66)

$$F_1 = \frac{m_1 v_1^{\,2}}{r_1}.$$

This force is directed toward the center of mass (or, what is the same thing, toward particle 2). The oppositely directed force $\mathbf{F}_2$ has magnitude

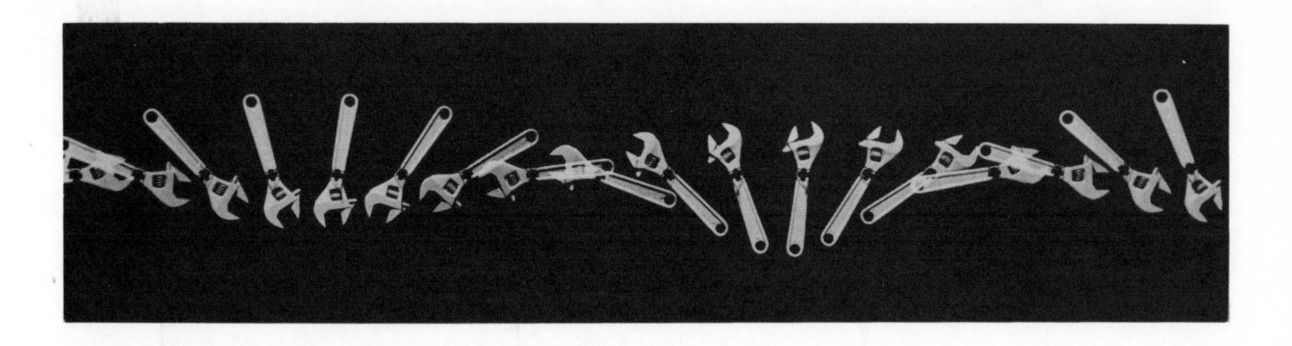

FIGURE 8.8 Motion of wrench without friction. Only the center of mass (black cross) moves with constant velocity. (Photograph from P.S.S.C. *Physics*, D. C. Heath & Co., 1965.)

$$F_2 = \frac{m_2 v_2^{\,2}}{r_2}.$$

Now use Equation 8.24 to substitute for the magnitude of $\mathbf{r}_2$, and use Equation 8.26 to substitute for v_2, producing

$$F_2 = \frac{m_1 v_1^{\,2}}{r_1} = F_1. \tag{8.31}$$

The magnitudes are equal. Since the directions are opposite,

$$\mathbf{F} = \mathbf{F}_1 + \mathbf{F}_2 = 0.$$

■

8.4 Newton's third law

A modern statement of Newton's third law is this: *If two bodies* A *and* B *interact, the force exerted by* B *on* A *is equal in magnitude and opposite in direction to the force exerted by* A *on* B. Mathematically, it is

Newton's third law

$$\mathbf{F}_{\mathrm{AB}} = -\mathbf{F}_{\mathrm{BA}}. \tag{8.32}$$

As a matter of convention, we shall let the first subscript designate the body *experiencing* the force and the second subscript designate the body *exerting* the force (see Figure 8.9). In somewhat more general terms, the law can be stated in this way: *All of nature's forces come in equal and opposite matched pairs.** Note that the third law requires for its application two (or more) interacting bodies. The first and second laws can be phrased in terms of single bodies.

Another statement of the third law

A most important aspect of Newton's third law to understand is that it is a law of force, not directly a law of motion. It is a statement about forces exerted and experienced by particles or objects regardless of how they might be moving. It states, for example, that if a book experiences a downward force of gravity of

It is a law of forces, not a law of motion

* It is a common practice in physics—one that we follow here—to call a pair of vectors whose sum is zero "equal and opposite." This means, of course, that they are equal in magnitude and opposite in direction.

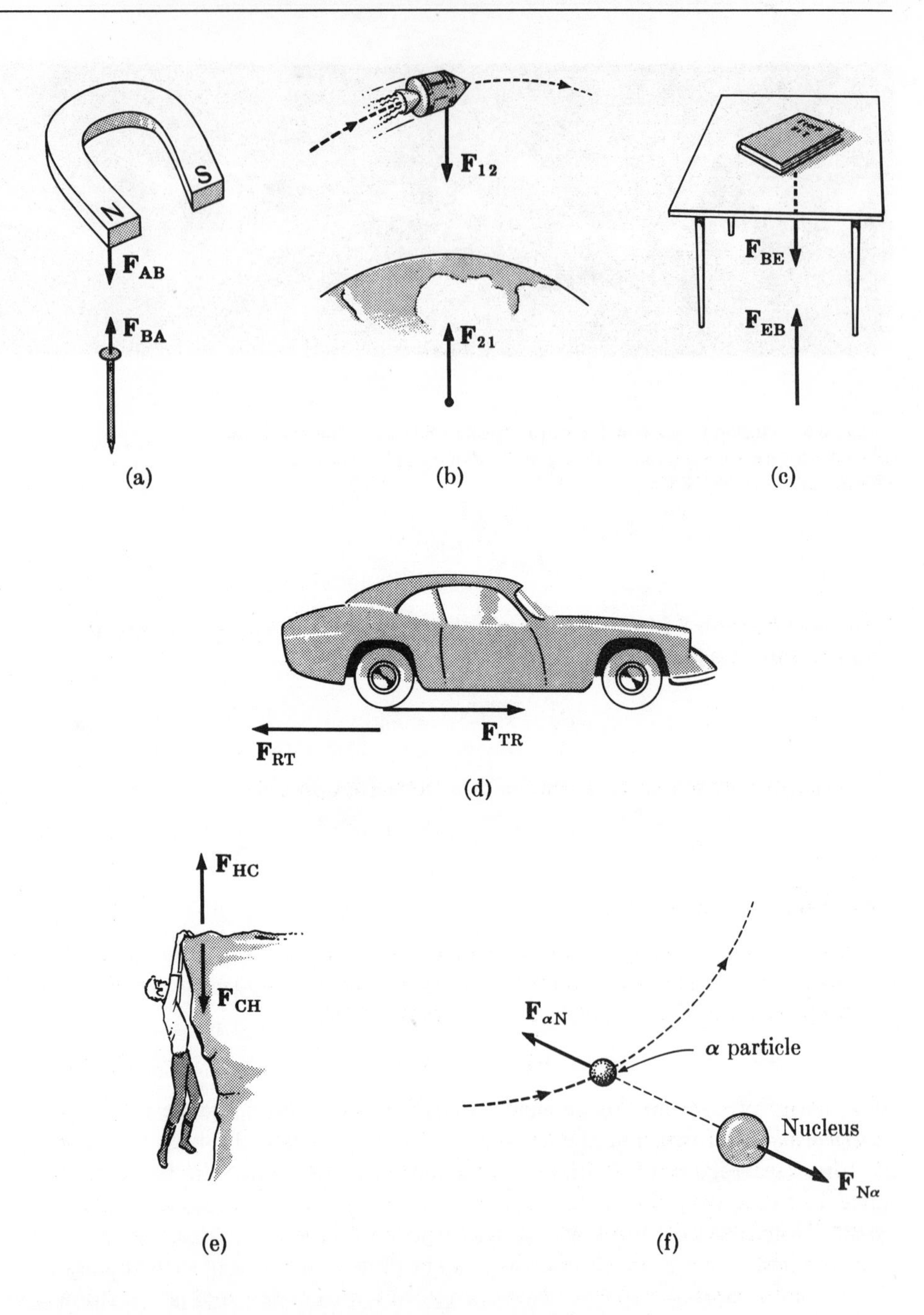

FIGURE 8.9 Examples of the action of Newton's third law: $\mathbf{F}_{AB} = -\mathbf{F}_{BA}$. (a) A nail pulls a magnet with a force equal and opposite to that of the magnet on the nail. (b) Satellite and earth, or (c) book and earth, pull equally on each other, regardless of their relative motion. (d) The force of a tire on the road is balanced by the propulsive force of the road on the tire. (e) The force of hands on cliff is equal and opposite to the force of cliff on hands. (f) An alpha particle and a nucleus repel one another with equal magnitudes of force.

10 N, the earth is pulled upward by the book with a net force of 10 N, regardless of whether the book is being lifted or rests on a table or falls to the floor and regardless of what other forces might be influencing the book and the earth. The motion of the earth and the motion of the book are determined by the *total* force acting on each (Newton's second law). Among all the forces possibly acting on the book, one is the earth's force of gravity. Among all the forces acting on the earth, one is the gravitational pull by the book. According to Newton's third law, these particular two forces are equal and opposite, a matched pair.

A second important point to appreciate about Newton's third law is its generality, a generality springing from what it does *not* say. It does not say anything about motion. It does not say anything about forces on object A arising from any source other than object B. It does not say anything about the kind of force acting between A and B. It is a completely general statement about forces in nature, independent of their strength, their source, their kind, or their effect. From Newton's third law one cannot learn the strength, for instance, or even the direction of the gravitational force exerted by the earth on a book. That knowledge comes from a particular law of force, the law of universal gravitation. The third law states only that the force of the earth on the book—whatever its strength and whatever its nature—is matched by an opposite force of the book on the earth. Moreover, it leaves open the possibility that as yet undiscovered forces may act between bodies, If they do, and if Newton's third law remains valid, the new forces too must come in equal and opposite pairs. When Newton formulated the third law, only the gravitational force was understood quantitatively. Later electric and magnetic forces were found to obey the third law. Ordinary contact forces, basically electric in nature, conform to the same law. In the subatomic domain, the strong and weak interactions have been found to be consistent with the law of momentum conservation, which may be considered to be a modern synthesis of Newton's first and third laws.

It governs all forces but specifies none

Although the idea of equal and opposite forces seems easy enough to grasp, it can sometimes be perplexing in practical application, and it has some unexpected implications. Consider the "cart and horse paradox." If a cart pulls backward on a horse with exactly the same magnitude of force that the horse pulls forward on the cart, how is it that cart and horse are able to move at all? Or in more general terms: If every force in nature is canceled by an opposite force, why is there any acceleration in the world? The resolution of the apparent paradox rests on one key idea: *The equal and opposite forces of a matched pair do not act on the same body.* The horse exerts *on the cart* a force $\mathbf{F}_{CH}$ (Figure 8.10). The force $\mathbf{F}_{HC}$ that is equal and opposite to $\mathbf{F}_{CH}$ does not act on the cart, but on the horse. The motion of the cart is determined only by the forces acting on it, not by any forces it may be exerting on other things. If we designate by $\mathbf{F}_C$ the total force acting on the cart (equal to $\mathbf{F}_{CH}$ plus frictional force from the ground plus the force of air resistance), we may apply Newton's second law to the cart alone:

The "cart and horse paradox"

$$\mathbf{F}_C = m_C \mathbf{a}_C. \tag{8.33}$$

Its mass times its acceleration is equal to the total force acting on it. The force

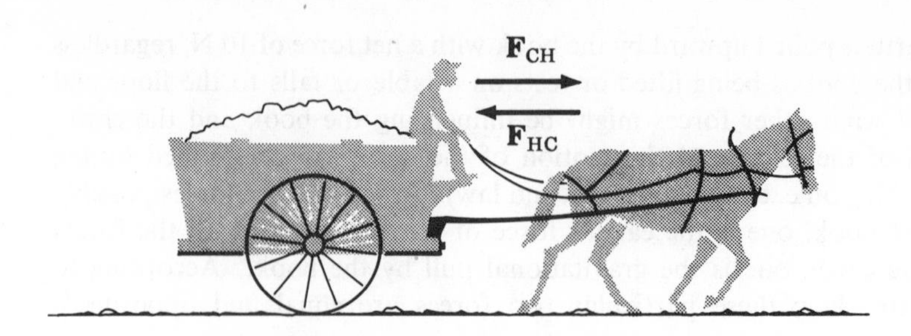

FIGURE 8.10 The "cart and horse paradox." Why does the system accelerate?

of cart on horse that "cancels" the force of horse on cart actually has nothing to do with the motion of the cart. Since Newton's third law does not relate the different forces acting *on* the cart, they need not cancel, and the cart may accelerate.

Propulsion of vehicles

Consider now the interesting application of Newton's third law where it does apply—to the interaction between ground and horse. Because the horse pushes backward on the ground, the ground pushes forward on the horse with an equal and opposite force. If the horse accelerates forward, it is because this force supplied by the ground is greater in magnitude than the rearward force supplied by the cart. Friction, far from being the retarding force it is usually assumed to be, is in this case the motive force driving the horse ahead. What drives an automobile ahead? Not its engine, but the road. Rails push a locomotive ahead; the air propels an airplane (even a jet). The engines of the automobile, locomotive, or airplane do not push the vehicle forward; they push some material (roadway, rails, or air) backward. The material, constrained by Newton's third law, must react to push the engines, and the vehicles attached to them, forward. Only a rocket, with nothing to push against, must use a different method of propulsion (Section 8.10).

Newton's version of his third law

Newton's original phrasing of his third law is of some interest, in part because the idea of "action and reaction" has entered the common language. Reference to it is often heard in settings far removed from physics. Newton wrote: "To every action there is always opposed an equal reaction; or the mutual actions of two bodies upon each other are always equal, and directed to contrary parts." In modern terminology, the same statement would be rendered as follows: For every force in nature there is always an equal and opposite force; that is, the mutual forces of two bodies upon each other are always equal in magnitude and opposite in direction. Obviously, the statements of the third law that appear at the beginning of this section do not differ in substance from Newton's statement. Why Newton chose to use the word "action" instead of the word "force"* is not certain—perhaps to let "action" summarize both the idea of "force" and the idea of "change of momentum," perhaps only for the

* In his original Latin, Newton refers to *vis* (usually translated "force") in his statements of the first and second laws, and to *actio* (usually translated "action") in his statement of the third law.

linguistic reason that "action" has a convenient opposite, "reaction," whereas "force" does not.

8.5 The cancellation of internal forces

No consequence of Newton's third law is more important than the cancellation of internal forces in every system. An *internal force* is one whose source and action (or whose cause and effect) are both within the system being considered. If cart and horse together are being considered as a single system, the force of the cart on the horse is an internal force. For any material object, the forces among its atomic constituents are internal forces. An *external force*, on the other hand, is one that acts on a system but whose source is outside the system. If the cart alone is treated as a system, both the force supplied by the horse and the force of friction supplied by the ground are external forces. The same force may, of course, be external for one system and internal for another. The sun's force on the earth is an external force if the earth alone is regarded as the system of interest. In the solar system as a whole, the same force is an internal force.

Definition of an internal force

Definition of an external force

For any system, then, the total force may be divided into an internal part and an external part:

$$\mathbf{F} = \mathbf{F}_{\text{int}} + \mathbf{F}_{\text{ext}}. \tag{8.34}$$

The definitions of internal and external forces together with Newton's third law at once imply that the sum of internal forces must be zero. For every force there is an equal and opposite force. For an internal force, the opposite force also acts within the system and is itself an internal force. In brief, all internal forces come in balanced pairs. Since the sum of each pair is zero, the sum of all pairs is zero. Thus we can write

$$\mathbf{F}_{\text{int}} = 0, \tag{8.35}$$

$$\mathbf{F} = \mathbf{F}_{\text{ext}}. \tag{8.36}$$

A vital implication of Newton's third law: Internal forces sum to zero

This means that Equations 8.29 and 8.30 can be rewritten

$$\mathbf{F}_{\text{ext}} = M\mathbf{a}_c, \tag{8.37}$$

$$\mathbf{F}_{\text{ext}} = \frac{d\mathbf{P}}{dt}. \tag{8.38}$$

Improved forms of Newton's second law for systems

These are the important final versions of Newton's second law for systems.

Equation 8.37 states that the center of mass of a system responds to external forces as if it were a single particle of mass M. This explains at once why the center of mass of the wrench pictured in Figure 8.8 moves with constant velocity (since, for the wrench, $\mathbf{F}_{\text{ext}} = 0$). It explains too why a billiard ball and an elementary particle follow the same laws of mechanics, despite the complexity of structure of the billiard ball [Figure 8.11(a)]. For the earth-moon system, it is the center of mass, not any other point, that follows a smooth elliptical path around the sun [Figure 8.11(b)]. The cancellation of internal forces brings to the description of complex systems the same simplicity as it does to the description of single particles.

Because $\mathbf{F}_{\text{int}} = 0$, complex systems have simple bulk properties

There exists, in fact, a complete correspondence between the motion of particles and the motion of systems. Here is a summary of the important equations:

Analogous equations govern systems and single particles

For a Single Particle	For a System
$\mathbf{p} = m\mathbf{v}$	$\mathbf{P} = M\mathbf{v}_c$
$\mathbf{F} = m\mathbf{a}$	$\mathbf{F}_{ext} = M\mathbf{a}_c$
$\mathbf{F} = \dfrac{d\mathbf{p}}{dt}$	$\mathbf{F}_{ext} = \dfrac{d\mathbf{P}}{dt}$.

The equations of the first line define momentum; the equations of the next two lines give alternative forms of Newton's second law.

THE CONNECTION OF NEWTON'S THIRD LAW TO NEWTON'S FIRST AND SECOND LAWS

The canceling of internal forces in every system makes for more than convenience in mechanics. One might almost say it makes mechanics possible. We are so accustomed to the fact that rigid objects accelerate only in response to external forces that we may easily overlook the importance of the canceling internal forces. If internal forces had any effect on the motion of macroscopic objects, Galileo would not have learned about uniform acceleration near the earth, and Newton would not have learned the nature of the gravitational force from the study of planetary motion. In our everyday world, objects on which no outside forces act would spontaneously accelerate. The fact that Newton's second law applies to "bodies" and "objects" (which are themselves complicated systems) depends directly on the validity of Newton's third law.

Newton's second law requires the third law

For an isolated body, $\mathbf{F}_{ext} = 0$. If such a body is to move without acceleration (Newton's first law), it is, of course, also necessary that $\mathbf{F}_{int} = 0$. This means that Newton's third law underlies the first law and the definition of an inertial frame of reference, just as it underlies the second law. This important connection was dealt with, in more qualitative terms, in Section 4.8.

The first law also requires the third law

These considerations show that Newton's first and second laws, as originally phrased, are not quite adequate. To say that a body moves with constant velocity or with some particular acceleration is ambiguous, since different parts of the body may move differently (see the wrench in Figure 8.8). To remove the ambiguity, these laws should refer to the velocity and the acceleration of the center of mass.

With the greater insight afforded by Newton's third law, we can now rephrase Newton's first and second laws in a more satisfactory and more general form:

Newton's first and second laws rephrased

1. The center of mass of an isolated system (one free of external influences) moves with constant velocity.
2. The net external force acting on a system of mass M is related to the acceleration of the center of mass of the system by $\mathbf{F}_{ext} = M\mathbf{a}_c$.

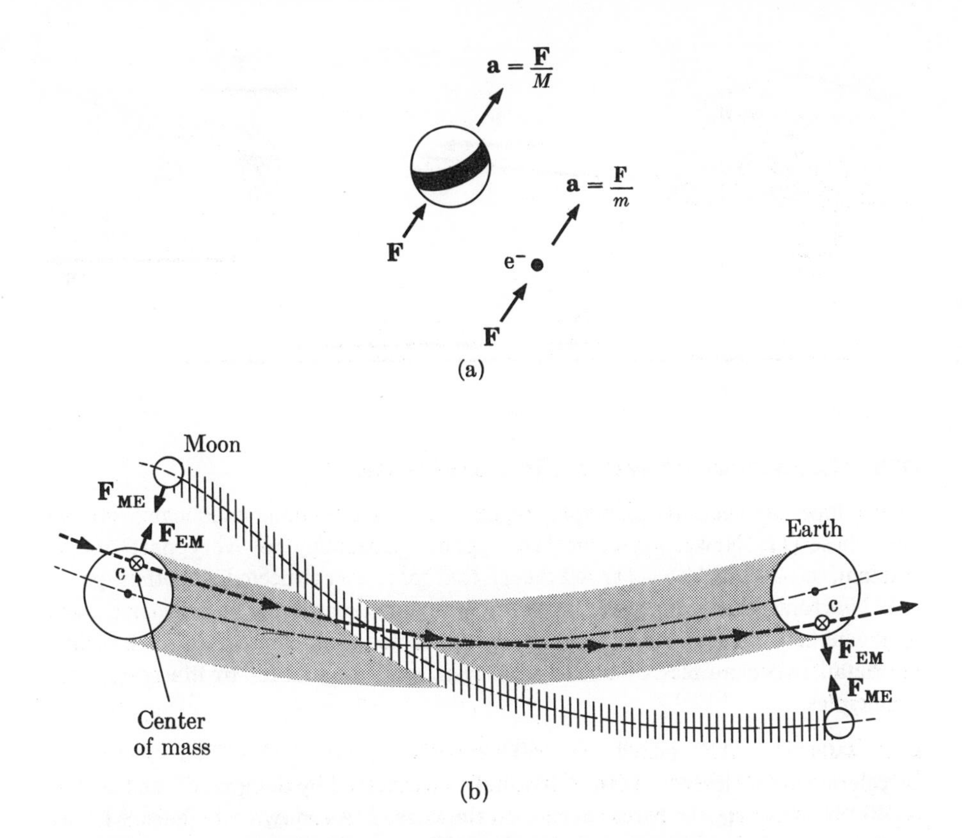

FIGURE 8.11 The cancellation of internal forces. (a) A billiard ball responds to an applied force as a single entity, in the same way as a single electron. (b) The forces of attraction between moon and earth are internal forces if the earth and moon together are regarded as a single system. These forces have no effect on the motion of the center of mass of the earth-moon system, which follows a smooth elliptical path about the sun. (This drawing is not to scale.)

We can see now that Newton's third law is *required* to validate the first and second laws.

Finally, let us examine the cart-and-horse paradox on a global scale. If every force in nature is canceled by an opposite force, why is there any acceleration in the world? The answer: For an isolated system, indeed there is no acceleration! At least there is no acceleration of the center of mass, the perfect representative point of the system. Individual parts of an isolated system experience acceleration, but that is because those parts are not themselves isolated. Anything in nature, to be accelerated, must be linked to something else outside itself (not necessarily connected by material bonds, of course, but linked through forces). We see acceleration and the interesting complexities of motion only when we fix attention on some part of a total system. If we were to take an Olympian view of nature, noting only the average motion (the center-of-mass motion) of isolated systems, we would see nothing but constant-velocity motion.

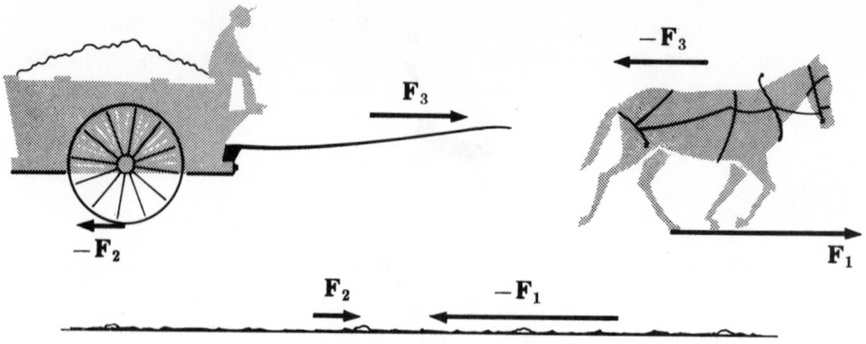

FIGURE 8.12 Cart, horse, and earth, shown as separate systems. The total force acting on any one of these systems need not be zero.

8.6 The motion of systems in one dimension

Two practical principles

Two important general principles of great utility in solving practical problems are these: (1) Newton's second law applies separately to every part of any system, no matter how (or whether) that part is connected to other parts. (2) Newton's third law applies to any two parts of a system, no matter how (or whether) those parts are physically connected. In this section we consider in detail two examples of one-dimensional motion in order to illustrate these principles.

■ EXAMPLE 1: A horse of mass 500 kg pulls a cart of mass 300 kg with an acceleration of 2 m/sec^2. If the frictional force exerted by the ground on the cart is 400 N, what are the forces acting on the horse? As shown in Figure 8.12, we call the magnitude of the forward frictional force on the horse F_1, the magnitude of the backward frictional force on the cart F_2, and the magnitudes of the equal and opposite forces acting between horse and cart F_3. By recognizing that two forces have the same magnitude F_3, we have already taken advantage of Newton's third law and the second of the two principles laid down in the preceding paragraph. We utilize the first principle by applying Newton's second law in three different ways: to the horse alone, to the cart alone, and to horse and cart together. The equations for horizontal components are

$$\text{horse:}\quad F_1 - F_3 = m_{\rm H}a; \tag{8.39}$$

$$\text{cart:}\quad F_3 - F_2 = m_{\rm C}a; \tag{8.40}$$

$$\text{horse and cart:}\quad F_1 - F_2 = (m_{\rm H} + m_{\rm C})a. \tag{8.41}$$

Since horse and cart have the same acceleration, the same symbol a appears in all three equations. It is very important to notice that if the horse is chosen as the system of interest, the force $-\mathbf{F}_2$ is irrelevant despite the physical connection that exists between the horse and cart. In thinking of the horse alone, one must mentally erase the cart and the earth from the picture. There stands the horse alone in space, accelerating under the combined action of the forces $\mathbf{F}_1$ and $-\mathbf{F}_3$ that he actually experiences. Similarly, the cart, despite its physical connection to the horse, can be regarded as a separate system accelerating under the action of forces $\mathbf{F}_3$ and $-\mathbf{F}_2$.

Not all three of Equations 8.39–8.41 are necessary, although it is a good idea to examine all three to decide which pair can yield the solution most easily. Since F_2 is known (400 N), it is simplest to use the second and third equations, 8.40 and 8.41, which contain F_2. Numerically, these two equations take the following form:

$$\text{cart:}\quad F_3 - 400\text{ N} = 300\text{ kg} \times 2\text{ m/sec}^2 = 600\text{ N};$$

$$\text{horse and cart:}\quad F_1 - 400\text{ N} = 800\text{ kg} \times 2\text{ m/sec}^2 = 1{,}600\text{ N}.$$

Their solutions provide the desired answers, the forces acting on the horse:

$$F_1 = 2{,}000\text{ N},$$

$$F_3 = 1{,}000\text{ N}.$$

Equation 8.39 may be used to check these answers. Its left and right sides are each found to be equal to 1,000 N, verifying the consistency of the answers obtained.

It is left as an exercise to calculate the leftward acceleration of the earth arising from the unbalanced pair of forces $-\mathbf{F}_1$ and $\mathbf{F}_2$ acting on the ground (Figure 8.12). ■

■ EXAMPLE 2: A three-link chain is being raised by a hook (Figure 8.13). Each link in the chain has a mass of 10 kg, and the upward force supplied by the hook is 414 N. What is the acceleration of the chain? What are the forces acting on the middle link? Included in Figure 8.13 are the directions of the various forces acting and the symbols for their magnitudes: F_H for the upward force of the hook, F_G for the downward force of gravity on each link, and F_1 and F_2 for the pairs of equal and opposite forces acting between the links. A problem of this kind requires a methodical approach, and it requires a decision—a decision about what to pick as the systems of interest. It can do no harm to pick more systems or parts than are actually needed. We may, for example, apply Newton's second law both to the chain as a whole and also to each link separately. The equations relating the vertical components of force and acceleration are

Every system or subsystem obeys Newton's second law

$$\text{whole chain:}\quad F_H - 3F_G = 3ma; \tag{8.42}$$

$$\text{upper link:}\quad F_H - F_G - F_1 = ma; \tag{8.43}$$

$$\text{middle link:}\quad F_1 - F_G - F_2 = ma; \tag{8.44}$$

$$\text{lower link:}\quad F_2 - F_G = ma. \tag{8.45}$$

The acceleration of the chain is represented by a, the mass of each link by m. In the first equation the total mass $3m$ appears; in each of the others the mass m of a single link appears. Upward-acting forces appear with plus signs, downward-acting forces with minus signs.

In this example, the knowns are F_H (414 N), m (10 kg), and F_G ($F_G = mg = 10 \times 9.8 = 98$ N). The unknowns are F_1, F_2, and a. From the first of the four equations, the acceleration may be found. That equation—Newton's second law applied to the chain as a whole—is, numerically,

$$414\text{ N} - 3 \times 98\text{ N} = 30\text{ kg} \times a.$$

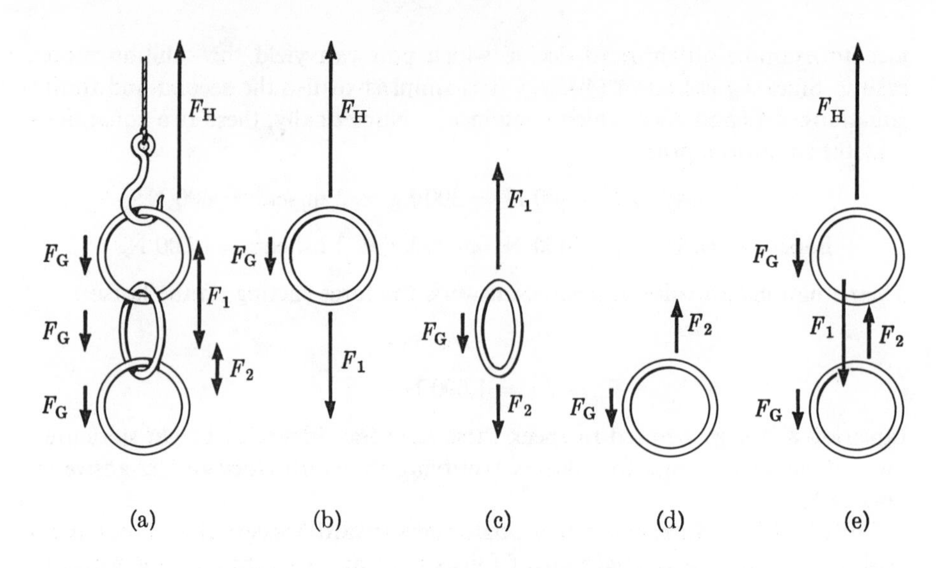

FIGURE 8.13 Acceleration of a three-link chain. (a) The whole chain is regarded as the system of interest. (b) The upper link alone is considered. (c) The middle link alone is considered. (d) The lower link alone is considered. (e) The upper and lower links together are treated as a single system. Newton's second law applies to every part or combination of parts.

Its solution gives the acceleration,

$$a = 4.0 \text{ m/sec}^2.$$

Now the upper-link and lower-link equations may be used to find the forces F_1 and F_2. In numerical form, they are

$$\text{upper link:} \quad 414 - 98 - F_1 = 10 \text{ kg} \times 4 \text{ m/sec}^2 = 40 \text{ N};$$

$$\text{lower link:} \quad F_2 - 98 = 10 \text{ kg} \times 4 \text{ m/sec}^2 = 40 \text{ N}.$$

Their solutions give the two interaction forces,

$$F_1 = 276 \text{ N},$$

$$F_2 = 138 \text{ N}.$$

The middle-link equation can serve as a check of these answers. Its left side is

$$F_1 - F_G - F_2 = 276 - 98 - 138 = 40 \text{ N}.$$

Its right side is

$$ma = 10 \text{ kg} \times 4 \text{ m/sec}^2 = 40 \text{ N}.$$

To show the power of the principle that Newton's second law may be applied to any part of a system and to obtain a further check on our answers, we may choose as the system of interest the upper link plus the lower link. In Figure 8.13(e), the hook and the middle link have been dissolved away. The remaining system has a mass of 20 kg and, according to our solution, both parts of it have the same upward acceleration, 4 m/sec². What are the forces acting on this system? Acting upward are the forces F_H on the upper link and F_2 on the lower link. Acting downward are the gravitational force $2F_G$ and the force F_1 on the upper link. The net upward force on this two-link system is

$$F_{up} = F_H + F_2 - 2F_G - F_1$$
$$= 414 + 138 - 196 - 276$$
$$= 80 \text{ N}.$$

This is equal, as it should be, to $2ma$:

$$2 \times 10 \text{ kg} \times 4 \text{ m/sec}^2 = 80 \text{ N}.$$ ■

8.7 Conservation of momentum

According to Equation 8.38 ($\mathbf{F}_{ext} = d\mathbf{P}/dt$), the total momentum of a system changes in response to external forces acting on the system. From this it follows at once that in the absence of external force (in particular for an isolated system), the total momentum is constant:

Conservation of momentum

$$\mathbf{P} = \textbf{constant} \qquad (\mathbf{F}_{ext} = 0). \tag{8.46}$$

This appears to be little more than a restatement of Newton's first law for a system. Its significance as a conservation law is clearer if we consider the individual parts of the system and write

Constancy during change: The sum of a set of variables is conserved

$$\mathbf{P} = \mathbf{p}_1 + \mathbf{p}_2 + \mathbf{p}_3 + \cdots = \textbf{constant.} \tag{8.47}$$

Each of the *individual* momenta may change in time, but their *sum* is conserved if the system is free of external force. Note that we are dealing here with a vector sum. The total momentum is constant in magnitude and in direction.

It is a vector law

DECAY OF UNSTABLE PARTICLES

One domain in which the conservation of momentum is of special importance is the submicroscopic realm of elementary particles, where almost all interaction events occur among a small number of particles that are effectively isolated from the rest of the universe. The example of the decay of an unstable particle into two other particles, the so-called two-body decay, was discussed in Section 4.5 (see Figure 4.1). The observation of two particles emerging from a decay process "back to back" with total momentum zero (in a frame of reference in which the parent particle was at rest) provides good evidence that no third particle escaped detection. If only one of the two product particles is seen, the law of momentum conservation can be used to help fix the properties of the unseen particle. In three-body decays of a particle at rest, the final momenta of the product particles must still sum to zero, but there are infinitely many ways in which the three particles may fan out without violating either the law of energy conservation or the law of momentum conservation. The three-body decay of a positive kaon, denoted by

Momentum conservation pins down unseen particles

$$K^+ \rightarrow \pi^+ + \pi^+ + \pi^-,$$

is illustrated in Figure 8.14.

MACROSCOPIC EXAMPLES: THE BURSTING OF SHELLS

An event in the macroscopic world not unlike the decay of an elementary particle is the bursting of an artillery shell into two or more fragments.

■ EXAMPLE 1: Suppose that a shell fired vertically upward explodes into two fragments just as it reaches its peak height. One fragment, with a mass of 5 kg, shoots downward with a speed of 600 m/sec. The mass of the other fragment is 10 kg. In what direction and with what speed does it fly off? We must first decide whether or not the shell is adequately approximated as an isolated system. In the explosion, the downward-going fragment is given an impulse $I = mv =$ 3,000 kg m/sec. If the duration of the explosion is 1 msec (a mere guess), the average explosive force experienced by this fragment is (from Equation 8.12)

$$\bar{F} = \frac{I}{\Delta t} = 3 \times 10^6 \text{ N}.$$

By comparison, the external force of gravity acting on this fragment is

$$F_{\text{grav}} = mg \cong 50 \text{ N}.$$

We may assume that the force of air resistance is also small in comparison with the explosive force, so for the time scale of the explosion, the external forces may be ignored.* This means that if the momentum just before the explosion is zero, it must also be zero just after the explosion. Here we see clearly the value of the principle of momentum conservation. Regardless of the complexity of the explosive internal forces and regardless of our ignorance of their exact nature (as true for elementary particles as for artillery shells), momentum conservation provides a simple link between "before" and "after." After the explosion,

$$\mathbf{p}_1 + \mathbf{p}_2 = 0,$$

so the 10-kg fragment must fly vertically upward with initial magnitude of momentum

$$p_2 = p_1 = 3{,}000 \text{ kg m/sec}.$$

Since $m_2 = 10$ kg, the initial speed of the second fragment is

$$v_2 = 300 \text{ m/sec}.$$ ■

■ EXAMPLE 2: An explosion (or decay) into two fragments results in equal and opposite momenta only if the shell (or particle) is initially at rest. Suppose instead that a shell of mass $m_0 = 10$ kg is moving horizontally with a speed $v_0 = 5 \times 10^2$ m/sec at the moment of its explosion into two equally massive fragments ($m_1 = m_2 = 5$ kg). If one fragment moves at 45 deg above the initial flight path and the other at 45 deg below the initial flight path (Figure 8.15),

---

* It is true, of course, that the *total* internal force is zero and, therefore, not larger than the external force. However, the part of the internal force acting on each fragment far exceeds the external force acting on that fragment. The same idea can be expressed this way: During the time of the explosion, the acceleration of the center of mass is much less than the accelerations of the fragments.

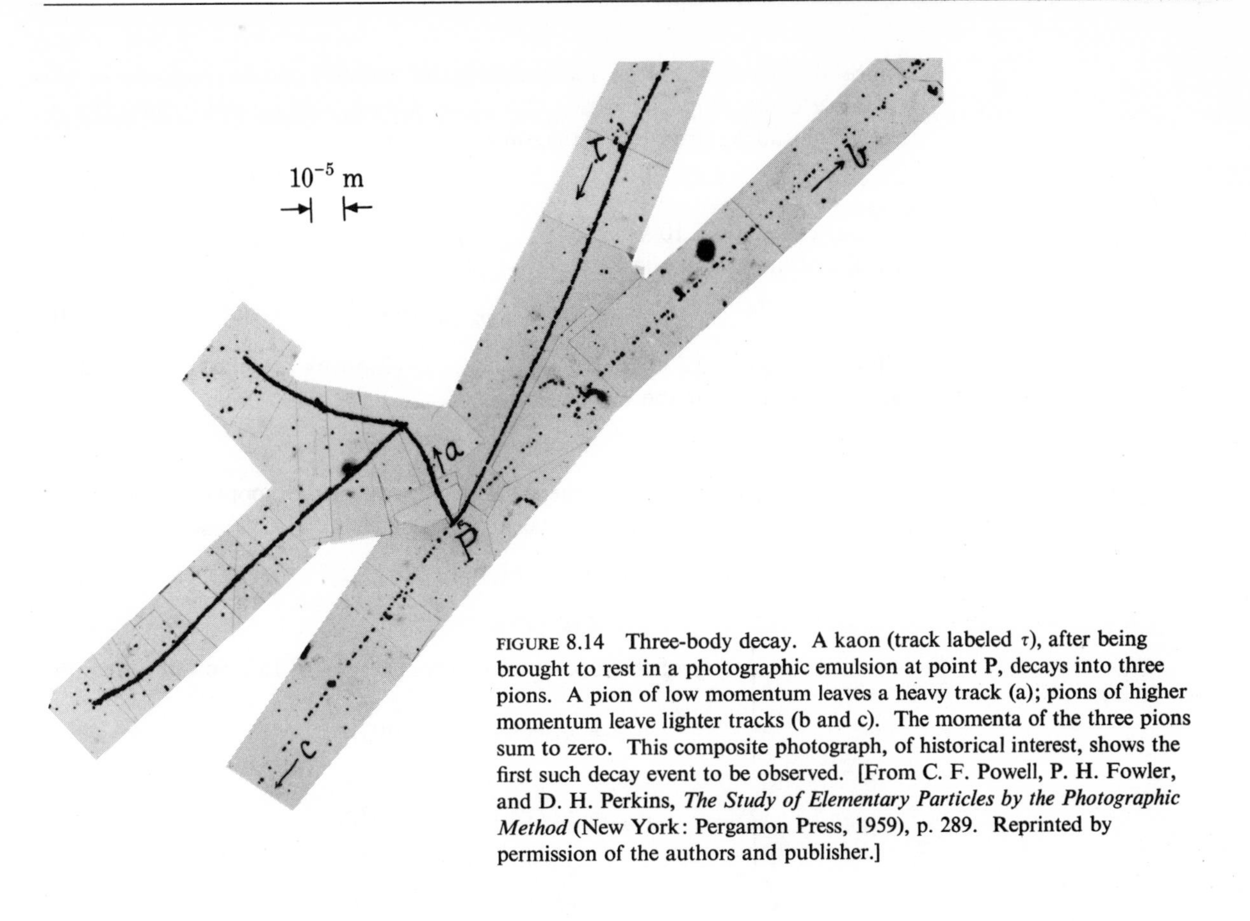

FIGURE 8.14 Three-body decay. A kaon (track labeled τ), after being brought to rest in a photographic emulsion at point P, decays into three pions. A pion of low momentum leaves a heavy track (a); pions of higher momentum leave lighter tracks (b and c). The momenta of the three pions sum to zero. This composite photograph, of historical interest, shows the first such decay event to be observed. [From C. F. Powell, P. H. Fowler, and D. H. Perkins, *The Study of Elementary Particles by the Photographic Method* (New York: Pergamon Press, 1959), p. 289. Reprinted by permission of the authors and publisher.]

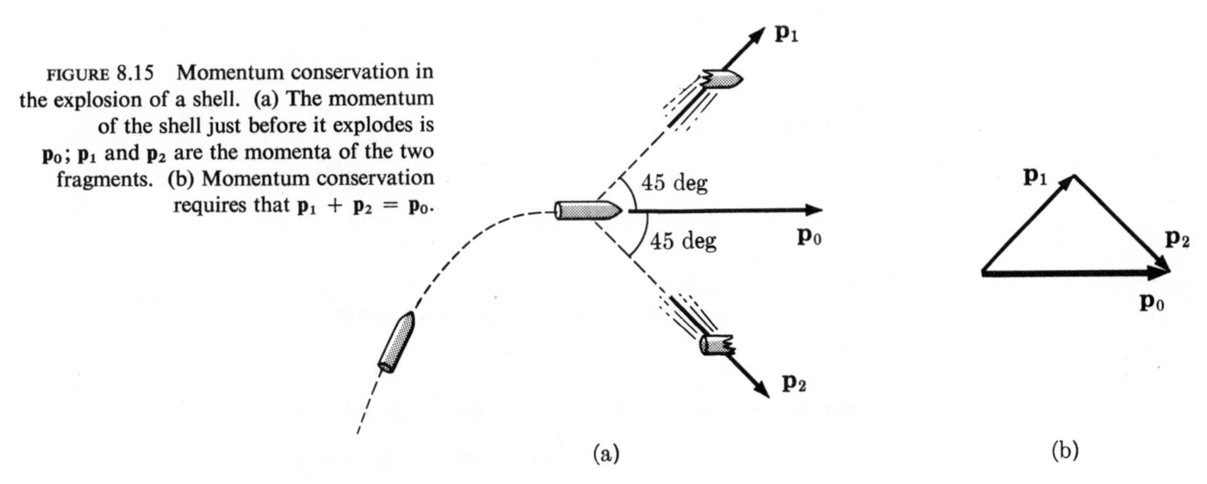

FIGURE 8.15 Momentum conservation in the explosion of a shell. (a) The momentum of the shell just before it explodes is $\mathbf{p}_0$; $\mathbf{p}_1$ and $\mathbf{p}_2$ are the momenta of the two fragments. (b) Momentum conservation requires that $\mathbf{p}_1 + \mathbf{p}_2 = \mathbf{p}_0$.

what is the speed of each? Application of the law of momentum conservation in this case requires the vector addition of the momenta $\mathbf{p}_1$ and $\mathbf{p}_2$ of the fragments, as shown in Figure 8.15(b). Momentum conservation is expressed by

$$\mathbf{p}_1 + \mathbf{p}_2 = \mathbf{p}_0. \tag{8.48}$$

Since the arrow diagram representing the vector addition happens in this example to be a right triangle, with equal sides p_1 and p_2 and hypotenuse p_0, the magnitudes satisfy the Pythagorean theorem

$$p_0^2 = p_1^2 + p_2^2. \tag{8.49}$$

Numerically, $p_0 = 10 \text{ kg} \times 5 \times 10^2 \text{ m/sec} = 5 \times 10^3 \text{ kg m/sec}$. Since $p_1 = p_2$, Equation 8.49 implies that $p_1^2 = \frac{1}{2}p_0^2$, or

$$p_1 = \sqrt{\tfrac{1}{2}}\, p_0 = 0.707 p_0. \tag{8.50}$$

Thus $p_1 = p_2 = 3.535 \times 10^3$ kg m/sec. Division by the mass of each, 5 kg, gives the sought-for speeds,

$$v_1 = v_2 = 707 \text{ m/sec}.$$

It is instructive to solve the same problem using components. Taking the x axis horizontal and the y axis vertical, we have

$$p_{1x} + p_{2x} = p_{0x} = p_0, \tag{8.51}$$

$$p_{1y} + p_{2y} = p_{0y} = 0. \tag{8.52}$$

From the geometry of the momentum vectors (Figure 8.16), we can write

$$p_{1x} = p_1 \cos\left(\frac{\pi}{4}\right) = 0.707 p_1,$$

$$p_{1y} = p_1 \cos\left(\frac{\pi}{4}\right) = 0.707 p_1,$$

$$p_{2x} = p_2 \cos\left(\frac{\pi}{4}\right) = 0.707 p_2,$$

$$p_{2y} = p_2 \cos\left(\frac{3\pi}{4}\right) = -0.707 p_2.$$

Then Equation 8.52 becomes

$$0.707(p_1 - p_2) = 0,$$

from which it follows that $p_1 = p_2$. Equation 8.51 becomes

$$0.707(p_1 + p_2) = p_0.$$

Since $p_1 = p_2$, this may also be written $1.414 p_1 = p_0$, or

$$p_1 = 0.707 p_0,$$

the same result as Equation 8.50, but obtained by a different method. ■

There is, of course, nothing fundamentally important about this particular example. It only illustrates a typical application of the law of momentum conservation in the macroscopic world, emphasizing the important facts that (1) details of internal forces are irrelevant to momentum conservation and (2) the vector nature of momentum plays an essential role in the law of momentum conservation.

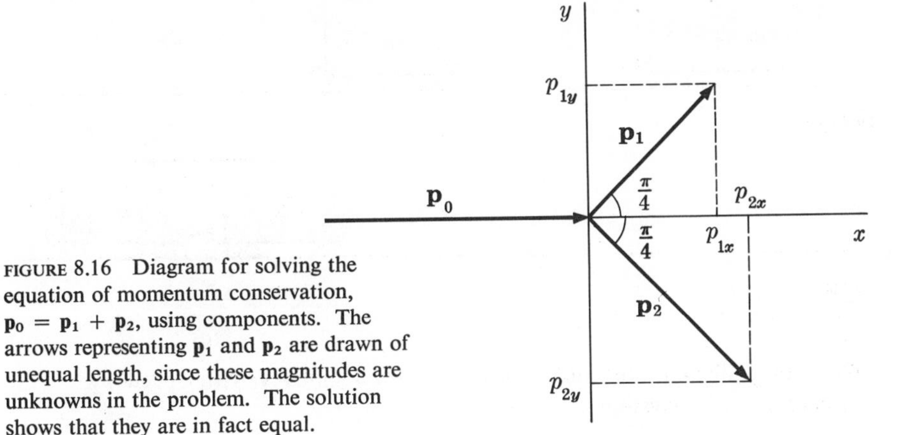

FIGURE 8.16 Diagram for solving the equation of momentum conservation, $\mathbf{p}_0 = \mathbf{p}_1 + \mathbf{p}_2$, using components. The arrows representing $\mathbf{p}_1$ and $\mathbf{p}_2$ are drawn of unequal length, since these magnitudes are unknowns in the problem. The solution shows that they are in fact equal.

8.8 Momentum conservation in systems that are not isolated

Even for systems not isolated from external influences, momentum conservation can play an important role. This can come about in two ways. (1) The external forces may add up to zero. Even though the system is not isolated, it experiences no net external force, and its momentum is therefore a constant. A freight car rolling on a straight horizontal track is such a system provided friction can be ignored. The downward pull of gravity is balanced by the upward push of the tracks; the total external force is zero. (2) The *component* of the net external force in some direction may vanish. Since momentum is a vector quantity, the law of momentum conservation for an isolated system is really three laws in one; each component is separately a constant. For a system that is not isolated, the components may be treated separately. It may then happen that only one or two components of momentum are conserved because only one or two components of the external force vanish. A projectile near the earth, for example, experiences only a vertically downward force (if air resistance is negligible). Therefore, its horizontal components of momentum remain constant. The earth in its orbit around the sun experiences no force perpendicular to the plane of its orbit. Therefore the component of its momentum in that direction, initially zero, remains zero, and the earth's orbit does not deviate from its original plane.

One component of momentum may be conserved and another not

■ EXAMPLE: A freight car of mass $m_1 = 1.5 \times 10^4$ kg rolling along a horizontal track at a speed $v_1 = 2.0$ m/sec collides and couples with another car of mass $m_2 = 1.0 \times 10^4$ kg, after which the pair continue together with a lesser speed v_f (Figure 8.17). What is the magnitude of v_f? Problems of this kind are best solved first algebraically, because the algebraic solution has a much wider applicability than the particular numerical solution. Before the collision, the second car has no momentum; the total momentum of the two-car system is that of the first car. Its horizontal component, the same as its total magnitude, is

$$p_1 = m_1 v_1. \tag{8.53}$$

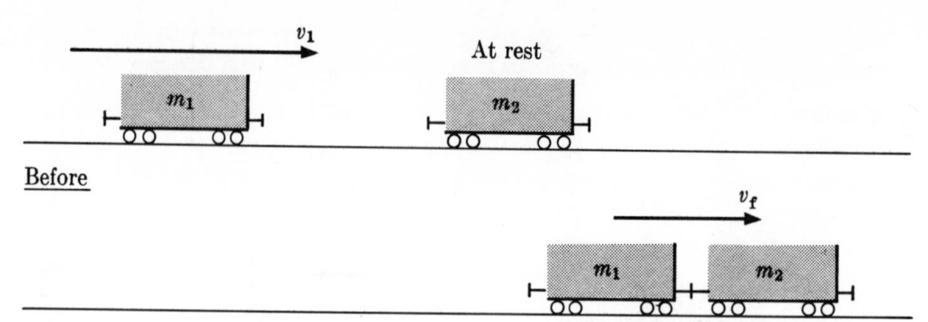

FIGURE 8.17 Momentum conservation in the collision and coupling of freight cars.

After the collision, the entire mass, $m_1 + m_2$, moves with speed v_2. The momentum of the system is

$$p_2 = (m_1 + m_2)v_f. \tag{8.54}$$

Because the forces acting during the collision and coupling are all internal forces for the chosen system of two cars, momentum is conserved: $p_1 = p_2$. The right sides of Equations 8.53 and 8.54 may be equated to give

$$m_1v_1 = (m_1 + m_2)v_f.$$

The algebraic solution for the final speed v_f is

Collision without rebound in one dimension

$$v_f = \frac{m_1}{m_1 + m_2} v_1. \tag{8.55}$$

For the particular numbers of this example,

$$v_f = \frac{1.5 \times 10^4}{(1.5 + 1.0) \times 10^4} \times 2.0 \text{ m/sec} = 1.2 \text{ m/sec}.$$

Even *with* frictional forces acting on the cars, this result is valid provided these frictional forces change the total momentum of the system by a small fraction during the short time of collision and coupling. ■

The mathematics of this example may be applied without change to other examples. Suppose that a bullet of mass m_1 and speed v_1 strikes a stationary wooden block of mass m_2 and is imbedded in the block. What speed will the block and bullet together acquire? Equation 8.55 provides the answer. This example has practical utility. It provides a convenient way to determine the speed of a bullet. The larger speed v_1 may be discovered by measuring the smaller speed v_f.

8.9 Momentum conservation and center of mass

From Equation 8.22 it follows that if $\mathbf{P}$ = **constant**,

If $\mathbf{F}_{ext} = 0$, then $\mathbf{v}_c$ = constant

$$\mathbf{v}_c = \textbf{constant}; \tag{8.56}$$

the center of mass moves with constant velocity. Also, if any component of the total momentum is conserved, the corresponding component of the center-of-mass velocity is constant. In solving problems concerned with momentum conservation, it is often convenient to use the center-of-mass concept.

Collision without rebound, using center-of-mass concept

■ EXAMPLE 1: What is the velocity of the center of mass of the freight cars considered in the preceding section (see Figure 8.17)? Choose an arbitrary origin, and let $\mathbf{r}_1$ and $\mathbf{r}_2$ be the position vectors of the centers of mass of the two cars. The center-of-mass vector for the combined system is then

$$\mathbf{r}_c = \frac{m_1\mathbf{r}_1 + m_2\mathbf{r}_2}{m_1 + m_2}. \tag{8.57}$$

Its velocity is

$$\mathbf{v}_c = \frac{m_1\mathbf{v}_1 + m_2\mathbf{v}_2}{m_1 + m_2}.$$

Before the cars collide, $\mathbf{v}_2 = 0$, so

$$\mathbf{v}_c = \frac{m_1}{m_1 + m_2}\mathbf{v}_1. \tag{8.58}$$

After they collide, $\mathbf{v}_c$ is unchanged and must be the same as $\mathbf{v}_f$ since all parts of the system then move with the same velocity. This result,

$$\mathbf{v}_f = \mathbf{v}_c = \frac{m_1}{m_1 + m_2}\mathbf{v}_1, \tag{8.59}$$

is equivalent to the previous result, Equation 8.55. ■

■ EXAMPLE 2: A boy of mass 75 kg and a girl of mass 50 kg face each other on ice skates on an ideally smooth horizontal ice surface. They are 4.0 m apart and are joined by a rope (Figure 8.18). Starting from rest, both the boy and the girl pull toward each other by pulling hand over hand on the rope. Where will they meet? It might seem at first sight that their meeting point would depend on which of them pulls harder on the rope. This is not the case, because boy, girl, and rope together form a system whose horizontal component of momentum must be conserved. All forces on the rope are internal forces that cannot influence their total momentum or the motion of their center of mass. The center-of-mass

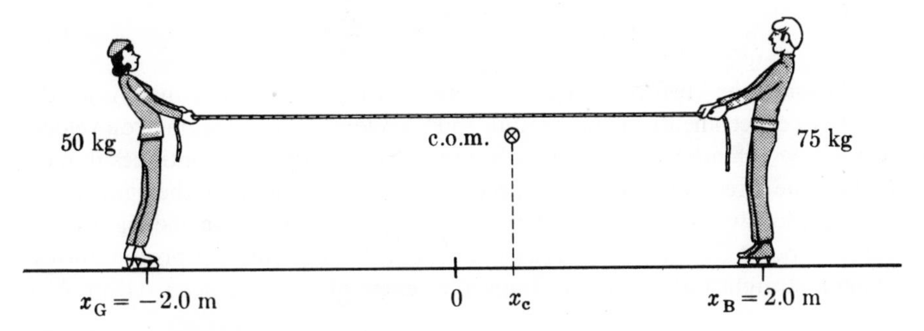

FIGURE 8.18 Conservation of the horizontal component of momentum in a system free of external horizontal force. The center of mass, initially at rest, moves neither to the left nor to the right.

concept provides the easiest solution to the problem. Since the center of mass is initially at rest, it remains at rest. It is a fixed point between the boy and the girl.* No matter how close together they come, that point remains between them and remains stationary. It must therefore be the point where they meet. Then the question to be answered is, Where is the center of mass? Choosing the line joining boy and girl as the x axis, we have

$$x_c = \frac{m_B x_B + m_G x_G}{m_B + m_G},$$

where subscripts B and G refer to boy and girl. If the origin is chosen midway between the two so that $x_B = 2.0$ m and $x_G = -2.0$ m, the center-of-mass coordinate is

$$x_c = \left(\frac{75 \text{ kg}}{125 \text{ kg}}\right) \times 2.0 \text{ m} - \left(\frac{50 \text{ kg}}{125 \text{ kg}}\right) \times 2.0 \text{ m}$$

$$= 0.4 \text{ m}.$$

The center of mass is 2.4 m from the girl and 1.6 m from the boy. That is the point where they meet. ■

★EXAMPLE OF A TWO-BODY COLLISION

Two-dimensional collision

A neutron moving at 4×10^6 m/sec strikes a proton that is initially at rest. As a result of the collision, the neutron is deflected through 30 deg, and the proton recoils at 60 deg to the initial neutron direction, that is, 90 deg to the final neutron direction (Figure 8.19). What are the final speeds of the two particles? For the purpose of this example, we may assume that the neutron and proton are equally massive (they actually differ in mass by less than one part in a thousand). There are various ways to solve this problem (all, of course, using the conservation of momentum). An algebraic approach, using components, is suggested in Exercise 8.38. Here let us use the center-of-mass concept and a geometrical approach.

Before the collision, the physical situation is the same as that of the freight cars in Example 1. The center-of-mass velocity, from Equation 8.58, is

$$\mathbf{v}_c = \frac{m_1}{m_1 + m_2} \mathbf{v}_1 = \tfrac{1}{2}\mathbf{v}_1. \tag{8.60}$$

The center-of-mass velocity is constant

As shown in the "before" diagram of Figure 8.19, the center of mass is located midway between neutron and proton, and its velocity is half the neutron velocity ($v_c = 2 \times 10^6$ m/sec). After the collision, the center of mass continues to move in the same direction at the same speed. The "after" diagram in the figure shows the situation when the center of mass has reached point C to the right of the collision point O. At this time (as at all times) the neutron and proton must lie along a straight line passing through the center of mass, such as AB or A′B′.

* More accurately stated, the center of mass does not move horizontally. Why can it move vertically up and down?

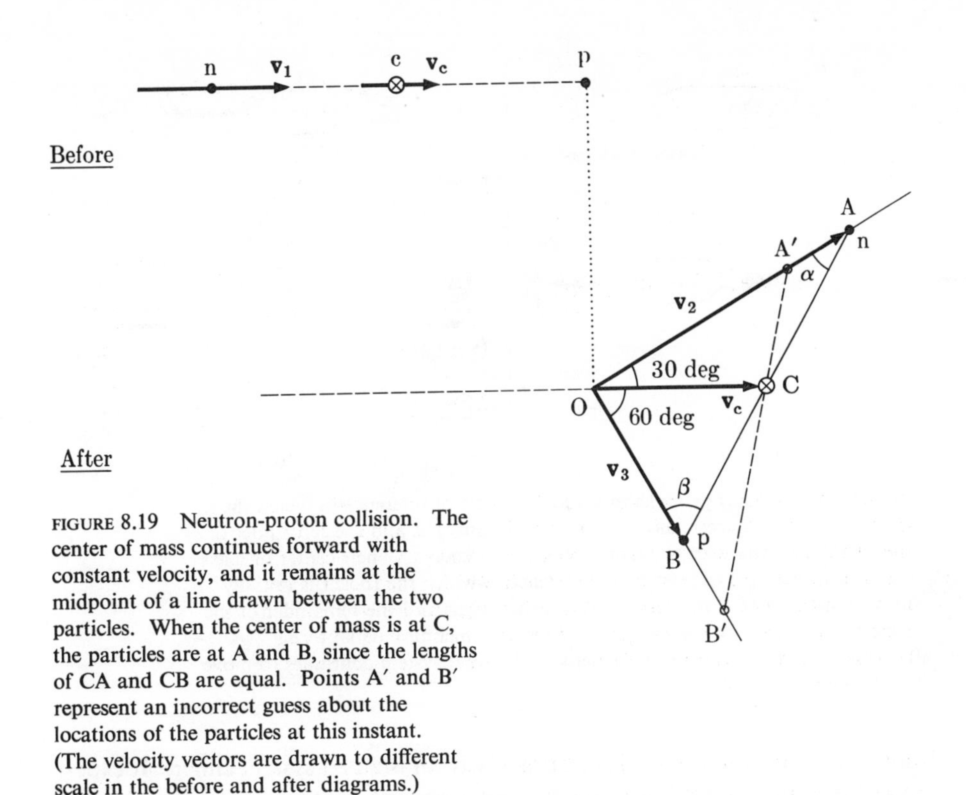

FIGURE 8.19 Neutron-proton collision. The center of mass continues forward with constant velocity, and it remains at the midpoint of a line drawn between the two particles. When the center of mass is at C, the particles are at A and B, since the lengths of CA and CB are equal. Points A′ and B′ represent an incorrect guess about the locations of the particles at this instant. (The velocity vectors are drawn to different scale in the before and after diagrams.)

Moreover, since the particles have equal mass, they must be equidistant from C. Let AB be the correct line with the property CA = CB. (A′B′ is an incorrect line, with CA′ < CB′.) Now we invoke a theorem of plane geometry. The triangle OAB is a right triangle, and C is the midpoint of its hypotenuse. The theorem states that the line OC divides the right triangle into two isosceles triangles, with OC = CA = CB. The angle α must therefore be 30 deg, and the angle β must be 60 deg. This completes the geometrical argument. The directed line segments OA, OB, and OC then represent respectively the final velocities of the two particles and the velocity of the center of mass. The sought-for speeds are

$$\begin{aligned} v_2 &= 2v_c \cos(30 \text{ deg}) \\ &= 2 \times 2 \times 10^6 \times 0.866 = 3.46 \times 10^6 \text{ m/sec}, \\ v_3 &= 2v_c \cos(60 \text{ deg}) \\ &= 2 \times 2 \times 10^6 \times 0.500 = 2 \times 10^6 \text{ m/sec}. \end{aligned}$$

8.10 Rocket propulsion

Ordinary propulsion on earth relies heavily on friction—on the fact that the vehicle, be it boat, locomotive, automobile, or airplane, is *not* an isolated system. For a rocket in space, the situation is entirely different. The rocket has neither earth nor air nor water to push against and must achieve its acceleration another

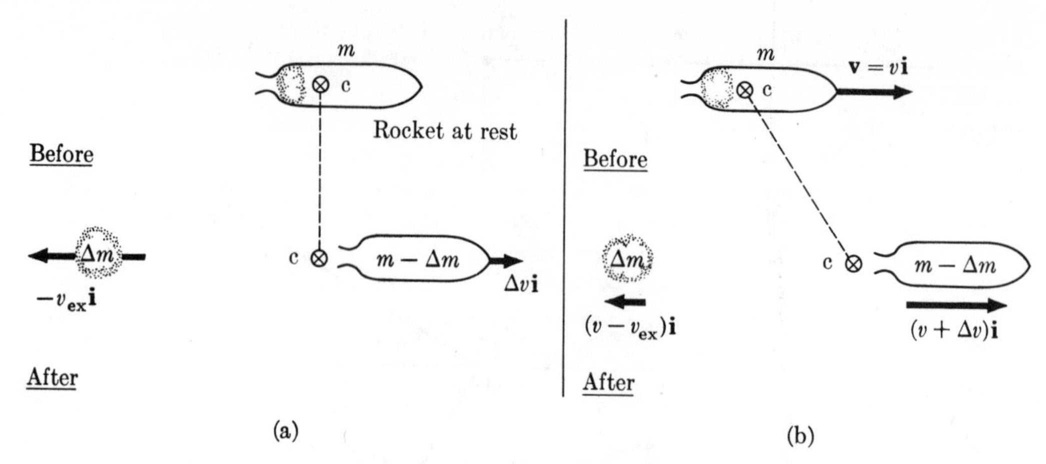

FIGURE 8.20 Rocket propulsion. (a) In a frame of reference in which the rocket is initially at rest, the increment of exhaust gas and the remainder of the rocket separate with equal and opposite momenta. Their center of mass remains at rest. (b) In a frame of reference in which the rocket is initially moving with speed v, the increment of exhaust gas and the remainder of the rocket separate in such a way that the total momentum, equal to mv, remains constant. The center of mass of the total system continues to move with constant velocity.

way. Of course, the rocket is not precisely an isolated system either. It experiences the gravitational force of sun, earth, moon, or other astronomical bodies and accelerates accordingly. During its coasting period, which is the major part of any rocket flight, the trajectory is determined by these external gravitational forces. However, there is nothing the rocket pilot—or the ground controller—can do about these. If an astronaut wishes to accelerate according to his own choice, he must find a substitute for the friction that provides the basis of propulsive control on earth. It is through rocket exhaust and the law of momentum conservation that he finds this control.

Imagine now a rocket in empty space that is truly an isolated system. Its center of mass moves with constant velocity. No internal motions or changes or forces whatever can prevent the center of mass of the system from continuing inexorably forward at constant velocity. But one part of the system can be caused to accelerate and therefore change its momentum provided that another part of the system experiences a compensating opposite change of momentum. The astronaut must be prepared literally to sacrifice some of his rocket ship for the benefit of the rest of it. *Only* by discarding some mass can he achieve an acceleration of what remains.

The principle of rocket propulsion: Mass must be thrown away

In simplest terms, we can say that the rocket ship is divided into two parts, the payload part and the fuel-exhaust part. Neither of these parts is an isolated system because each interacts with the other part. Therefore, on each part considered alone an external force can act to accelerate that part. What happens is that the payload part pushes the exhaust backward, and the exhaust pushes the payload forward with an equal and opposite force. The center of mass of the entire system continues uniformly, unaffected by all this activity. The payload part does, however, achieve the desired acceleration. Usually, the payload is a small part of the original total mass of the rocket. For it to have acquired a high velocity in one direction, a large mass of exhaust must have been accelerated in the opposite direction.

To give these ideas mathematical expression, we consider a rocket moving along the x axis in some "stationary" frame of reference. At a particular moment, it has mass m and velocity $\mathbf{v} = v\mathbf{i}$. A brief interval of time later, it has lesser mass $m - \Delta m$ and greater speed $v + \Delta v$ (Figure 8.20). During this interval, a mass of exhaust Δm has been discarded with velocity $-v_{ex}\mathbf{i}$ relative to the payload part of the rocket, or velocity $(v - v_{ex})\mathbf{i}$ in the stationary frame. At the start of the interval the momentum of the system is

$$\mathbf{p}_1 = mv\mathbf{i}. \tag{8.61}$$

At the end of the interval it is

$$\mathbf{p}_2 = (m - \Delta m)(v + \Delta v)\mathbf{i} + \Delta m(v - v_{ex})\mathbf{i}. \tag{8.62}$$

Since no external forces acted on the system, momentum is conserved:

For rocket plus exhaust, momentum is conserved

$$\mathbf{p}_1 = \mathbf{p}_2.$$

The x components on the right sides of Equations 8.61 and 8.62 may therefore be equated. Algebraic manipulation yields a formula for the amount of mass Δm that must be thrown away to achieve the increment of speed Δv:

Speed gain Δv requires mass loss Δm

$$\Delta m = \left(\frac{\Delta v}{v_{ex} + \Delta v}\right) m. \tag{8.63}$$

Suppose, for example, that the gain of speed is $\Delta v = 100$ m/sec (about 225 mile/hr) and that the exhaust speed is 900 m/sec. For these speeds, Equation 8.63 gives

$$\Delta m = 0.10m;$$

one-tenth of the mass must be discarded as exhaust.

Actually Equation 8.63 is valid only if $\Delta v \ll v_{ex}$. (Why is this?) It becomes perfectly accurate, and also most useful, if the increments Δm and Δv are allowed to approach zero. To achieve this limit, first rewrite Equation 8.63 in the form

$$\frac{\Delta v}{\Delta m} = \frac{v_{ex} + \Delta v}{m}. \tag{8.64}$$

Now let Δm (and, with it, Δv) approach zero. On the right side, Δv vanishes, leaving v_{ex}/m. On the left side, there is a slight subtlety associated with sign. We defined Δv to be $v_2 - v_1$ (the later minus the earlier speed) but Δm to be $m_1 - m_2$ (the earlier minus the later mass). In the limit, the ratio is therefore the negative of a derivative:

$$\lim_{\Delta m \to 0} \frac{\Delta v}{\Delta m} = \lim_{m_2 \to m_1} \frac{v_2 - v_1}{m_1 - m_2} = -\frac{dv}{dm}.$$

So Equation 8.64 leads to the differential equation

The fundamental rocket equation in differential form

$$\frac{dv}{dm} = -\frac{v_{ex}}{m}. \tag{8.65}$$

This is a fundamental equation of rocket engineering. It is strictly valid only for a rocket far removed from all centers of gravitational attraction. However, it is approximately valid whenever the momentum change produced by the rocket exhaust is much greater than the momentum change produced by gravity in the same time interval.

Equation 8.65 is a simple form of differential equation that can readily be integrated if we write

$$v = \int \left(\frac{dv}{dm}\right) dm = -v_{ex} \int \frac{dm}{m}. \tag{8.66}$$

In particular, consider some finite interval in which the mass of the payload part of the rocket changes from m_1 to m_2 (a decrease) and its speed changes from v_1 to v_2 (an increase). The integral in Equation 8.66 becomes a definite integral, and we have

$$v_2 - v_1 = -v_{ex} \int_{m_1}^{m_2} \frac{dm}{m}$$

$$= -v_{ex}(\ln m_2 - \ln m_1).$$

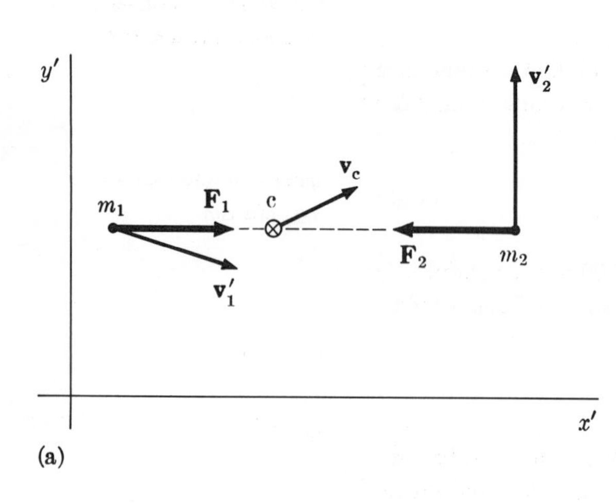

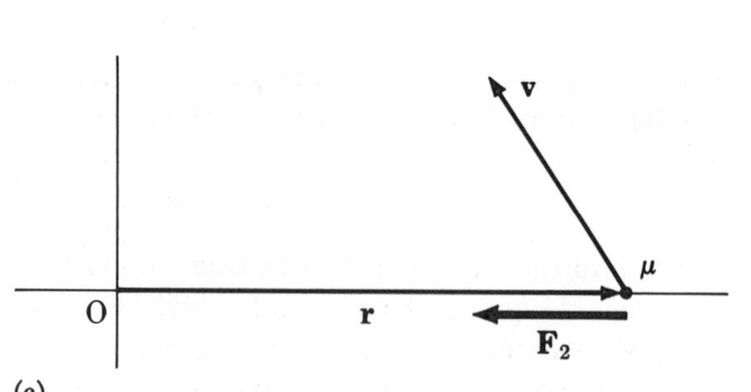

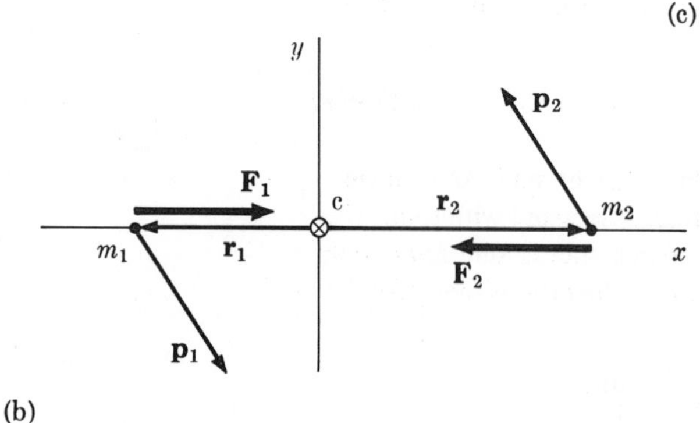

FIGURE 8.21 Isolated system of two particles. (a) In any frame of reference, the forces are equal and opposite ($\mathbf{F}_1 = -\mathbf{F}_2$), and the center-of-mass velocity $\mathbf{v}_c$ is constant. (b) In a frame of reference in which the center of mass is at rest, the momenta are also equal and opposite ($\mathbf{p}_1 = -\mathbf{p}_2$). (c) The motion of the two particles can be found by first solving an effective one-body problem. A single particle of mass μ (Equation 8.74) moves under the action of force $\mathbf{F}_2$. (Central forces are shown here. However, the treatment of the two-body problem in the text requires only that $\mathbf{F}_1 + \mathbf{F}_2 = 0$, not that $\mathbf{F}_1$ and $\mathbf{F}_2$ act along the same line.)

Written most compactly, this is

$$v_2 - v_1 = v_{ex} \ln\left(\frac{m_1}{m_2}\right). \tag{8.67}$$

The integrated rocket equation in algebraic form

Suppose a velocity change equal to v_{ex} were desired. This would require $\ln(m_1/m_2) = 1$, or $m_1/m_2 = 2.72$; nearly two-thirds of the mass would have to be exhausted. To obtain a velocity change of $3v_{ex}$ would require $\ln(m_1/m_2) = 3$, or $m_1/m_2 \cong 20$; about 95 percent of the mass would have to be exhausted. It is evident that the exhaust velocity should be as large as possible in order to minimize the mass thrown away as exhaust.

It is instructive to look at Equation 8.65 in another way. It may be re-expressed in terms of time derivatives:

$$-v_{ex}\frac{dm}{dt} = m\frac{dv}{dt}. \tag{8.68}$$

Rocket thrust

(Here we have used the equality $(dv/dt)/(dm/dt) = dv/dm$.) On the right side of this equation is mass times acceleration. The left side may therefore be interpreted as the force on the rocket produced by its exhaust. This force is called thrust:

$$\text{thrust} = v_{ex}\left|\frac{dm}{dt}\right|. \tag{8.69}$$

Not surprisingly, thrust is proportional both to mass flow rate and to exhaust speed. Equation 8.69 accurately gives the thrust even if other forces, such as gravity and air friction, are also at work.

★8.11 Systems of two particles; reduced mass

The earth and the moon, the sun and Mercury, a proton and an electron, an alpha particle and a nucleus—all these are two-body systems. In this section, we will demonstrate how the two-body problem in classical mechanics can be simplified to an effective one-body problem if the system is isolated.

Consider two particles of mass m_1 and m_2 experiencing forces $\mathbf{F}_1$ and $\mathbf{F}_2$. If the system is isolated, these forces must cancel ($\mathbf{F}_1 = -\mathbf{F}_2$) so that the center of mass moves with constant velocity $\mathbf{v}_c$ [Figure 8.21(a)]. One way to simplify the description of the motion is to choose a frame of reference in which the center of mass is at rest at the origin. Then, as shown in Figure 8.21(b), $\mathbf{r}_1$ and $\mathbf{r}_2$ lie along the same straight line, and the momenta are equal and opposite ($\mathbf{p}_1 = -\mathbf{p}_2$). Even greater simplification results if we introduce the relative position vector,

In an arbitrary frame of reference, the motion seems complex

The center-of-mass frame is simpler

$$\mathbf{r} = \mathbf{r}_2 - \mathbf{r}_1, \tag{8.70}$$

and the relative velocity and acceleration,

$$\mathbf{v} = \mathbf{v}_2 - \mathbf{v}_1, \tag{8.71}$$

$$\mathbf{a} = \mathbf{a}_2 - \mathbf{a}_1. \tag{8.72}$$

On the right side of Equation 8.72, we may use Newton's second law to replace

$\mathbf{a}_2$ by $\mathbf{F}_2/m_2$ and $\mathbf{a}_1$ by $\mathbf{F}_1/m_1$; the relative acceleration is then

$$\mathbf{a} = \frac{\mathbf{F}_2}{m_2} - \frac{\mathbf{F}_1}{m_1}.$$

From Newton's third law for this isolated system, $\mathbf{F}_1 = -\mathbf{F}_2$, so

$$\mathbf{a} = \mathbf{F}_2\left(\frac{1}{m_2} + \frac{1}{m_1}\right).$$

With some slight rearrangement, this relation can be written

Equation of motion for the effective one-body problem

$$\mathbf{F}_2 = \mu\mathbf{a}, \tag{8.73}$$

where μ, the so-called *reduced mass*, is defined by

$$\mu = \frac{m_1 m_2}{m_1 + m_2}. \tag{8.74}$$

Equation 8.73 evidently has the form of Newton's second law for a particle of mass μ. It defines an effective one-body problem. A particle of mass μ (not a real particle) has position vector $\mathbf{r}$, velocity $\mathbf{v}$, acceleration $\mathbf{a}$, and experiences force $\mathbf{F}_2$ [Figure 8.21(c)]. If the motion of the effective single particle can be found, the motion of the two actual particles can be deduced from it. In the frame of reference of Figure 8.21(b), where the center of mass is fixed at the origin, the connecting equations are (we introduce M for $m_1 + m_2$)

$$\mathbf{r}_1 = -\left(\frac{m_2}{M}\right)\mathbf{r}, \tag{8.75}$$

$$\mathbf{r}_2 = \left(\frac{m_1}{M}\right)\mathbf{r}, \tag{8.76}$$

Relations among kinematic variables

$$\mathbf{v}_1 = -\left(\frac{m_2}{M}\right)\mathbf{v}, \tag{8.77}$$

$$\mathbf{v}_2 = \left(\frac{m_1}{M}\right)\mathbf{v}, \tag{8.78}$$

$$\mathbf{a}_1 = -\left(\frac{m_2}{M}\right)\mathbf{a}, \tag{8.79}$$

$$\mathbf{a}_2 = \left(\frac{m_1}{M}\right)\mathbf{a}. \tag{8.80}$$

The proof of the first two of these equations is left as an exercise. The other four follow directly from the first two. If the properties of the effective single particle ($\mathbf{r}$, $\mathbf{v}$, and $\mathbf{a}$) are substituted in the right sides of Equations 8.75–8.80, the properties of the actual particles are determined on the left sides (for the chosen frame of reference).

A momentum,

The one-particle momentum

$$\mathbf{p} = \mu\mathbf{v}, \tag{8.81}$$

may be defined for the effective single particle. However, it is *not* a relative momentum, $\mathbf{p}_2 - \mathbf{p}_1$. On the right side of Equation 8.81, use terms from Equations 8.74 and 8.71 to get

$$\mathbf{p} = \frac{m_1 m_2}{M}(\mathbf{v}_2 - \mathbf{v}_1)$$

$$= \frac{m_1}{M}\mathbf{p}_2 - \frac{m_2}{M}\mathbf{p}_1.$$

In the chosen frame of reference, where $\mathbf{p}_1 = -\mathbf{p}_2$, this can be written

$$\mathbf{p} = \frac{m_1 + m_2}{M}\mathbf{p}_2 = \mathbf{p}_2. \qquad (8.82)$$

It is equal to the momentum of an actual particle

The effective single-particle momentum is in fact equal to an actual particle momentum.* In magnitudes,

$$p = p_1 = p_2. \qquad (8.83)$$

■ EXAMPLE: Two equal masses, each of 1 kg, are connected by a string 3 m long and rotate about their stationary center of mass on a nearly frictionless ice surface. Each mass exerts on the other a force of magnitude 14 N. What is the reduced mass of the system? What is the motion of the effective single particle? For the reduced mass, Equation 8.74 gives

$$\mu = \tfrac{1}{2}\text{m} = 0.5 \text{ kg}.$$

(Note that for any combination of masses, μ is always less than the smaller of m_1 and m_2.) The position vector $\mathbf{r}$ of the effective single particle, being equal to $\mathbf{r}_2 - \mathbf{r}_1$, has magnitude 3 m. Since it is a fixed distance from the origin and experiences a centripetal force of fixed magnitude, this particle executes uniform circular motion. From the formula for centripetal force, $F = \mu v^2/r$, we get

$$v = \sqrt{\frac{Fr}{\mu}} = \sqrt{\frac{14 \text{ N} \times 3 \text{ m}}{0.5 \text{ kg}}}$$

$$= 9.17 \text{ m/sec}.$$

The acceleration of the effective single particle is

$$a = \frac{F}{\mu} = 28 \text{ m/sec}^2.$$

Equations 8.77 and 8.78 then show that

$$v_1 = v_2 = \tfrac{1}{2}v = 4.58 \text{ m/sec}.$$

From Equations 8.79 and 8.80,

$$a_1 = a_2 = \tfrac{1}{2}a = 14 \text{ m/sec}^2.$$

* An effective single-particle angular momentum can also be defined and proves to be equal to the *total* angular momentum of the two actual particles.

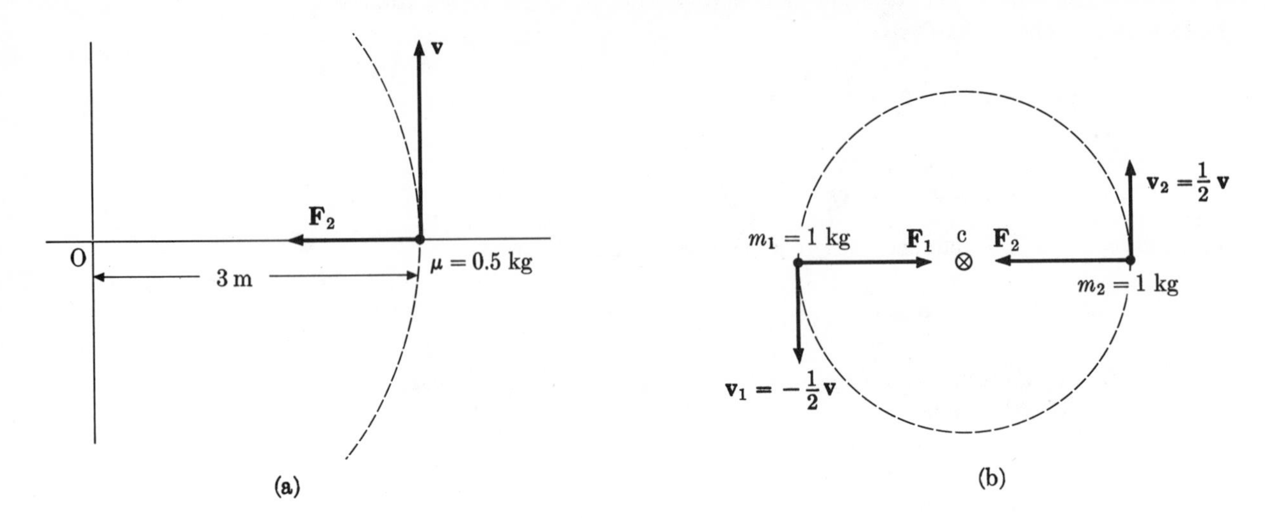

FIGURE 8.22 Example of equal masses rotating about a fixed point. (a) The effective single particle moves in a circle of radius 3 m with speed v. (b) Each actual particle moves in a circle of radius 1.5 m with half the speed and half the acceleration of the effective single particle. Forces and momenta have the same magnitudes in the two diagrams.

The momenta of the effective single particle and of each of the actual particles are equal, as they should be:

$$p = \mu v = 4.58 \text{ kg m/sec},$$

$$p_2 = m_2 v_2 = 4.58 \text{ kg m/sec}.$$

Figure 8.22 compares the motion of the effective single particle and the motion of the actual particles. ■

8.12 The significance of momentum conservation

Consider a system of two particles interacting with each other but isolated from all outside influences. Two fundamental laws of their interaction and motion have been the themes of this chapter. One is Newton's third law of equal and opposite forces, expressed mathematically by

$$\mathbf{F}_{12} = -\mathbf{F}_{21}.$$

The other is the law of momentum conservation, expressed by

$$\mathbf{p}_1 + \mathbf{p}_2 = \textbf{constant}.$$

Is one of these laws more fundamental than the other? There is no unambiguous logical way to answer this question. However, experience can tend to show that one law is more general, more useful, or apparently deeper than another. Originally, Newton chose his three laws as the most suitable axioms for the development of mechanics. From the third law (together with the second) the law of momentum conservation was deduced. We have followed the classical approach in this chapter so far: Momentum conservation is a *consequence* of Newton's laws. However, it is important to be aware that the law of momentum conservation could supplant Newton's third law as one of the basic foundation stones of mechanics. If it did, the third law would become secondary, a law that could be deduced mathematically from momentum conservation and Newton's second law. The evolution of physics since Newton, particularly the developments of the twentieth century, suggest that the law of momentum conservation indeed has a good claim to be regarded as a more fundamental statement about nature than does Newton's third law.

Momentum conservation may be derived from Newton's laws

It now seems preferable to choose it as a basic law

Among the reasons for attaching a special significance to the law of momentum conservation is its simplicity. Every conservation law, by picking from the chaos of activity and change in nature a single constant quantity, has a special appeal. Because momentum is a constant for every isolated system or interaction, it is naturally looked upon as a fundamental concept, and its conservation as a particularly important law.

1. *Momentum conservation is particularly simple*

As mathematicians of the eighteenth and nineteenth centuries studied and reformulated mechanics, the concept of force was eclipsed in importance by the concepts of energy and momentum. This was, so to speak, accidental. It simply turned out that the equations with the greatest power for solving mechanical problems were equations in which momentum and energy appeared explicitly and in which force did not. The demotion of force from its central position in mechanics was completed by the twentieth-century revolutions of relativity and quantum mechanics. In these theories, force may be introduced, although it need not be. Momentum, on the other hand (suitably redefined), remains a central concept in these modern theories and still obeys a conservation law.

2. *Momentum appears naturally in advanced formulations of mechanics*

Force (and with it, Newton's third law) has encountered two main difficulties in modern physics. The first is that in the catastrophic events of annihilation and creation that govern the behavior of elementary particles, we usually do not know what forces act. The conservation of momentum from before to after the event can be verified, and the probability of the event can be determined; but exactly what forces are at work during the event (if indeed the event has any duration) cannot be determined. The second difficulty connected with force is more of an inconvenience than a fundamental problem. It has to do with the idea of action at a distance. Since forces do not act instantaneously across space, being rather propagated with finite speed, Newton's third law must be examined with new care. We must ask, *When* are the pair of forces equal and opposite? It takes some time for particle number 1 to react to the presence of particle number 2 at a distance, and in that time the distance between the particles and the force between them may have changed. The result of this is that Newton's third law is greatly complicated, for it is no longer true that a pair of forces need be precisely equal and opposite at a given instant of time. On the other hand, the law of momentum conservation retains a simple form even when the time lag of propagating forces is taken into account. It is still true that the system has a constant total momentum at each instant of time. The only added feature of the new view is that only part of the total momentum is contributed by the matter in the system. The rest comes from photons or other

3. *Force is not a useful concept in submicroscopic physics*

massless messengers that transmit force, energy, and momentum from one material part to another.

4. Momentum is related to symmetry in nature

In some ways the most compelling of the reasons for regarding momentum conservation as a profound law of nature is that regarding the connection between this conservation law and the uniformity of space. The principle of the uniformity of space might be called the principle of the sameness of nothingness. How a law as general and important as the law of momentum conservation can rest on a principle apparently so innocuous was explained in Chapter 4; as emphasized there, the true beauty of every conservation law may reside in the symmetry principle upon which it is founded. The symmetries of space and time underlie the laws of conservation of energy, momentum, and angular momentum. The search for subtler symmetries of nature underlying some of the other conservation laws is one of the most exciting endeavors of physics.

Summary of ideas and definitions

The momentum of a particle (nonrelativistically) is defined by $\mathbf{p} = m\mathbf{v}$.

Impulse is defined by $\mathbf{I} = \int_{t_1}^{t_2} \mathbf{F}\, dt$.

Newton's second law in terms of momentum is $\mathbf{F} = d\mathbf{p}/dt$. In terms of impulse, it is $\Delta\mathbf{p} = \mathbf{I}$.

The center-of-mass vector of a system is the weighted average position of all the mass in the system: $\mathbf{r}_c = (m_1\mathbf{r}_1 + m_2\mathbf{r}_2 + m_3\mathbf{r}_3 + \cdots)/M$. The quantities $\mathbf{v}_c$ and $\mathbf{a}_c$ are similarly weighted averages of velocities and accelerations.

The momentum of a system is $\mathbf{P} = M\mathbf{v}_c$.

Newton's third law is written $\mathbf{F}_{AB} = -\mathbf{F}_{BA}$: If two bodies A and B interact, the force exerted by B on A is the negative of the force exerted by A on B.

An internal force is one whose source and whose action (or whose cause and effect) are both within the system being considered. In any system, the total internal force vanishes.

An external force is one whose source is outside the system.

Newton's second law for systems is $\mathbf{F}_{ext} = M\mathbf{a}_c$, or $\mathbf{F}_{ext} = d\mathbf{P}/dt$.

The center of mass of a system behaves in every way like a single particle of mass M experiencing force $\mathbf{F}_{ext}$. Analogous equations govern systems and single particles (see page 282).

Newton's first and second laws are valid for a material object only because of the action of Newton's third law and the cancellation of internal forces within the object.

To avoid ambiguity, Newton's first and second laws for systems should be restated in terms of the center of mass (page 282).

Newton's second law applies separately to every part of a system or any combination of parts.

The conservation of momentum: If $\mathbf{F}_{ext} = 0$, then $\mathbf{P} =$ **constant**. If any component of the external force vanishes, the corresponding component of total momentum is constant.

If momentum is conserved, the center-of-mass velocity is constant.

Rocket propulsion utilizes momentum conservation. To accelerate the payload, mass must be thrown away.

In the absence of gravity or air friction, the change of speed of a rocket is given by $v_2 - v_1 = v_{ex} \ln (m_1/m_2)$. Rocket thrust is equal to $v_{ex}|dm/dt|$.

The problem of an isolated two-body system can be replaced by an effective one-body problem whose equation of motion is $\mathbf{F}_2 = \mu\mathbf{a}$, where the reduced mass is defined by $\mu = m_1 m_2/M$.

Developments in physics since the time of Newton make the law of momentum conservation appear to be more fundamental than Newton's third law.

QUESTIONS

Section 8.1

Q8.1 (1) Can you run fast enough to have the same momentum as an automobile rolling at 1 mile/hr? (2) About how fast must a proton move in order to have the same momentum as a gently falling feather?

Q8.2 Does Equation 8.1 provide an operational definition of momentum? Does Equation 8.7 provide an operational definition of impulse?

Q8.3 The concept of impulse is particularly useful when a large force acts for a short time. Give several examples of phenomena to which the concept of impulse might be applied.

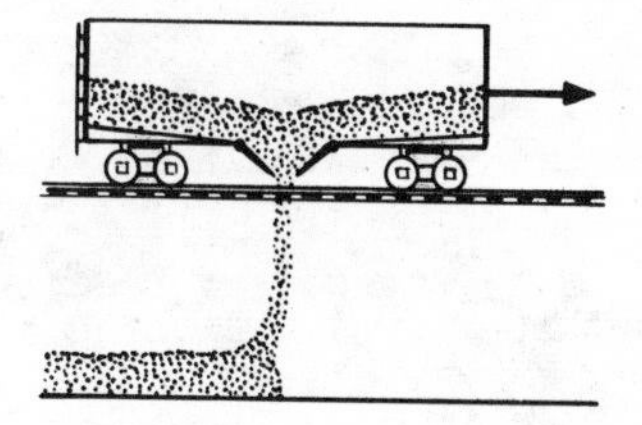

Q8.4 As the freight car in the figure rolls across an elevated level track, its load of sand is being dumped at a steady rate ($dm/dt = constant$). If the car rolls without friction, does its speed change? Does the momentum of the car and its load change? Give reasons for your answers.

Section 8.2

Q8.5 Give an example of a solid object whose center of mass is outside the object.

Q8.6 Where is the center of mass of each of the following objects: (a) a homogeneous sphere, (b) a sphere whose density is a function of radius ($\rho = \rho(r)$), (c) a homogeneous cylinder, (d) a doughnut, and (e) a water glass?

Q8.7 What is the approximate location of the center of mass of the earth's atmosphere?

Q8.8 Nathan and Samuel float in space near their orbiting laboratory. Nathan tosses a wrench to Samuel; Samuel catches it and throws it back to Nathan. Describe the subsequent motion of the two men. Discuss the motion of their center of mass.

Section 8.4

Q8.9 Explain why the phrase "equal and opposite," taken literally, does not make sense for vectors. (The phrase is nevertheless a convenient contraction.)

Q8.10 For each of the following forces, there is an equal and opposite matching force required by Newton's third law. State the nature of the matching force and the body on which it acts. (a) A force of magnitude mg on a baseball, (b) a force of frictional air drag on a baseball, (c) the force of a hammer on a nail, (d) the tidal force of the moon on the oceans of the earth, (e) the frictional force on a skidding automobile, and (f) the thrust propelling a rocket outside the atmosphere.

Q8.11 A short-order cook slides a plate of spaghetti down a counter. He misjudges the stopping power of the counter; the plate continues off the edge of the counter, arcs through the air, and crashes on the floor. Describe *all* the forces exerted by the plate on its environment in each part of its motion.

Q8.12 (1) What gravitational force do you exert on the earth (a) when you are standing motionless on the ground? and (b) when you are leaping through the air? (2) What *total* force do you exert on the earth in each of these cases?

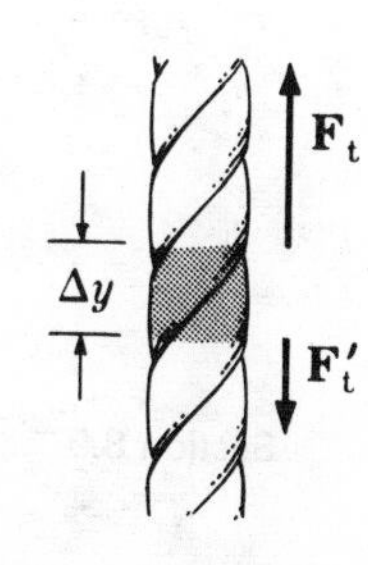

Q8.13 The forces of tension acting on a section of cable (the shaded portion of the figure) are $\mathbf{F}_t$, exerted by the cable above the section, and $\mathbf{F}_t'$, exerted by the cable below the section. (1) These forces are not of equal magnitude ($\mathbf{F}_t + \mathbf{F}_t' \neq 0$). Why does this not violate Newton's third law? (2) As Δy approaches zero, the magnitudes F_t and F_t' approach equality. Is this because of Newton's third law or for some other reason?

Q8.14 Speculate briefly on possible aspects of the behavior of matter in a world where Newton's third law is not valid.

Q8.15 Two weights joined by a light, flexible cord are sent flying through the air. Discuss the motion of their center of mass (a) if the cord is taut, allowing the weights to exert forces on each other, and (b) if the cord is slack so that each weight experiences only a gravitational force.

Section 8.5

Q8.16 For what system is your weight an external force? For what system is it an internal force?

Q8.17 The velocity of the center of mass of a pair of particles is constant before the particles collide. Discuss the motion of the center of mass during the collision of the particles and after their collision.

Q8.18 Explain in what sense Newton's third law is implied by Newton's first law. Show, for instance, that if the third law failed, the first law would fail.

Q8.19 Suppose that a yet-to-be-discovered force within the atomic nucleus does *not* obey Newton's third law. Explain why this is very unlikely. (Discussions in both Sections 8.5 and 4.8 are relevant.)

Section 8.7

Q8.20 Two particles are observed to fly with equal and opposite momenta from the point where an unstable particle decayed at rest. Does this prove that no third particle was created in the decay event?

Q8.21 An experimenter studying bubble-chamber photographs finds a single track coming from a point where he is quite certain a stationary particle decayed. Should he announce his discovery of a violation of the law of momentum conservation?

Q8.22 As an automobile rolls down a hill with its engine off, the magnitude of its momentum increases. Reconcile this fact with the law of momentum conservation.

Q8.23 Two identical spheres approach one another with equal and opposite velocity (not necessarily directly head on). (1) What exactly does the law of momentum conservation say about their motion after the collision? What does it not say? (2) Are there any circumstances in which momentum is not conserved in such a collision?

Section 8.8

Q8.24 Explain why the law of momentum conservation can be applied to a collision that takes place (a) in air, where external forces of gravity act on the system, and (b) on a road surface, where external forces of friction act on the system. (HINT: Consider the impulses contributed by both internal and external forces.)

Q8.25 A body is acted on by a force that is always directed toward a fixed center. Why does the body move in a plane?

Q8.26 A skydiver is falling straight down at terminal speed. (1) Is his momentum constant because of the law of momentum conservation or for some other reason? (2) A camera slips out of the skydiver's pocket and moves upward relative to him (i.e., it falls more slowly). Does the total momentum of the skydiver and his camera remain constant? Why or why not?

Section 8.9

Q8.27 By moving back and forth in a certain way, a man with neither oars nor motor can cause a boat to be propelled in one direction. By repeating the action, he can move the boat as far as he likes. (1) Explain why this method of propelling a system by actions taken within the system does not violate the law of momentum conservation. (2) Can astronauts change the velocity of their spacecraft in a similar way?

Q8.28 A strip of opaque material hides part of a billiard table (see the figure). One white billiard ball disappears under the strip from the left side; then a click is heard and two white balls appear on the right side. Why is it impossible to learn from velocity measurements which ball on the right was the "projectile" and which one was the "target"?

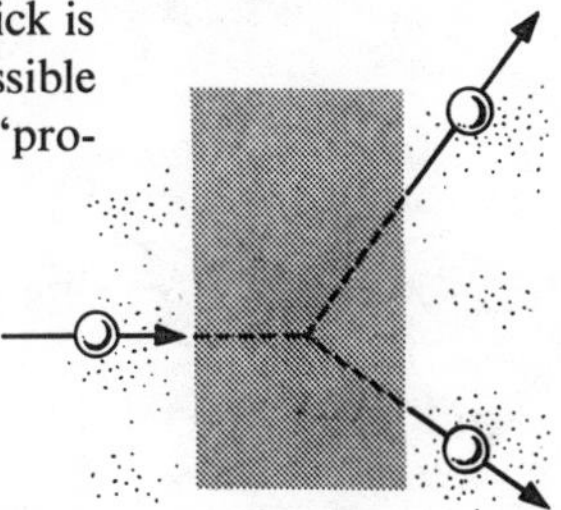

Section 8.10

Q8.29 A rocket engine on a spacecraft burns for a few seconds to accelerate the craft to a higher speed. Does the change in speed of the rocket depend on its speed before the engine burns? Does the *fractional* change in speed depend on the initial speed?

Q8.30 Are there any feasible methods of propulsion in space other than rocket thrust?

Section 8.12

Q8.31 Explain why the law of momentum conservation could replace Newton's third law as a basic law of mechanics. In particular, show that Newton's third law would then be a *consequence* of the basic laws.

EXERCISES

Section 8.1

E8.1 A body falls with constant acceleration. Write expressions for its momentum as a function of (a) time and (b) distance.

E8.2 A proton of mass 1.7×10^{-27} kg is moving with a speed of 10^7 m/sec. (1) What is the magnitude of its momentum? (2) It is acted upon by a force of 5×10^{-15} N directed parallel to its velocity. What is the rate of change of its momentum? (3) How much time is required for the momentum of the proton to double?

E8.3 An electron moves horizontally to the right with momentum $p_0 = 3 \times 10^{-24}$ kg m/sec. It then experiences a steady force of magnitude 1.5×10^{-18} N. Find its momentum (magnitude and direction) after 1 μsec if the force acts (a) horizontally to the right, (b) horizontally to the left, or (c) vertically upward.

E8.4 (1) Verify that the dimension of force $\times$ time is the same as the dimension of momentum. (2) Verify that the dimension of force $\times$ distance $\times$ mass is the same as the dimension of $(\text{momentum})^2$.

E8.5 A man pushes his stalled automobile along a level stretch of highway. To keep it moving forward with a constant momentum of 1.5×10^3 kg m/sec, he must exert a steady force of 120 N. (1) What force of friction acts? (2) If he stops pushing and the same force of friction continues to act, how long a respite will he have until the momentum of the automobile is halved?

E8.6 A marble of mass m and velocity $\mathbf{v}$ strikes a stationary marble of equal mass. After the collision, the incident marble is at rest and the struck marble is moving with velocity $\mathbf{v}$. (1) What impulse is delivered to each marble? (2) What total impulse is delivered to both marbles?

E8.7 A car with a mass of 1,200 kg and a speed of 5 m/sec is initially traveling northward. After completing a turn of 90 deg to the right in a period of 5 sec, the daydreaming driver runs into a telephone pole. (1) What impulse was delivered to the car (a) during the turn and (b) during the crash? (2) What average force $\overline{\mathbf{F}}$ acted on the car during its turn?

E8.8 (1) For a body of fixed mass,

$$\frac{d\mathbf{p}}{dt} = m\frac{d\mathbf{v}}{dt}.$$

What is the corresponding expression for the rate of change of momentum of a body of variable mass? (2) Name a commonly encountered object or system that experiences a change of momentum with no change of velocity.

E8.9 Show that the definition of average velocity,

$$\overline{\mathbf{v}} = \frac{\Delta\mathbf{r}}{\Delta t} = \frac{\mathbf{r}_2 - \mathbf{r}_1}{t_2 - t_1},$$

where $\mathbf{r}$ is a position vector, follows from the general definition of an average over time given by Equation 8.11.

E8.10 Use Equation 8.10 to find the average position and the average x component of velocity of a particle moving along the x axis with constant acceleration. Give the averages for (a) an arbitrary time interval and arbitrary initial conditions and (b) the time interval 0 to T with the initial conditions $x = 0$ and $v_x = 0$ at $t = 0$.

Section 8.2

E8.11 In the xy plane a 10-kg mass is placed at the point $x = 5$, $y = 1$ and a 5-kg mass at the point $x = 5$, $y = 4$. (1) Where is their center of mass? (2) A third mass of 30 kg is added at the point $x = 2$, $y = 2$. Where is the center of mass of the combination of three masses?

E8.12 The distance between the centers of the carbon atom and the oxygen atom in a molecule of carbon monoxide (CO) is 1.13×10^{-10} m (1.13 Å). (1) Where is the center of mass of the molecule? (2) If at a given instant the carbon nucleus is moving downward with a speed of 3×10^2 m/sec and the center of mass is at rest, in what direction and with what speed is the oxygen nucleus moving? (Ignore any contribution of electrons to the molecular mass.)

E8.13 Verify by calculation the statement made in Section 8.2 that the center of mass of the earth-moon system is 2,900 miles from the center of the earth.

E8.14 A projectile particle of mass m_1 moves with speed v_1 toward a stationary target particle of mass m_2. Find the velocity and the acceleration of the center of mass of the two-particle system.

E8.15 The position vector of a marble of mass $2m$ is $\mathbf{r}_1 = v_0t\mathbf{i}$; the position vector of another marble, of mass m, is $\mathbf{r}_2 = -v_0t\mathbf{j}$. (1) Sketch the paths of the marbles and show their positions at a particular instant. (2) Find (a) the position, (b) the velocity, and (c) the acceleration of their center of mass.

E8.16 A thrown ball follows a parabolic path described by $\mathbf{r}_1 = v_0 t\mathbf{i} + (v_1 t - \frac{1}{2}gt^2)\mathbf{j}$. Beneath it rolls a ball of equal mass, its position given by $\mathbf{r}_2 = v_0 t\mathbf{i}$. Find the velocity and acceleration of the center of mass of this pair of balls.

E8.17 (1) For the system of two balls described in the preceding exercise, find (a) the total momentum and (b) the rate of change of momentum. (2) What total force acts on this system?

Section 8.3

E8.18 At one edge of a spacecraft of mass 2,000 kg a vernier-thrust rocket fires accidentally, setting the craft into a twisting and gyrating motion. The rocket pushes the craft for 2 sec with a force of 400 N. (1) Describe in words the motion of the center of mass of the craft. (2) Assuming that the direction of the thrust changes very little during the 2 sec in which the rocket fires, calculate the change of velocity of the craft.

E8.19 At $t = 0$, a stone of mass m_1 is dropped from the top of a tower; it falls with constant acceleration g. At $t = t_1$, a second stone, of mass m_2, is dropped from the same point. The initial speed of each is zero. (1) For this system of two stones after time t_1, find (a) the velocity of the center of mass, (b) the acceleration of the center of mass, (c) the total momentum of the system, and (d) the total force acting on the system. (2) Verify explicitly that Equations 8.22, 8.29, and 8.30 are satisfied. (3) Explain why the two stones may be considered to form a system, even though neither one has any effect on the other.

Section 8.4

E8.20 A chain hoist lifts a mass of 10^4 kg at a constant speed of 0.1 m/sec. What downward force does the chain hoist exert on the girder that supports it? (Neglect the mass of the chain hoist in comparison with the mass of its load.)

E8.21 A 10-kg mass falls from rest through a distance of 3 m. It strikes the ground and experiences an acceleration of $200g$ as it is being brought to rest. What total force does it exert on the earth (a) just after it starts to fall, (b) just before it hits the ground, (c) as it is being stopped, and (d) after it has been stopped?

Section 8.6

E8.22 Referring to Figure 8.12 and Example 1 in Section 8.6, calculate the acceleration imparted to the whole earth by the net leftward force, $F_1 - F_2$, exerted by the horse and cart on the ground. If this acceleration persisted for 1 hr, how far would the earth move? Compare this distance with the size of an atom, a nucleus, or an elementary particle.

E8.23 Show that Equations 8.42–8.45 are partially redundant: derive one of them algebraically from the other three.

E8.24 A three-link chain, each link of mass 10 kg, hangs from a hook. The chain is drawn upward with a constant acceleration of 2 m/sec. Find all the forces acting on the center link.

E8.25 A locomotive pulls N cars on a level track with acceleration a. Car 1 is nearest the locomotive; car N is at the rear of the train. Each car has the same mass m and is acted on by the same frictional force F_f. (1) What force does car n exert on car $n + 1$? (2) What force does car $n + 1$ exert on car n? (3) What force does car 1 exert on the locomotive?

E8.26 Exerting a horizontal component of force of 20 N, a boy pulls an 8-kg wagon at constant speed. (1) What force does the wagon exert on the ground? (2) As the wagon rolls onto a smoother surface, the frictional force is halved.

If the boy continues to pull with the same force, what is the acceleration of the wagon? (3) What horizontal component of force does the wagon exert on the boy (a) as it rolls with constant speed and (b) as it is accelerated?

Section 8.7

E8.27 An alpha particle with an initial momentum of 7×10^{-20} kg m/sec approaches a stationary uranium nucleus. In the encounter, the alpha particle is turned through 90 deg and emerges with nearly the same magnitude of momentum. What is the recoil velocity (magnitude and direction) of the uranium nucleus, whose mass is 4×10^{-25} kg?

E8.28 A neutral pion with a momentum of 5.10×10^{-20} kg m/sec decays into two photons, one of which has a momentum of 2.94×10^{-20} kg m/sec and flies away on a line perpendicular to the original path of the pion. (1) What is the momentum vector of the other photon? Include a diagram with your answer. (2) If the momentum of the pion, whose mass is 2.4×10^{-28} kg, were correctly given by the nonrelativistic formula, $p = mv$ (it is not), what would be its speed before it decayed? *Optional:* Use Equation 21.25 to find the speed of the pion before its decay.

E8.29 An artillery shell that is traveling horizontally at a speed of 800 m/sec explodes into three fragments. The first fragment has a mass of 15 kg and flies forward in a horizontal direction with a speed of 1,000 m/sec. The second fragment, with a mass of 10 kg, flies forward in a horizontal direction with a speed of 900 m/sec. The third fragment has a mass of 5 kg. What is its velocity a moment after the explosion?

E8.30 The recoil definition of mass is given by Equation 2.10. (1) Derive this equation from the law of momentum conservation. (2) Under what conditions is the equation valid? (3) It can be applied to carts on an air track. Why?

E8.31 Two docked space vehicles, one of mass 1,500 kg, the other of mass 2,500 kg, are caused to separate by means of a small explosive charge located between them. The explosion gives an impulse of 1,000 N sec to the less massive vehicle. With what relative speed do the two vehicles drift apart? (Assume that they separate along a line joining their centers of mass.)

Section 8.8

E8.32 A head-on collision takes place between a 10,000-kg truck traveling 60 mile/hr and a 1,000-kg passenger car traveling 80 mile/hr in the opposite direction. If the car and truck lock together at the instant of collision, in what direction and at what speed is the wreckage traveling immediately after the collision? What is the change of velocity of the passenger car? What is the change of velocity of the truck? (NOTE: This exercise can be solved without units conversion.)

E8.33 A 50-kg girl on ice skates catches a 0.5-kg ball that was moving at an angle of 30 deg to the horizontal with a speed of 15 m/sec just before she caught it. If friction with the ice can be neglected, what is her speed after catching the ball?

E8.34 A certain rifle fires a bullet of mass 0.025 kg with a muzzle velocity of 700 m/sec (about Mach 2). The bullet strikes a 1-kg wooden block. The block, with the bullet imbedded in it, acquires a speed of 14.8 m/sec. (1) How much loss of momentum did the bullet suffer between the time it left the rifle barrel and the time it struck the block? Express your answer both in kg m/sec and in percentage of initial momentum. (2) What happened to the "lost" momentum?

E8.35 A pickup truck with a mass of 2,000 kg including its load is coasting at 6 m/sec. A man standing in the rear throws out a 50-kg bag of sand with a speed of 1 m/sec relative to the truck. What is the speed of the truck after the bag is thrown out if the bag is thrown (a) directly to the rear or (b) directly to the side?

E8.36 A rocket sled with a mass of 3,000 kg moves with negligible friction at a constant speed of 250 m/sec over a set of rails. At a certain point a scoop on the sled dips into a trough of water located between the rails and scoops water into an empty tank on the rocket sled. After 1 m^3 of water has entered the tank, what is the speed of the sled?

Section 8.9

E8.37 A collision slightly less catastrophic than the one described in Exercise 8.32 occurs between a pair of automobiles, each of mass 2×10^3 kg, one traveling north at 15 mile/hr, the other traveling east at 20 mile/hr. What is the velocity of the center of mass of the system (magnitude and direction) before, during, and after the collision?

E8.38 As shown in Figure 8.19, a neutron is deflected by 30 deg when it strikes a stationary proton, and the proton recoils at an angle of 60 deg to the initial direction of the neutron. The initial speed of the neutron is $v_1 = 4 \times 10^6$ m/sec; the masses of the two particles are nearly equal. (1) Express the law of momentum conservation in component form and solve the two resulting equations to find the final speeds v_2 and v_3. (2) Is kinetic energy conserved in this collision?

Section 8.10

E8.39 (1) Working from the equations of momentum conservation, carry out the algebraic manipulations needed to derive Formula 8.63 for the mass of exhaust gas needed to change the velocity of a rocket. (2) Why is this formula valid only if $\Delta v \ll v_{ex}$?

E8.40 During a lunar mission it is necessary to make a midcourse correction of 2 m/sec in the speed of the spacecraft. If the speed of the exhaust gases is $v_{ex} = 10^3$ m/sec, what percentage of the mass of the craft must be discarded as exhaust?

E8.41 Rearrange Equation 8.67 to express the final mass m_2 of a rocket as a function of the initial mass m_1 and the change of speed, $v_2 - v_1$. Sketch a graph that illustrates the nature of this function. Discuss it in two limits: (a) $v_2 - v_1 \ll v_{ex}$ and (b) $v_2 - v_1 \gg v_{ex}$.

E8.42 A rocket with a mass of 10^5 kg is on the launching pad pouring out exhaust whose speed is $v_{ex} = 2.5 \times 10^3$ m/sec. (1) What is the rate of flow of exhaust (in kg/sec) required to start the rocket upward with an acceleration of $0.5g$? (2) After 1 min, what is the remaining mass of the rocket? What is its acceleration at this time? (Neglect the change of gravitational acceleration with altitude.)

Section 8.11

E8.43 The center of mass of a two-body system is fixed at the origin. Show that the position vectors of the two particles are given by $\mathbf{r}_1 = -(m_2/M)\mathbf{r}$ and $\mathbf{r}_2 = (m_1/M)\mathbf{r}$, where $M = m_1 + m_2$ and $\mathbf{r} = \mathbf{r}_2 - \mathbf{r}_1$.

E8.44 (1) For two particles with masses m_1 and m_2 prove that the reduced mass μ is less than the smaller of the two masses. (2) Prove that the reduced mass of a two-particle system of fixed total mass is greatest if the two masses m_1 and m_2 are equal. (3) If $m_1 \ll m_2$, show that $\mu \cong m_1(1 - (m_1/m_2))$.

PROBLEMS

Momentum and force

P8.1 At a certain instant the momentum of an electron is given by $\mathbf{p} = p_0\mathbf{i}$, and it is experiencing a force given by $\mathbf{F} = F_0\mathbf{j}$ ($\mathbf{i}$ and $\mathbf{j}$ are unit vectors in the x and y directions respectively). (1) What is the rate of change of momentum of the electron? (2) What is the rate of change of the angle that defines the direction of the momentum? (3) What is the rate of change of the magnitude of the momentum? (4) If $\mathbf{F}$ remains constant, which of these three rates of change will remain constant? Which, if any, will change?

Momentum and impulse

P8.2 A fastball approaches the plate at 45 m/sec. The batter connects; the ball leaves his bat at an angle of 45 deg to the horizontal and is caught in deep center field 120 m from the plate. Assume the mass of the baseball to be 0.145 kg and ignore air friction. (1) What is the momentum of the ball immediately before it is struck by the bat? (2) What is its momentum immediately after it has been struck? (3) What impulse is delivered (a) by the bat to the ball and (b) by the ball to the bat? (4) Estimate the length of time the ball is in contact with the bat and calculate the average force $\bar{\mathbf{F}}$ exerted by the bat on the ball.

Center of mass of a composite system

P8.3 A complex system of many particles is divided into two parts: a part of mass M_1 with center-of-mass vector $\mathbf{R}_1$ and a part of mass M_2 with center-of-mass vector $\mathbf{R}_2$. Prove that the center-of-mass vector of the whole system is given by

$$\mathbf{r}_c = \frac{M_1\mathbf{R}_1 + M_2\mathbf{R}_2}{M_1 + M_2}.$$

Three-body system

P8.4 A system is composed of three objects that may be treated as if they were particles. At a particular instant, object 1, whose mass is 3 kg, is located at the origin and has an acceleration given by $\mathbf{a}_1 = (2\mathbf{i} + 2\mathbf{j})$ m/sec^2; object 2, whose mass is 1 kg, is located at $x = 5$, $y = 0$ and has an acceleration given by $\mathbf{a}_2 = 4\mathbf{i}$ m/sec^2; object 3, whose mass is 2 kg, is located at $x = 0$, $y = 3$. The total external force on the system is known to be zero. State as much as possible about (a) the location of the center of mass, (b) the velocity of the center of mass, (c) the acceleration of the center of mass, (d) the velocities of all three objects, and (e) the acceleration of object 3.

Forces on a system

P8.5 A man lifts a 10-kg box with the help of a pulley of mass 2 kg and a rope of negligible mass (see the figure). (1) What upward force is exerted by the supporting hook if (a) the box is stationary, (b) the box rises at a constant speed of 0.5 m/sec, (c) the box experiences an upward acceleration of

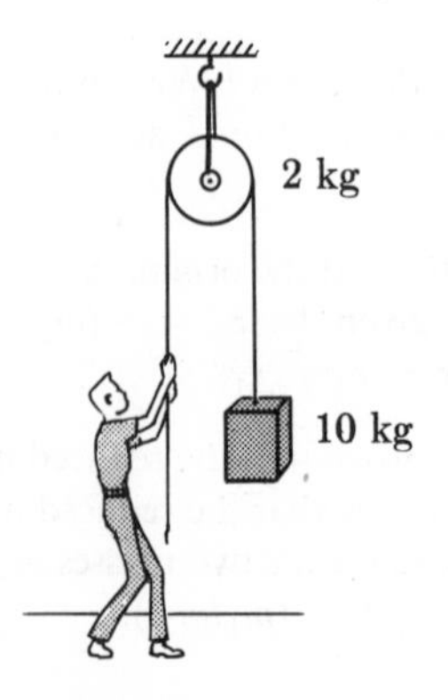

1.2 m/sec^2, and (d) the man lets go of the rope? (2) With what force does the rope pull on the man in cases (a), (b), and (c)? (Treat the pulley as frictionless.)

Systems with variable mass

P8.6 Beginning at $t = 0$, sawdust falls at a steady rate onto a moving conveyer belt (see the figure). The mass of sawdust on the belt is $m(t) = \alpha t$, where α is a constant. (1) What force is required to keep the belt moving at a steady speed v? (2) What is the integral $\int_0^t F\,dt$? What is its significance?

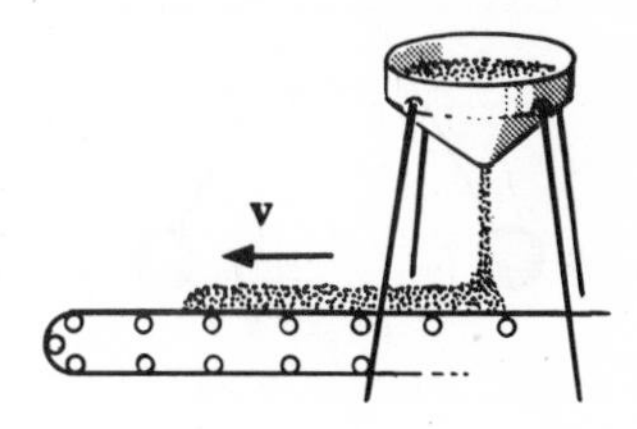

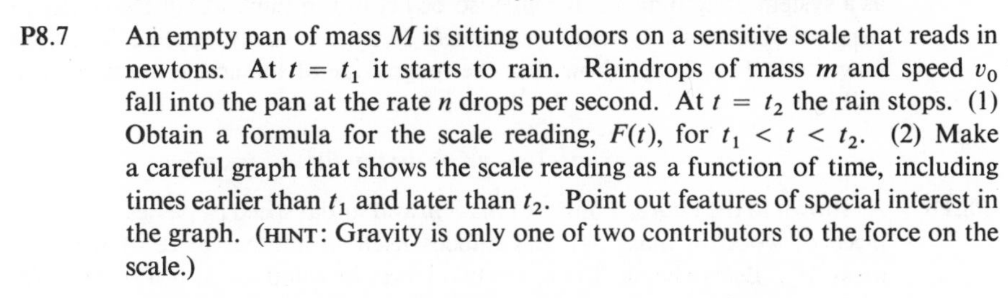

P8.7 An empty pan of mass M is sitting outdoors on a sensitive scale that reads in newtons. At $t = t_1$ it starts to rain. Raindrops of mass m and speed v_0 fall into the pan at the rate n drops per second. At $t = t_2$ the rain stops. (1) Obtain a formula for the scale reading, $F(t)$, for $t_1 < t < t_2$. (2) Make a careful graph that shows the scale reading as a function of time, including times earlier than t_1 and later than t_2. Point out features of special interest in the graph. (HINT: Gravity is only one of two contributors to the force on the scale.)

Motion of a system accelerated by gravity

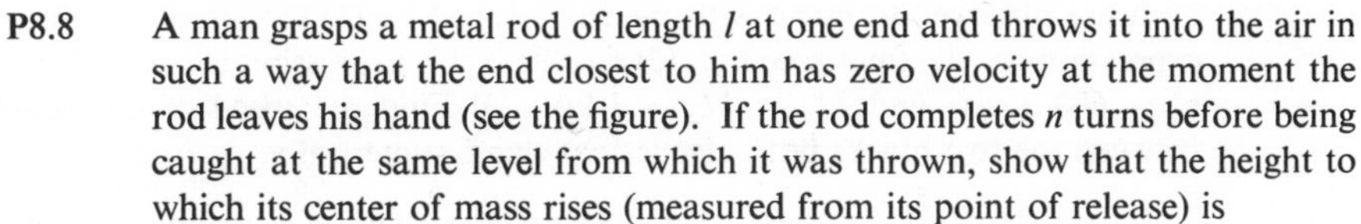

P8.8 A man grasps a metal rod of length l at one end and throws it into the air in such a way that the end closest to him has zero velocity at the moment the rod leaves his hand (see the figure). If the rod completes n turns before being caught at the same level from which it was thrown, show that the height to which its center of mass rises (measured from its point of release) is

$$h = \frac{\pi}{4} nl.$$

(HINT: The motion of the rod relative to its center of mass is the same as if no gravitational force acted.)

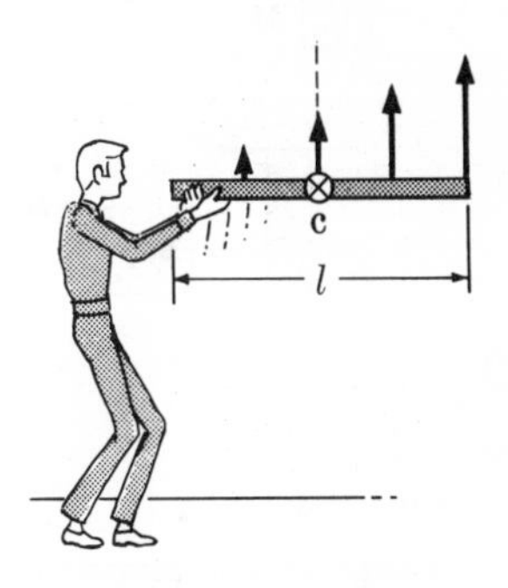

P8.9 Weights of mass m_1 and m_2 are joined by a light rod of length 0.8 m and of negligible mass. As this system rotates and falls in the earth's gravitational field, its center of mass follows a straight vertical path (see the figure). At time t, the velocities of the weights are directed as shown and have magnitudes $v_1 = 5.4$ m/sec and $v_2 = 10.6$ m/sec. At time $t' = t + 1.41$ sec, the velocities of the weights are directed as shown and have magnitudes $v_1' = 28.5$ m/sec and $v_2' = 12.5$ m/sec. (1) If $m_1 + m_2 = 8$ kg, what are the separate masses m_1 and m_2? (2) Through how many revolutions has the system turned between times t and t'?

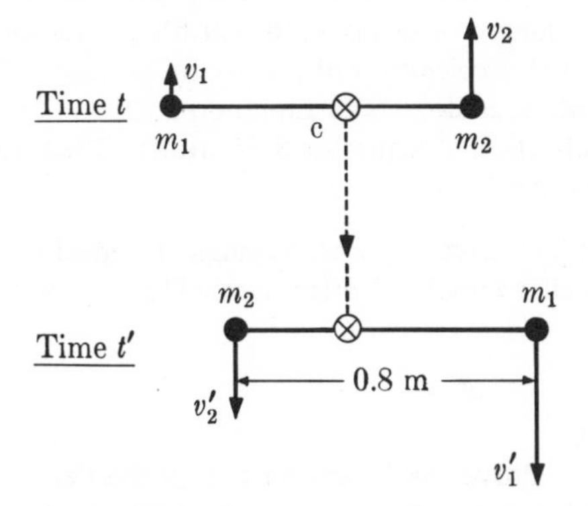

Plastic balls on a string

P8.10 A child's toy consists of two plastic balls, each of mass m, joined by strings of equal length l to a small ring (see the figure). (1) Prove that if a child applies a

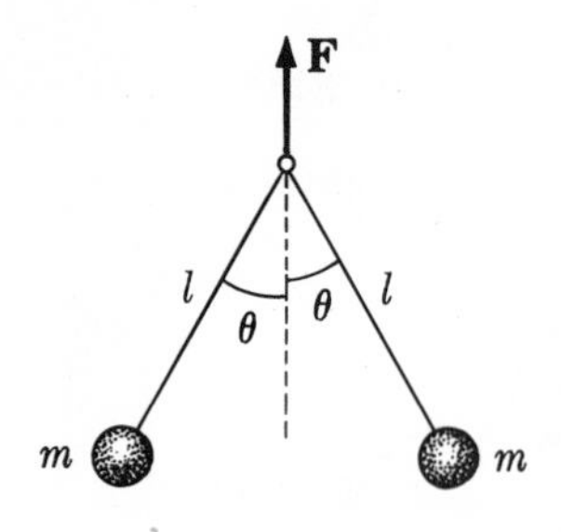

vertical force **F** of magnitude $2mg$ to the ring and the strings make equal angles θ with the vertical, the magnitude of the acceleration of each ball is $a = g \tan \theta$. (2) What is the direction of the acceleration of each ball? (3) If the strings are initially at angle θ_0 to the vertical, through what vertical distance does the ring move until the balls come together? (Ignore the finite radius of the balls for this calculation) (HINTS: The ring alone may be treated as a system of zero mass. It may also be helpful to think about the center of mass of the whole system.) *Optional:* If the vertically applied force has magnitude $F = 2\gamma mg$, show that the magnitude of the acceleration of each ball is

$$a = g\sqrt{(1 - \gamma)^2 + (\gamma \tan \theta)^2}.$$

Conservation of momentum in one dimension

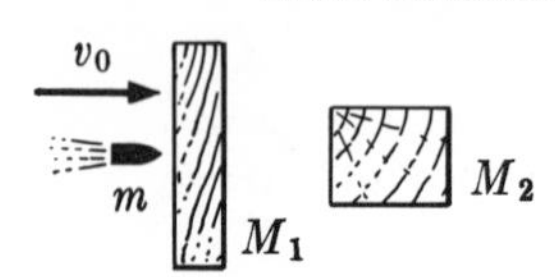

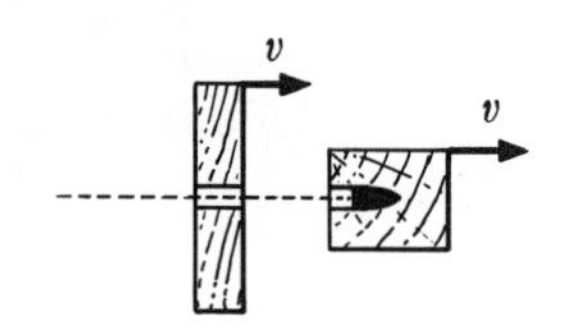

P8.11 As shown in the figure, a bullet of mass m and initial speed v_0 passes through a piece of wood of mass M_1 and imbeds itself in another piece of wood of mass M_2. Before being struck, the two pieces of wood are at rest; afterward, both are observed to move with the same speed v. (1) Obtain an algebraic expression for the final speed v. (2) Evaluate this expression for $m = 2$ gm, $v_0 = 600$ m/sec, and $M_1 = M_2 = 1$ kg. (3) Find the speed of the bullet between the two blocks both algebraically and numerically.

P8.12 Two baseball players attempt to practice on the surface of a frozen lake where friction is negligible. The pitcher throws the ball at a speed of 40 m/sec relative to himself. The batter hits a line drive back to the pitcher, the batted ball moving at a speed of 40 m/sec relative to the batter. The mass of the baseball is 0.14 kg and the mass of each player is 70 kg. (1) With what speed does the pitched ball approach home plate? (2) What is the speed of the pitcher (a) while the ball is in the air and (b) after he catches the batted ball? (Ignore the effects of gravity, air friction, and ice friction.)

Conservation of momentum in two dimensions

P8.13 When a moving billiard ball (A) strikes a stationary billiard ball (B) in any way other than a head-on collision, the velocities of the two balls after the collision are nearly perpendicular. (1) Use this fact together with the law of momentum conservation to derive an expression for the speed of ball A after the collision as a function of its initial speed v_0 and its angle of deflection θ.

Rocket in a gravitational field

P8.14 Equations 8.65 and 8.67 were derived for a rocket free of gravitational forces. Consider instead a rocket moving vertically in the earth's gravitational field. (Ignore the dependence of the acceleration of gravity on altitude.) Derive a differential equation of motion analogous to Equation 8.65, and integrate it in order to obtain a generalization of Equation 8.67. (HINT: These equations will involve time as well as speed and mass.)

Specific impulse

P8.15 A quantity frequently used by rocket engineers to gauge the effectiveness of a certain rocket fuel is "specific impulse," often symbolized by $\mathscr{I}$. Specific impulse is defined by

$$\mathscr{I} = \frac{v_{ex}}{g}.$$

It is measured in seconds. Prove the following two properties of specific impulse. (1) Rocket thrust is equal to $\mathscr{I}$ times the *weight* of fuel exhausted per second. (2) If a rocket just clear of the launching pad barely supports its own weight (thrust $= mg$), $\mathscr{I}$ is the time required for the rocket weight to drop to $1/e$ of its initial value.

P8.16 The center of mass of two-body system has position vector $\mathbf{r}_c$ and moves with constant velocity $\mathbf{v}_c$. (1) Generalize Equations 8.75–8.80 to cover this case. Which of these equations require no change? (2) If the system is isolated so that $\mathbf{F}_1 + \mathbf{F}_2 = 0$, show that $\mathbf{F}_2 = d\mathbf{p}/dt$, where $\mathbf{p}$ is the momentum of the effective single particle (see Equation 8.81).

Two-body system

P8.17 Two unequal masses m_1 and m_2, joined by a spring of equilibrium length s_0 and spring constant k, vibrate along a straight line. As shown in the figure, their separation is $s = s_1 + s_2$, where s_1 and s_2 are their respective distances from their fixed center of mass. (1) Set up and solve the effective one-body problem to find $s(t)$. (2) What is the period of vibration? (3) What are the separate position functions $s_1(t)$ and $s_2(t)$? (HINT: The quantity $s - s_0$ is a useful variable.)

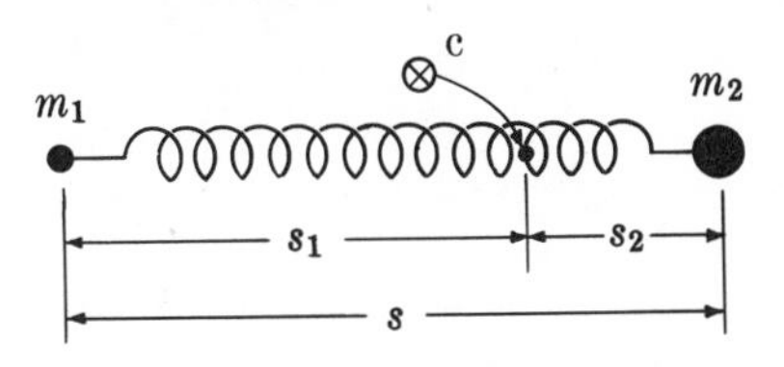

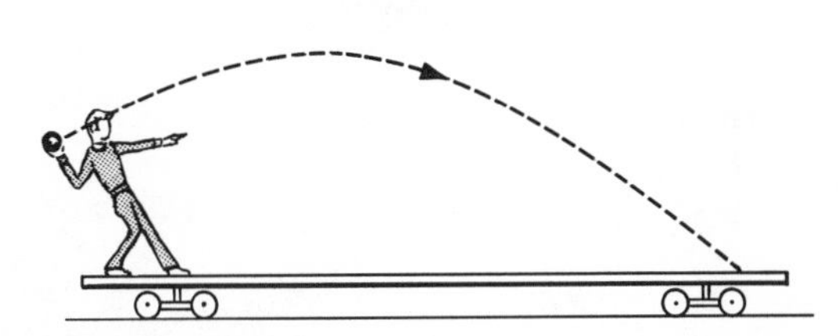

P8.18 A shot-putter standing at one end of a light flatcar (see the figure) puts the shot along the length of the car twice, first with the car locked in position, then with the car free to roll on horizontal rails with negligible friction. On both trials the shot, just after it leaves the hand of the shot-putter, has the same velocity (both magnitude and direction) *relative* to him. On which trial, if either, does the shot cover the greater distance measured along the flatcar? Carefully justify your answer.

Shot-putter on a movable car

P8.19 (NOTE: In this problem "velocity" means x component of velocity.) Ball 1, of mass m_1, moves along the x axis with positive velocity v_0 toward ball 2, which has mass m_2 and is initially stationary. After their collision, the two balls move along the x axis with velocities v_1 and v_2 respectively. (1) Write an equation that expresses momentum conservation in the collision. (2) Is this equation valid if the balls are made of steel? if they are made of putty? (3) Considering v_0 to be a fixed constant in the momentum-conservation equation, prepare a graph of v_2 vs v_1. Label the points where your curve intercepts the v_1 and v_2 axes. (4) Each point on the curve in the v_1v_2 plane represents a pair of values of v_1 and v_2 that satisfy the condition of momentum conservation. If the balls are impenetrable, there is another condition on v_1 and v_2. What is this condition? Show how it divides the curve into parts representing physical and nonphysical final states of motion. (5) With the condition of impenetrability, what are the maximum and minimum possible values of v_1? of v_2? (NOTE: Energy considerations are added to this problem in Problems 10.26 and 10.27.)

Two-body collision in one dimension

P8.20 In one frame of reference, a set of n particles with masses $m_1, m_2, m_3, \ldots$ have velocities $\mathbf{v}_1, \mathbf{v}_2, \mathbf{v}_3, \ldots$. Another frame of reference moves with constant velocity $\mathbf{v}_0$ with respect to the first frame. (1) Prove that the total momentum of the system is constant in one frame if it is constant in the other frame. (2) Is the physical *quantity* momentum invariant (the same in different inertial frames of reference)? Is the *law* of momentum conservation invariant?

Invariance of a conservation law

9 Angular Momentum

Angular momentum is one of those basic concepts of mechanics that has proved to be a hardy perennial, even more deeply rooted in contemporary physics than in the classical physics that gave it birth. In this chapter we are concerned with its definition, its application to mechanical problems, and its conservation.

9.1 The concept of angular momentum

Angular momentum is a vector quantity.* As a derived concept, it is defined in terms of the already understood concepts of mass, velocity, and distance. The usual symbol for angular momentum is **L**. Its defining equation is

The angular momentum of a particle

$$\mathbf{L} = m\mathbf{r} \times \mathbf{v}, \tag{9.1}$$

where m is the mass of a particle, **r** is its position vector, and **v** is its velocity. More compactly, angular momentum may be expressed in terms of the particle momentum **p**:

An equivalent definition

$$\mathbf{L} = \mathbf{r} \times \mathbf{p}. \tag{9.2}$$

L is defined with respect to an arbitrary origin

In Equations 9.1 and 9.2, **r** is the position vector of the particle *with respect to an arbitrary origin*. This means that angular momentum, both in magnitude and in direction, may depend upon the choice of origin. As we shall discover in this chapter, it is nevertheless a very useful concept. As shown in Figure 9.1, angular momentum is directed perpendicular to the plane containing **r** and **v**, with the right-hand rule of the vector product determining its direction along this perpendicular line. The magnitude of **L** is

$$L = mrv \sin\theta = rp \sin\theta, \tag{9.3}$$

* More exactly, it is an axial vector quantity (see Section 6.8).

where θ is the smaller angle between **r** and **v**. The choice $\mathbf{r} \times \mathbf{p}$ rather than $\mathbf{p} \times \mathbf{r}$ to define **L** is an arbitrary convention, similar to the standard choice of a right-handed coordinate system.

If the velocity **v** is perpendicular to the position vector **r**, as it would be for circular motion about the reference point [Figure 9.2(a)], the magnitude of the angular momentum is

$$L = mrv = rp \qquad (\mathbf{r} \perp \mathbf{v}). \tag{9.4}$$

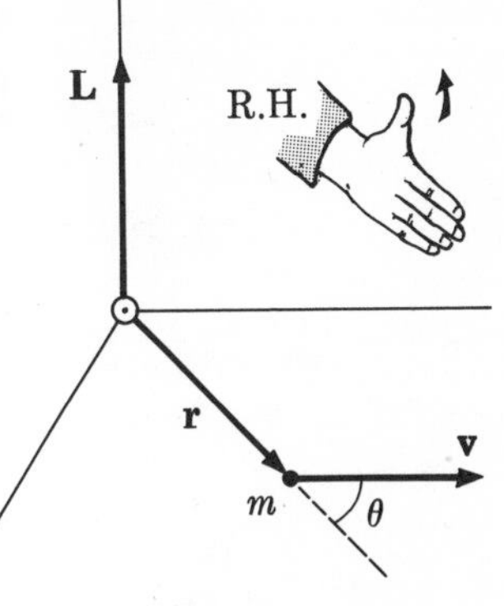

FIGURE 9.1 The angular momentum of a particle is defined with respect to an arbitrary origin by the vector-product formula $\mathbf{L} = m\mathbf{r} \times \mathbf{v}$.

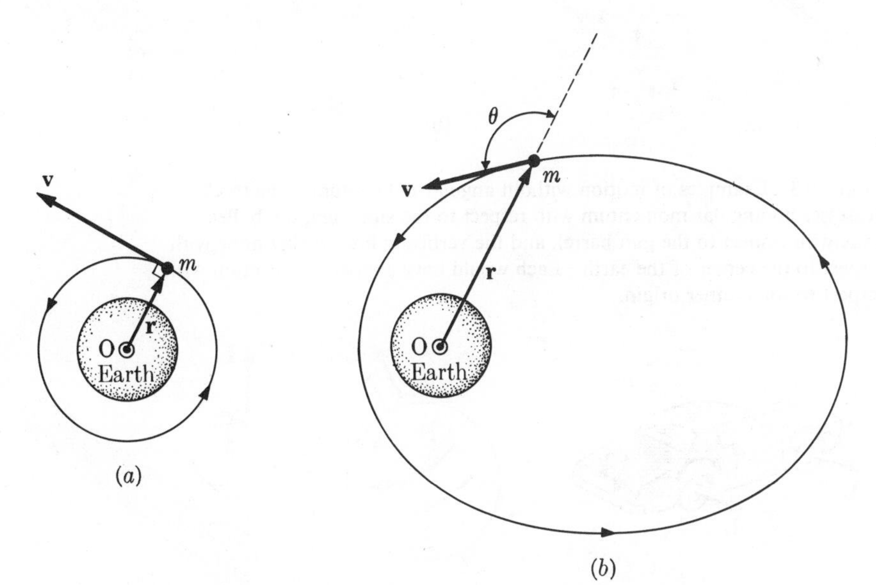

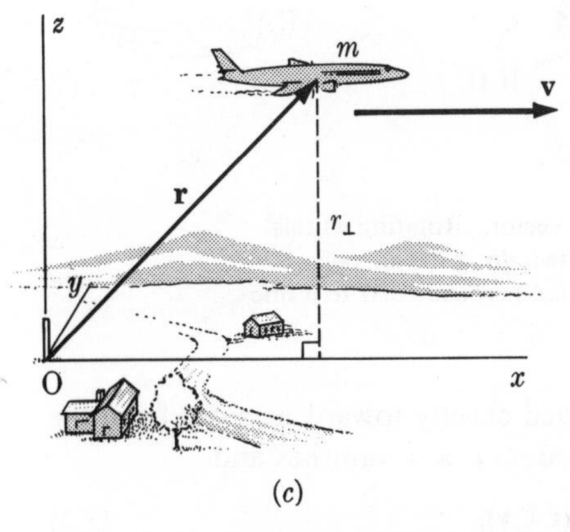

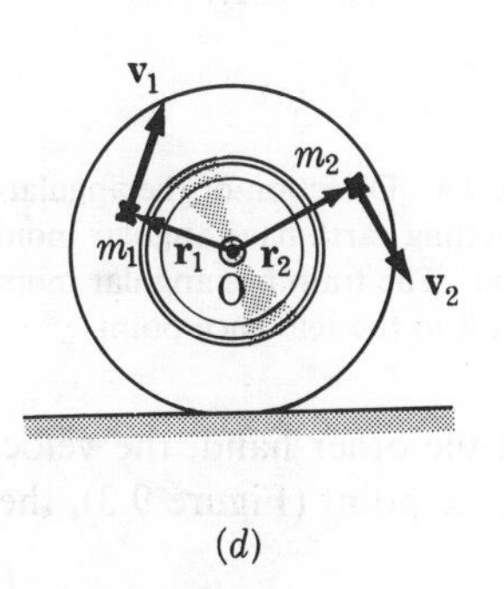

FIGURE 9.2 Examples of motion with angular momentum. (a) A satellite in a circular orbit, and (b) a satellite in an elliptical orbit. For both, the center of the earth is the most convenient reference point. (c) An airplane flying straight possesses angular momentum with respect to a point on the earth. (d) For a rotating wheel, each bit of mass contributes some angular momentum to the total. In each example, the reference point is arbitrary. See also Figure 2.5.

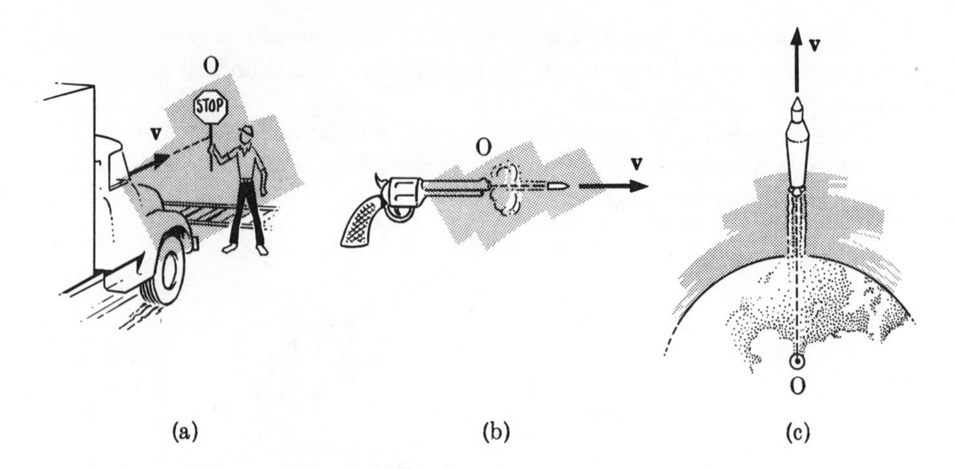

FIGURE 9.3 Examples of motion without angular momentum. The truck possesses no angular momentum with respect to the stop sign, the bullet none with respect to the gun barrel, and the vertically fired rocket none with respect to the center of the earth. Each would have angular momentum with respect to some other origin.

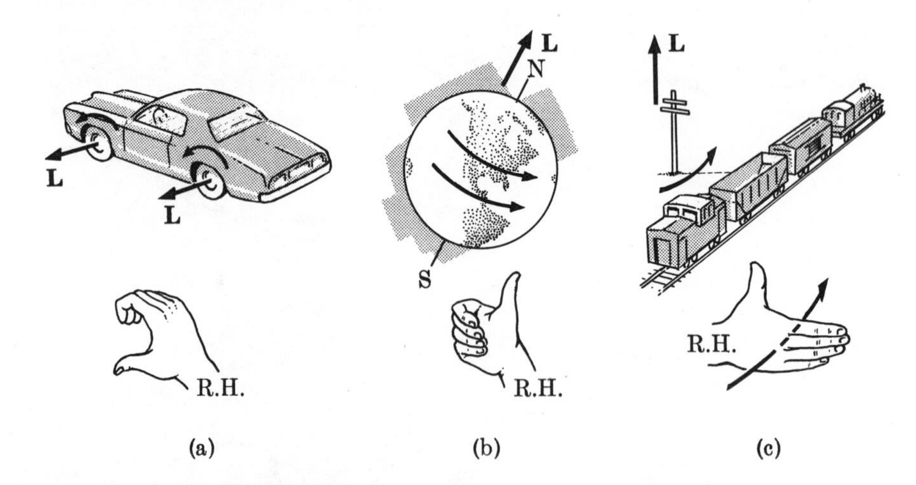

FIGURE 9.4 Direction of the angular-momentum vector. Rotating wheels and rotating earth have angular momentum directed along the axis of rotation. The train has angular momentum directed perpendicular to a line joining it to the reference point.

If, on the other hand, the velocity is directed exactly toward or away from the reference point (Figure 9.3), the vector product $\mathbf{r} \times \mathbf{v}$ vanishes and

$$\mathbf{L} = 0 \qquad (\mathbf{r} \parallel \mathbf{v}). \tag{9.5}$$

For intermediate directions of the velocity [Figures 9.2(b) and 9.2(c)], the factor $\sin \theta$ in Equation 9.3 causes the magnitude of $\mathbf{L}$ to be intermediate between 0 and rp. The important directional property of angular momentum is emphasized in Figure 9.4.

Alternative ways of writing Equation 9.3 are occasionally useful. Note

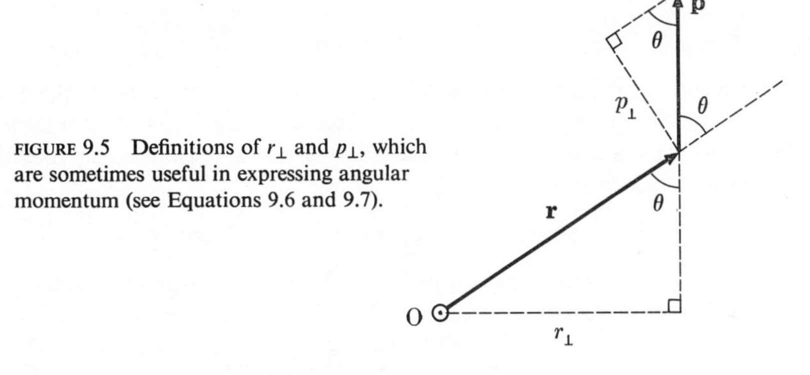

FIGURE 9.5 Definitions of $r_\perp$ and $p_\perp$, which are sometimes useful in expressing angular momentum (see Equations 9.6 and 9.7).

that $r \sin\theta$ is the same as $r_\perp$, the component of **r** perpendicular to **p** (Figure 9.5). Therefore, the magnitude of **L** may be written

$$L = r_\perp p. \tag{9.6}$$

Other expressions for L

Another equivalent expression is

$$L = rp_\perp, \tag{9.7}$$

in which $p_\perp$ is the component of **p** perpendicular to **r** (Figure 9.5). Equation 9.7 can be given an interesting physical interpretation. Angular momentum measures a *rotational* property of motion. If a particle flies straight toward (or away from) a point, it has no tendency to rotate about that point. Accordingly, it has no angular momentum with respect to that particular point. If the particle moves transverse to the line joining it to the reference point, it has the maximum tendency to rotate about that point. In that case, the perpendicular component $p_\perp$ takes on its maximum value. Between these two extremes, a certain part of the momentum, its perpendicular component, contributes to the rotational tendency while another part, its component parallel to **r**, does not. The easiest way to recognize what we here call a rotational tendency is to picture the motion of an imaginary line drawn from the reference point to the particle. If this line rotates as the particle moves, the particle possesses angular momentum. If the line does not rotate but only stretches or shrinks as the particle moves, the particle possesses no angular momentum.

As the definition makes clear, the physical dimension of angular momentum is the product of the dimensions of mass, distance, and velocity. Its SI unit is kg m^2/sec. This particular combination of units has received no new name of its own (unlike, for example, 1 kg m/sec^2, which is 1 newton). A qualitative discussion of angular momentum is to be found in Section 2.11. We can best think of the "meaning" of angular momentum as the strength of rotational motion, a measure of the effort required to start or stop such motion.

A particle moves with constant velocity

That a particle can possess angular momentum without actually rotating around a point is made clear through the example of straight-line motion at constant speed. Imagine a particle moving vertically upward with speed v past a point P; at its point of nearest approach, it passes within a distance b of point P (Figure 9.6). At this point of nearest approach, labeled A, **r** is perpendicular to **v**, and, according to Equation 9.3 or Equation 9.6, the angular momentum has magnitude

$$L_A = mvb. \tag{9.8}$$

The vector $\mathbf{L}_A$ is directed upward from the page. Later, at point B, the particle is at a greater distance r from point P. However, the factor $\sin\theta$ is smaller, and the product $r \sin\theta$ remains equal to b. By this reasoning, or directly from Equation 9.6, it follows that

$$L_B = mvb, \tag{9.9}$$

the same as L_A. The direction of $\mathbf{L}_B$ is also upward from the page, so we may write the vector equality

It has constant angular momentum

$$\mathbf{L}_A = \mathbf{L}_B. \tag{9.10}$$

Since point B represents any point at all along the particle trajectory, the angular momentum is constant. In terms of the defining Equation 9.1, this constancy comes about because the increase in the magnitude of $\mathbf{r}$ as the particle moves upward from point A is exactly compensated by a decrease in the angle between $\mathbf{r}$ and $\mathbf{v}$ such that the vector product $\mathbf{r} \times \mathbf{v}$ remains unchanged.

This example of simple constant-velocity motion reveals two important aspects of angular momentum. First, it shows that the definition of angular momentum, despite its complexity, can be useful because it leads to the result that in undisturbed inertial motion, the angular momentum is constant. Notice that the angular momentum of the uniformly moving particle is different with respect to different reference points but that for any particular reference point, it does not change in time. Second, the example shows that angular momentum can accompany nonrotational motion. *Something* is rotating, however, and that something is the imaginary line joining the reference point to the particle. This line sweeps counterclockwise in Figure 9.6 as the particle moves. Without such rotation, there would be no angular momentum.

A particle moves uniformly in a circle

Among examples of motion with angular momentum, none is math-

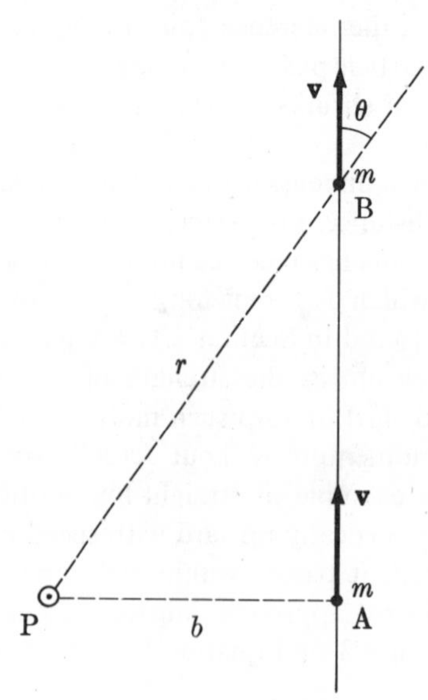

FIGURE 9.6 Straight-line motion with constant velocity. The particle possesses constant angular momentum with respect to the point P.

ematically simpler than the uniform rotation of a particle in a circle [Figure 9.2(a)]. This simplicity requires that the center of the circle be chosen as the reference point. With this choice, the angular momentum is directed along the axis of the circle and has constant magnitude

It has constant angular momentum with respect to the center

$$L = mvr = pr, \tag{9.11}$$

where r is the radius of the circle. This equation could be used to calculate with good accuracy the angular momentum of the earth about the sun, an astronaut in a circular orbit about the earth, or a stone swung on the end of a rope.

■ EXAMPLE 1: A typical electron near the outer part of an atom is about half an angstrom (0.5×10^{-10} m) from the atomic nucleus and moves at about 1 percent of the speed of light, or 3×10^6 m/sec. What is its angular momentum with respect to the nucleus? Since the methods of classical mechanics are not generally valid in the submicroscopic world governed by quantum mechanics,* Equation 9.11 can provide at best only a rough estimate for the electron's angular momentum. It gives, for assumed circular motion,

$$L = 9.1 \times 10^{-31} \text{ kg} \times 3 \times 10^6 \text{ m/sec} \times 0.5 \times 10^{-10} \text{ m}$$

$$= 1.37 \times 10^{-34} \text{ kg m}^2\text{/sec.}$$

Actually, in the submicroscopic world (and in the large-scale world as well), all orbital angular momenta come in units of Planck's constant divided by 2π (whose magnitude is $\hbar = 1.05 \times 10^{-34}$ kg m^2/sec). An electron could possess an angular momentum equal to $\hbar$, or $2\hbar$, or $3\hbar$ but not in fact equal to the calculated value, which is about $1.3\hbar$. Nevertheless, the calculation does provide the correct order of magnitude. ■

The significance of $\hbar$

■ EXAMPLE 2: The airplane shown in Figure 9.2(c) has a mass of 10^5 kg and a speed of 250 m/sec. If its altitude is 8 km and its horizontal distance from the chosen origin is also 8 km, what is its angular momentum with respect to this origin? From Equation 9.6, the magnitude of **L** is

$$L = r_{\perp}p = 8 \times 10^3 \text{ m} \times 10^5 \text{ kg} \times 250 \text{ m/sec}$$

$$= 2 \times 10^{11} \text{ kg m}^2\text{/sec.}$$

(The horizontal distance does not matter.) Its direction is horizontal—to the north, for instance, if the airplane flies to the east. The same example can be handled more formally using vectors. Choose an x axis directed parallel to the velocity **v** and a z axis directed vertically upward, as shown in the figure. Then

$$\mathbf{L} = m\mathbf{r} \times \mathbf{v} = m(x\mathbf{i} + z\mathbf{k}) \times v\mathbf{i},$$

where $x = z = 8$ km. Equation 6.68 permits us to write

$$\mathbf{L} = mxv\mathbf{i} \times \mathbf{i} + mzv\mathbf{k} \times \mathbf{i}.$$

* Equation 9.1, for instance, is totally invalid for massless particles, which can possess angular momentum despite the fact that, for them, $m = 0$.

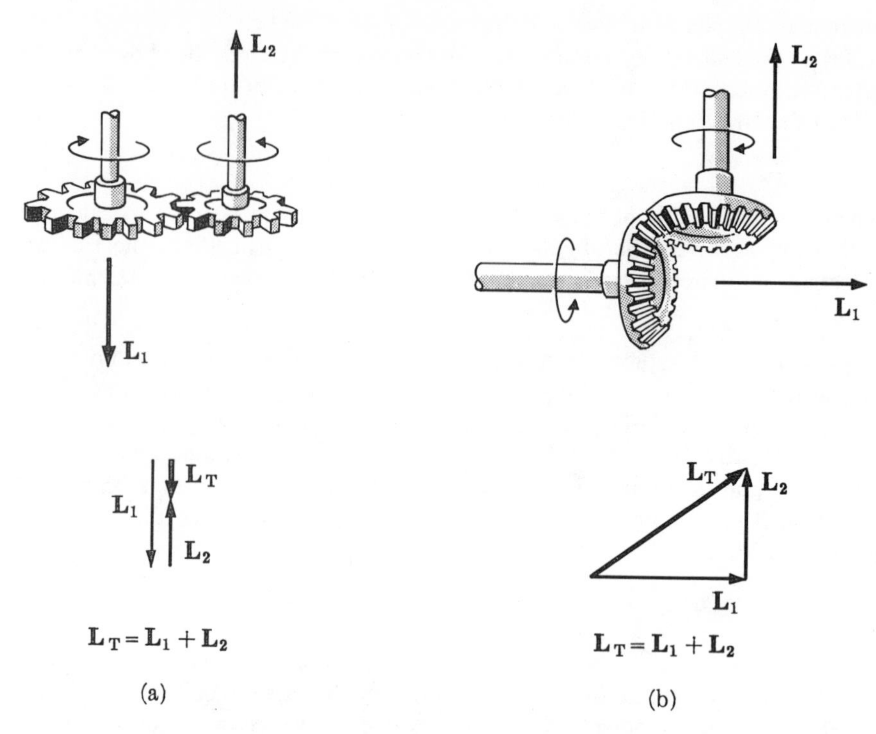

FIGURE 9.7 The vector addition of angular momenta. (a) The total angular momentum of this pair of gears is small in magnitude and directed downward. (b) The total angular momentum of this pair of gears is not directed along either axis.

Since the vector products of the unit vectors are $\mathbf{i} \times \mathbf{i} = 0$ and $\mathbf{k} \times \mathbf{i} = \mathbf{j}$, the final answer is

$$\mathbf{L} = mzv\mathbf{j},$$

with direction along the y axis and magnitude 2×10^{11} kg m^2/sec. ■

Angular momenta add as vectors

The vector character of angular momentum is important. In particular, two or more angular momenta add as vectors. Such addition is illustrated in Figure 9.7. Two equal and opposite angular momenta, just like two equal and opposite forces or momenta, can add to give zero (see Figure 4.3).

9.2 The angular momentum of systems

An important theorem

Most examples of rotational motion encountered in the large-scale world are not orbital motions of particles. They are rotations of objects, systems of many particles or constituent parts. Even in the small-scale world, the spin of a single particle must be viewed as the rotation of a structure. In this section we will consider the angular momentum of systems in quite general terms. In so doing, we can prove the following theorem: *If the center of mass of a system of particles is at rest, the angular momentum of the system is the same with respect to all reference points.* For a single moving particle the angular momentum always depends on the choice of reference point and might even be variable with respect to one reference point and constant with respect to another. For two or more particles, on the other hand, the angular momentum can be uniquely defined—the same for all reference points—in a frame of reference in which the center of mass is stationary. The proof of this statement, although somewhat complicated, provides an interesting example of the power of vector calculus and vector algebra.

Consider a set of particles with masses $m_1, m_2, m_3, \ldots$ and position vectors $\mathbf{r}_1, \mathbf{r}_2, \mathbf{r}_3, \ldots$. As shown in Figure 9.8, each position vector may be expressed as the sum of the center-of-mass position vector, $\mathbf{r}_c$, and the position vector of the particle relative to the center of mass. For the ith particle, we write

$$\mathbf{r}_i = \mathbf{r}_c + \mathbf{r}'_i. \tag{9.12}$$

With respect to the arbitrarily chosen origin, the angular momentum of the system of particles can be written as the sum

$$\mathbf{L} = \sum_i m_i \mathbf{r}_i \times \mathbf{v}_i\,, \tag{9.13}$$

where $\mathbf{v}_i$ is the velocity of the ith particle. Substitution of Equation 9.12 in Equation 9.13 gives

$$\mathbf{L} = \sum_i m_i(\mathbf{r}_c + \mathbf{r}'_i) \times \mathbf{v}_i\,. \tag{9.14}$$

This formula for $\mathbf{L}$ can be rewritten as

Total angular momentum can be divided into two parts

$$\mathbf{L} = \mathbf{L}_1 + \mathbf{L}_2, \tag{9.15}$$

where

$$\mathbf{L}_1 = \mathbf{r}_c \times \sum_i m_i \mathbf{v}_i\,,$$

$$\mathbf{L}_2 = \sum_i m_i \mathbf{r}'_i \times \mathbf{v}_i\,.$$

Notice that the vector $\mathbf{r}_c$ can be taken outside the sum in the expression for $\mathbf{L}_1$. To further simplify $\mathbf{L}_1$, we use Equation 8.18, which is equivalent to

$$\sum_i m_i \mathbf{v}_i = M\mathbf{v}_c\,. \tag{9.16}$$

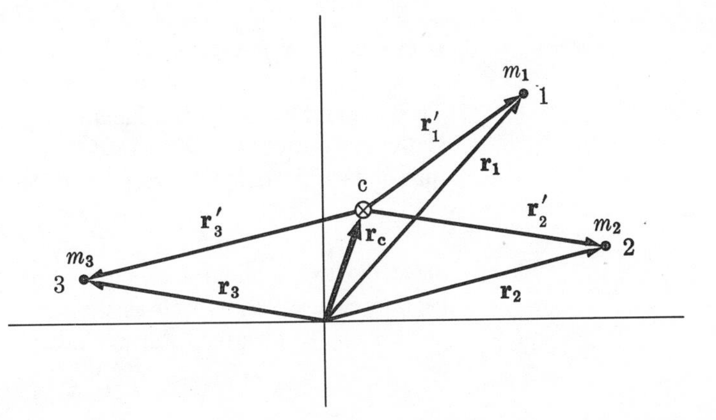

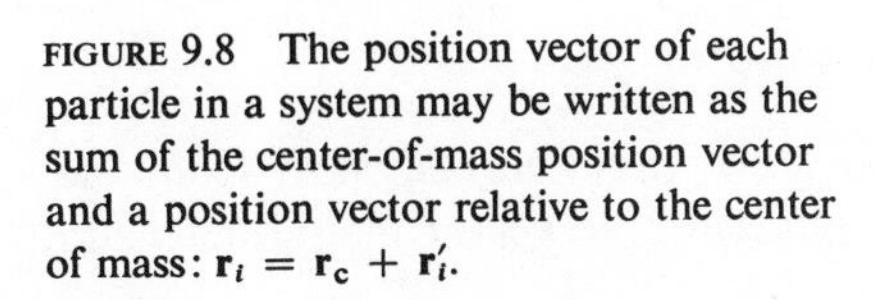

FIGURE 9.8 The position vector of each particle in a system may be written as the sum of the center-of-mass position vector and a position vector relative to the center of mass: $\mathbf{r}_i = \mathbf{r}_c + \mathbf{r}'_i$.

Here M is the total mass ($M = \sum_i m_i$) and $\mathbf{v}_c$ is the velocity of the center of mass. Therefore,

The first part is expressed with center-of-mass variables

$$\mathbf{L}_1 = M\mathbf{r}_c \times \mathbf{v}_c. \tag{9.17}$$

To deal with $\mathbf{L}_2$, we note that

$$\mathbf{v}_i = \frac{d}{dt}(\mathbf{r}_c + \mathbf{r}_i') = \mathbf{v}_c + \mathbf{v}_i', \tag{9.18}$$

so we can write

$$\mathbf{L}_2 = \sum_i m_i\mathbf{r}_i' \times (\mathbf{v}_c + \mathbf{v}_i')$$

$$= \left(\sum_i m_i\mathbf{r}_i'\right) \times \mathbf{v}_c + \sum_i m_i\mathbf{r}_i' \times \mathbf{v}_i'.$$

What is the quantity $\sum_i m_i\mathbf{r}_i'$? If divided by M, it defines the center-of-mass vector relative to the center of mass! It must therefore be zero. (An alternative way to show that it vanishes is suggested in Exercise 9.11.) The quantity $\mathbf{L}_2$ is, therefore,

The second part is expressed relative to the center of mass

$$\mathbf{L}_2 = \sum_i m_i\mathbf{r}_i' \times \mathbf{v}_i'. \tag{9.19}$$

The total angular momentum of the system, $\mathbf{L}_1 + \mathbf{L}_2$, is then

$$\mathbf{L} = M\mathbf{r}_c \times \mathbf{v}_c + \sum_i m_i\mathbf{r}_i' \times \mathbf{v}_i'. \tag{9.20}$$

Now consider what happens if the origin of the coordinate system is changed. It is evident from Figure 9.8 that the center-of-mass vector $\mathbf{r}_c$ changes but that the relative position vectors $\mathbf{r}_i'$ do not. Neither do their time derivatives $\mathbf{v}_i'$. Therefore, in the general expression for angular momentum, Equation 9.20, the first term depends on the location of the origin (the reference point for defining angular momentum), but the second term does not. To complete the proof of the theorem stated at the beginning of this section, we need only consider the special case in which the center of mass is stationary. Then $\mathbf{v}_c = 0$ and the first term in Equation 9.20 vanishes. The total angular momentum is equal to the second term, which is independent of the choice of origin. To phrase it differently: If the center of mass is at rest, the angular momentum can be calculated with respect to *any* reference point and the result is the same.

If $\mathbf{v}_c = 0$, then $\mathbf{L}$ is independent of the choice of origin

■ EXAMPLE: A simple system can serve to illustrate this general result. A pair of equally massive particles rotate about their fixed midpoint (Figure 9.9). Their separation is d; each has mass m, speed v, and is at distance $\frac{1}{2}d$ from the stationary center of mass. Accordingly, the magnitude of the angular momentum of each particle with respect to the midpoint is

$$L_1 = L_2 = \tfrac{1}{2}mvd.$$

Since the two angular momenta have the same direction (outward from the page in Figure 9.9) as well as the same magnitude, the vector magnitudes add numerically to give the total magnitude:

$$L = L_1 + L_2 = mvd. \tag{9.21}$$

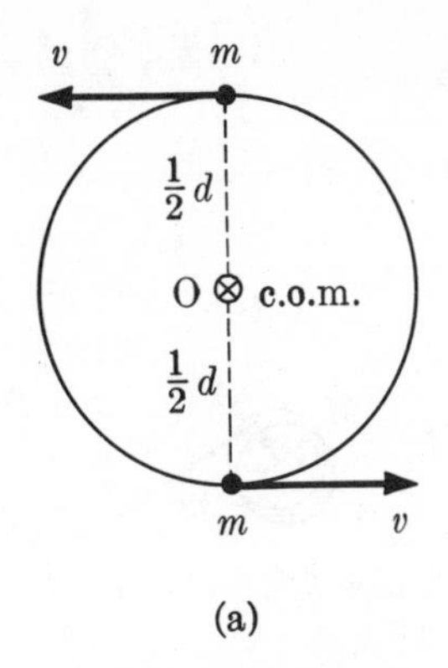

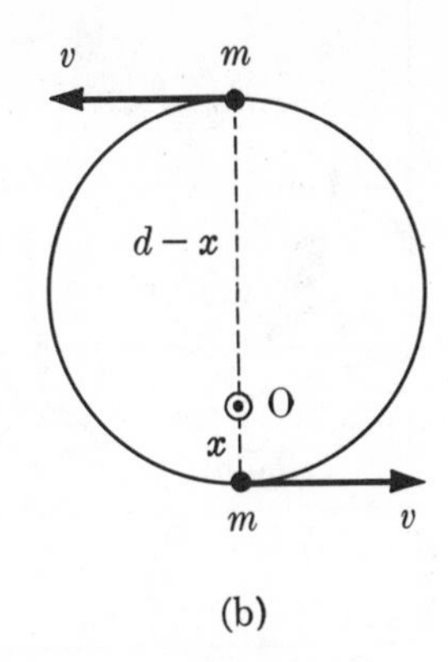

FIGURE 9.9 Equally massive particles rotating about their stationary center of mass have the same total angular momentum with respect to any point. This illustrates a general theorem: If $\mathbf{v}_c = 0$, $\mathbf{L}_{\text{total}}$ is independent of the choice of origin.

This is the total angular momentum of the pair with respect to the point halfway between them. Consider now a reference point located at the position of one of the particles. This particle has no angular momentum with respect to its own location ($\mathbf{r} = 0$). The other particle, a distance d away, has angular momentum mvd. The sum of the two is again mvd. Alternatively, pick an intermediate reference point a distance x from one particle and a distance $d - x$ from the other [Figure 9.9(b)]. With respect to this reference point, the two angular momenta are

$$L_1 = mvx,$$

$$L_2 = mv(d - x).$$

Once again the sum has the same value,

$$L = L_1 + L_2 = mvd.$$ ■

9.3 Orbital angular momentum and spin angular momentum

According to Equation 9.20, the total angular momentum of a system can be split into two terms of quite different form. The first contains only the total mass and properties of the center of mass. It is called *orbital angular momentum*. The second is the angular momentum of the system relative to its center of mass. This is called *spin angular momentum*, or often simply *spin*. To summarize results of the previous section, using these names:

New names for the two kinds of angular momentum

$$\mathbf{L} = \mathbf{L}_{\text{orbital}} + \mathbf{L}_{\text{spin}}; \tag{9.22}$$

$$\mathbf{L}_{\text{orbital}} = M\mathbf{r}_c \times \mathbf{v}_c, \tag{9.23}$$

$$\mathbf{L}_{\text{spin}} = \sum_i m_i \mathbf{r}'_i \times \mathbf{v}'_i. \tag{9.24}$$

$\mathbf{L}_{\text{spin}}$ is independent of the choice of origin

Spin is the angular momentum of a system calculated with respect to the center of mass, whether or not the center of mass is moving. It is an intrinsic property of the system, independent of the observer's point of reference. If

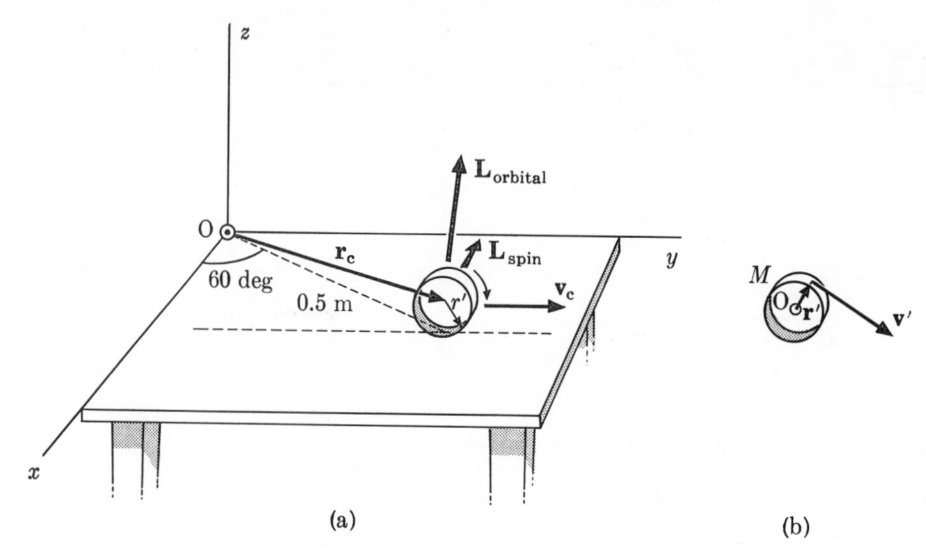

FIGURE 9.10 Hoop rolling on a table. (a) In the laboratory frame of reference, with the origin at a corner of the table, $\mathbf{L}_{spin}$ is directed along the axis of the hoop, and $\mathbf{L}_{orbital}$ is slightly inclined away from the vertical (it is perpendicular to the plane defined by $\mathbf{r}_c$ and $\mathbf{v}_c$). (b) The spin angular momentum is calculated by considering the motion in a frame of reference whose origin is at the center of mass of the hoop.

some object—for example, a top or the idling propeller of a parked airplane—is rotating about a stationary axis passing through its center of mass, its spin angular momentum is the same as its total angular momentum. More generally, the center of mass may be moving, in which case spin is only part of the total. The spin itself, however, is independent of the bulk motion. The spin angular momentum of the earth, for instance, does not depend on the orbital speed of the earth. It is calculated as if the earth's axis were stationary.

To calculate $\mathbf{L}_{orbital}$, think of the system as a point at the center of mass

The analogy between Equation 9.23, defining the orbital angular momentum of a system, and Equation 9.1, defining the angular momentum of a single particle, is evident. We can say, therefore, that the *orbital* angular momentum of a system is the same as the angular momentum of a particle of mass M located at the center of mass. To calculate the orbital angular momentum of the earth with respect to the sun, for example, replace the earth by a particle of mass M located at its center. Since the total momentum $\mathbf{P}$ of a system is equal to $M\mathbf{v}_c$ (Equation 8.22), orbital angular momentum may also be defined by

$$\mathbf{L}_{orbital} = \mathbf{r}_c \times \mathbf{P}, \tag{9.25}$$

which is closely analogous to Equation 9.2 for a particle. Orbital angular momentum, like the angular momentum of a particle, does depend on the chosen reference point.

■ EXAMPLE: A hoop of radius 6 cm and mass 0.3 kg rolls across a table with a speed of 40 cm/sec. As shown in Figure 9.10(a), its motion is referred to a Cartesian coordinate system with origin at the left rear corner of the table. A line drawn from the origin to the point of contact of the hoop with the table has a length of 50 cm and makes an angle of 60 deg with the x axis. For this choice of origin, what are the orbital and spin angular momenta? What is the total angular momentum? At first glance, the problem appears to be complicated, but a methodical approach will show it to be rather simple. We may first specify the properties of the center of mass (in mks units):

$$\mathbf{r}_c = 0.5 \cos(60 \text{ deg})\,\mathbf{i} + 0.5 \sin(60 \text{ deg})\,\mathbf{j} + 0.06\mathbf{k},$$

$$\mathbf{v}_c = 0.4\mathbf{j} \text{ m/sec}.$$

The orbital angular momentum is therefore

$$\mathbf{L}_{\text{orbital}} = M\mathbf{r}_c \times \mathbf{v}_c$$
$$= 0.3(0.25\mathbf{i} + 0.433\mathbf{j} + 0.06\mathbf{k}) \times 0.4\mathbf{j}.$$

Since $\mathbf{i} \times \mathbf{j} = \mathbf{k}$, $\mathbf{j} \times \mathbf{j} = 0$, and $\mathbf{k} \times \mathbf{j} = -\mathbf{i}$, the result is

$$\mathbf{L}_{\text{orbital}} = (0.030\mathbf{k} - 0.0072\mathbf{i}) \text{ kg m}^2/\text{sec}.$$

This is a vector of magnitude 0.0309 kg m^2/sec, tilted slightly away from the vertical.

To obtain $\mathbf{L}_{\text{spin}}$, we consider the motion relative to the center of mass. This is uniform circular motion. Every element of mass is the same distance from the center of mass, $r' = 0.06$ m; and, relative to the center of mass, every element of mass moves with the same speed $v' = 0.4$ m/sec, with $\mathbf{v}'$ and $\mathbf{r}'$ perpendicular [Figure 9.10(b)]. In Equation 9.24, $\mathbf{r}_i' \times \mathbf{v}_i'$ is the same for each bit of mass, so

$$\mathbf{L}_{\text{spin}} = M\mathbf{r}' \times \mathbf{v}'.$$

This is a vector in the negative x direction, of magnitude $Mr'v'$:

$$\mathbf{L}_{\text{spin}} = -0.0072\mathbf{i} \text{ kg m}^2/\text{sec}.$$

This is, of course, independent of the location of the origin. The total angular momentum, $\mathbf{L}_{\text{orbital}} + \mathbf{L}_{\text{spin}}$, is

$$\mathbf{L} = (0.030\mathbf{k} - 0.0144\mathbf{i}) \text{ kg m}^2/\text{sec}. \quad \blacksquare$$

The division of angular momentum into orbital and spin parts has found great usefulness in the world of particles. Most of the elementary particles possess a spin that is an invariable property of the particle. In addition, any particle may possess orbital angular momentum with respect to any reference point (see Figure 4.4). In an interesting class of particle collisions called spin-flip collisions, the direction of the spin angular momentum changes during the collision. Because of the vector nature of angular momentum and the law of conservation of angular momentum that applies in these collisions, a spin flip must be accompanied by a change in orbital angular momentum—in magnitude or direction or both—in order that the total angular momentum remain unchanged.

When $\mathbf{L}_{\text{total}} = $ **constant,** $\mathbf{L}_{\text{spin}}$ *and* $\mathbf{L}_{\text{orbital}}$ *may undergo compensating changes*

The earth-moon system provides another interesting example of interchange between spin and orbital angular momentum. According to present theory, the moon once possessed more spin and less orbital angular momentum with respect to the earth than at present. Now it spins on its axis only once in each 27-day trip around the earth, and its orbital angular momentum with respect to the earth far exceeds its spin angular momentum.

9.4 Moments of inertia

Any solid object—a billiard ball, a top, an automobile wheel—may be looked upon as a system of particles. If the object rotates about an axis, each microscopic speck of matter in the object is a particle tracing out a circular orbit and contributing its microscopic bit to the total angular momentum of the whole object [Figure 9.2(d)]. Integral calculus, concerned with summing an infinite number of infinitesimal contributions, is the mathematical tool required to calculate the total angular momentum of an extended object of any shape.

ROTATION ABOUT AN ARBITRARY FIXED AXIS

If a rigid body rotates about an arbitrary fixed axis, which we may call the z axis, and if the reference point for defining angular momentum lies on this axis, the component of angular momentum along the axis is proportional to the angular speed of the body:

Moment of inertia defined

$$L_z = I\omega. \tag{9.26}$$

The constant of proportionality I is called the *moment of inertia* of the body. The moment of inertia depends on the size and shape and mass of the body and on the location of the axis of rotation, but it is independent of the rate of rotation. The SI unit of I is kg m^2.

We shall first illustrate Equation 9.26 with a simple example and then carry out a derivation that will (1) prove the proportionality of L_z to ω, and (2) provide a general formula for I.

■ EXAMPLE 1: The mass of an idealized bicycle wheel or hoop is concentrated in a circle of radius R. The hoop rotates about an axis through its center, perpendicular to its plane (Figure 9.11). What is the moment of inertia of the hoop? To obtain I, we must develop a formula for L_z and divide it by ω. A convenient choice of origin is the center of the hoop, which is, of course, also the center of mass. Let τ designate the mass per unit distance around the rim so that in the increment of distance ds there is contained mass

$$dm = \tau\, ds.$$

This increment of mass, moving with speed v, possesses angular momentum $d\mathbf{L}$, directed along the z axis. Its magnitude is

$$dL = dL_z = vR\, dm = vR\tau\, ds.$$

The total component of angular momentum, L_z, is the sum of such infinitesimal contributions—that is, an integral:

$$L_z = \int dL_z = vR\tau \int_0^{2\pi R} ds. \tag{9.27}$$

None of the quantities v, R, and τ depends on s, the distance measured around the rim. The integral on the right gives simply $2\pi R$ so that

$$L_z = 2\pi v R^2 \tau.$$

Since the mass M of the wheel is equal to $2\pi R\tau$, the quantity L_z—which, in this example, is the same as the magnitude of the total angular momentum—can be

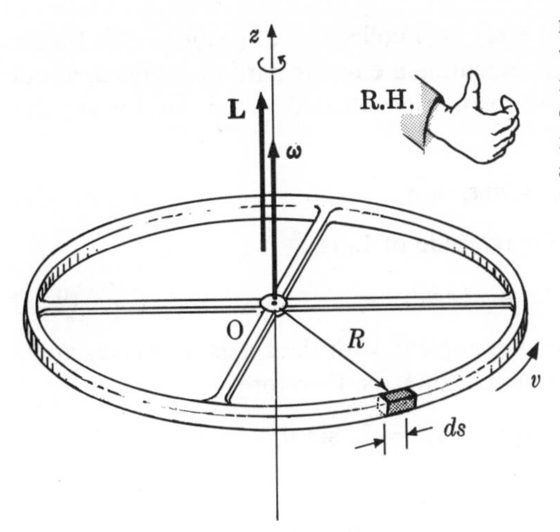

FIGURE 9.11 A rotating wheel, almost all of whose mass lies at a fixed distance R from its center. Mass dm lies in the increment of distance ds. For the indicated direction of rotation, $\mathbf{L}$ is directed upward parallel to the axis.

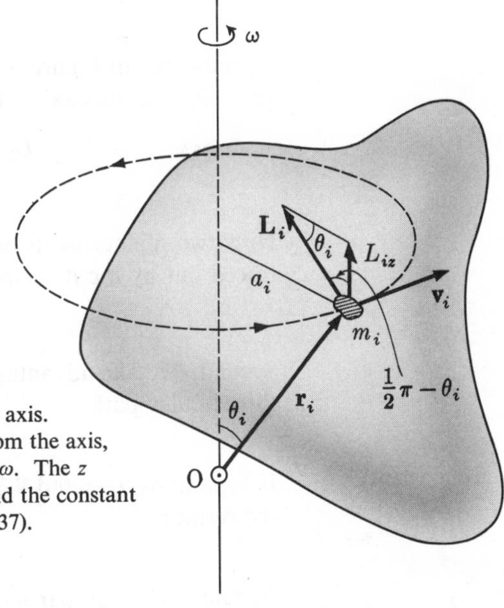

FIGURE 9.12 A rigid body rotates about an arbitrarily chosen z axis. A particular element of mass m_i in the body, at a distance a_i from the axis, has a z component of angular momentum given by $L_{iz} = m_i a_i^2 \omega$. The z component of total angular momentum is proportional to ω, and the constant of proportionality is the moment of inertia (Equation 9.36 or 9.37).

written, not surprisingly,

Spin angular momentum of a hoop

$$L_z = L = MvR. \tag{9.28}$$

This is the same result that was obtained by slightly different arguments and was used in the example of the preceding section. To bring in ω, we note that the rim speed v is related to the angular speed by

$$v = \omega R.$$

Substitution of this formula in Equation 9.28 gives

$$L_z = MR^2\omega, \tag{9.29}$$

which demonstrates the proportionality of L_z to ω and (by comparison with Equation 9.26) shows the moment of inertia to be

Moment of inertia of a hoop about its axis of cylindrical symmetry

$$I = MR^2 \qquad \text{(thin wheel or hoop).} \tag{9.30}$$

■

A hoop is a special case in which all the mass moves at the same speed. More generally, different parts of a rotating body move with different speed. It is the angular speed ω, not the speed v, that is constant for all parts of the body.

The general case: any rigid body and any fixed axis

Consider now the general case: A rigid body with any distribution of mass rotates about an arbitrarily chosen z axis (Figure 9.12). In treating this motion, it is convenient to regard the rigid body as a collection of particles. In Figure 9.12, the ith particle, of mass m_i, is executing a circular path of radius a_i about the axis as the body rotates. Relative to an origin located on the axis, the angular momentum of this particle is

$$\mathbf{L}_i = m_i\mathbf{r}_i \times \mathbf{v}_i \,.$$

Since $\mathbf{v}_i$ is perpendicular to $\mathbf{r}_i$, the magnitude of $\mathbf{L}_i$ is

$$L_i = m_i r_i v_i \,. \tag{9.31}$$

As indicated in Figure 9.12, if $\mathbf{r}_i$ makes angle θ_i with the z axis, $\mathbf{L}_i$ makes angle $\frac{1}{2}\pi - \theta_i$ with this axis. The z component of $\mathbf{L}_i$ is, therefore,

$$\begin{aligned} L_{iz} &= L_i \cos\left(\tfrac{1}{2}\pi - \theta_i\right) = L_i \sin\theta_i \\ &= m_i r_i v_i \sin\theta_i \,. \end{aligned} \tag{9.32}$$

Now two replacements are useful. First, we note that the radius a_i of the circle traced out by the ith particle is

$$a_i = r_i \sin\theta_i \,. \tag{9.33}$$

Second, we take advantage of the relation between speed and angular speed for this circular path,

$$v_i = a_i\omega. \tag{9.34}$$

If Equations 9.33 and 9.34 are used to eliminate r_i and v_i from Equation 9.32, the result is

$$L_{iz} = m_i a_i^2 \omega.$$

Summing over all particles gives the total z component of angular momentum,

$$L_z = \left(\sum_i m_i a_i^2\right)\omega. \tag{9.35}$$

Proof that $L_z \sim \omega$

Two features of this equation are important. First, it shows that L_z does not depend on the location of the origin, which may lie anywhere on the axis of rotation. Second, it proves the proportionality of L_z to ω since the quantity in parentheses depends only on the distribution of mass relative to the axis and not on the angular speed. This quantity is, by definition, the moment of inertia of the body:

$$I = \sum_i m_i a_i^2. \tag{9.36}$$

Formulas for moment of inertia

If a continuous distribution of mass is considered, this sum becomes an integral:

$$I = \int a^2\, dm. \tag{9.37}$$

(This is a definite integral, whose limits of integration must encompass the body.) In different notation, these two important formulas are saying the same thing: To calculate a moment of inertia, multiply each element of mass by the square of its distance from the axis, and sum over all parts of the body.

■ EXAMPLE 2: Show that the moment of inertia of a hoop, as given by Equation 9.30, follows directly from the general formula for I, Equation 9.36. Since all of the mass in the hoop is at the same distance from the axis, $a_i = R$, and the factor a_i^2 can be taken outside the sum in Equation 9.36. This step gives

$$I = R^2 \sum_i m_i .$$

The remaining sum is the total mass of the hoop, so $I = MR^2$, in agreement with Equation 9.30. ■

■ EXAMPLE 3: A thin rod, pivoted at one end, swings along the surface of a cone (Figure 9.13). What is its moment of inertia? Let us call the length of the rod l, its mass M, and its angle to the axis α. An increment of mass dm at distance s from the pivot point can be written

$$dm = \tau \, ds,$$

where τ is the mass per unit length (equal to M/l). The distance of this increment of mass from the axis is

$$a = s \sin \alpha.$$

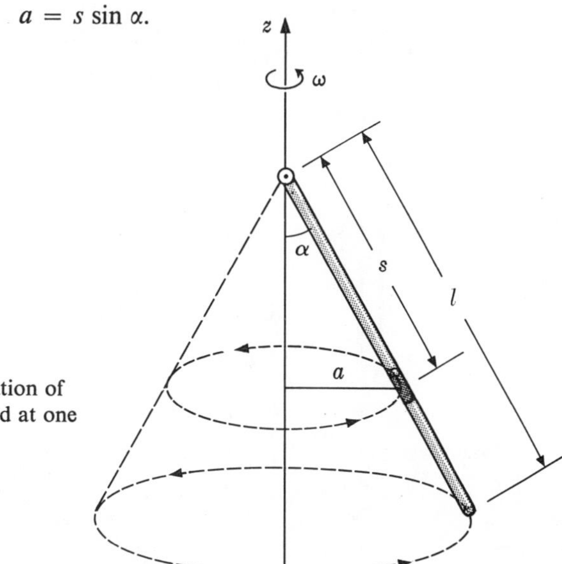

FIGURE 9.13 Diagram for the calculation of the moment of inertia of a rod pivoted at one end and swinging with fixed angle α.

Using these expressions in Equation 9.37 gives

$$I = \tau \sin^2 \alpha \int_0^l s^2 \, ds. \tag{9.38}$$

The integration extends over the length of the rod, from $s = 0$ to l. The integral is $l^3/3$, so

$$I = \tfrac{1}{3}\tau \sin^2 \alpha \, l^3.$$

Finally, substitution of M/l for τ gives

$$I = \tfrac{1}{3}Ml^2 \sin^2 \alpha. \tag{9.39}$$

Moment of inertia of a rod pivoted at one end

The moment of inertia is zero for $\alpha = 0$ (since the cross-sectional area of the rod has been assumed negligible), and has a maximum value of $\frac{1}{3}Ml^2$ for $\alpha = \frac{1}{2}\pi$. Note the characteristic appearance in Equation 9.39 of mass multiplied by the square of a distance. ■

ROTATION ABOUT A SYMMETRY AXIS

Equation valid for special axes

For special choices of axes, Equation 9.26 can be replaced by the vector equation

$$\mathbf{L} = I\boldsymbol{\omega}. \tag{9.40}$$

Here $\boldsymbol{\omega}$ is the angular *velocity*, a vector whose magnitude is the angular speed ω and whose direction is along the axis of rotation, as defined by a right-hand rule (Figure 9.11). When Equation 9.40 is valid, as it is for the hoop in Figure 9.11, the angular momentum $\mathbf{L}$ is also directed along the axis, parallel to $\boldsymbol{\omega}$. The moment of inertia has the same meaning in Equation 9.40 as in Equation 9.26, and is still given by either Equation 9.36 or 9.37.

Definition of symmetry axis

A body is said to have a symmetry axis if the hypothetical operation of reflecting every mass point through the axis leaves the body unchanged. The hoop in Figure 9.11 has a symmetry axis (in fact more than one). Other examples appear in Figure 9.14. One situation correctly described by Equation 9.40 is rotation of a symmetric body about its axis of symmetry. For such motion, the center of mass is at rest. The total angular momentum $\mathbf{L}$ is therefore the same as $\mathbf{L}_{\text{spin}}$ and is independent of the choice of origin. However, it is usually convenient to place the origin at the center of mass.

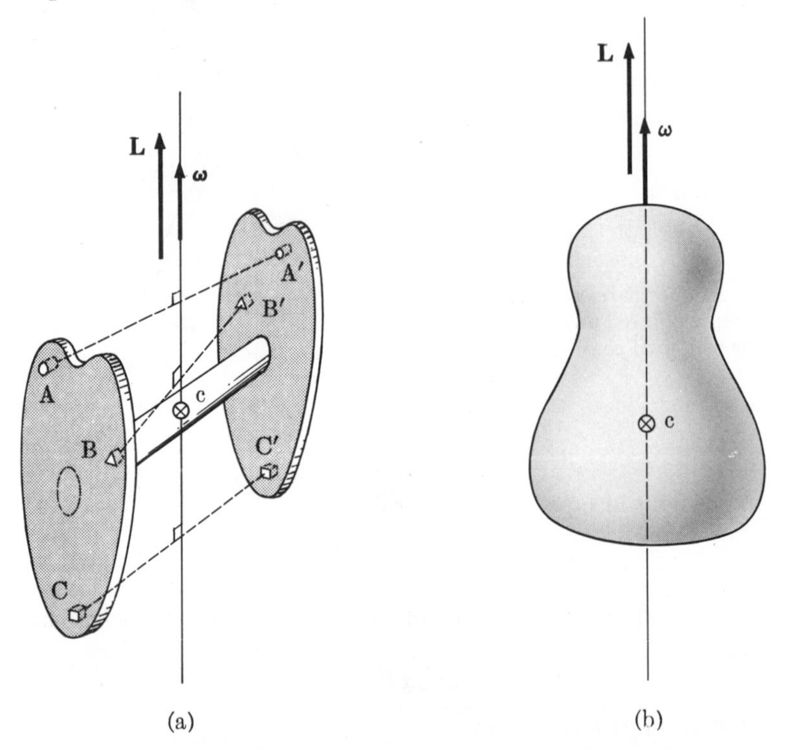

FIGURE 9.14 Symmetry axes. (a) For every element of mass A, B, or C in this body, there is a corresponding element A′, B′, or C′ reflected through the symmetry axis. The body is unchanged by rotation about its symmetry axis through 180 deg. (b) This body is called cylindrically symmetric. It is unchanged by rotation about its symmetry axis through *any* angle. For both bodies, angular momentum and angular velocity are parallel (Equation 9.40). Note that the centers of mass lie on the symmetry axes.

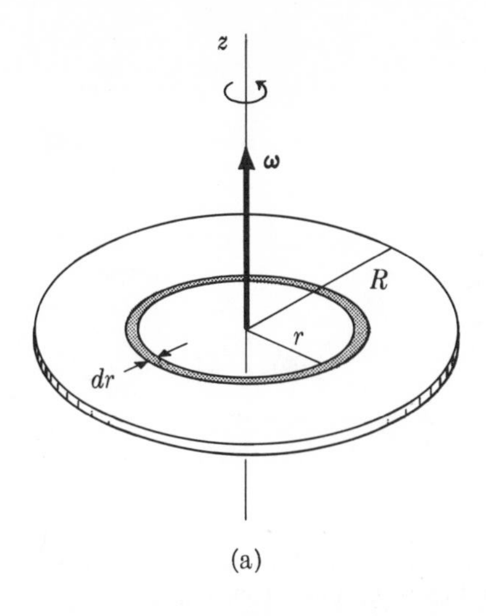

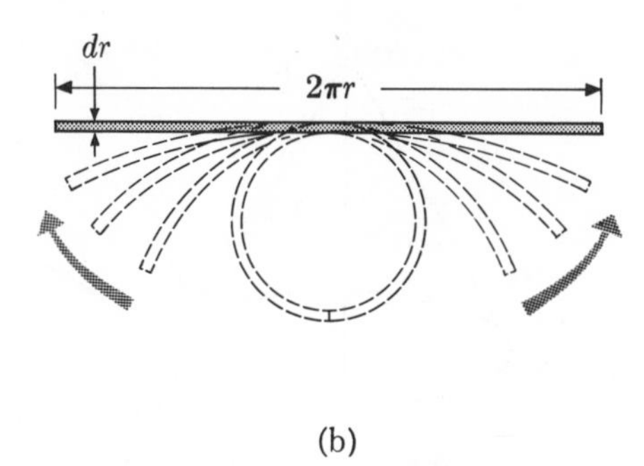

FIGURE 9.15 A uniform disk rotating about its axis of cylindrical symmetry. (a) The disk may be treated as a series of hoops with radii varying from $r = 0$ to $r = R$. (b) "Straightening" a hoop of infinitesimal width shows its area to be $2\pi r\, dr$.

■ EXAMPLE 4: What is the moment of inertia of a uniform disk of radius R rotating about its axis of cylindrical symmetry (Figure 9.15)? The disk can be thought of as a series of hoops of varying radius r, each of infinitesimal thickness dr. If σ designates the mass per unit area in the disk (such that $M = \pi R^2\sigma$), the mass in a hoop contained between r and $r + dr$ is

$$dm = 2\pi r\, dr \times \sigma.$$

The distance of this mass from the axis is $a = r$. Substituting for a and dm in Equation 9.37 (with appropriate limits of integration added) gives

$$I = 2\pi\sigma \int_0^R r^3\, dr.$$

Integration yields

$$I = \frac{2\pi\sigma}{4} r^4 \Big|_0^R = \tfrac{1}{2}\pi\sigma R^4. \tag{9.41}$$

As a final step, it is convenient to express I in terms of the total mass M rather than the mass per unit area σ, these being related by

$$M = \pi R^2\sigma. \tag{9.42}$$

The result is

$$I = \tfrac{1}{2}MR^2 \qquad \text{(disk).} \tag{9.43}$$

Moment of inertia of a disk about its axis of cylindrical symmetry

Because this is the moment of inertia about a symmetry axis, the vector equation $\mathbf{L} = I\boldsymbol{\omega}$ is valid. ■

We state without proof another useful result. The moment of inertia of a uniform sphere of mass M and radius R rotating about an axis through its center is

$$I = \tfrac{2}{5}MR^2 \qquad \text{(sphere).} \tag{9.44}$$

A similar result for a sphere

One way to appreciate the significance of the concept of moment of inertia is to compare the formula for momentum, $\mathbf{p} = m\mathbf{v}$, and the formula for angular

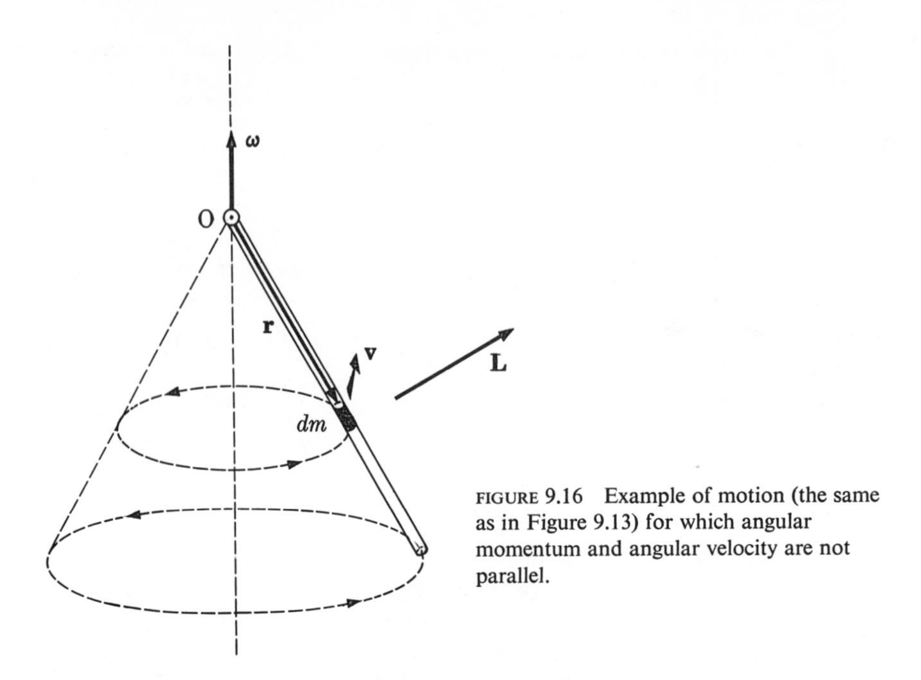

FIGURE 9.16 Example of motion (the same as in Figure 9.13) for which angular momentum and angular velocity are not parallel.

momentum about a symmetry axis, $\mathbf{L} = I\boldsymbol{\omega}$. Momentum varies in proportion to velocity $\mathbf{v}$, and angular momentum varies in proportion to angular velocity $\boldsymbol{\omega}$. Thus the moment of inertia I is, like the mass m, an inertial property of a body. One measures the resistance of the body to a change of velocity; the other measures the resistance of the body to a change of angular velocity.

Moment of inertia is the rotational analog of mass

★ROTATION ABOUT AN AXIS PARALLEL TO A SYMMETRY AXIS

$\mathbf{L}$ is not always parallel to $\boldsymbol{\omega}$

To see that the vectors $\mathbf{L}$ and $\boldsymbol{\omega}$ are not always parallel, it is only necessary to examine the pivoted rod of Example 3. As shown in Figure 9.16, the angular velocity is directed vertically, whereas the angular momentum of every part of the rod—and therefore of the whole rod—is directed at an angle to the vertical, and changes its direction as the rod swings.

A full discussion of the rotation of rigid bodies is beyond the scope of this text. In general, the moment of inertia is what is called a tensor quantity. Only for certain axes of rotation, called *principal axes*, does it behave as a scalar quantity. Then Equation 9.40 is valid. In Equation 9.26, which is valid for *any* fixed axis, I is a numeric but need not be a scalar.

$\mathbf{L} = I\boldsymbol{\omega}$ for principal axes

Examples of principal axes

Any symmetry axis of a body is a principal axis. Examples appear in Figures 9.11, 9.14, 9.15, and 9.17(a). Any axis perpendicular to a plane of symmetry is also a principal axis* provided the origin is chosen at the intersection of the axis and the symmetry plane [Figures 9.17(b) and 9.18]. Figure 9.18 also illustrates the fact that an axis *parallel* to a symmetry axis is a principal axis if the origin is located on the axis at the point closest to the center of mass.

---

* Actually three (or sometimes more) principal axes are associated with every point in a rigid body. We mention only certain special cases here.

Let us study in detail the example illustrated in Figure 9.18. A disk of mass M and radius R is attached at a point on its edge to an axis that is perpendicular to the plane of the disk. The disk rotates about this axis with angular velocity $\boldsymbol{\omega}$. We wish to find the angular momentum with respect to the point where the axis touches the disk and the moment of inertia for this same choice of origin. Although Equation 9.37 is applicable, it would be unwieldy in this example. Instead of using it, we return to Equation 9.22 and write, for the angular momentum of the disk,

$$\mathbf{L} = \mathbf{L}_{\text{orbital}} + \mathbf{L}_{\text{spin}}.$$

(1) *The orbital angular momentum* is defined by Equation 9.23. Inspection of Figure 9.18 shows that $\mathbf{r}_c \times \mathbf{v}_c$ is directed vertically upward, parallel to $\boldsymbol{\omega}$. Therefore, $\mathbf{L}_{\text{orbital}}$ has this direction. Its magnitude is

$$L_{\text{orbital}} = Mr_c v_c \sin(90 \text{ deg}) = Mr_c v_c.$$

Since $r_c = R$ and since $v_c = R\omega$, this can be written

$$L_{\text{orbital}} = MR^2\omega.$$

As a vector, it is

$$\mathbf{L}_{\text{orbital}} = MR^2\boldsymbol{\omega}. \tag{9.45}$$

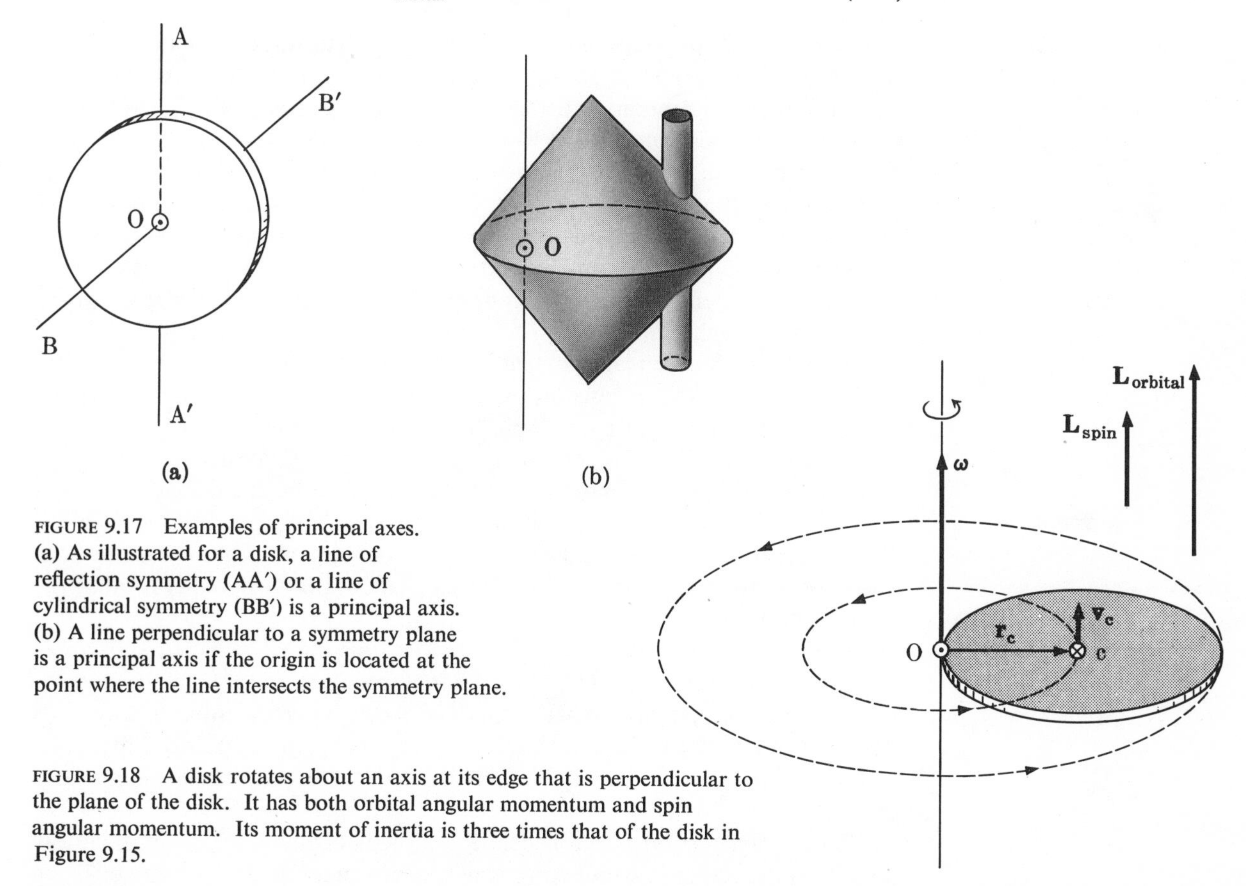

FIGURE 9.17 Examples of principal axes.
(a) As illustrated for a disk, a line of reflection symmetry (AA′) or a line of cylindrical symmetry (BB′) is a principal axis.
(b) A line perpendicular to a symmetry plane is a principal axis if the origin is located at the point where the line intersects the symmetry plane.

FIGURE 9.18 A disk rotates about an axis at its edge that is perpendicular to the plane of the disk. It has both orbital angular momentum and spin angular momentum. Its moment of inertia is three times that of the disk in Figure 9.15.

(2) *The spin angular momentum* is the angular momentum of the body relative to its center of mass. Since it rotates about its center of mass once each time it turns about the axis, its angular velocity relative to the center of mass is also $\boldsymbol{\omega}$. Its spin angular momentum is therefore the same as that of a disk rotating about a fixed center. According to Equation 9.43, this is

$$\mathbf{L}_{\text{spin}} = \tfrac{1}{2}MR^2\boldsymbol{\omega}. \tag{9.46}$$

Adding Equations 9.45 and 9.46 gives the total angular momentum of the disk,

$$\mathbf{L} = \tfrac{3}{2}MR^2\boldsymbol{\omega}. \tag{9.47}$$

This has the form of Equation 9.40 and shows that the axis is indeed a principal axis for the origin at the edge of the disk. The moment of inertia is

Moment of inertia of the disk in Figure 9.18

$$I = \tfrac{3}{2}MR^2. \tag{9.48}$$

Moving the axis from the center to the edge of the disk increases the moment of inertia by a factor of 3.

9.5 Torque and the law of angular momentum change

A review of key points so far

In the preceding four sections, the concept of angular momentum has been defined and elaborated but not yet placed in the context of a physical law. To summarize the main points emphasized so far: (1) The angular momentum of a particle with respect to a point, a measure of the strength of its rotational tendency about that point, is defined in terms of the mass and velocity of the particle and its position relative to the reference point. Even more simply, it may be defined in terms of the particle's momentum and its position. (2) A particle moving at constant velocity in a straight line has constant angular momentum with respect to any fixed point. A particle moving at fixed speed in a circle has constant angular momentum with respect to the center of the circle. (3) Angular momentum is a vector concept. Its direction is the axial direction as defined by the right-hand rule. (4) The angular momentum of a system is the vector sum of the angular momenta of the constituent parts of the system. (5) It is convenient to separate the angular momentum of a material object into a spin part and an orbital part. Spin is the object's angular momentum with respect to its center of mass. Orbital angular momentum, the additional angular momentum arising from motion of the center of mass, is calculated as if all the mass of the object were concentrated at the center of mass. (6) For rotation about certain axes called principal axes, the angular momentum of a body is proportional to its angular velocity. The constant of proportionality is called the moment of inertia.

Still to be answered are the important questions, Under what circumstances is angular momentum conserved? What is required to change the angular momentum of a system? What is the law of angular momentum change? If the third question can be answered, the answers to the first two will follow readily. In this section we will explore the classical law of angular-momentum change.

Since angular momentum is defined mathematically in terms of previously established concepts, it is not surprising to learn that the law of its change can be derived mathematically from Newton's laws and need not have a new experimental foundation. We proceed by differentiating both sides of Equation 9.1 with respect to time:

$$\frac{d\mathbf{L}}{dt} = m\frac{d\mathbf{r}}{dt} \times \mathbf{v} + m\mathbf{r} \times \frac{d\mathbf{v}}{dt}. \tag{9.49}$$

Since $\mathbf{v} = d\mathbf{r}/dt$, the first term on the right contains $\mathbf{v} \times \mathbf{v}$, which is zero. Since $\mathbf{a} = d\mathbf{v}/dt$, the second term may be written $m\mathbf{r} \times \mathbf{a}$, which is the same as

$$\mathbf{r} \times (m\mathbf{a}).$$

From Newton's second law, mass times acceleration may be replaced by force so that Equation 9.49 yields

$$\frac{d\mathbf{L}}{dt} = \mathbf{r} \times \mathbf{F}. \tag{9.50}$$

A consequence of Newton's second law

This is the law of angular-momentum change for a particle experiencing force **F**. The combination $\mathbf{r} \times \mathbf{F}$ is sufficiently important to deserve a name of its own. It is called *torque*, for which the usual symbol is **T**:

$$\mathbf{T} = \mathbf{r} \times \mathbf{F}. \tag{9.51}$$

The definition of torque

This is the *definition* of torque: It is a vector quantity with the dimension of force times distance and its SI unit is the newton meter. In terms of torque, the law of angular momentum change is simply

$$\frac{d\mathbf{L}}{dt} = \mathbf{T} \tag{9.52}$$

The law of change of angular momentum

(here **L** and **T** must be referred to the same origin). This equation is valid for extended objects as well as particles if **L** is the total angular momentum of the object and if **T** is the total torque acting on it.

Figure 9.19 indicates geometrically the definition of torque. Like angular momentum, torque is defined relative to an arbitrarily chosen origin. Some practical examples of torque, emphasizing the vector character of this concept, are shown in Figure 9.20.

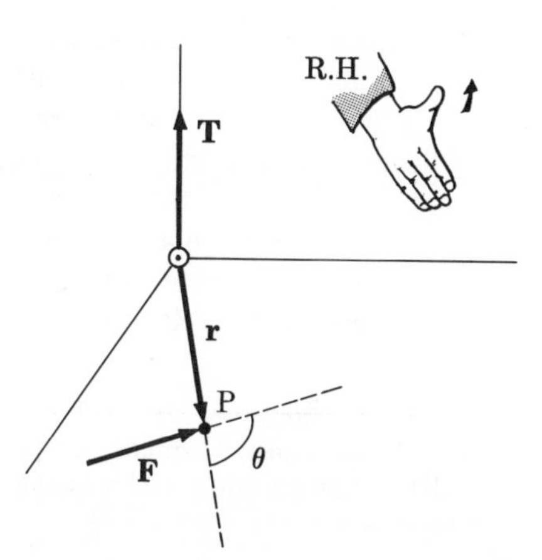

FIGURE 9.19 Definition of torque. A force **F** acts at point P. With respect to the origin, it produces a torque $\mathbf{T} = \mathbf{r} \times \mathbf{F}$.

■ EXAMPLE: In Figure 9.20(b), the length of the flagpole is 3 m, and the force, of magnitude 200 N, acts at an angle of 30 deg to the flagpole. What is the torque with respect to point O where the flagpole touches the building? Why does this torque not change the angular momentum of the flagpole? From the definition of Equation 9.51, the magnitude of the torque is

$$T = 3\text{ m} \times 200\text{ N} \times \sin(150\text{ deg})$$
$$= 300\text{ N m}.$$

Its direction is perpendicular to the plane containing **r** and **F**, as shown in the figure. If the flagpole is to remain stationary, one or more other forces must also be acting in order to make the *total* torque vanish. Gravity, acting downward on the pole and the flag, provides the required opposite torque. ■

Two ways to have force without torque

Normally, if a force acts on a body, it produces a torque. However, there are two ways in which a force may act without producing a torque: (1) The force may act at the chosen origin. Then, since $\mathbf{r} = 0$, Equation 9.51 says that $\mathbf{T} = 0$. In Figure 9.20(a), for example, if the rider pushed down on the sprocket axle rather than the pedal, there would be no torque applied to the sprocket and no change of angular momentum. (2) The force may act directly toward or away from the origin. Then **r** is parallel to **F**, and $\mathbf{r} \times \mathbf{F} = 0$. As illustrated in Figure 9.20(d), for instance, the sun exerts no torque on a planet. The angular momentum of the planet is therefore constant relative to the center of the sun.*

(1) $\mathbf{r} = 0$

(2) **r** *parallel to* **F**

A way to have torque without force

If a body is isolated and free of all forces, it is also free of torque. If two or more forces act on the body, however, these forces may sum to zero and yet produce a net torque. An example of torque without any net force is shown in Figure 9.21. When two forces, equal and opposite, do not act along the same line, they produce a torque of magnitude

$$T = lF, \tag{9.53}$$

where F is the magnitude of each force and where l is the perpendicular distance between the lines of action of the forces. This result is evidently correct for the example in Figure 9.21. With respect to the central axis, each force produces an upward torque of magnitude $\frac{1}{2}lF$. A more general proof is left as a problem. A torque of this kind, arising from a pair of equal and opposite forces, is sometimes called a *couple*. It is also left as a problem to show that a couple is independent of the choice of origin.

Just as force is the idea of a push or pull made into a quantitative concept, torque is the idea of a twist rendered scientifically precise and useful. Torque bears the same relationship to angular momentum that force bears to ordinary momentum. Force is that which changes momentum; torque, similarly, is that which changes angular momentum. In the absence of force, momentum is conserved; in the absence of torque, angular momentum is conserved. These and other related parallels in mechanics are summarized in Table 9.1.

---

* For a spherical planet, the orbital and spin angular momenta are separately constant. For a nonspherical planet orbital and spin angular momenta each change slowly, but their sum remains constant (see Section 11.5).

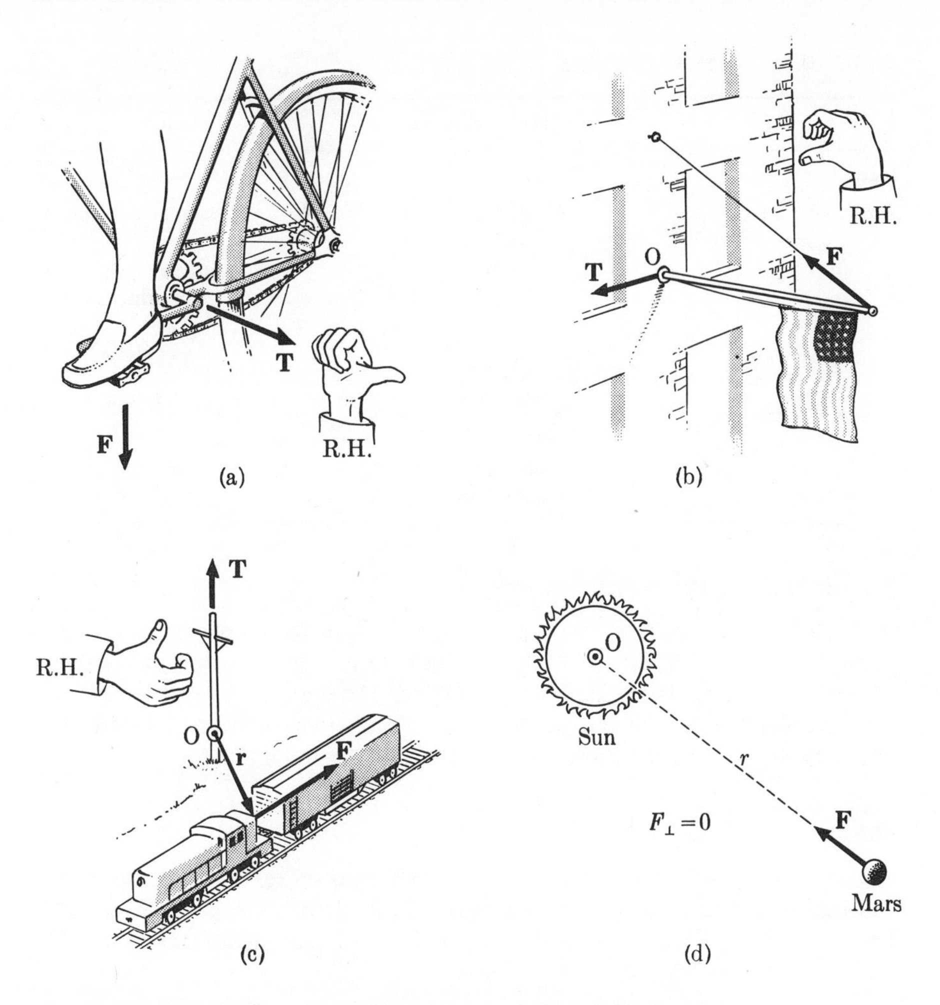

FIGURE 9.20 The vector nature of torque. (a) Force on a bicycle pedal produces an axial torque to the left. (b) With respect to a point at the base of the flagpole, the force of the guy wire produces a horizontal torque as shown. (c) An engine pushing a freight car produces a vertical torque if the origin is chosen to one side of the track. (d) With respect to the center of the sun, the gravitational force of the sun on Mars produces no torque.

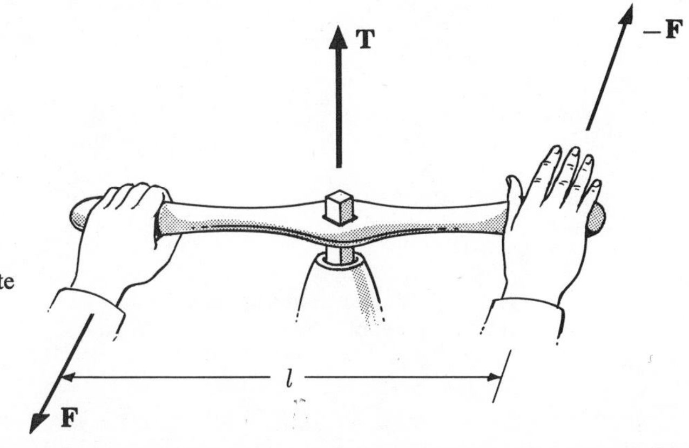

FIGURE 9.21 A pair of equal and opposite forces, not acting along the same line, produce a torque called a couple.

TABLE 9.1 SOME PARALLELS IN MECHANICS

Feature	Translation	Rotation
Inertial property	Mass, M	Moment of inertia, I
Aspect of motion	Center-of-mass velocity, $\mathbf{v}_c$	Angular velocity, $\boldsymbol{\omega}$
	Momentum, $\mathbf{P} = M\mathbf{v}_c$	Angular momentum about principal axis, $\mathbf{L} = I\boldsymbol{\omega}$
Impetus of change	Force, $\mathbf{F}$	Torque, $\mathbf{T}$
Law of change	$\mathbf{F} = \frac{d\mathbf{P}}{dt}$	$\mathbf{T} = \frac{d\mathbf{L}}{dt}$
Requirement for conservation	Absence of external force	Absence of external torque

9.6 Rotation about a fixed axis

An important special case: $\boldsymbol{\omega}$, $\mathbf{T}$, and $\mathbf{L}$ lie along the same line

In many practical devices, involving wheels, pulleys, gears, or shafts, rotation occurs about a fixed axis. We limit consideration here to rotation about a principal axis, such as a symmetry axis. Then the important vectors—angular velocity, torque, and angular momentum—all lie along the same line, the axis of rotation, and we need only consider components along this line.

CONSTANT ANGULAR VELOCITY

The rotation of a body with constant angular velocity ω_0 is so similar to the uniform motion of a particle in a circle that little further discussion is required. The angle through which the body turns increases linearly with time:

$$\theta = \theta_0 + \omega_0 t. \tag{9.54}$$

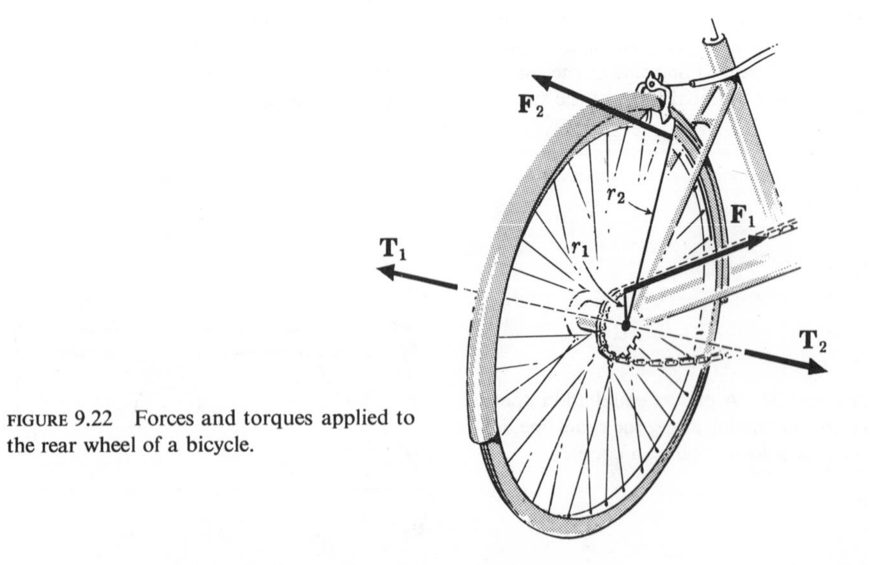

FIGURE 9.22 Forces and torques applied to the rear wheel of a bicycle.

The time derivative of the angle is, of course, the constant angular speed ω_0:

$$\frac{d\theta}{dt} = \omega_0. \tag{9.55}$$

The angular momentum of the body is constant:

$$\mathbf{L} = I\omega_0; \tag{9.56}$$

and the torque acting on it must be zero (since $\mathbf{T} = d\mathbf{L}/dt$).

CONSTANT ANGULAR ACCELERATION

Suppose that the angular speed of a body changes with time at a constant rate. It can be written

$$\omega = \omega_0 + \alpha t. \tag{9.57}$$

The quantity α is the angular acceleration:

Definition of angular acceleration

$$\alpha = \frac{d\omega}{dt} = \frac{d^2\theta}{dt^2}. \tag{9.58}$$

Its unit is radian/sec^2. (Since α and ω are actually components of vectors and not magnitudes, they may be positive or negative.) For such motion, the angle as a function of time is

$$\theta = \theta_0 + \omega_0 t + \tfrac{1}{2}\alpha t^2. \tag{9.59}$$

The analogy with uniformly accelerated motion in a straight line is obvious (see Equations 5.55–5.57).

The angular momentum of the body, being proportional to angular velocity, changes linearly with time. Its component is given by

$$L = I\omega = I\omega_0 + I\alpha t. \tag{9.60}$$

Finally, differentation of this expression with respect to time gives the component of the torque,

Torque related to angular acceleration

$$T = I\alpha. \tag{9.61}$$

Here is an equation with an important similarity to Newton's second law, $F = ma$. Just as constant force produces constant acceleration, constant *torque* produces constant *angular acceleration*. To the parallels in Table 9.1 can be added acceleration $\mathbf{a}$ and angular acceleration $\boldsymbol{\alpha}$. We remark, incidentally, that for any motion about a principal axis, whether constantly accelerated or not, the equation

A close similarity to $\mathbf{F} = M\mathbf{a}$

$$\mathbf{T} = I\boldsymbol{\alpha} \tag{9.62}$$

Angular acceleration is a vector quantity

is valid. The vector $\boldsymbol{\alpha}$ points along the axis of rotation, parallel to $\boldsymbol{\omega}$ if the angular speed is growing and opposite to $\boldsymbol{\omega}$ if the angular speed is shrinking.

As an example for study, consider a bicycle suspended so that its wheels can spin freely without touching the ground. We fix attention on the rear wheel as the system of interest (Figure 9.22). A force $\mathbf{F}_1$ can be applied at the rear sprocket by the chain, and a force $\mathbf{F}_2$ can be applied at the rim by the brakes. The frictional forces at the axle are assumed to be small enough to ignore.

With respect to the center of the axle chosen as a reference point, each of the forces $\mathbf{F}_1$ and $\mathbf{F}_2$ gives rise to a torque. Since the direction of each force is perpendicular to a line drawn from the axis to the point of application of the force, the torques have magnitudes

Torques if $\mathbf{r} \perp \mathbf{F}$

$$T_1 = F_1 r_1, \tag{9.63}$$

$$T_2 = F_2 r_2. \tag{9.64}$$

These torques are directed axially, $\mathbf{T}_1$ to the left and $\mathbf{T}_2$ to the right, as indicated in Figure 9.22. This means that any change in the wheel's angular momentum produced by the torque $\mathbf{T}_1$ must be directed axially to the left; any change in its angular momentum produced by the torque $\mathbf{T}_2$ must be directed axially to the right.

■ EXAMPLE 1: Suppose that the physical parameters of the wheel are

$$r_1 = 0.040 \text{ m},$$

$$r_2 = 0.316 \text{ m},$$

$$M = 2.5 \text{ kg},$$

and that it can be approximated as a hoop of radius r_2 so that its moment of inertia is

$$I = Mr_2^2 = 0.250 \text{ kg m}^2.$$

If the wheel starts from rest and the chain applies a force $F_1 = 100$ N for 3 sec, what angular momentum and what angular velocity does the wheel acquire? Through what angle does it turn as it is being accelerated? First, the magnitude of the torque T_1 must be calculated. From Equation 9.63, we get

$$T_1 = 100 \text{ N} \times 0.04 \text{ m} = 4 \text{ N m}.$$

This torque is equal to the rate of change of angular momentum (Equation 9.52). In terms of magnitudes,

$$\frac{dL}{dt} = T_1 = 4 \text{ kg m}^2/\text{sec}^2.$$

For constant torque, the change of angular momentum is

$$\Delta L_1 = T_1 \int dt = T_1 \, \Delta t. \tag{9.65}$$

In this example, with $\Delta t = 3$ sec,

$$\Delta L_1 = 12 \text{ kg m}^2/\text{sec}.$$

Since the angular momentum was initially zero, this *change* of angular momentum is the same as the angular momentum of the wheel after 3 sec. It is directed to the left, corresponding correctly to clockwise rotation if the wheel is viewed from the right side. The final angular speed is found from Equation 9.40:

$$\omega = \frac{L}{I} = \frac{12 \text{ kg m}^2/\text{sec}}{0.25 \text{ kg m}^2}$$

$$= 48 \text{ radian/sec}.$$

To get the angular rotation of the wheel, we need its acceleration. From Equation 9.62, we get

$$\alpha_1 = \frac{T_1}{I} = 16 \text{ radian/sec}^2.$$

Equation 9.59, with $\theta_0 = \omega_0 = 0$, gives the angle:

$$\theta = \tfrac{1}{2}\alpha_1 t^2 = \tfrac{1}{2}(16)(3)^2$$
$$= 72 \text{ radians},$$

or about 11.5 revolutions. To check the final angular speed, we may also use Equation 9.57, which gives

$$\omega = \alpha_1 t = (16)(3) = 48 \text{ radian/sec}.$$ ■

■ EXAMPLE 2: The wheel now spins freely for some time with constant angular momentum, after which time the brakes are applied with constant force to halve the angular momentum of the wheel in 2 sec. What brake force must be applied to achieve this result? What is the angular acceleration? Following the generally preferable procedure, we solve the problem first algebraically, then numerically. In general, for *constant* torque, the change of angular momentum is

$$\Delta\mathbf{L} = \mathbf{T}\,\Delta t. \tag{9.66}$$

In the preceding example,

$$\Delta\mathbf{L}_1 = \mathbf{T}_1\,\Delta t_1,$$

and in this example,

$$\Delta\mathbf{L}_2 = \mathbf{T}_2\,\Delta t_2.$$

Since $\Delta\mathbf{L}_2 = -\tfrac{1}{2}\,\Delta\mathbf{L}_1$, these equations yield

$$\mathbf{T}_2\,\Delta t_2 = -\tfrac{1}{2}\mathbf{T}_1\,\Delta t_1.$$

The required braking torque is

$$\mathbf{T}_2 = -\frac{\mathbf{T}_1\,\Delta t_1}{2\,\Delta t_2}.$$

With the help of Equation 9.64, the magnitude of the braking force can therefore be written

$$F_2 = \frac{T_1\,\Delta t_1}{2r_2\,\Delta t_2}.$$

From Equation 9.62, the magnitude of the acceleration during braking is $\alpha_2 = T_2/I$, or

$$\alpha_2 = \frac{F_2 r_2}{I}.$$

In these last two equations, we may substitute numerical values: $T_1 = 4$ N m, $\Delta t_1 = 3$ sec, $\Delta t_2 = 2$ sec, $r_2 = 0.316$ m, and $I = 0.250$ kg m^2; and we obtain

$$F_2 = 9.49 \text{ N},$$
$$\alpha_2 = 12.0 \text{ radian/sec}^2.$$

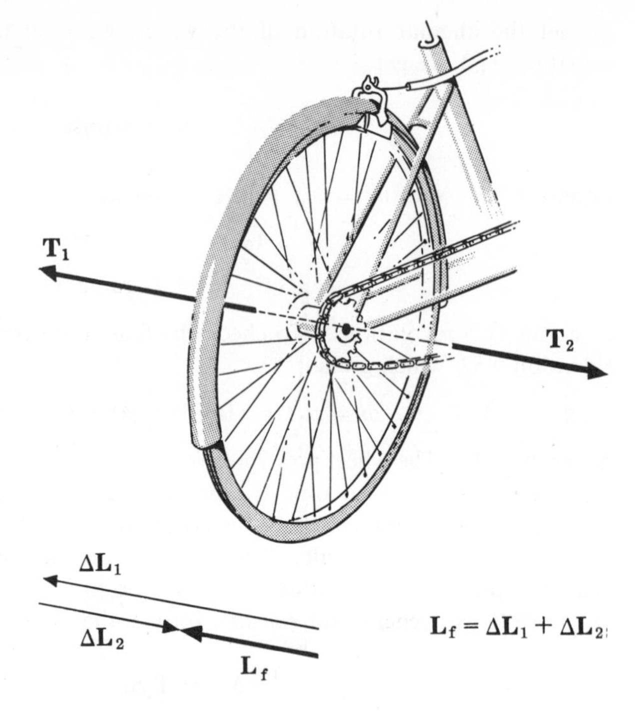

FIGURE 9.23 Changes of angular momentum produced by torques. Set spinning with torque $\mathbf{T}_1$, the wheel has angular momentum $\Delta\mathbf{L}_1$. A braking torque $\mathbf{T}_2$ then contributes a change of angular momentum $\Delta\mathbf{L}_2$. The final angular momentum $\mathbf{L}_f$ is the vector sum of $\Delta\mathbf{L}_1$ and $\Delta\mathbf{L}_2$.

As illustrated in Figure 9.23, the change of angular momentum produced by the braking torque is directed to the right, so the final angular momentum is the vector *sum* of the two changes of angular momentum:

$$\mathbf{L}_{\text{final}} = \Delta\mathbf{L}_1 + \Delta\mathbf{L}_2 . \quad \blacksquare$$

★9.7 Precession

In the examples in the preceding section, the vector nature of angular momentum played a rather minor role since all torques and angular momenta were directed left or right along the same fixed axis. When this is not the case—if, for instance, a torque acts upon a spinning object in a direction other than the direction of its spin—new effects arise that at first seem rather startling despite the fact that they are direct consequences of Equation 9.52.

The fall of a wheel, viewed in a new way

Consider a bicycle wheel mounted on a pair of handles (Figure 9.24), the kind often used for lecture demonstrations. If the wheel is not spinning and if one end of one handle is placed on a pointed support, as shown in Figure 9.24, the wheel does what is expected of it; it falls off the support. Its fall can be easily understood in terms of torque and angular momentum vectors. To be precise, suppose that the handle is initially horizontal in the east-west direction with its west end resting on the support. The downward force of gravity, which can be supposed to act at the center of the wheel, gives rise to a torque with respect to the point of support. The direction of this torque is horizontal to the north. It produces a twist about a north-south axis. Accordingly, the wheel acquires rotational motion about this axis. The mass of the wheel starts vertically downward. With respect to the point of support, it gains angular momentum northward. As the wheel falls from the support, it rotates, not about its own axle, but about a north-south axis dictated by the direction of the gravitationally produced torque.

This description of the simple fall of the wheel from its support may seem unnecessarily roundabout, but it is a very useful preamble to the discussion of an altered situation. Suppose that the wheel is set spinning about its own axle before it is placed, as before, with the west end of its handle on a point of support. Viewed from the east, if the wheel is rotating counterclockwise, its angular-momentum vector is directed eastward. The same gravitational force acts as before, producing a northward torque (Figure 9.25). Now we have an interesting new situation in which torque and angular momentum are not parallel. However, the rate of *change* of angular momentum is parallel to the torque (in fact equal to the torque), northward in this case. If the vector $\mathbf{L}_1$ designates the initial angular momentum and if the vector $\Delta\mathbf{L}$ designates the change of

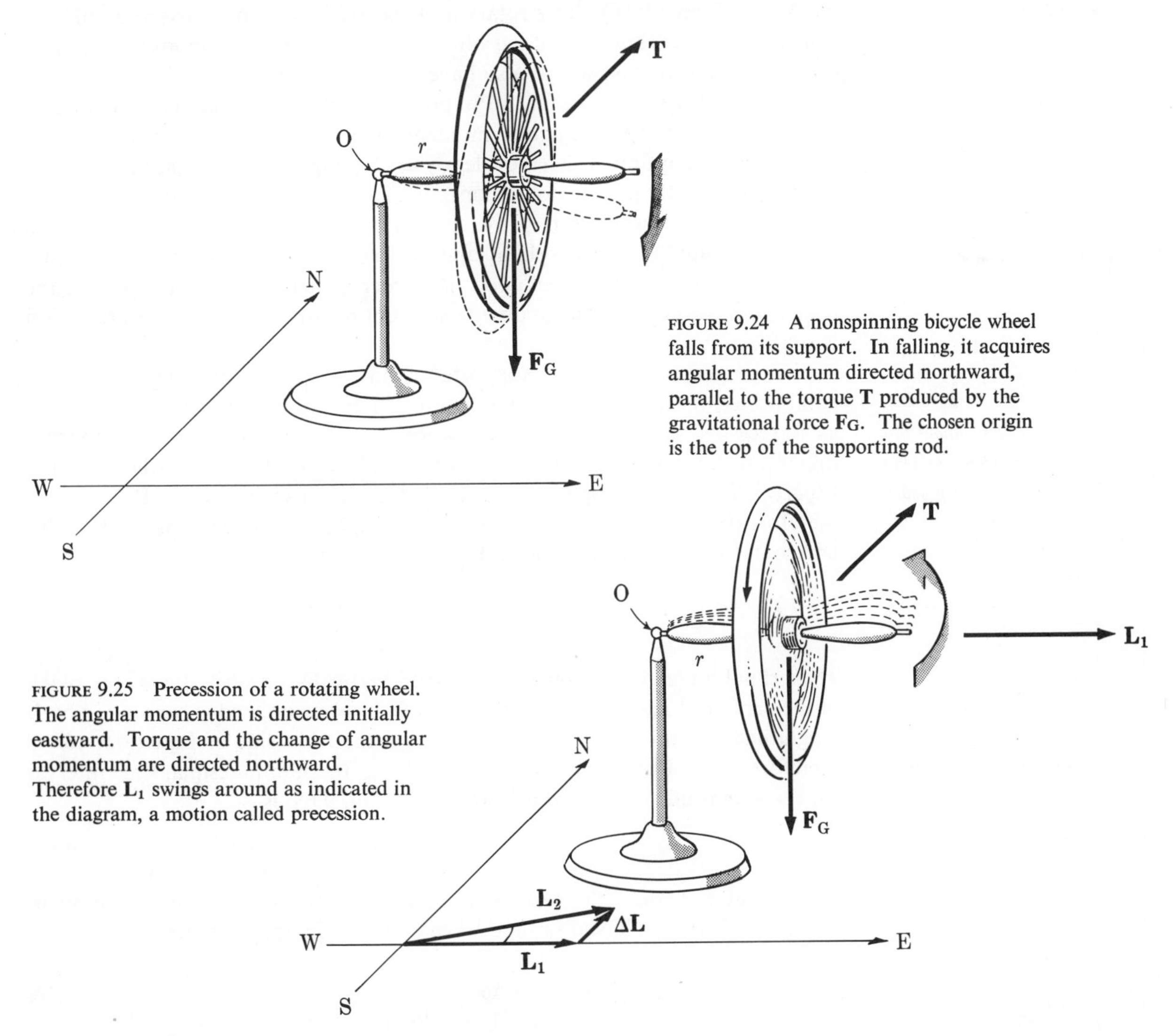

FIGURE 9.24 A nonspinning bicycle wheel falls from its support. In falling, it acquires angular momentum directed northward, parallel to the torque **T** produced by the gravitational force $\mathbf{F}_G$. The chosen origin is the top of the supporting rod.

FIGURE 9.25 Precession of a rotating wheel. The angular momentum is directed initially eastward. Torque and the change of angular momentum are directed northward. Therefore $\mathbf{L}_1$ swings around as indicated in the diagram, a motion called precession.

angular momentum in a brief interval of time ($\Delta\mathbf{L} = \mathbf{T}\,\Delta t$), their sum $\mathbf{L}_2$ will designate the angular momentum of the wheel after this time interval:

$$\mathbf{L}_2 = \mathbf{L}_1 + \Delta\mathbf{L}. \tag{9.67}$$

This vector addition—which is strictly valid only in the limit $\Delta t \to 0$—is diagrammed in Figure 9.25. The result of the gravitational torque is not to topple the wheel from the support; rather it shifts the wheel's axle around to a new direction, in this case slightly north of east. Once the wheel has so shifted, the gravitational torque has also shifted, to a direction slightly west of north. Consequently, the new total angular momentum $\mathbf{L}_2$ is pulled further counterclockwise (as seen from above) by a new change $\Delta\mathbf{L}$. The process continues indefinitely, or until friction puts a stop to it, with the angular-momentum vector of the wheel swinging through the points of the compass. Physically, this means that the axle of the wheel, as seen from above, rotates counterclockwise (Figure 9.26). This rotation of the axle, which proceeds at a uniform rate, is called *precession*. Had the wheel been spinning in the opposite direction, its precession direction would also have been opposite.

A spinning wheel precesses

Although unstated so far, an important assumption underlies this discussion. It is that the total angular momentum of the wheel is directed along its axle. This assumption, which is evidently used in Figure 9.25, means that changes in the direction of angular momentum can be equated to changes in the direction of the axle. The assumption is in fact valid only if the spin angular momentum is much greater than the orbital angular momentum, which means roughly that the rotation of the wheel about its center of mass is much faster than the rotation of the center of mass about the point of support. Then the precession is simple and uniform.

An approximation used in Figure 9.25: $L_{\text{spin}} \gg L_{\text{orbital}}$

More generally, a wheel supported as in Figure 9.25 may undergo rather complicated motion, in which the axle oscillates up and down (or "nutates") as it precesses. Among the possible motions, however, even without the approximation of large spin, is pure precession, which is easy to treat mathematically. Figure 9.26 shows the top view of a wheel engaged in such motion. Its axle, and therefore its spin angular momentum, remains horizontal. The torque $\mathbf{T}$ is also horizontal and is perpendicular to $\mathbf{L}_{\text{spin}}$. The equation of motion of the wheel is

Even without this approximation, pure precession is possible

$$\mathbf{T} = \frac{d\mathbf{L}}{dt} = \frac{d\mathbf{L}_{\text{orbital}}}{dt} + \frac{d\mathbf{L}_{\text{spin}}}{dt}. \tag{9.68}$$

The orbital angular momentum is directed vertically upward, out of the plane of the paper in Figure 9.26. It arises from the precession, and is *constant* in both magnitude and direction. The first term on the right side of Equation 9.68 is therefore zero. The torque changes only $\mathbf{L}_{\text{spin}}$. In an infinitesimal time interval Δt, the magnitude of the change, from Equation 9.68, is

Then only L_{spin} changes

$$\Delta L_{\text{spin}} = T\,\Delta t. \tag{9.69}$$

In this interval, the axle swings through angle $\Delta\theta$. Reference to the small triangle in Figure 9.26 shows that $\Delta\theta$ can be written approximately as

$$\Delta\theta = \frac{\Delta L_{\text{spin}}}{L_{\text{spin}}}. \tag{9.70}$$

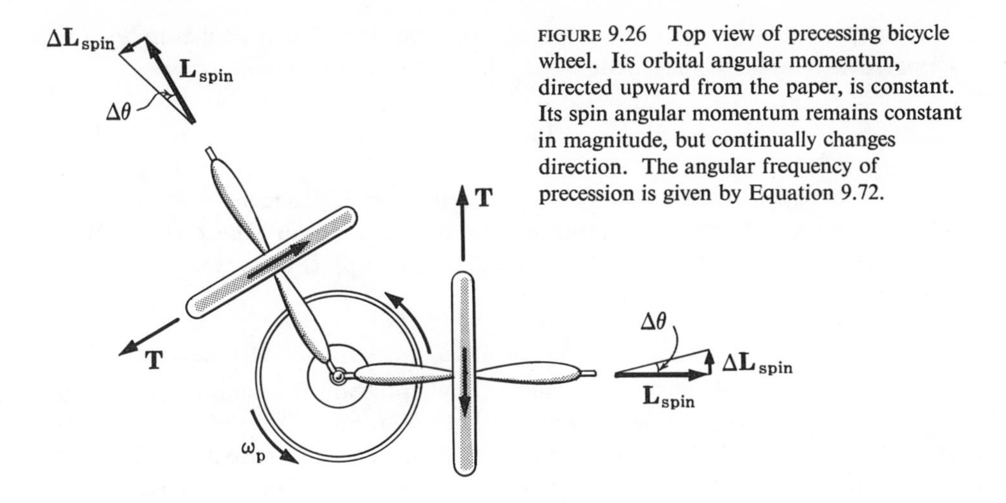

FIGURE 9.26 Top view of precessing bicycle wheel. Its orbital angular momentum, directed upward from the paper, is constant. Its spin angular momentum remains constant in magnitude, but continually changes direction. The angular frequency of precession is given by Equation 9.72.

Equations 9.69 and 9.70 combine to give

$$\frac{\Delta\theta}{\Delta t} = \frac{T}{L_{\text{spin}}}.$$

In the limit of small Δt, the ratio on the left becomes the angular speed of precession,

$$\omega_p = \lim_{\Delta t \to 0} \frac{\Delta\theta}{\Delta t}. \tag{9.71}$$

We have, therefore, obtained a very simple formula for the rate of precession,

$$\omega_p = \frac{T}{L_{\text{spin}}}, \tag{9.72}$$

A simple formula for the rate of precession

where T is the torque acting on the wheel with respect to its point of support and where L_{spin} is the spin angular momentum, which is independent of the choice of origin.

Actually watching the bicycle wheel, supported at the end of one of its handles, precess is a worthwhile experience. Notwithstanding our ability to understand what is happening, it seems rather surprising that the wheel does not fall from the support. The beauty of this example is the way it shows the power of a mathematical law to easily penetrate a new range of experience. The law connecting torque and the rate of change of angular momentum, based on observations in full accord with common sense, readily accounts for new phenomena whose explanations would otherwise strain common sense.

9.8 The conservation of angular momentum

Under what circumstances does the angular momentum of a system remain constant? Equation 9.52 provides an immediate answer. If the total torque on a system is zero, the total rate of change of its angular momentum is zero; therefore, its angular momentum is constant. As it stands, Equation 9.52 applies to any particle or any part of a system. As a reminder that it applies to an entire system as well, we may write

$$\mathbf{T}_{\text{total}} = \frac{d\mathbf{L}_{\text{total}}}{dt}. \tag{9.73}$$

Actually, a more important and more useful statement than this about torque and angular momentum can be made. Consider the division of the total torque into an internal part and an external part:

$$\mathbf{T}_{\text{total}} = \mathbf{T}_{\text{int}} + \mathbf{T}_{\text{ext}}. \tag{9.74}$$

An important new principle: Internal torques sum to zero

External torques arise from outside the system; internal torques arise from within the system. A skydiver falling through the air, for example, is subjected to external torques from the air and internal torques from his own muscles. According to excellent experimental evidence—evidence based on the validity of the law of angular-momentum conservation—the *internal torques in fact always add up to zero*. This is a new and important statement about nature. It might be called the rotational equivalent of Newton's third law. Because of Newton's third law, which leads directly to the conclusion that internal *forces* always add up to zero, the *momentum* of an isolated system is conserved. Similarly, the vanishing of total internal *torque* leads to the conservation of *angular momentum* for isolated systems. For central forces—those forces acting along the lines joining particles—Newton's third law leads to the vanishing of total internal torque as well as total internal force. In general, however, the requirement that total internal torque vanish is a more powerful restriction on the forces of nature than is Newton's third law.

Since internal torques sum to zero, Equation 9.73 may be replaced by

The rotational equation of motion for a system

$$\mathbf{T}_{\text{ext}} = \frac{d\mathbf{L}_{\text{total}}}{dt}. \tag{9.75}$$

Then the law of angular momentum conservation can be written

The conservation of angular momentum

$$\mathbf{L}_{\text{total}} = \textbf{constant} \qquad \text{if} \qquad \mathbf{T}_{\text{ext}} = 0. \tag{9.76}$$

The total angular momentum of a system changes only in response to an external torque. If after falling for a while in a swan dive, the skydiver begins to rotate, it can only be because of torques delivered by the air. Falling through a vacuum, he would be unable to change his angular momentum by any contortions whatever.* His own muscles, by changing his shape, can trigger torques in the air that impart angular momentum. No pushing, pulling, or twisting of the muscles, however, can lead to a net, internally produced torque.

For the same reason, nothing done internally in a spacecraft can stop or start a uniform rotational motion. This can be accomplished only in the same

* He could, however, change his orientation without changing his angular momentum. A cat, usually able to land on its feet, is very clever at doing exactly that. For the falling cat, the air contributes nothing important to the turning maneuver, which could, in principle, be performed in a vacuum. The explanation of the cat's feat is left to the reader (see Question 9.27).

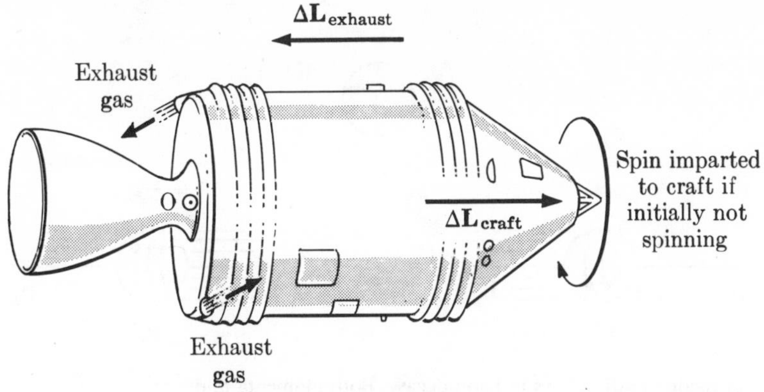

FIGURE 9.27 Application of torque to spacecraft by tangential rocket thrust. For the total system, spacecraft plus exhaust, angular momentum is conserved, so $\Delta\mathbf{L}_{\text{exhaust}} + \Delta\mathbf{L}_{\text{craft}} = 0$.

way that its momentum is changed—by throwing away mass. The exhaust from a tangential nozzle (Figure 9.27) possesses angular momentum with respect to the center of the spacecraft. The craft and what remains in it thereby acquire an equal and opposite change of angular momentum. Spacecraft plus exhaust viewed together as a single system suffer no change of angular momentum since they experience no external torque. If the exhaust is *not* regarded as part of the system, it is part of the external world delivering a torque and a change of angular momentum to what is left behind. What about the exhaust by itself? Is it acted upon by an external torque?

L = constant *for spacecraft plus exhaust*

■ EXAMPLE: The spacecraft in Figure 9.27, whose moment of inertia is $I = 2{,}000 \text{ kg m}^2$, is rotating freely about its symmetry axis with an angular speed $\omega = 0.2$ radian/sec. The astronauts wish to stop this motion by using two control jets. Each is located $r = 1.5$ m from the axis, and together they shoot out exhaust at the rate $dm/dt = 2$ kg/sec, with exhaust speed $v_{\text{ex}} = 50$ m/sec. For how long a time should these tangential control jets be turned on? We can proceed approximately, noting that the peripheral speed of the rocket, ωr, is much less than v_{ex} and assuming that the exhaust mass is much less than the spacecraft mass. Initially, the angular momentum of the system is

$$L_{\text{before}} = I\omega.$$

Finally, with the spacecraft brought to rest, the only angular momentum is that of the exhaust gas, whose mass is Δm:

$$L_{\text{after}} = \Delta m\, v_{\text{ex}} r.$$

These two angular momenta must be equal, so

$$\Delta m = \frac{I\omega}{v_{\text{ex}} r}. \tag{9.77}$$

Since dm/dt is constant, we can also write

$$\Delta m = \int \left(\frac{dm}{dt}\right) dt = \left(\frac{dm}{dt}\right) \Delta t.$$

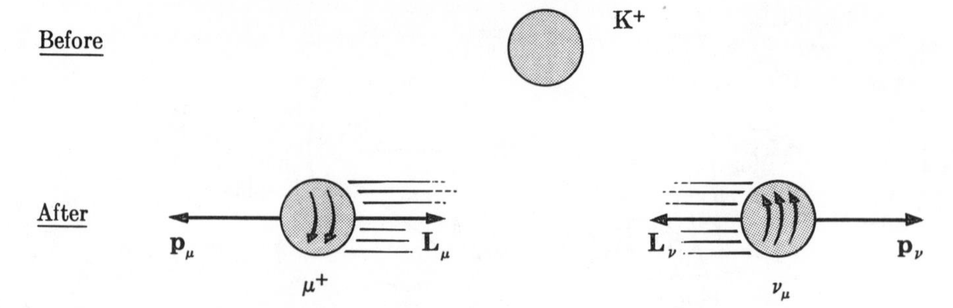

FIGURE 9.28 Two conservation laws in kaon decay. Both momenta and angular momenta of the two product particles add to zero. Although the neutrino normally escapes detection, $\mathbf{p}_\nu$ and $\mathbf{L}_\nu$ can be deduced from measurements of $\mathbf{p}_\mu$ and $\mathbf{L}_\mu$.

The required time is, therefore,

$$\Delta t = \frac{I\omega}{v_{ex}r(dm/dt)}.$$

Substitution of the given numbers yields

$$\Delta m = 5.33 \text{ kg},$$

$$\Delta t = 2.67 \text{ sec.}$$

■

L = constant *for an isolated system*

The easiest way to rid a system of external torque is to rid it of all external forces. An *isolated* system experiences no external force and therefore no external torque. Consequently, we have as a fundamental law of nature: *The total angular momentum of an isolated system is conserved.* The skydiver without any air and the spacecraft outside the atmosphere illustrated the principle. Examples are easy to find also in the world of atoms and particles. Indeed it is in the submicroscopic world, where true isolation of an individual system from the influence of all other systems is very nearly a reality and not just an idealized abstraction, that the law of angular-momentum conservation can be most precisely tested. Despite the fact that the *quantization* of angular momentum is significant in the submicroscopic world and irrelevant in the macroscopic world, the *conservation* of angular momentum takes exactly the same form and has the same significance in both domains.*

Most particle processes are isolated

Examples of angular-momentum conservation in particle transformations appear in Chapter 4. As another example, consider the decay of a kaon into two pions, denoted, for example, by

$$K^+ \rightarrow \pi^+ + \pi^0.$$

All three particles are spinless. With respect to the point occupied initially by the kaon, the total angular momentum is zero. After the decay, angular-

* It should be borne in mind, however, that the law of *change*, Equation 9.75, is a classical law that loses its validity in the quantum world.

momentum conservation requires that the two pions fly apart with no orbital angular momentum. Another mode of decay of the positive kaon is into muon and neutrino:

$$K^+ \rightarrow \mu^+ + \nu_\mu.$$

Again the total angular momentum is zero, but this time each final particle has one-half quantum unit of spin. The spinning neutrino has the characteristic property of being "left-handed." This means that its angular momentum is always directed opposite to its momentum, or equivalently, opposite to its direction of flight (Figure 9.28). To conserve angular momentum in this mode of kaon decay, the muon must also emerge in a left-handed state of motion. Both the momenta and the angular momenta of the two product particles cancel.

9.9 The law of areas

For an isolated system upon which no external forces act, both momentum and angular momentum are conserved. However, isolation is not required for angular-momentum conservation; all that is required is the vanishing of external torque. It is quite possible for a system to be acted upon by an outside force yet be free of torque. For one special yet very important kind of force—a *central* force—the angular momentum of a system is conserved even though the system is not isolated and its momentum is not constant.

L = constant *for central forces*

Consider Mars, the planet studied most intensively by Kepler, and choose the sun as a reference point. (More precisely, the reference point should be the center of mass of the Mars-sun system, which is not far from the center of the sun.) With respect to this point, Mars possesses orbital angular momentum. However, with respect to the same point, Mars experiences no torque [Figure 9.20(d)]. Therefore, Mars, as it moves around the sun, retains constant angular momentum. Actually the spin and orbital angular momenta of the planet are separately constant since it experiences no torque with respect to its own center and none with respect to the sun. We are here concerned with the constancy of its orbital angular momentum.

A central force is defined as a force acting along the line joining a particle or object to a center of force. With an origin at the force center, the position vector and the force vector are either parallel or antiparallel. In either case,

The reason: Central forces produce no torque

$$\mathbf{T} = \mathbf{r} \times \mathbf{F} = 0 \qquad \text{(central force).} \tag{9.78}$$

Therefore, with respect to the force center, angular momentum is constant. Uniform motion in a circle, which requires a central force, is a special and often cited example of accelerated motion with constant angular momentum. However, an object acted upon by a central force need not move in a circle, or even move around the center of force. A few kinds of motion in central force fields are shown schematically in Figure 9.29. For each, angular momentum with respect to the force center is conserved. Also each such orbit or trajectory has in common an interesting geometric property called the "law of areas."

Figure 9.30 illustrates the geometric construction that makes clear the meaning of the law of areas. At a particular moment a particle acted upon by a central force is at point A, a distance r from the force center C, and is moving with speed v. After the lapse of a time interval Δt it has moved ahead to

Derivation of the law of areas

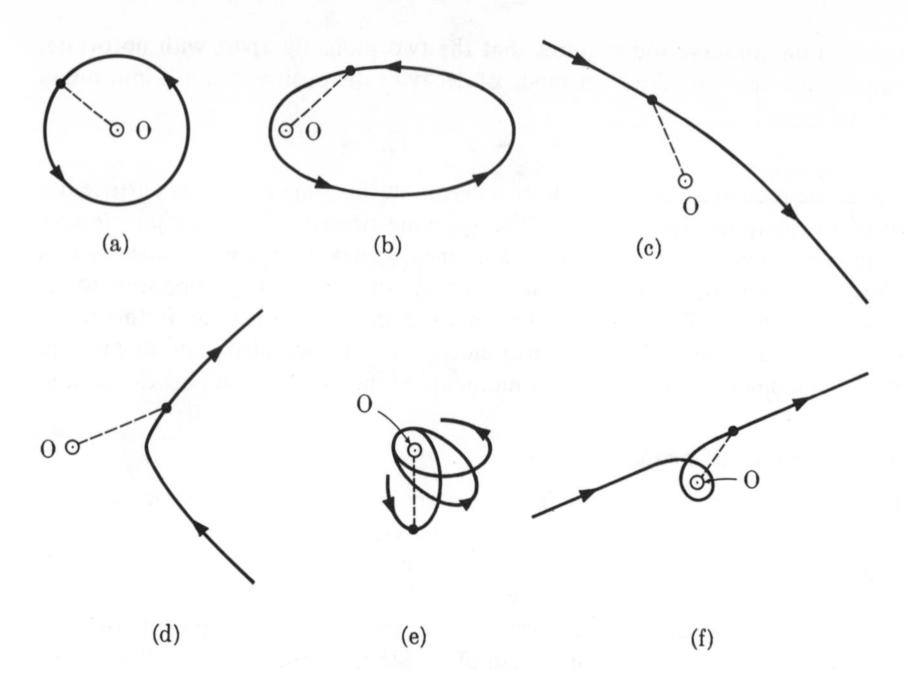

FIGURE 9.29 Possible motions in central force fields. Diagrams (a) through (d) show circular, elliptical, and hyperbolic paths that are possible when the force varies inversely with the square of the distance ($F \sim 1/r^2$). If the force varies in other ways with distance, an infinite variety of other trajectories are possible. Two such are shown in diagrams (e) and (f).

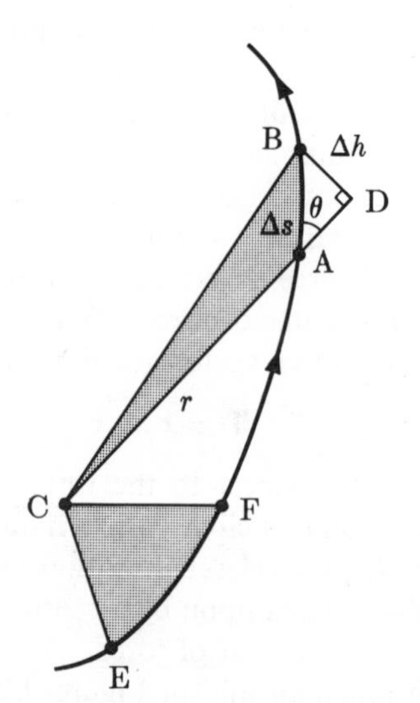

FIGURE 9.30 Geometric construction for the law of areas. Points E, F, A, and B lie along the particle trajectory. Point C is the force center.

point B. If the interval Δt is sufficiently brief, the short segment of orbit AB may be approximated as a straight line traversed at constant speed v. The length of the segment AB may be called Δs. It is equal to the product of speed and time:

$$\Delta s = v\,\Delta t.$$

Next a line BD perpendicular to the radial line CA is drawn. Its length Δh may be written

$$\Delta h = \Delta s \sin\theta = v\,\Delta t \sin\theta, \tag{9.79}$$

where θ is the angle between AD and AB, which is the same as the angle between the vectors $\mathbf{r}$ and $\mathbf{v}$. We now fix attention on the shaded area ΔA of triangle ABC. This is referred to as the area "swept out" by the particle in the time interval Δt. It is equal to half the product of the base r and the height Δh:

$$\Delta A = \tfrac{1}{2} r\,\Delta h. \tag{9.80}$$

Replacing Δh in Equation 9.80 by $v\,\Delta t \sin\theta$ (from Equation 9.79) yields

$$\Delta A = \tfrac{1}{2} rv \sin\theta\,\Delta t.$$

The reason for these manipulations now begins to appear, for on the right side of this equation is the same combination $rv \sin\theta$ that appears in the definition of angular momentum (Equation 9.3). The combination is equal to the angular momentum L with respect to the force center divided by the particle mass m. The area ΔA may therefore be written

$$\Delta A = \frac{1}{2}\frac{L}{m}\,\Delta t.$$

As a final step, the factor Δt is moved to the left side of the equation, which becomes

$$\frac{\Delta A}{\Delta t} = \frac{L}{2m}.$$

In the limit $\Delta t \to 0$,

$$\frac{dA}{dt} = \frac{L}{2m}. \tag{9.81}$$

Area is swept out at a constant rate if **L = constant**

The rate of change of area—or the rate at which area is being swept out—is equal to the angular momentum of the particle divided by twice the particle mass.* If the force is central, this is a constant for any particular particle, the same at one time as at another. In Figure 9.30, for instance, the areas CAB and CEF are equal if the time intervals for the particle to traverse the orbit segments AB and EF are the same. As the figure makes clear, this means that the particle speed must be greater from E to F than from A to B.

The fact that a line drawn from the force center to the particle (or object) sweeps out area at a constant rate is what is called the law of areas. We have derived it as a geometric consequence of the law of angular-momentum conservation. Historically, the law of areas was not a consequence but rather a precursor of angular-momentum conservation. Johannes Kepler in 1609—before the concept of angular momentum had been invented, before the idea of a

---

* Area may be defined as a vector quantity, the direction of the vector being perpendicular to the plane of the area. The vector area of triangle CAB, for instance, is $\Delta\mathbf{A} = \tfrac{1}{2}\mathbf{r} \times \mathbf{v}\,\Delta t$. The vector form of Equation 9.81 is $d\mathbf{A}/dt = \mathbf{L}/2m$.

Kepler discovered the law of areas for Mars

central force existed, even before it was understood that planetary motion had anything to do with force—discovered as an empirical fact that the law of areas correctly describes the motion of Mars around the sun. This was one of three important properties of planetary motion discovered by Kepler. Kepler's laws, which tied together a myriad of careful observations of the sun and planets into three neat packages, will be discussed further in Chapter 11. His law of areas, many decades after its discovery, provided strong support for Newton's contention that the planets were drawn toward the sun by a central force.

At its perigee (nearest point) and at its apogee (farthest point), an earth satellite is moving perpendicular to the line joining it to the center of the earth (Figure 9.31). At these particular points, its angular momentum can be simply expressed (see Equation 9.4):

$$L_{\rm p} = mr_{\rm p}v_{\rm p}, \tag{9.82}$$

$$L_{\rm a} = mr_{\rm a}v_{\rm a} \tag{9.83}$$

(the subscripts p and a refer to perigee and apogee). Since the earth's gravitational force acting on the satellite is very nearly a central force, angular momentum changes very little, and $L_{\rm p}$ may be set equal to $L_{\rm a}$. This equality leads to a particularly simple relation between speeds and distances at perigee and apogee.

Simple ratio for perigee and apogee

$$\frac{v_{\rm a}}{v_{\rm p}} = \frac{r_{\rm p}}{r_{\rm a}}. \tag{9.84}$$

Minimum speed occurs at apogee, maximum speed at perigee.

Orbits of some communications satellites are very eccentric

The decrease of speed at apogee dictated by the conservation of angular momentum has been used to good advantage in the choice of orbit for some communications satellites. The Soviet *Molniya* satellites have been launched into orbits with apogee about seven times as distant (from the earth's center) as perigee. They are used to relay messages principally when they are near apogee and moving slowly enough to allow uninterrupted use for many hours. As they

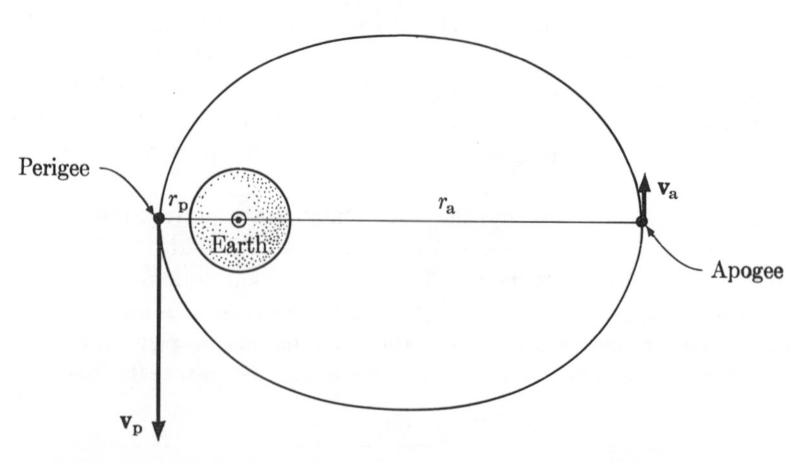

FIGURE 9.31 A satellite at perigee and apogee. The ratio of its speeds at these two points is the inverse ratio of its distances from the center of the earth (Equation 9.84).

TABLE 9.2 SOME PROPERTIES OF PLANETARY ORBITS KNOWN TO KEPLER*

Planet	Average Distance from Sun in Astronomical Units†	Ratio of Aphelion Distance to Perihelion Distance	Ratio of Longest Diameter of Orbit to Shortest Diameter
Mercury	0.387	1.518	1.022
Venus	0.723	1.014	1.000023
Earth	1.000	1.034	1.00014
Mars	1.524	1.206	1.0044
Jupiter	5.20	1.102	1.0012
Saturn	9.54	1.118	1.0016

* The data in the table are current values of these quantities.
† The astronomical unit (A.U.) is defined to be the average distance of the earth from the sun. It is equal to 1.496×10^{11} m, or about 93 million miles.

disappear over the horizon and approach perigee, they speed up so they waste less time getting around the other side of the earth and back to a useful position than would be the case if they were in nearly circular orbits. Man cannot repeal the law of angular-momentum conservation, but he can use it to his advantage.

Planetary orbits do not differ greatly from circles

Kepler, in studying the orbits of the six planets then known, was not favored with the wide variety of orbital shapes now exhibited by artificial earth satellites. As shown in Table 9.2, none of these planets has an orbit markedly different from circular. Mercury, whose orbit is farthest from circular among these six, has an aphelion distance 52 percent greater than its perihelion distance (although the longest diameter of its orbit is only 2 percent greater than the shortest diameter—see Figure 9.32). The orbit of Venus is nearly a perfect circle. Mars, whose orbit is flattened by less than 0.5 percent, comes about 20 percent closer to the sun at perihelion than at aphelion. The data of Tycho de Brahe available to Kepler were of sufficient accuracy to allow him to discover the law of areas for the motion of Mars and to verify it for the other known planets.

The precision tracking of modern satellites provides further tests of the law of areas every day. Of course, for satellites the law of areas is a tool, not a test. At one extreme are the elongated orbits of some of the Explorer satellites, which brushed the fringes of the atmosphere at little more than 100 miles above the earth at perigee and cruised out to distances of more than 100,000 miles—halfway to the moon—at apogee. Explorer 18 rounded its apogee more than 126,000 miles from the center of the earth at a leisurely pace of about 780 mile/hr as compared with its speed of over 24,000 mile/hr at perigee. At the other extreme of orbit shapes is the almost perfect circle of a synchronous communications satellite. In 1969, Intelstat 3D, over 26,000 miles from the center of the earth, had an apogee and a perigee that differed by less than 1 mile.*

* Since the communications satellites carry small rockets controllable from earth, the orbit of Intelstat 3D may now be slightly different. Because it is entirely free of frictional force, it will in any case be a companion of the earth indefinitely.

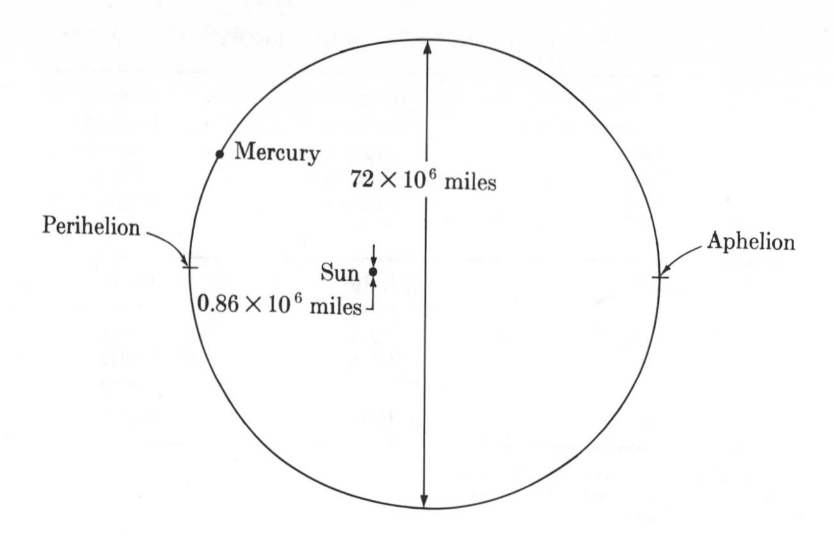

FIGURE 9.32 Orbit of Mercury drawn to scale. The size of the sun is also drawn to scale, but Mercury is greatly enlarged. The speed of Mercury at its perihelion is about 52 percent greater than its speed at aphelion.

Fortunately for Kepler and the advance of science in the seventeenth century, the law of areas is very nearly an exact law for planetary motion. However, we now know—and indeed Newton knew—that no planet experiences precisely a central force. Mars, for example, although predominantly influenced by the sun, is weakly pulled in other directions by Jupiter, the earth, and every other planet. With respect to the sun's center as a reference point, these weaker forces, known as perturbations, are not central forces. Because of the perturbations, Mars experiences some torque, its angular momentum changes slightly in time, and for it the law of areas is not quite an exact law. Because the perturbations have only slight effects, the solar system has a pleasing simplicity that it would otherwise lack, a piece of good luck for man in his early efforts to comprehend the world beyond the earth. On the other hand, the perturbations proved eventually to serve science, not impede it. They made possible sensitive tests of the law of gravitational force; and perhaps most significant, they led to the discovery of two new planets—Neptune and Pluto. How this came about will be described in Chapter 11.

Perturbations cause the law of areas to be inexact

For satellites circling the earth, the law of areas is also imperfect, an imperfection that can be put to good use in learning more about the earth. If the earth were a perfect sphere with its mass arranged in uniform shells, the force experienced by a satellite would be exactly a central force. But the earth is not a perfect sphere. It is flattened at the poles, it is very slightly pear-shaped (a fact first revealed by satellites), and its surface layers are rather lumpy. For all these reasons, the earth can exert small torques on a satellite, thereby changing its angular momentum and causing it to deviate slightly from the law of areas. These orbital deviations have been measured with enough precision for some satellites to reveal the earth's pear shape and other finer details of its mass distribution.

The earth exerts small torques on satellites

★9.10 The isotropy of space

The concept of angular momentum has undergone an interesting evolution in physics. Kepler discovered the law of areas as an empirical fact long before angular momentum entered the vocabulary or the tool kit of physics. Still without the help of the angular-momentum concept, Newton related the law of areas to the action of a central force. Not until the eighteenth century was angular momentum defined and used in mechanics. And not until the nineteenth, with the development of alternative and more powerful mathematical formulations of mechanics, did angular momentum come to be regarded as one of the most fundamental concepts of mechanics. Finally, in the twentieth century, angular momentum has joined momentum and energy as one of the preeminent mechanical concepts.* There are several reasons for its present station. One reason is its conservation; another is its quantization; a third is its relation to a simple symmetry of empty space.

Why angular momentum is now regarded as a key concept

As momentum conservation is related to, and indeed can be founded upon, the *homogeneity* of space (the indistinguishability of one point in space from another), angular-momentum conservation is similarly tied to the *isotropy* of space (the indistinguishability of one direction from another). An isolated object at rest in space is not expected to be self-accelerating in some direction, for that would imply an inhomogeneity of space. Nor is it expected to set itself spontaneously into rotation, for that would imply an anisotropy of space. The absence of spontaneous rotation requires the absence of any net internal torque, which in turn implies that the angular momentum of an isolated system is conserved. The bland sameness of space is at the root of both momentum conservation and angular-momentum conservation (see Section 4.8).

The chain of argument for angular momentum is worth reviewing in somewhat more detail. Consider a wheel that is uniform except for a weight placed at one point of its periphery (Figure 9.33). If suspended near the earth and initially at rest, it will begin to rotate "spontaneously" if the weight is placed at any point other than the lowest point. There is nothing surprising about this. An external force, the force of gravity, acts on the wheel, and it produces an external torque. (Another external force, supporting the wheel at its axle, contributes no torque with respect to the axle.) Because of the external torque, the wheel's angular momentum changes. If initially at rest, it begins to rotate. For the wheel, angular momentum is not conserved. Another way to describe this situation is to say that in the neighborhood of the earth there exists a preferred direction, the vertical direction of the earth's gravitational force. The preferred spoke—the one connecting the hub to the weight on the rim—moves in such a way as to align itself with the preferred direction or to oscillate equally about the preferred direction. In empty space, on the other hand, far from the earth or other external influences, there should be no preferred direction and no spontaneous rotation. The same wheel, placed at rest in an ideally remote location, should remain at rest. The "should" in these sentences is based on the

Near the earth, there is a preferred direction

In remote space, all directions are equivalent

* A reminder: All three of these concepts have required new definitions in the modern theories, definitions that encompass the old.

fundamental postulate of the isotropy of space. If space possesses the indistinguishability of direction called isotropy, no isolated object should spontaneously begin to rotate. This is the first key step in the argument. The second is to note that no rotation means no torque. Torques within the isolated object, if any, must cancel exactly. Having reached the conclusion that total internal torque must equal zero, we may for the final step in the argument allow our isolated object to be rotating initially rather than being stationary. Since it experiences neither external torque nor internal torque, its angular momentum remains constant.

Because of this isotropy, internal torques must sum to zero

This argument, intended only to be indicative of the existence of a link between the isotropy of space and the conservation of angular momentum, is less rigorous than it may appear. It provides a tight, logical link only for rigid objects such as the wheel of Figure 9.33. For looser systems whose parts are in relative motion, the connection between spatial isotropy and angular-momentum conservation is more subtle (but just as real). The absence of any preferred direction leads only to the conclusion that *some* rotational property should be conserved. Angular momentum is defined in just such a way that it is the conserved quantity. It is not hard to think of other rotational quantities—angular velocity, for instance—that are not conserved.

Is angular momentum conserved in the universe at large?

Angular-momentum conservation has not been put to the test over domains of space larger than the solar system. It remains a question for the future, and a most intriguing question, whether or not this conservation law will fail in the galactic and intergalactic domains. If it does, man will have learned that space in the large is not perfectly isotropic, a discovery that would have most important bearing on the structure of the universe as a whole and also on the question of whether the universe is finite or infinite.

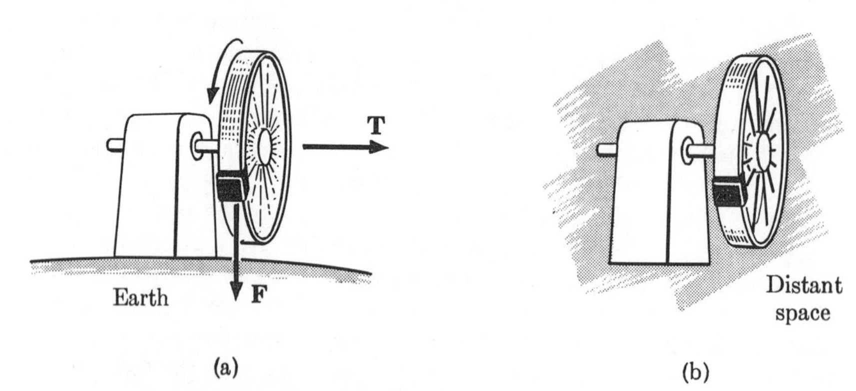

FIGURE 9.33 Angular-momentum conservation and the isotropy of space. (a) An unbalanced wheel near the earth rotates "spontaneously." Its angular momentum is not constant. (b) The same wheel in the depth of empty space does not start to rotate spontaneously if space is isotropic. This means that it experiences no net torque, which in turn implies that angular momentum is conserved for an isolated system.

Summary of ideas and definitions

Angular momentum is an axial vector quantity, defined for a particle by $\mathbf{L} = m\mathbf{r} \times \mathbf{v} = \mathbf{r} \times \mathbf{p}$. It is defined with respect to an arbitrary origin.

If a line drawn from the origin to a particle rotates, the particle possesses angular momentum.

A particle moving uniformly in a straight line has constant angular momentum with respect to any origin.

A particle moving uniformly in a circle has constant angular momentum with respect to the center of the circle.

The total angular momentum of a system can be written $\mathbf{L} = \mathbf{L}_{\text{orbital}} + \mathbf{L}_{\text{spin}}$.

$\mathbf{L}_{\text{orbital}}$ (given by Equation 9.23 or 9.25) arises from the motion of the center of mass.

$\mathbf{L}_{\text{spin}}$ (given by Equation 9.24) is the angular momentum relative to the center of mass.

If the center of mass of a system is at rest, then $\mathbf{L} = \mathbf{L}_{\text{spin}}$, which is independent of the choice of origin.

If a rigid body rotates about certain axes called principal axes, its angular momentum is proportional to its angular velocity: $\mathbf{L} = I\boldsymbol{\omega}$.

The quantity I is called the moment of inertia of the body. It depends on the choice of axis. It is measured in kg m^2. It is an inertial property, analogous to mass.

Any line of cylindrical symmetry or reflection symmetry is a principal axis. Any line perpendicular to a plane of symmetry is a principal axis if the origin is at the intersection of the line and the symmetry plane.

For rotation about an arbitrary z axis, the moment of inertia is defined by $L_z = I\omega$; Equations 9.36 and 9.37 provide formulas for I.

Torque is an axial vector quantity defined with respect to an arbitrary origin by $\mathbf{T} = \mathbf{r} \times \mathbf{F}$.

A force produces no torque if it acts at the origin or along a line that passes through the origin.

If two forces equal in magnitude and opposite in direction do not act along the same line, they produce a torque called a couple. A couple is independent of the choice of origin.

For a particle, the law of angular momentum change is $d\mathbf{L}/dt = \mathbf{T}$. For a system, it is $d\mathbf{L}/dt = \mathbf{T}_{\text{ext}}$. These laws are analogous to $d\mathbf{p}/dt = \mathbf{F}$ and $d\mathbf{P}/dt = \mathbf{F}_{\text{ext}}$.

For a rigid body rotating about a principal axis, the angular acceleration is related to torque by $\mathbf{T} = I\boldsymbol{\alpha}$. This is analogous to $\mathbf{F} = M\mathbf{a}$.

Zero torque implies constant angular velocity. Constant torque implies constant angular acceleration.

If a torque acts on a spinning wheel in a direction not along its axis, the wheel precesses. Uniform precession of a wheel with a horizontal axis occurs at the rate $\omega_{\text{p}} = T/L_{\text{spin}}$.

The total internal torque in any system is equal to zero. This principle is related to Newton's third law and the vanishing of internal force, but it is a distinct principle.

The total angular momentum of a system is conserved if the total external torque is zero. In particular, the angular momentum of an isolated system is conserved.

A particle acted upon by a central force has constant angular momentum with respect to the force center. The central force exerts no torque.

A particle moving in a central field of force with angular momentum L sweeps out area at the rate $dA/dt = L/2m$.

The constancy of dA/dt is called the law of areas. It is very nearly a valid law for planets and satellites.

Angular momentum is a fundamental concept because of its conservation, its quantization, and its relation to a simple symmetry principle.

The isotropy of space implies that internal torques must sum to zero. Otherwise an isolated body would experience spontaneous angular acceleration. Angular-momentum conservation is founded on spatial isotropy.

QUESTIONS

Section 9.1

Q9.1 In a baseball game a pitcher throws a baseball directly to the catcher. Discuss the angular momentum of the moving ball relative to the pitcher, the catcher, the first baseman, and the third baseman.

Q9.2 Seen from above, the main rotor of a helicopter rotates counterclockwise. A small tail rotor mounted on a horizontal axis rotates counterclockwise when seen from the right side of the craft. What are the directions of the angular momenta of the two rotors? What is the approximate direction of their total angular momentum?

Q9.3 Can a particle moving along a curved path have zero angular momentum with respect to a point not on that path (a) at one instant of time? (b) for a finite span of time? Justify both answers.

Q9.4 Under what conditions, if any, can a particle at one instant of time have (a) zero angular momentum with respect to *all* reference points? (b) nonzero angular momentum with respect to all reference points?

Q9.5 A particle *accelerates* in a straight line. (1) Specify (a) an origin with respect to which its angular momentum **L** is constant and (b) an origin with respect to which **L** is not constant. (2) Is there any inertial frame of reference in which the particle's momentum **p** is constant?

Q9.6 Some vector quantities are defined with respect to an origin, and some are independent of the choice of origin. In which of these two categories does each of the following vector quantities belong: (a) position, (b) displacement, (c) velocity, (d) momentum, (e) angular momentum, and (f) force?

Section 9.3

Q9.7 (1) Can the spin angular momentum of a system be zero with respect to one reference point and nonzero with respect to another? (2) Can the orbital angular momentum of a system have different values with respect to different reference points?

Q9.8 A car is driven counterclockwise around a circular track. (1) Give the directions of (a) the orbital angular momentum of the car with respect to the center of the track and (b) the spin angular momentum of one of the wheels. (2) Which wheels possess the greater magnitude of spin angular momentum?

Q9.9 (1) A bicycle is held stationary in the air and its front wheel is set spinning. Is the angular momentum of the wheel orbital or spin angular momentum or a combination of both? (2) The bicycle is then placed on the ground and is ridden in a straight line. Is the angular momentum of its front wheel now orbital or spin angular momentum or a combination of both?

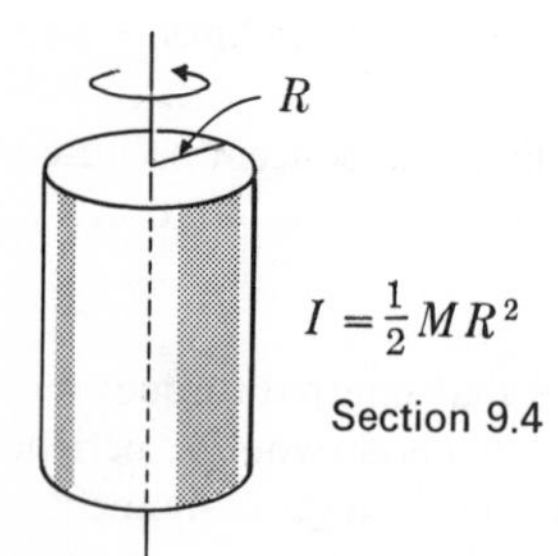

Section 9.4

Q9.10 A thin, uniform disk rotating about its axis of cylindrical symmetry has a moment of inertia given by $I = \frac{1}{2}MR^2$ (Equation 9.43). As indicated in the figure, the moment of inertia of an extended cylindrical body of uniform density rotating about its axis of cylindrical symmetry is given by the same formula. Explain the reason for this "equivalence" of disk and cylinder.

Q9.11 Cylindrical container A can pass through hoop B. The two objects are equally massive and rotate about the axis shown in the figure. Without any knowledge of the distribution of mass within container A, why can you be certain that their moments of inertia satisfy the inequality $I_A < I_B$?

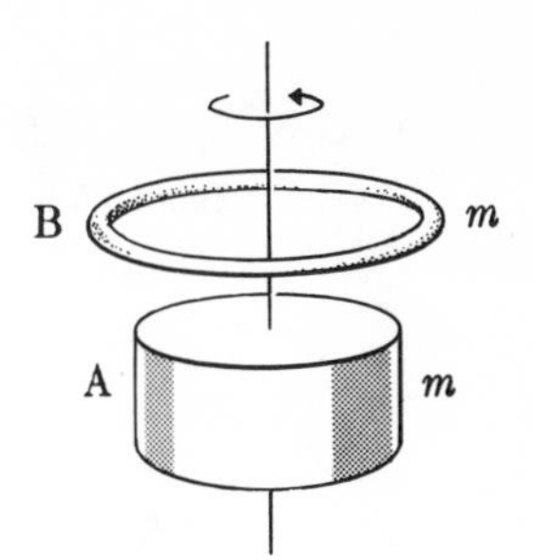

Q9.12 Explain why an axis of symmetry of a rigid body necessarily passes through the center of mass of the body.

Q9.13 How many symmetry axes has a thin, circular hoop? Describe them.

Q9.14 How many symmetry axes has a cylindrical tin can? Describe them.

Q9.15 (1) The angular momentum **L** of a spiraling football on its way to a receiver is approximately constant. Why? (2) Nevertheless, the axis of spin usually wobbles; the angular velocity vector ω is not constant. Interpret this fact. (3) Occasionally, a passer throws a spinning football that does not wobble. What does this mean?

Section 9.5

Q9.16 (1) Verify that the two sides of Equation 9.52 have the same physical dimension. (2) Express the SI unit of torque in terms of kg, m, and sec. What other important physical quantity has the same dimension as torque?

Q9.17 Many scales (such as those found in doctors' offices) balance a large weight with a small weight. Explain the operation of such a scale in terms of torque.

Q9.18 A particle moving uniformly in a circle under the action of a centripetal force experiences no torque and no change of angular momentum. If you tie a weight to a string, you are able to exert on the weight only an inward force along the string toward your hand. How, then, are you able to set the weight into rotational motion—that is, to impart to it some angular momentum?

Q9.19 Let **F** be the total external force and **T** be the total external torque acting on a system. Give examples of systems for which (a) $\mathbf{F} = 0$ and $\mathbf{T} \neq 0$, (b) $\mathbf{F} \neq 0$ and $\mathbf{T} = 0$, (c) $\mathbf{F} \neq 0$ and $\mathbf{T} \neq 0$, and (d) $\mathbf{F} = 0$ and $\mathbf{T} = 0$.

Q9.20 Divide the force of the sun on the earth into many forces acting on different parts of the earth. Explain why the total torque with respect to the center of the sun is precisely zero. (NOTE: This demonstration requires neither the approximation of a spherical earth nor the approximation that all parts of the earth are equally distant from the sun.)

Q9.21 If a body rotates about an axis that is not a principal axis, a torque must be applied to the axis to keep it aligned in a fixed direction. If the rotation takes place about a symmetry axis, no such torque is required. Explain these facts.

Section 9.6

Q9.22 Suggest a simple experimental procedure to measure the moment of inertia of the man shown in Exercise 9.45 with respect to the axis of the turntable on which he stands.

Section 9.7

Q9.23 Two wheels of equal mass and equal physical dimensions are mounted on identical handles supported as shown in Figure 9.25. The wheels differ in that the mass of one is concentrated near its rim; the mass of the other is distributed more uniformly. (1) If the wheels rotate with equal angular momentum, which one, if either, precesses more rapidly? (2) If the wheels rotate with equal angular velocity, which one, if either, precesses more rapidly?

Q9.24 Discuss the fact that it is easy to stay on a bicycle that is rolling but difficult to balance on one that is not rolling.

Section 9.8

Q9.25 While waiting for a traffic light to change, a driver "revs" up his engine and notices that his car lurches slightly to one side as he does so. (1) What causes this effect? (2) Is the angular momentum of the car, including its engine and driver, conserved? Does any external torque act on the car? (3) Would it be possible to build an engine that did not exhibit this lurching effect?

Q9.26 Why do divers and trapeze artists double up in a ball when they want to make several turns in the air? Why does a diver completing a double somersault seem to stop turning as he straightens out just before entering the water? Does he, in fact, stop turning?

Q9.27 A falling cat with zero angular momentum about its own center of mass can turn and land on its feet. It changes its angular position without changing its angular momentum. How can it do so? (HINT: Think of the front and rear halves of the cat as two connected parts of a single system. Only the sum of the angular momenta of the two parts need remain zero.)

Q9.28 An artificial satellite mounted atop a rocket on the launching pad at Cape Kennedy has a small angular momentum (owing to the fact that it is attached to the rotating earth). After it is injected into orbit, it has a considerably larger angular momentum. Carefully reconcile the two statements, "Angular momentum is conserved," and "The angular momentum of the satellite is changed."

Q9.29 When a boatload of ore is transported from Alaska to Peru, does the length of the day change? Why or why not?

Q9.30 For exercise, a crew of astronauts take along a bicycle. One of the astronauts gets on the bicycle. Initially, he and the bicycle float in midcabin, motionless with respect to the spaceship. Then he starts to pedal. Describe the resulting motion. Is angular momentum conserved? What happens when he brakes and the wheels stop turning?

Q9.31 A student on a nearly frictionless turntable holds a spinning wheel as shown. Initially, the turntable is not rotating. What happens as the student does each of the following things in turn? (1) He moves the axle of the wheel farther from the axis of the turntable and then brings it back. (2) He turns the wheel over. (3) He turns it back to its original position. (4) He touches the wheel to his jacket and stops it with friction.

Q9.32 Can an isolated body change its angular velocity? Illustrate the possibility or impossibility of such a change with an example.

Q9.33 The spin of a proton is $\frac{1}{2}\hbar$. If two protons approach one another with relative orbital angular momentum zero, what is the greatest magnitude of relative orbital angular momentum with which they can separate?

Q9.34 A girl holds a spinning wheel by handles attached to its axles as shown. She pushes forward with her left hand and pulls back with her right hand. The handles respond by exerting forces on her hands—but not horizontal forces. One handle pushes up, the other down. (1) Why is this? (2) Does this violate Newton's third law? (3) If the wheel is spinning as shown, which handle seemingly "wants" to go up? which "wants" to go down?

Section 9.9

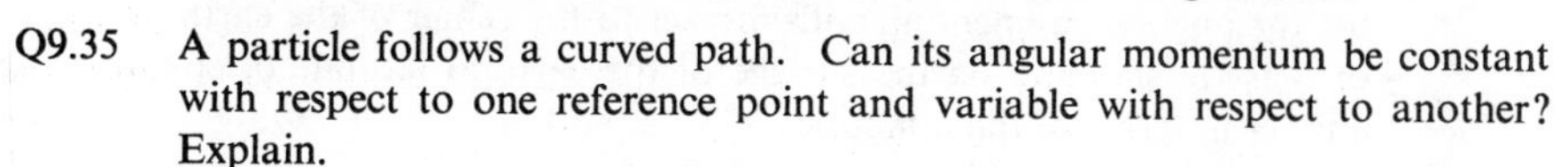

Q9.35 A particle follows a curved path. Can its angular momentum be constant with respect to one reference point and variable with respect to another? Explain.

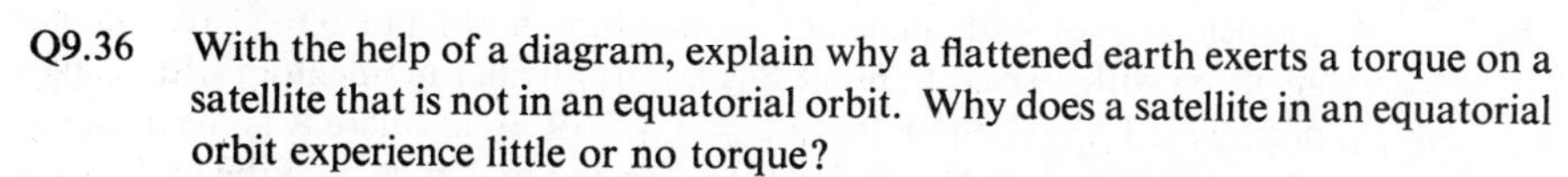

Q9.36 With the help of a diagram, explain why a flattened earth exerts a torque on a satellite that is not in an equatorial orbit. Why does a satellite in an equatorial orbit experience little or no torque?

Section 9.10

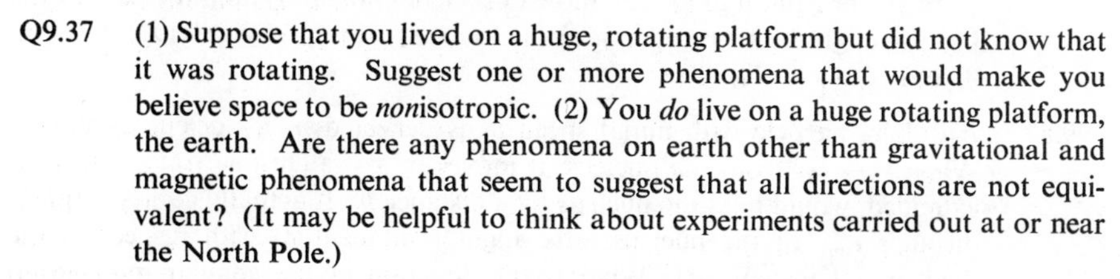

Q9.37 (1) Suppose that you lived on a huge, rotating platform but did not know that it was rotating. Suggest one or more phenomena that would make you believe space to be *non*isotropic. (2) You *do* live on a huge rotating platform, the earth. Are there any phenomena on earth other than gravitational and magnetic phenomena that seem to suggest that all directions are not equivalent? (It may be helpful to think about experiments carried out at or near the North Pole.)

Q9.38 An experimenter in a windowless room finds that if he places an object on a frictionless surface, the object "spontaneously" starts to rotate. Suggest one radical and one conservative interpretation for this behavior.

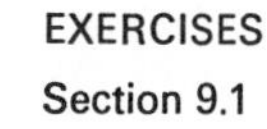

EXERCISES

Section 9.1

E9.1 A particle moves in the xy plane. (1) Show that its z component of angular momentum is expressed in terms of its coordinates (x and y) and its components of velocity (v_x and v_y) by

$$L_z = m(xv_y - yv_x).$$

(2) How do L_z and the magnitude L differ, if at all?

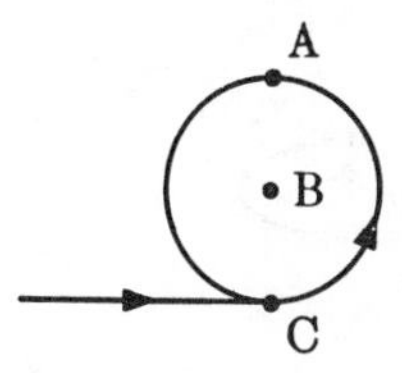

E9.2 In the diagram a proton ($m = 1.67 \times 10^{-27}$ kg) is injected into a circular synchrotron at point C with a speed of 10^7 m/sec. The radius of the synchrotron is 50 m. What is the initial angular momentum of the proton (a) with respect to point A? (b) with respect to point B? (c) with respect to point C? (Give directions as well as magnitudes.)

E9.3 A particle with position vector **r** and velocity **v** has angular momentum **L**. (1) Show that $\mathbf{r} \cdot \mathbf{L} = 0$ and that $\mathbf{r} \times \mathbf{L}$ is a vector in the plane defined by **r** and **v**. (2) Is $\mathbf{r} \times \mathbf{L}$ a polar vector or an axial vector?

E9.4 Name systems that have magnitudes of angular momenta given very roughly by (a) 10^{-34} kg m^2/sec, (b) 1 kg m^2/sec, and (c) 10^{+34} kg m^2/sec. Justify your answers with approximate calculations.

E9.5 (1) Demonstrate the equivalence of Equations 9.6 and 9.7 ($L = r_\perp p$ and $L = rp_\perp$). (2) Which among the quantities r, p, $r_\perp$, and $v_\perp$ are constant for (a) motion with constant velocity, such as that shown in Figure 9.6, (b) motion with constant speed in a circle about the origin, and (c) motion with variable speed in a circle about the origin?

E9.6 Let θ represent the latitude angle of a point on the earth (zero at the equator, +90 deg at the North Pole, −90 deg at the South Pole), R the radius of the earth, and T the period of rotation of the earth. Find an algebraic expression for the angular momentum with respect to the center of the earth of a man of mass m standing on the surface of the earth at latitude θ. Express the answer in terms of the quantities θ, R, T, and m.

E9.7 A particle moves with constant momentum $\mathbf{p}$ (Figure 9.6). Its angular momentum with respect to point P is $\mathbf{L}_P$; its angular momentum with respect to point Q is $\mathbf{L}_Q$. (1) Show that $\mathbf{L}_P = \mathbf{L}_Q + \mathbf{R} \times \mathbf{p}$, where $\mathbf{R}$ is the displacement vector from P to Q. (2) Show in sketches how P, Q, and the path of the particle might be arranged to give (a) $L_P > L_Q$, (b) $L_P < L_Q$, (c) $L_P = L_Q \neq 0$, and (d) $L_P = L_Q = 0$.

E9.8 An alpha particle with initial speed v_0 is deflected by a nucleus as shown. When very far from the nucleus, it moves in a straight-line track, which, if extended, would miss the nucleus by a distance b. It actually comes within a distance r_{min} of the nucleus. Its angular momentum with respect to the nucleus is constant. (1) What is the direction of the angular momentum vector? (Be careful.) (2) What is its magnitude? (3) If $r_{min} = 2b$, what is the ratio of the speed v at the point of closest approach to the speed v_0 at great distance?

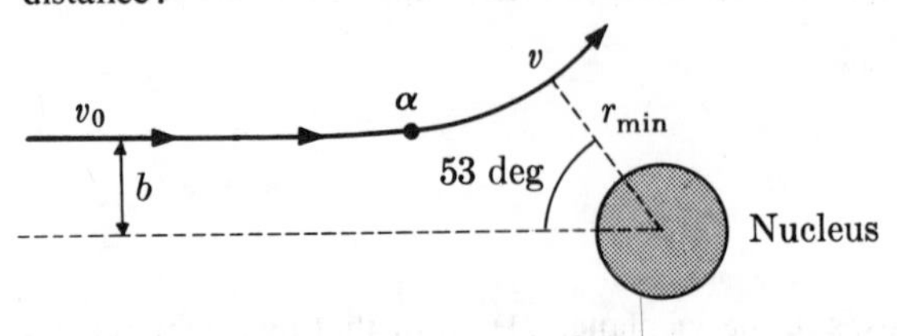

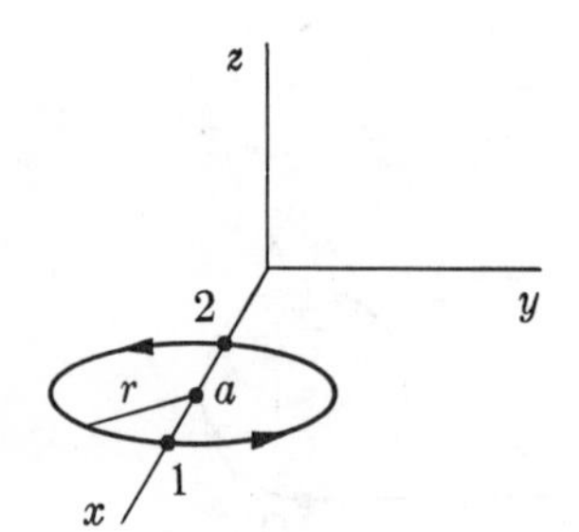

E9.9 A particle of mass m moves with constant speed v in a circle of radius r. The circle lies in the xy plane and, as shown in the fig, its center is at $x = a$. The angular momentum of the particle with respect to the origin is $\mathbf{L}_1$ when the particle crosses the x axis at point 1 and $\mathbf{L}_2$ when the particle crosses the x axis at point 2. Prove that $\mathbf{L}_1 - \mathbf{L}_2 = 2mva\mathbf{k}$. Be sure that your proof is valid for all values of a.

E9.10 The coordinates of a particle of mass m moving in the xy plane are given as a function of time by $x = v_0 t$ and $y = A \sin \omega t$, where v_0, A, and ω are constants. (1) Sketch the path of the particle. (2) Obtain an expression for L_z, the z component of the particle's angular momentum. (3) At what times and at what positions does $dL_z/dt = 0$?

E9.11 As shown in Figure 9.8, the position vectors of a set of particles can be written $\mathbf{r}_i = \mathbf{r}_c + \mathbf{r}'_i$ (the notation is defined in Section 9.2). By making use of this vector equality, show that

Section 9.2

$$\sum_i m_i \mathbf{r}'_i = 0.$$

Express in words the meaning of this result.

E9.12 Four particles of equal mass m have the positions and velocities shown in the following table.

Particle	x	y	v_x	v_y
1	0	0	0	$-v_0$
2	a	a	$-v_0$	0
3	$2a$	0	0	v_0
4	a	$-a$	v_0	0

(1) Where is the center of mass of this system of particles? (2) What is the center-of-mass velocity? (3) With respect to the point $x = -b$, $y = 0$, calculate the angular momentum of each particle and the angular momentum of the system. (4) Comment on the relationship of these results to the theorem proved in Section 9.2

E9.13 Imagine an observer A stationed on one of the two bodies shown in Figure 9.9 and shown here to larger scale. (1) If A considers himself to be stationary at the center of the universe, what speed does he assign to body B? (2) What angular momentum does he assign to body B? Note that this is *not* the same as the angular momentum $L = mvd$ calculated in the example in Section 9.2. (3) Reconcile this result with the theorem in Section 9.2. (HINT: The reconciliation does depend on the fact that A is moving, but it does not depend on the fact that he is in an accelerated frame of reference.)

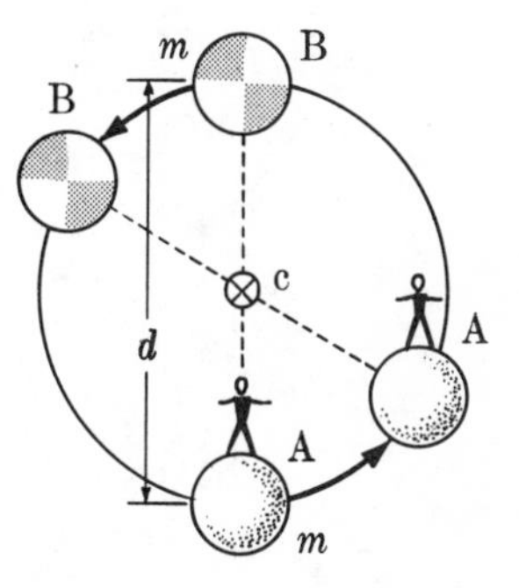

E9.14 At a particular instant, two particles of equal mass m have the following positions and velocities.

Section 9.3

Particle	x	y	v_x	v_y
1	0	0	$-v_0$	$-v_0$
2	0	d	v_0	$3v_0$

(1) Find the position and velocity of the center of mass of this two-particle system. (2) Find the orbital, spin, and total angular momenta of the system.

E9.15 A dumbbell consisting of two 5-kg weights joined by a light rod of length 0.4 m is sent spinning across the surface of a frozen lake. In a coordinate system on the surface of the lake, its center-of-mass vector is $\mathbf{r}_c = v_0 t\mathbf{i} + y_0\mathbf{j}$, with $v_0 = 2$ m/sec and $y_0 = 1$ m. The dumbbell spins clockwise about its center of mass with angular speed $\omega = 50$ radian/sec. Find its orbital, spin, and total angular momenta.

E9.16 The earth and the moon circle about their center of mass, which is 4,660 km from the center of the earth. Which of the two bodies has the greater orbital angular momentum with respect to their center of mass?

E9.17 A bug of mass m is clinging to the edge of a wheel of radius a that is rotating at angular speed ω about a fixed axle. (1) With respect to a reference point located at the axle, what are (a) the orbital angular momentum and (b) the spin angular momentum of the bug? (NOTE: The bug alone is the "system" of interest. For calculational purposes, treat the bug as a point particle.) (2) A second bug alights on the wheel at the opposite end of a diameter from the first bug. Keeping the origin at the axle, find the orbital and spin angular momenta of the two-bug system.

E9.18 Assume the approximate validity of classical mechanics in describing particle motion, and consider a neutron moving along a straight line with a speed of 10^6 m/sec past a stationary proton located at the origin of a coordinate system. At its point of closest approach, the neutron comes to within 3.78×10^{-13} m of the proton. (1) What is the orbital angular momentum of the neutron with respect to the proton? Express your answer in quantum units of $\hbar$. (2) Show in a sketch how the spins of neutron and proton must be aligned to maximize the total angular momentum of this system with respect to the chosen origin. With this alignment, what is the magnitude of the total angular momentum?

E9.19 A girl is rolling a plastic hoop along the sidewalk at a speed of 1.5 m/sec. The hoop has a diameter of 1 m and a mass of 0.2 kg. (1) What is the angular speed in radian/sec of the hoop about its center? Give both an algebraic and a numerical answer. (2) What is the angular momentum of the hoop with respect to its center?

E9.20 A cylinder of radius r rolls without slipping on a flat surface; the speed and acceleration of its center of mass are v_c and a_c. (1) Show that the angular acceleration of the cylinder with respect to its center of mass is $\alpha = a_c/r$. (2) What are its angular speed and angular acceleration with respect to its point of contact with the ground?

E9.21 A hoop of radius 0.5 m and mass 1 kg rolls on level ground around a circular path of radius 3 m, completing one circuit in 10 sec. The hoop remains vertical as it rolls. (1) What is the orbital angular momentum of the hoop with respect to the center of its circular path? (2) What is the spin of the hoop? Give directions as well as magnitudes.

Section 9.4

E9.22 A rigid body of arbitrary shape rotates with angular velocity ω about a fixed z axis. A point in the body with position vector $\mathbf{r}_i$ moves with velocity $\mathbf{v}_i$ (see Figure 9.12). Show that $\mathbf{v}_i = \boldsymbol{\omega} \times \mathbf{r}_i$. Does this result remain valid if the z axis does not pass through the body?

E9.23 Show that the formula for the moment of inertia of a thin hoop, $I = MR^2$ (Equation 9.30), follows from the general formula $I = \int a^2\, dm$ (Equation 9.37). What are the appropriate limits of integration?

R

M

E9.24 What is the moment of inertia of a circular ribbon rotating about the axis shown in the figure?

E9.25 A pair of parallel rods of mass m, length l, and separation $2a$ rotate about an axis parallel to the rods and half way between them (see the figure). What is the moment of inertia of the combination?

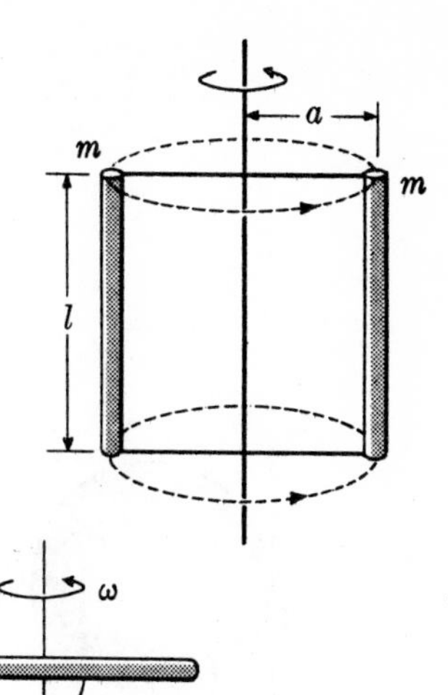

E9.26 A thin rod of mass M and length l rotates about an axis that is perpendicular to the rod and passes through the center of the rod. Show that the moment of inertia of the rod is $I = \frac{1}{12}Ml^2$.

E9.27 Two thin rods of equal mass and equal length rotate about an axis. As shown in the figure, the axis passes through the center of rod A and through the end of rod B; rod A makes a fixed angle $\pi/2$ with the axis and rod B makes a fixed angle α with the axis. The two rods have identical moments of inertia. (1) What is the angle α? (2) Which rod extends a greater distance from the axis? (Use the results of Exercise 9.26 and of Example 3 in Section 9.4.)

E9.28 A wheel with a mass of 10 kg and an outer radius of 0.3 m has a moment of inertia of 0.25 kg m^2. (1) If the same mass were distributed uniformly in a disk with the same moment of inertia, what would be the radius of the disk? (2) If the same mass were concentrated at the rim of a wheel with the same moment of inertia, what would be the radius of that wheel? (3) What sort of distribution of mass must the original wheel have?

E9.29 (1) Ten kilograms of aluminum are fashioned into a solid sphere. (a) What is the radius of the sphere? (b) What is its moment of inertia about an axis through its center? (The density of Al is $\rho = 2.7$ gm/cm^3 $= 2.7 \times 10^3$ kg/m^3.) (2) The same quantity of Al is formed into a circular disk 1 cm thick. What is its radius? What is its moment of inertia about its axis of cylindrical symmetry?

E9.30 Prove the following additive property of moments of inertia: If object 1 has moment of inertia I_1 for rotation about a particular axis and object 2 has moment of inertia I_2 for rotation about the same axis, the combined system rotating about that axis has moment of inertia $I = I_1 + I_2$. (HINT: Look at Equation 9.36.)

E9.31 What is the moment of inertia of a uniform rectangular plate of sides a and b and mass M rotating about the axis shown in the figure? Why is the dimension b "irrelevant"?

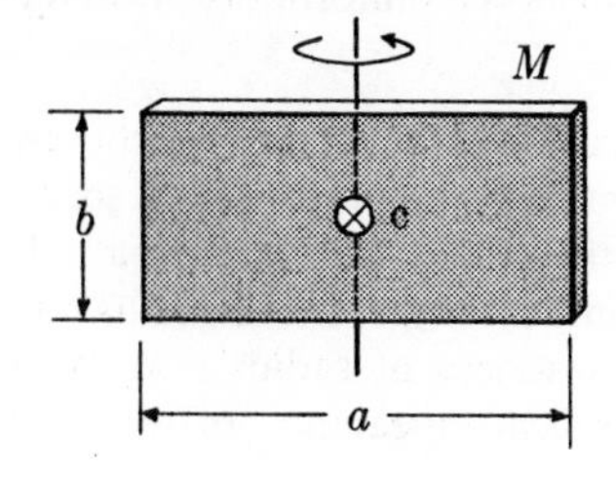

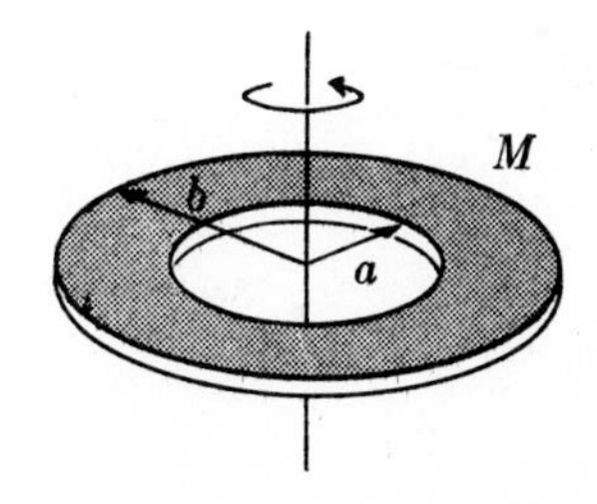

E9.32 An annular disk of mass M and uniform density, with inner radius a and outer radius b, rotates about its axis of cylindrical symmetry (see the figure). Find its moment of inertia. (HINT: Refer to Example 4 in Section 9.4.)

E9.33 Two weights of equal mass m are held a fixed distance apart by a light rod of length l. (1) An axis of rotation passes through the center of mass of this system. What is the total angular momentum of the rotating weights? What is their moment of inertia? (2) The axis of rotation is displaced so that it passes through one of the weights. Evaluate the total angular momentum of the system as the sum of a spin part and an orbital part. What is the moment of inertia of the system for rotation about this displaced axis? (In both parts, take the axis to be perpendicular to the road.)

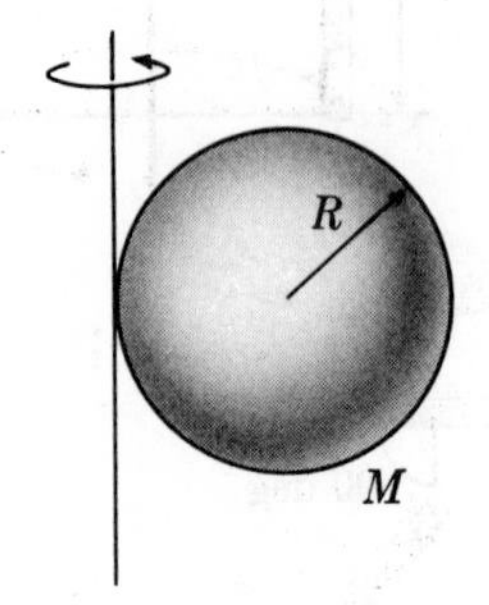

E9.34 The spin angular momentum of a uniform sphere of mass M and radius R rotating with angular velocity ω is $\mathbf{L} = \frac{2}{5}MR^2\omega$ (see Equation 9.44). (1) What is the orbital angular momentum of a sphere rotating about an axis tangent to its surface (see the figure)? What is its total angular momentum? (2) What is the moment of inertia of the sphere rotating about this axis? By what factor does it differ from the moment of inertia of a sphere rotating about an axis through its center?

Section 9.5

E9.35 A wheel of radius 0.5 m requires a minimum torque of 2 N m to set it in motion. At what angle should a force of 6 N be applied at the rim of the wheel in order to provide this minimum torque?

E9.36 In Figure 9.21, let $l = 0.5$ m and $F = 250$ N. (1) If $\mathbf{F}$ and $-\mathbf{F}$ are perpendicular to the length of the handle, what is the magnitude of the torque $\mathbf{T}$ being exerted by the man? (2) If he is unable to turn the handle because of a frictional force that acts 1.5 cm from the axis, what is the magnitude of this frictional force? (3) Would the man stand a better chance of turning the handle if he moved his hands or changed the directions of the forces he applies?

E9.37 A boy tosses a stone of mass 0.08 kg into the air at an angle of 45 deg to the horizontal with an initial speed of 14 m/sec. It lands 20 m away, again with a speed of 14 m/sec and at an angle of 45 deg to the horizontal. (1) What is the angular momentum of the stone with respect to the boy at the moment it leaves his hand? (2) What is its angular momentum with respect to the boy just before it lands? (3) Why does the angular momentum of the stone change?

E9.38 A heavy flywheel with negligible friction has a radius of 0.3 m. In order to start the wheel spinning, a man wraps a rope around its periphery and pulls on the rope for 3 sec with a force of 1.5 N. (1) What is the final angular momentum of the flywheel? (2) If the wheel is a uniform disk with a mass of 40 kg, what is its final angular speed?

E9.39 The *radius of gyration* of an object of mass m with respect to a chosen axis is defined to be a distance k such that the moment of inertia with respect to that axis can be written $I = mk^2$. (1) Give the radii of gyration of (a) a hoop of radius r about its axis of cylindrical symmetry, (b) a disk of radius r about its axis of cylindrical symmetry, and (c) a sphere of radius r about an axis through its center. (2) Explain the physical "meaning" of the concept of radius of gyration.

Section 9.6

E9.40 For motion with constant angular acceleration, obtain a formula relating angle and angular velocity. Like Equation 5.58, this formula should not contain time.

E9.41 The wheels of a large airplane are set spinning before the plane lands in order to minimize tire wear. Suppose that the touchdown speed of a plane is 60 m/sec and that each wheel, including the tire, has an outer radius of 0.8 m, a mass of 400 kg, and a moment of inertia of 180 kg m^2. Suppose further that the wheel is uniformly accelerated from rest to the "correct" angular speed in 25 sec. Find each of the following quantities: (a) the final angular speed of the wheel, (b) its angular acceleration, (c) the number of revolutions of the wheel as it is being accelerated, (d) its final angular momentum, (e) the torque applied to it, and (f) the force needed to produce this torque if the force is applied 0.2 m from the axis of rotation.

Section 9.7

E9.42 The bicycle wheel pictured in Figure 9.25 has a mass of 2.5 kg. The distance from its support point to its center is $r = 0.2$ m. (1) With respect to its support point, what is the torque produced by gravity? (2) If the magnitude of the spin angular momentum of the wheel is $L_{spin} = 12$ kg m^2/sec, through what angle does the wheel precess in 1 sec? (3) Is the approximation $L_{spin} \gg L_{orbital}$ well satisfied (with the origin at the support point)?

Section 9.8

E9.43 (1) Find the angular momentum of the exhaust gas in the example in Section 9.8. Verify that it is the same as the initial angular momentum of the spacecraft. (2) Does a torque act on this gas as it is being shot out?

E9.44 Disk A, of radius r and mass m, is spinning initially at angular speed ω about the axis shown; disk B, of radius $2r$ and mass $4m$, is initially stationary. The disks are then brought together and they start to rotate as a unit at angular speed ω_f. (1) Assuming angular momentum to be conserved, find ω_f. (2) Will frictional forces between the surfaces of the disks in fact prevent the total angular momentum of this system from being conserved? Why or why not?

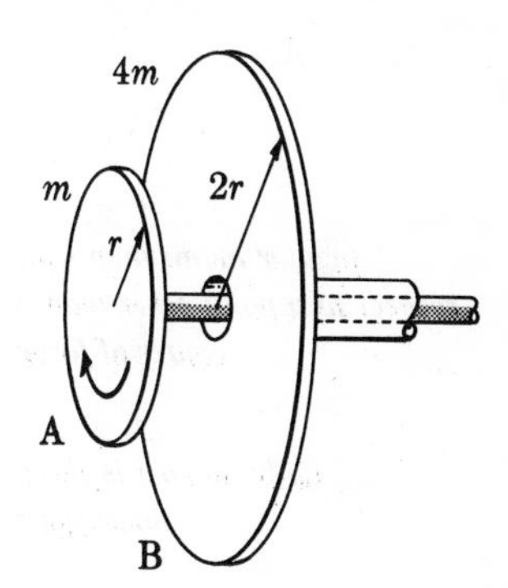

E9.45 Standing at rest on a nearly frictionless turntable, an instructor hurls an eraser tangentially, as shown in the figure. The 0.1-kg eraser leaves his hand with a speed of 12 m/sec at a distance of 0.8 m from the axis of the turntable. (1) What angular momentum does the system of instructor plus turntable acquire? (2) Estimate roughly the instructor's resulting angular speed. (Make reasonable guesses for any required magnitudes.)

E9.46 A student standing on a small turntable with his arms held straight out is turning at 5 radian/sec. He then brings his arms in to minimize his moment of inertia. (1) Make realistic estimates of all needed magnitudes and calculate his new angular speed. (2) Would the calculation be significantly affected if the student were holding 2-kg weights in each hand? (If the opportunity presents itself, check your conclusions experimentally.)

Section 9.9

E9.47 A comet moves in an orbit of eccentricity $e = 0.90$. If its speed at its greatest distance from the sun is v_0, what is its speed at its point of nearest approach to the sun? (Refer ahead to Figure 11.12 and Equation 11.36.)

E9.48 An earth satellite has a perigee that is 110 miles above the earth's surface and an apogee that is 220 miles above the surface. What is the ratio of the velocity at perigee to that at apogee?

E9.49 The velocity of a satellite at apogee is perpendicular to the radial line joining the satellite to the center of the earth. This seemingly obvious fact (see Figure 9.31) requires proof. To prove it, assume that $\mathbf{v}$ is *not* perpendicular to $\mathbf{r}$ and show that this assumption is incompatible with the given fact that the satellite is at apogee.

PROBLEMS

Angular momentum with respect to a center of force

P9.1 Equations describing the trajectory of a particle of mass m moving in the xy plane are $x = a \cos \omega t$, $y = b \sin \omega t$. (1) Sketch the path of the particle. (2) Write vector expressions for the position and velocity of the particle. (3) Show that its angular momentum is a constant given by $\mathbf{L} = mab\omega\mathbf{k}$. (4) Use Equation 9.81 to find an expression for the area enclosed by the trajectory of the particle. (SUGGESTION: Integrate Equation 9.81 over an appropriate time interval.)

Angular momentum with respect to a point other than a center of force

P9.2 (1) Find the angular momentum of the particle described in the preceding problem with respect to the point $x = -d$, $y = 0$. (2) With respect to this point, what torque acts on the particle? (3) For $d = a$, identify the points of maximum and minimum L and the points of maximum and minimum T.

Body moving in three dimensions

P9.3 A boy on a merry-go-round follows a path specified by the equations $x = a \cos \omega t$, $y = a \sin \omega t$, and $z = d \sin 5\omega t$. The origin is on the central axis of the merry-go-round, and $d \ll a$. (1) Give a brief description in words of the boy's path in space. (2) Obtain an expression for his angular momentum $\mathbf{L}$. (3) Show that the magnitude of his angular momentum varies between maximum and minimum values given by $L_{\text{max}} = m\omega a^2\sqrt{1 + (5d/a)^2}$ and $L_{\text{min}} = m\omega a^2\sqrt{1 + (d/a)^2}$, where m is the mass of the boy. (HINT: The quantity L^2 is extremal when L is extremal. It is easier to work with L^2 than with L.)

Two-body systems

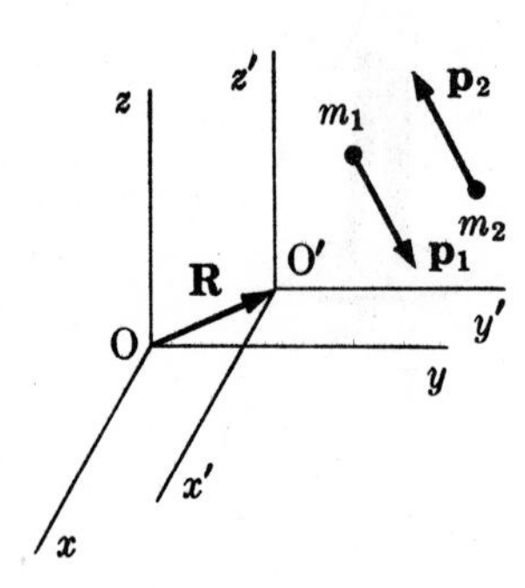

P9.4 As shown in the figure, two coordinate systems are separated by a fixed displacement $\mathbf{R}$. Two particles have angular momenta $\mathbf{L}_1$ and $\mathbf{L}_2$ with respect to origin O and angular momenta $\mathbf{L}_1'$ and $\mathbf{L}_2'$ with respect to origin O′. The particles move with equal and opposite momenta ($\mathbf{p}_1 = -\mathbf{p}_2$). (1) Show that $\mathbf{L}_1 + \mathbf{L}_2 = \mathbf{L}_1' + \mathbf{L}_2'$ even though $\mathbf{L}_1$ and $\mathbf{L}_1'$ are not equal in general and $\mathbf{L}_2$ and $\mathbf{L}_2'$ are not equal in general. (2) What can you say about the center of mass of this system?

P9.5 As shown in the figure, two particles with unequal masses m_1 and m_2 execute circular orbits about their stationary center of mass. (1) Obtain expressions for their angular momenta $\mathbf{L}_1$ and $\mathbf{L}_2$ and for the ratio L_1/L_2. (2) For the effective one-body problem, angular momentum is defined by $\mathbf{L} = \mu\mathbf{r} \times \mathbf{v}$, where $\mathbf{r} = \mathbf{r}_2 - \mathbf{r}_1$, $\mathbf{v} = \mathbf{v}_2 - \mathbf{v}_1$, and μ is the reduced mass, equal to $m_1m_2/(m_1 + m_2)$. Show that $\mathbf{L} = \mathbf{L}_1 + \mathbf{L}_2$. (It may be helpful to refer to Section 8.11.) *Optional:* Establish the equality $\mathbf{L} = \mathbf{L}_1 + \mathbf{L}_2$ for noncircular motion of the particles about their stationary center of mass.

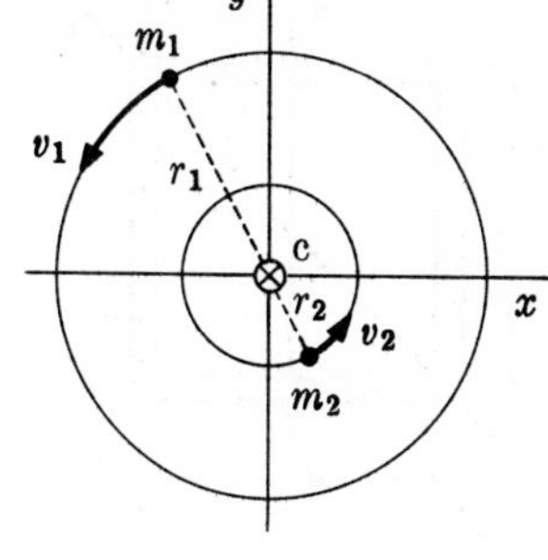

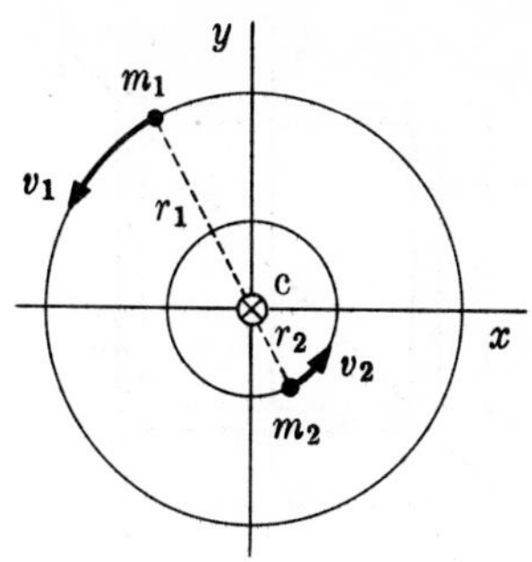

Orbital and spin angular momentum

P9.6 The ends of two light rods of equal length r are pivoted at the origin. Attached to the other ends of the rods are weights of equal mass m. The rods rotate at angular speed ω about the origin and maintain a fixed angular separation θ as they do so (see the figure). Obtain expressions for the orbital, spin, and total angular momentum of the system. Discuss the two limits $\theta = 0$ and $\theta = \pi$.

Moments of inertia

P9.7 (1) As shown in the figure, a sphere of diameter d and mass M rotates about an axis that passes through the sphere. Show that the moment of inertia of the sphere must have the form $I = \gamma Md^2$, with $\gamma < 1$. (2) If the axis is outside the sphere, can γ be greater than 1? Can it be equal to 1?

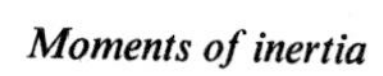

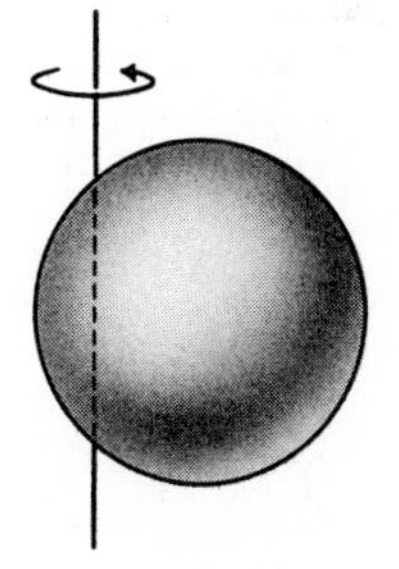

P9.8 A thin rod of mass M and length l rotates about an axis perpendicular to its length (see the figure). The point of intersection of the axis and the rod is at distance x from the center of the rod; this point of intersection is chosen to be the origin. Find the moment of inertia of the rod as a function of x for $0 \le x \le \frac{1}{2}l$. Sketch a graph of I vs x.

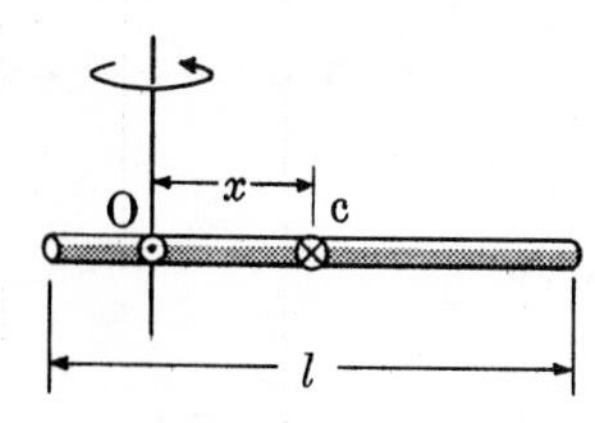

P9.9 A cylindrical can half full of water is placed on its side in a freezer. When it is later put in a warm room, a thin layer of ice inside the can melts, permitting the slug of ice to rotate within the can (see the figure). Show that the moment of inertia of the ice with respect to the center of the can C is $I = \frac{1}{2}MR^2$. (M is the mass of the ice and R is the inner radius of the can.)

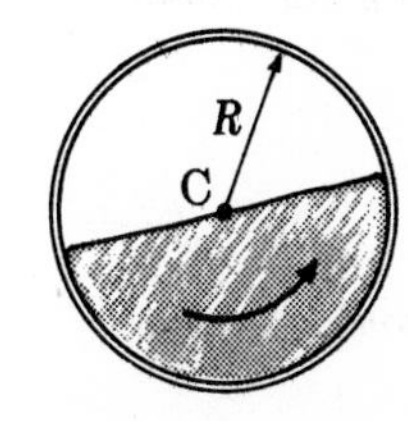

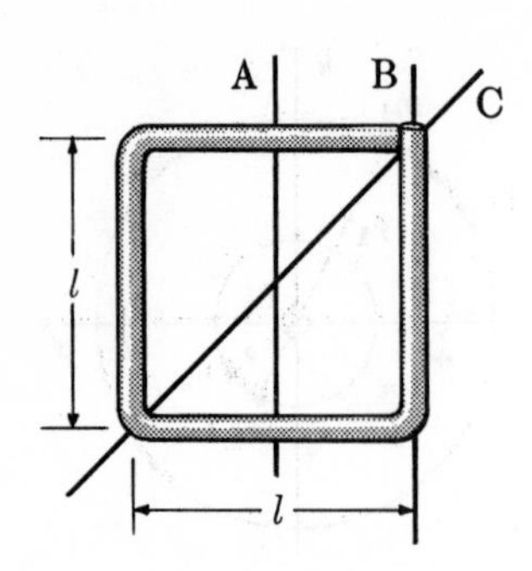

P9.10 A uniform metal rod of mass M and length $4l$ is bent into the shape of a square (see the figure). What is its moment of inertia for rotation about (a) axis A? (b) axis B? (c) axis C?

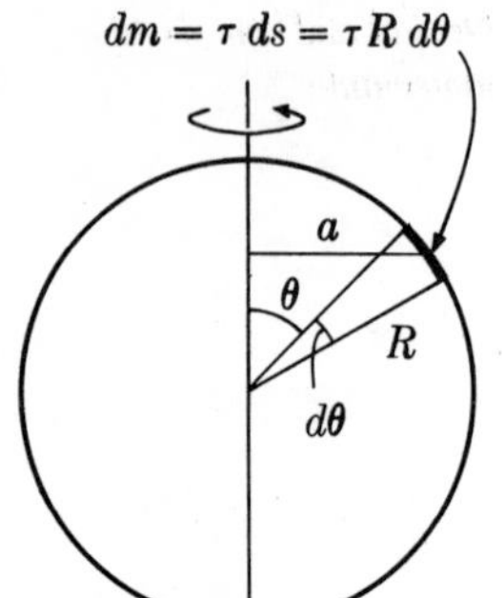

P9.11 A thin hoop of radius R and mass M rotates about an axis in its plane that passes through its center. The mass per unit length around its circumference is designated by τ ($\tau = M/2\pi R$). Prove that its moment of inertia is

$$I = \tfrac{1}{2}MR^2.$$

(SUGGESTION: Use Equation 9.37. An expression for the increment of mass dm is given in the figure. It may be necessary to consult a table of definite integrals.)

P9.12 A thin, uniform disk of radius R and mass M rotates about an axis in its plane that passes through its center. The mass per unit area in the plane of the disk is designated by σ ($\sigma = M/\pi R^2$). Prove that its moment of inertia is

$$I = \tfrac{1}{4}MR^2.$$

(SUGGESTION: Use Equation 9.37. An expression for an increment of mass dm at distance a from the axis is given in the figure. It may be necessary to consult a table of definite integrals.)

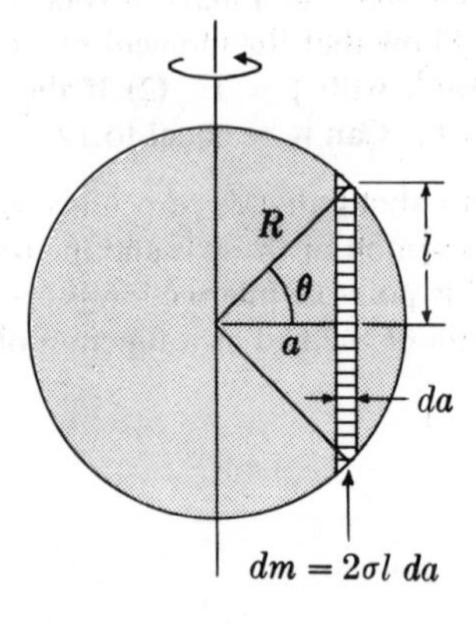

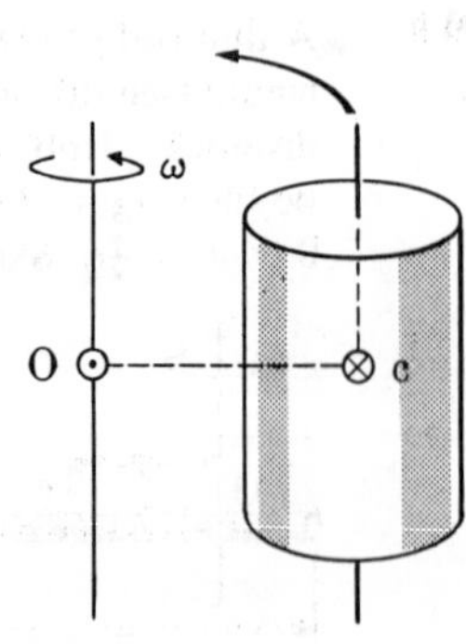

A principal-axis theorem

P9.13 As shown in the figure, a body rotates about an axis parallel to one of its symmetry axes. The origin O lies on the axis of rotation. Prove that this axis of rotation is a principal axis (one for which **L** is parallel to $\boldsymbol{\omega}$) if a line joining O and the center of mass of the body is perpendicular to the axis. (HINT: Use $\mathbf{L} = \mathbf{L}_{\text{orbital}} + \mathbf{L}_{\text{spin}}$.)

The parallel-axis theorem

P9.14 A body of mass M rotates about an arbitrary fixed axis A. Prove that its moment of inertia is given by

$$I = I_0 + Mb^2,$$

where I_0 is its moment of inertia about a parallel axis B passing through the center of mass and b is the distance between these axes. (Note the implication that $I \geq I_0$.) *Method*: Work from Equation 9.36. As noted in the figure, it is convenient to introduce displacement vectors $\mathbf{a}_i$ from axis A to an element of mass m_i, $\mathbf{a}_i'$ from axis B to the element of mass, and $\mathbf{b}$ from axis A to axis B. Recall that $a_i^2 = \mathbf{a}_i \cdot \mathbf{a}_i$. *Optional:* Repeat the proof using Equation 9.37.

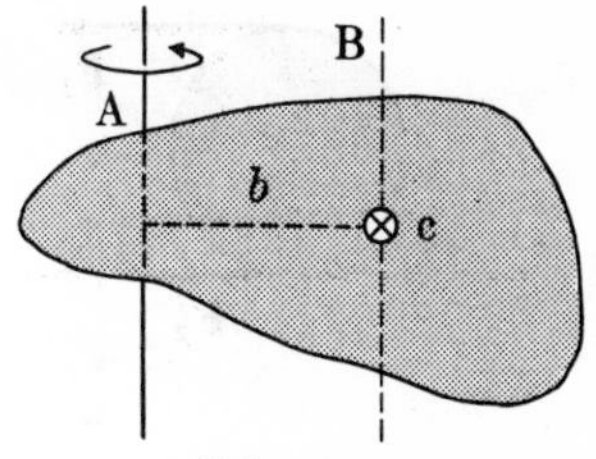

Side view

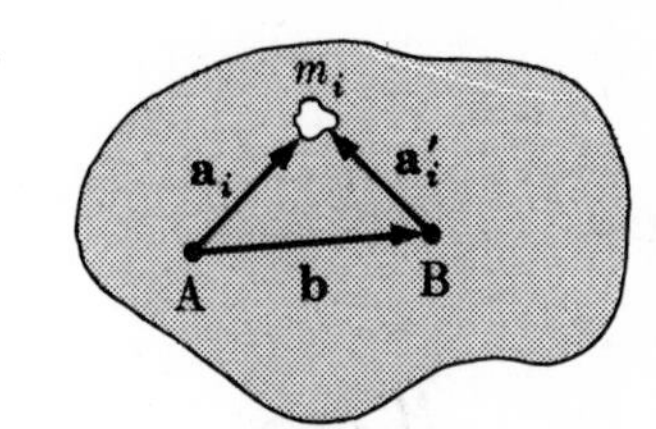

Top view

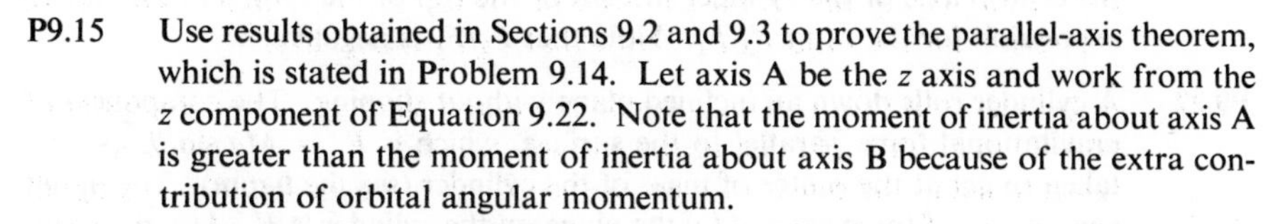
P9.15 Use results obtained in Sections 9.2 and 9.3 to prove the parallel-axis theorem, which is stated in Problem 9.14. Let axis A be the z axis and work from the z component of Equation 9.22. Note that the moment of inertia about axis A is greater than the moment of inertia about axis B because of the extra contribution of orbital angular momentum.

Torque of a couple

P9.16 (1) Prove that the torque of a "couple" (two forces equal in magnitude and opposite in direction) is independent of the choice of origin. (2) Show that the magnitude of the torque is $T = lF$, where F is the magnitude of each force and l is the distance between the lines along which they act (see the figure).

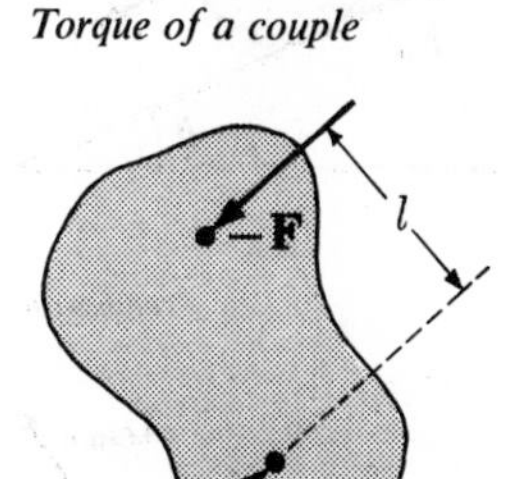

Angular momentum of a projectile

P9.17 A ball of mass m is thrown with initial speed v_0 at an angle θ to the horizontal. Let the origin of a coordinate system be located at the point where the ball is released, with the x axis horizontal, the y axis vertical, and the path of the ball in the xy plane. Neglect air resistance. (1) Write expressions for $x(t)$, $y(t)$, $v_x(t)$, and $v_y(t)$. (2) Obtain a formula for the angular momentum of the ball, $\mathbf{L} = m\mathbf{r} \times \mathbf{v}$. (3) The gravitational force on the ball produces a torque. Find this torque. Note that it is time-dependent. (4) Integrate the equation $\mathbf{T} = d\mathbf{L}/dt$ with respect to time in order to obtain an expression for $\mathbf{L}$. Compare the answers to parts 2 and 4.

Rotational version of Newton's second law

P9.18 (1) Derive the equation relating torque and angular acceleration, $\mathbf{T} = I\boldsymbol{\alpha}$. Explain why it is valid only for rotation about a principal axis. (2) What closely related equation is valid for rotation about any fixed axis? (HINT: Compare Equations 9.26 and 9.40.)

Impulse that produces rotation and translation

P9.19 A pole that stands vertically on the ground is struck an impulsive horizontal blow at its upper end. If there is no friction, what is the direction of motion of the lower end of the pole an instant after the blow is struck? Carefully justify your answer.

Rolling cylinder

P9.20 A cylinder of mass M and radius r rolls without slipping on a horizontal surface. A horizontal force of magnitude F_1 is applied at the top of the cylinder; a force with horizontal component F_2 acts on the cylinder at its

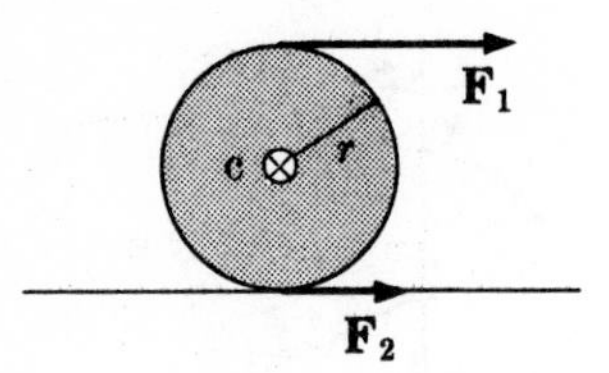

point of contact with the surface (see the figure). (1) Obtain expressions for the acceleration of the center of mass and for the angular acceleration of the cylinder about its center of mass. (2) Taking into account the condition that the cylinder does not slip, show that the ratio of the horizontal components of force is

$$\frac{F_2}{F_1} = \frac{(Mr^2/I) - 1}{(Mr^2/I) + 1},$$

where I is the moment of inertia of the cylinder. (3) Express the same result in terms of the radius of gyration of the cylinder, which is defined in Exercise 9.39.

P9.21 Suppose that the horizontal force $\mathbf{F}_1$ in the preceding problem is applied at the central axis of the cylinder instead of the top of the cylinder. Obtain an expression for the ratio F_2/F_1. Note that F_2/F_1 is negative.

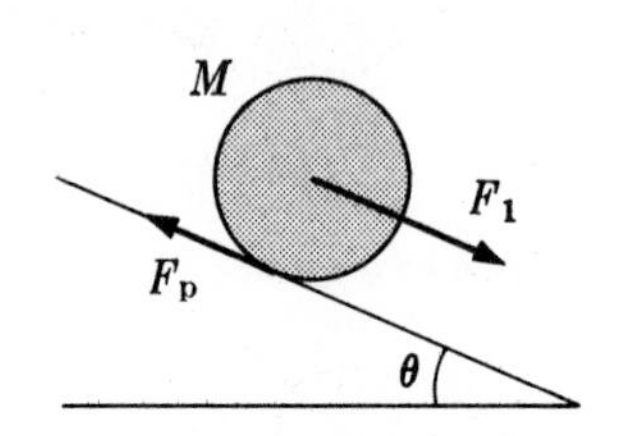

P9.22 A cylinder rolls down an inclined plane without slipping. The component of gravitational force parallel to the surface, which is $F_1 = Mg \sin\theta$, may be taken to act at the center of mass of the cylinder (see the figure). The uphill component of force exerted by the plane on the cylinder is F_p. Use the result obtained in the preceding problem to find the net downhill component of force, $F_1 - F_p$, and the acceleration of the center of mass of the cylinder. Contrast this acceleration with the acceleration of a body sliding without friction down the same plane.

Precession

P9.23 With its axis inclined at a constant angle α to the vertical, the spinning gyroscope shown in the figure precesses about a vertical axis at a constant rate. (1) Prove that the rate of change of its spin angular momentum is given by

$$\frac{d\mathbf{L}_{\text{spin}}}{dt} = \boldsymbol{\omega}_p \times \mathbf{L}_{\text{spin}},$$

where ω_p is the angular velocity of precession. (SUGGESTION: Work separately with magnitudes and directions.) (2) Prove that an equation of identical form,

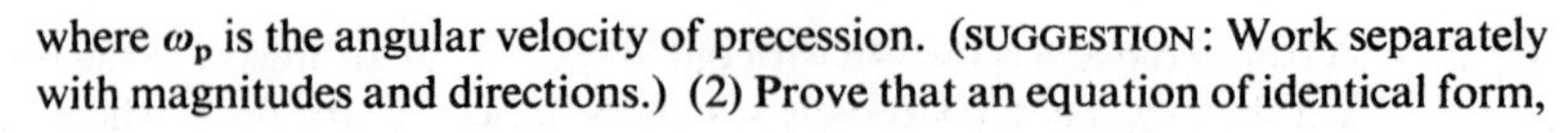

$$\frac{d\mathbf{L}}{dt} = \boldsymbol{\omega}_p \times \mathbf{L},$$

is valid for the total angular momentum $\mathbf{L}$.

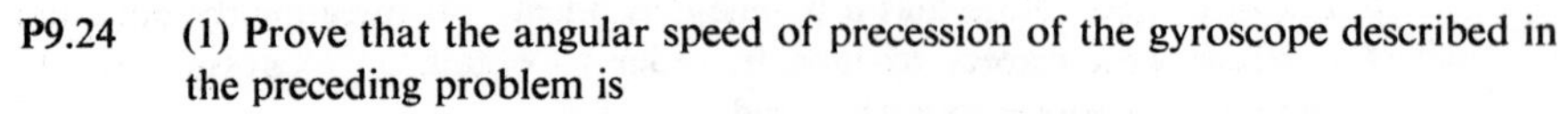

P9.24 (1) Prove that the angular speed of precession of the gyroscope described in the preceding problem is

$$\omega_p = \frac{T}{L_{\text{spin}} \sin\alpha},$$

where T is the magnitude of the gravitational torque applied to the gyroscope. (This result generalizes Equation 9.72.) (2) How does T depend on the angle α? What is the net dependence of ω_p on α?

Rotating rocket

P9.25 A cylindrically symmetric body (it could be a spacecraft or a Fourth of July pinwheel) is caused to rotate by the flow of exhaust gas from a pair of tangential nozzles located at opposite ends of a diameter (see the figure). The distance from the center of the body to each nozzle is r and the exhaust speed relative to the nozzle is v_{ex}. Develop an equation that expresses the conservation of angular momentum for the brief time from t to $t + \Delta t$. From it derive the differential equation

$$\frac{d\omega}{dm} = -\frac{v_{ex}r}{I},$$

in which m is the mass of the body, I is its moment of inertia (a variable), and ω is its angular speed. Assume that the exhaust material is stored at the outer edge of the body before it is ejected. Comment on the connection between this equation and Equation 9.77. (NOTE: The analogous development in Section 8.10 should be helpful.)

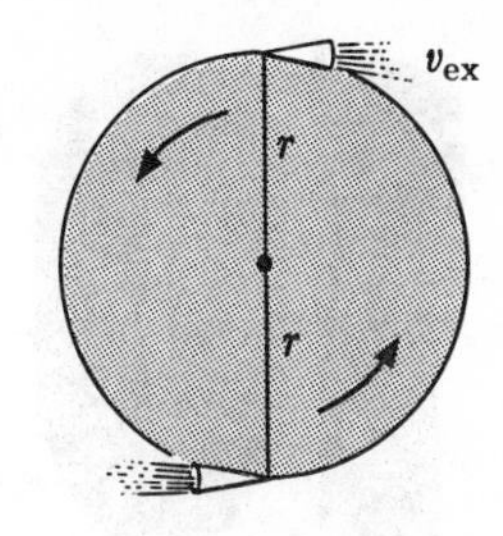

Contraction of a nebula

P9.26 A spherical mass of gas of astronomical dimensions, rotating with initial angular speed ω_0, contracts slowly under the mutual gravitational attraction of its constituent parts. (1) When the mass has contracted to half its original radius, what is its angular speed? Make the simplifying assumption that the mass of gas has uniform density and rotates as a solid body (all parts having the same angular velocity at any one time). (2) Show that the centripetal acceleration of an element of mass near the outer edge of the gas varies in proportion to $1/r^3$, where r is the radius of the sphere. The gravitational force on the same element of mass varies in proportion to $1/r^2$. Discuss the implication of the fact that these two quantities depend on the radius in different ways.

Monkey on a rope

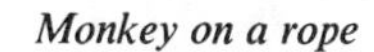

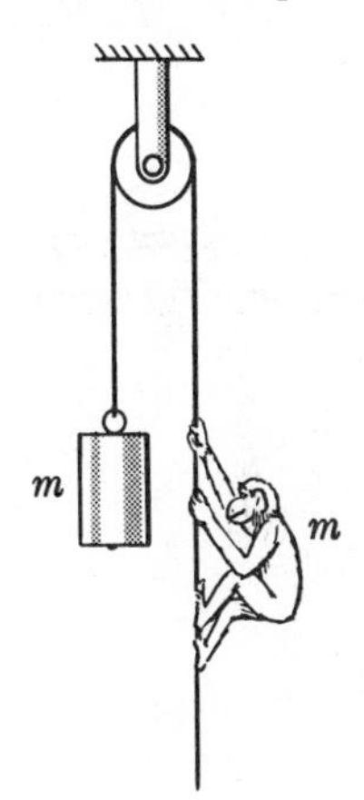

P9.27 A rope goes over a pulley and hangs down on either side. On one end of the rope is fastened a counterweight of mass m; clinging to the rope on the other side is a monkey of equal mass m (see the figure). The masses of the pulley and rope are negligible. (1) Let the system of interest consist of the pulley, rope, counterweight, and monkey. What external forces act on this system? Can these forces sum to zero? Do they necessarily sum to zero? (2) With respect to the axis of the pulley, what is the total external torque? (3) If the monkey starts climbing upward, making good a speed v relative to the pulley, how fast does the rope move, and in which direction?

P9.28 The monkey in the preceding figure is initially at rest exactly opposite the counterweight. Then, after climbing upward for a few seconds, he lets go and falls. (1) What force is needed to support the pulley (a) before the monkey starts to climb, (b) as he is climbing at a constant speed v relative to the pulley, and (c) after he lets go? (2) Which reaches the floor first, the monkey or the counterweight? Assume that the monkey had an upward component of velocity when he left the rope.

P9.29 The monkey of mass m shown in Problem 9.27 is replaced by a monkey of mass $2m$. Initially, the counterweight is held stationary. Then it is released. (1) With respect to the axis of the pulley, what is the total external torque acting on the system? (2) Integrate the equation $d\mathbf{L}/dt = \mathbf{T}$ (with an appropriate initial condition) in order to find $\mathbf{L}(t)$, the angular momentum of the system. (3) Can the monkey move in such a way that the counterweight remains stationary? If so, how does he move? (4) Can the monkey reach the pulley? Explain your reasoning.

Angular momentum of a planet

P9.30 Consider an idealized solar system containing a fixed center of attractive force and a single spherical planet spinning on its axis and rotating about the force center. (1) Explain why the spin angular momentum and orbital angular momentum of the planet are separately constant. (2) Now imagine that the planet is deformed into a nonspherical shape (very nonspherical if you like). Explain why its spin angular momentum need not be constant.

10 Energy

Energy conservation unites the many forms of energy

Energy is a central idea in physics, chemistry, biology—indeed in every area of natural science, as well as in engineering and practical affairs.* It is marked by variety of forms and by a conservation law that unites its many forms. From its humble beginning as a secondary concept in mechanics, energy has grown into what is perhaps the most important unifying idea of natural science.

Figure 10.1 shows schematically some of the interconnections that exist among different forms of energy. Mass energy and kinetic energy in particle transformations were considered in Chapter 4. In this chapter the primary emphasis will be on the mechanical aspects of energy, encircled by a dashed line in Figure 10.1: work, kinetic energy, and potential energy. Energy in these and other forms will recur often in later chapters.

10.1 Work done by a constant force

We begin by defining the concept of work for the special case that a *constant* force acts on a body. The utility of the concept will be made clear in the next section, where a link between work and kinetic energy is established. In Section 10.3, the definition of work will be generalized to encompass arbitrary forces. In these first three sections, we limit our attention to the motion of single particles or of rigid bodies undergoing translational motion without rotation (these two kinds of motion being mechanically equivalent). Systems of particles will be taken up in Section 10.4, and rotational motion will be treated in Section 10.9.

---

* An example: The gross national product of a country is, to a good approximation, proportional to its "consumption" of energy. See S. Fred Singer, "Human Energy Production as a Process in the Biosphere," *Scientific American*, September, 1970.

If a body experiencing a constant force $\mathbf{F}$ suffers a displacement $\mathbf{s}$,* the work done on the body by the force $\mathbf{F}$ is defined to be the scalar product of $\mathbf{F}$ and $\mathbf{s}$:

A definition of work

$$W = \mathbf{F} \cdot \mathbf{s} \qquad (\text{constant } \mathbf{F}). \tag{10.1}$$

This definition at once reveals several significant things about work. It is a scalar quantity. Its SI unit is the newton meter, which is the same as kg m^2/sec^2; as noted previously, this combination is called the joule (J):

$$1 \text{ J} = 1 \text{ N m} = 1 \text{ kg m}^2/\text{sec}^2. \tag{10.2}$$

Work is defined over an interval

Work is defined not at an instant or at a point but over an interval. It involves both the force acting on a body and the motion of the body.† Using the definition of the scalar product, we can write

$$W = Fs \cos \theta \qquad (\text{constant } \mathbf{F}). \tag{10.3}$$

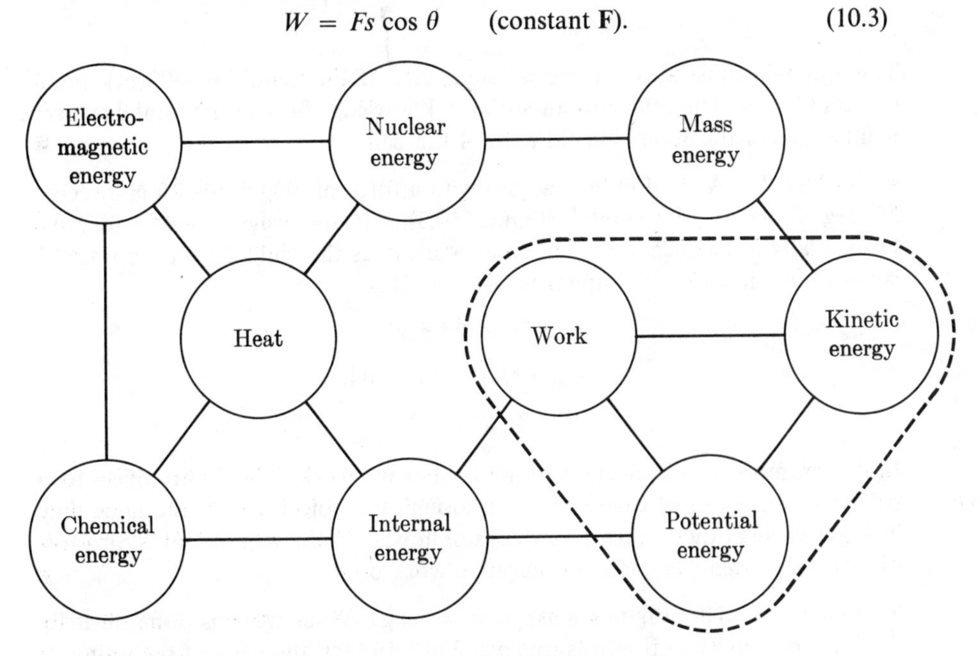

FIGURE 10.1 Some forms of energy, and some of their interconnections. Not all of these forms are clearly distinct: Internal energy is in part molecular kinetic energy; stored chemical energy can be considered to be atomic potential energy; and changes of mass accompany almost all other energy changes. The dashed line encircles the forms of energy relevant to macroscopic mechanics. Heat and work, in the center of the diagram, are modes of energy transfer.

---

* Since we postulate only translational motion, the displacement $\mathbf{s}$ is the same for every point of the body. If we were considering more general kinds of motion, $\mathbf{s}$ would be the displacement of the point at which the force is applied.

† There is an interesting analogy between the scalar quantity *work* and the vector quantity *impulse*. Recall that for constant force, impulse is defined by $\mathbf{I} = \mathbf{F}\,\Delta t$, the product of force and a time interval. As we shall see, the effect of work is to change energy, whereas the effect of impulse is to change momentum.

The significance of positive and negative work

If the angle θ between **F** and **s** is less than $\pi/2$, then W is positive; the external force does positive work on the body [Figure 10.2(a)]. If θ is greater than $\pi/2$, the force is generally opposing the motion, and W is negative [Figure 10.2(b)]. We describe negative W either by saying that the force does negative work on the body or by saying that the body does work on its environment.

■ EXAMPLE 1: The displacement vector **s** of the basketball in Figure 10.2(b) has magnitude 4 m and makes an angle of 45 deg with the horizontal. If the mass, m, of the basketball is 0.6 kg, what is the work done on it by the gravitational force, mg? We may use Equation 10.3 and write

$$\begin{aligned} W &= mgs \cos (135 \text{ deg}) \\ &= 0.6 \text{ kg} \times 9.8 \text{ m/sec}^2 \times 4 \text{ m} \times (-0.707) \\ &= -16.6 \text{ J}. \end{aligned}$$

This example illustrates the special simplicity of the definition of work for a constant force. The calculation requires a knowledge only of the total displacement **s**, not of the actual curved path of the ball. ■

■ EXAMPLE 2: A child pulls a wagon with a force of magnitude 25 N directed 35 deg above the horizontal (Figure 10.3). If the wagon moves forward horizontally a distance $d = 2$ m, what work does the child do on the wagon? Again, Equation 10.3 is simplest to use. It gives

$$\begin{aligned} W &= Fd \cos (35 \text{ deg}) \\ &= 25 \text{ N} \times 2 \text{ m} \times 0.819 \\ &= +41 \text{ J}. \end{aligned}$$

Work may be defined for each force separately

In this example, it is important to notice that the work done on an object by a particular force can be calculated even though the object may at the same time be experiencing other forces, known or unknown. As the wagon moves forward, for example, *frictional* forces do negative work on it. ■

■ EXAMPLE 3: The wagon's mass is $m = 5$ kg. What work is done on it by the force of gravity as it moves forward 2 m? In fact, the mass of the wagon is irrelevant, for gravity does no work. Since $\mathbf{F}_{\text{grav}}$ is perpendicular to **s**,

$$W_{\text{grav}} = \mathbf{F}_{\text{grav}} \cdot \mathbf{s} = 0.$$

Three ways to have zero work

Work vanishes under three conditions: if there is no force (as in the free motion of an isolated body); if there is no displacement (as when opposing forces cancel and a body remains at rest); and if the force is perpendicular to the displacement (as in this example of the gravitational force on the wagon). ■

As noted in the caption for Figure 10.3, the work done by a constant force may also be written

$$W = F_{\parallel} d, \tag{10.4}$$

where d is the magnitude of the displacement and $F_{\parallel}$ is the component of **F** in the direction of **s**. Expressing it in words, we can say that only that part of the force acting in the direction of the displacement contributes to the work.

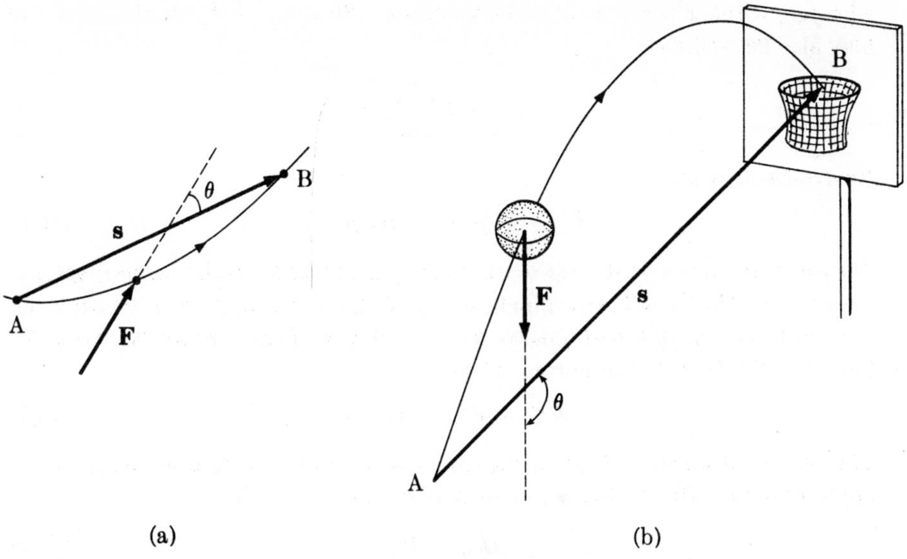

FIGURE 10.2 Positive and negative work. (a) A particle moves from A to B along a curved path under the action of a constant force **F**. The displacement **s** makes angle θ with **F**. Work, defined by $\mathbf{F} \cdot \mathbf{s}$, is positive, since $\cos\theta > 0$. (b) A basketball follows the curve AB under the action of the constant gravitational force **F**. Since $\cos\theta < 0$, the work done on the basketball is negative.

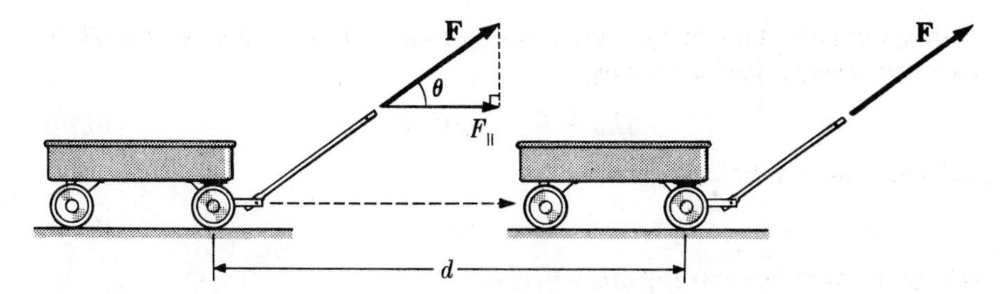

FIGURE 10.3 The work performed by one force may be calculated even if it is not the only force acting on the system. In this example, the work is $Fd\cos\theta = F_{\|}d$.

10.2 Work and kinetic energy for one-dimensional motion

Consider the following idealized experiment of great simplicity, which provides evidence that work is a useful concept. On a frictionless horizontal track, a cart is pushed with a constant force $\mathbf{F}_A$ for a time, then allowed to slide freely for a time, then decelerated and brought to rest with a constant force $\mathbf{F}_C$, whose magnitude differs from the magnitude of the accelerating force $\mathbf{F}_A$. For purposes of discussion, we shall divide the motion into three parts, A, B, and C, as illustrated in Figure 10.4.

In part A of the motion, the cart is uniformly accelerated, starting from rest at the origin. Its velocity and position may therefore be related by Equation 5.54:

$$v^2 = 2a_A x, \tag{10.5}$$

where a_A is the x component of acceleration. Since $a_A = F_A/m$, this equation may also be written

$$v^2 = 2\left(\frac{F_A}{m}\right)x.$$

Rearrangement gives

$$\tfrac{1}{2}mv^2 = F_A x \qquad \text{(part A)}. \tag{10.6}$$

The quantity on the right is the work done by the force $\mathbf{F}_A$ as the cart is displaced from 0 to x. On the left is the kinetic energy K of the cart at point x, which is the same as its *change* of kinetic energy ΔK from 0 to x. Therefore we can write, for the interval 0 to x, the important result,

Change of kinetic energy equals net work done

$$\Delta K = W \qquad \text{(part A)}. \tag{10.7}$$

The change of kinetic energy of the cart is equal to the work done on it. If we apply Equation 10.7 to the whole of part A, it can be written

$$\Delta K_A = W_A, \tag{10.8}$$

where $\Delta K_A = \tfrac{1}{2}mv_1{}^2$, the kinetic energy at x_1 (see Figure 10.4), and $W_A = F_A x_1$, the total work done in part A.

In part B of the motion, no force acts in the direction of motion, so no work is done:

$$W_B = 0. \tag{10.9}$$

During this part of the motion, the cart slides with constant velocity. Therefore its kinetic energy does not change:

$$\Delta K_B = 0 \qquad \text{(part B)}. \tag{10.10}$$

Evidently this can be written

$$\Delta K_B = W_B, \tag{10.11}$$

having the same form as Equation 10.8.

To relate velocity and position in the decelerating phase, part C, we need Equation 5.58. It gives

$$v^2 - v_2{}^2 = 2a_C(x - x_2). \tag{10.12}$$

Proceeding as before, we can replace a_C by $-F_C/m$ (the x component of acceleration is negative), and multiply both sides of Equation 10.12 by $\tfrac{1}{2}m$ to obtain

$$\tfrac{1}{2}m(v^2 - v_2{}^2) = -F_C \cdot (x - x_2). \tag{10.13}$$

Let us set $x = x_3$ and consider the whole of part C. The right side of this equation becomes $-F_C \cdot (x_3 - x_2)$, which is equal to the work in part C:

$$W_C = \mathbf{F}_C \cdot \mathbf{s}_C = -F_C \cdot (x_3 - x_2). \tag{10.14}$$

The left side becomes the total change of kinetic energy in part C (which, since $v_3 = 0$, is $-\tfrac{1}{2}mv_2{}^2$). Therefore,

$$\Delta K_C = W_C. \tag{10.15}$$

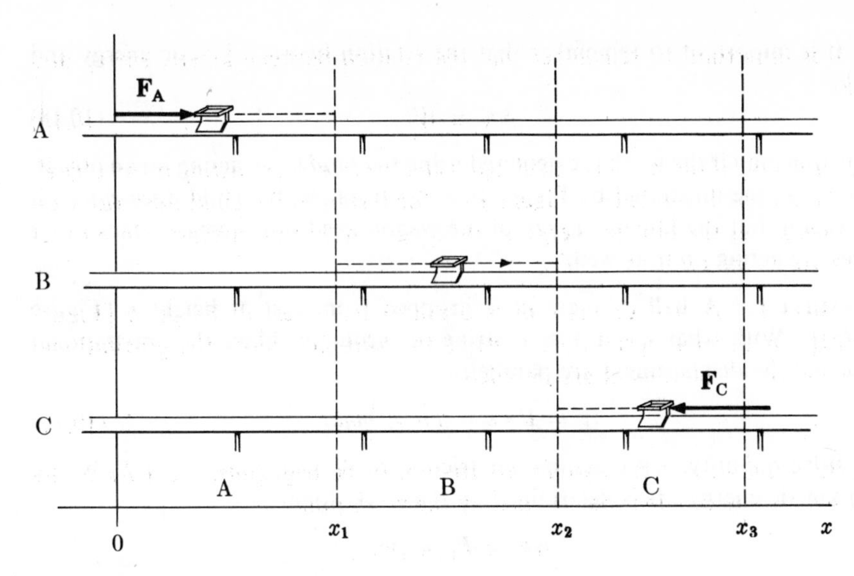

FIGURE 10.4 Example of motion used to establish the connection between work and kinetic energy. A cart is accelerated with a constant force $\mathbf{F_A}$ from $x = 0$ to x_1, then allowed to coast from x_1 to x_2, then brought to rest by a constant force $\mathbf{F_C}$ applied between x_2 and x_3.

The equality of work and change of kinetic energy is valid for every part of the motion—indeed, it is valid for every subpart and for the motion as a whole. For the entire motion, which begins and ends at rest,*

$$\Delta K_T = \Delta K_A + \Delta K_B + \Delta K_C = 0. \tag{10.16}$$

The total work must also vanish:

$$W_T = W_A + W_B + W_C = W_A + W_C = 0. \tag{10.17}$$

The positive work W_A is exactly balanced by the negative work W_C.

From this simple example, we can draw several conclusions: (1) Work is a useful measure of the effort required to start or stop an object. The work required to give the cart a certain velocity in part A is equal and opposite to the work done in stopping the cart in part C, regardless of the relative magnitudes of the forces $\mathbf{F_A}$ and $\mathbf{F_C}$. (2) Kinetic energy is a scalar property of motion that is very simply related to work. The change of kinetic energy of a rigid body in any interval is equal to the work done on the body in that interval by the *net* force acting on the body. (3) In this example, we see the hint of a conservation law. The work done in part A is fully recoverable in part C. The image suggested by the mathematics is that the positive work in part A adds something to the cart, something we can call energy; that this energy is preserved during the coasting motion in part B; and that the negative work in part C subtracts this energy from the cart.

The hint of a conservation law: Work is recoverable

* To see that the sum in Equation 10.16 vanishes, use

$$\Delta K_A = \tfrac{1}{2}mv_1^2 = \tfrac{1}{2}mv_2^2 \qquad \Delta K_B = 0 \qquad \text{and} \qquad \Delta K_C = -\tfrac{1}{2}mv_2^2.$$

$\Delta K = W$ for any part of the motion provided W is the total work from all forces

It is important to remember that the relation between kinetic energy and work,

$$\Delta K = W, \tag{10.18}$$

holds true only if the work is calculated using the *total* force acting on an object. In the example illustrated by Figure 10.3, for instance, the child does work on the wagon, but the kinetic energy of the wagon need not increase, since other forces are acting on it as well.

■ EXAMPLE 1: A ball of mass m is dropped from rest at height h [Figure 10.5(a)]. With what speed does it strike the ground? Since the gravitational force and the displacement are parallel,

Work done by the gravitational force

$$W = \mathbf{F} \cdot \mathbf{s} = Fh = mgh, \tag{10.19}$$

a positive quantity. (We assume air friction to be negligible.) Let K_f be the final kinetic energy. It is determined by the work done:

$$\Delta K = K_f = W.$$

Put $K_f = \frac{1}{2}mv_f^2$ and $W = mgh$; then solve to find the final speed,

$$v_f = \sqrt{2gh}. \tag{10.20}$$

This result could also be obtained directly from Newton's second law. However, Equation 10.18 probably provides the simplest way to get it. For a fall of 1 m, $v_f = 4.43$ m/sec, independent of mass. For other heights, v_f scales as $\sqrt{h}$. ■

■ EXAMPLE 2: A ball is thrown downward from the same height h, with initial speed v_1 [Figure 10.5(b)]. With what speed does the ball strike the ground? The work, and therefore the *change* of kinetic energy, is the same as in the preceding example. We can write

$$\tfrac{1}{2}mv_f^2 - \tfrac{1}{2}mv_1^2 = mgh.$$

The solution for v_f is

$$v_f = \sqrt{v_1^2 + 2gh}. \tag{10.21}$$

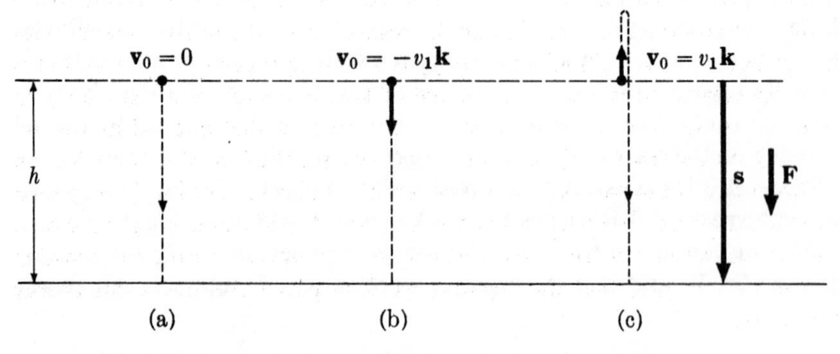

FIGURE 10.5 (a) A ball is dropped from rest at height h. (b) A ball is thrown downward from the same height. (c) A ball is launched upward from the same height. In each example, the change of kinetic energy is the same.

Suppose that $h = 1$ m and that $v_1 = 10$ m/sec. Then $v_f = 10.9$ m/sec. The work adds relatively little to the speed. ■

■ EXAMPLE 3: Starting from the same height h, a ball is thrown upward [Figure 10.5(c)]. What must be its initial speed v_1 in order that it strikes the ground with speed $2v_1$? Despite the longer path of the ball, the net displacement is the same as in the preceding examples,

$$\mathbf{s} = -h\mathbf{k}$$

(where $\mathbf{k}$ is a unit vector directed upward). The work and change of kinetic energy are therefore also the same. Equation 10.18 takes the form

$$\tfrac{1}{2}m(2v_1)^2 - \tfrac{1}{2}m(v_1)^2 = mgh.$$

The solution for v_1 is

$$v_1 = \sqrt{\tfrac{2}{3}gh}.$$

It is an easy matter, again using Equation 10.18, to find to what maximum height this ball rises. ■

10.3 Work and kinetic energy for the general motion of a particle

The work done by a force that is not constant is defined for an infinitesimal displacement $d\mathbf{s}$ by

$$dW = \mathbf{F} \cdot d\mathbf{s}. \tag{10.22}$$

The general definition of work

This is the most general definition of work in physics. It implies that a finite (and therefore measurable) increment of work must be expressed as an integral.

ARBITRARY MOTION IN ONE DIMENSION

Before examining the implications of Equation 10.22 in three dimensions, let us consider one-dimensional motion in which a particle moves along the x axis. (As in the previous sections, the analysis will apply also to the translational motion of a rigid body.) Then, since $d\mathbf{s} = dx\,\mathbf{i}$,

$$dW = F_x\,dx. \tag{10.23}$$

Note that this is valid even if F_y and/or F_z are not zero (refer to Equation 6.61). Of course, if a particle moves only along the x axis, the *total* force acting on it must be in the x direction. However, we might wish to find the work done by one among several forces, and that one need not be parallel to the x axis.

To obtain from Equation 10.23 the work over a finite interval, we must sum the infinitesimal contributions—that is, perform an integral. For displacement from x_1 to x_2, the work is

$$W = \int_{x_1}^{x_2} F_x\,dx. \tag{10.24}$$

Work in one dimension

Graphically, work is the area under a curve in a graph of F_x vs x (Figure 10.6). To calculate it no details of the actual motion, such as speed or the time required to move from x_1 to x_2, need be known. Figure 10.6(a) shows an arbitrary

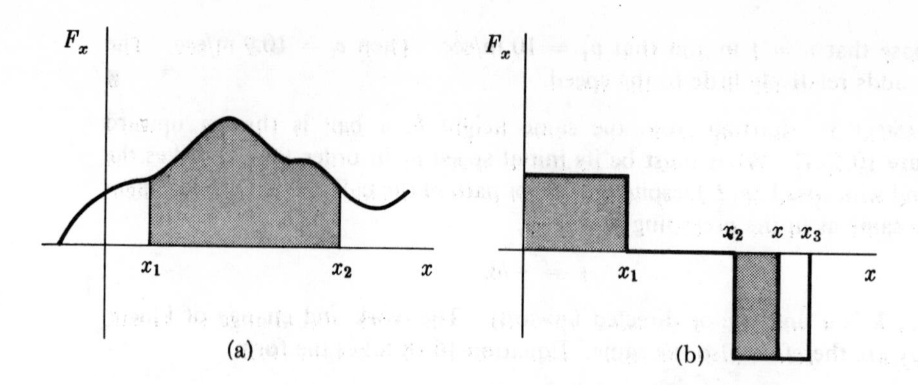

FIGURE 10.6 For one-dimensional motion, work is equal to the area under a curve in a graph of F_x vs x. (a) An arbitrary force. The work done during a displacement from x_1 to x_2 is $\int_{x_1}^{x_2} F_x\, dx$. (b) The force associated with the motion depicted in Figure 10.4. The total area for displacement from 0 to x_3 is zero.

force function. Figure 10.6(b) shows the force function describing the example illustrated in Figure 10.4. The shaded area (partly positive, partly negative) gives the work done from $x = 0$ to some value of x between x_2 and x_3.

■ EXAMPLE 1: A rocket of mass 10^4 kg is launched vertically upward. The upward component of the gravitational force acting on it is given by

$$F_r = -mg_0\left(\frac{R}{r}\right)^2, \tag{10.25}$$

where r is the distance of the rocket from the center of the earth, R is the radius of the earth, and mg_0 is the weight of the rocket at the surface of the earth. If the variation of mass of the rocket is ignored and if the rocket escapes from the earth, how much work is done on it by the gravitational force (see Figure 10.7)? The calculation requires evaluation of the work integral (Equation 10.24) for the particular law of force (Equation 10.25):

$$W = \int_R^\infty F_r\, dr$$

$$= -mg_0R^2\int_R^\infty \frac{dr}{r^2}.$$

The indefinite integral, $\int (1/r^2)\, dr$, is $-1/r$ (if this is not a familiar result, verify it by differentiation). Evaluated between the limits R and ∞, it is $+1/R$. The work is therefore

$$W = -mg_0R.$$

In particular, for $m = 10^4$ kg, $g_0 = 9.8$ m/sec^2, and $R = 6.37 \times 10^6$ m,

$$W = -6.24 \times 10^{11}\text{ J}.$$

The work needed to escape from the earth

This is the negative work done by gravity. In order for the rocket to escape the earth, at least this same magnitude of positive work must be done by the rocket thrust. ■

Returning to Equation 10.24, we can derive a general connection between work and kinetic energy in one dimension. If F_x is now the x component of the *total* force, it is related to the acceleration through Newton's second law:

$$F_x = m\frac{dv_x}{dt}.$$

Since the motion proceeds in only one dimension, v_x may be expressed either as a function of t or as a function of x. If it is considered as a function of x, the chain rule of differentiation gives for the acceleration

$$\frac{d}{dt}[v_x(x)] = \frac{dv_x}{dx}\cdot\frac{dx}{dt}. \tag{10.26}$$

The second factor on the right is just v_x, so Newton's second law can be re-expressed as

$$F_x = mv_x\frac{dv_x}{dx}. \tag{10.27}$$

The right side of this equation can be written as a total derivative, which leads to the following interesting form of Newton's second law (in one dimension):

A new form of Newton's second law

$$F_x = \frac{d}{dx}(\tfrac{1}{2}mv_x^2). \tag{10.28}$$

The factor in parentheses is the familiar expression for kinetic energy. Equation 10.24 therefore takes the form

$$W = \int_{x_1}^{x_2}\frac{dK}{dx}\,dx.$$

The integral is K itself, so

$$W = K_2 - K_1 = \Delta K. \tag{10.29}$$

The result of the last section—work equals change of kinetic energy—is therefore valid not just for a constant force but for *any* force in one dimension. A reminder: The validity of this expression requires that the work be calculated using the *total* force.

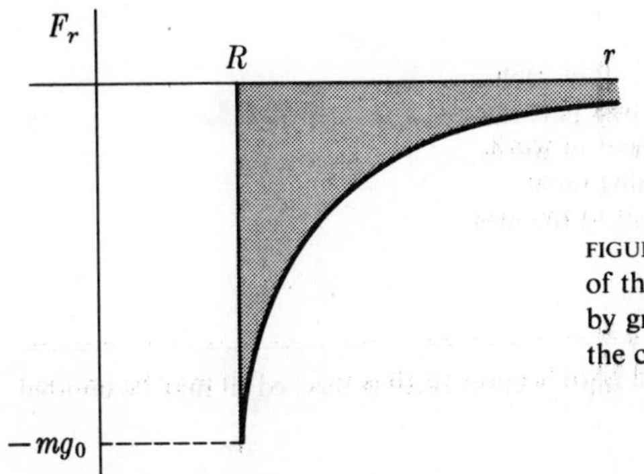

FIGURE 10.7 A rocket flies from the surface of the earth to infinity. The work done on it by gravity (negative) is the shaded area under the curve of the inverse-square force.

★ARBITRARY MOTION IN THREE DIMENSIONS*

Equation 10.22 leads formally to the integral expression for work,

$$W = \int_{\mathbf{r}_1}^{\mathbf{r}_2} \mathbf{F} \cdot d\mathbf{s}. \tag{10.30}$$

The idea of a line integral

This is called a *line integral* because it must be evaluated along the actual path, or line in space (not necessarily a straight line), followed by the particle under study as it is displaced from $\mathbf{r}_1$ to $\mathbf{r}_2$. Physically, this is not a difficult idea. For each increment of displacement $\Delta\mathbf{s}$ along the path, the increment of work, $\Delta W = \mathbf{F} \cdot \Delta\mathbf{s}$, is calculated, and these scalar quantities are summed to give the total work (Figure 10.8). Mathematically, however, the line integral is a somewhat difficult concept; a complete discussion of it is beyond the scope of this text. Special examples of line integrals that are simple to handle appear in Section 10.10 and later in Chapters 15 and 16. Here we wish to show that the previously discussed connection between work and kinetic energy is a general consequence of Equation 10.30.

Consider a particle following some trajectory through space (Figure 10.8). In an infinitesimal time interval dt its displacement is simply the product of its velocity and the time interval:

$$d\mathbf{s} = \mathbf{v}\, dt. \tag{10.31}$$

The increment of work is therefore

$$dW = \mathbf{F} \cdot \mathbf{v}\, dt. \tag{10.32}$$

If $\mathbf{F}$ is the *total* force acting on the particle, we can use Newton's second law and replace $\mathbf{F}$ by $m\, d\mathbf{v}/dt$ to give

$$dW = m \frac{d\mathbf{v}}{dt} \cdot \mathbf{v}\, dt.$$

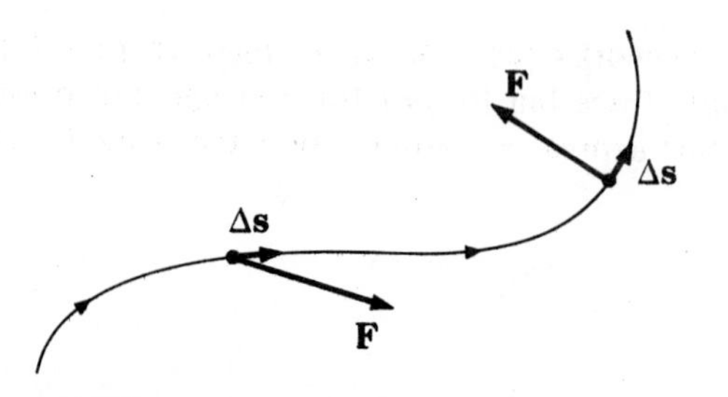

FIGURE 10.8 Work in general. For each infinitesimal displacement $\Delta\mathbf{s}$ of a particle along its trajectory, an increment of work, $\Delta W = \mathbf{F} \cdot \Delta\mathbf{s}$, is done. Summing these infinitesimal contributions leads to the idea of a line integral.

* Study of this subsection may be postponed until Section 10.10 is reached; it may be omitted if Section 10.10 is omitted.

The quantity multiplying dt proves to be the time derivative of the kinetic energy. To prove this write $K = \frac{1}{2}mv^2 = \frac{1}{2}m\mathbf{v} \cdot \mathbf{v}$. Its time derivative is

$$\frac{d}{dt}\left(\tfrac{1}{2}m\mathbf{v} \cdot \mathbf{v}\right) = \tfrac{1}{2}m\left(\frac{d\mathbf{v}}{dt} \cdot \mathbf{v} + \mathbf{v} \cdot \frac{d\mathbf{v}}{dt}\right).$$

Since the scalar product is commutative, the two terms within the parentheses are equal, and

$$\frac{dK}{dt} = m\frac{d\mathbf{v}}{dt} \cdot \mathbf{v}. \tag{10.33}$$

The increment of work is therefore

$$dW = \frac{dK}{dt}\,dt; \tag{10.34}$$

we have transformed the expression $\mathbf{F} \cdot d\mathbf{s}$ in such a way that Equation 10.30 can be written

$$W = \int_{t_1}^{t_2} \frac{dK}{dt}\,dt,$$

Work as an integral over time

just as for one-dimensional motion. From this it follows that the change of kinetic energy is equal to the work done,

$$\Delta K = W, \tag{10.35}$$

a result that we now see to be generally valid for any force and for motion in three dimensions, provided the following conditions are met: (1) The work is calculated from the total force. (2) The system is a single particle, or a rigid body executing translational motion without rotation. In some practical examples, the apparent simplicity of Equation 10.35 may be illusory, for the calculation of W need not be simple.

Conditions under which $\Delta K = W$ is valid

■ EXAMPLE 2: A particle moves in a circle under the action of a force directed toward the center of the circle. By how much can its speed change during one revolution? Since the force is at all times perpendicular to the displacement, $\mathbf{F} \cdot d\mathbf{s} = 0$, and the work is zero for any displacement. Therefore $\Delta K = 0$. The kinetic energy does not change, so the speed is constant. Circular motion under the action of a central force can only be *uniform* circular motion. ■

Proof that v = constant for circular motion under the action of a central force

■ EXAMPLE 3: A satellite, when at its apogee 10,000 miles from the center of the earth, has kinetic energy K_a. How much work is done on it by the earth's gravitational force as it moves to its perigee 5,000 miles from the center of the earth? In this case we can work "backwards" from kinetic energy to work. The law of areas (see Equation 9.84) tells us that this satellite's speed at perigee is twice its speed at apogee. Its kinetic energy at perigee is therefore $K_p = 4K_a$. The work done by the gravitational force must be $\Delta K = K_p - K_a$, or

$$W = 3K_a,$$

a positive quantity. As the satellite returns to apogee, the gravitational work is $-3K_a$, making the total work during a full orbit zero. ■

★10.4 Work and kinetic energy for a system of particles

We learned in Chapters 8 and 9 that it is fruitful, and in some ways surprisingly easy, to investigate the properties of an arbitrary system of any number of particles moving under the action of arbitrary forces (see Figure 8.2 or 8.7). Let us take this approach now with the concepts of work and kinetic energy.

Consider an increment of time dt in which the ith particle in the system experiences total force $\mathbf{F}_i$ (the sum of both internal and external forces acting on it) and suffers displacement $d\mathbf{s}_i$. The increment of work done on this particle is

$$dW_i = \mathbf{F}_i \cdot d\mathbf{s}_i. \tag{10.36}$$

Since $d\mathbf{s}_i = \mathbf{v}_i\, dt$ and since, from Newton's second law, $\mathbf{F}_i = m_i\, d\mathbf{v}_i/dt$, this increment of work can be written

$$dW_i = m_i \frac{d\mathbf{v}_i}{dt} \cdot \mathbf{v}_i\, dt.$$

The treatment of many particles parallels the treatment of a single particle

The derivation following Equation 10.32 can be applied again, leading to the equivalent of Equation 10.34:

$$dW_i = \frac{dK_i}{dt}\, dt. \tag{10.37}$$

The *total* work done on the system during this time interval (still an infinitesimal quantity) is the sum

$$dW_{\mathrm{T}} = \sum_i dW_i\,. \tag{10.38}$$

Summing both sides of Equation 10.37 over all the particles in the system ($i = 1$ to N) leads to

$$dW_{\mathrm{T}} = \frac{dK_{\mathrm{T}}}{dt}\, dt, \tag{10.39}$$

where K_{T} is the total kinetic energy,

$$K_{\mathrm{T}} = \sum_i K_i\,. \tag{10.40}$$

Finally (paralleling the reasoning below Equation 10.34), the total work done in a finite time, t_1 to t_2, is

$$W_{\mathrm{T}} = \int_{t_1}^{t_2} \frac{dK_{\mathrm{T}}}{dt}\, dt, \tag{10.41}$$

which leads to

Work equals change of kinetic energy for the total system

$$\Delta K_{\mathrm{T}} = W_{\mathrm{T}}. \tag{10.42}$$

The change in the total kinetic energy of the system is equal to the total work done on all parts of the system during any interval.

The result (Equation 10.42) seems beautifully simple and general. In this form, however, it is less useful than might at first appear. One reason is that the total kinetic energy of a system does not have such simple properties as the total momentum. There is no frame of reference, for instance, in which the kinetic energy of a general system is zero. Another reason is that internal forces can do work on a system. Even if the system is isolated, work may be done and the total kinetic energy may change. This fact can be illustrated with a two-particle system. Let the forces acting on the particles be $\mathbf{F}_1$ and $-\mathbf{F}_1$. In time dt, the work done on the system is

Internal forces can do work

$$dW_T = \mathbf{F}_1 \cdot d\mathbf{s}_1 - \mathbf{F}_1 \cdot d\mathbf{s}_2.$$

Since the displacements need not be equal, the work need not be zero. (For example, think of a pair of charged particles attracting each other. Each may do positive work on the other, and their kinetic energy grows in time.) A new concept, potential energy (Section 10.7), is needed in order to put Equation 10.42 in a more useful form.

TRANSLATIONAL MOTION OF A RIGID BODY

We may think of a rigid body as a collection of many particles in a fixed spatial relationship. If the body undergoes translational motion, every particle in it experiences the same displacement:

$$d\mathbf{s}_1 = d\mathbf{s}_2 = d\mathbf{s}_3 = \cdots. \tag{10.43}$$

Each displacement $d\mathbf{s}_i$ may therefore be set equal to $d\mathbf{s}_c$, the center-of-mass displacement. The total increment of work on the body is

$$\begin{aligned} dW_T &= \sum_i \mathbf{F}_i \cdot d\mathbf{s}_i \\ &= \sum_i \mathbf{F}_i \cdot d\mathbf{s}_c \\ &= \mathbf{F}_T \cdot d\mathbf{s}_c, \end{aligned}$$

where F_T is the total force. Since internal forces vanish, $\mathbf{F}_T = \mathbf{F}_{ext}$, and the increment of work is

$$dW_T = \mathbf{F}_{ext} \cdot d\mathbf{s}_c. \tag{10.44}$$

This is precisely equivalent to Equation 10.22, which defines the work on a single particle.

Since each part of the rigid body has the same displacement, each part has the same velocity, $\mathbf{v}_c$, and the total kinetic energy is

$$K_T = \tfrac{1}{2} \sum_i m_i v_c^2,$$

$$K_T = \tfrac{1}{2} M v_c^2. \tag{10.45}$$

This too has the same form as the kinetic energy of a single particle. The results expressed by Equations 10.44 and 10.45 justify the examples in the preceding sections in which a rigid body moving without rotation was treated as a particle.

Translation of a rigid body is equivalent to the motion of a particle

A solid piece of matter actually differs in an important way from an idealized rigid body: At the atomic level, its constituent atoms vibrate and possess a great deal of kinetic energy, even when $\mathbf{v}_c = 0$, so that Equation 10.45 gives only part of the total kinetic energy—in practice, only a small part.

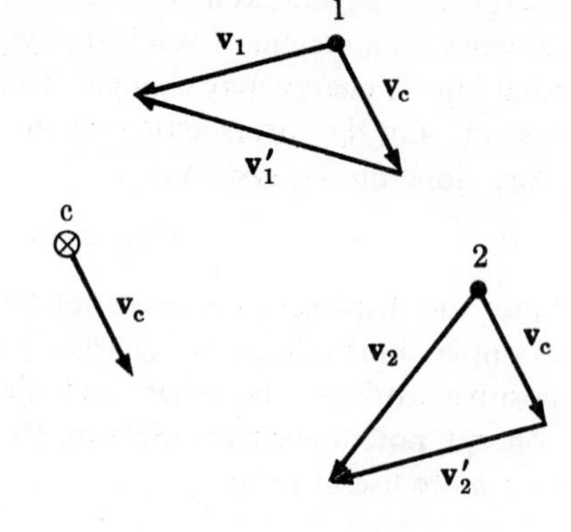

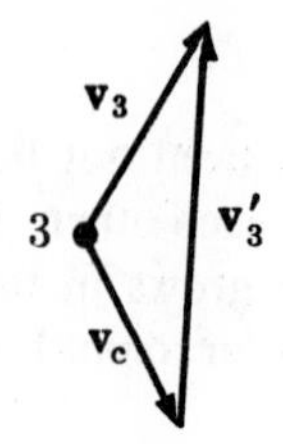

FIGURE 10.9 The velocity of any particle in a system can be expressed as the sum of the center-of-mass velocity and a relative velocity (Equation 10.47). The relative velocity $\mathbf{v}_i'$ is the velocity that would be measured in a frame of reference moving with the center of mass.

Nevertheless, if deformation and compression of the body are negligible, the work done by external forces goes primarily into what we may call the "translational kinetic energy," $\frac{1}{2}Mv_c^2$, not into the atomic vibrational energy. Internal energy may then be ignored, and the solid object is successfully treated as an ideal rigid body. For real solids, Equations 10.44 and 10.45 are approximations justified by experiment.*

KINETIC ENERGY IN GENERAL

The kinetic energy of a system is the sum of the kinetic energies of its constituent parts. Since kinetic energy is a scalar quantity, not a vector quantity, and since it can never be negative, any two kinetic energies always combine to give a larger kinetic energy. A system can have zero kinetic energy only if *all* of its constituent parts are at rest.

The kinetic energy of a system does have one simplifying feature. It can be separated into two parts, an internal kinetic energy and a bulk kinetic energy. As indicated in Figure 9.8, we can write the position of each particle as the position of the center of mass plus the position of the particle relative to the center of mass:

$$\mathbf{r}_i = \mathbf{r}_c + \mathbf{r}_i'. \tag{10.46}$$

The time derivative of this equation gives a similar relation for velocity (see Figure 10.9),

$$\mathbf{v}_i = \mathbf{v}_c + \mathbf{v}_i'. \tag{10.47}$$

The total kinetic energy of the system is

$$\begin{aligned} K_T &= \tfrac{1}{2}\sum_i m_i v_i^2 \\ &= \tfrac{1}{2}\sum_i m_i(\mathbf{v}_c + \mathbf{v}_i')\cdot(\mathbf{v}_c + \mathbf{v}_i'). \end{aligned} \tag{10.48}$$

* Corrections to this approximation, involving the influence of external work on internal energy at the atomic level, carry one away from the subject of mechanics to the subject of thermodynamics.

The scalar product that appears here can be written as a sum of three terms, $v_c^2 + 2\mathbf{v}_c \cdot \mathbf{v}_i' + v_i'^2$, so that the kinetic energy becomes

$$K_T = \tfrac{1}{2}\sum_i m_i v_c^2 + \sum_i m_i \mathbf{v}_c \cdot \mathbf{v}_i' + \tfrac{1}{2}\sum_i m_i v_i'^2.$$

In the first of these three terms, only the masses are summed. This term is the same as that found earlier for a rigid body:

$$K_I = \tfrac{1}{2}Mv_c^2. \tag{10.49}$$

We call this the *translational kinetic energy*. The second term proves to be zero. It may be written

$$K_{II} = \mathbf{v}_c \cdot \sum_i m_i \mathbf{v}_i'.$$

The sum can be recognized as the velocity of the center of mass relative to the center of mass, which is necessarily zero:

$$K_{II} = 0. \tag{10.50}$$

The third term,

$$K_{III} = \tfrac{1}{2}\sum_i m_i v_i'^2, \tag{10.51}$$

is the kinetic energy relative to the center of mass, which is a combination of *rotational kinetic energy* (Section 10.9) and *internal kinetic energy* (such as molecular vibration). In summary,

Kinetic energy can be separated into a translational part and a part relative to the center of mass

$$K_T = \tfrac{1}{2}Mv_c^2 + \tfrac{1}{2}\sum_i m_i v_i'^2. \tag{10.52}$$

For an ideal rigid body undergoing translational motion, the second term is zero (compare Equation 10.45).

A two-particle system provides an interesting special case (Figure 10.10). We define the relative position vector and relative velocity vector by

$$\mathbf{r} = \mathbf{r}_2 - \mathbf{r}_1, \tag{10.53}$$

$$\mathbf{v} = \mathbf{v}_2 - \mathbf{v}_1. \tag{10.54}$$

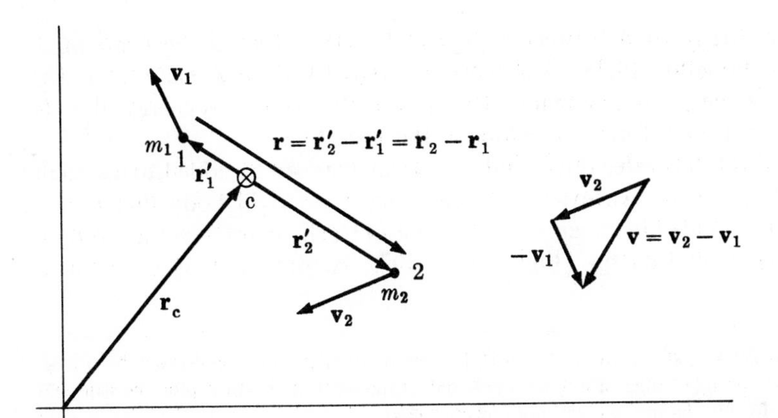

FIGURE 10.10 Two-particle system. The relative position vector $\mathbf{r}$ is equal to $\mathbf{r}_2 - \mathbf{r}_1$. The relative velocity is $\mathbf{v} = \mathbf{v}_2 - \mathbf{v}_1$. The kinetic energy of the system is given by Equation 10.58.

With a minor change of notation, Equations 8.77 and 8.78 provide the velocities relative to the center of mass,

$$\mathbf{v}_1' = -\frac{m_2}{M}\mathbf{v}, \tag{10.55}$$

$$\mathbf{v}_2' = +\frac{m_1}{M}\mathbf{v}. \tag{10.56}$$

Substituting these expressions in

$$K_{\text{III}} = \tfrac{1}{2}m_1 v_1'^2 + \tfrac{1}{2}m_2 v_2'^2$$

leads to

$$K_{\text{III}} = \tfrac{1}{2}\mu v^2,$$

where μ is the reduced mass (Equation 8.74):

$$\mu = \frac{m_1 m_2}{M}. \tag{10.57}$$

Kinetic energy of a two-body system

The total kinetic energy of the two-particle system is therefore

$$K_{\text{T}} = \tfrac{1}{2}Mv_c^2 + \tfrac{1}{2}\mu v^2. \tag{10.58}$$

The internal kinetic energy is the same as that of a single particle of mass μ moving with the relative velocity $\mathbf{v}$.

10.5 Work as a mode of energy transfer

A definition of energy

One way to *define* energy in general is as that which is capable of doing work. Not all forms of energy are directly or easily converted to work, but through some suitable chain of transformation, all can manifest themselves as work. If work is done on an object (or a system), it acquires an amount of energy equal to the work done (possibly, but not necessarily, in the form of kinetic energy). This relation of energy change and work can be expressed symbolically by an equation that generalizes Equation 10.35:*

Work changes energy

$$\Delta E = W. \tag{10.59}$$

In this form, the relation between energy and work is free of the restrictions stated below Equation 10.35. The work associated with any *single* force, for example, adds energy to a system (if W is positive) or subtracts energy (if W is negative) even if other forces are acting at the same time.

Figure 10.11 illustrates three kinds of energy that can be added to a system by doing work on it. An external force acting on a rigid body that moves without friction adds kinetic energy [Figure 10.11(a)]. A force compressing a spring adds potential energy [Figure 10.11(b)]. A force moving a piston to

* A still more general statement is the first law of thermodynamics, expressed by $\Delta E = W + Q$: The energy change of a system is equal to the work done on it plus the heat (Q) added to it. In this chapter, we will not consider heat.

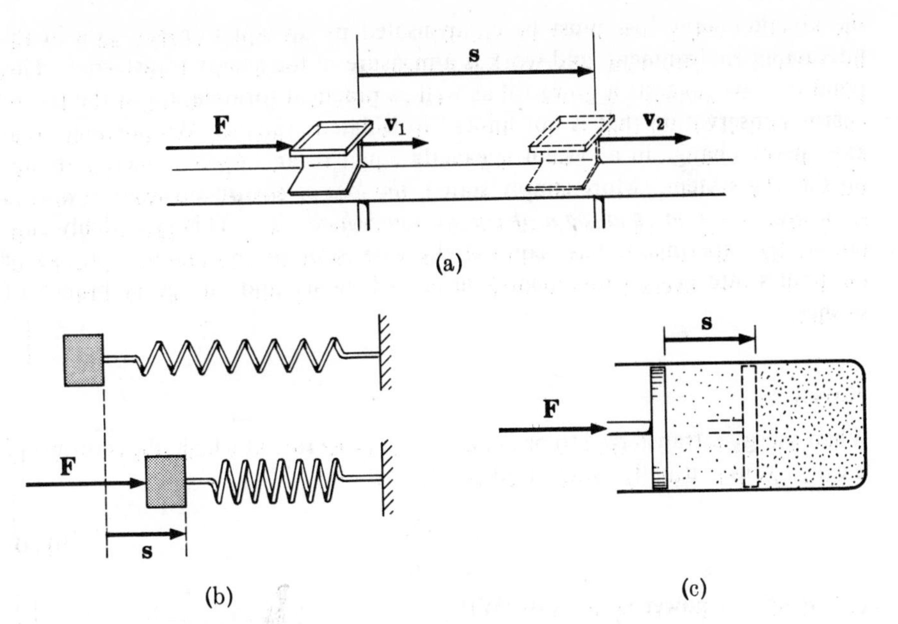

FIGURE 10.11 Work and energy change. (a) Work adds kinetic energy to a sliding cart. (b) Work adds potential energy to a spring. (c) Work adds internal energy to a gas.

compress a gas adds what is called internal energy—actually a combination of kinetic and potential energy at the molecular level [Figure 10.11(c)]. In all three examples, the energy can manifest itself again as work.

Ways of looking at energy

The fact that we speak of energy being added to or taken away from a system does not mean that energy is in any sense regarded as an actual material substance. It can be called at best an attribute or property of the system. Nevertheless, in order to form a mental picture of the difference between energy and work, it is helpful to think of energy as substance and work as action. A system *possesses* energy; it can *do* work. At any instant a system has a certain energy *content*. Part or all of this energy can be transformed into the *activity* of work. In one sense work is just another form of energy because it can be transformed without loss to and from energy. In another and deeper sense, work is only the active measure of energy and not itself a form of energy. It is best regarded as a mode of transfer of energy from one form to another.

Work as a medium of exchange

The role of work as a medium of exchange can be elucidated by considering again the sliding cart shown in Figure 10.4. Initially, work is done on it and its kinetic energy accordingly increases. This work must have been supplied by something else. Whatever the source of the work, we may assume that somewhere else a loss of energy occurred to make this work available. This unidentified other loss of energy must be equal in magnitude to the work done, which in turn is equal to the kinetic energy gained by the cart. Jumping over the intermediate idea of work, we can say simply that the cart gains kinetic energy because energy is transferred from outside the system to inside the system (the "system" here is the sliding cart alone). Looked at this way work is a transmitter of energy, a medium of exchange. Similarly, in part C of the motion, the kinetic energy loss must be compensated by an equal energy gain in the interacting environment, and work is a measure of the energy transferred. This point of view suggests a powerful as well as practical formulation of the law of energy conservation that is not limited to isolated systems. We postulate that any energy change in a system is exactly equal to an opposite energy change outside the system. More simply stated, *the loss of energy anywhere is always compensated by an equal gain of energy somewhere else*. This law, richly supported by experiment, has required the extension of the energy concept of mechanics into every subsequently developed theory and into every branch of science.

The law of energy conservation

POWER

When energy is transferred to or from a system, the rate at which it is transferred is called *power*, usually symbolized by P:

Definition of power

$$P = \frac{dE}{dt}. \tag{10.60}$$

The SI unit of power is the watt (W):

$$1 \text{ W} = 1 \text{ J/sec} = 1 \text{ kg m}^2/\text{sec}^3. \tag{10.61}$$

This unit is familiar from the rating of light bulbs and appliances. A 100-watt light bulb is one that (when connected to the proper voltage) transforms electrical energy into heat and light energy at the rate of 100 J/sec.

When a force acts on a particle, the power expended on the particle is

$$P = \frac{dW}{dt} = \mathbf{F} \cdot \frac{d\mathbf{s}}{dt},$$

or

Power expended on a moving particle

$$P = \mathbf{F} \cdot \mathbf{v}. \tag{10.62}$$

This follows directly from the basic definition of work given by Equation 10.22.

■ EXAMPLE 1: A certain horse exerts a horizontal component of force of 2,000 N. How fast must the horse move in order to supply 1 horsepower (746 W)? Since **F** and **v** are parallel, Equation 10.62 can be written, for this example,

$$746 \text{ W} = 2{,}000 \text{ N} \times v.$$

The required speed is, therefore,

$$v = 0.373 \text{ m/sec},$$

or about 0.83 mile/hr. ■

■ EXAMPLE 2: An electric locomotive can use its electric motors for braking. The motors then act as generators and feed electric power back into the electric line. If a locomotive of mass 150,000 kg moving at a speed of 60 mile/hr starts to decelerate at a rate of 0.5 m/sec^2, what is the maximum power that it can supply to the electric line? Its mass times acceleration gives the force that must be acting on it:

$$F = 150{,}000 \text{ kg} \times 0.5 \text{ m/sec}^2$$
$$= 75{,}000 \text{ N}.$$

Since **F** is directed opposite to **v**, Equation 10.62 gives

$$P = -75{,}000 \text{ N} \times 60 \text{ mile/hr} \times 0.447 \frac{\text{m/sec}}{\text{mile/hr}}$$
$$= -2.01 \times 10^6 \text{ W}.$$

The negative sign means that negative work is being done on the locomotive, another way of saying that the locomotive is doing work on its environment. If all of this work went back to the electric line, it would be at the rate of about 2 megawatts (MW), or about 2,700 horsepower. ■

10.6 Conservation of work

Certain mechanical devices are transformers of work without being intermediary storers of energy. Work is done on the device and simultaneously work is done by the device. If the heat energy created by frictional forces is sufficiently small, the output work will, to good approximation, be equal to the input work. The study of levers and pulleys, which are devices of this kind, provided the earliest hints of the idea of energy conservation.

A simple lever consists of a rigid bar pivoted at some point, not necessarily its center (Figure 10.12). Let us call the distances from the pivot to the two ends of the lever L_1 and L_2. If a force F_1 is applied to move the left end of the lever through a distance d_1, an amount of work F_1d_1 is done on the lever. If the lever is so arranged that its other end applies a force F_2 to some other object as it moves through a distance d_2, the output work is equal to F_2d_2. (We assume that the forces act parallel to the direction of motion.) In the absence of frictional energy dissipation, input work and output work are equal:

For a lever, work in ≅ work out

$$F_1d_1 = F_2d_2. \tag{10.63}$$

The input and output *forces*, on the other hand, are unequal. Instead, their ratio is the inverse ratio of the distances moved:

$$\frac{F_1}{F_2} = \frac{d_2}{d_1}. \tag{10.64}$$

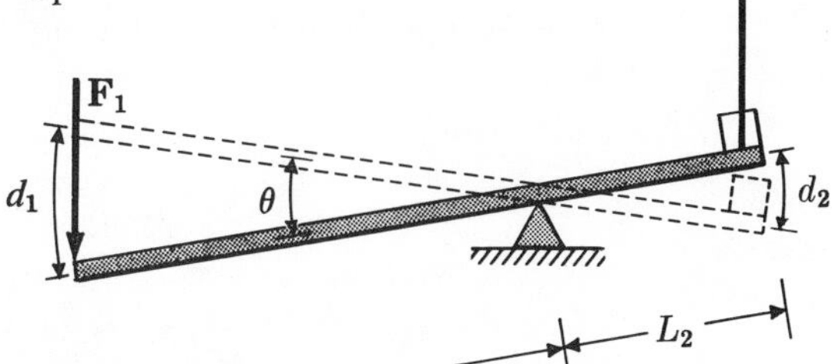

FIGURE 10.12 The simple lever. The input work is equal to the output work.

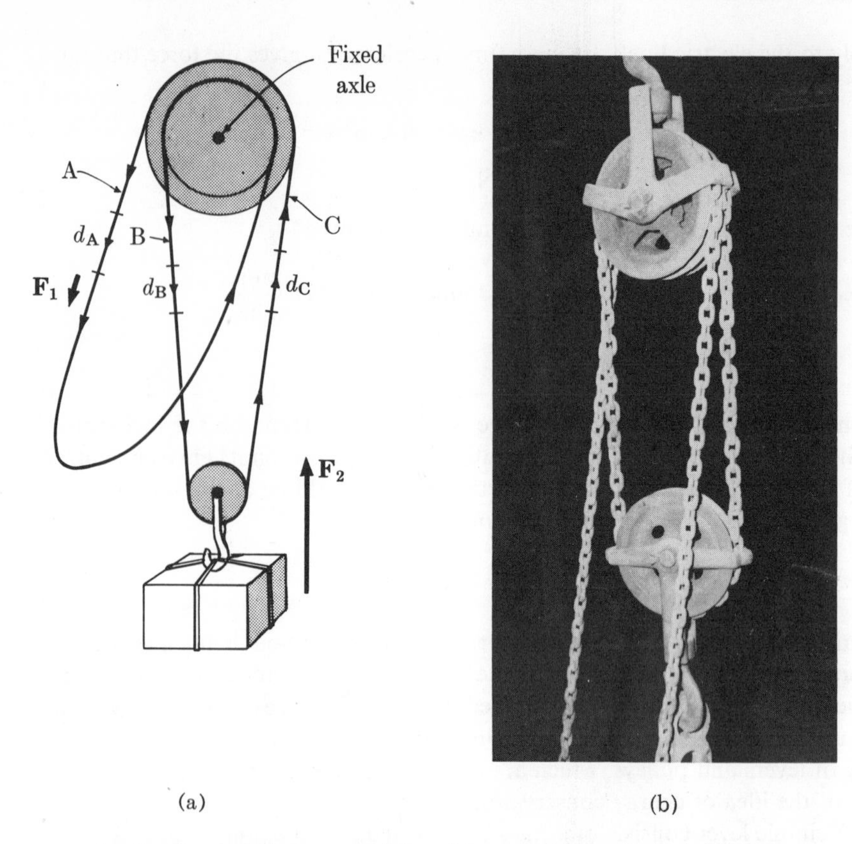

(a) (b)

FIGURE 10.13 The chain hoist, a multiplier of force. (a) The principal of its operation. While force $\mathbf{F}_1$ acts through distance d_A, the much greater force $\mathbf{F}_2$ lifts the load through a much smaller distance, $\frac{1}{2}(d_C - d_B)$. (b) A chain hoist in action. (Photograph courtesy of L. L. Baumunk & Son.)

The distances moved, in turn, are proportional to the lengths of the arm of the lever. If the lever turns through angle θ (radians), its left end moves a distance given by

$$d_1 = \theta L_1,$$

and its right end moves a distance given by

$$d_2 = \theta L_2.$$

Therefore the ratio d_2/d_1 is the same as the ratio L_2/L_1. The normal purpose of a lever is to gain an output force greater than the input force. The ratio of output force to input force is called the *mechanical advantage* (*M.A.*) of the lever. In symbols,

Mechanical advantage of a lever

$$M.A. = \frac{F_2}{F_1} = \frac{L_1}{L_2}. \tag{10.65}$$

Evidently, because of the conservation of work, the greater force must act through the smaller distance. The lever pictured in Figure 10.12, with the length L_1 twice the length L_2, has a mechanical advantage of 2. Archimedes, who first discovered the law of the lever represented by Equation 10.65 around 250 B.C., is said to have been so impressed by the unlimited possible magnitude of the mechanical advantage that he declared, "Give me a fulcrum on which to rest, and I will move the earth." It is entirely believable that Archimedes did say this, for the statement reveals two common attributes of the creative scientist: his enthusiasm for his subject and his ability to extrapolate from the familiar realm in which a law is discovered to a new domain where the law could have new power or receive new tests.

Archimedes' law of the lever can be founded on either the idea of balancing torque or the idea of equal input and output work. For the lever, these two approaches are equally simple. For understanding most other devices, however, the idea of the conservation of work is decidedly the simpler, more useful, and more general approach. Consider, for example, its application to the operation of a chain hoist. In the chain hoist, a pair of pulleys of slightly different circumference are attached to the same fixed shaft. Below them, as shown in Figure 10.13, is connected a movable pulley which supports the weight to be raised. The input force F_1 is applied to the chain leaving the larger upper pulley. The output force F_2 is supplied by the lower pulley to the load being raised. Of interest is the mechanical advantage, which is the ratio of forces, F_2/F_1. In Figure 10.13, three chain segments are labeled A, B, and C. As segment A is pulled down a distance d_A, segment B moves down a distance d_B, segment C moves up a distance d_C, and the lower pulley moves up a distance $\frac{1}{2}(d_C - d_B)$. Segments B and C, supporting the lower pulley, suffer a net decrease of length $d_C - d_B$, which is equally divided between the two sides—thus the factor of $\frac{1}{2}$. The conservation of work may be expressed by the equality

$$F_1 d_A = \tfrac{1}{2} F_2 (d_C - d_B),$$

so the mechanical advantage is

$$M.A. = \frac{F_2}{F_1} = \frac{2d_A}{(d_C - d_B)}.$$

Now consider the motions that take place during one full revolution of the upper shaft. Segment C moves up and segment A moves down through the same distance L_1, the circumference of the outer pulley. At the same time segment B moves down through distance L_2, the circumference of the inner pulley. Therefore, the mechanical advantage may be re-expressed in terms of these two circumferences:

Mechanical advantage of a chain hoist

$$M.A. = \frac{2L_1}{(L_1 - L_2)}. \tag{10.66}$$

If the *difference* of the two circumferences, $L_1 - L_2$, is much less than the outer circumference L_1, the mechanical advantage will be very large. The next time that you are in an automobile garage, watch a chain hoist and notice the large movement of chain segment A required to produce a small movement of the lower pulley and its load.

Among other devices illustrating the principle of conservation of work is the vise that is so common to home workbenches. The explanation of its action is left as an exercise. For all such multipliers of force, the principle is the same: The input force multiplied by the distance through which it acts is equal to the output force multiplied by the distance through which it acts. This conservation of work is a very special case of the general law of energy conservation. To be valid, it requires that the device store no energy and that it transform no work (or, in practice, very little work) into nonrecapturable energy, such as heat energy, so work input appears immediately as equal work output.

10.7 Potential energy in one dimension

When you ride in an elevator, you experience two quite different kinds of forces. One is the force of gravity, which depends very slightly on your altitude above the surface of the earth but does not depend either on the velocity or acceleration of the elevator or on any other features of the motion. The other is the force exerted by the floor of the elevator. This is determined by the acceleration of the elevator, which may depend on the age and condition of the elevator motor, the number of people on the elevator, and whether someone has pushed a particular button. These two forces could be called fundamental and nonfundamental, but a more useful distinction is the following:

Two types of forces

> Forces of type 1: those that depend only on position—the position of a system as a whole or the positions of its parts.
>
> Forces of type 2: those that depend on other variables (including possibly the whim of a human being or the vagaries of a machine).

The force of gravity and the electrical force between two charged particles are forces of type 1. Most of the pushes and pulls produced by humans and by machines are forces of type 2. The reason for drawing the distinction between these two kinds of force is that a form of energy—potential energy—can usually be associated with forces of type 1. These forces therefore deserve to be segregated and studied separately.*

As another illustration of the distinction between these two types of forces, consider a block attached to a spring, as shown in Figure 10.11(b). If the block is pushed to the right by hand, it experiences a force of type 2. At the same time, it experiences an opposite force from the compressed spring. This is a force of type 1 because it depends only on the displacement of the block away from its equilibrium position.

In one dimension, potential energy is associated with forces of type 1

For motion confined to one dimension, potential energy can *always* be associated with a position-dependent force of type 1. (The somewhat more complicated situation in three dimensions will be discussed in Section 10.10.) In this section we will define one-dimensional potential energy in a general way, then show how the idea applies to gravity near the earth and to the harmonic oscillator.

---

* A force of type 2 may be reducible in principle to forces of type 1. A push by hand or a force of friction, for example, finds its ultimate explanation in terms of electrical forces between atoms.

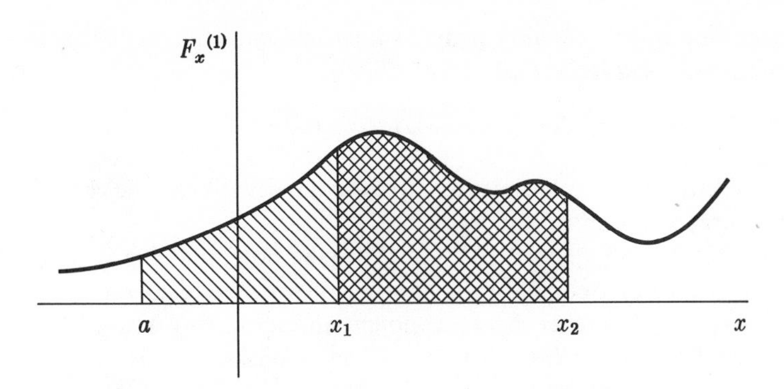

FIGURE 10.14 For a force in one dimension that depends only on position x, potential energy is defined by $U(x) = -\int_a^x F_x^{(1)}(x')\,dx'$, the negative of the area under the force curve between a and x. The location a is the arbitrarily chosen point where $U = 0$. The cross-hatched area between x_1 and x_2 is the *difference* of the area between a and x_2 and the area between a and x_1.

Consider a particle moving along the x axis; it is acted upon by a force $F_x^{(1)}(x)$, of type 1, depending only on the position x of the particle, and acted upon by a force $F_x^{(2)}$, of type 2, depending on other variables. In undergoing a displacement from x_1 to x_2, the particle experiences a change of kinetic energy equal to the work done on it (Equation 10.29):

$$\Delta K = W = W_1 + W_2. \tag{10.67}$$

The two contributions to the work are

$$W_1 = \int_{x_1}^{x_2} F_x^{(1)}(x)\,dx, \tag{10.68}$$

$$W_2 = \int_{x_1}^{x_2} F_x^{(2)}\,dx. \tag{10.69}$$

To calculate W_2, we would need to know more than has been specified so far about the nature of $F_x^{(2)}$ and the manner in which the particle moved from x_1 to x_2. To calculate W_1, however, no details of the motion need be known. Once the force function, $F_x^{(1)}(x)$, is given, W_1 is determined as a simple integral (see Figure 10.6). Think of an arbitrary but, for the moment, definite force, such as the one graphed in Figure 10.14. The area under this graph from a fixed lower limit, $x = a$, to a variable upper limit x, may be used to define a new function, $-U(x)$:

Work done by forces of type 1 is easy to calculate

$$-U(x) = \int_a^x F_x^{(1)}(x')\,dx'. \tag{10.70}$$

Potential energy defined

The integral is, of course, the same as the work associated with a displacement from a to x. The reason for the choice of negative sign will be made clear later. The quantity U is called the *potential energy* associated with the force $F_x^{(1)}(x)$. Next we make use of the equality, valid for any integrand $f(x)$,

$$\int_{x_1}^{x_2} f(x)\,dx = \int_a^{x_2} f(x)\,dx - \int_a^{x_1} f(x)\,dx.$$

The graphical meaning of this equality is made clear in Figure 10.14. Because of it, we can write (using Equations 10.68 and 10.70)

Work related to change of potential energy

$$W_1 = -[U(x_2) - U(x_1)] = -\Delta U. \qquad (10.71)$$

The work associated with the displacement x_1 to x_2 is the negative of the change of the potential energy between x_1 and x_2.

The definition given by Equation 10.70 might make potential energy seem to be nothing more than a new name for the negative of work. Actually, there is an important conceptual difference between potential energy and work. The work W_1 is defined for an *interval*, x_1 to x_2. The potential energy $U(x)$ is defined at a *point*.* From one point to another, we have a quantity of work, but we have a *change* of potential energy.

Potential energy can be considered to exist at a point

If W_1 is replaced by $-\Delta U$ (from Equation 10.71), Equation 10.67 can be written

$$\Delta K = -\Delta U + W_2.$$

Moving the ΔU term from the right side to the left side gives

$$\Delta K + \Delta U = W_2,$$

or

$$\Delta(K + U) = W_2. \qquad (10.72)$$

This makes clear the reason for the choice of a negative sign in the definition of U. On the left side of Equation 10.72 appears the sum of two energy terms:

Kinetic energy, K: energy of motion;
Potential energy, U: energy of position.

The sum of these two energies we may call the *mechanical energy*, E:

Definition of mechanical energy

$$E = K + U. \qquad (10.73)$$

(The same symbol E is often used for other kinds of energy too.) Equation 10.72, written

Work done by forces of type 2 changes mechanical energy

$$\Delta E = W_2, \qquad (10.74)$$

states, therefore, that the change of mechanical energy of a system (kinetic plus potential) is equal to the work done on the system by any forces whose effect is not already taken into account by the potential energy. In case there are no forces of type 2 contributing such additional work, we have a law of *conservation of mechanical energy*,

When all forces are described by potential energy, E is conserved

$$\Delta E = 0, \qquad (10.75)$$

or

$$E = K + U = \textit{constant}. \qquad (10.76)$$

A direct connection between potential energy and force can be obtained by

---

* However, see the discussion on the next page.

considering an infinitesimal displacement dx. Then Equation 10.71 becomes

$$dW_1 = -dU.$$

From Equation 10.23, $dW_1 = F_x\,dx$, so

$$dU = -F_x\,dx. \qquad (10.77)$$

If U is a specific function of x, this differential statement is equivalent to

$$\frac{dU}{dx} = -F_x. \qquad (10.78)$$

Force is the negative gradient of potential energy

Since a derivative with respect to a position coordinate is often called a gradient,* we may say that the force is the negative gradient of the potential energy. This kind of force, with which a potential energy can be associated, is called a *conservative* force. When the only forces that act are conservative forces, the law of energy conservation, Equation 10.76, is valid.

Conservative force defined

Finally, the role of the arbitrary limit of integration a in Equation 10.70 requires comment. Any choice within the range where $F_x^{(1)}$ is defined is permissible. This means that the magnitude of U depends on the choice of a. If a is changed, U changes by a constant amount. The potential energy U is therefore arbitrary to within an additive constant. Another function,

$$U' = U + A,$$

where A is a constant, is also an acceptable potential-energy function. This arbitrariness is consistent with Equation 10.78 since $dU'/dx = dU/dx$. The significance of a is that it is the value of x where U is defined to be zero. In physical problems, this arbitrary zero has no effect because only *changes* of potential energy are significant. The fundamental Equation 10.72, for instance, refers only to change of energy. If mechanical energy is conserved (Equation 10.76), the magnitude of the constant E can be adjusted to reflect the arbitrary zero of U.

The location of the zero of potential energy is arbitrary

This discussion has been rather formal and general. To make its meaning clear, we turn to specific examples.

GRAVITY NEAR THE EARTH

Consider a rocket fired vertically upward (Figure 10.15). If a z axis is chosen to point upward, the gravitational force acting on the rocket is

$$\mathbf{F}_1 = -mg\mathbf{k}.$$

Gravitational force is of type 1

This is a force of type 1. From it, a potential energy may be derived. Choose the zero of potential energy to be at $z = 0$. Then (from Equation 10.70)

$$U(z) = -\int_0^z F_{1z}\,dz' = +mg\int_0^z dz' = mgz. \qquad (10.79)$$

* A more general definition of a gradient appears in the footnote on page 418.

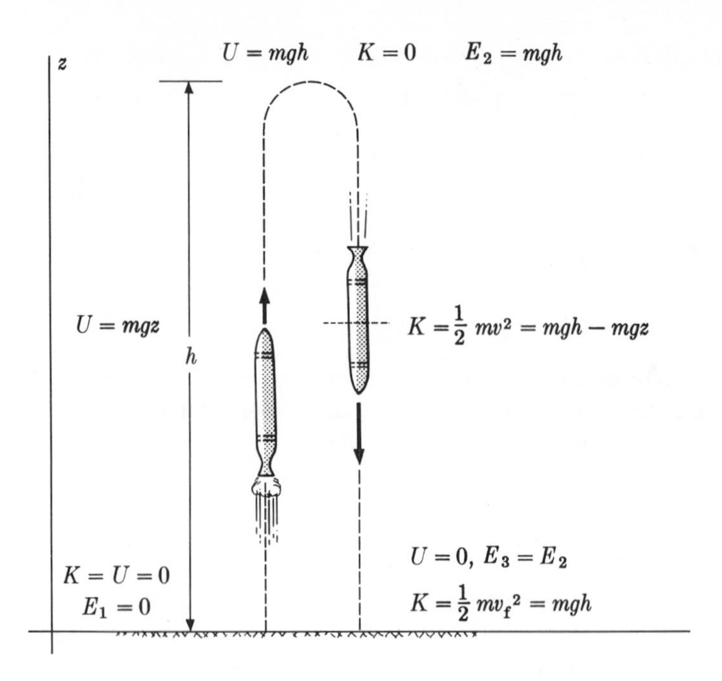

FIGURE 10.15 Gravitational potential energy. Lifted very slowly a distance h, a rocket gains potential energy equal to mgh, which in turn is equal to the work done in lifting it. In free fall through the same distance with the engine off, the rocket gains kinetic energy equal to its loss of potential energy in the fall.

If the zero of potential energy were located instead at $z = a$, the potential energy would be

Gravitational potential energy

$$U' = mg(z - a). \tag{10.80}$$

This is an important potential-energy function, which is actually valid for two- and three-dimensional motion near the earth, as well as for one-dimensional motion.

The thrust of the rocket is a force of type 2:

Rocket thrust is a force of type 2

$$\mathbf{F}_2 = F_2\mathbf{k}. \tag{10.81}$$

As the rocket is displaced upward from $z = 0$ to $z = h$, this force does work on the rocket given by

$$W_2 = \int_0^h F_2\,dz. \tag{10.82}$$

This work is "put into" the mechanical energy of the rocket, kinetic plus potential. For this example, if we use $U = mgz$, Equation 10.72 takes the form

$$\Delta E = \Delta(\tfrac{1}{2}mv^2 + mgh) = W_2. \tag{10.83}$$

■ EXAMPLE 1: Owing to a faulty engine, the rocket thrust is barely sufficient to lift the rocket. The rocket staggers upward, reaching a height h with negligible kinetic energy (Figure 10.15). How much work is done? What energy transformation takes place? Initially, at ground level, $K = 0$ and $U = 0$, so

$$E_1 = 0 \qquad (z = 0).$$

At height h, K is still zero, and $U = mgh$, so

$$E_2 = mgh \qquad (z = h).$$

Therefore $\Delta E = E_2 - E_1 = mgh$, which must be the same as the work done by the rocket thrust:

$$W_2 = mgh.$$

This can be checked using Equation 10.82. If the thrust is just equal to the weight (or slightly greater), $F_2 = mg$, and the integral defining W_2 yields mgh. More generally, however, F_2 might have been sometimes greater and sometimes less than mg. The work W_2 would still have to be mgh. In this example, the work is transformed entirely into potential energy. By virtue of its change of position, the rocket has gained energy even though its speed is the same at height h as it was at the ground.

Note that gravity has done negative work on the rocket during its ascent. The *total* work, $W_1 + W_2$, contributed by gravity plus the thrust, is zero, which accounts for the zero change of kinetic energy. It is important to avoid "double counting." If the force of gravity is already taken into account by means of a potential-energy function, it should not be counted again in computing the work that it does. ■

The danger of double counting

■ EXAMPLE 2: The rocket engine cuts off at height h, and the rocket falls back to earth (Figure 10.15). With what speed does it strike the earth if air friction is negligible? No forces of type 2 act during this part of the motion. Mechanical energy is therefore conserved. At the start of the fall,

$$E_2 = K + U = 0 + mgh = mgh.$$

At the end of the fall,

$$E_3 = K + U = \tfrac{1}{2}mv_f^2 + 0 = \tfrac{1}{2}mv_f^2.$$

Since $E_2 = E_3$, these two expressions may be equated, giving the final speed,

$$v_f = \sqrt{2gh}.$$

This is the same as Equation 10.20, which was derived in a slightly different way. In this example, potential energy is transformed into kinetic energy. ■

This consideration of gravity suggests that the conservation of energy does not require that a system be isolated. The rocket alone may be taken to be the system of interest, in which case the gravitational force is an *external* force. Alternatively, rocket plus earth may together be regarded as a single system. Then the gravitational force is an *internal* force. In either case, the law of energy conservation takes the same form. This is a somewhat subtle point. Actually, energy is strictly conserved only for an isolated system. The reason that the earth can be left out of the system is that the energy associated with its response to the rocket's motion is negligible. If rocket and earth were of more nearly equal mass, the motion of one would significantly affect the motion of the other, and it would be essential to keep both in the system in order to apply the law of energy conservation.

Energy is sometimes approximately conserved in a nonisolated system

FIGURE 10.16 An idealized harmonic oscillator.

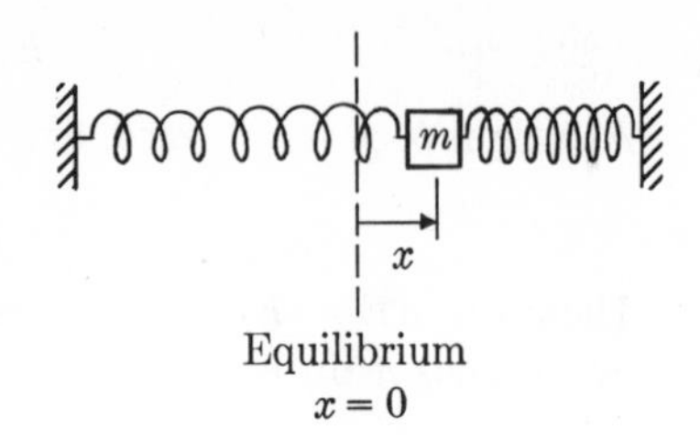

THE HARMONIC OSCILLATOR

An idealized harmonic oscillator is shown in Figure 10.16. The system includes the springs, but the mass of the springs is assumed negligible in comparison with the mass m of the block. The total kinetic energy is therefore $K = \frac{1}{2}mv^2$, where v is the speed of the block. The restoring force of the springs is

A conservative force

$$F_x = -kx. \tag{10.84}$$

Since this is a position-dependent force (type 1) in one dimension, it is a conservative force. The zero of potential energy may conveniently be located at $x = 0$. Then, from Equation 10.70,

$$U(x) = -\int_0^x (-kx')\,dx'$$

Potential energy of the harmonic oscillator

$$U(x) = +\tfrac{1}{2}kx^2. \tag{10.85}$$

The potential energy is a quadratic function of x. The overall work-energy relation (Equation 10.72) for the harmonic oscillator is therefore

$$\Delta E = \Delta(\tfrac{1}{2}mv^2 + \tfrac{1}{2}kx^2) = W_2, \tag{10.86}$$

where W_2 is the input work from any source other than the springs.

■ EXAMPLE 3: The block in Figure 10.16 starts from rest at a displacement $+x_0$ from its equilibrium position and oscillates freely. What maximum speed does it achieve, and where does this occur? How much external work is required in order to bring the block to rest at $x = 0$? During its free oscillation, the block has constant energy, which must be equal to its initial energy,

$$E = K_0 + U_0 = \tfrac{1}{2}kx_0^2.$$

At any later time, energy conservation is expressed by

$$\tfrac{1}{2}mv^2 + \tfrac{1}{2}kx^2 = \tfrac{1}{2}kx_0^2. \tag{10.87}$$

Both terms on the left side are positive or zero, and their sum is constant. Maximum kinetic energy must therefore occur at the point of minimum potential energy, which is $x = 0$. At this point,

$$\tfrac{1}{2}mv_{\max}^2 = \tfrac{1}{2}kx_0^2.$$

The maximum speed (at $x = 0$) is

Speed is maximum where U is minimum

$$v_{\max} = \sqrt{\frac{k}{m}}\,x_0. \tag{10.88}$$

If the block is then to be brought to rest at the origin ($v = 0$ and $x = 0$), its final energy is zero, and its change of energy is

$$\Delta E = E_2 - E_1 = 0 - \tfrac{1}{2}kx_0^2.$$

This, in turn, is the work that must be done to bring it to rest (Equation 10.74):

$$W_2 = -\tfrac{1}{2}kx_0^2.$$

The negative sign means that energy is transferred from the oscillator to its environment. ■

10.8 Conservation of mechanical energy in one dimension; energy diagrams

In the previous section, the conservation of mechanical energy was introduced:

$$E = K + U = \textit{constant},$$

valid when all the forces that act are accounted for by the potential energy U. Here we wish to expand on this subject, paying particular attention to the role of the energy diagram.

GRAVITY NEAR THE EARTH

In an energy diagram for one-dimensional motion, energy is plotted vertically, distance horizontally. For motion near the earth, let us again choose a z axis directed vertically upward, with its origin at ground level, and write for the potential energy

$$U = mgz.$$

In the energy diagram, first plot the potential-energy function $U(z)$, in this case a straight line of slope mg passing through the origin (Figure 10.17). This slope, according to Equation 10.78, gives the magnitude of the force. Next draw a horizontal line whose height is equal to the total energy E. It is horizontal because the energy E is the same at all distances z. Finally, a vertical line may be drawn connecting any point on the potential-energy curve (or in this case, the potential-energy line) with the point on the energy line directly above it. The length of this vertical line gives the magnitude of the kinetic energy since it is the difference between total energy and potential energy.

Potential energy curve: Its slope gives the force

The energy diagram is considerably more abstract than a physical picture of the actual motion, in part because vertical distances in the diagram are used to measure more than one quantity, in part because a single diagram represents more than one example of motion. Nevertheless, the energy diagram has a simplicity of its own and will repay careful study. It is in its own way a picture of motion. If a baseball is thrown vertically upward with some total energy E, this energy is initially all in the form of kinetic energy, which gradually is converted to potential energy as the ball rises. The rise of the ball corresponds to moving to the right in the energy diagram until the potential energy line and the total energy line meet (from A to B to C). This meeting point, where $K = 0$, is called a turning point of the motion. It locates the maximum height possible for the given value of energy. Another motion described by identically the same energy diagram would be that of a ball dropped from this maximum

The energy diagram: a useful aid to analysis

Turning point for $K = 0$

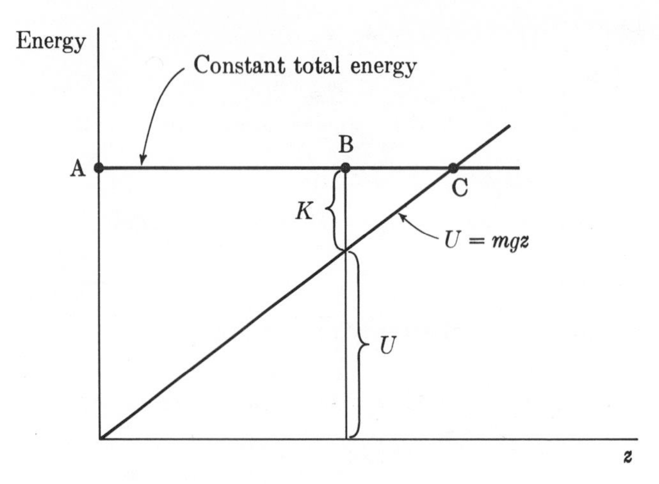

FIGURE 10.17 Energy diagram for vertical motion near the earth. Point C locates a turning point of the motion.

height. It moves to the left in the diagram, from C to B to A. Alternatively, a ball with the same total energy (and therefore the same energy diagram) might start at a lower height and be thrown downward with an initial velocity so that it starts out with some kinetic energy and some potential energy. Its progress in the energy diagram is only from B to A.

The different motions just described have different initial conditions but the same energy. For a given energy, the maximum possible height is defined by the intersection of the energy line and the potential energy line, although this maximum height need not always be reached. For a different energy, the diagram requires but little modification. The horizontal energy line is simply slid up or down to a new level. For greater energy, the maximum height increases in proportion to the energy. To summarize: The advantages of the energy diagram are that it shows at a glance the limit of the motion (in this example the maximum height); it simultaneously describes motions with different initial conditions; it gives pictorially the relative magnitudes of the kinetic energy and the potential energy at any point; from the slope of the potential-energy curve, it gives the force; and it reveals simply the effect of changing the energy. One disadvantage is that it does not directly show how fast the motion occurs from one part of the diagram to another. However, it does give the kinetic energy directly, which is easily related to speed.

■ EXAMPLE 1: A body falls from rest. Can the conservation-of-energy principle be used to determine all of the details of its motion? Yes. The energy of the body, if it starts from rest at $z = z_0$, is $E = mgz_0$. The conservation-of-energy equation, $K + U = E$, then reads

E = constant gives a differential equation

$$\tfrac{1}{2}m\left(\frac{dz}{dt}\right)^2 + mgz = mgz_0. \tag{10.89}$$

This is a *first-order differential equation* whose solution gives $z(t)$, the position as a function of time. From this equation,

$$\frac{dz}{dt} = -\sqrt{2g(z_0 - z)}.$$

(Why is the negative square root chosen in this example?) The derivative of time with respect to the z coordinate is, therefore,

$$\frac{dt}{dz} = -\frac{1}{\sqrt{2g(z_0 - z)}}.$$

This derivative can be substituted in the equation

$$t = \int \left(\frac{dt}{dz}\right) dz$$

to give

$$t = -\frac{1}{\sqrt{2g}} \int \frac{dz}{\sqrt{z_0 - z}}. \qquad (10.90)$$

The integral on the right can be treated as an indefinite integral, the constant of integration being adjusted later to fit the boundary condition. A change of variable to $s = z_0 - z$ (which is the distance of fall) gives

$$t = \frac{1}{\sqrt{2g}} \int s^{-1/2}\, ds$$

$$= \frac{2}{\sqrt{2g}} (\sqrt{s} + C),$$

where C is the constant of integration. To match our initial condition—which is $s = 0$ at $t = 0$—the constant C must be set equal to zero. The displacement as a function of time then takes a familiar form:

A familiar result reached by a new route

$$s = z_0 - z = \tfrac{1}{2}gt^2.$$

For this particular example, direct application of Newton's second law is somewhat simpler than using the law of energy conservation. In other examples, energy conservation may provide the easiest route to a solution. ■

THE HARMONIC OSCILLATOR

An energy diagram for the simple harmonic oscillator appears in Figure 10.18. The heavy curve is a graph of the potential-energy function

$$U = \tfrac{1}{2}kx^2.$$

This is a parabola. Its increasing positive slope for $x > 0$ indicates a force directed to the left, whose magnitude grows in proportion to x. Its negative slope for $x < 0$ indicates a force directed to the right. At $x = 0$, where the force is zero, the function $U(x)$ has zero slope.

For zero energy, the oscillator is quiescent at its midpoint (energy line A). For any greater energy, it moves back and forth between its turning points in the

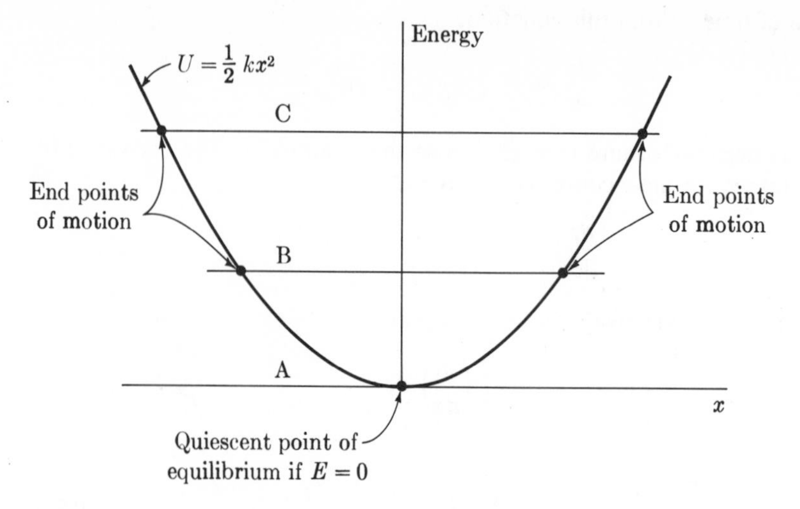

FIGURE 10.18 Energy diagram for the harmonic oscillator. The potential energy curve is a parabola. Lines A, B, and C designate three different magnitudes of total energy.

symmetrical energy diagram. At these limits, its total energy E and its potential energy are equal:

$$\tfrac{1}{2}k{x_{\text{limit}}}^2 = E.$$

The limiting displacement is therefore given by

$$x_{\text{limit}} = \pm\sqrt{\frac{2E}{k}}.$$

The amplitude may be defined as the magnitude of x_{limit},

Amplitude of the harmonic oscillator

$$A = |x_{\text{limit}}| = +\sqrt{\frac{2E}{k}}. \tag{10.91}$$

Doubling the energy (energy line B to energy line C) multiplies the amplitude of oscillation by $\sqrt{2}$.

The quantity k is a "stiffness" constant. A soft spring has a small value of k; a stiffer spring has a larger value of k. According to Equation 10.85, a larger value of k produces a more rapidly rising potential-energy curve. This in turn produces, for the same energy, a smaller amplitude of oscillation (Equation 10.91).

For the simple harmonic oscillator, the energy conservation equation reads

$$\tfrac{1}{2}mv^2 + \tfrac{1}{2}kx^2 = E.$$

The substitution $v = dx/dt$ converts this to a differential equation,

Again a first-order differential equation from E = constant

$$\tfrac{1}{2}m\left(\frac{dx}{dt}\right)^2 + \tfrac{1}{2}kx^2 = E. \tag{10.92}$$

Just by looking at the energy diagram, we can conclude that the oscillator must execute periodic motion that is symmetrical about $x = 0$. It is left as a problem to prove that the solution to Equation 10.92 is

$$x = A \sin(\omega t + \varphi), \tag{10.93}$$

where A is given by Equation 10.91, ω is given by

$$\omega = \sqrt{\frac{k}{m}},$$

and φ is an arbitrary constant. This is the same solution given by Equation 7.35. Recall that it was obtained previously from the second-order differential equation that expressed Newton's second law,

$$m \frac{d^2x}{dt^2} = -kx.$$

The energy equation, by contrast, leads to a first-order differential equation.

★SMALL OSCILLATIONS IN GENERAL

It has been emphasized previously that many systems executing oscillatory motion behave as simple harmonic oscillators. The energy diagram helps to explain why. Consider an arbitrary potential-energy function $U(x)$ (Figure 10.19). If this function has a minimum at some point x_0 and if the second derivative, d^2U/dx^2, does not vanish at that point, the function $U(x)$ near x_0 will resemble the parabolic function of the harmonic oscillator. If the system described by this function has an energy E not too much greater than the minimum potential energy U_0 (see the figure), it will oscillate harmonically about x_0.

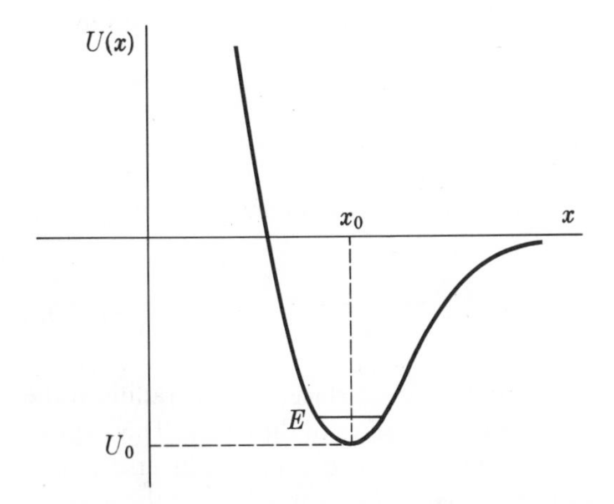

FIGURE 10.19 A potential energy function has a minimum, U_0, near which the function is well approximated by a parabola. For an energy E not much greater than U_0, the system executes approximately simple harmonic motion. The particular function shown here represents the potential energy within a diatomic molecule: x is the separation of the two nuclei, and x_0 is their equilibrium separation. The nuclei can vibrate harmonically.

We can make these ideas quantitative with the help of a Taylor series expansion of the potential energy U about the point x_0:

Taylor series for U near its minimum

$$U(x) = U(x_0) + \left(\frac{dU}{dx}\right)_{x_0} (x - x_0) + \frac{1}{2}\left(\frac{d^2U}{dx^2}\right)_{x_0} (x - x_0)^2 + \cdots.$$

Since x_0 has been chosen to be a point where $U(x)$ has a minimum, then $(dU/dx) = 0$ at x_0. The Taylor series thus gives, as an approximation to $U(x)$ for x near x_0,

$$U(x) \cong U_0 + \frac{1}{2}\left(\frac{d^2U}{dx^2}\right)_{x_0} (x - x_0)^2. \tag{10.94}$$

This is exactly equivalent to the harmonic oscillator form,

$$U = \tfrac{1}{2}kx^2,$$

differing only in having the zeros of U and of x shifted. The force constant for the arbitrary potential is

$$k = \left(\frac{d^2U}{dx^2}\right)_{x_0}. \tag{10.95}$$

If the energy difference, $E - U_0$, is sufficiently small, the motion will be confined to a narrow range of x near x_0 (see Figure 10.19), and the next term in the Taylor series,

$$\frac{1}{6}\left(\frac{d^3U}{dx^3}\right)_{x_0} (x - x_0)^3,$$

Oscillation near a potential energy minimum is harmonic

will be negligible. Then the system, whatever it may be, will execute simple harmonic motion. If the cubic term in the potential is *not* negligible, the vibration is said to be *anharmonic*. It is then more complicated and no longer described by a sine function.

GRAVITY IN THE LARGE

Outside the earth, a body of mass m experiences a radial component of gravitational force given by

Gravitational force outside the earth

$$F_r = -\frac{mgR^2}{r^2}, \tag{10.96}$$

where g is the acceleration of gravity at the earth's surface, R is the radius of the earth, and r is the distance of the body from the center of the earth. This expression correctly gives the force $F_r = -mg$ for $r \cong R$ and the inverse-square dependence on r for larger r. (A more fundamental expression for the gravitational force is given by Equation 11.10.) If we consider motion along a radial line, we can write for the potential energy,

$$U = -\int_a^r F_r(r')\, dr'.$$

Substituting the right side of Equation 10.96 for F_r gives

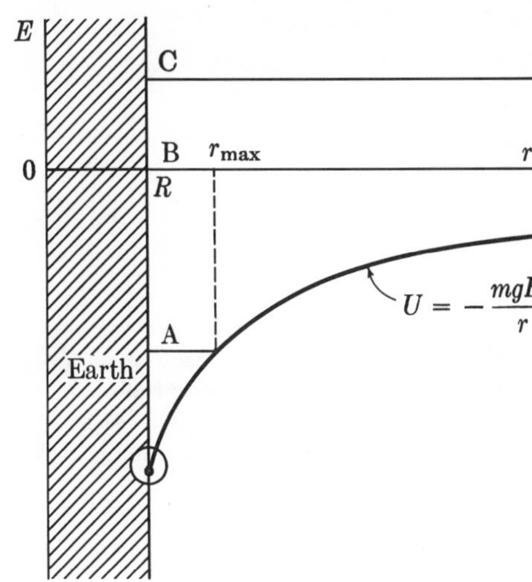

FIGURE 10.20 Energy diagram for an object influenced by the earth's gravity. The vertical scale is chosen to place zero potential energy at infinite distance. The horizontal scale measures distance from the center of the earth. The shaded region represents the interior of the earth. Energy line A is associated with a vertically fired rocket which rises to a maximum radial distance r_{max}. Energy line B ($E = 0$) is associated with a rocket that is barely able to escape the earth. Energy line C describes a rocket that escapes with energy to spare.

$$U = +mgR^2 \int_a^r \frac{dr'}{r'^2} = -mgR^2 \frac{1}{r'}\bigg|_a^r = -mgR^2\left(\frac{1}{r} - \frac{1}{a}\right).$$

It is conventional to set $a = \infty$, which places the zero of potential energy at infinity. With this choice, the function $U(r)$ takes on a simple form:

Gravitational potential energy outside the earth

$$U(r) = -\frac{mgR^2}{r}. \tag{10.97}$$

A graph of this potential-energy function appears in Figure 10.20 (it is valid only for $r \geq R$). The small circle in the diagram near $r = R$ encloses the region where the linear approximation

$$U \cong mg(r - R) + constant \tag{10.98}$$

is valid. (To show the equivalence of Equations 10.97 and 10.98 for $r - R \ll R$ is left as a problem.) In this circled region, the straight-line graph in Figure 10.17 adequately approximates the hyperbola graphed in Figure 10.20.

For a rocket fired vertically, the energy equation (during its coasting period) is

$$\tfrac{1}{2}mv^2 - \frac{mgR^2}{r} = E.$$

For $E < 0$ (energy line A in the figure), there is a maximum radius, found by setting $v = 0$:

$$-\frac{mgR^2}{r_{max}} = E,$$

which gives

$$r_{max} = -\frac{mgR^2}{E}. \tag{10.99}$$

$E < 0$: bounded motion

This evidently requires negative energy to be meaningful. As the energy diagram makes clear, there is no upper limit on the radial motion if $E \geq 0$. For positive

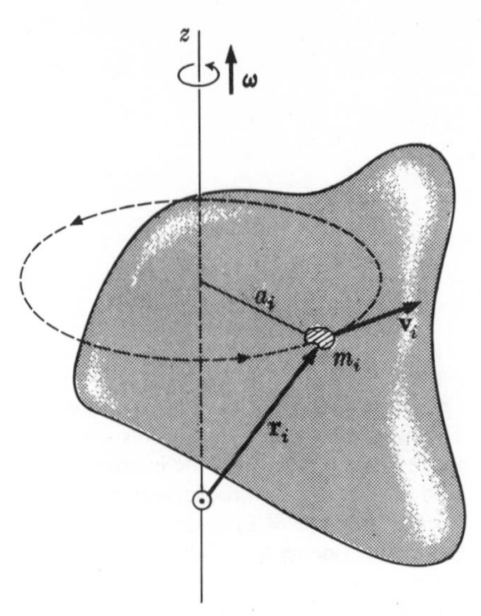

FIGURE 10.21 Rotational energy. As in Figure 9.12, a rigid body rotates about an arbitrary fixed axis. An element of mass m_i has kinetic energy $K_i = \frac{1}{2}m_i v_i^2 = \frac{1}{2}\omega^2 m_i a_i^2$. The total kinetic energy of the body is given by Equation 10.105.

$E \geq 0$: unbounded motion

energy (energy line C), the rocket coasts away to infinity with final potential energy approaching zero, and final kinetic energy approaching E, so that its final speed is

$$v_f = \sqrt{\frac{2E}{m}}. \qquad (10.100)$$

■ EXAMPLE 2: A rocket is given a short burst of acceleration near the earth and then coasts vertically upward. What is the minimum speed that it must have as it starts its coasting motion in order that it escape from the earth? This is called its *escape speed.* It is clear from the energy diagram that the required condition is $E = 0$ (energy line B). For any lesser E, the rocket does not escape. For any greater E, it escapes with kinetic energy to spare. If the vertical ascent of the rocket during its acceleration phase is ignored, the energy equation at $r \cong R$ can be written

$$\tfrac{1}{2}mv_{esc}^2 - mgR = 0.$$

The solution is

$$v_{esc} = \sqrt{2gR}. \qquad (10.101)$$

Escape speed from the earth

Substitution of $g = 9.80$ m/sec and $R = 6.37 \times 10^6$ m gives

$$v_{esc} = 1.12 \times 10^4 \text{ m/sec}. \qquad (10.102)$$

This speed is about 34 times the speed of sound in air, or about 25,000 mile/hr. Comparison of Equations 10.101 and 7.71 shows that the escape speed exceeds the orbital speed at low altitude by a factor of $\sqrt{2}$. ■

10.9 Rotational energy

Consider a rigid body rotating about a fixed axis. As in previous discussions, we may think of the body as being composed of a set of particles of masses m_i, position vectors $\mathbf{r}_i$, and velocities $\mathbf{v}_i$ (Figure 10.21). Its kinetic energy is

$$K = \tfrac{1}{2} \sum_i m_i v_i^2.$$

The speed v_i may be expressed in terms of the angular speed ω and the distance a_i of the mass m_i from the axis of rotation:

$$v_i = \omega a_i. \tag{10.103}$$

The kinetic energy may therefore be written

$$K = \tfrac{1}{2}\omega^2 \sum_i m_i a_i^2. \tag{10.104}$$

We recognize the sum as the moment of inertia of the body for the chosen axis of rotation (see Equation 9.36), so the kinetic energy takes the simple form

Kinetic energy of a rigid body rotating about a fixed axis

$$K = \tfrac{1}{2} I \omega^2. \tag{10.105}$$

For a set of discrete particles, we have

$$I = \sum_i m_i a_i^2. \tag{10.106}$$

For a continuous distribution of matter, the sum becomes an integral,

$$I = \int a^2 \, dm. \tag{10.107}$$

(These formulas repeat Equations 9.36 and 9.37.)

■ EXAMPLE 1: A flywheel in the form of a uniform disk of radius 15 cm rotates with angular speed ω = 3,000 rpm. Its mass is 25 kg. What is the kinetic energy of the flywheel? How fast would it have to move in a straight line in order to have an equal translational kinetic energy? First, let us express its angular speed in radian/sec:

$$\omega = \frac{3{,}000 \text{ rev/min} \times 2\pi \text{ radian/rev}}{60 \text{ sec/min}}$$
$$= 314 \text{ radian/sec}.$$

According to Equation 9.43, the moment of inertia of a disk is $I = \tfrac{1}{2}MR^2$. Putting this expression for I into Equation 10.105 gives for the rotational kinetic energy of the disk

$$K = \tfrac{1}{4}MR^2\omega^2$$
$$= \tfrac{1}{4} \times 25 \text{ kg} \times (0.15 \text{ m})^2 \times (314 \text{ radian/sec})^2$$
$$= 1.39 \times 10^4 \text{ J}.$$

To find the equivalent translational speed, equate the expressions for translational and rotational kinetic energies:

$$\tfrac{1}{2}Mv^2 = \tfrac{1}{4}MR^2\omega^2.$$

The solution is $v = R\omega/\sqrt{2}$, or

$$v = 33.3 \text{ m/sec}.$$

■

A derivation in Section 10.4 led to Equation 10.52, which expresses the kinetic energy of a system as the sum of a translational kinetic energy associated with motion of the center of mass and a kinetic energy relative to the center of mass. For an ideally rigid body, the latter term is the same as a rotational kinetic energy since in a frame of reference in which the center of mass is at rest, the only possible motion of a rigid body is rotation about an axis passing through the center of mass. For any motion of a rigid body, the total kinetic energy may therefore be written

Kinetic energy of a rigid body with arbitrary motion

$$K = \tfrac{1}{2}Mv_c^2 + \tfrac{1}{2}I\omega^2. \tag{10.108}$$

The moment of inertia and the angular speed in the second term both refer to the center-of-mass frame of reference, with an axis of rotation passing through the center of mass.

■ EXAMPLE 2: A hoop rolls down a plane inclined at angle θ to the horizontal (Figure 10.22). If the work done by frictional forces is negligible, what is the acceleration of the center of mass of the hoop? Interestingly, we need to know neither the mass nor the radius of the hoop. Let s be the displacement of the center of mass of the hoop down the plane; then $v_c = ds/dt$ is the center-of-mass speed [Figure 10.22(a)]. Superimposed on this translational motion is the rotational motion of the hoop about its center of mass. In the center-of-mass frame, it is easy to see that $\omega = v_c/R$ [Figure 10.22(b)]. In this frame, the hoop, at its point of contact with the plane, has velocity $-\mathbf{v}_c$ directed uphill. This velocity, added to the velocity $\mathbf{v}_c$ of the center of mass, gives the required zero velocity of the hoop at its point of contact with the plane in the laboratory frame of reference. Since for a hoop $I = MR^2$ (Equation 9.30), Equation 10.108 gives for the total kinetic energy

$$K = \tfrac{1}{2}Mv_c^2 + \tfrac{1}{2}MR^2\left(\frac{v_c}{R}\right)^2,$$

or

$$K = Mv_c^2. \tag{10.109}$$

The potential energy of the hoop is $U = Mgy$. As shown in Figure 10.22(a), y can be written

$$y = y_0 - s\sin\theta.$$

The equation of energy conservation, $K + U = E$, then takes the form

$$Mv_c^2 + Mg(y_0 - s\sin\theta) = E. \tag{10.110}$$

To get the acceleration, dv_c/dt, we may differentiate this equation with respect to time. Only v_c and s are variables. The result is

$$2Mv_c\frac{dv_c}{dt} - Mg\sin\theta\frac{ds}{dt} = 0.$$

Since $v_c = ds/dt$, the factors v_c and ds/dt cancel, as do the factors M. The center-of-mass acceleration is therefore

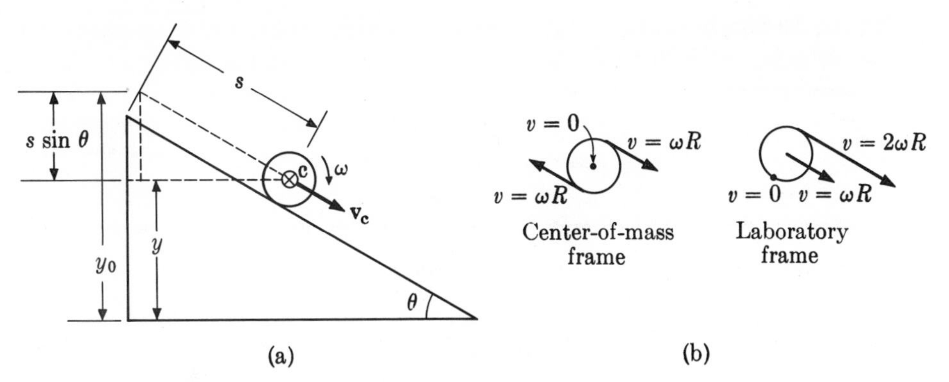

FIGURE 10.22 (a) A hoop rolls down a plane. Its potential energy is $U = Mgy$. (b) Comparison of the hoop's motion in the center-of-mass frame of reference and in the laboratory frame.

$$a_c = \frac{dv_c}{dt} = \tfrac{1}{2}g \sin \theta. \tag{10.111}$$

This is half the acceleration of a body sliding without friction down a plane. ■

Returning to rotation about a fixed axis, we may derive an interesting connection between torque and the rate of change of kinetic energy. A starting point is the law of angular-momentum change, Equation 9.52:

$$\frac{d\mathbf{L}}{dt} = \mathbf{T}.$$

Let the z axis be the axis of rotation. Then $L_z = I\omega$, and the z component of this equation is

$$\frac{d}{dt}(I\omega) = T_z. \tag{10.112}$$

Multiplying both sides by ω gives

$$I\omega \frac{d\omega}{dt} = T_z\omega.$$

The left side is evidently the rate of change of rotational kinetic energy. This equation may be written

Rate of change of rotational energy

$$\frac{d}{dt}(\tfrac{1}{2}I\omega^2) = T_z\omega. \tag{10.113}$$

Since rate of change of energy is power, we have an expression for the power delivered by a torque to a rotating body:

Power delivered by a torque

$$P = T_z\omega. \tag{10.114}$$

Note the close analogy of Equation 10.114 to Equation 10.62. This and other correspondences between rotational motion and translational motion are summarized in Table 10.1 (see also Table 9.1).

TABLE 10.1 CORRESPONDING FORMULAS FOR TRANSLATIONAL MOTION AND ROTATIONAL MOTION

Physical Quantity	Formula for Translational Motion	Formula for Rotational Motion about z axis
Acceleration	$F = m\mathbf{a} = m\dfrac{d\mathbf{v}}{dt}$	$T_z = I\alpha = I\dfrac{d\omega}{dt}$
Kinetic energy	$K = \frac{1}{2}mv^2$	$K = \frac{1}{2}I\omega^2$
Work	$dW = \mathbf{F} \cdot d\mathbf{s}$	$dW = T_z\, d\theta$
Power	$P = \mathbf{F} \cdot \mathbf{v}$	$P = T_z\omega$

★10.10 Conservative forces and potential energy in general

In this section we will restrict attention to particle mechanics and consider the general definition of a conservative force in three dimensions. If a particle is acted upon by a force **F** (which need not be constant) as it moves from point A to point B, this force does a certain amount of work on the particle, which may be called W_{AB}:

$$W_{AB} = \int_A^B \mathbf{F} \cdot d\mathbf{s} = \int_A^B F_s\, ds, \tag{10.115}$$

where F_s is the component of **F** parallel to the increment of displacement $d\mathbf{s}$. In general, this work might depend upon the particular path followed by the particle as it moves from A to B. For example, the work done on an airplane by drag forces depends on the path followed by the plane through the air—on whether it flies high or low. If, however, the work does *not* depend on the path, the force is said to be a *conservative force* (Figure 10.23).* Several important forces in nature are conservative forces. The path-independence of the work for such forces means that W_{AB} depends only on the end points, A and B, not on points in between. In fact, as we may easily demonstrate, W_{AB} depends only on the *difference* of two functions, one depending on A and one on B. Let us choose a fixed origin O and define the potential-energy function U as follows:

The work done by a conservative force is path-independent

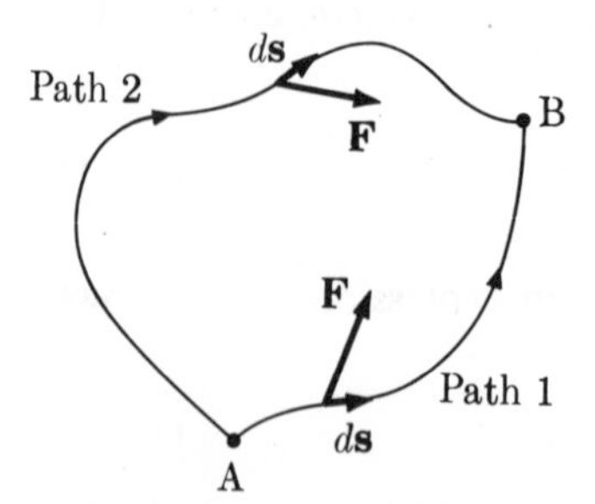

FIGURE 10.23 Definition of a conservative force: If the work W_{AB} done by the force in a displacement from point A to point B depends only on the end points, not on the path, the force is conservative. Potential energy is associated with a conservative force.

* For one-dimensional motion, *every* force depending only on position is a conservative force because there is only one path and one possible magnitude of work for displacement from one point to another.

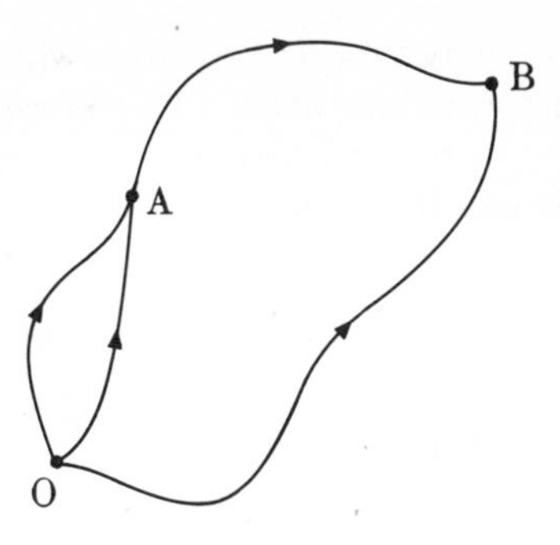

FIGURE 10.24 Definition of potential energy. An arbitrary zero is chosen at point O. Then $U(\mathrm{A}) = -W_{\mathrm{OA}}$, which is independent of the path from O to A; and $U(\mathrm{B}) = -W_{\mathrm{OB}}$, independent of the path from O to B. Consideration of a path from O through A to B leads to Equation 10.118.

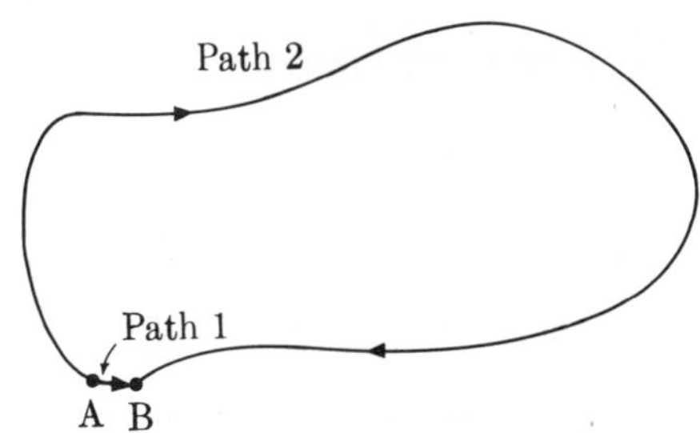

FIGURE 10.25 Diagram used to illustrate the fact that a conservative force does zero net work in a displacement around a closed loop.

$$U(\mathbf{r}_\mathrm{B}) = -W_{\mathrm{OB}} = -\int_\mathrm{O}^\mathrm{B} \mathbf{F} \cdot d\mathbf{s}. \qquad (10.116)$$

Potential energy

Since $\mathbf{F}$ is conservative, U depends only on the position of the end point, not on the path. Suppose we choose a particular path from O to B that passes through a point A (Figure 10.24). Then the integral in Equation 10.116 is

$$\int_\mathrm{O}^\mathrm{B} \mathbf{F} \cdot d\mathbf{s} = \int_\mathrm{O}^\mathrm{A} \mathbf{F} \cdot d\mathbf{s} + \int_\mathrm{A}^\mathrm{B} \mathbf{F} \cdot d\mathbf{s}. \qquad (10.117)$$

The first term on the right defines $-U(\mathbf{r}_\mathrm{A})$; the second term is W_{AB}, the work done as the particle moves from A to B. With a slight rearrangement of terms, Equation 10.117 can therefore be written

$W = -\Delta U$ in three dimensions

$$W_{\mathrm{AB}} = U(\mathbf{r}_\mathrm{A}) - U(\mathbf{r}_\mathrm{B}). \qquad (10.118)$$

The work done by a conservative force on a particle that moves from A to B—by any path—is the negative of the change of potential energy, $-[U(\mathbf{r}_\mathrm{B}) - U(\mathbf{r}_\mathrm{A})]$. This generalizes Equation 10.71 to three dimensions.

A simple consequence of Equation 10.118 is the fact that a conservative force does no net work on a particle that executes a closed path and returns to its starting point. Formally, from Equation 10.118, we can write

$$W_{\mathrm{AA}} = U(\mathbf{r}_\mathrm{A}) - U(\mathbf{r}_\mathrm{A}) = 0. \qquad (10.119)$$

To see the reason for this result more clearly, consider end points A and B separated by an infinitesimal distance (Figure 10.25). As B approaches A, the work done on the shortest direct path from A to B must approach zero since $d\mathbf{s} \to 0$. Since the work is the same for all paths, it must also approach zero for an arbitrary circuitous path from A to B.*

The general connection between force and potential energy in three dimensions is important in more advanced treatments of mechanics, and it is worth mentioning here. It involves the idea of the *partial derivative*. If a function

* A common notation for a line integral around a closed path is $\oint$. For a conservative force, $\oint \mathbf{F} \cdot d\mathbf{s} = 0$.

Partial derivative defined

depends on more than one variable, we may take its derivative with respect to one of its variables, holding the other variables fixed. This is called partial differentiation. The notation $\partial f/\partial x$ means the derivative of f with respect to x, other variables being treated as constants. If, for instance,

$$f = x^2 y,$$

its partial derivatives are

$$\frac{\partial f}{\partial x} = 2xy,$$

$$\frac{\partial f}{\partial y} = x^2.$$

With this understanding of the meaning of a partial derivative, we may state the relations between force and potential energy:

Components of a conservative force

$$F_x = -\frac{\partial U}{\partial x}, \tag{10.120}$$

$$F_y = -\frac{\partial U}{\partial y}, \tag{10.121}$$

$$F_z = -\frac{\partial U}{\partial z}. \tag{10.122}$$

We shall not attempt to derive these equations. Note that they generalize Equation 10.78, derived previously for one-dimensional motion.*

To make clear the concept of path independence and conservative forces, we consider two important laws of force.

A CONSTANT FORCE

A body that is near the earth or a charged particle that is in a uniform electric field experiences a constant force. To be precise, suppose that the force acts in the negative y direction (Figure 10.26) so that it can be written

$$\mathbf{F} = -F\mathbf{j} \tag{10.123}$$

Let us consider the line-integral definition of potential energy, Equation 10.116, for the three paths shown in the figure. Remembering that $\mathbf{F} \cdot d\mathbf{s} = F_s\, ds$, for path 1 we can write

$$U(\mathbf{r}_P) = -\int_0^a F_x(x, 0)\, dx - \int_0^b F_y(a, y)\, dy. \tag{10.124}$$

Along the first segment of the path, represented by the first integral, x varies from 0 to a, and $y = 0$. Along the second segment, represented by the second

* In vector form, Equations 10.120–10.122 are sometimes written $\mathbf{F} = -\nabla U$ or $\mathbf{F} = -\mathbf{grad}\ U$, which are shorthand expressions for $\mathbf{F} = -[(\partial U/\partial x)\mathbf{i} + (\partial U/\partial y)\mathbf{j} + (\partial U/\partial z)\mathbf{k}]$. The quantity in brackets defines the *gradient* in three dimensions. A conservative force is the negative gradient of a potential-energy function.

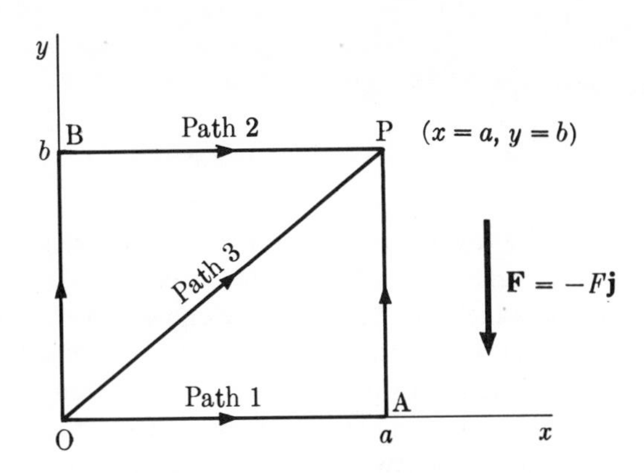

FIGURE 10.26 Three paths in a constant force field. The work is independent of the path.

integral, $x = a$, and y varies from 0 to b. Since $F_x = 0$ everywhere, the first integral vanishes. Since $F_y = -F$, a constant, the second integral is easily evaluated, leading to

$$U(r_P) = +Fb.$$

(If $F = mg$, the right side is mgb, the familiar expression for gravitational potential energy at height b.) Similar reasoning for path 2 leads to

$$U(\mathbf{r}_P) = -\int_0^b F_y(0, y)\,dy - \int_0^a F_x(x, b)\,dx.$$

This time the second integral vanishes. The first leads again to

$$U(r_P) = Fb.$$

Path 3 is only slightly more difficult to handle. For any increment of displacement along this path, write

$$d\mathbf{s} = dx\mathbf{i} + dy\mathbf{j}.$$

The scalar product $\mathbf{F} \cdot d\mathbf{s}$ is

$$\mathbf{F} \cdot d\mathbf{s} = F_x\,dx + F_y\,dy. \tag{10.125}$$

This expression would be true for *any* path. What distinguishes a particular path is the choice of arguments x and y for the force components. If we pretend for the moment that F_x and F_y *do* depend on x and y, for path 3 we would write

$$\mathbf{F} \cdot d\mathbf{s} = F_x(x, x\tan\theta)\,dx + F_y\left(\frac{y}{\tan\theta}, y\right)dy$$

since $y = x\tan\theta$ along this path (Figure 10.26). Actually, in this case, $F_x = 0$ and $F_y = -F$, so for path 3, or for *any* path,

Path independence for a constant force

$$U(\mathbf{r}_P) = F\int_0^b dy = Fb. \tag{10.126}$$

The initial assumption is verified: A constant force is a conservative force. Thus the expression for gravitational potential energy near the earth, $U = mgy$, is shown to be correct for two- and three-dimensional motion, as well as for one-dimensional motion.

INVERSE-SQUARE LAW OF FORCE

Consider a radially directed force given by

An inverse-square central force

$$\mathbf{F} = \frac{C\mathbf{i}_r}{r^2}, \tag{10.127}$$

where $\mathbf{i}_r$ is a unit vector directed radially away from the origin and C is a constant—positive for a repulsive force, negative for an attractive force. This is the so-called inverse-square law of force, which, with appropriate choices of C, can describe both gravitational and electrical forces. Suppose that a particle in such a field of force is displaced along the path shown in Figure 10.27 from point A, a distance r_A from the force center, to point B, a distance r_B from the force center. From point A, it moves radially outward to a distance R, then swings in a circular arc, then continues radially outward to point B. How much work is done on it by the given force? In the first segment of its path, $\mathbf{F}$ is parallel to $d\mathbf{s}$, so

$$W_I = C\int_{r_A}^{R}\frac{dr}{r^2}$$

$$= C\left(\frac{1}{r_A} - \frac{1}{R}\right). \tag{10.128}$$

In the second segment, $\mathbf{F}$ is perpendicular to $d\mathbf{s}$; this means that $\mathbf{F}\cdot d\mathbf{s} = 0$, and

$$W_{II} = 0. \tag{10.129}$$

In the third segment, $\mathbf{F}$ is again parallel to $d\mathbf{s}$. The work is

$$W_{III} = C\int_{R}^{r_B}\frac{dr}{r^2}$$

$$= C\left(\frac{1}{R} - \frac{1}{r_B}\right). \tag{10.130}$$

The sum of these three contributions is the total work:

$$W_{AB} = W_I + W_{II} + W_{III}$$

$$= \frac{C}{r_A} - \frac{C}{r_B}. \tag{10.131}$$

We may at once notice two things about this expression. First, it is independent

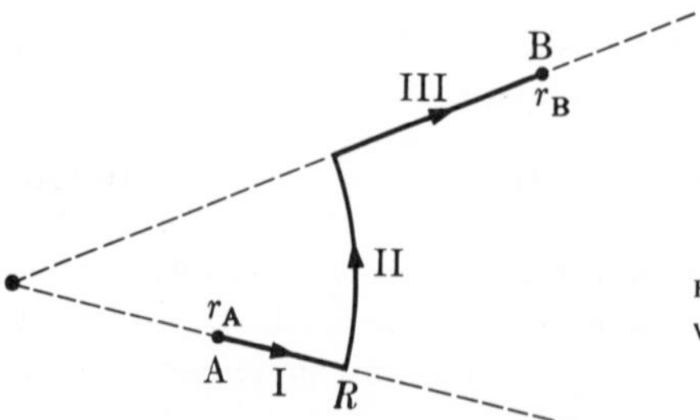

FIGURE 10.27 Path used to calculate the work done by an inverse-square central force.

of R; the work does not depend on the particular circular arc chosen to move from one radial line to the other. This is a partial proof of path independence. A full proof is left as a problem. Second, the expression has the form of Equation 10.118, the difference of a function evaluated at two positions. Comparison of Equations 10.131 and 10.118 shows that the potential-energy function may be chosen to be

$$U(r) = \frac{C}{r} + D,$$

where D is an arbitrary constant. Usually, D is chosen to be zero. This is equivalent to placing the zero of potential energy at $r = \infty$. Then

$$U = \frac{C}{r}. \qquad (10.132)$$

Its potential energy if $U(\infty) = 0$

This agrees with Equation 10.97, derived for one-dimensional motion. (In that equation, the constant C was $-mgR^2$.) Equations 10.127 and 10.132 illustrate a relationship between force and potential energy that is more generally valid. It is

$$F_r = -\frac{\partial U}{\partial r}. \qquad (10.133)$$

(When U depends only on r, the partial derivative is the same as an ordinary, or total, derivative.)

■ EXAMPLE: How much work must be done on a satellite by forces other than gravity in order to put it into a circular orbit a distance $2R$ from the center of the earth (R is the radius of the earth)? To what height could it have been raised if the same work had been done in launching it vertically upward? According to Equation 10.72, the work done by forces of type 2 is equal to the change in mechanical energy:

$$\Delta(K + U) = W_2.$$

Equation 10.72 remains valid in three dimensions

This equation remains valid in three dimensions. In moving from R to $2R$, the change of potential energy of the satellite is

$$\begin{aligned} U &= U(2R) - U(R) \\ &= -\frac{mgR^2}{2R} + \frac{mgR^2}{R} \\ &= +\tfrac{1}{2}mgR. \end{aligned} \qquad (10.134)$$

Its kinetic energy at the earth's surface is zero. To determine its kinetic energy in a circular orbit, we may use Newton's second law and equate its mass times its centripetal acceleration ($m \times v^2/r$) to the magnitude of the gravitational force acting on it (mgR^2/r^2), evaluated at $r = 2R$:

$$\frac{mv^2}{2R} = \frac{mgR^2}{4R^2}.$$

This gives $\frac{1}{2}mv^2 = \frac{1}{4}mgR$. Since the kinetic energy in orbit is the same as the change of kinetic energy, we have

$$\Delta K = \tfrac{1}{4}mgR. \qquad (10.135)$$

The required work, equal to $\Delta K + \Delta U$, is, therefore,

$$W_2 = \tfrac{3}{4}mgR,$$

two-thirds of which goes into the change of potential energy and one-third into kinetic energy. Had the satellite been launched vertically upward and allowed to coast to a maximum altitude, its final kinetic energy would have been zero so that

$$\Delta K = 0.$$

In that case,

$$\Delta U = W_2 = \tfrac{3}{4}mgR.$$

This means

$$-\frac{mgR^2}{r_{max}} + \frac{mgR^2}{R} = \tfrac{3}{4}mgR.$$

The solution for r_{max} is

$$r_{max} = 4R,$$

which means an altitude of $3R$ above the surface of the earth, three times higher than its orbital altitude. ■

10.11 The simple pendulum

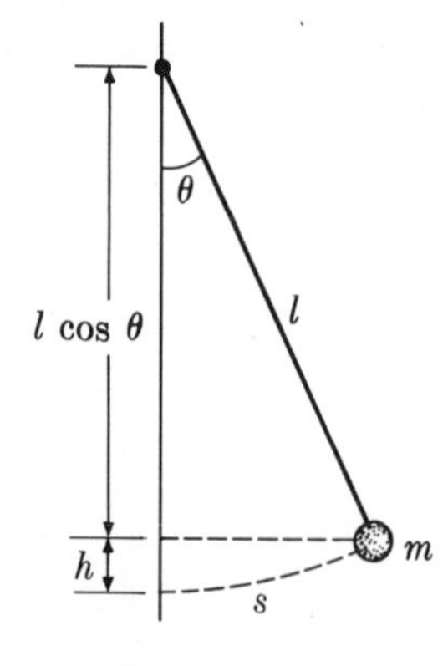

FIGURE 10.28 The simple pendulum. Its potential energy is $U = mgh$.

One of the most commonly encountered oscillators is the pendulum. The ideal simple pendulum consists of a bob of mass m suspended by a rod of length l, which is rigid but of negligible mass (Figure 10.28). In Section 7.6, the simple pendulum was studied using ideas of force and Newton's second law. Here we wish to present the much simpler analysis of its motion using the idea of gravitational potential energy.

As the rod swings away from its equilibrium position through angle θ, the pendulum bob rises through a vertical height h given by

$$h = l - l\cos\theta, \qquad (10.136)$$

and it gains potential energy

$$U = mgh. \qquad (10.137)$$

For small angles θ the cosine is approximated* by

$$\cos\theta \cong 1 - \tfrac{1}{2}\theta^2.$$

* If this approximation is unfamiliar, you may obtain it from $\sin^2 \tfrac{1}{2}\theta = \tfrac{1}{2}(1 - \cos\theta)$, using $\sin x \cong x$ for small x.

This means that

$$h \cong \tfrac{1}{2}l\theta^2,$$

or, in terms of the displacement s (which is equal to $l\theta$),

$$h \cong \frac{s^2}{2l}. \qquad (10.138)$$

Substitution of this small-angle approximation for h in Equation 10.137 gives for the potential energy of the pendulum

Potential energy of pendulum for small displacement

$$U = \frac{mg}{2l} s^2 \qquad (s \ll l), \qquad (10.139)$$

which has the same mathematical form as Equation 10.85. The kinetic energy of the pendulum is

$$K = \frac{1}{2} m \left(\frac{ds}{dt}\right)^2.$$

Its equation of energy conservation is, therefore,

It is approximately a harmonic oscillator

$$\frac{1}{2} m \left(\frac{ds}{dt}\right)^2 + \frac{1}{2}\left(\frac{mg}{l}\right) s^2 = E, \qquad (10.140)$$

equivalent to Equation 10.92 for the harmonic oscillator, with force constant k given by

$$k = \frac{mg}{l}. \qquad (10.141)$$

For small θ, the pendulum vibrates harmonically, and any previous result for the harmonic oscillator may be used. The period of the pendulum, for instance, is

$$T = 2\pi \sqrt{\frac{m}{k}} = 2\pi \sqrt{\frac{l}{g}},$$

independent of mass, as one expects for motion in a gravitational field. The proportionality of T to $\sqrt{l}$ was probably discovered first by Galileo.

10.12 Assessment of energy conservation

$K + U =$ constant is of limited validity

Within the framework of the theory of mechanics alone, the conservation of energy (kinetic plus potential) is often nearly correct and often useful, but it is not universally valid. Mechanical energy can be lost (via frictional forces, for instance) or gained (via mass to energy conversion, for instance).

$E_{\text{total}} =$ constant appears to be universally valid

Yet the very limitations of energy conservation as a law of mechanics have proved to be the sources of its power and significance as a more general principle of nature. The transformation of mechanical energy via heat to internal energy proved to be basic to the development of the theory of thermodynamics. Concern about the consistent transfer of energy from mechanical form to the form of electromagnetic radiation was an important ingredient of the foundation of relativity theory, and it led in turn to Einstein's deep insight into the equivalence of mass and energy.

Convertibility of energy unifies nature and unifies the description of nature

As nature's most convertible and many-faceted concept, energy plays a special role in science. It prevents man from compartmentalizing the phenomena of nature into rigid, well-separated disciplines. Mechanics cannot be separated from thermodynamics or from electromagnetism, nor can any of these subjects ignore relativity. As a common concept of every theory describing nature, energy joins together the parts of science. As the principal medium of exchange among the phenomena of nature, it literally joins the world together. The sea, the sky, and the land are joined by energy transfer, some of it as subtle as the evaporation of dew, some as violent as a summer thunderstorm. Through nuclear, electromagnetic, and gravitational energy, the galaxies join to form a whole. Through nature's most elaborate and wonderful sequence of energy conversions, the thermonuclear reactions of the sun energize the human body.

Because of its conservation, and only because of its conservation, energy is the central unifying concept of science. Without conservation, the disappearance of energy in one place could not be linked to the appearance of energy in different places or in different guises. Without conservation, two different manifestations of energy would instead be two distinct concepts. Without conservation, man could not so easily buy and sell and ship energy like a commodity.

Energy conservation is related to the uniformity of time

In this century, man's view of energy has deepened in two important ways, first through its connection with mass and second through its connection with the time symmetry of nature. Although it is beyond the scope of this text to discuss in detail this latter connection, it is so fundamental and at the same time so startling that it must at least be mentioned. The conservation of energy has been linked to the uniformity of time, more particularly to the invariance of the laws of nature with respect to change of time. Energy is conserved, according to the modern argument, because the laws of nature are the same yesterday, today, and tomorrow. The enterprising student in search of a hint about the reason for this remarkable connection should look first at Section 21.5 and then at Section 4.7 again.

Summary of ideas and definitions

Definitions of work:

for a constant force, $W = \mathbf{F} \cdot \mathbf{s}$; (10.1)

in one dimension, $W = \int F_x\, dx$; (10.24)

in general, $dW = \mathbf{F} \cdot d\mathbf{s}$, $W = \int \mathbf{F} \cdot d\mathbf{s}$. (10.22, 10.30)

Work is a mode of energy transfer; it is defined over an interval, not at a point.

Energy may be defined as that which can do work (either directly or via a chain of transformation).

Positive work: The system acted upon by the force gains energy. Negative work: The system loses energy.

Work done on a particle or a rigid body changes its kinetic energy: $\Delta K = W$ if W is the total work from all external forces (Equation 10.35).

For an arbitrary system, $\Delta K_{\mathrm{T}} = W_{\mathrm{T}}$ if W_{T} is the total work done by internal and external forces (Equation 10.42).

In one dimension, Newton's second law may be transformed to $F_x = dK/dx$ (Equation 10.28).

The kinetic energy of a system is,

$$K_T = \tfrac{1}{2}Mv_c^2 + \tfrac{1}{2}\sum_i m_i v_i'^2. \qquad (10.52)$$

The kinetic energy of a two-body system is

$$K = \tfrac{1}{2}Mv_c^2 + \tfrac{1}{2}\mu v^2. \qquad (10.58)$$

The kinetic energy of a rigid body is

$$K = \tfrac{1}{2}Mv_c^2 + \tfrac{1}{2}I\omega^2. \qquad (10.108)$$

Power is defined as the rate of change of energy, $P = dE/dt$; power expended by a force is $P = \mathbf{F} \cdot \mathbf{v}$; power expended by a torque is $P = T_z\omega$ (Equations 10.60, 10.62, and 10.114).

For certain mechanical devices, such as levers, pulleys, chain hoists, and vises, work in $\cong$ work out.

Mechanical advantage is defined as the ratio of output force to input force.

A conservative force is one for which the work done depends only on the end points of a displacement, not on the path followed.

In one dimension, any force that depends only on position is conservative.

In three dimensions, a constant force and an inverse-square central force are examples of conservative forces.

A potential energy may be associated with a conservative force. It is defined as the negative of the work done by the force in a displacement from an arbitrarily chosen reference point to any other point (Equation 10.70 or 10.116).

Mechanical energy is governed by Equation 10.72, $\Delta E = \Delta(K + U) = W_2$, where W_2 is the work done by nonconservative forces.

If all the forces that act are conservative forces, mechanical energy is conserved:

$$E = K + U = \textit{constant}. \qquad (10.76)$$

A conservative force is the negative gradient of potential energy (Equations 10.78, 10.133, and 10.120–10.122).

The location of the zero of potential energy is arbitrary.

Some potential energy functions:

- gravity near the earth, $U = mgz$; (10.79)
- gravity outside the earth, $U = -mgR^2/r$; (10.97)
- the harmonic oscillator, $U = \tfrac{1}{2}kx^2$; (10.85)
- the simple pendulum, $U \cong \tfrac{1}{2}(mg/l)s^2$. (10.139)

Any system with a minimum in its potential-energy function can execute approximately simple harmonic motion (provided the second derivative of U does not vanish at the minimum).

In the gravitational field of the earth, negative energy corresponds to bounded motion and positive energy to unbounded motion.

Energy is a unifying concept in science because of its convertibility into many forms and because of its conservation.

QUESTIONS

Section 10.1

Q10.1 Work and torque have the same dimension and both can be measured in newton meters. Does this mean that torque is a form of energy? Explain.

Q10.2 A ball is thrown straight up and is then caught at the same level from which it was thrown. (1) Why does the gravitational force do no net work on the ball? (2) Suppose that air friction is important enough so that the speed of the ball is considerably less when it is caught than when it was thrown. Why does gravity still do no net work on the ball? (3) In the latter case, is the work done on the ball by the air positive, negative, or zero?

Q10.3 Why is it tiring to support a heavy load even though it takes no work to do so?

Q10.4 (1) Does Equation 10.1 provide an operational definition of work? (2) Is work a scalar quantity by definition or by experimental confirmation or by both? (3) Is work a true scalar or a pseudoscalar?

Section 10.2

Q10.5 Express the kinetic energy of a particle in terms of (a) its momentum and its mass and (b) its momentum and its speed. (The former expression is frequently useful, the latter rarely so.)

Q10.6 If a ball falls 1 m from rest, it gains a speed of 4.4 m/sec. If it starts downward with a speed of 10 m/sec, its speed increases by only 0.9 m/sec in a fall of 1 m. Why the difference? How do these numbers indirectly support the usefulness of the energy concept?

Q10.7 Two observers measure the mass, the kinetic energy, and the x component of velocity of a particle. For which of the three quantities will they obtain the same numerical value if they use Cartesian coordinate systems that (a) have different orientations and no relative velocity? (b) have different orientations and are in relative motion?

Section 10.3

Q10.8 A planet executes a circular orbit about a star. (1) Explain why the star does no work on the planet. (2) Name two quantities that remain constant during the course of this motion and two that change.

Q10.9 A 10-MeV alpha particle approaches a nucleus head-on and is turned straight back. (1) Up to the time that the alpha particle reaches its turning point (point of closest approach), what work does the nucleus do on the alpha particle? (2) During the complete motion of the alpha particle toward and away from the nucleus, what total work does the nucleus do on the alpha particle?

Q10.10 During a complete space mission, from liftoff to touchdown back on earth, gravitational forces do no net work on an astronaut. Does rocket thrust do any net work on him? Does frictional air drag do any net work on him? Do all forces combined do any net work on him?

Section 10.4

Q10.11 Describe a system composed of two particles that has zero linear momentum and zero angular momentum but nonzero kinetic energy.

Q10.12 In a sample of gas at constant temperature, the total kinetic energy of the molecules remains constant. What does this imply about the work done on the molecules by the forces acting among them?

Q10.13 A man squeezes a spring between his hands in such a way that the center of mass of the spring remains at rest. Is any net work done on the spring? Give a reason for your answer.

Q10.14 Two bodies A and B collide elastically (with no change in their total kinetic energy). This means that the total work done on A and B is zero. (1) Is the work done on A alone zero? (2) Is the total work done on A and B zero at every instant during the collision, or only from before to after the collision?

Q10.15 A battery-powered machine hangs by a cord from a fixed point of support in an evacuated region. If the machine is initially motionless, can it start itself swinging?

Section 10.5

Q10.16 In what form is energy manifested as a result of each of the following examples of work: (a) the work done by a pitcher on a baseball, (b) the work done by a baseball on a catcher's mitt, (c) the work done in pushing a tire pump, (d) the work done by an elevator in lifting passengers, and (e) the work done by steam turning turbine blades?

Q10.17 A book called *Ecology at Home* (J. Killeen, ed. [San Francisco: 101 Productions, 1971]) states, "The wattage rating of an appliance is the total watts used in one hour." Criticize this statement, and correct it.

Section 10.6

Q10.18 One end of a lever is pushed down briskly so that the load on the other end is accelerated upward. (1) Does this require more force than a slow, steady push? Why or why not? (2) Is output work still equal to input work?

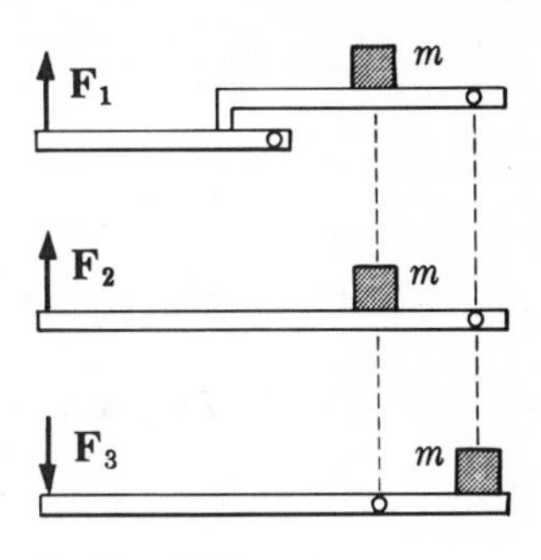

Q10.19 Why can it be particularly damaging to have a finger pinched in the crack of a door on the hinged side of the door?

Q10.20 Name two quantities other than force that can be "multiplied" by mechanical devices without violating the law of energy conservation. Describe simple devices that could be used to increase each of the quantities you have named.

Q10.21 Forces F_1, F_2, and F_3 are used to support equal weights, using the levers shown in the figure (small circles show the pivot points). Arrange the three forces in order of increasing magnitude. Are any of the forces equal?

Section 10.7

Q10.22 (1) Give an example in which kinetic energy supplies work—that is, the decrease of kinetic energy of a system is equal to the work done by the system. (2) Give an example in which potential energy supplies work with no change of kinetic energy.

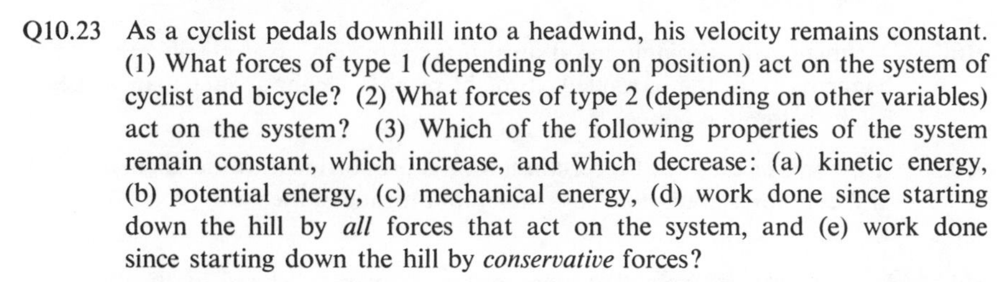

Q10.23 As a cyclist pedals downhill into a headwind, his velocity remains constant. (1) What forces of type 1 (depending only on position) act on the system of cyclist and bicycle? (2) What forces of type 2 (depending on other variables) act on the system? (3) Which of the following properties of the system remain constant, which increase, and which decrease: (a) kinetic energy, (b) potential energy, (c) mechanical energy, (d) work done since starting down the hill by *all* forces that act on the system, and (e) work done since starting down the hill by *conservative* forces?

Q10.24 Is the magnetic force on a moving charged particle a force of type 1 (depending only on the position of the particle) or a force of type 2 (depending on other variables)?

Q10.25 A worker on an assembly line performs a repetitive operation that requires him to push on a clip with a force that depends only on the position of the clip. Can a potential energy be associated with this force?

Section 10.8

Q10.26 Give a qualitative description of the motion of a particle whose energy diagram is the one shown here. Point out any features of interest about the motion when the particle is at the points x_1, x_2, x_3, x_4, and x_5.

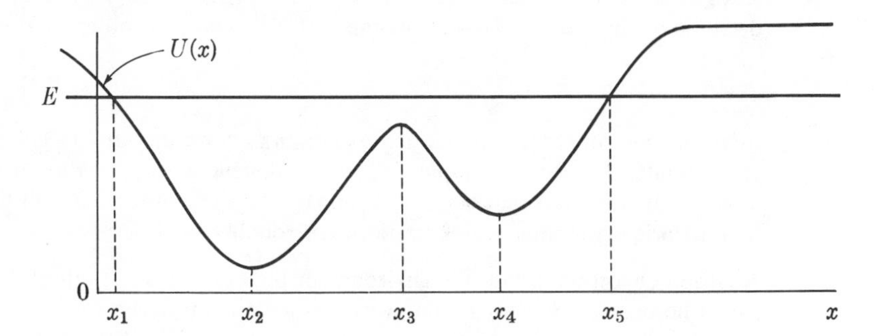

Q10.27 How do qualitative features of the motion of the particle considered in the preceding question change if the energy E is (a) halved or (b) doubled?

Q10.28 How is an ideally rigid wall represented in an energy diagram? How could a realistic wall—one that yields slightly when it is struck and that can exert only a finite force—be represented in an energy diagram?

Q10.29 (1) How would a satellite in a circular orbit be represented in the energy diagram of Figure 10.20? (2) How would a satellite in an elliptical orbit be represented?

Q10.30 Is there a significant advantage in launching orbital spacecraft from terrain 1 to 2 miles above sea level instead of launching them from sea level? Consider both gravity and air resistance.

Section 10.9

Q10.31 Why was the wheel such an important invention?

Q10.32 A hoop, a solid cylinder, and a solid sphere start rolling down a plane together. In what order do they reach the bottom?

Q10.33 A race is arranged between a block sliding without friction down one plane and a cylindrical body rolling down an adjacent parallel plane. Why does the cylinder have no chance to win, no matter what its mass, its radius, or its internal distribution of mass?

Q10.34 A rubber ball is spinning as it arcs through the air. Is it possible that after bouncing from the pavement, it might reach a height greater than its peak height before the bounce?

Q10.35 If the power output of a certain motor is proportional to the angular speed of the motor's shaft, how does the output torque of the motor depend on the angular speed?

Section 10.10

Q10.36 Explain why a force that does no work over a closed path is called "conservative."

Q10.37 How can a force be nonconservative without violating the general law of energy conservation? Name one force that is nonconservative.

Energy transformations

Q10.38 Two identical safety pins are completely dissolved in two identical acid baths. One safety pin was placed in the acid unfastened and the other was fastened. What happened to the extra potential energy of the fastened safety pin?

Q10.39 When a sugar cube is held so that it just touches the coffee in a cup, some of the coffee travels up the sugar cube. Work is done on the coffee to raise it in the earth's gravitational field. What is the source of this work? That is, what decrease of energy compensates for the coffee's increase of potential energy?

EXERCISES

Energy comparisons

E10.1 In an outside source find the annual per capita "consumption" of energy in your country. Calculate a typical figure for the energy supplied to an adult in a year by the food that he eats. Compare the two figures. (Recall that a food calorie is the same as a kilocalorie, or roughly 4,000 J.)

E10.2 A citizen concerned about the environment buys a 7-W electric toothbrush, which he plans to use for 4 min every day. To compensate for this use of power, he decides to cut down on the use of his automobile. If the automobile delivers about 50 horsepower, how much driving should he forego in a year?

E10.3 Calculate and compare the following energies in units of eV: (a) kinetic energy of a hydrogen atom with speed $v = 2 \times 10^3$ m/sec, typical of normal thermal motion; (b) energy of a photon of red light emitted by hydrogen with frequency $\nu = 4.6 \times 10^{14}$ Hz ($E = h\nu$); (c) energy released per H atom in an H-bomb if 1 part in 200 of the mass is converted to energy; and (d) kinetic energy of a 0.1-gm flea lazing along at 0.1 cm/sec.

E10.4 Energy released in nuclear explosions is commonly measured in kilotons or megatons. The "ton" is an energy unit equal to 4.2×10^9 J; it is the energy released in the chemical explosion of 1 ton of TNT. (1) Express in kilotons the kinetic energy of 10 thousand km^3 of air (the air over a small county) moving at 5 m/sec (a light breeze). Take the average air density to be 1 kg/m^3. (2) If the energy of a 10-megaton H-bomb could be harnessed to lift an ocean liner with a mass of 10^7 kg, could the ship be put into orbit? (A useful number is provided by Equation 10.102).

E10.5 (1) How many joules of energy are supplied by a 50-megawatt nuclear generating station in a year? (2) How much mass is converted to energy at this station in a year if in addition to 50 megawatts of useful power it generates 100 megawatts of waste heat? (3) What is the annual conversion of mass to energy at a coal-fired generating station of the same capacity and efficiency?

E10.6 Above the earth's atmosphere the flux of energy arriving from the sun is 1.39×10^3 J/m^2 sec. (1) Assuming isotropic radiation by the sun, calculate the total rate of energy radiation by the sun in J/sec. (2) What is the rate of conversion of mass to energy in the sun in kg/sec? (3) How long does it take the sun to transform one earth-mass to energy?

Section 10.1

E10.7 (1) A particle is acted on by two constant forces, $\mathbf{F}_1$ and $\mathbf{F}_2$. Show that the work done on the particle is the sum of the works that would be done by the forces acting separately. (2) A particle moves from $\mathbf{r}_1$ to $\mathbf{r}_2$, then from $\mathbf{r}_2$ to $\mathbf{r}_3$. Show that the total work done on the particle by a constant force is the sum of the works done in the two displacements. (3) A particle executes a closed path. Why does a constant force do no net work on it?

E10.8 A raindrop of mass m is falling vertically at a constant terminal speed v_0. As it falls through a distance l, what work is done on it (a) by the force of gravity and (b) by the force of air resistance?

E10.9 A lightweight ball follows the path through the air shown in the figure. What work is done by gravity on the ball as it moves (a) from A to B? (b) from B to C? (c) from A to C?

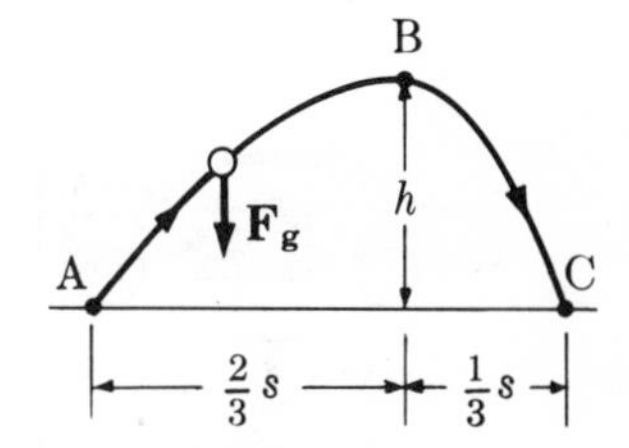

E10.10 Refer to Example 2 in Section 10.1 and to Figure 10.3. (1) In the 2-m displacement, what work does the wagon do on the boy who is pulling it? (2) While the wagon is moving with constant speed through a 1-m displacement, what is the net work done by all the forces that act on the wagon?

E10.11 A particle with position coordinates x, y in a plane executes the following straight-line displacements: (a) from 0, 0 to 0, $2a$; (b) from 0, $2a$ to $2a$, $2a$; (c) from $2a$, $2a$ to $2a$, 0; and (d) from $2a$, 0 to a, a. (1) Find the work done on the particle in each part of its motion by the force $\mathbf{F} = F_0(\mathbf{i} + \mathbf{j})$. (2) What total work is done on the particle? Compare this with the work that would be done in a straight-line displacement from 0, 0 to a, a.

E10.12 During a sports-car rally, a man drives a 1,000-kg car around a circular path of radius 100 m that has been marked on the side of a hill. The hill can be approximated as a plane inclined at an angle of 10 deg to the horizontal. What work is done by gravity on the car as it moves (a) from the high point to the low point of the circle? (b) from the low point to the high point? (c) once around the circle?

Section 10.2

E10.13 An automobile with a mass of 10^3 kg is stalled on a horizontal roadway. Starting from rest, a man pushes the car forward with a steady force of 200 N for 5 sec. Assuming that the car rolls with negligible friction, calculate the following quantities: (a) the speed of the automobile after 5 sec, (b) the distance covered, (c) the kinetic energy of the automobile after 5 sec, and (d) the work done by the man.

E10.14 The ball in Figure 10.5(c) starts upward with speed v_1 and hits the ground with speed $2v_1$. What maximum distance from the ground does it reach?

E10.15 Within a television tube, an electron acquires a kinetic energy of 4×10^{-16} J in a distance of 2 cm. (1) What is the approximate speed of an electron with this kinetic energy (the nonrelativistic formula suffices)? (2) What average force must have acted on the electron to give it this energy? (3) What average force acts on the electron while it is being brought to rest in the tube face in a distance of 10^{-6} m? How does this force compare with the weight of an electron?

E10.16 Suppose that when an automobile is being braked there is a fixed maximum horizontal component of force exerted on the tires by the road, independent of speed. Show that the minimum stopping distance is then proportional to the square of the initial speed. Is this result approximately valid in practice?

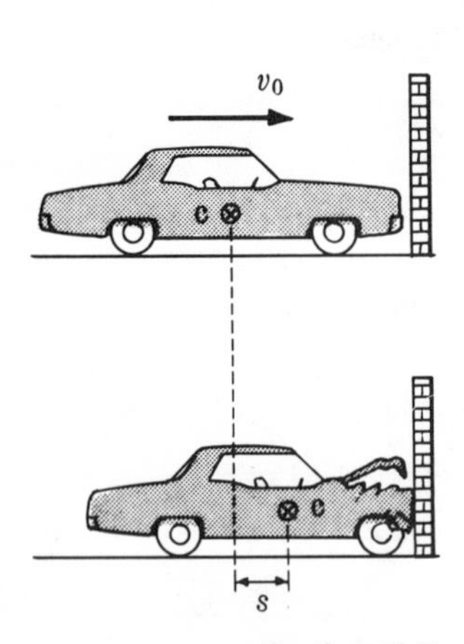

E10.17 The kinetic energy of a spacecraft re-entering the atmosphere is given approximately by $K = K_0 - As$, when K_0 is the craft's kinetic energy at some initial time, s is the distance it moves after that time, and A is a constant. (1) What is the physical significance of the constant A? (2) How does K depend on time? (3) How does the acceleration of the spacecraft depend on distance and on time?

E10.18 Studies of low-speed collisions of automobiles with rigid walls (see the figure) reveal the following two facts: (1) The acceleration of the center of mass of a car as it is being stopped by a wall is approximately proportional to the initial speed v_0. (2) The cost of repair is approximately proportional to the initial kinetic energy K_0. (1) How does the "crumpling distance" s (the distance moved by the center of mass of the car during the collision) depend on v_0? (2) How does the cost of repair depend on the crumpling distance?

Section 10.3

E10.19 A 70-kg man walks from the third to the first floor of a building, then rides an elevator to the fourteenth floor. If the distance between floors is 3.5 m, what work is done on the man by the earth's gravitational force? What work is done on him by the elevator? Explain why the work done by the elevator does not depend on its speed or its acceleration as it moves between floors.

E10.20 Two boys standing a fixed distance R apart are throwing a ball of mass m back and forth. (1) Show that if the ball is thrown at angle θ to the horizontal, its initial kinetic energy should be

$$K_0 = \tfrac{1}{4}mgR \csc 2\theta.$$

(2) Sketch a graph of K_0 vs θ. For what angle θ is K_0 minimum?

E10.21 A body moving along a line parallel to the x axis experiences the force $F_x = -kx$. (1) Sketch a graph of F_x vs x. (2) Evaluate the integral $\int_0^{x_1} F_x \, dx$.

Show the graphical meaning of this integral in your sketch, and state its physical meaning.

E10.22 Two particles move in the xy plane from the origin to point P at $x = a$, $y = a$; one moves along path A and one along path B (see the figure). Both particles experience a force given by $\mathbf{F} = \beta y\mathbf{i}$, where β is a constant. Find the work done by this force on each particle. Is it possible that this is the only force acting on the particles? *Optional:* Find the work done by the same force on a third particle that moves in a straight line from O to P.

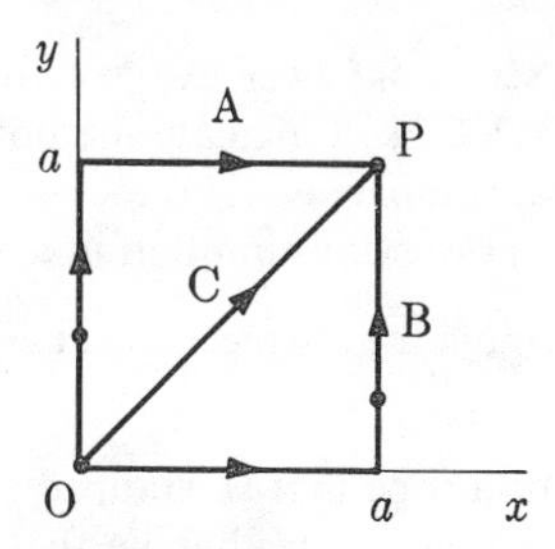

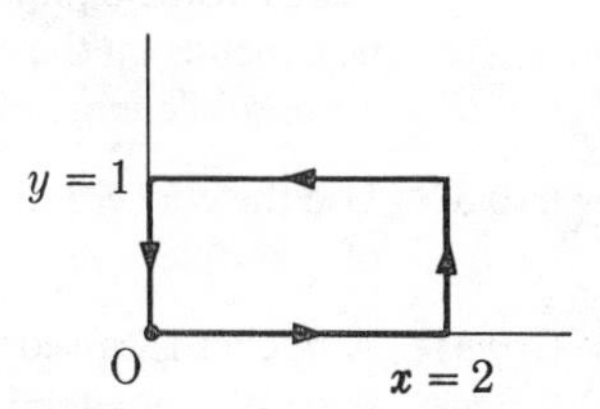

E10.23 Starting at the origin, a particle follows the rectangular path shown in the figure. What is the net work done on it by the force $\mathbf{F} = 3y\mathbf{i} + 2x\mathbf{j}$? Distance is measured in meters, force in newtons.

E10.24 The figure shows the side view of two roller-coaster tracks. (1) Prove that the work done by gravity is the same for cars rolling from A to B as for cars rolling from A to C. (2) A car passes point A with a speed of 6 m/sec and passes point B at 20 m/sec. Points A and D are at the same elevation, both 20 m higher than point B. Is the car likely to roll past point D? (Do not ignore friction.)

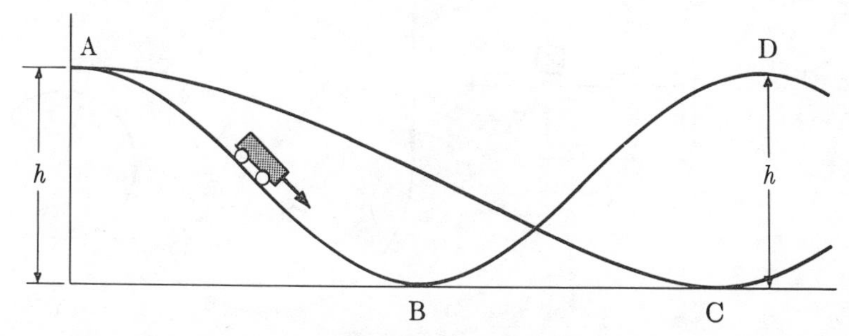

E10.25 Two apples of equal mass are dropped simultaneously from the same height h above the ground on opposite sides of the earth. (1) What total force acts on the system of two apples? (2) What is the change of momentum of the system as the apples fall through distance h? (3) What total work is done on the apples as they fall? (4) What is the change of kinetic energy of the system?

Section 10.4

E10.26 Two bodies of unequal mass move about their stationary center of mass. Show that their kinetic energies are inversely proportional to their masses:

$$\frac{K_1}{K_2} = \frac{m_2}{m_1}.$$

Point out the significance of this result for simplifying the description of systems such as the earth and the moon or a proton and an electron.

E10.27 Two carts of equal mass m on a linear air track are joined by a light spring. The x components of velocity of the carts are given by $v_1 = v_c + v_0 \cos \omega t$ and $v_2 = v_c - v_0 \cos \omega t$. (1) Describe the motion in words. (2) Obtain an expression for the total kinetic energy and identify its translational part and its internal part. (2) Find the reduced mass of this two-body system and give an exact description of the effective one-body motion.

Section 10.5

E10.28 A weight lifter raises 180 kg a vertical distance of 2 m in 3 sec. What is his average power output during this time?

E10.29 After reaching a terminal speed of 60 m/sec, a skydiver exerts on the air a steady force equal to his own weight $F = 700$ N. Calculate the power that he expends on the air. How does this power compare with (a) the power of an automobile engine? (b) the typical electric power consumption in a home?

Section 10.6

E10.30 Use the concept of torque to derive the formula for the mechanical advantage of a simple lever (Equation 10.65).

E10.31 A bucket is raised from a well by means of a rope that is wrapped around a shaft of diameter 15 cm. The shaft is turned by a crank handle that extends 30 cm from the axis of the shaft. What force on the handle is required to raise a bucket of water weighing 80 N?

E10.32 (1) What is the mechanical advantage of the single lever shown in the figure? (2) What is the mechanical advantage of the combination of two levers shown in the figure?

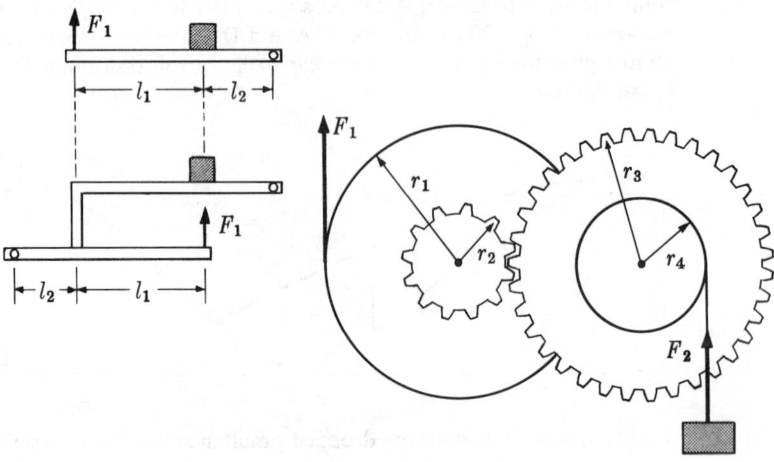

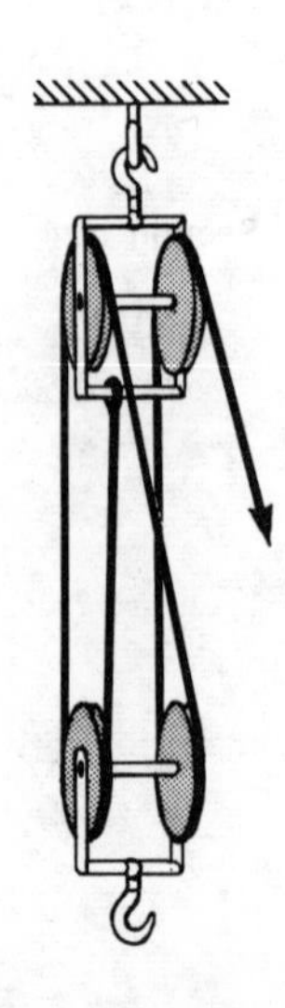

E10.33 What is the mechanical advantage of the wheel and gear combination shown in the figure? Assume that each wheel is rigidly attached to its coaxial gear.

E10.34 A common predecessor of the chain hoist was the block and tackle, a combination of multiple pulleys. The combination with two double pulleys (pictured in the figure) has a mechanical advantage of 4. Explain why.

E10.35 The upper pulleys of a chain hoist consist of two sprockets, one with 15 teeth and the other with 14 teeth. (To engage the same chain, the two sets of teeth must, of course, have the same spacing.) This chain hoist is used to lift half the weight of a 1,500-kg automobile. (1) If 80 percent of the input work appears as output work, what force must be applied to the chain? (2) Could a

32-kg child apply this force (a) by pulling down while standing on the floor? (b) by hanging from the chain with his feet off the floor? (c) by jumping from a ladder and clutching the chain to stop his fall?

E10.36 The input force on the vise shown in the figure is applied a distance R from the axis of the screw. In one turn, the screw thread advances a distance d. The output force is applied at the jaws of the vise. (1) Assuming conservation of work, derive a formula for the mechanical advantage of the vise. (2) If half of the input work is transformed into heat and half appears as output work, what is the mechanical advantage?

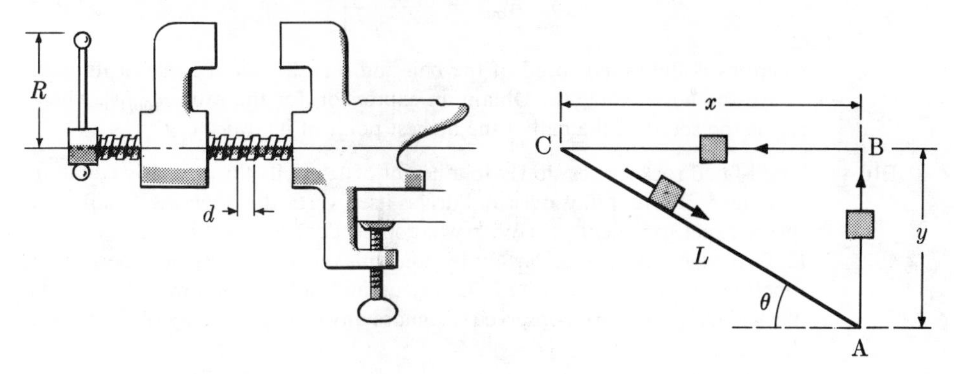

Section 10.7

E10.37 The block of mass m shown in the figure is lifted vertically a distance y from A to B, which requires an input work equal to mgy. It is then moved horizontally a distance x from B to C, which requires no expenditure of work. Then it slides back to A along a frictionless track. Calculate its final kinetic energy in two ways. (1) Find the component of gravitational force along its direction of motion, then its acceleration, and finally its speed at the bottom. (2) Apply energy conservation directly. Comment on the relative merits of these two approaches to the problem.

E10.38 An automobile whose mass is $m = 1{,}500$ kg coasts with negligible friction on a horizontal road at a speed $v_0 = 20$ mile/hr. (1) What is the automobile's kinetic energy in joules? (2) What is the greatest vertical height of a hill over which it can coast? Demonstrate that this height does not depend on the mass of the automobile.

E10.39 A shell of mass 10 kg is fired vertically upward with an initial speed of 600 m/sec. (1) If there were no air friction, what would be the maximum height of the shell? (2) If on its way up the shell dissipates 8×10^5 J in the form of heat, what height does it reach? (In both parts, ignore the variation of gravitational force with height.)

E10.40 As a girl lifts a 1-kg book through 1 m, she does 12 J of work on the book. What is its change of (a) potential energy U? (b) kinetic energy K? (c) mechanical energy, $K + U$?

E10.41 A 75-gm plastic flying saucer (frisbee) is thrown from a height of 1 m above the ground with a speed of 12 m/sec. When it is at a height of 2 m above the ground, its speed is measured to be 10.5 m/sec. (1) How much work has been done on the flying saucer by the air? (2) How much work has been done on it by gravity?

E10.42 (1) Write the equation of energy conservation of a projectile moving without friction in two dimensions; express it in terms of the projectile's horizontal component of velocity v_x, its vertical component of velocity v_y, and its distance y above a reference plane. How many adjustable constants does this equation contain? (2) Is it possible to infer from this equation alone that v_x is constant and v_y is variable? Why or why not?

E10.43 (1) Use energy considerations to show the maximum height of a thrown baseball is

$$y_{\max} = \frac{v_0^2 \sin^2 \theta}{2g},$$

where v_0 is the initial speed of the ball and θ is the initial angle of its path above the horizontal. (2) Obtain an expression for the ratio v_{top}/v_0, where v_{top} is the speed of the ball at the highest point of its trajectory.

E10.44 An ill-fated rocket rises slowly to a height of 1 km, its thrust nearly equaling its weight. There it flips over and drives itself vertically downward, still with thrust equal to weight. (1) With what speed does it crash into the ground? (2) Give algebraic expressions for its pontential energy and kinetic energy as a function of vertical distance on its way up and on its way down. Why is its mechanical energy not conserved? (Ignore any change of mass of the rocket during its trip.)

E10.45 An electron of charge $-e$ in a uniform electric field E experiences a constant force of magnitude $F = eE$. Suppose that this force is directed to the left and the positive x axis is directed to the right ($F_x = -eE$). (1) Obtain an expression for the potential energy of the electron in this field of force. (2) Is the location of the zero of potential energy arbitrary in this example?

E10.46 A particle moving along the x axis experiences the force $F_x = -k_1x + k_2x^3$. (1) What is its potential-energy function $U(x)$ if $U(0)$ is chosen to be zero? (2) What work does this force do on the particle as the particle moves from $-x_0$ to $+x_0$, where $x_0 = \sqrt{k_1/k_2}$? (3) What is the change of kinetic energy of the particle as it moves from $x = x_0$ to $x = 0$?

E10.47 The potential energy of a particle moving in one dimension is

$$U = \tfrac{1}{2}k(x - a)^2 + mg(x - a).$$

(1) What force acts on the particle? (2) If the particle moves with constant energy E ($K + U = E$), at what point is its kinetic energy maximum? What are the values of K and U at this point?

E10.48 The mechanical energy of a harmonic oscillator decreases by 1 percent in each cycle of oscillation so that its energy is $E_1 = 0.99E_0$ after one cycle, $E_2 = (0.99)^2E_0$ after two cycles, and $E_n = (0.99)^nE_0$ after n cycles. (1) Show that the energy of the oscillator is given approximately by

$$E \cong E_0e^{-t/\tau},$$

where $\tau \cong 100T$ (T is the period of the oscillator). (2) What power is transferred from the oscillator to its environment as a function of time?

Section 10.8

E10.49 A certain ball loses 20 percent of its kinetic energy when it bounces from a hard surface. Its energy loss through air friction is negligible. (1) With what initial speed must the ball be thrown downward from height h in order to bounce back to the same height? (2) In an energy diagram show the history of the ball from the moment it is released until it returns to its starting point.

E10.50 A tennis racket moving with speed v_1 hits a tennis ball moving with speed v_0. Show that the maximum speed of the rebounding ball is $v_0 + 2v_1$.

E10.51 For a body moving vertically near the earth, the energy conservation equation can be written

$$\tfrac{1}{2}m\left(\frac{dz}{dt}\right)^2 + mgz = E.$$

Verify by substitution that $z = z_0 + v_0t - \tfrac{1}{2}gt^2$ is a solution to this equation. How is the energy E related to the constants z_0 and v_0?

E10.52 If displaced from its equilibrium position ($x = 0$), an atom in a crystal experiences a restoring force proportional to its displacement (the force is given by Equation 10.84). Consider an atom whose force constant is $k = 800$ N/m, vibrating with an energy of 1 eV. (1) What is its maximum displacement, x_{limit}? (2) If the mass of the atom is 4×10^{-26} kg, what are (a) its maximum speed and (b) its maximum acceleration? (3) At what values of x are the maximum speed and maximum acceleration attained?

E10.53 An object moving along the x axis is acted on by the force $F_x = -ax^2 - bx$, with $a = 6$ N/m^2 and $b = 4$ N/m. (1) Sketch a graph of F_x vs x. (2) Obtain a potential-energy function $U(x)$, and sketch a graph of U vs x. (3) For small x, a good approximation is $U(x) = \tfrac{1}{2}bx^2$. Why? (4) For what values of x is this harmonic-oscillator approximation for $U(x)$ accurate to within 10 percent? For what values of x is it accurate to within 1 percent?

E10.54 A potential-energy function for the one-dimensional motion of a particle is

$$U(x) = A(1 - \cos \alpha x).$$

(1) Sketch a graph of U vs x, paying attention to its form near $x = 0$. (2) For small x ($x \ll 1/\alpha$) and small energy ($E \ll A$), the particle executes approximately harmonic motion. What is the force constant k in the small-x approximation $U(x) \cong \tfrac{1}{2}kx^2$? (3) What is the angular frequency ω for vibrations of small amplitude?

E10.55 A rocket is given a burst of acceleration near the earth's surface until it reaches a speed $v = 2\sqrt{gR}$ (compare this with Equation 10.101). It then coasts vertically upward. Find its speed, both algebraically and numerically, as it drifts away at great distance from the earth. At this speed, how long would it take to reach the sun?

E10.56 What is the escape speed from the moon? Compare this number with the speed of sound in air.

Section 10.9

E10.57 Two bodies of unequal mass circle about their fixed center of mass. Obtain a formula for the moment of inertia I of this system and verify explicitly that $\tfrac{1}{2}I\omega^2 = \tfrac{1}{2}m_1{v_1}^2 + \tfrac{1}{2}m_2{v_2}^2$.

E10.58 A rolling hoop and a sliding block of wood are observed to keep pace as they move down an inclined plane. If the energy loss of the hoop is negligible, what frictional force acts on the block?

E10.59 The moment of inertia of a certain cylinder of radius r and mass M is $I = Mk^2$ (k is called its radius of gyration). (1) Show that the kinetic energy of the cylinder when it rolls without slipping is $K = \frac{1}{2}mv_c^2[1 + (k^2/r^2)]$, where v_c is the speed of its center of mass. (2) Show that the acceleration of the cylinder when it rolls down a plane inclined at an angle θ to the horizontal is

$$a_c = \frac{g \sin \theta}{1 + (k^2/r^2)}.$$

E10.60 The body of a cart has mass M; each of its four wheels has radius r, mass m, and moment of inertia $I = \frac{1}{2}mr^2$. What is the acceleration of the cart as it rolls without internal friction down a plane inclined at angle θ to the horizontal? What are the limiting values of its acceleration for $M \gg 4m$ and $M \ll 4m$?

E10.61 (1) If the maximum torque that an automobile engine can deliver at 2,000 rpm is 214 N m, what is its maximum power at this angular speed? (2) If the maximum engine power at 4,000 rpm is 100 horsepower, by what percentage does the torque decrease between 2,000 and 4,000 rpm?

Section 10.10

E10.62 As shown in the figure, a projectile is launched horizontally from height h with initial speed v_0; it hits the ground with speed v_1 after covering horizontal distance d. (1) Express v_1 as a function of d. (2) Obtain approximate expressions for $v_1(d)$ for (a) $d \ll h$ and (b) $d \gg h$.

E10.63 A small object slides without friction in a trough whose cross section is given by $y = x^2/a$ (a parabola). If the object, while at rest at the lowest point of the trough, is given a blow that imparts to it a velocity $\mathbf{v} = v_0\mathbf{i}$, what will be its greatest straight-line distance from its starting point? Obtain approximate expressions for this distance if $v_0 \ll \sqrt{2ga}$ and if $v_0 \gg \sqrt{2ga}$.

E10.64 A section of roller-coaster track is described by the equation $z = \frac{1}{2}h[1 + \sin(\pi x/a)]$, in which x is horizontal distance and z is vertical distance. (1) What is the physical significance of the constant h? the constant a? (2) Find the speed of a coasting car at any x if its speed at the top of a "hill" is v_0 (ignore friction). (3) Sketch a graph of z vs x and below it a graph of v vs x.

E10.65 In terms of Cartesian coordinates, the potential energy associated with an inverse-square central force is written

$$U(x, y, z) = \frac{C}{\sqrt{x^2 + y^2 + z^2}}.$$

Take the partial derivatives of this function with respect to x, y, and z to find the Cartesian components of force (see Equations 10.120–10.122), and verify that $F_x\mathbf{i} + F_y\mathbf{j} + F_z\mathbf{k} = C\mathbf{r}/r^3$, which is equivalent to Equation 10.127.

E10.66 A spacecraft is in a circular orbit at a distance from the center of the earth equal to two earth radii ($r = 2R$). (1) Show that the speed of the satellite is $v_1 = \sqrt{\frac{1}{2}gR}$ (g is the acceleration of gravity at the surface of the earth). (2) The craft is given a burst of acceleration that increases its speed to v_2 and sends it into a parabolic orbit, one that barely escapes from the earth. What work is done on the spacecraft during the burst of acceleration? (3) What is v_2?

Section 10.11

E10.67 (1) For the simple pendulum, what is the exact equation of energy conservation valid for all angles? (2) Sketch a graph of the pendulum's potential energy as a function of either its angular displacement θ or its displacement s measured along the arc of its circle. (3) Based on this graph, give a qualitative explanation of the fact that the period of a pendulum increases as its amplitude increases.

E10.68 Alfred, whose mass is 35 kg, is sitting motionless on a swing. Brenda tosses him a 10-kg watermelon. He catches it as it is moving horizontally with a speed of 1 m/sec; then he and the watermelon swing to a maximum angle of 2 deg. "Just as I thought," says Brenda. "The ropes supporting your swing are 4.1 meters long." Is Brenda right? (A cautionary question: Is mechanical energy conserved in all parts of this activity?)

PROBLEMS

Work and change of speed

P10.1 A small amount of work W is done on a body of mass m moving at high speed v. Show that the change of speed of the body is given approximately by

$$\Delta v \cong \frac{W}{mv}.$$

What condition is required to validate this approximation?

Rate of change of kinetic energy

P10.2 (1) A particle experiences a *total* force $\mathbf{F}$. Prove that the rate of change of kinetic energy of the particle is

$$\frac{dK}{dt} = \mathbf{F} \cdot \mathbf{v},$$

where $\mathbf{v}$ is the velocity of the particle. (2) If $\mathbf{F}'$ is only one among several forces acting on a particle, what is the significance of $\mathbf{F}' \cdot \mathbf{v}$?

Force proportional to $\cos(2\pi x/l)$

P10.3 A particle moves in a field of force given by $\mathbf{F} = F_0 \cos(2\pi x/l)\mathbf{i}$. (1) If the particle is released from rest at the origin, describe its subsequent motion in qualitative terms, answering such questions as, Does it move in 1, 2, or 3 dimensions? Is the motion bounded within a finite region? Is it oscillatory? Does it continue indefinitely? (2) Where, besides the origin, does the particle have zero velocity?

Kinetic energy of a three-body system

P10.4 Three particles have the following masses, positions, and velocities.

Particle	Mass	x	y	v_x	v_y
1	m	0	0	$3v_0$	0
2	m	0	a	0	$3v_0$
3	$2m$	a	a	?	?

The center-of-mass velocity of this system is $\mathbf{v}_c = v_0\mathbf{i}$. Find each of the following quantities: (a) the location of the center of mass, (b) the position vector of each particle relative to the center of mass, (c) the velocity of particle 3, (d) the velocity of each particle relative to the center of mass, (e) the translational kinetic energy of the system, and (f) the kinetic energy relative to the center of mass.

Accelerated lever

P10.5 Suppose that the magnitude of the downward force $\mathbf{F}_1$ applied to the lever in Figure 10.12 is greater than the minimum needed to lift the load. Use both angular momentum and energy concepts to find the acceleration and kinetic energy of the load. Neglect the mass of the lever in comparison with the mass of the load, and treat the problem only for a narrow range of angles within which vertical components of force and components perpendicular to the lever may be equated.

Equal pay for unequal work

P10.6 Two men lift a 60-kg piece of metal cut in the shape of an equilateral triangle through a vertical distance of 1.2 m. One man lifts at a corner of the triangle; the other man lifts at the middle of the opposite side (see the figure). (1) What is the total work done by the two men? (2) What is the work done by each man?

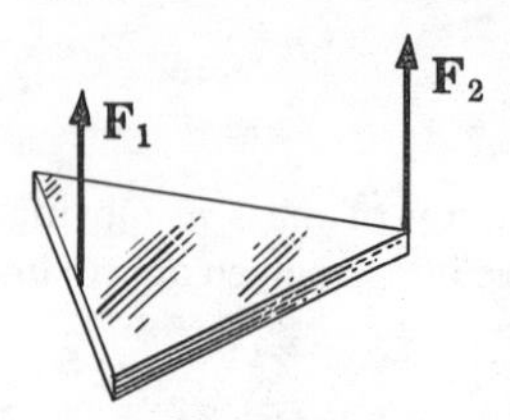

Time related to energy

P10.7 In one dimension, a particle whose mechanical energy is conserved moves from x_1 to x_2 without retracing any part of its path (this means that v_x does not change sign). (1) Prove that the time required for this part of its motion is

$$t_2 - t_1 = \int_{x_1}^{x_2} \frac{dx}{\sqrt{\dfrac{2}{m}\,[E - U(x)]}}$$

if $x_1 < x_2$. What is the slight change required in this formula if $x_1 > x_2$? (2) Use this formula to derive the familiar result $x = v_0 t + \frac{1}{2}a_0 t^2$ for a particle starting from the origin with speed v_0 and acted on by the constant force $F = ma_0$.

P10.8 Use the formula stated in the preceding problem to find the period of a harmonic oscillator whose potential-energy function is $U = \frac{1}{2}kx^2$. (SUGGESTION: Find the time required for the particle to move from $x = 0$ to one turning point; multiply this by 4 to get the period.)

"Triangular" potential

P10.9 The potential energy of a particle moving in one dimension is $U(x) = K|x|$ (see the figure). Find the period of the oscillatory motion as a function of the particle energy E. (SUGGESTION: Consider positive x and negative x separately.)

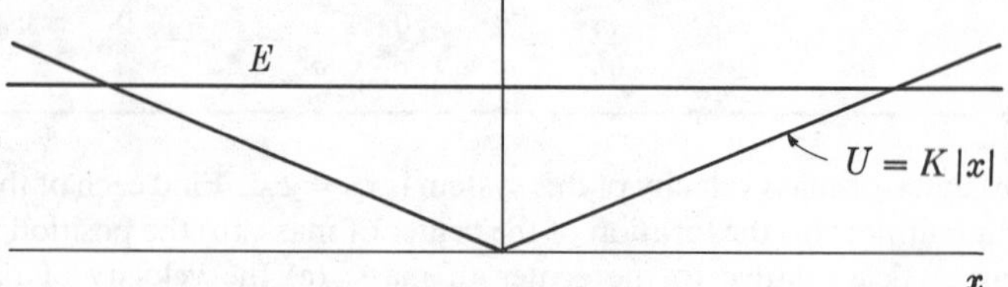

x(t) for harmonic oscillator

P10.10 Use two methods to demonstrate that $x = A \sin(\omega t + \varphi)$ (see Equation 10.93) is a solution to the energy-conservation equation of the harmonic oscillator (see Equation 10.92). (1) Treat Equation 10.93 as a "guess" and verify its correctness by substituting from it in Equation 10.92. (2) Set up the integral $t = \int (dt/dx)\,dx$; find dt/dx from Equation 10.92 and carry out the integration. (3) Explain why only two of the four constants A, E, ω, and φ are arbitrary.

P10.11 *Hooke's law in reverse*

A potential-energy function for an object moving in one dimension is $U(x) = -\frac{1}{2}kx^2$, in which k is a positive constant. (1) Sketch an energy diagram and discuss the motion qualitatively for $E > 0$, $E = 0$, and $E < 0$. (2) Set up the energy-conservation equation as a differential equation analogous to Equation 10.92, and solve the equation to find $x(t)$ for the special case $E = 0$. Discuss the motion near $x = 0$.

P10.12 *Linear approximation for gravitational potential energy*

An accurate expression for the gravitational potential energy of a body of mass m acted on by the earth is $U = -mgR^2/r$ (Equation 10.97). Show that the formula $U = mg(r - R) + C$ (Equation 10.98) is a good approximation if $r - R \ll R$. What is the constant C? *Optional:* Develop three or four terms of the Taylor series for $U(r)$ near $r = R$.

P10.13 *Harmonic oscillator approximation*

The potential-energy function of a body moving in one dimension is $U(x) = A(e^{\alpha x} + e^{-\alpha x} - 1)$. (1) Locate the minimum of this function and verify that $U(x)$ can be approximated by $U(x) \cong \frac{1}{2}k(x - a)^2$ near the minimum. What is k? What is a? (2) If $\alpha = 3\ \text{m}^{-1}$, $A = 1$ J, the energy of the body is $E = 0.04$ J, and its mass is $m = 2$ kg, what is its approximate period? What is its approximate amplitude? (3) For small energy, the period is nearly independent of energy; for larger energy, the period decreases with increasing energy. Explain these two facts.

P10.14 *Electrostatic potential energy*

An alpha particle of charge $2e$ moving toward a nucleus of charge Ze experiences a force given by

$$F_r = \frac{2Ze^2}{4\pi\varepsilon_0} \cdot \frac{1}{r^2}.$$

(Values of e and ε_0 can be found in Appendix 3A.) (1) Find the potential energy $U(x)$ with $U(\infty) = 0$. (2) Draw an energy diagram for the head-on approach of an alpha particle to a nucleus. (3) Obtain a formula for the distance of closest approach. Evaluate this distance for a 1-MeV alpha particle approaching a gold nucleus ($Z = 79$).

P10.15 *Bead on a circular wire*

A bead of mass m slides without friction on a wire bent into a circle of radius a. The circle is in a vertical plane. (1) With what speed does the bead pass the bottom of the circle if it exerts no force on the wire at the top of the circle? (2) For motion of the bead with arbitrary energy, obtain an expression for the change in magnitude of the *total* force acting on the bead as it moves from the top to the bottom of the circle.

P10.16 *Hooke's law in Hollywood*

A Hollywood stunt man experiments with a rubber rope fastened at its upper end to a rafter. First he climbs slowly down the rope, hangs from its lower end, and finds that it stretches by 1 m. Then he drops from the rafter, grabs the lower end of the rope as he passes it, and finds that he continues downward another 4.3 m before the rope stops his fall. He wonders what he might be able to calculate if he assumes that the stretched rope obeys Hooke's law (restoring force is proportional to displacement). (1) Can he calculate his mass? If so, what is it? (2) Can he calculate the acceleration of gravity? If so, what does he find it to be? (3) Can he calculate the original length of the rope? If so, what is it?

Hoop rolling down an inclined plane

P10.17 As shown in Figure 10.22, a hoop rolls down a plane inclined at angle θ to the horizontal. Let P be the point of contact of the hoop with the plane. At any instant, the hoop is rotating with angular speed ω about point P. (1) Find the orbital angular momentum of the hoop with respect to P, its spin angular momentum, and its total angular momentum with respect to P. (2) With respect to P, what total external torque acts on the hoop? (3) With the help of the answers to parts 1 and 2, use the equation $d\mathbf{L}/dt = \mathbf{T}$ to find the angular acceleration $\alpha = d\omega/dt$; then use α to find a_c, the acceleration of the center of mass. Check your answer against Equation 10.111.

Accelerated rolling wheel

P10.18 A wheel of radius r, mass m, and moment of inertia $I = mk^2$ (k is its radius of gyration—see Exercise 9.39) is pulled along a level surface by the application of a horizontal force $\mathbf{F}$ to a rope unwinding from an axle of radius b (see the figure). (1) Explain why the frictional force $\mathbf{F}_f$ at the surface does no work if the wheel does not slip. (2) Show that the work done by the force $\mathbf{F}$ as the wheel's center of mass moves through distance x_c is $W = Fx_c[1 + (b/r)]$. (HINT: How far does the rope move in the same time?) (3) Express the energy equation $\Delta K = W$ in terms of the variables m, k, b, r, F, x_c, and dx_c/dt. (4) Differentiate this equation with respect to time in order to prove that the acceleration of the wheel is

$$a_c = \frac{d^2x_c}{dt^2} = \frac{F}{m}\cdot\frac{1 + (b/r)}{1 + (k^2/r^2)}.$$

Mechanics of a yo-yo

P10.19 A yo-yo of mass m and moment of inertia I about its spin axis is allowed to fall from rest. As it does so, the string unwinds from the inner shaft of radius b (see the figure). Mechanical energy is conserved as the yo-yo falls. (1) Working from the equation of energy conservation, find expressions for the speed of the center of mass, v_c, and the angular speed of the yo-yo, both as functions of its distance of fall, $z_0 - z$. (2) What is the acceleration of the center of mass? (3) If the yo-yo is approximated as a uniform disk of radius r and if $r = 10b$, find (a) the ratio a_c/g and (b) the ratio of rotational kinetic energy to translational kinetic energy.

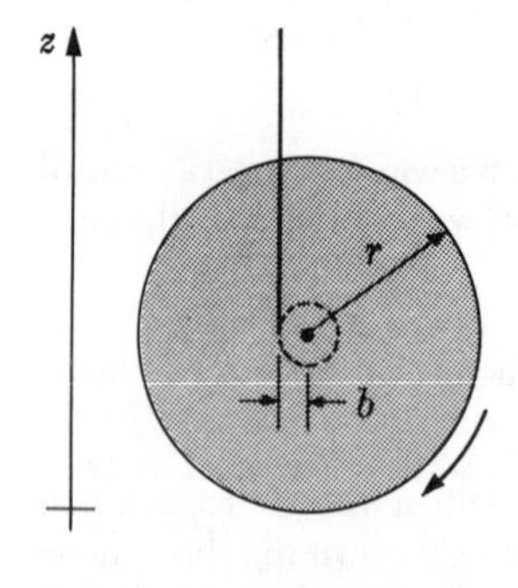

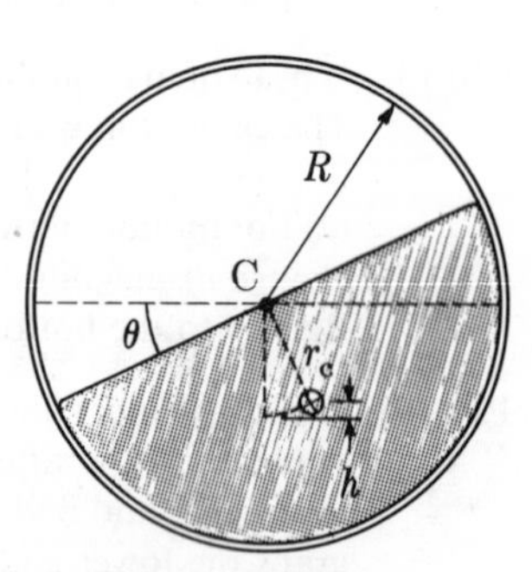

Harmonic vibration of solid body

P10.20 The figure shows a slug of ice of mass M, which half fills a cylindrical can of inner radius R. As stated in Problem 9.9, the moment of inertia of the ice with respect to the center of the can C is $I = \frac{1}{2}MR^2$. The center of mass of the ice is at a distance $r_c = (4/3\pi)R$ from C. Suppose that a melted layer of ice permits the slug to rotate within the can with negligible friction. Show that the slug's period of small vibration with the can held motionless is

$$T = 2\pi \sqrt{\frac{3\pi}{8}\frac{R}{g}}.$$

(SUGGESTION: Seek to express the kinetic energy of the ice in terms of ω [equal to $d\theta/dt$] and its potential energy in terms of θ. With a small-angle approximation, the energy equation should resemble Equation 10.92 or 10.140.) *Optional:* Verify the formula given above for r_c.

P10.21 A particle of mass m moves in an elliptical orbit in an attractive inverse-square central field of force; its potential energy is $U = -C/r$. Apply both energy conservation and angular-momentum conservation at points 1 and 2 shown in the figure—the points of greatest and least distance of the particle from the force center. Show that to satisfy these conservation laws, the energy of the particle must be

Energy in an inverse-square force field

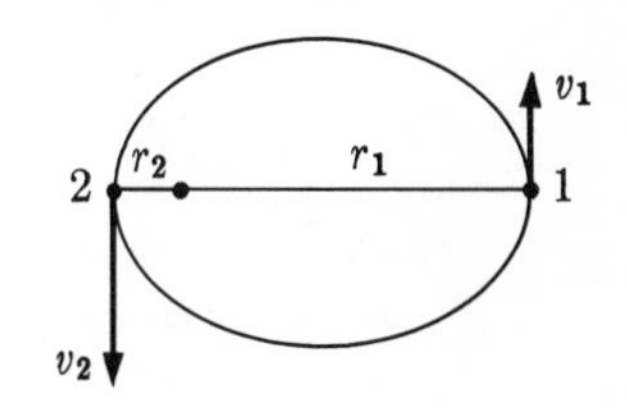

$$E = -\frac{C}{r_1 + r_2}.$$

This important result applies to satellites and planets. It says that the energy of the body depends only on the longest diameter of its elliptical orbit.

P10.22 For the inverse-square law of force defined by $\mathbf{F} = (C/r^2)\mathbf{i}_r$ ($\mathbf{i}_r$ is a unit vector directed radially), prove that the line integral $\int_A^B \mathbf{F} \cdot d\mathbf{s}$ is independent of the path from A to B and has the value given by Equation 10.131. There are various ways to carry out this proof. Here is one method. Form the scalar product $\mathbf{F} \cdot d\mathbf{s}$ and prove that $\mathbf{i}_r \cdot d\mathbf{s} = dr$ (refer to the figure and keep in mind that the magnitude of $\mathbf{i}_r$ is 1). The line integral then becomes an ordinary integral over r that can easily be evaluated.

Proof that inverse-square central force is conservative

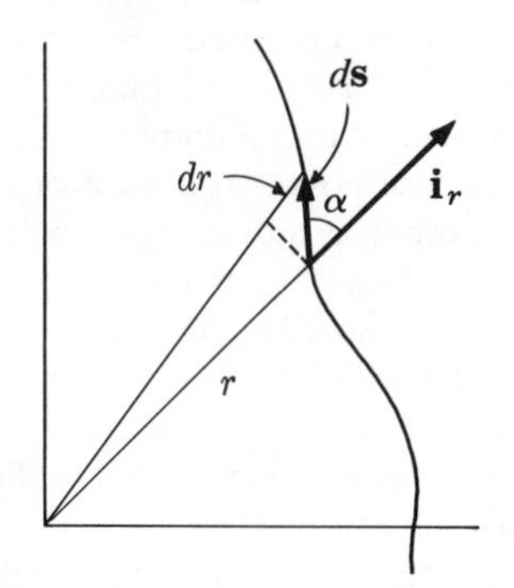

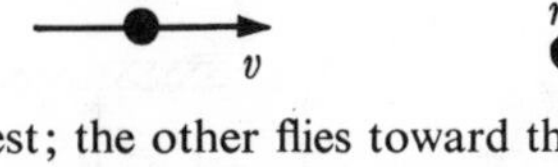

P10.23 Two particles have equal mass m. One is at rest; the other flies toward the first with speed v (see the figure). They collide elastically, which means that their total kinetic energy is the same after the collision as before. For convenience, define the x direction to be the flight direction of the incident particle. If the target particle flies off at 45 deg to the x axis, what are the final velocities of both particles? (SUGGESTION: Assign symbols to unknown quantities and write before-and-after conservation equations for total kinetic energy and for x and y components of total momentum.)

Two-body collision

P10.24 As in the preceding problem, a projectile particle of mass m strikes a stationary target particle of the same mass m. The collision is elastic. (1) Prove that the path of the projectile particle after the collision is perpendicular to the path of the recoiling target particle whenever both final velocities are nonzero. (2) What is the situation if one or the other final velocity is zero?

Two-body collision in one dimension; energy conserved

P10.25 Solve Problem 8.19 if you have not done so already. Then continue with this problem. (1) Assume the collision to be elastic and write the equation of energy conservation. (2) Think of this equation as a relation between the variables v_1 and v_2 for fixed masses and fixed initial speed v_0. Graph the equation in the v_1v_2 plane. What is the name of the resulting curve? (3) How many points of intersection are there between the energy-conservation curve and the momentum-conservation curve in the v_1v_2 plane? Identify the one physically acceptable intersection point if the balls are impenetrable. (4) Solve for the final velocities v_1 and v_2, which are the coordinates of this intersection point.

Two-body collision in one dimension; energy not conserved

P10.26 This problem is a continuation of Problems 8.19 and 10.25. Now assume the collision is inelastic (mechanical energy is not conserved). The energy after the collision can be written $\frac{1}{2}m_1v_1^2 + \frac{1}{2}m_2v_2^2 = \frac{1}{2}\gamma m_1v_0^2$, where $\gamma \leq 1$. (1) Carry out a graphical analysis like the one in the preceding problem, and identify the physically acceptable intersection point in the v_1v_2 plane if the balls are impenetrable. (2) How does the graphical analysis change as the "inelasticity factor" γ changes? Show that there is a minimum physically acceptable value of γ. Obtain an expression for γ_{min}. (3) What is the *graphical* situation when γ takes on its minimum value? What is the *physical* situation when $\gamma = \gamma_{\text{min}}$?

Multiple collisions in one dimension

P10.27 An "executive toy" consists of five metal balls suspended by strings of equal length. To idealize the toy's action, suppose that the balls are not quite in contact when they hang stationary and that collisions between the balls are elastic. (1) Show that if ball 1 is pulled aside and released, it will transfer all of its energy to ball 5. (2) What will happen if balls 1 and 2 are pulled aside and released in rapid succession (so that balls 1 and 2 do not make contact until after ball 2 strikes ball 3)? (3) Suppose now that a drop of solder is placed between balls 1 and 2 so that they move as a unit. What will happen when ball 5 is pulled aside and released? (4) Now the soldered pair of balls is pulled aside; they swing in to strike ball 3 with speed v. Show that ball 5 will swing out with speed $\frac{4}{3}v$, followed by ball 4 with speed $\frac{4}{9}v$, then ball 3 with speed $\frac{4}{27}v$; balls 1 and 2 together will follow with speed $\frac{1}{27}v$. (This problem shows that momentum and energy conservation alone are not sufficient to account for the simple behavior usually observed when the balls start in contact. For an informative discussion, see T. A. Walkiewicz and N. D. Newby, Jr., *American Journal of Physics* **40** (1972).)

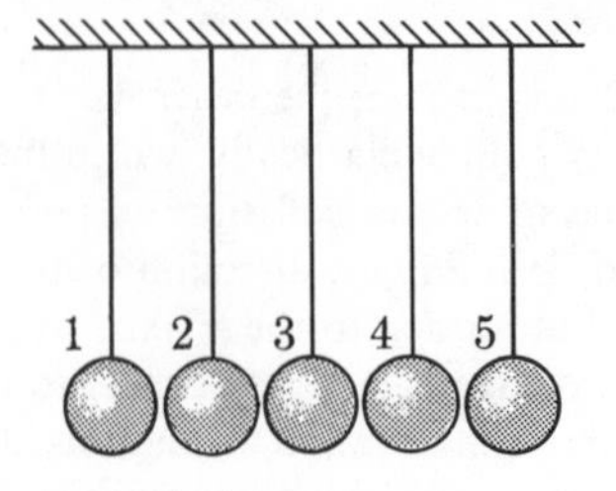

11 Gravitation

To bring mechanics to bear on nature requires more than Newton's laws; specific laws of force are also required. One force in particular—gravitation—is singled out for attention in this chapter.

11.1 Laws of force in mechanics

Consider the vast generality of Newton's three laws alone: The first specifies the nature of undisturbed motion for *all* bodies and defines inertial frames of reference. The second describes how an object reacts to *any* force. The third states that *all* nature's forces come in matched pairs. These laws encompass both the experimentally circumscribed domain of physics and the mathematical domain of unfettered abstraction. Newton himself studied the properties of motion associated with some hypothetical laws of force. Students of physics ever since have been solving problems of motion for arbitrary forces—forces that may not resemble any of those actually encountered in nature.

Specific laws of force supplement Newton's laws

Scientists study nature as it might be only to the extent that such study can teach something about nature as it is. To narrow mechanics to actuality, Newton's laws must be supplemented by laws of force governing natural phenomena, in particular the laws of those fundamental forces that act over macroscopic distances*—the gravitational, electric, and magnetic forces.

The fact that Newton's laws do not distinguish real forces from hypothetical forces is not a weakness of mechanics. In fact it is a strength, for it means that mechanics is open, ready and able to accommodate new discovery. The

* The short-range nuclear forces and the weak interactions act in the subatomic domain, where classical mechanics is not valid.

mechanics of Newton accurately describes the deflection of an electron by an electric force, for instance, although neither the existence of the electron nor the law of electric force was known to Newton.

THE ROLE OF GRAVITY

Gravity, the universal force

Among the fundamental forces of nature, gravity is of special interest for several reasons. It is, first of all, the only truly universal force. It acts on every material thing from electron to galaxy, and as we have learned in this century, it acts even on immaterial things—photons, neutrinos, or energy in any form. Second, a practical reason: Gravity is the most immediately evident force, affecting man and his environment in many ways every day. Indeed, we are so used to gravity that the weightless condition of astronauts strikes us as exceptional and fascinating.

Gravity's historic role

Gravity has played a uniquely important role in the development and growth of mechanics. In his definitive formulation of mechanics, Newton drew upon past studies of motion near the earth, influenced by local gravity, and of planetary motion far from earth, influenced by the sun's gravity. Since Newton's time, motion governed by gravitational force has provided the most stringent tests of mechanics, has served as a stimulus for much of the mathematical elaboration of the theory of mechanics, has led to the discovery of distant new planets, and in our own era of artificial satellites has revealed new details of the shape and structure of the earth. Through the study of the orbit of Mercury came the first hint of an imperfection in Newtonian mechanics. Mercury's refusal to follow precisely the laws of classical mechanics stands now as one of the experimental supports of the new mechanics of Einstein's general relativity.

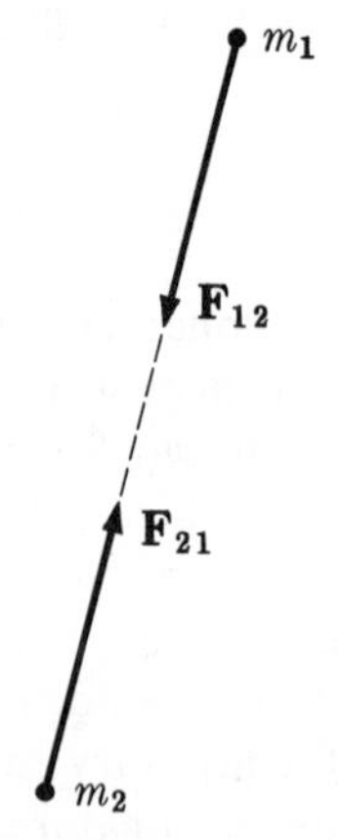

11.2 The law of universal gravitation

Consider a pair of particles separated by a distance d, one with mass m_1 and the other with mass m_2. The gravitational force exerted by each on the other is attractive and central; that is, it acts along the line joining the particles. The magnitude of these equal and opposite forces is proportional to the product of the two masses and inversely proportional to the distance separating them. It may be written

The law of gravitational force

$$F = G\frac{m_1 m_2}{d^2}\,; \tag{11.1}$$

the constant of proportionality G is called the gravitational constant. The *law of universal gravitation* states that this specific law of gravitational force applies to every pair of particles in the universe, always with the same fixed gravitational constant G. It was Isaac Newton who first advanced this grand generalization (his line of reasoning will be plausibly reconstructed in Section 11.7). We now recognize Equation 11.1 as a law richly supported by experiment and valid over an enormous domain, but inadequate under certain extreme conditions such as might exist in or near a neutron star. (A neutron star is a postulated star that has collapsed to a density more than 10^9 times the density of the sun.)

In SI units, the numerical value of the gravitational constant G is

$$G = 6.67 \times 10^{-11} \text{ N m}^2/\text{kg}^2. \quad (11.2)$$

The gravitational constant

This small number reflects the extraordinary weakness of the gravitational force. Two equal masses of 1 kg separated by 1 m attract each other with a force of 6.7×10^{-11} N, less than the weight of a man by a factor of 10^{13}. It is only because of the enormous mass of the earth that its gravitational force seems strong. The action of gravity between any pair of ordinary-sized objects on earth is negligibly small. So far as we know, gravity is also of no consequence in the world of particles. (In the yet-to-be-explored subparticle domain, it is conceivable that the exceedingly small distances may more than compensate for the small masses and small value of G and cause gravity again to become important.)

Gravity is the weakest fundamental force

An important thing to notice about Equation 11.1 is that it is a law of force, not a law of motion. It specifies the magnitude of the gravitational force independent of how the pair of particles are moving or even *whether* they are moving. The masses that appear in the numerator are *gravitational* masses. They are measures of the *strength of interaction*. As noted earlier (Section 7.7), gravitational mass and inertial mass are found experimentally to be equal.

If the factors m_1 and m_2 on the right side of Equation 11.1 are interchanged, the force is unchanged. This mass symmetry is no coincidence. Were it not true, the gravitational-force law would violate Newton's third law. Suppose, for example, we started with the fact that the gravitational force *experienced* by an object is proportional to its own mass. According to Newton's third law, the force exerted by the object is equal in magnitude to the force it feels. Therefore, the force it *exerts* is also proportional to its mass. This means that for any pair of objects, the gravitational force must be proportional to both the attracting mass and the attracted mass. The logic of this argument can be expressed more clearly in a diagram (Figure 11.1).

$m_1 \leftrightarrow m_2$: mass symmetry and Newton's third law

Implicit in Equation 11.1, but not obvious, is a superposition principle for gravitational forces. There are two aspects of the superposition principle that can best be explained by example. Consider three particles of different mass, arranged as shown in Figure 11.2. The superposition principle states first that the force exerted on particle 1 by particle 2 ($\mathbf{F}_{12}$) is independent of the position or mass of particle 3, indeed totally independent of the presence of particle 3. This means that Equation 11.1 remains valid for any pair of particles, regardless of their environment. Second, the superposition principle states that the total force acting on particle 1 is the vector sum of the separate contributing forces, in this example $\mathbf{F}_{12} + \mathbf{F}_{13}$. Obviously the superposition principle is a principle of simplicity. It is hard to conceive of any easier way to extend a law of force from two particles to more than two particles.

Superposition

1. *$\mathbf{F}_{12}$ is independent of 3, 4, ...*

2. $\mathbf{F}_1 = \mathbf{F}_{12} + \mathbf{F}_{13} + \cdots$

GRAVITY OF A SPHERICAL SHELL

As it stands, Equation 11.1 applies only to idealized point particles—or, in practice, to objects whose size is much less than their separation. To find the gravitational force exerted by an extended body, it is necessary to sum the contributions from all parts of the body. The problem of the gravitating

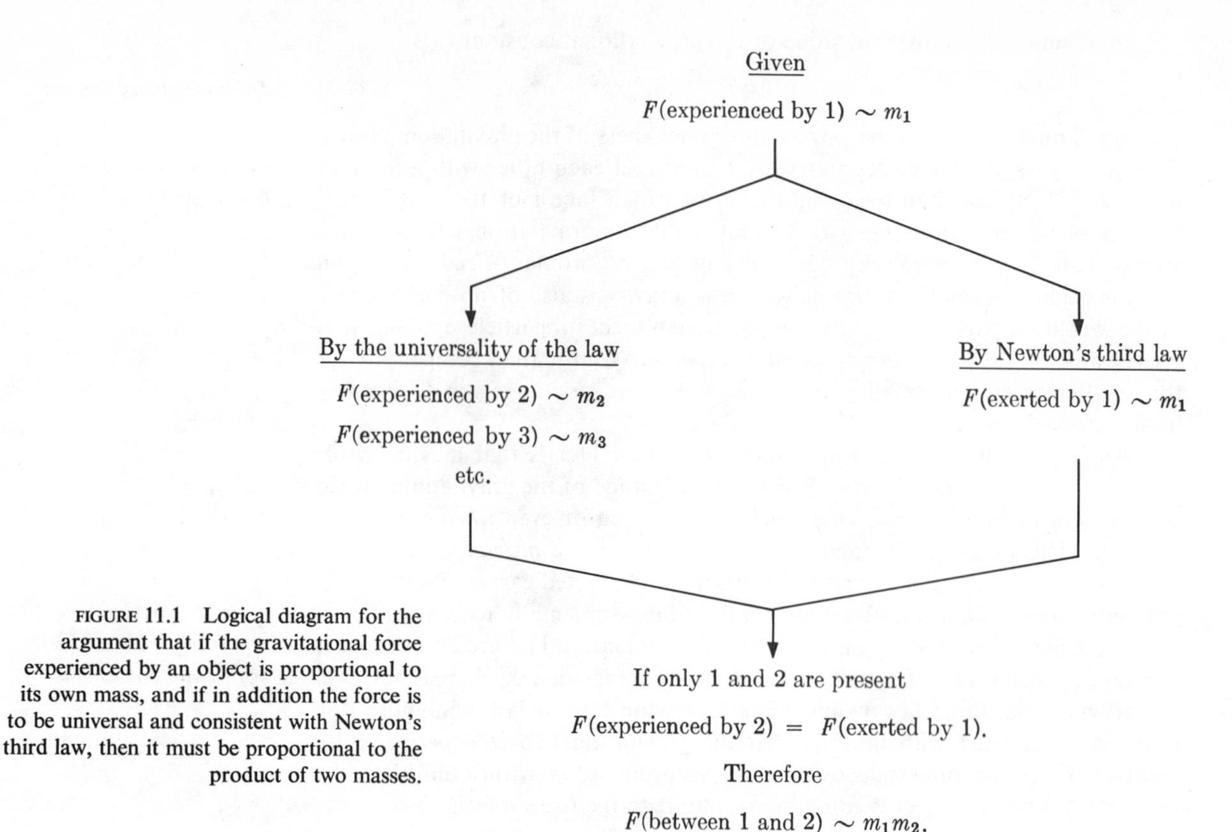
Given
F(experienced by 1) $\sim m_1$
By the universality of the law
F(experienced by 2) $\sim m_2$
F(experienced by 3) $\sim m_3$
etc.
By Newton's third law
F(exerted by 1) $\sim m_1$
If only 1 and 2 are present
F(experienced by 2) $=$ F(exerted by 1).
Therefore
F(between 1 and 2) $\sim m_1 m_2$.

FIGURE 11.1 Logical diagram for the argument that if the gravitational force experienced by an object is proportional to its own mass, and if in addition the force is to be universal and consistent with Newton's third law, then it must be proportional to the product of two masses.

FIGURE 11.2 The superposition principle for gravitational forces.

Gravity outside a shell: The shell acts as a point

sphere was solved first by Newton. His exact results are these: (1) A particle outside a uniform, spherical shell is drawn toward the shell exactly as if all the mass of the shell were concentrated at its center. (2) A particle inside the shell is drawn equally in all directions and experiences no force at all. The proof of the first result is slightly complicated; a method of carrying it out is suggested in Problem 11.14. We give here a proof of the second result.

Consider a point mass m located anywhere within a uniform spherical shell (Figure 11.3). From the point mass, extend a cone of infinitesimal angle in any direction to intercept area dA_1 on the surface of the sphere. The reflection of the same cone intercepts area dA_2 on another part of the sphere. Contained within these intercepted areas are masses dm_1 and dm_2 at distances r_1 and r_2 from mass m. To complete the notation, let dA_1' and dA_2' designate the areas of the two perpendicular bases of the cones at distances r_1 and r_2 (see Figure 11.3), and let $d\mathbf{F}_1$ and $d\mathbf{F}_2$ designate the forces exerted on the mass m by the increments of mass dm_1 and dm_2.

The forces $d\mathbf{F}_1$ and $d\mathbf{F}_2$ are oppositely directed. Their magnitudes are

$$dF_1 = G\frac{m\,dm_1}{{r_1}^2}, \qquad dF_2 = G\frac{m\,dm_2}{{r_2}^2}.$$

The ratio of their magnitudes is

$$\frac{dF_1}{dF_2} = \frac{dm_1/{r_1}^2}{dm_2/{r_2}^2} = \left(\frac{r_2}{r_1}\right)^2\frac{dm_1}{dm_2}. \tag{11.3}$$

Since the shell is uniform, the mass in any increment of area is proportional to the area so that the ratio dm_1/dm_2 is equal to dA_1/dA_2. Now we take advantage of geometrical aspects of the diagram in Figure 11.3. First, the angle between areas dA_1 and dA_1' is the same as the angle between dA_2 and dA_2'. Calling this angle θ, we can write

$$\begin{aligned} dA_1' &= dA_1\cos\theta, \\ dA_2' &= dA_2\cos\theta, \end{aligned} \tag{11.4}$$

so that the ratios dA_1/dA_2 and dA_1'/dA_2' are equal. Second, the areas dA_1' and dA_2' are proportional to the squares of their respective distances from the apex of the cones so that

$$\frac{dA_1'}{dA_2'} = \frac{{r_1}^2}{{r_2}^2}.$$

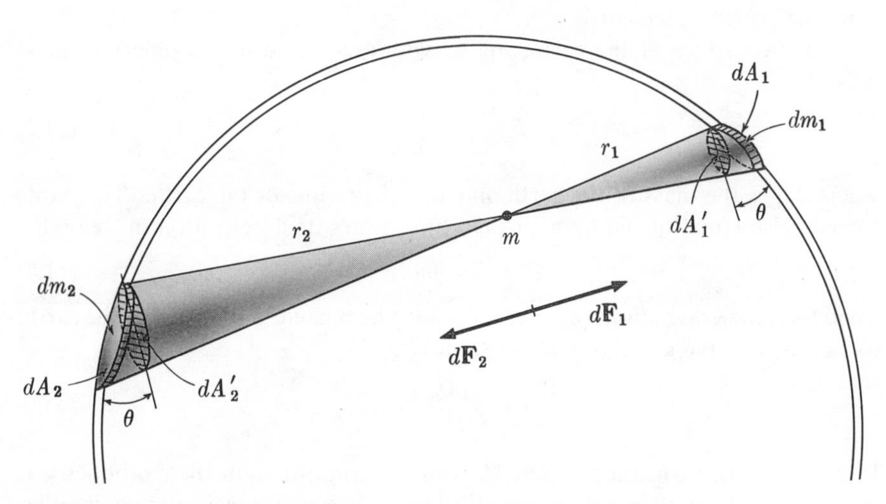

FIGURE 11.3 Geometrical construction used to demonstrate that the net gravitational force on a mass m within a uniform spherical shell is zero.

Putting these facts together, we can write

$$\frac{dm_1}{dm_2} = \frac{dA_1}{dA_2} = \frac{dA'_1}{dA'_2} = \frac{r_1^{\,2}}{r_2^{\,2}}. \tag{11.5}$$

Substituting from Equation 11.5 in Equation 11.3 yields

$$\frac{dF_1}{dF_2} = \left(\frac{r_2}{r_1}\right)^2 \left(\frac{r_1}{r_2}\right)^2 = 1. \tag{11.6}$$

The oppositely directed forces are equal in magnitude. Therefore, their vector sum is zero.

Inside a shell, the gravitational force vanishes

To complete the proof, imagine such cones extending in all directions from the mass m to cover the whole surface of the sphere. For each double cone, the opposite pulls from two parts of the sphere exactly cancel. The total force on the interior mass m is zero. This beautifully simple result is a unique attribute of the inverse-square law of force. For any other dependence of force on distance, the net force within a spherical shell would not vanish. The same uniqueness applies to a particle outside a spherical shell. Only for the inverse-square law is the force outside the spherical shell the same as if all the mass were concentrated at its center. For such reasons do we marvel at the simplicity of nature's fundamental laws.

11.3 Gravity of the earth

The earth can be likened to an onion of successive spherical shells. A man on the surface is pulled downward by each shell as if all its mass were at the center of the earth. The force is then calculable by Equation 11.1. Mass m_1 is the mass of the man, mass m_2 is the mass of the earth, and distance d is the distance from the man to the center of the earth. The force is the same as if the man were left suspended in space while the earth collapsed to a tiny ball four thousand miles beneath his feet. For calculating the force of gravity on a satellite, the radius of the earth is irrelevant. All that matters is the distance of the satellite from the center of the earth.

At the surface of the earth, the gravitational force on an object of mass m is

$$F = G\frac{mM_\text{E}}{R^2}, \tag{11.7}$$

where M_E is the mass of the earth and R is the radius of the earth. This same force is also equal to the mass of the object times its acceleration in free fall:

$$F = mg. \tag{11.8}$$

Therefore, the gravitational acceleration must be related to the mass of the earth, its radius, and the gravitational constant G by

The acceleration of gravity for an ideal spherical earth

$$g = \frac{GM_\text{E}}{R^2}. \tag{11.9}$$

In Newton's time, neither G nor M_E was known, although their product was well-known. Even today, the product is known to higher precision than is either factor. Outside the earth, Equation 11.7 is replaced by

$$F = G\frac{mM_E}{r^2}, \tag{11.10}$$

where r is the distance from the center of the earth. Because of Equation 11.9, this expression for extraterrestrial weight can also be written

Gravity outside the earth

$$F = mg\left(\frac{R}{r}\right)^2, \tag{11.11}$$

a form that was used in the last chapter (see Equation 10.96).

Imagine now a hole bored straight through the center of the earth. How would a man's weight vary as he was lowered into the hole? At a depth of 100 m, the outermost spherical shell of the earth, 100 m thick, would cease to exert any net force on him. All the rest of the earth below would still be pulling him downward. At every depth the earth above would be irrelevant, and his weight would be the same as if he were on the surface of a smaller planet. At the center his weight would be zero. To express his weight mathematically, we may assume that the density of the earth is constant (not, in fact, a very good assumption). If this were true, volume and mass would vary in proportion to the cube of the radius. Inside radius r would be mass M given by

$$M = M_E\left(\frac{r}{R}\right)^3. \tag{11.12}$$

The weight of the man would then be

$$F = G\frac{mM}{r^2} = G\frac{mM_E r}{R^3}. \tag{11.13}$$

With the help of Equation 11.9, this expression simplifies to

Gravity inside the earth

$$F = mg\left(\frac{r}{R}\right). \tag{11.14}$$

Within the earth, weight varies approximately in proportion to distance from the center.

Gravitational force inside and outside the earth is shown graphically in Figure 11.4. Pursuing the discussion of the imaginary hole drilled through the center of the earth, we may ask about the fate of a man who fell into such a hole. Equation 11.13 or 11.14 shows that if air friction could be neglected, he would execute simple harmonic motion because he is drawn toward the center with a force proportional to distance. Comparison of Equation 11.13 or 11.14 with Equation 7.21 shows that the force constant is

$$k = \frac{mg}{R} = \frac{GmM_E}{R^3}. \tag{11.15}$$

Any previous result for simple harmonic motion applies, with appropriate substitution for the constant k. The period, for example, is

$$T = 2\pi\sqrt{\frac{m}{k}} = 2\pi\sqrt{\frac{R}{g}}. \tag{11.16}$$

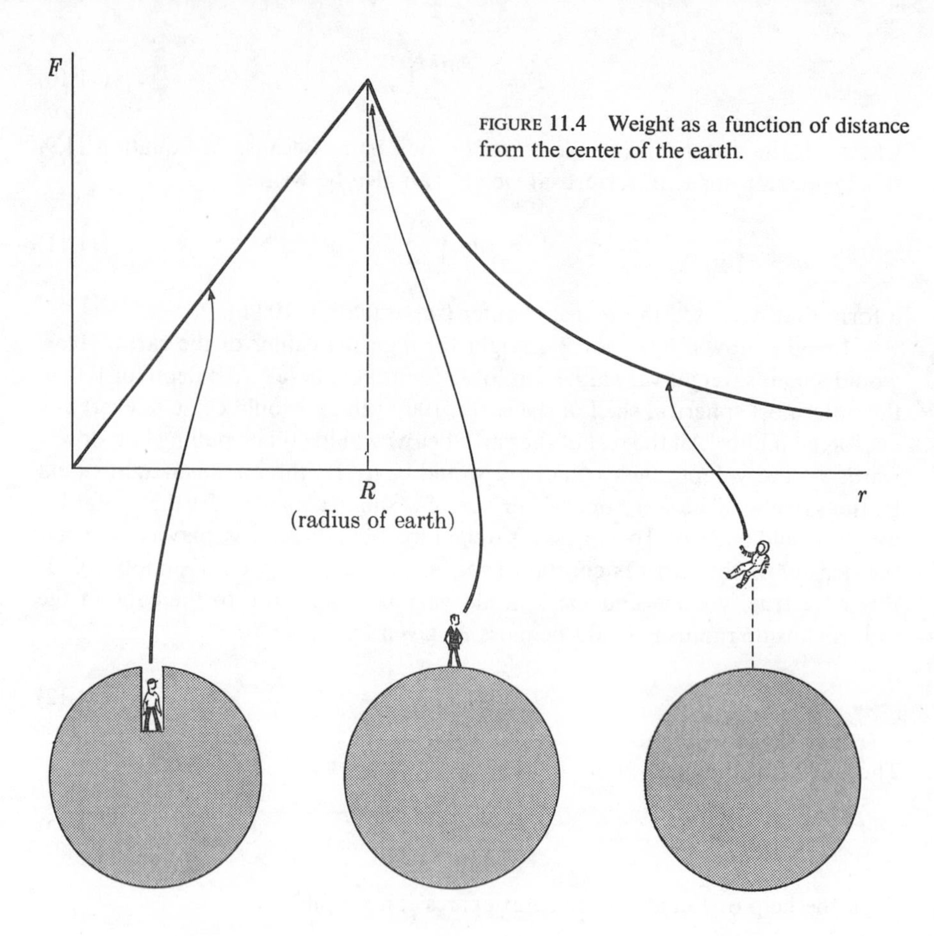

FIGURE 11.4 Weight as a function of distance from the center of the earth.

Except for such problems as air resistance, friends and relatives at the edge of the hole would have to wait just 84 min for the man who fell in to return from his 16,000-mile round trip through the earth. This happens to be also exactly the time required for a low-altitude satellite to circle the earth.

11.4 Gravitational potential energy: escape speed

If the zero of potential energy is chosen at infinite distance from the earth, an object of mass m has negative potential energy in the gravitational force field of the earth, given by Equation 10.97, which we repeat here:

$$U = -\frac{mgR^2}{r} \quad (r \geq R). \tag{11.17}$$

Gravitational potential energy outside the earth

This can also be written, in more fundamental form,

$$U = -G\frac{mM_E}{r} \quad (r \geq R). \tag{11.18}$$

Because the inverse-square central force is a conservative force, this potential-energy function is valid for arbitrary three-dimensional motion of an object influenced by the earth.

At the surface of the earth, where $r = R$, the potential energy of a body of mass m is

$$U_{\text{surface}} = -mgR = -G\frac{mM_{\text{E}}}{R}. \tag{11.19}$$

This may be called the "depth of the potential well." It is equal in magnitude to the least energy required to remove the body completely from the influence of the earth's gravitational forces. In Section 10.8, we defined *escape speed* to be the speed that a rocket must have near the surface of the earth in order (barely) to coast to infinite distance. Putting $\frac{1}{2}mv_{\text{esc}}^2 = |U_{\text{surface}}|$ led to the formula

$$v_{\text{esc}} = \sqrt{2gR}. \tag{11.20}$$

Terrestrial escape speed

In terms of the gravitational constant G, the same quantity is

$$v_{\text{esc}} = \sqrt{\frac{2GM_{\text{E}}}{R}}. \tag{11.21}$$

Its magnitude is given by Equation 10.102.*

Since the gravitational-force law inside the earth is that of a harmonic oscillator, the potential-energy function inside the earth is

$$U = \tfrac{1}{2}kr^2 + C, \tag{11.22}$$

where k is given by Equation 11.15 and where C is a constant. For an oscillator, C is usually set equal to zero, but in this case it is preferable to adjust C so that Equation 11.22 gives $U(R) = U_{\text{surface}}$. Then the potential energy suffers no discontinuity at $r = R$.† Use of Equation 11.15 in Equation 11.22, together with the requirement that U be continuous, leads to

$$C = -\frac{3}{2}mgR = -\frac{3}{2}G\frac{mM_{\text{E}}}{R}, \tag{11.23}$$

and

$$U = mgR\left[-\frac{3}{2} + \frac{1}{2}\left(\frac{r}{R}\right)^2\right] \quad (r \le R), \tag{11.24}$$

Gravitational potential energy inside the earth

or, equivalently,

$$U = G\frac{mM_{\text{E}}}{R}\left[-\frac{3}{2} + \frac{1}{2}\left(\frac{r}{R}\right)^2\right] \quad (r \le R). \tag{11.25}$$

Figure 11.5 shows a graph of $U(r)$ both inside and outside the earth.

---

* In the units preferred in the United States space program, $v_{\text{esc}} = 36{,}700$ ft/sec.

† A discontinuity in U is physically permissible but is normally quite inconvenient because it invalidates the simple connection between work and potential energy (Equation 10.71) and it means that the total energy E is not independent of position (see Question 11.10).

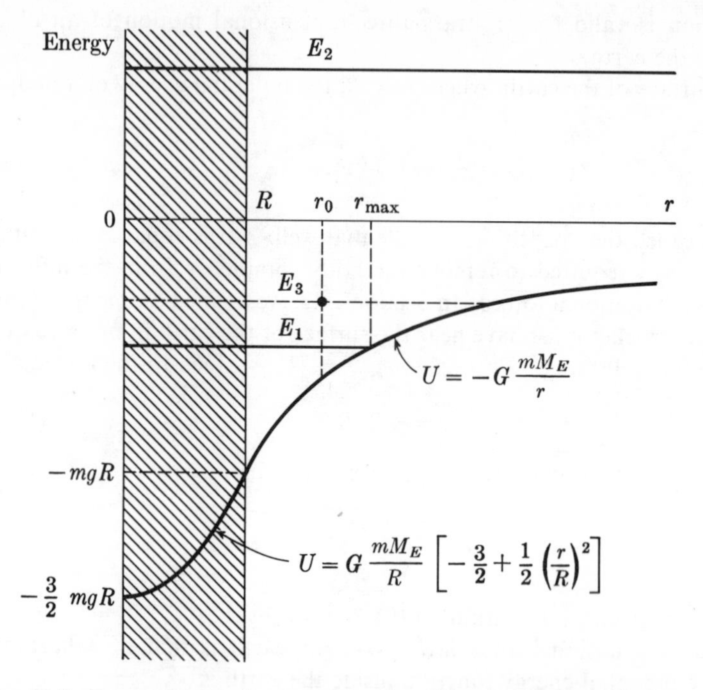

FIGURE 11.5 Energy diagram for rocket or other object interacting with the earth. *Example* 1: A vertically fired rocket with energy E_1 reaches a peak height $r_{max} - R$. *Example* 2: A rocket with energy E_2 escapes from the earth. *Example* 3: A satellite in a circular orbit, with energy E_3 and radial distance r_0, occupies a fixed point in the diagram. At this point, $E_3 = \frac{1}{2}U$.

■ EXAMPLE 1: A vertically fired rocket starts upward with total energy $E_1 = -\frac{1}{2}mgR$ (see Figure 11.5). What is its maximum height? At the peak of its motion, $K = 0$, and $U = E_1$, or

$$-G\frac{mM_E}{r_{max}} = -\frac{1}{2}mgR.$$

On the right, g can be replaced by GM_E/R^2 (Equation 11.9), so

$$-G\frac{mM_E}{r_{max}} = -\frac{1}{2}G\frac{mM_E}{R}.$$

The solution to this equation is

$$r_{max} = 2R.$$

The peak distance above the earth is

$$h_{max} = r_{max} - R = R \cong 4{,}000 \text{ miles.}$$ ■

■ EXAMPLE 2: After exhausting its fuel in a very short climb, a vertically fired rocket of mass 10^5 kg has a kinetic energy of 10^{13} J. Does it escape? If so, what is its final speed at great distance from the earth? First, determine its total energy. According to Equation 11.19, its potential energy at the start of its trip is

$$U_0 = -mgR = -10^5 \text{ kg} \times 9.80 \text{ m/sec}^2 \times 6.37 \times 10^6 \text{ m}$$

$$= -6.24 \times 10^{12} \text{ J}.$$

Its total energy, $E_2 = K_0 + U_0$, is

$$E_2 = 10^{13} \text{ J} - 0.624 \times 10^{13} \text{ J}$$

$$= 3.76 \times 10^{12} \text{ J}.$$

Since this energy is positive, the rocket does escape (Figure 11.5). At infinite distance, $U = 0$, and $K = E_2$, or

$$\tfrac{1}{2}mv_{\text{final}}{}^2 = E_2 = 3.76 \times 10^{12} \text{ J}.$$

The solution is

$$v_{\text{final}} = \sqrt{\frac{2E_2}{m}} = 8.67 \times 10^3 \text{ m/sec}.$$

■

■ EXAMPLE 3: A satellite is in a circular orbit at distance r_0 from the center of the earth. What is the ratio of its kinetic energy to its potential energy? Let us call the total energy of the satellite E_3. As shown in Figure 11.5, the circling satellite is represented by a stationary point in an energy diagram, since both r_0 and E_3 are constant. The force acting on the satellite is

$$F = G\frac{mM_E}{r_0{}^2},$$

so its acceleration is

$$a = \frac{GM_E}{r_0{}^2}.$$

This must be equal to v^2/r_0, the centripetal acceleration of any object moving uniformly in a circle. From this equality, the kinetic energy is derived:

$$K = \frac{1}{2}mv^2 = \frac{1}{2}G\frac{mM_E}{r_0}. \tag{11.26}$$

The potential energy at the same radial distance is

$$U = -G\frac{mM_E}{r_0}. \tag{11.27}$$

Therefore, the ratio K/U is simply

$$\frac{K}{U} = -\frac{1}{2}, \tag{11.28}$$

Simple properties of circular motion in an inverse-square field of force

independent of r_0. The sum $K + U$ gives the total energy:

$$E = -\frac{1}{2}G\frac{mM_E}{r_0}, \tag{11.29}$$

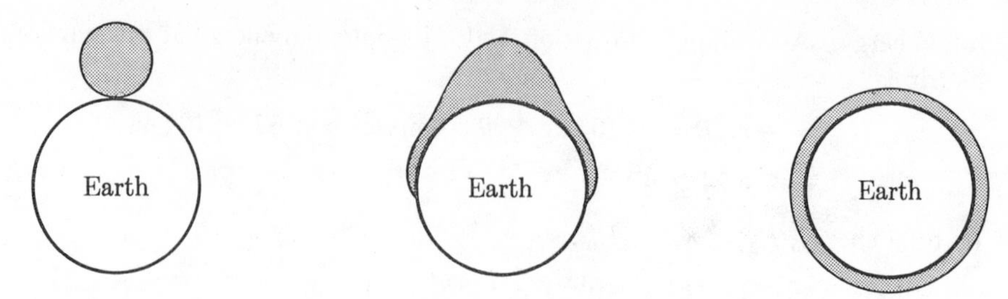

FIGURE 11.6 Imaginary spread of a giant water drop over the surface of the earth. It seeks the almost spherical form of lowest gravitational potential energy.

which is the same as $-K$ or $\frac{1}{2}U$. These simple relationships mean that the point in the energy diagram representing circular motion lies halfway between the potential-energy curve and the zero-energy line. ■

11.5 The shape of the earth

The fact that the earth, moon, sun, and planets are spheres—or very nearly spheres—was accepted for millennia without ever really being explained. This was hardly a scientific puzzle, for it is unlikely that more than a handful of people before Newton thought that any explanation was required. Yet he realized that an explanation was called for and that gravity could provide it. Using modern terminology, we can say that the spherical form is the shape of least potential energy. Imagine, for instance, a giant sphere of water suddenly released at one point on the earth (Figure 11.6). We pretend for the moment that the earth is a perfect sphere. Each part of the liquid is attracted to the center of the earth, and consequently flows until it is as close as it can get to the earth's center. It establishes a spherical shell surrounding the earth. The whole earth was probably once hot enough to flow in this way. Even if it were not, after the passage of enough millions of years, solid rock too can "flow" to establish the spherical form. Our older mountain ranges are gradually being worn away and their mass redistributed over the earth. The eventual fate of a nonrotating body would be, in principle, an ideal sphere. In practice, of course, surface irregularities might never be totally erased.

A stationary earth would be a sphere

What if the sphere is rotating? Then a "stationary" bit of matter on the surface of the earth is in fact experiencing acceleration as it circles the axis of the earth once each day. Consider a book on a supposedly frictionless table [Figure 11.7(a)]. The contact force $\mathbf{F}_c$ exerted by the table on the book acts vertically upward. Usually, we say that the gravitational force on the book, $\mathbf{F}_G$, has equal magnitude and acts vertically downward, so the book experiences no net force. Actually, however, these two forces cannot quite cancel. They must add to give a small net force directed toward the axis of the earth ($\mathbf{F}_{net} = \mathbf{F}_c + \mathbf{F}_G$). Since the book is experiencing a net acceleration, it must be experiencing a net force. This means that the gravitational force cannot act vertically downward (except at pole and equator). As indicated in Figure 11.7(b), the gravitational force acts approximately toward the center of the earth, a direction inclined slightly away from the vertical. If the earth remained precisely spherical,

A "motionless" object on earth is accelerated

the vertical direction (perpendicular to the surface) and the direction of the center of the earth would be the same, and it would be impossible to combine a vertical contact force and a gravitational force to produce the proper net force toward the axis.

A rotating earth is an ellipsoid

Newton was able to prove that the equilibrium shape of a rotating body, if it is capable eventually of flowing, is an ellipsoid, flattened at its poles, and bulging at its equator. He confidently predicted such a flattening of the earth, calculated the magnitude of the effect, and produced some striking indirect evidence in support of it. But fifty years went by before direct measurements confirmed the prediction. Imagine east-west circles of latitude drawn on the earth at 1-deg intervals from the equator (0 deg) to the North Pole (90 deg). Each circle of latitude is labeled by the angle that a line pointing toward the North Star makes with the local horizontal. Experimentally, the distance between these circles of latitude can be determined by measuring the distance that must be traveled northward or southward to change the angular position of the North Star in the sky by 1 deg. On a spherical earth, this distance is the same on all parts of the earth ($\frac{1}{360}$ × circumference of earth). On a flattened earth, changing the latitude angle by 1 deg requires greater travel near either pole than near the equator. In the years 1735–1737 expeditions went out from France to Peru and Lapland to make the measurements and settle the question of the shape of the earth. Both returned with convincing evidence in support of Newton's prediction. Pierre de Maupertuis, leader of the Lapland expedition, was henceforth known as "the grand flattener."

Satellites reveal new details of the earth's shape

No essentially new evidence on the shape of the earth came to light until the age of artificial earth satellites. Precise tracking of satellites can reveal in great detail the deviations of the earth's mass distribution from perfect sphericity,

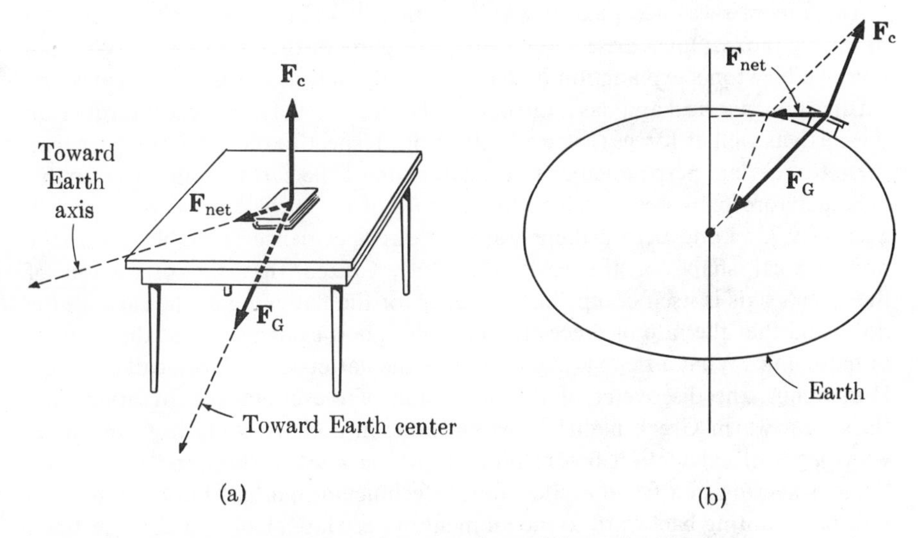

FIGURE 11.7 Explanation of the nonspherical shape of the earth. For a stationary bit of matter at the earth's surface, the gravitational force $\mathbf{F}_G$ and the contact force $\mathbf{F}_c$ must add to give a small net force $\mathbf{F}_{net}$ toward the earth's axis. (The amount of the earth's flattening is greatly exaggerated in the drawing.)

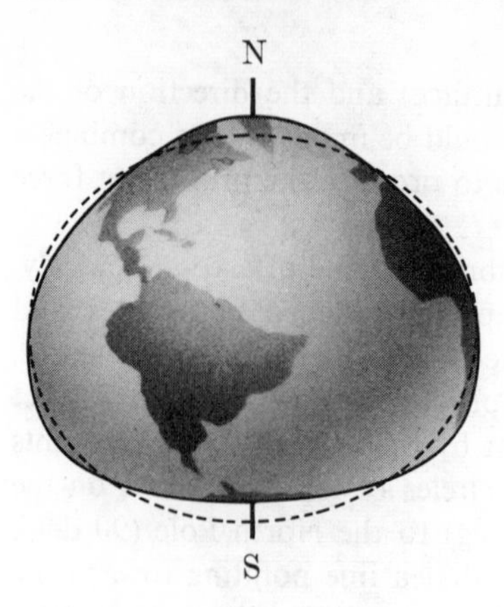

FIGURE 11.8 A highly exaggerated view of the shape of the earth. Besides being flattened at the poles, the earth is very slightly distorted into a pear shape.

since any such deviation alters slightly the inverse-square law of force outside the earth and influences the acceleration of the satellite.* In 1959 it was discovered that the earth, besides being flattened at the poles, has a very slight pear shape, a minute tendency toward a pointed North-Pole head and the barest hint of jowls about the southern hemisphere (Figure 11.8).

PRECESSION OF THE EQUINOXES

Probably the most dramatic and beautiful aspect of Newton's work on the nonspherical earth was his explanation of the precession of the equinoxes. Since at least 200 B.C., it had been known that the point in the sky to which the earth's pole points moves gradually among the stars. After the Copernican view of the universe was accepted, it was clear that this effect, called the precession of the equinoxes, must arise from a slow wobbling of the earth's axis. We shall not give Newton's explanation in detail but only indicate its basis. Because the earth is nonspherical and has a tilted axis, the sun and the moon exert torques on the earth as well as forces (Figure 11.9). With respect to the center of the earth, these torques act perpendicular to the direction of the earth's spin. The earth's axis therefore precesses, much in the manner of the bicycle wheel discussed in Section 9.7. Long before there was any direct evidence available about the nonspherical shape of the earth, Newton realized that the precession of the earth's axis is itself compelling evidence for the flattening of the earth at the poles and that the rate of precession provides good evidence about the amount of flattening. What a rich variety of images the flattened earth can call to mind: Hipparchus, the discoverer of the precession of the equinoxes, observing the stars on a warm Greek night; Newton, alone in a chilly study in Cambridge, with pages of calculation before him; de Maupertuis, a thousand miles from home, shivering in a frigid arctic wind; a technician in a half-buried bunker in Florida, counting backward to the moment of fiery launch of a satellite destined to reveal new details of the earth's shape.

Torques act on the earth; its spin precesses

* See Desmond King-Hele, "The Shape of the Earth," *Scientific American*, October, 1967.

THE ACCELERATION OF GRAVITY

For a stationary and perfectly spherical earth, the acceleration of gravity would be precisely constant over the face of the earth, and it is given by Equation 11.9. Not so for a rotating earth. At the equator, the earth's surface is itself accelerated toward the center of the earth so that a falling object has to overtake a downward accelerating surface. Thus the measured value of the acceleration g *relative to the surface* is less at the equator than at the pole. The surface acceleration may be calculated from the formula $a = v^2/r$. At the equator, $v = 465$ m/sec (about 1,040 mile/hr), and

$$a_{\text{equator}} = \frac{(4.65 \times 10^2 \text{ m/sec})^2}{6.37 \times 10^6 \text{ m}} = 0.034 \text{ m/sec}^2. \tag{11.30}$$

Two effects cause g to vary with latitude

This is the amount by which the value of g would diminish in going from pole to equator if the earth were spherical. Because it is not spherical, the force of gravity is actually slightly greater at the pole than at the equator, enough greater to account for an additional change in the value of g of 0.018 m/sec^2. Altogether, the measured value of g is less at the equator than at the pole by an amount Δg equal to the sum of the two contributions:

$$\Delta g = 0.052 \text{ m/sec}^2. \tag{11.31}$$

The modern values for g at pole and equator and their difference are

$$g_{\text{pole}} = 9.8322 \text{ m/sec}^2, \tag{11.32}$$

$$g_{\text{equator}} = 9.7804 \text{ m/sec}^2, \tag{11.33}$$

$$\Delta g = 0.0518 \text{ m/sec}^2. \tag{11.34}$$

The approximate value of g commonly employed, 9.80 m/sec^2, is correct to within 0.3 percent everywhere on earth.

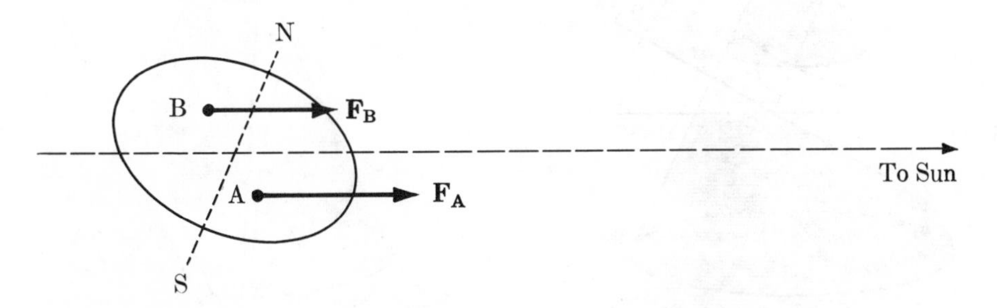

FIGURE 11.9 Explanation of the torque exerted by the sun on the earth. The dashed line represents the ecliptic plane (plane of earth's orbit). The center of mass of the part of the earth below this plane (point A) is slightly closer to the sun than the center of mass of the part of the earth above this plane (point B). Accordingly the sun exerts a force $\mathbf{F_A}$ on the lower half that is slightly greater than the force $\mathbf{F_B}$ which it exerts on the upper half. Because these forces are not precisely equal there is a net torque exerted with respect to the center of the earth (directed out of the page in this diagram). The scale of the figure is of course highly distorted.

11.6 Kepler's laws

Newton's laws are statements about motion in general and forces in general. Kepler's laws are of a quite different kind; they are statements about the motion of one single system, the planetary system. Newton's and Kepler's laws differ also in the number of concepts they draw together. In modern terminology, Newton's laws are *dynamic*, connecting mass, force, distance, and time. Kepler's laws are *kinematic*, concerning only distance and time. Kepler's laws are best looked upon as summarized observation. They distill and neatly package a myriad of observations, converting into a few beautifully simple relationships what would otherwise be long tables of numbers with much substance and no form. Newton's laws, rather than *summarizing* a particular set of observations, *generalize* from observation. They connect a few basic concepts for all motion and all systems. Despite the fact that they are both called "laws," the nature and intent of Newton's laws and Kepler's laws are quite different.

Kepler's laws summarize planetary data

Laboring for many years over the astronomical data of Brahe, Kepler was rewarded by the discovery of the three laws of planetary motion that now bear his name. The first two, published in 1609, describe the shape of planetary orbits and the variations of speed of a planet as it executes its orbit. The third, published in 1619, relates one planetary orbit to another. Kepler's laws remain as valid today as when they were discovered.

KEPLER'S FIRST LAW

The geometry of planetary orbits

The path traced out by each planet is an ellipse with the sun at one focus.

An ellipse is a conic section. It may be defined in several ways: (1) If a

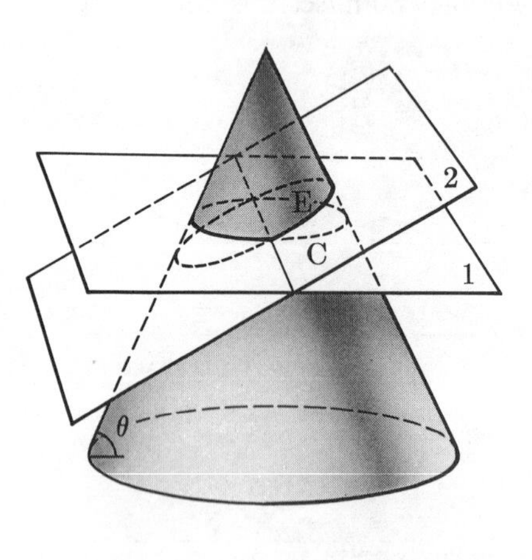

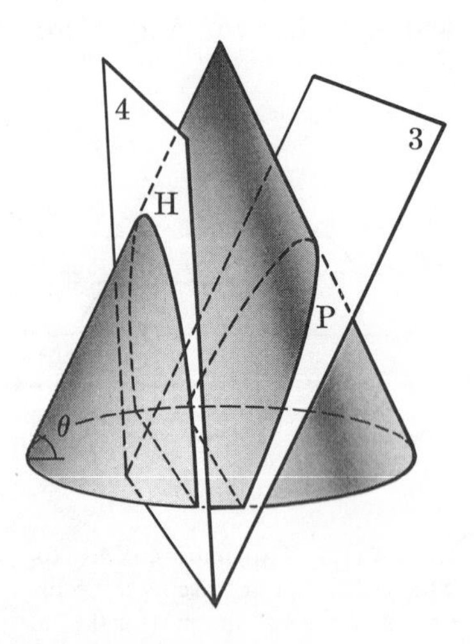

FIGURE 11.10 Conic sections. The edge of the cone makes an angle θ with the horizontal. The intersection of a horizontal plane (1) with the cone defines a circle (C). A plane inclined to the horizontal at less than the angle θ (2) intersects the cone in an ellipse (E). Planes inclined to the horizontal at angles equal to θ (3) and greater than θ (4) define the parabola (P) and the hyperbola (H).

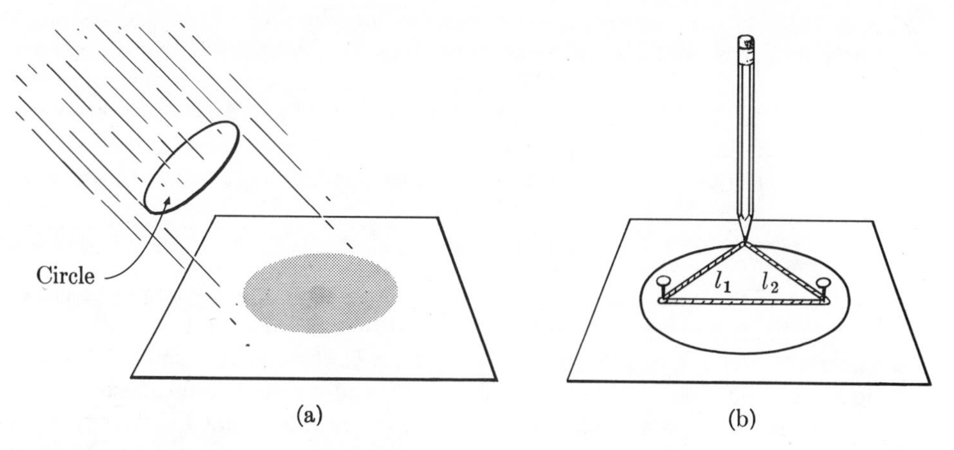

FIGURE 11.11 Other definitions of the ellipse. (a) The shadow of a circle intercepting parallel beams of light is an ellipse. (b) Points, the sum of whose distances, l_1 and l_2, from fixed points is constant, define an ellipse.

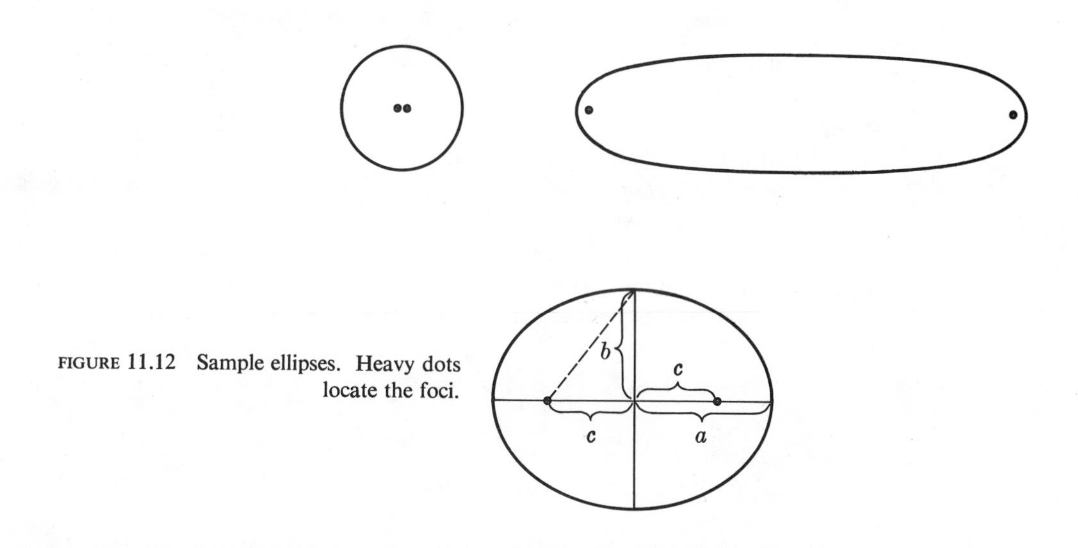

FIGURE 11.12 Sample ellipses. Heavy dots locate the foci.

plane intersects a circular cone in such a way that the intersection is a closed curve, that curve is an ellipse (Figure 11.10). Other conic sections, as shown in Figure 11.10, are the parabola, the hyperbola, and the circle. The parabola and hyperbola are open curves. The circle is a special ellipse. Other geometrical definitions of the ellipse include the following: (2) The shadow cast on a flat surface by a circle exposed to parallel rays of light is an ellipse [Figure 11.11(a)]. (3) If a pencil point moves in a plane in such a way that the sum of its distances from two fixed points in the plane is constant, the curve it traces out is an ellipse [Figure 11.11(b)]. These two fixed points within the ellipse are called its foci.

Properties of ellipses

Figure 11.12 shows an ellipse with various points and distances labeled. The longest straight line through the center of the ellipse joining its opposite ends is called the major axis. Either half of the major axis is called a semimajor axis. The shortest line through the center of the ellipse joining its sides is called the minor axis, and half of it the semiminor axis. Major and minor axes are perpendicular to each other. The two foci lie on the major axis equidistant from its center. In the diagram, the length of the semimajor axis is called a; the length of the semiminor axis, b; and the distance from the center of the ellipse to either focus, c. For every ellipse, there is a simple algebraic connection among a, b, and c:

$$c^2 = a^2 - b^2. \tag{11.35}$$

Because of this relation, the distance from one end of the minor axis to either focus (the dashed line in Figure 11.12) is also equal to a. For a "fat" ellipse (nearly a circle), a is only slightly larger than b, and c is very small. The two foci are close together. (For a circle, $a = b$ and $c = 0$; the two foci coalesce at the center.) For a "thin" ellipse, whose major axis is much longer than its minor axis ($a \gg b$), the foci are widely separated near the ends of the ellipse. One other parameter is used to characterize an ellipse. It is the *eccentricity* e, defined by

Eccentricity

$$e = \frac{c}{a}. \tag{11.36}$$

The planetary orbits all have rather small eccentricities. Some comets have large eccentricities (that is, e nearly equal to 1, which is the maximum value), and they sweep very close to the sun at one end of their orbit, very far from it at the other end. What is the eccentricity of a circle?

We add one algebraic definition of the ellipse, not because it contributes more to the understanding of Kepler's first law but because it illustrates the variety of approaches to this interesting curve: (4) If an ellipse is placed with its center at the origin of a coordinate system and with its major and minor axes along the x and y axes respectively, the curve is defined by the equation

Equation of an ellipse

$$\frac{x^2}{a^2} + \frac{y^2}{b^2} = 1. \tag{11.37}$$

In this equation, a and b are fixed constants, as defined earlier, characterizing the particular ellipse. The variable coordinates x and y locate any point on the curve.

Newton, using his second law and the law of gravitational force, first proved that if a planet happened to be started in the right way, it could move along an elliptical orbit of any shape or size whatever. In fact, he proved even more—that all of the possible trajectories are conic sections. A sufficiently energetic planet or comet could move along a parabolic or hyperbolic path and escape from the solar system. Now that man has created a new planetary system of satellites about the earth, he has created a much wider range of elliptical shapes than were originally recognized in the solar system. He has also sent vehicles into escaping orbits along hyperbolic paths relative to the earth.*

* No space vehicle has so far escaped from the solar system. Once free of the earth, the interplanetary vehicles settle into elliptical orbits about the sun.

Kepler's first law, it should be noted, is purely a geometrical law. It refers only to the spatial aspect of the planetary orbit, not to time or speed or any other concept. Yet what a great step it was after two thousand years of circles. With a single statement that planets move along ellipses, Kepler vastly simplified the picture of the solar system, and at the same time he achieved better concordance between observation and calculation than had the most elaborate previous scheme.

KEPLER'S SECOND LAW

The law of areas

Kepler's second law is the *law of areas*, discussed in Section 9.9. As a planet moves along its elliptical orbit, its speed varies in such a way that the radial line connecting the sun to the planet sweeps out area at a constant rate. This means that the planet moves more rapidly when close to the sun, more slowly when far from the sun. The variation is such as to preserve the constancy of orbital angular momentum.

As pointed out in Section 9.9, the application of Kepler's second law is especially simple at the points of greatest and least distance from the sun (or from any other center of force). At these two points, the products of speed and radial distance are equal:

$$v_a r_a = v_p r_p. \tag{11.38}$$

The subscripts a and p refer to aphelion and perihelion. For the earth, $r_a =$ 152 million km and $r_p =$ 147 million km. The earth's greatest speed (at perihelion) divided by its least speed (at aphelion) is

$$\frac{v_p}{v_a} = \frac{r_a}{r_p} = \frac{1.52 \times 10^{11} \text{ m}}{1.47 \times 10^{11} \text{ m}} = 1.034. \tag{11.39}$$

Kepler's second law deals only in ratios of speeds. It does not reveal the speed at any point along the orbit unless the speed at another point is already known. The measured minimum speed of the earth is

$$v_a = 2.93 \times 10^4 \text{ m/sec}. \tag{11.40}$$

(This is about 90 times the speed of sound in air and nearly 4 times the speed of a low-altitude satellite relative to the earth.)

In modern notation, Kepler's second law may be written

A conservation law

$$\mathbf{r} \times \mathbf{v} = \textbf{constant}. \tag{11.41}$$

Like the first law, it is valid for each orbit separately but establishes no connection among different orbits. Although Kepler did not recognize it as such, his second law of planetary motion is one of the great conservation laws of mechanics.

KEPLER'S THIRD LAW

If a denotes the length of the semimajor axis of a planetary orbit and if T denotes the time required for the planet to complete one revolution about the sun, then

$$T^2 \sim a^3. \tag{11.42}$$

Kepler's third law links different orbits

The square of the period is proportional to the cube of the length of the semimajor axis. Alternatively, Kepler's third law may be written

$$\frac{T^2}{a^3} = \kappa. \tag{11.43}$$

The quantity κ is a constant, the same for every planet.

For many years, Kepler had sought to supplement his first two laws with a third that would relate one planetary orbit to another. Since the time of Copernicus, it had been recognized that planets successively more distant from the sun had successively longer periods. Kepler was the first to establish the quantitative relationship between times and distances. Both the form of the law and the numerical value of its constant of proportionality were for Kepler empirical discoveries, unrelated to any deeper rules of planetary motion. Now, with the help of the inverse-square law of gravitational force, we may readily derive Kepler's third law. To do this most easily, let us specialize to circular orbits. We may then make use of the simple connection between kinetic energy and potential energy given by Equation 11.28:

$$U = -2K.$$

For the planets, this equality can be written

$$-G\frac{mM_S}{r} = -mv^2, \tag{11.44}$$

where m is the mass of a planet, r is its distance from the sun, v is its speed, and M_S is the mass of the sun. The mass of the planet cancels. Since its period is given by

$$T = \frac{2\pi r}{v},$$

we may replace v in Equation 11.44 by $2\pi r/T$, then rearrange factors to give the form of Kepler's third law,

$$\frac{T^2}{r^3} = \frac{4\pi^2}{GM_S}. \tag{11.45}$$

The right side is a constant, the same for all planets. We may call this κ_S:

$$\kappa_S = \frac{4\pi^2}{GM_S}. \tag{11.46}$$

Theory relates κ to the mass of the force center

Clearly, the constant κ in Kepler's third law is fixed only for a particular center of force. It has one value for all the planets and comets rotating around the sun, another value for all of the moons of Jupiter ($\kappa_J = 4\pi^2/GM_J$), and another value,

$$\kappa_E = \frac{4\pi^2}{GM_E}, \tag{11.47}$$

for all satellites of the earth, including our moon. As Newton first demonstrated, Equations 11.43 and 11.46 are valid for elliptical orbits, not just circular orbits.

The principal properties of the sun's nine known planets are summarized in Table 11.1 (see also Table 9.2).

TABLE 11.1 PRINCIPAL PROPERTIES OF THE SUN'S NINE KNOWN PLANETS*

Name of Planet	Mean Distance from Sun†		Period		Eccentricity of Orbit	Mass of Planet		Mean Radius of Planet		Acceleration of Gravity at Planet Surface‡
	m	A.U.§	sec	Earth years		kg	Earth masses	m	Earth radii	m/sec^2
Mercury	5.79×10^{10}	0.387	7.60×10^{6}	0.241	0.2056	3.30×10^{23}	0.055	2.43×10^{6}	0.38	3.7
Venus	1.08×10^{11}	0.723	1.94×10^{7}	0.615	0.0068	4.87×10^{24}	0.815	6.06×10^{6}	0.95	8.8
Earth	1.496×10^{11}	1.000	3.156×10^{7}	1.000	0.0167	5.97×10^{24}	1.000	6.37×10^{6}	1.00	9.80
Mars	2.28×10^{11}	1.524	5.94×10^{7}	1.881	0.0934	6.42×10^{23}	0.107	3.37×10^{6}	0.53	3.76
Jupiter	7.78×10^{11}	5.204	3.74×10^{8}	11.86	0.0485	1.90×10^{27}	317.9	6.99×10^{7}	10.97	24.
Saturn	1.43×10^{12}	9.58	9.35×10^{8}	29.6	0.055	5.69×10^{26}	95.2	5.85×10^{7}	9.18	10.
Uranus	2.86×10^{12}	19.14	2.64×10^{9}	83.7	0.047	8.7×10^{25}	14.5	2.33×10^{7}	3.66	10.
Neptune	4.52×10^{12}	30.2	5.22×10^{9}	165.4	0.008	1.03×10^{26}	17.2	2.21×10^{7}	3.47	14.
Pluto	5.90×10^{12}	39.4	7.82×10^{9}	248	0.249	$\sim 5 \times 10^{23}$	~ 0.08	3×10^{6}	0.5	~ 3

* References for planetary data: W. M. Kaula, *An Introduction to Planetary Physics*, (New York: John Wiley and Sons, 1968) and Michael E. Ash, Irwin I. Shapiro, and William B. Smith, "The System of Planetary Masses," *Science* **174**, 551 (1971). Copyright 1971 by the American Association for the Advancement of Science.
† The planet's mean distance from the sun is equal to its semimajor axis.
‡ Numbers in this column give average acceleration relative to rotating planet surface.
§ Earth's mean distance from the sun is defined to be one astronomical unit (A.U.).

★11.7 Deduction of the law of gravitational force

One of the greatest dramas in the history of science was Newton's deduction of the law of gravitational force and his synthesis of celestial and terrestrial motion. From a utilitarian point of view, a knowledge of this great generalization is irrelevant to the student's task of understanding and applying the law of gravitational force. Yet it is both interesting and instructive to see how Newton was able to build the edifice of universal gravitation on the foundation of Kepler's laws. That development is our concern in this section.

An important thing to understand first is why Newton was convinced that a mechanical explanation of planetary motion was needed. For Aristotle and Copernicus, circular motion was the "natural" motion in the heavens. It was the way stars or planets moved when left to themselves (and obviously man had no opportunity to do otherwise than leave them to themselves). It required no further explanation. Kepler and his English contemporary William Gilbert were perhaps the first to think seriously about a force to explain planetary motion. Both imagined that the force might be magnetic. But their speculations could not lead far, for neither understood the principle of inertia, that an undisturbed object continues to move uniformly in a straight line. They imagined that a force should act along the planetary orbit to impel the planet continually along its path through space. Galileo's attitude toward planetary motion was most interesting, for while he took giant strides forward in understanding mechanical principles and in understanding gravity, he reverted to a view of natural circular motion scarcely more sophisticated than Aristotle's. He was the first to state that undisturbed motion ("natural" motion) is motion in a straight line with constant velocity. This sounds like Newton's first law. However, Galileo believed the validity of this law to be limited to the macroscopic, human-sized domain. He pointed out that an object set sliding on an enormous frictionless plane in contact with the earth at one point (Figure 11.13) would not continue indefinitely at constant velocity. As it moved further from the center of the earth it would be decelerated; what started out as horizontal motion would become uphill motion. On the other hand, on an imaginary perfectly smooth sphere surrounding the earth, an object set sliding would continue at constant speed in a great circle around the earth. Because he could not free himself mentally from the shackles of the earth's gravitational field, Galileo failed to extend the principle of inertia to the heavens.

Newton's imagination was able to cut loose from the bonds of earthly gravity that had restrained Galileo. He unhesitatingly extended Galileo's principle of inertial motion to infinity and stated it as his first law of mechanics. This was a necessary first step to the discovery of the law of universal gravitation. If the "natural" motion of a planet in the absence of force is straight-line motion, the actual orbital motion requires an explanation. A force must be acting to deflect the planet from its otherwise straight course into its curved path around the sun.

The importance of Newton's first law: planetary motion requires an explanation

In the five subsections below, we will trace a line of reasoning from Kepler's laws to the law of gravitational force. This should be taken simply as a plausible reconstruction of Newton's work and not as authoritative history. There is some evidence about the early development of Newton's ideas in his later corre-

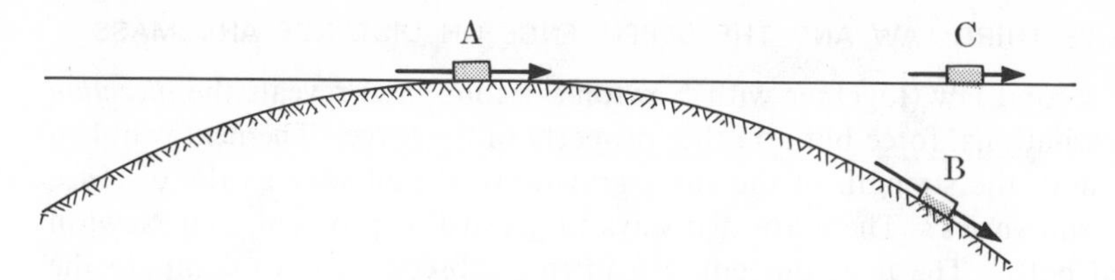

FIGURE 11.13 Galileo's view of frictionless undisturbed motion. An object sliding without friction around the surface of the earth (A to B) maintains constant speed, whereas an object sliding without friction in a straight line (A to C) is decelerated.

spondence, but Newton himself may have been rewriting the events. Insights in science are much more likely to occur in a patchwork of intuition, deduction, guesswork, and calculation than in any logical, orderly chain. Not the least important element is the preconception, the belief that the solution must be found along a certain path.

1. KEPLER'S SECOND LAW AND THE DIRECTION OF THE FORCE

Before the time of Newton, the kinematics of uniform circular motion has not been understood. The essential facts are that the acceleration is directed inward toward the center of the circle and has constant magnitude given by

$$a = \frac{v^2}{r},$$

where v is the speed of the object and r is the radius of the circle. These facts about uniform circular motion were first published by Christiaan Huyghens in 1673, but they were probably known to Newton in 1666. They mean that if an object moves uniformly in a circle, it must experience a force directed toward the center of the circle. (Here we use the proportionality of force and acceleration.) This conclusion was all-important to the progress in astronomy because it suggested that the planets move as they do as a result of a force pulling them toward the sun. Newton put this reasonable guess on a firm footing by proving a theorem not just for uniform circular motion but for any orbital motion at all that is executed in a plane. In Section 9.9 we proved that an object acted upon by a central force obeys Kepler's second law (the law of areas). Because it experiences no torque with respect to the center of force, it has constant angular momentum with respect to this point, and constant angular momentum in turn implies that the radial line is sweeping out area at a constant rate. To put it in planetary terms: If the sun exerts a central force on a planet, the planet moves in such a way as to satisfy Kepler's second law. This is the theorem that Newton proved, along with its converse: If a line drawn from the sun to an accelerated planet sweeps out area at a constant rate, the force acting on the planet must be directed toward the sun. The essential point is the connection between the central force and Kepler's second law. Using only one of Kepler's laws of planetary motion, Newton could prove the vastly important result that the planets are all acted upon by a central force directed toward the sun.

A theorem: The law of areas is valid if and only if the force is central

2. KEPLER'S THIRD LAW AND THE DEPENDENCE ON DISTANCE AND MASS

To learn radial dependence of force, compare different planets

Kepler's second law (together with Newton's second law) reveals the *direction* of the gravitational force but no other property of the force. The next problem is: How does the strength of the sun's gravitational pull vary as the distance from the sun varies? There are two ways to get at this problem, and Newton employed both. The first, our concern in this subsection, is to compare the motion of different planets at different distances from the sun. The second, discussed in the next subsection, is to study the motion of a single planet as its distance from the sun varies during the course of its orbit. For comparing the motion of different planets, Newton had Kepler's third law at hand; this law states that the squares of the periods of the planets are proportional to the cubes of the semimajor axes of their elliptical orbits (Equation 11.45). In the previous section, we showed that for circular orbits, this law can be derived if the inverse-square law of force is assumed. Here let us reverse the argument and show that the inverse-square law of force is implied by Kepler's third law. We again approximate the planetary orbits as circles, as Newton undoubtedly did in his earliest calculations. Introducing a subscript p to designate a particular planet, we can write, for the acceleration of the planet,

Treat orbits as circles

$$a = \frac{v_p^2}{r_p} .$$

(Here a designates acceleration and is not to be confused with the distance a in Equation 11.43.) In this formula, we may replace the speed v_p by the orbital circumference divided by the period:

$$v_p = \frac{2\pi r_p}{T_p} . \tag{11.48}$$

The formula for the planet's acceleration then becomes

$$a = \frac{1}{r_p}\left(\frac{2\pi r_p}{T_p}\right)^2 = \frac{4\pi^2 r_p}{T_p^2} . \tag{11.49}$$

So far this is a kinematic statement about uniform circular motion in general. Kepler's third law, which can be written

$$T_p^2 = \kappa_S r_p^3, \tag{11.50}$$

is, on the other hand, an observational fact about the planets. Substitution from Equation 11.50 in the denominator of Equation 11.49 gives a formula for planetary acceleration:

An implication of Kepler's third law: $a \sim 1/r^2$

$$a = \frac{4\pi^2}{\kappa_S}\frac{1}{r_p^2} . \tag{11.51}$$

This states that the acceleration of a planet toward the sun is inversely proportional to the square of its distance from the sun. Multiplication of the acceleration of planet by its mass gives the force acting on it:

$$F_p = \frac{4\pi^2}{\kappa_S}\frac{m_p}{r_p^2} . \tag{11.52}$$

Compare this with the law of gravitational force stated by Equation 11.1. Here, derived from Kepler's third law, is a major part of the final form of the law of force. Equation 11.52 states that the gravitational force experienced by a planet is proportional to its mass and inversely proportional to the square of its distance from the sun. All that is missing is the proportionality of this force to the mass of the sun. (That proportionality, as shown in Figure 11.1, is required by Newton's third law.)

3. KEPLER'S FIRST LAW: MORE EVIDENCE ON THE RADIAL DEPENDENCE

The idea of an inverse-square law of gravitational force acting on the planets was in the air around Newton's time. His genius lay not so much in thinking of it as in demonstrating its validity mathematically and weaving it into a coherent theory of universal gravitation. The really crucial test of the inverse-square law is provided by Kepler's first law, the statement that planets move in elliptical orbits with the sun at one focus. In any orbit that is not a circle, a planet periodically alters its distance from the sun, sampling stronger and weaker regions of the sun's gravitational field as it moves around. The precise form of its orbit, therefore, depends on the law of force, on exactly how the force weakens as the distance increases. In London in 1684, Edmund Halley, Robert Hooke, and Sir Christopher Wren worked at the problem of connecting the elliptical orbits of the planets to the law of force emanating from the sun. Although they believed in the inverse-square law, they failed to connect it to Kepler's first law of elliptical motion. Finally, Halley journeyed up to Cambridge to ask Newton about the problem. Here is an account of their meeting, written soon afterward by John Conduitt:

> Without mentioning either his own speculations, or those of Hooke and Wren, he [Halley] at once indicated the object of his visit by asking Newton what would be the curve described by the planets on the supposition that gravity diminished as the square of the distance. Newton immediately answered, *an Ellipse*. Struck with joy and amazement, Halley asked him how he knew it? Why, replied he, I have calculated it.*

Three years later, in 1687, under the auspices of Halley, Newton's monumental *Principia* was published.

Before Newton, the inverse-square law of gravitational force had been an unsupported hypothesis with little more weight than Democritus' belief in atoms. In Newton's hands it became an established law. Not only was he able to carry out the complicated calculation deriving the planetary motion from his law of gravitational force, but he correctly realized that Kepler's first law provided the most sensitive test of the inverse-square law. If the force diminished in some slightly different way than inversely with the square of the distance, the orbits would not be ellipses, nor would they close on themselves (Figure 11.14). Instead a planet would follow a somewhat different path on each successive trip around the sun. Any small deviation of the true law of force from the

Elliptical orbits provide the most sensitive test of $F \sim 1/r^2$

* See Charles C. Gillispie, *The Edge of Objectivity* (Princeton, New Jersey: Princeton University Press, 1960), p. 137.

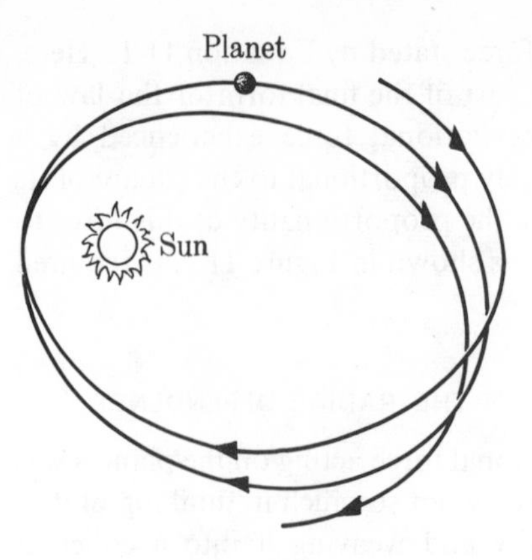

FIGURE 11.14 Hypothetical planetary orbit if the gravitational force depended on distance in some way other than inversely proportional to the square of the distance (see also Figure 9.29). In fact very small deviations of planetary orbits from ellipses do occur, because each planet experiences small forces contributed by the other planets besides the dominant force contributed by the sun.

inverse-square law would show itself in the geometry of planetary orbits before it would significantly alter Kepler's third law.

4. NEWTON'S THIRD LAW AND THE UNIVERSALITY OF GRAVITATION

Insofar as there was thinking about the mechanics of celestial motion before Newton, it was in terms of a force exerted *by* the sun acting *on* the planets. Newton assumed that the planets equally well must pull on the sun and on each other and indeed that every massive object must exert a gravitational force on every other. This idea of the universality of gravitation rested basically on the principle of action and reaction that we now call Newton's third law. The existence of reaction forces opposing applied forces is obvious in many commonplace examples, but before Newton no one had recognized the general significance of this pairing of forces. Newton obviously grasped its significance, for he adopted as his third fundamental law of mechanics the principle that every force in nature is opposed by an equal and opposite force. Moreover, he generalized this principle straight from the force of a man lifting a weight or a horse pulling a cart to the sun pulling the earth. He argued that if the planets experience forces to hold them in their orbits, they must also exert forces, forces equal and opposite to those they experience. If the sun attracts the earth, the earth equally attracts the sun. But if the earth attracts the sun, it must attract other objects as well—other planets, the moon, or an apple falling from a tree. By this sort of reasoning, Newton was led to think of the *universality* of gravitation—that the local gravity accounting for weight on earth and the cosmic gravity accounting for the orbit of the earth around the sun were all one.

If planets experience forces, they must exert forces

Because of the universality of gravitation, Newton realized that the sun need not be the only center of a planetary system. He noted that the moons of Jupiter obey Kepler's three laws, as do the moons of Saturn. These moons must therefore be held in their orbits by an inverse-square central force exactly as are the planets. The only difference is in the total intensity of the force, a difference that shows itself in a different proportionality constant in Kepler's third law for each center of force. The earth, no less than Jupiter and Saturn, could be—and now is—the center of a planetary system, a situation that

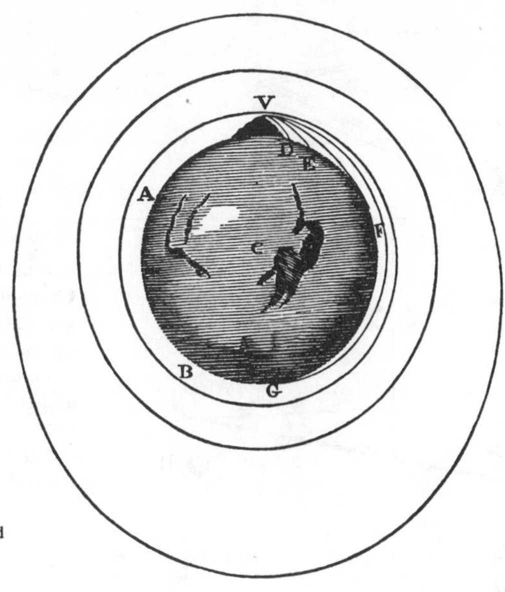

From Newton's *Principia*, 1687

Newton correctly foresaw, although he surely could not have imagined the hundreds of satellites now circling the earth.

Equation 11.52 contains the proportionality

$$F_p \sim \frac{m_p}{r_p^{\,2}}.$$

Because of Newton's third law (recall the argument in Section 11.2), the force must also be proportional to the sun's mass:

$$F_p \sim \frac{m_S m_p}{r_p^{\,2}}.$$

Newton postulated that *the force depends on nothing else.* If this is true, the proportionality may be converted to an equation with a single universal constant of proportionality:

The law of gravitational force determined from Kepler's empirical laws, Newton's theoretical laws, and assumptions of simplicity

$$F_p = G\frac{m_S m_p}{r_p^{\,2}}. \tag{11.53}$$

So far as we know, the gravitational constant G is indeed universal, the same for all matter in all places at all times.*

The measurement of G

In order to determine the constant G, it is necessary to measure the force between a pair of objects of manageable size whose individual masses can be separately measured. This experiment, first carried out by Henry Cavendish in the 1790s, is sometimes referred to theatrically as "weighing the earth." Indeed this is what it achieves indirectly, although it could just as well be called weighing

* Some physicists have speculated that the magnitude of G may decrease gradually as the universe expands. There is as yet no experimental evidence in support of this view (there is some indirect evidence against it).

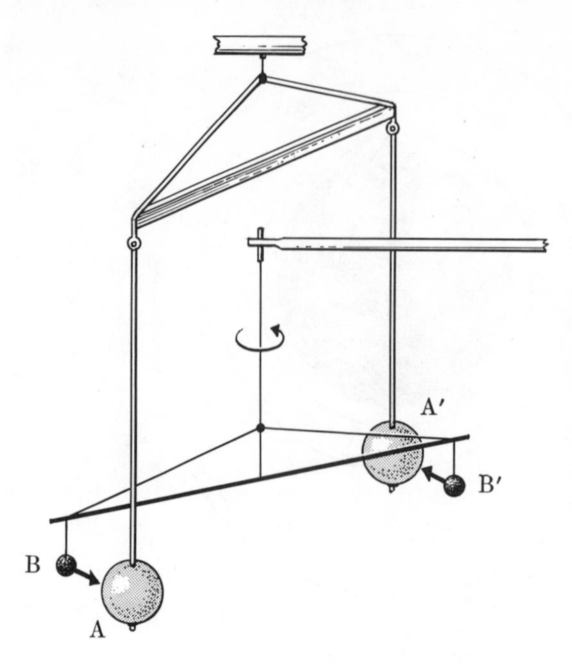

FIGURE 11.15 Schematic diagram of the Cavendish apparatus to measure the gravitational force between objects of ordinary size. Massive lead spheres (A and A′) are placed close to smaller spheres (B and B′) attached to the ends of a horizontal rod. In response to the torque produced by the gravitational forces, the fiber supporting the rod twists through a measurable angle.

the sun. Since the products Gm_S and Gm_E are known, the masses of sun and earth become known as soon as the constant G is separately determined. The difficulty of the Cavendish experiment springs from the extreme weakness of the gravitational force. In order that it be detectable for a pair of human-sized objects, it is necessary to use a highly sensitive balance to measure the force, and it is necessary to get rid of all electric forces, which are intrinsically far stronger. The balance used by Cavendish was a torsion balance (Figure 11.15), a long wire whose twist can respond to a very weak force.

5. THE LINK BETWEEN THE TERRESTRIAL AND THE CELESTIAL

Looking back on his earliest discoveries, Newton later wrote:

> And the same year I began to think of gravity extending to the orb of the Moon, and . . . from Kepler's Rule [Kepler's third law] . . . I deduced that the forces which keep the Planets in their Orbs must [vary] reciprocally as the squares of their distances from the centers about which they revolve: and thereby compared the force requisite to keep the Moon in her Orb with the force of gravity at the surface of the earth, and found them [to] answer pretty nearly.*

The moon linked earth's gravity and sun's gravity

The moon proved to be the vital link between terrestrial gravity and the cosmic gravitational force. The motion of the moons of Jupiter and Saturn proved that these planets attracted their moons in exactly the same way that the sun attracts the planets. But to our own moon fell the job of proving that this

* This passage is quoted in Gillispie, *The Edge of Objectivity*, pp. 119–120.

inverse-square force that reaches out from every massive body is exactly the same as the familiar force of gravity that we all experience on earth.

To establish this link, Newton had to use the fact that a sphere acts gravitationally just as if all its mass were concentrated at its center. The problem of the spherical mass provides a beautiful example of a self-consistent circle of reasoning. The equivalence of sphere and point was used in helping to establish the correctness of the inverse-square law of force. Yet the sphere is in fact equivalent to a point mass *only* for the inverse-square force. Had the gravitational force turned out to have a different form, it would not have been proper to replace the earth in calculations by a point mass at its center. This is only one of the simple properties of the inverse-square law that makes it seem so uniquely right.

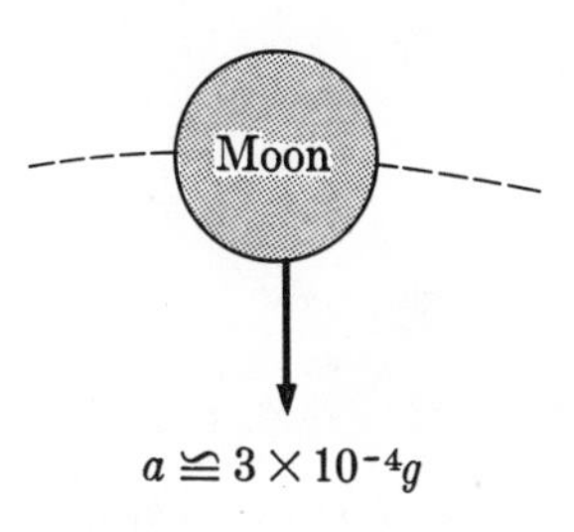

Newton knew that near the surface of the earth all objects fall with the same acceleration. Therefore, every object experiences a gravitational force proportional to its own mass. The moon, in its nearly circular orbit, is also "falling" toward the earth, with an easily calculable acceleration ($a = v^2/r$). Since the force acting on the moon is presumed also to be proportional to its mass, its inward acceleration is independent of its mass. To find how the gravitational force weakens with increasing distance, it is necessary only to compare the acceleration of the falling moon with the acceleration of a falling apple, both directed toward the center of the earth.

The moon is 384,000 km distant from the center of the earth, and the apple is 6,370 km from the center of the earth. If the earth's gravity extends on out into space, diminishing as the inverse square of the distance from its center, it is a simple matter to calculate how much less should be the moon's acceleration than the apple's acceleration. The ratio of these accelerations should be

$$\frac{a_{\text{moon}}}{g} = \left(\frac{6{,}370 \text{ km}}{384{,}000 \text{ km}}\right)^2 = 2.75 \times 10^{-4}.$$

Since the moon's speed is known to be 1.02×10^3 m/sec, its acceleration is

$$a_{\text{moon}} = \frac{v^2}{r} = \frac{(1.02 \times 10^3 \text{ m/sec})^2}{3.84 \times 10^8 \text{ m}} = 2.71 \times 10^{-3} \text{ m/sec}^2.$$

The actual ratio of the accelerations of moon and apple is

$$\frac{a_{\text{moon}}}{g} = \frac{2.71 \times 10^{-3} \text{ m/sec}^2}{9.82 \text{ m/sec}^2} = 2.76 \times 10^{-4}.$$

Using the best data available to him, Newton made a similar comparison of the theoretical and actual acceleration ratios and found them to "answer pretty nearly."

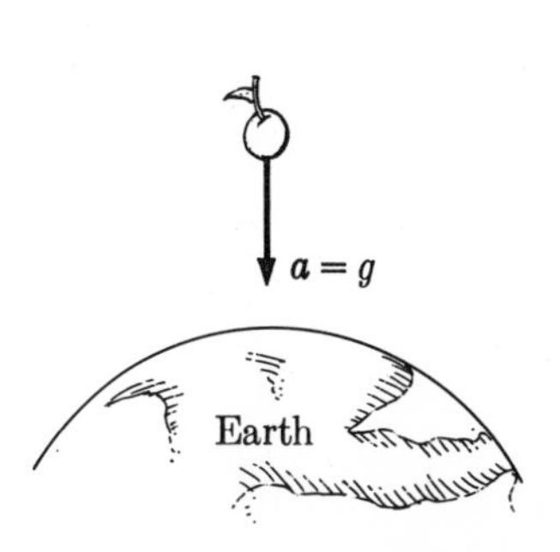

The universal law of gravitational force was the product of Newton's genius and nature's kindness. Nature was kind in arranging for a law of force of such magnificent simplicity. And nature was also kind in providing a solar mass so much greater than any planetary mass that the solar force on any planet greatly exceeds the force produced by any neighboring planet. To good approximation every planet responds to the sun and ignores its fellow planets. Only for this reason can planetary motion be well described by Kepler's laws.

11.8 The discovery of Neptune

Perhaps the most dramatic success of mechanics after Newton was the discovery of the planet Neptune in 1846. Newton had recognized that the forces acting directly between one planet and another, although small, would be significant in causing small deviations from the perfect elliptical orbit that would be followed if only the force of the sun were acting. These extra disturbing forces, which can be ignored for most purposes, are called perturbations. Throughout the eighteenth century, perturbations were studied with increasing care as astronomical observations became more precise and calculational methods more refined. The culminating work on perturbations was Pierre Simon Laplace's *Celestial Mechanics*, published early in the nineteenth century. The perturbing interplanetary forces had been identified and measured, resulting in values of planetary masses that could not be ascertained from the action of the sun's force alone. By this time it was clear that gravitation was indeed a universal force acting among all of the bodies in the solar system. The world system of Laplace was a beautifully self-consistent system of mutually interacting bodies, each influencing and influenced by all the others.

Uranus discovered by chance

Six planets were known to Newton. In 1781, in England, William Herschel, after a telescopic survey of the whole sky, had turned up a seventh, Uranus, about twice as distant from the sun as the sixth planet, Saturn. The full machinery of observation and calculation was turned upon Uranus and before long its future orbit was mapped out. With a period of 84 years, Uranus was in no hurry to get on with its job of tracing out the predicted orbit. Nearly 50 years went by before it became clear that Uranus was, in fact, not following precisely the orbit that should have resulted from the action of the sun and other planets. Either Newton's law of gravitation was failing at this great distance from the sun, or still another planet, more remote and still unseen, was perturbing Uranus. As happens so often in science, the more conservative possibility proved to be the correct one. In the 1840s two young men, John Adams in England and Urbain Leverrier in France, unknown to each other, decided to assume that Newton's law of gravitational force was perfectly correct and to calculate where an eighth planet should be located to account for the unexpected perturbation of Uranus. Both completed their calculations at nearly the same time, and both predicted nearly the same position in the sky for the new planet. But Leverrier had the greater luck. While astronomers at Greenwich politely ignored Adams's suggestion to search in a certain area of the sky, the director of the Berlin Observatory acted at once upon receipt of a letter from Leverrier and discovered Neptune the very same day.

Neptune predicted to account for perturbation of Uranus's orbit

Pluto also predicted

A somewhat similar but less dramatic chain of events led to the discovery of the ninth and, so far, the last planet, Pluto. Early in this century, an American astronomer, Percival Lowell, completed calculations on the orbits of Neptune and Uranus that suggested that there should exist yet another planet beyond Neptune. It was not so easy to predict the exact location of this new planet, and it escaped detection for 25 years. Finally, in 1930, Pluto was definitely identified at the Lowell Observatory in Arizona. Not until the year 2178 will it have completed one trip around the sun and be back at the point in the sky where it was first seen.

Summary of ideas and definitions

To describe nature, Newton's law must be supplemented by specific laws of force.

Gravity, although the weakest known force, is the only universal force, and it played a vital role in the development of mechanics.

The gravitational force is a central attractive force with magnitude given by

$$F = G\frac{m_1 m_2}{d^2}. \quad (11.1)$$

For more than two bodies, it satisfies a superposition principle.

Outside a spherical shell or any set of shells, the gravitational force is the same as if all the mass were concentrated at the center of the shell(s).

Inside a spherical shell, the net gravitational force is zero.

Inside a uniform sphere, the gravitational force is proportional to radius, the same dependence on distance as for a harmonic oscillator (Equation 11.13 or 11.14).

The acceleration of gravity at the earth's surface is

$$g = \frac{GM_{\mathrm{E}}}{R^2}. \quad (11.9)$$

The gravitational potential energy of a body influenced by a uniform sphere is proportional to $1/r$ outside the sphere (Equation 11.17 or 11.18) and is parabolic inside the sphere (Equation 11.24 or 11.25).

Escape speed from the earth is

$$v_{\mathrm{esc}} = \sqrt{2gR} = \sqrt{\frac{2GM_{\mathrm{E}}}{R}}. \quad (11.20, 11.21)$$

For circular motion in an inverse-square field of force, potential energy has twice the magnitude of kinetic energy ($U = -2K$), and

$$E = -K = \tfrac{1}{2}U. \quad (11.28, 11.29)$$

The rotation of the earth causes its shape to be ellipsoidal.

Because it is nonspherical, the earth experiences a torque, and its spin angular momentum precesses.

The acceleration of gravity is influenced by the shape of the earth and by its rotation; g decreases by 0.052 m/sec^2 in going from pole to equator.

Kepler's first law: Planets follow elliptical paths with the sun at a focus. This provides the most stringent proof of the radial dependence of the force, $F \sim 1/r^2$.

Kepler's second law is the law of areas: Planets sweep out area at a constant rate. This is equivalent to the conservation of angular momentum and proves that the gravitational force is a central force.

Kepler's third law: Periods and semimajor axes of planetary orbits are related by $T^2 = \kappa_{\mathrm{S}} a^3$. This law is related to the inverse-square dependence of the force.

The product $m_1 m_2$ in the law of gravitational force is a reflection of Newton's third law. Every body experiences gravitational forces and exerts gravitational forces.

The acceleration of the moon showed that the earth's gravity reaches out into space, decreasing as the square of the distance from the center of the earth.

The eighth and ninth planets, Neptune and Pluto, were predicted theoretically because they perturbed the motion of other planets.

QUESTIONS

Section 11.1

Q11.1 "Unreal" examples in physics are of two kinds: idealizations that can be approached but never fully realized in practice, and hypothetical situations quite different from anything known in nature. Give one example of each kind, and explain why each may help to provide insight into properties of the real world.

Q11.2 How could an astronaut in orbit, although "weightless," deduce that he is being influenced by gravity?

Section 11.2

Q11.3 Equation 11.2 states the value of the gravitational constant G in N m^2/kg^2. Re-express this SI unit (a) in kg, m, and sec and (b) in J, m, and kg.

Q11.4 The text states that the superposition principle is "implicit in Equation 11.1." Explain how this is so.

Q11.5 At noon, the sun's gravity pulls an object on earth upward, opposite to the pull of the earth's gravity; at midnight, the pull of the sun on the object is downward, parallel to the earth's gravity. Do objects on earth therefore weigh slightly more at midnight than at noon? Why or why not? (In answering this question, it is permissible to make the approximation that all parts of the earth are equally distant from the sun.)

Q11.6 One possible way to define the concept of mass is to make use of the law of gravitational force between all pairs of material objects. Explain how this law might be used to define mass. (Any measurements you suggest should be possible in principle but need not be very practical.) Would this method define gravitational mass or inertial mass?

Q11.7 The electrical force between two charged particles varies inversely as the square of the distance between them. Does a charged particle inside a spherical shell experience a net force if the shell is (a) uniformly charged? (b) not uniformly charged?

Section 11.3

Q11.8 One man is located at the bottom of a shaft 2,000 miles deep, another is on the surface of the earth, another is on top of a tower 2,000 miles high, and another is on the surface of the moon. List the four men in order of increasing weight if all are of equal mass.

Q11.9 The earth is denser near its core than near its surface. Qualitatively, how does this cause the law of force within the earth to deviate from the linear dependence on r represented by Equation 11.14 and Figure 11.4?

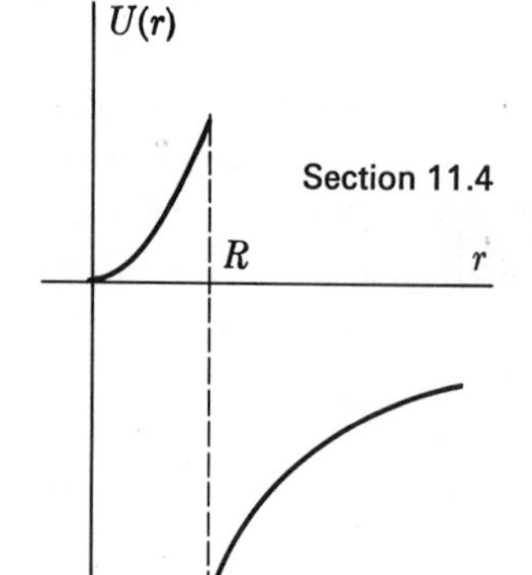

Section 11.4

Q11.10 The figure shows a permissible but inconvenient potential-energy function for a body moving inside and outside the earth. If a ball is dropped from above the earth into a hole so that its radial coordinate r passes through the value R where U is discontinuous, which, if any, of the following quantities change discontinuously: (1) the kinetic energy of the ball, (2) its total energy, and (3) the work done on the ball by gravity?

Q11.11 In an energy diagram, sketch the "paths" of (a) a satellite raised from the surface of the earth and put into an elliptical orbit and (b) a meteorite approaching the earth from a great distance, losing most of its kinetic energy in the atmosphere, and striking the earth.

Q11.12 Two satellites pass close together several hundred miles above the earth's surface. One is moving parallel to the earth's surface; one is moving away from the earth at an angle of 45 deg to the vertical. Their speed is the same. Which one is more likely to escape from the earth?

Q11.13 Why do astronauts experience greater acceleration returning to the earth than leaving the earth?

Section 11.5

Q11.14 Why do new mountain ranges come into existence even though their formation requires that work be done against gravity?

Q11.15 Latitude may be defined as the angle that the vertical to the earth's surface makes with a line drawn from the center of the earth to the equator. Draw an exaggerated diagram of the flattened earth, and with its help explain the following facts: (1) The latitude angle of a point halfway from equator to pole is greater than 45 deg. (2) The distance of a 1-deg change of latitude measured along a north-south meridian line is greater near a pole than near the equator.

Q11.16 A trader buys coffee beans in Colombia and weighs them out into 1-lb bags. When he gets ready to sell them in an Eskimo village, he weighs them again and finds that each bag still weighs exactly 1 lb on his scale. Assuming his scale to be perfectly reliable, explain why it does not show an increase in weight of the bags from the tropics to the arctic.

Q11.17 Discuss the accuracy required in the construction of a pendulum and in measurements of its length and period in order to obtain values of g as accurate as those quoted in Formulas 11.32 and 11.33.

Section 11.6

Q11.18 (1) Why do we speak of Kepler's "laws" instead of Kepler's "theory"? (2) Why are Kepler's laws applicable and useful in the description of the motion of artificial earth satellites?

Q11.19 (1) What is the eccentricity of a circle? (2) What is the appearance of an ellipse whose eccentricity is equal to 1?

Q11.20 If you were given the distance of an asteroid from the sun at a particular time and its velocity at that time, how would you decide whether it was following an elliptic, parabolic, or hyperbolic orbit?

Q11.21 Why can a satellite "hover" over a point on the equator but not over a point north or south of the equator?

Q11.22 A satellite brushing the fringes of the atmosphere may *gain* average kinetic energy as a result of atmospheric friction. Explain.

Q11.23 In an outside reference, read about the Ptolemaic system of the world (a geocentric, or earth-centered, system) and the Copernican system (a heliocentric, or sun-centered, system). (1) Briefly describe the motions of the planets envisioned by Ptolemy and by Copernicus. (2) In what sense are the ellipses of Kepler "simpler" than the circles of Copernicus?

Q11.24 (1) In orbit an astronaut wants to turn his vehicle around without in any way disturbing the motion of its center of mass. Explain how he can do this. (2) Later he wants to change the orbit in order to descend into the atmosphere. Explain how a retrorocket achieves this aim. At what point in his elliptical orbit should the retrorocket be fired to get the desired effect most easily?

Q11.25 Two spacecraft are "flying formation" in a circular orbit about the earth. The rocket engine of one is turned on briefly to accelerate it in its direction of motion. Then both spacecraft coast. Are they likely ever to be close together again? Why or why not?

Section 11.7

Q11.26 Does a body on which no force acts sweep out area at a constant rate? If so, with respect to what point?

Q11.27 A hypothetical central force varies in proportion to $1/r^3$. Which, if any, of Kepler's laws remain valid?

Q11.28 Aristotle argued that if the earth rotated, the natural motion of matter in the earth must be in circles, in contradiction to the observed natural motion straight downward of a piece of solid matter that is dropped. Carefully criticize this reasoning.

Q11.29 Which two planets can never be seen at midnight? How does the heliocentric theory account for this fact? Illustrate your answer with a diagram. Can the geocentric theory offer an equally plausible explanation?

Q11.30 (1) What is meant by the retrograde motion of planets? (Consult an outside reference if necessary.) (2) With the help of a diagram, explain the reason for retrograde motion.

Q11.31 A simple calculation can show that the force exerted by the sun on the earth's oceans is much greater than the force exerted by the moon on the oceans. Why then is it the moon that is primarily responsible for the earth's tides? (HINT: The *change* of force from one part of the earth to another is relevant.)

Section 11.8

Q11.32 Deviations of the earth (more exactly, of the earth-moon system) from an elliptical path are caused principally by Jupiter. Why?

EXERCISES

Section 11.2

E11.1 A 2,000-kg spacecraft is located 10^7 m from the center of the earth at a particular instant. (1) What force of gravity does it experience? (2) What gravitational force does it exert on the earth? (3) What is the acceleration of the spacecraft? (4) By how much is the earth accelerated because of the pull of the spacecraft?

E11.2 A 75-kg man stands 2 m from the center of mass of a 2,000-kg automobile. (1) Compute the gravitational force of attraction exerted by the car on the man, and then find the ratio of this force to the weight of the man. (2) Estimate roughly, by any means you can think of, the force exerted on the man by a light breeze and compare this with the gravitational attraction of the car.

E11.3 Let $\mathbf{r}_{21}$ be the displacement vector from particle 1, whose mass is m_1, to particle 2, whose mass is m_2. Write a vector expression for $\mathbf{F}_{21}$, the force exerted by particle 1 on particle 2.

E11.4 The force of electrical attraction between a proton and an electron separated by 1 Å (10^{-10} m) is 2.3×10^{-8} N. By what factor is this force greater than the gravitational force between the same two particles? What is the implication of this result for atomic structure?

E11.5 Among the forces acting on a man on the earth's surface are gravitational forces of the earth, sun, moon, and Venus. Make approximate calculations of the relative magnitudes of these forces (to one significant figure) and show that $F_{\text{earth}} \gg F_{\text{sun}} \gg F_{\text{moon}} \gg F_{\text{Venus}}$. Give approximate values for the ratios of each of the latter three forces to the weight of the man. (Note that neither the man's weight nor the value of G need enter into your calculations. Why not?)

E11.6 Prove that the two angles designated θ in Figure 11.3 are equal in the limit that dA_1 and dA_2 approach zero. (The equality of these angles is used in the proof that the net force on a particle within the sphere vanishes.)

E11.7 (1) What is your weight in newtons at the surface of the earth (absolute honesty not required)? (2) What would be your weight 3,960 miles above the surface of the earth? (3) At what distance from the center of the earth would your weight be 10 N?

E11.8 A space traveler on an alien planet is placed on a horizontal and truly frictionless surface by his unfriendly hosts and left there to flounder. To add to his torment, they pile good things to eat on a table that is out of his reach. But they forget that gravity will pull him to the food. If the table with its load has a mass of 100 kg and if it is rigidly attached to the ground 3 m from the earthling, what will be his initial acceleration toward his next meal? Estimate roughly the time required for him to reach the food. Is there any danger that he will starve to death on the way?

Section 11.3

E11.9 (1) Verify numerically that $GM_E/R^2 = 9.8\ \text{m/sec}^2$. (2) Verify that GmM_E/R^3 has the same dimension as the force constant k of a harmonic oscillator.

E11.10 (1) How does the period of a small pendulum vary as it is (a) lowered into a deep mine shaft and (b) raised above the earth's surface? (2) How far must the pendulum be moved in each direction to suffer a 1 percent change in period? *Optional:* Discuss the relative merits of a pendulum and a barometer for measuring altitude.

E11.11 An object oscillates in an imaginary hole drilled through the center of the earth. Let z measure the position of the object relative to the center of the earth. (1) Give expressions for its velocity component v_z (a) as a function of time and (b) as a function of z. (2) What is its speed, both algebraically and numerically, at the center of the earth?

E11.12 An engineer uninhibited by practical considerations designs a straight-line tunnel about 1,000 miles long to connect Los Angeles and Seattle. At both terminals the tunnel is inclined at an angle of about 7 deg to the horizontal. (1) By means of an *order-of-magnitude calculation*, estimate the time required for a train to roll without friction through the tunnel. You might, for example, consider the approximate average acceleration of the train during each half of a one-way trip. (As stated in Problem 11.7, the exact answer is 42 min.) (2) In a similar way, estimate the maximum speed achieved by the train at the midpoint of the tunnel. (This exercise is intended to encourage order-of-magnitude reasoning. Do *not* submit exact solutions unless they are to be compared with approximate answers.)

Section 11.4

E11.13 (1) At what point between earth and moon is an object attracted equally to these two bodies? (2) Is its potential energy at this point positive, negative, or zero (if its potential energy at infinity is zero)? Why?

E11.14 If the potential energy of a body of mass m is set equal to zero at the center of the earth and has no discontinuities anywhere, what is its potential energy (a) at the surface of the earth and (b) at infinite distance from the earth?

E11.15 (1) Imagine that the earth collapses, with no change in its mass, until the work required to remove an object of mass m from the surface of the earth is equal to mc^2, the rest energy of the object. What, then, would be the radius of the earth (give both an algebraic and a numerical answer)? This is called the *gravitational radius* of the earth. (2) What is the gravitational radius of the sun? (It is hypothesized that some burned-out stars may collapse to their gravitational radius or less.)

E11.16 Find the gravitational potential energy of a body of mass m inside and outside a thin, hollow shell of radius R and mass M. Adjust $U(r)$ to have no discontinuity at $r = R$. Sketch a graph of U vs r.

E11.17 A satellite moves in a circular orbit at a distance r from the center of the earth. Let R and M_E be the radius and mass of the earth respectively; g is the acceleration of gravity at the surface of the earth, and G is the gravitational constant. (1) Obtain formulas for the speed of the circling satellite (a) as a function of r, R, and g and as a function of r, G, and M_E. (2) At what radial distance r is the speed equal to 330 m/sec (in air this would be Mach 1)?

E11.18 Show that the work required to put a satellite into a low earth orbit is slightly more than half the work required to send it to the vicinity of the moon.

E11.19 (1) Show that the work required to put a satellite into a circular orbit at a distance of 200,000 miles from the earth is only 1 percent greater than the work required to send the satellite along a straight vertical path to that distance. (2) Show that the work required to put a satellite into a circular orbit 100 miles above the earth is 20 times greater than the work required to send it vertically to that altitude. Include energy diagrams with your answers.

E11.20 A body leaves the earth moving vertically upward with 99 percent of escape speed. What height does it reach? (Ignore air friction.)

E11.21 When 200,000 miles from the earth, the kinetic energy of a spacecraft returning from the moon is negligible in comparison with its kinetic energy when it nears the earth. (1) What is the approximate total energy of the spacecraft if its zero of potential energy is at the earth's surface? (2) What is its approximate total energy if its zero of potential energy is at infinity? (3) What is its approximate speed as it nears the earth (before air friction starts to decelerate it)? (Answer parts 1 and 2 algebraically and part 3 both algebraically and numerically.)

E11.22 (1) What is the escape speed of a body from the moon? Give a numerical answer. (2) What is the ratio of escape speed from the earth to escape speed from the moon?

E11.23 (1) Two planets of equal density have radii R_1 and R_2. What is the ratio of escape speeds from their surfaces? (2) Two planets of equal radius have masses M_1 and M_2. What is the ratio of escape speeds from their surfaces?

E11.24 An asteroid has a radius of 4 miles and a density equal to the average density of the earth. (1) By flexing his knees and jumping, could a man escape from this asteroid? (2) Could he throw a baseball into orbit around this asteroid?

E11.25 A space traveler in interstellar space is working near his craft when his safety line breaks. At that moment he is 3 m from the center of mass of the craft and drifting away from it at a speed of 1 mm/sec. If the mass of the craft is 10,000 kg, will he reach a maximum distance and be drawn back, or will he drift away indefinitely?

E11.26 A vertically fixed rocket is accelerated upward until its distance from the center of the earth is r; then it coasts. What must be its minimum speed at the moment its engine is turned off in order that it escape from the earth? Discuss the two limits of impulsive firing (r only slightly greater than the radius of the earth R) and slow firing ($r \gg R$).

E11.27 A spacecraft is in a circular orbit moving with speed v at distance r from the center of the earth. Obtain a formula for Δv, the increase of speed required to bring the energy of the spacecraft up to zero and allow it to escape from the earth. What is the ratio $\Delta v/v$?

E11.28 Let r represent the radius of the earth's orbit (approximated as a circle), M_E the mass of the earth, and M_S the mass of the sun. In terms of these quantities and the gravitational constant G, answer all parts of this exercise algebraically. (1) What is the potential energy of the earth in the gravitational field of the sun? (2) What is the kinetic energy of the earth? (3) How much additional energy would the earth require in order to escape from the solar system? (4) Show that the escape speed of the earth from the solar system is

$$V_{esc} = \sqrt{\frac{2GM_S}{r}}.$$

E11.29 (1) Using the formula given in the preceding exercise, compute the escape speed of the earth from the solar system. Note that this escape speed does not depend on the mass of the escaping body. (Once having "escaped" from the earth, a spacecraft would need this much speed in order to continue on to escape from the sun.) (2) What is the ratio of the solar escape speed to the average orbital speed of the earth given in Appendix 3D? (3) What is the ratio of the solar escape speed to the terrestrial escape speed given by Equation 11.21? What does this imply about the relative difficulty of getting to the moon and getting to Neptune?

Section 11.6

E11.30 When launched in 1969, a Soviet communications satellite, *Molniya* 1L, had a perigee 323 miles above the earth's surface and an apogee 24,570 miles above the earth's surface. (1) Calculate the three distances a, b, and c characterizing its elliptical orbit (see Figure 11.12). Slide-rule accuracy suffices. (2) What is the eccentricity of this ellipse?

E11.31 Intelsat 2A, a commercial communications satellite launched in 1966, was intended for a near-circular orbit (eccentricity near zero). Instead it went into orbit with eccentricity $e = 0.634$. At apogee it is 27,000 miles from the center of the earth. (1) How far above the surface of the earth is it at perigee? (2) What is the ratio of its speed at perigee to its speed at apogee? (3) Show that its period is closer to 12 hr than to the intended 24 hr.

E11.32 With distances measured in astronomical units (1 A.U. = average earth-sun distance) and with a suitably chosen coordinate system, the equation of the orbit of Halley's comet can be written $(x/17.8)^2 + (y/4.6)^2 = 1$. (1) Using graph paper, plot the orbit. Label the axes and locate the positions of the foci. (2) Compare the dimensions of your plotted ellipse with the dimensions of planetary orbits given in Table 11.1, and state where in the solar system Halley's comet is to be found when it is at perihelion and when it is at aphelion. (This comet, studied by Edmund Halley in 1682, is next expected at perihelion in 1986.)

E11.33 If r and θ are polar coordinates, show that

$$r = \frac{b}{\sqrt{1 - e^2 \cos^2 \theta}}$$

is the equation of an ellipse of eccentricity e and semiminor axis b.

E11.34 The equations

$$x = a \cos u, \qquad y = b \sin u$$

are called parametric equations because x and y are expressed in terms of the variable parameter u. Show that these parametric equations represent an ellipse. Through what range must u vary to move once around the ellipse?

E11.35 The equation

$$\frac{x^2}{a^2} - \frac{y^2}{b^2} = 1$$

describes a conic section. Which one? Sketch a graph of this equation for $a = 1, b = 2$. (Note that the graph has two branches, that is, two distinct curves.) What is the behavior of the curves as $x^2 \to \infty$?

E11.36 What is the ratio of the maximum speed to the minimum speed of Halley's comet (the equation of its orbit is given in Exercise 11.32)?

E11.37 An unseen planet called Vulcan was once hypothesized to exist in an orbit smaller than the orbit of Mercury. If Vulcan's mean distance from the sun were 0.2 A.U., what would be its period in earth years?

E11.38 Working from the moon's kinematic data given in Appendix 3D, scale downward using Kepler's third law to find (a) the distance in miles from the center of the earth of a satellite whose period is 24 hr and (b) the period in minutes of a low-altitude satellite whose distance from the center of the earth is 4,100 miles. Assume circular orbits for both satellites.

E11.39 Assume that a synchronous communications satellite is in a perfect circular orbit. (1) In order that its period be within 30 sec of its "nominal" period of 23 hr 56 min, how accurately must its altitude be established? Express the answer in miles. (2) If its period is 15 sec less than its nominal period, what is the "drift speed" of the satellite, that is, the speed of the point on the equator that lies directly below the satellite? Express this answer in miles per day.

E11.40 During part of their Gemini 12 flight in 1966, astronauts James Lovell and Edwin Aldrin were in an orbit whose distance from the center of the earth varied between 4,060 and 4,135 miles. The period of this orbit was 89 min. (1) What was the ratio of their maximum speed to their minimum speed in orbit? (2) Scaling from the period of Gemini 12, find the period of a satellite in an orbit with a semimajor axis of 8,200 miles. (3) If the gravitational force varied not as $1/r^2$ but as $1/r^3$, one of the two preceding answers would be changed, the other unchanged. Which would be unchanged, and why?

E11.41 Express the constant κ_S in Kepler's third law (see Equation 11.46) numerically in terms of astronomical units and years.

E11.42 Use Equations 11.46 and 11.47, together with kinematic data (distances and times) on planets and satellites, to find the ratio of the mass of the sun to the mass of the earth. Check your answer against the masses given in Appendix 3.

E11.43 Consider planets in circular orbits about a central sun. How are their orbital speeds related to their orbital radii?

E11.44 A satellite with a mass of 10^4 kg in a circular orbit 8×10^6 m from the center of the earth turns on its rocket engine for 2 min, during which time it experiences a force of 10^4 N in its direction of motion. (1) What is the applied torque relative to the center of the earth? (2) What is the change in its angular momentum? (3) What were its speed, its angular momentum, and its kinetic energy before the impulse? (4) What are each of these same three quantities after the impulse? (5) Qualitatively, how does its orbit change? (6) How would its orbit have changed had an impulse of the same magnitude been directed (a) opposite to its velocity? (b) vertically upward away from the earth?

PROBLEMS

P11.1 *Net force within a sphere if gravity is not an inverse-square force*

(1) With the help of a diagram like that of Figure 11.3 and reasoning similar to that in Section 11.2, demonstrate that the net gravitational force on a particle within a spherical shell does not vanish if the force between material particles does not vary in proportion to $1/r^2$. (2) If the force varies in proportion to $1/r^n$, give the direction of the net force on the particle for (a) n slightly larger than 2 and (b) n slightly smaller than 2.

P11.2 *Motion toward a center of force*

Exercise 11.8 describes the plight of a stranded earthling. (1) Develop an exact formula for the time required for him to reach the food. (HINT: For a conservative force, the energy equation is usually a good starting point for finding a relationship between time and distance. One integration is then required.) (2) Evaluate the result numerically and compare it with the time that would be calculated assuming a constant force equal to the initial force. (3) Give a qualitative explanation of the fact that the exactly calculated time and the crudely calculated time do not differ greatly.

P11.3 *Small-angle deflection by a gravitational force*

A boy skating straight down a frozen river is attracted (gravitationally) to a girl sitting on the bank. By how much can her attraction deflect his path? (1) Use *order-of-magnitude* reasoning to show that an approximate answer for his angle of deflection is

$$\theta \cong \frac{Gm}{bv^2},$$

where G is the gravitational constant, m is the mass of the girl, b is the distance of closest approach of the boy to the girl, and v is the boy's speed. (HINT: The concept of impulse may be useful.) (2) Insert reasonable values of m, b, and v in this formula and evaluate θ. (Besides re-emphasizing the weakness of gravity, this problem shows the usefulness of order-of-magnitude reasoning.)

P11.4 In the preceding problem, it is a very crude approximation to attribute either constant magnitude or constant direction to the force, but it is an excellent approximation to assume that the skater moves in a straight line with constant speed (if the weak force of gravity is the only force that acts horizontally on him). (1) Take advantage of the near-constancy of his velocity in order to evaluate the impulse $\mathbf{I} = \int \mathbf{F}\, dt$ ($\mathbf{F}$ is the force exerted by the girl on the boy). (2) Using this result, show that the boy's angle of deflection is

$$\theta = \frac{2Gm}{bv^2}$$

(the notation is given in the preceding problem). (3) Express this angle in terms of a potential energy and a kinetic energy.

Gravitational acceleration near the earth

P11.5 Suppose that a crank was convinced that the gravitational force of the earth should depend in a simple way not on distance from the center of the earth but on distance from the surface of the earth. For low-altitude satellites he could successfully fit the facts with this "law" of acceleration:

$$a = g - \frac{2gh}{R},$$

in which g is the acceleration of gravity at the earth's surface, h is the distance above the surface, and R is the radius of the earth. (1) Prove that $-2g/R$ is the correct coefficient of h to give agreement with Newton's law of gravitational force at low altitude. (2) Suggest an experiment or observation that could easily demonstrate the limitations of this hypothesis for the gravitational acceleration. (3) What is the fractional error in this linear-in-h approximation at an altitude of 200 miles? *Optional:* Derive a more accurate formula for a at low altitude that contains both linear and quadratic terms in h.

Harmonic oscillation through the earth

P11.6 (1) What is the average over time of the speed of an object oscillating in an imaginary hole drilled through the center of the earth? (2) What is the average over distance of the speed of the object? Compare both of these average speeds with the speed of a low-altitude satellite. (All answers may be given algebraically.)

Hypothetical rapid-transit system

P11.7 In a hypothetical rapid-transit system, a train rolls without friction through a straight tunnel joining two cities (see the figure). Prove that the transit time between *any* pair of cities is $\pi\sqrt{R/g}$, or about 42 min. (SUGGESTION: Introduce a coordinate s that measures the position of the train relative to the midpoint of the tunnel.)

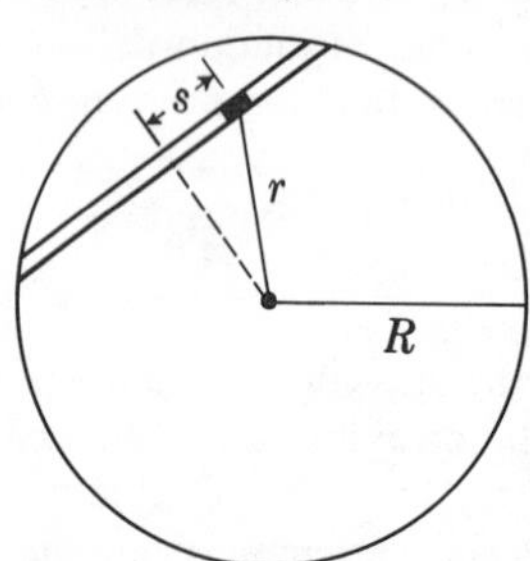

Gravity of a hollow planet

P11.8 A peculiar planet has the same mass and radius as the earth, but has all its mass concentrated in a 10-mile-thick outer shell. Assuming the density of this shell to be constant, discuss in detail the motion of an object as it falls through a 10-mile hole and continues into the hollow interior of the planet (ignore friction). Include sample numerical calculations.

Satellite in an elliptical orbit

P11.9 A certain satellite of mass m in an elliptical orbit has a kinetic energy that is four times greater at perigee than at apogee. (1) Prove that its total energy is

$$E = -\frac{1}{3}\frac{GmM_{\rm E}}{r_{\rm p}} = -\frac{2}{3}\frac{GmM_{\rm E}}{r_{\rm a}},$$

where $r_{\rm p}$ and $r_{\rm a}$ are the satellite's radial distances from the center of the earth at perigee and apogee respectively. (2) Show how this satellite is represented in an energy diagram.

P11.10 A rocket of mass m has energy E and angular momentum L (with respect to the center of the earth) as it coasts in the earth's gravitational field. Close to the earth, its path makes angle θ with the vertical. (1) Show that its equation of energy conservation may be written

Escape along a curved path

$$\frac{1}{2}mv_{\parallel}^2 + \frac{L^2}{2mr^2} - \frac{GmM_E}{r} = E,$$

where $v_{\parallel}$ is the component of $\mathbf{v}$ parallel to $\mathbf{r}$ (see the figure). (2) Show that the rocket escapes from the earth if $E \geq 0$, no matter what the value of θ. (3) Show that its escape speed is given by Equation 11.21, independent of θ.

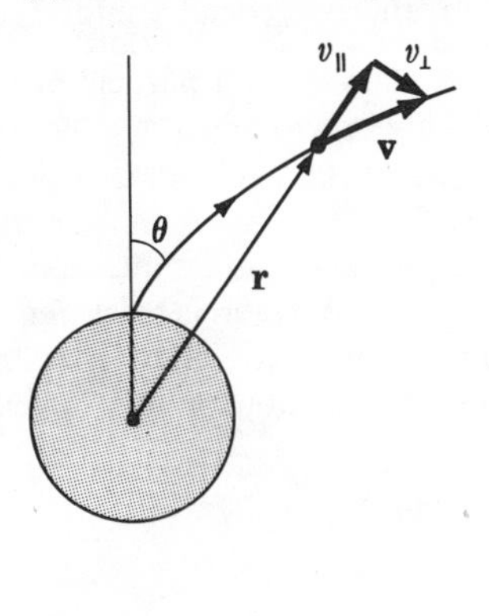

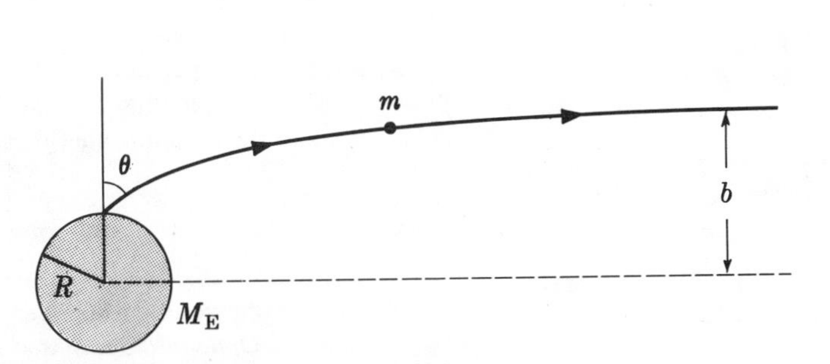

P11.11 A rocket starts near the earth with positive energy E and with its velocity vector inclined at angle θ to the vertical. It coasts to great distance, where its path approaches a straight line a distance b from a parallel line through the center of the earth (see the figure). (1) Using both energy conservation and angular-momentum conservation, show that

$$b = \sqrt{\frac{E + B}{E}}\, R \sin\theta,$$

where $B = GmM_E/R$. (HINT: The distance b enters into a simple expression for angular momentum.) (2) Discuss the two limits $E \to 0$ and $E \to \infty$.

P11.12 (1) Prove that $v_{\parallel}$, as defined in Problem 11.10, is the same as dr/dt, where r is the radial distance from the center of the earth to the rocket. (2) The energy-conservation equation stated in Problem 11.10 can be written

Radial component of two-dimensional motion

$$\frac{1}{2}m\left(\frac{dr}{dt}\right)^2 + U_{\text{eff}}(r) = E.$$

Write an expression for the effective potential energy, U_{eff}, and sketch a graph of U_{eff} vs r for (a) $L = 0$ and (b) $L \neq 0$. Such a graph provides an energy diagram for the radial component of the two-dimensional motion. (3) Show both graphically and algebraically that r has both a maximum and a minimum value if $E < 0$ and only a minimum value if $E \geq 0$ (assume $L \neq 0$). Give expressions for r_{min} and r_{max}. (4) How must energy and angular momentum be related in order that $r_{\text{min}} = r_{\text{max}}$? Verify this relationship by working from previously derived properties of uniform circular motion.

Potential energy en route to the moon

P11.13 Consider a body of mass m moving along a straight line between the earth and the moon. Overlook any effects associated with the orbital motion of the moon. (1) Write a formula for the potential energy U of the body as a function of its distance r from the center of the earth. (*Further notation:* The earth has radius R and mass M_E; the moon has radius R' and mass M_M; and the distance from the center of the earth to the center of the moon is r_m.) (2) Sketch a graph, approximately to scale, of U vs r for $R \le r \le r_m - R'$. (3) For what value of r is U maximum? What is U_{max}? Compare U_{max} with U at the earth's surface. (4) Let v'_{esc} be the minimum speed a body must have close to the earth in order to coast over the potential-energy "hill" and reach the moon. Both algebraically and numerically, find $v_{esc} - v'_{esc}$.

Potential energy outside a spherical shell

P11.14 Outside a spherical shell of radius R and mass m_1 is a particle of mass m_2 at distance r from the center of the shell. By proceeding along the following lines, show that the potential energy of interaction of the particle with the spherical shell is Gm_1m_2/r, the same as if the shell were concentrated at a point. The shaded region in the diagram represents an infinitesimal area dA of radius b at distance l from the particle. In this region is mass $dm = \sigma\, dA$; σ is the mass per unit area of the shell ($\sigma = m_1/4\pi R^2$) and $dA = 2\pi b\, ds = 2\pi bR\, d\theta$. Note that $b = R \sin\theta$. The contribution of this increment of mass to the potential energy is

$$dU = \frac{Gm_2\, dm}{l}.$$

The quantities on the right may be expressed in terms of fixed quantities and the variable quantity θ. Then an integration over θ yields the total potential energy. *Optional:* Show that if the particle is inside the shell, its potential energy is constant.

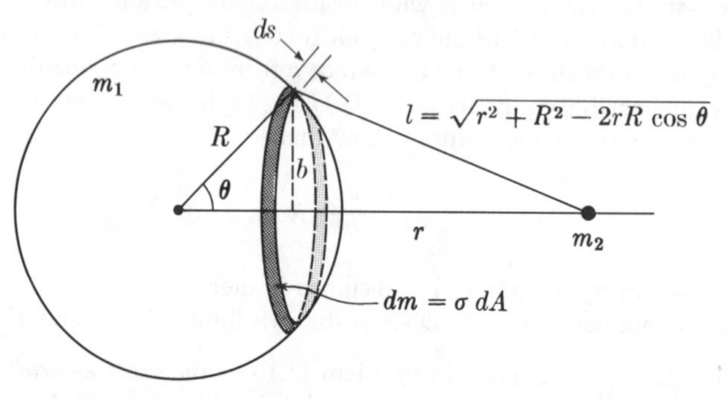

Angle of earth's gravitational force to the vertical

P11.15 (1) For a point on the earth's surface at 45 deg N latitude, calculate the inward acceleration produced by the rotation of the earth. Compare this acceleration with the free-fall acceleration g. Is it a legitimate approximation for most purposes to say that an object at rest on the surface of the earth experiences no net force? (2) The gravitational force at this latitude is inclined slightly away from the vertical (Figure 11.7). By approximately what angle?

Latitude angle

P11.16 If the origin of a coordinate system is placed at the center of the earth, with its x axis extending to the equator and its y axis extending to a pole, the surface of the earth can be described by the equation

$$\frac{x^2}{a^2} + \frac{y^2}{b^2} = 1.$$

(1) What are the magnitudes of a and b (in miles or kilometers)? (2) Prove that the latitude angle λ and the polar angle θ (defined in the figure) are related by the equation

$$\tan \lambda = \left(\frac{a}{b}\right)^2 \tan \theta.$$

(HINT: Tan $\psi = dy/dx$.)

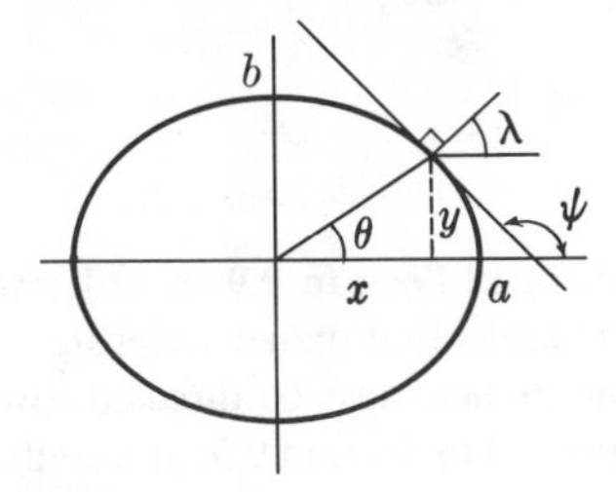

P11.17 As shown in the figure, the trajectory of a projectile on earth is part of an ellipse. (1) Show that for the choice of coordinates indicated in the figure, the part of the ellipse outside the earth can be represented by the parabolic approximation

$$z \cong -\frac{a}{2b^2}x^2,$$

where a is the semimajor axis of the ellipse and b is its semiminor axis. (SUGGESTION: Write an equation for an ellipse centered at the origin of a coordinate system; then shift the origin to the end of the major axis.) (2) State a condition that ensures the accuracy of the parabolic approximation. (3) For projectiles of short range and low altitude, show that

$$a \cong \frac{1}{2}R, \qquad b \cong v_x\sqrt{\frac{R}{2g}},$$

where R is the radius of the earth, g is the acceleration of gravity, and v_x is the horizontal component of velocity of the projectile. (4) A baseball is thrown with $v_x = 15$ m/sec. What is the semiminor axis of its "orbit"?

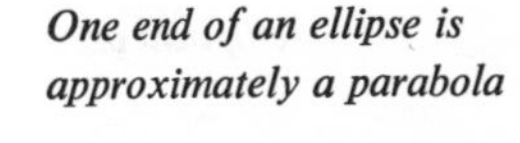
One end of an ellipse is approximately a parabola

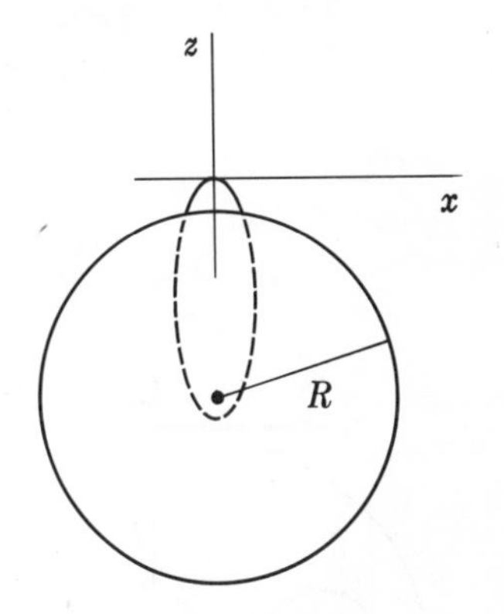

P11.18 Points C and D in the figure are the foci of an ellipse; point A is a point on the ellipse. Prove that the angles θ_1 and θ_2 are equal. This property means that sound waves or light waves emanating from one focus of an elliptical chamber will be reflected and converge at the other focus. (SUGGESTION: Consider a point B separated by a short distance s from A. The distance l_1' may be expressed as a function of l_1, s, and θ_1; and l_2' may be expressed as a function of l_2, s, and θ_2. For an ellipse, $l_1' + l_2' = l_1 + l_2 = constant$ [see Figure 11.11(b)]. Therefore $d(l_1' + l_2')/ds = 0$. In the limit of small s, the segment AB lies along the ellipse.)

Equal-angle property of an ellipse

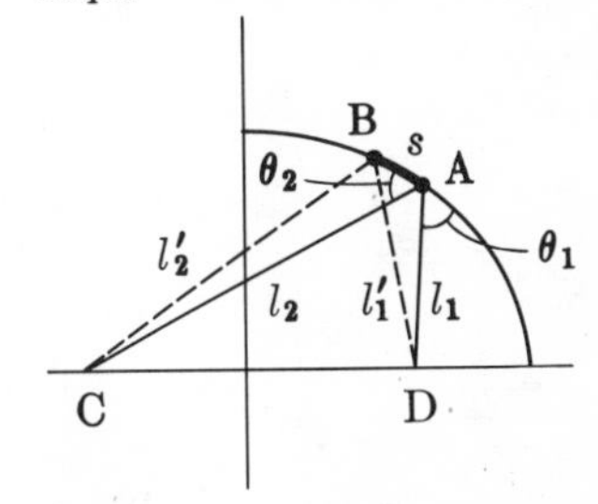

Hypothetical planetary system

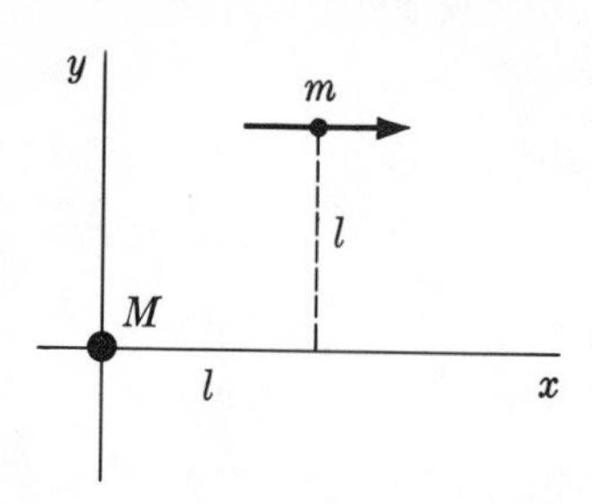

P11.19 In a certain mythical universe, a goddess places a star at the origin of a coordinate system and then throws planets from the point $x = l$, $y = l$; she throws them with different initial speeds but starts all of them moving parallel to the x axis (see the figure). Each planet follows an elliptical path with the star at one focus. Using the equal-angle property of ellipses given in the preceding problem, show that all of the other foci lie along a single curve. Give the equation of this curve. Sketch several of the planetary orbits. (Problem of Edward Kasner [1878–1955].)

The law of areas implies a central force

P11.20 Reconstruct the logic of the development in Section 9.9 in order to prove that if a line drawn from the sun to an accelerated planet sweeps out area at a constant rate, the force acting on the planet must be directed toward (or away from) the sun. (The converse is proved in Section 9.9: If a central force acts, area is swept out at a constant rate.)

Speed of a satellite in an elliptical orbit

P11.21 As shown in the figure, a satellite in an elliptical orbit has speed v_a at apogee, speed v_p at perigee, and speed v_b at the ends of the minor axis of its orbit. (1) Using angular momentum conservation, derive the following formulas for speed ratios in terms of eccentricity e:

$$\frac{v_b}{v_a} = \sqrt{\frac{1+e}{1-e}},$$

$$\frac{v_p}{v_a} = \frac{1+e}{1-e}.$$

(2) Evaluate these ratios numerically for an orbit with $a = 11{,}000$ miles, $b = 8{,}800$ miles, and $c = 6{,}600$ miles. (The labeled ellipse in Figure 11.12 happens to have these relative dimensions.) (3) Does the satellite gain more speed in going from A to B or in going from B to P?

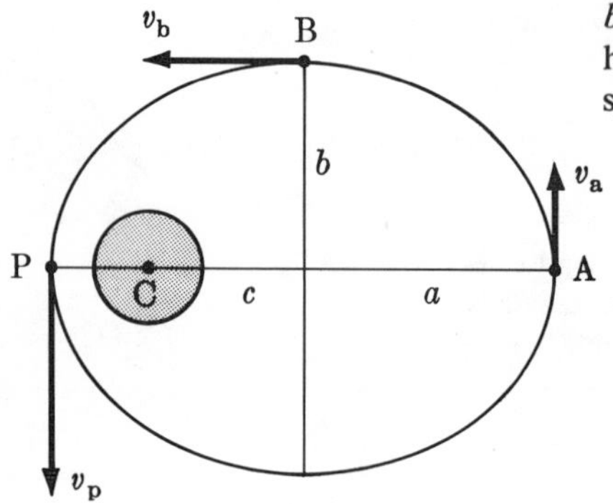

P11.22 In the previous problem, speed ratios were obtained from the law of angular-momentum conservation. Now add energy conservation to the analysis. (1) Show that the speed of the satellite at point B is given by

$$v_b = \sqrt{\frac{GM_E}{a}}$$

(M_E is the mass of the earth). (2) Find the ratio of the satellite's kinetic energy to its potential energy at points A, B, and P. Express the ratios in terms of the eccentricity of the orbit. (Take the zero of potential energy to be at $r = \infty$.)

P11.23 If the force on a planet were proportional to $1/r^n$, with n slightly different from 2, the orbit of the planet could be approximated as an ellipse whose major axis rotated slowly around the sun. Explain in qualitative terms why this is so. For n slightly greater than 2, what is the direction of rotation of the axis of the orbit relative to the direction of rotation of the planet in its orbit? Answer the same question for n slightly less than 2.

Kepler's first law in other worlds

P11.24 (1) Imagine a solar system in which planets circling around a central star have periods proportional to the cubes of their distances from the star ($T \sim r^3$). How does the force of gravity depend on distance in this world? (2) Is there any force law for which the period of a body in a circular orbit is independent of its distance from the force center? If so, what is the force law?

Kepler's third law in other worlds

P11.25 Let the planets be numbered sequentially, $n = 1$ for Mercury to $n = 9$ for Pluto, and let r_n be the mean distance of the nth planet from the sun, measured in A.U. (1) On semilog paper, plot $r_n - 0.4$ vs n, omitting $n = 1$. Note that if a missing planet is postulated and the more distant planets are renumbered accordingly, six planets give points that lie approximately on a straight line. What is the number of the missing planet? If this mathematical regularity has any physical significance, what should be the mean distance of the missing planet from the sun? (2) Show that the line describing the planets Venus through Uranus (after renumbering) can be represented by $r_n = 0.4 + B\,2^n$. What is the numerical value of B in A.U.? (3) Many asteroids are known to occupy a belt whose mean distance from the sun is about 2.8 A.U. Does their existence support the missing-planet hypothesis? (The regularity described here is usually known as Bode's law. When it was publicized by Johann Bode in 1772 [following its original suggestion by Johann Titius], neither Uranus nor any of the asteroids were known. Bode's law still remains an empirical regularity without a sure theoretical foundation.)

Bode's law

P11.26 Consider a satellite consisting of two parts of equal mass m separated by distance $2a$. Its center of mass is at distance r_0 from the center of the earth. For this calculation the earth may be regarded as a spherically symmetric body. (1) Calculate the gravitational potential energy of the satellite for two orientations: (a) parallel to a radial line (solid line in figure) and (b) perpendicular to a radial line (dashed line in figure). Define ε to be the small quantity a/r_0 and expand both expressions in powers of ε, dropping powers higher than the second. (2) Which orientation has the lower potential energy? What is the difference in potential energy for the two orientations? (3) What is the average over angle of the torque that acts on the satellite as it swings from one of these orientations to the other? (4) The moon always keeps one face to the earth. How is that fact related to the ideas in this problem? *Optional:* Carry out the preceding development for all angles of orientation of the satellite.

Torque on a satellite

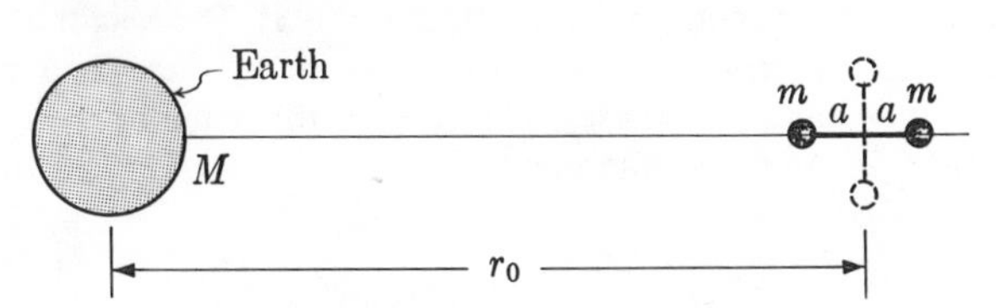

Satellites flying formation

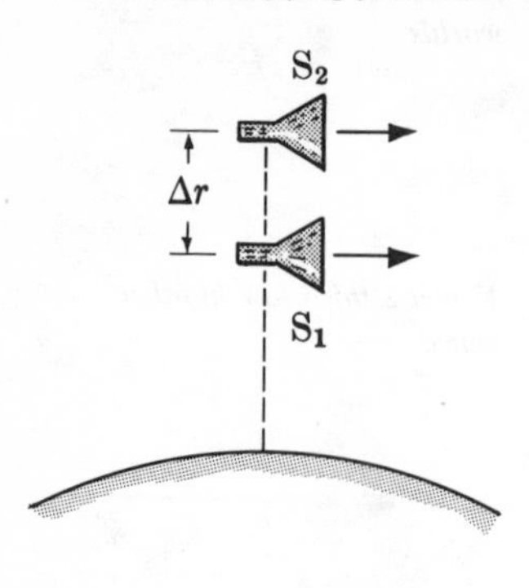

P11.27 Satellites S_1 and S_2 are in *circular* orbits. At a given instant, S_2 is a distance Δr vertically above S_1, and their velocities are parallel (see the figure). Their separation Δr is very much less than the distance r_1 of S_1 from the center of the earth. (1) Ignoring the direct gravitational attraction between these two bodies, derive a formula that gives the distance of lag of one of them behind the other after one revolution. Which one lags? (2) Find this distance of lag if $r_1 = 6{,}600$ km and $\Delta r = 4$ m. (3) If the mass of each satellite is 1,000 kg and r_1 and Δr are as given, does the direct gravitational attraction between the two satellites have a significant effect during the 90 min of a single revolution?

P11.28 Satellite S_1 is in a circular orbit at distance r_1 from the center of the earth. At a given instant, satellite S_2 is a distance Δr vertically above S_1, and the two satellites have the *same velocity* (see the figure of the preceding problem). (1) Prove that S_2 is at the perigee of an elliptical orbit at this moment. (2) Derive a formula valid for $\Delta r \ll r_1$ that gives the spacing of the two satellites when S_2 is at apogee. (3) Evaluate this formula for $r_1 = 6{,}600$ km and $\Delta r = 4$ m.

Tidal effect of the sun on the earth

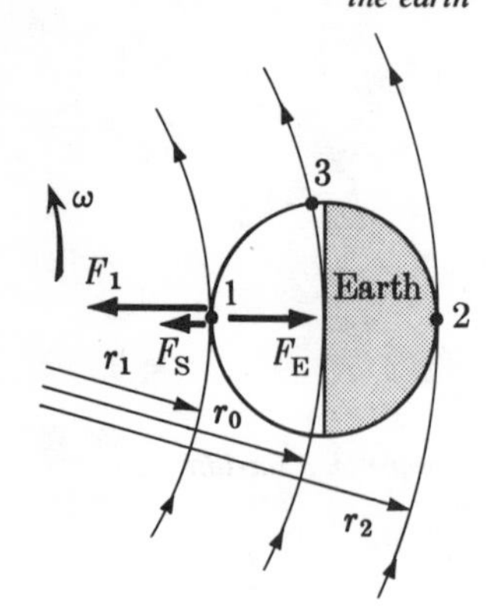

P11.29 To avoid unnecessary complication in this problem, ignore the rotation of the earth about its axis and the effect of the moon. Consider bodies of equal mass m at points 1, 2, and 3 on the surface of the earth (see the figure). Their distances from the sun are r_1, r_2, and r_0 respectively, r_0 also being the distance to the center of the earth. For convenience, define the small quantity $\varepsilon = R/r_0$, where R is the radius of the earth; then $r_1 = r_0(1 - \varepsilon)$ and $r_2 = r_0(1 + \varepsilon)$. The angular speed of the earth in its orbit is

$$\omega = \sqrt{\frac{GM_S}{r_0^3}}$$

(M_S is the mass of the sun). Acting on the body at point 1 are the gravitational force of the earth, which is $F_E = mg$, the gravitational force of the sun, which is $F_S = GmM_S/r_1^2$, and a contact force F_1, which is the same as the weight of the body. (1) Taking account of the acceleration toward the sun of the body at point 1, show that its weight is

$$F_1 = mg\left(1 - \frac{3R\omega^2}{g}\right)$$

(work only to first order in ε). (2) Apply similar reasoning (net force = mass × acceleration) at point 2 and show that $F_2 = F_1$. (3) Explain why the weight of the body at point 3 is, to first order in ε, the same as on a stationary earth: $F_3 = mg$. (These calculations show that equal masses weigh slightly less at the points nearest and farthest from the sun than at intermediate points. These differences give rise to tidal bulges.) *Optional:* Consider the work required to move a particle from the center of the earth to point 1 and from the center of the earth to point 3. The height of the tide can be estimated by equating the work done along these two paths—a condition of equal potential energy. Show that this equality requires that point 1 be about 25 cm higher than point 3.

12 Further Applications of Mechanics

The applications of mechanics extend almost without limit in all directions. In this chapter we will select for attention a few additional mechanical topics, with emphasis on the practical.

12.1 Surface friction

Friction may be a nuisance to designers of some machines, but it is a necessity for everyday life. Without it we could not walk, sit, write, or eat. Every living creature uses frictional forces in some way for locomotion and sustenance.

At the microscopic level, friction is a highly complicated phenomenon. Even though frictional forces are ultimately related to electromagnetic forces, there is no way to derive simple basic laws of friction. Nevertheless, there are approximate laws governing friction, which are useful under certain circumstances. Here we will consider surface friction between solids, and in Section 12.5 we will deal with frictional air drag.

If a pebble is placed on a plank and one end of the plank is gradually raised, we know from experience that the pebble will remain stationary on the plank for a time; then, when a certain angle of inclination is reached, it will start to slide. At angles less than this critical angle, the net force on the pebble is zero. This means that the plank must be exerting on the pebble a vertically upward force $\mathbf{F}_p$ equal in magnitude to the downward force of gravity, $\mathbf{F}_g$ [Figure 12.1(a)]. The vector equation is

$$\mathbf{F}_p + \mathbf{F}_g = 0, \tag{12.1}$$

with $F_p = F_g = mg$. It is convenient to write $\mathbf{F}_p$ as the sum of two other forces: $\mathbf{F}_n$, perpendicular (or normal) to the surface of the plank and $\mathbf{F}_f$, parallel to the

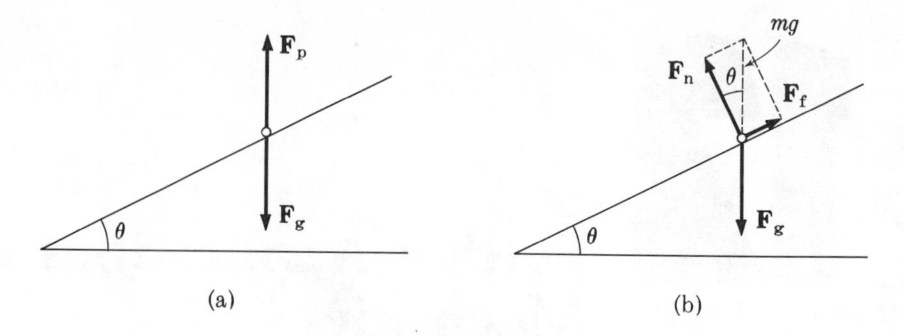

FIGURE 12.1 A pebble in equilibrium on a plank. (a) The total force exerted by the plank, $\mathbf{F}_p$, must be equal to $-\mathbf{F}_g$. (b) The component of $\mathbf{F}_p$ parallel to the surface of the plank is defined to be the frictional force $\mathbf{F}_f$.

surface [Figure 12.1(b)]. The force $\mathbf{F}_f$ is, by definition, the frictional force. The total force acting on the pebble is

$$\mathbf{F}_T = \mathbf{F}_n + \mathbf{F}_f + \mathbf{F}_g. \tag{12.2}$$

In the absence of motion, $\mathbf{F}_T = 0$, and

$$F_f = mg \sin \theta, \tag{12.3}$$

where θ is the angle of the plank to the horizontal. Equation 12.3, which follows from the geometry of the arrow diagram in Figure 12.1(b), is a statement about the frictional force but not a law of friction. The approximate law of frictional force between solid surfaces is

An approximate law of static friction

$$(F_f)_{max} = \mu_s F_n. \tag{12.4}$$

In words: The maximum frictional force is proportional to the normal force. The constant of proportionality, μ_s, called the *coefficient of static friction*, depends on the nature of the surfaces in contact, but for any pair of substances it is nearly independent of the size and mass of either object and of the area of the surfaces in contact. For dry surfaces, values of μ_s lie typically in the range 0.2 to 1.

From Equations 12.3 and 12.4 it follows that

$$mg \sin \theta \leq \mu_s F_n. \tag{12.5}$$

At some critical angle θ_c, the maximum frictional force is achieved, and

$$mg \sin \theta_c = \mu_s F_n. \tag{12.6}$$

At this angle,

$$F_n = mg \cos \theta_c. \tag{12.7}$$

Combining Equations 12.7 and 12.6 yields

$$\sin \theta_c = \mu_s \cos \theta_c,$$

or

Beyond a critical angle, sliding takes place

$$\tan \theta_c = \mu_s. \tag{12.8}$$

■ EXAMPLE 1: The sides of a sawdust pile are observed to make an angle of 38 deg with the horizontal (Figure 12.2). What is the coefficient of static friction for two bits of sawdust in contact? If the angle of the sides of the pile were greater than the critical angle θ_c, a piece of sawdust would slide down the pile, and the angle would readjust itself to a smaller value. If the angle were less than θ_c, sawdust would pile up on top, tending to increase the angle. The observed angle must therefore be the critical angle (for a pile fed from the top). For this example,

$$\mu_s = \tan(38 \text{ deg}) = 0.78. \tag{12.9}$$

The angle of repose

For a pile of loose material, such as sawdust, the critical angle θ_c is known as the angle of repose. ■

For one object sliding over another, the frictional force is generally less than the maximum force of static friction. The approximate law of sliding friction is

An approximate law of sliding friction

$$F_f = \mu_k F_n. \tag{12.10}$$

The frictional force is again proportional to the normal force. The constant of proportionality, μ_k, called the *coefficient of kinetic friction*, is, for a particular pair of surfaces, approximately independent of the normal force, of the area of contact, and of the speed of the motion.

■ EXAMPLE 2: A plate is sent sliding down a counter with an initial speed of 2.5 m/sec. If the coefficient of kinetic friction between plate and counter is $\mu_k = 0.3$, how far does the plate slide? The normal force on the plate is equal to its weight, mg. The horizontal frictional force is therefore

$$F_f = \mu_k mg.$$

The acceleration of the plate has magnitude F_f/m, or

$$a = \mu_k g.$$

Since this acceleration is directed opposite to the velocity, the appropriate equation of uniformly accelerated motion (Equation 5.58) is

$$v^2 = v_0^2 - 2as.$$

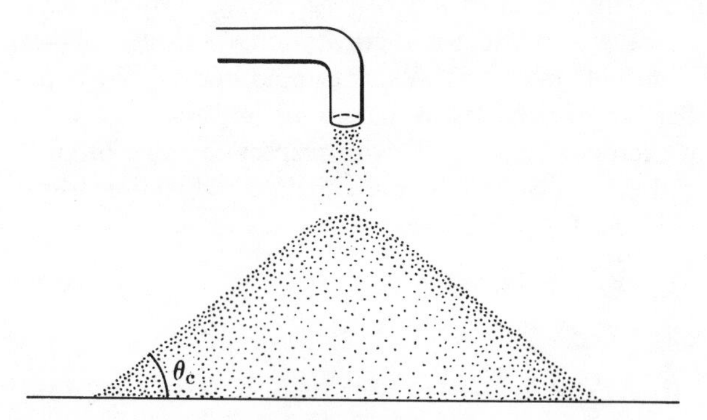

FIGURE 12.2 A sawdust pile, fed from the top, has sides sloping at angle θ_c to the horizontal. This critical angle is also known as the angle of repose.

The distance s is found by setting $v = 0$. Then

$$s = \frac{v_0^2}{2a} = \frac{v_0^2}{2\mu_k g}. \tag{12.11}$$

For this example, the numerical calculation is

$$s = \frac{(2.5 \text{ m/sec})^2}{2 \times 0.3 \times 9.8 \text{ m/sec}^2} = 1.06 \text{ m}.$$

■

12.2 Statics of rigid bodies

Civil engineers and architects are concerned with building structures that are stationary and remain that way. For them, the subject of statics is the most vital part of mechanics. The physicist is more likely to be interested in motion, but he too can find a certain pleasure in solving some of the problems of static equilibrium.

If a particle at rest is to remain at rest, the net force acting on it must be zero. For a rigid body, this condition of zero force is not enough. For instance, if two forces equal in magnitude and opposite in direction do not act along the same line, they create a torque and can set an object into rotational motion. The conditions for the static equilibrium of a rigid body are zero net force *and* zero net torque:

Conditions for static equilibrium

$$\mathbf{F}_T = 0, \tag{12.12}$$

$$\mathbf{T}_T = 0. \tag{12.13}$$

Expressed in terms of components, these equations may give rise, in the most general case, to six equations to be solved simultaneously. If all the forces act in a plane, all the torques act along a line perpendicular to that plane. Then only three simultaneous equations need to be solved (two of force, one of torque).

Equilibrium problems frequently involve the gravitational force, which acts not at any single point but at all points throughout a body. For handling gravity near the earth, the following theorem is useful: *The sum of all gravitational forces acting on a system is equivalent to a single force acting at the center of mass of the system.* "Equivalent" means that the single force accounts for both the total gravitational force and the total gravitational torque. The theorem actually applies to nonrigid systems as well as to rigid bodies. We can prove it for an arbitrary collection of particles of masses $m_1, m_2, m_3, \ldots$. Let these particles have position vectors $\mathbf{r}_1, \mathbf{r}_2, \mathbf{r}_3, \ldots$ with respect to some origin (Figure 12.3). If the gravitational force acts in the negative z direction, the individual forces are $\mathbf{F}_1 = -m_1 g\mathbf{k}$, $\mathbf{F}_2 = -m_2 g\mathbf{k}$, etc., or

A theorem for gravity near the earth

$$\mathbf{F}_i = -m_i g\mathbf{k}. \tag{12.14}$$

The total gravitational force is

The total force

$$\mathbf{F}_g = \sum_i \mathbf{F}_i = -Mg\mathbf{k}, \tag{12.15}$$

where $M = m_1 + m_2 + m_3 + \cdots$, the total mass of the system. With respect

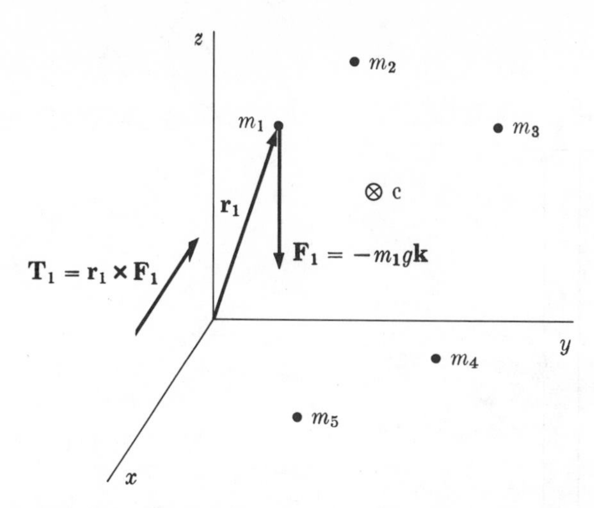

FIGURE 12.3 In a uniform gravitational field, both the total force and the total torque on a system are the same as if all of the mass were concentrated at the center of mass (Equations 12.15 and 12.17).

to the chosen origin, the gravitational force on the ith particle contributes a torque

$$\mathbf{T}_i = \mathbf{r}_i \times \mathbf{F}_i = -m_i g \mathbf{r}_i \times \mathbf{k}.$$

In summing these torques, we note that g and $\mathbf{k}$ are constants, the same for all the particles. Therefore the total gravitational torque is

$$\mathbf{T}_g = \sum_i \mathbf{T}_i = -g\left(\sum_i m_i \mathbf{r}_i\right) \times \mathbf{k}. \tag{12.16}$$

The sum within parentheses occurs in the definition of the center-of-mass position vector $\mathbf{r}_c$ (Equation 8.17). The relationship is

$$\sum_i m_i \mathbf{r}_i = M\mathbf{r}_c.$$

Therefore, the gravitational torque is

The total torque

$$\mathbf{T}_g = -gM\mathbf{r}_c \times \mathbf{k} = \mathbf{r}_c \times \mathbf{F}_g. \tag{12.17}$$

A single force, equal to $-Mg\mathbf{k}$, acting at the center of mass, gives the correct total gravitational force (Equation 12.15) and the correct total gravitational torque (Equation 12.17). This conclusion is independent of the choice of origin. Note, however, that it does depend on the uniformity of the gravitational field, as expressed by Equation 12.14.

■ EXAMPLE 1: A door of width w, height h, and mass m is supported in equilibrium by two hinges at points A and B, as shown in Figure 12.4. The force $\mathbf{F}_1$ at point A is known to act horizontally. What is the magnitude of $\mathbf{F}_1$? What is the direction of $\mathbf{F}_2$? According to the theorem proved above, the effect of gravity is accounted for by a single force $\mathbf{F}_g$ acting vertically downward at the center of mass of the door, here assumed to be at its geometrical center. To apply the condition of zero torque (Equation 12.13) requires that an origin be

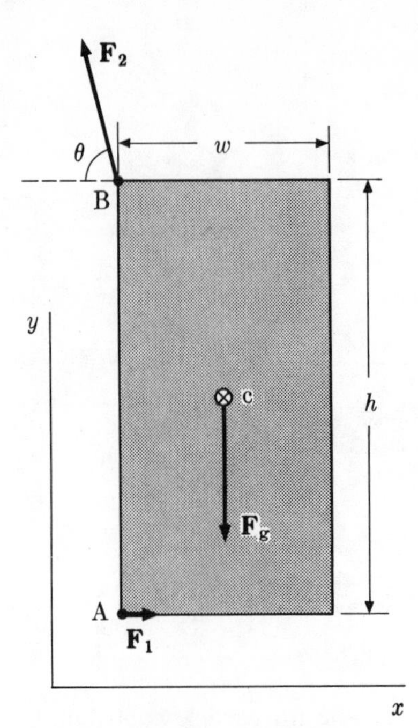

FIGURE 12.4 A door in equilibrium, supported by hinges at A and B.

chosen. Point B is a convenient choice because with respect to this point, the force $\mathbf{F}_2$ produces no torque. The forces $\mathbf{F}_g$ and $\mathbf{F}_1$ produce torques into and out from the page. Define a z axis outward from the page. The torque components along this axis are

$$T_1 = hF_1,$$

$$T_2 = 0,$$

$$T_g = -\tfrac{1}{2}wF_g.$$

The equation of zero torque (with F_g set equal to mg) is

$$hF_1 - \tfrac{1}{2}wmg = 0.$$

The magnitude of $\mathbf{F}_1$ is, therefore,

$$F_1 = \frac{w}{2h}\,mg, \qquad (12.18)$$

which provides the answer to the first question. The answer to the second question is provided by the condition of zero force,

$$\mathbf{F}_1 + \mathbf{F}_2 + \mathbf{F}_g = 0.$$

With axes chosen as in Figure 12.4, the component equations are

$$x: \; F_1 - F_2 \cos\theta = 0,$$

$$y: \; F_2 \sin\theta - mg = 0.$$

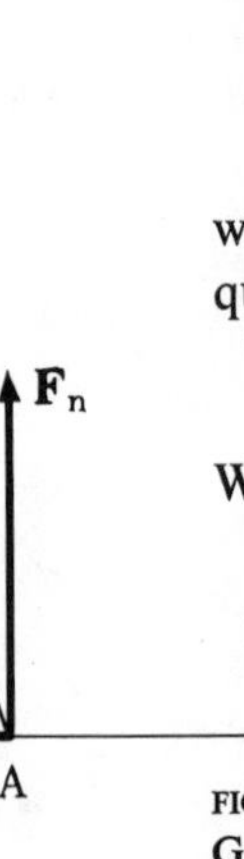

FIGURE 12.5 A man, represented by the point M, is near the top of a ladder. Gravity exerts force $\mathbf{F}_g$, the wall exerts force $\mathbf{F}_c$, and the floor exerts forces $\mathbf{F}_f$ and $\mathbf{F}_n$.

Elimination of F_2 between these two equations yields

$$\tan\theta = \frac{mg}{F_1}, \tag{12.19}$$

an equation that determines the angle θ (Figure 12.4). This result is better expressed by substituting the right side of Equation 12.18 for F_1 in Equation 12.19. Then

$$\tan\theta = \frac{2h}{w}. \tag{12.20}$$

■

■ EXAMPLE 2: A 100-kg man climbs a lightweight aluminum ladder of length 3 m (Figure 12.5). The ladder makes an angle of 70 deg with the floor. If the frictional force at the wall is negligible, what is the minimum coefficient of friction between ladder and floor that enables the man to ascend safely to the top of the ladder? Again, forces and torques must be balanced (if the ladder is not to fall). When the man has just reached the top of the ladder, the forces acting on the ladder are: $\mathbf{F}_c$, the contact force at the wall, acting horizontally to the right (since friction at the wall is negligible); $\mathbf{F}_g$, the weight of the man, acting vertically downward; $\mathbf{F}_n$, the normal force at the floor, acting vertically upward; and $\mathbf{F}_f$, the frictional force at the floor, acting horizontally to the left. We neglect the weight of the ladder. Such a problem is best solved algebraically first. As indicated in Figure 12.5, we call the distances from the corner to the bottom and top of the ladder x and y, the length of the ladder L, and its angle to the horizontal θ. The bottom of the ladder (point A) is a convenient origin. With respect to it, the torques $\mathbf{T}_c$ and $\mathbf{T}_g$ must be equal in magnitude:

$$F_c y = F_g x. \tag{12.21}$$

The condition of zero total force on the ladder yields the two equations

$$F_g = F_n,$$

$$F_c = F_f.$$

These equations imply that $F_f/F_n = F_c/F_g$. From Equation 12.21, it follows that $F_c/F_g = x/y$. Therefore,

$$\frac{F_f}{F_n} = \frac{x}{y} = \frac{1}{\tan\theta}. \tag{12.22}$$

For a coefficient of friction μ_s at the floor,

$$\frac{F_f}{F_n} \le \mu_s$$

(see Equation 12.4). This relation can also be written

$$(\mu_s)_{\min} = \frac{F_f}{F_n}, \tag{12.23}$$

an expression for the minimum coefficient of friction necessary to maintain equilibrium. Substituting from Equation 12.22 in Equation 12.23 yields

$$(\mu_s)_{min} = \frac{1}{\tan\theta} = \text{ctn}\,\theta. \tag{12.24}$$

Note that the mass of the man is irrelevant. For the given angle, $\theta = 70$ deg, $\tan\theta = 2.747$, and the answer to the question posed is

$$(\mu_s)_{min} = 0.364.$$

In what direction does the force $\mathbf{F}_n + \mathbf{F}_f$ act when the ladder is just on the point of slipping? ■

12.3 Forces in fluids

Even when the external forces acting on a rigid body are simple, the internal forces within the rigid body may be complex. Figure 12.6 illustrates this fact. A bar rests on two points of support. As shown in Figure 12.6(a), its equilibrium requires consideration of just three forces, two acting vertically upward at the points of support and one acting vertically downward at the center of mass. But if we look in greater detail at a part of the rod, such as its right end [Figure 12.6(b)], other forces are important. The rest of the rod, through interatomic forces, exerts on this part (1) forces of *tension*, F_t; (2) forces of *compression*, F_c; and (3) a force of *shear*, F_s. The directions of these forces are shown in the diagram. Together, these three kinds of force add up to give a total force and total torque sufficient to balance the effect of the gravitational force F_g acting on this segment of the rod.

Forces within a solid: tension, compression, and shear

We draw attention to these internal forces in solid matter in order to make clear the practical difference between a fluid and a solid. A fluid in equilibrium (either liquid or gas) cannot sustain shear forces, and it can sustain only very small forces of tension. For most purposes, only compressional forces are important in a fluid at rest.* These forces are measured by the *fluid pressure*. Pressure is defined through the relation

Force within a fluid at rest: compression

Definition of pressure

$$d\mathbf{F} = p\,d\mathbf{A}. \tag{12.25}$$

An infinitesimal area dA, either within the fluid or at its surface, may be represented by a vector $d\mathbf{A}$ directed perpendicular to the surface area (Figure 12.7). The *compressional* force $d\mathbf{F}$ that the fluid exerts on its surroundings across this area is proportional to the magnitude of $d\mathbf{A}$ and is in the direction of $d\mathbf{A}$ (it must have this direction, for any component of force perpendicular to $d\mathbf{A}$ would be a shear force). Therefore, $d\mathbf{F} \sim d\mathbf{A}$. The constant of proportionality (a scalar quantity) is called pressure; it is force per unit area. In terms of magnitudes,

$$p = \frac{dF}{dA}. \tag{12.26}$$

An act as simple as sticking your finger into a pan of water shows the

* In a moving fluid, shear forces are possible and are called viscous forces. Viscosity can be described as the result of sliding friction within a fluid as one part of the fluid moves relative to another part.

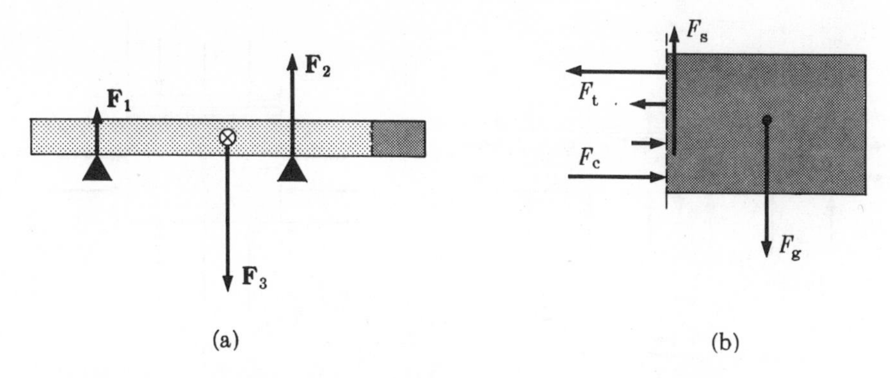

FIGURE 12.6 (a) For a bar resting on two points of support, three forces are sufficient to describe its equilibrium: forces $\mathbf{F}_1$ and $\mathbf{F}_2$ exerted by the points of support, and the gravitational force $\mathbf{F}_3$. (b) One end of the bar, examined in more detail. Forces of tension (F_t), compression (F_c), and shear (F_s) act within the bar, in addition to the force of gravity (F_g).

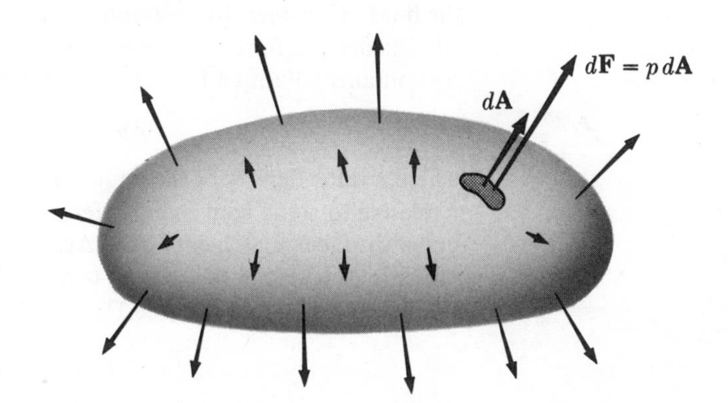

FIGURE 12.7 Fluid pressure. The compressional force, or pressure force, acts at every point in a direction perpendicular to the surface (parallel to the area vector $d\mathbf{A}$).

negligible effects of shear forces in a fluid. Shear forces act to stop your finger when it is pressed against a solid. In a fluid, they do not stop you. Now imagine the somewhat absurd situation of a "rod" of water placed on supports like those holding up the solid rod in Figure 12.6(a). The water would, of course, immediately run onto the table or floor. Since the water cannot support shear forces, nothing prevents gravity from pulling any unsupported part of the water downward. It is a familiar fact that fluids require containers.

Consider a pan of water at rest. The external forces acting on the outside surfaces of the water are the force of air pressure, acting downward, and forces of support by the pan, acting inward and upward. As indicated in Figure 12.8(a), these forces are all perpendicular to surfaces of the fluid. Within the fluid, too, the compressional, or pressure, forces act perpendicular to any area that is considered. Figure 12.8(b) shows the forces acting on a part of the fluid at the right end of the pan. The situation is evidently simpler than that prevailing in a rigid body [Figure 12.6(b)].

Pressure is a scalar measure of compressional force

The internal stress of a fluid in equilibrium, then, is characterized by a single scalar quantity, the pressure p. For a fluid acted upon by gravity, we

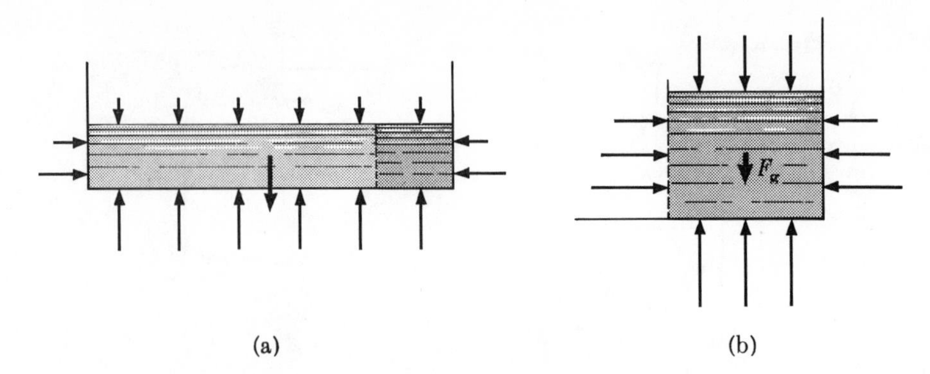

FIGURE 12.8 (a) External forces on a pan of water all act perpendicular to the surface of the water. (b) Internal forces also act perpendicular to a surface separating one part of the fluid from another.

can derive an important differential equation connecting pressure and depth in the fluid. Consider, for example, a cylindrical container of liquid (Figure 12.9). Results derived for it will apply as well to a gas. A slice of area A and thickness Δy contains volume $\Delta V = A\ \Delta y$ and mass

$$\Delta m = \rho\ \Delta V = \rho A\ \Delta y, \tag{12.27}$$

where ρ is the density of the liquid. Let the lower surface of the slice be at height y relative to some convenient reference plane, such as the bottom surface. The upper surface is at height $y + \Delta y$. If the pressure at height y is p and the pressure at height $y + \Delta y$ is $p + \Delta p$, the three external forces acting vertically on the slice of liquid are (see Figure 12.9)

$$\mathbf{F}_1 = pA\mathbf{j},$$

$$\mathbf{F}_2 = -(p + \Delta p)A\mathbf{j},$$

$$\mathbf{F}_g = -(\Delta m)g\mathbf{j} = -\rho Ag\ \Delta\ y\mathbf{j}.$$

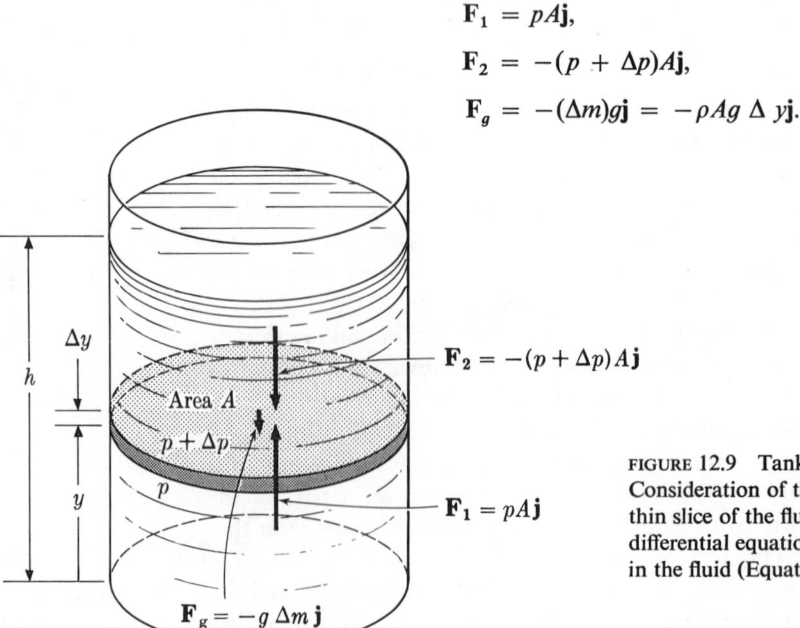

FIGURE 12.9 Tank of fluid in equilibrium. Consideration of the balancing of forces on a thin slice of the fluid leads to a simple differential equation governing the pressure in the fluid (Equation 12.29).

Since the vertical walls of the container contribute no vertical component of force, these three forces must add to zero:

$$[pA - (p + \Delta p)A - \rho Ag\,\Delta y]\mathbf{j} = 0.$$

This equation reduces to

$$A\,\Delta p = -\rho Ag\,\Delta y.$$

In the limit that Δy approaches zero, the corresponding differential equation is (after cancellation of the factors A)

$$dp = -\rho g\,dy, \tag{12.28}$$

or, equivalently,

A differential equation governing the static equilibrium of a fluid

$$\frac{dp}{dy} = -\rho g. \tag{12.29}$$

This result does not in fact depend upon the choice of a container with vertical sides. Sloping sides do contribute a vertical component of force, but as Δy approaches zero, their contribution vanishes, and the same result, Equation 12.29, is obtained.

INCOMPRESSIBLE FLUIDS

Most liquids, including water, are nearly incompressible. This means that their density ρ is a constant, or very nearly so, independent of the pressure. For a liquid, it is a good approximation to call the right side of Equation 12.29 a constant. Then the derivative, dp/dy, is constant, and the pressure is a linear function of vertical distance y:

Pressure in an incompressible fluid

$$p = p_0 - \rho gy \qquad (\rho = \text{constant}). \tag{12.30}$$

Figure 12.10 shows a graph of pressure vs height in a liquid, with pressure p_0 at its bottom surface and pressure $p_s = p_0 - \rho gh$ at its top surface, where h is the total depth of the liquid. If the liquid is exposed to air, the surface pressure p_s

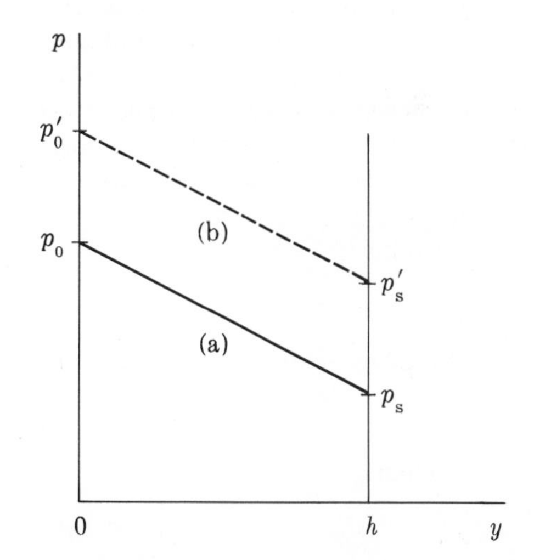

FIGURE 12.10 Pressure vs height in an incompressible fluid. (a) Pressure at the bottom of a tank is p_0, and pressure at the liquid surface is p_s. The slope of the line is $-\rho g$. (b) If the surface pressure is raised from p_s to p_s' the pressure at the bottom and throughout the fluid is raised by the same increment, an illustration of what is called Pascal's principle.

is normal air pressure, p_a, whose average value at sea level is

Normal air pressure at sea level

$$p_a = 1.013 \times 10^5 \text{ N/m}^2. \quad (12.31)$$

A pressure of 10^5 N/m^2 (or 10^6 dyne/cm^2) is called one *bar*, so 1 atmosphere (1 atm) of pressure is nearly equal to 1 bar. It is sometimes convenient to re-express Equation 12.30 in terms of surface pressure p_s and depth $d = h - y$. It then reads

$$p = p_s + \rho g d. \quad (12.32)$$

Water: 1 atm in about 10 m

For water, $\rho g = 10^3 \text{ kg/m}^3 \times 9.8 \text{ m/sec}^2 = 0.98 \times 10^4 \text{ N/m}^3$. Roughly, water pressure increases by 0.1 atm for each meter of depth. At a depth of 10.34 m (33.9 ft), it is increased by 1 atm.

■ EXAMPLE 1: The density of blood is about 1.06 gm/cm^3. What is the blood-pressure difference between head and toe of a person of height 175 cm (69 in.)? From Equation 12.30 or 12.32,

$$\begin{aligned}\Delta p &= \rho g h \\ &= 1.06 \text{ gm/cm}^3 \times 980 \text{ cm/sec}^2 \times 175 \text{ cm} \\ &= 1.82 \times 10^5 \text{ dyne/cm}^2 = 1.82 \times 10^4 \text{ N/m}^2 \\ &= 0.179 \text{ atm.}\end{aligned}$$ ■

If the surface pressure p_s is increased in some way—by a change in atmospheric conditions, by pumping more air into an enclosure containing the liquid, or in any other way—the pressure throughout the liquid will change by the same amount. At a particular depth d, the term $\rho g d$ in Equation 12.32 is unchanged, so

$$\Delta p = \Delta p_s. \quad (12.33)$$

Pascal's principle

This is a special case of what is known as *Pascal's principle*, a statement that a change of pressure in one part of a fluid in equilibrium is transmitted equally to all parts of the fluid. The effect is illustrated graphically in Figure 12.10. A change of surface pressure raises the whole pressure graph by a fixed amount.

The mercury barometer [Figure 12.11(a)] is a simple device whose action can be easily understood with the help of Equation 12.30. Atmospheric pressure p_a acts on its surface at distance d_1 above the bottom of the container. A higher surface at distance d_2 above the bottom is exposed only to mercury vapor, whose pressure is negligible compared with atmospheric pressure. Figure 12.11(b) shows a graph of pressure vs vertical distance y in the liquid. At $y = d_1$, Equation 12.30 gives

$$p_a = p_0 - \rho g d_1.$$

At the higher surface, $y = d_2$,

$$0 = p_0 - \rho g d_2.$$

Elimination of p_0 between these two equations gives

$$p_a = \rho g(d_2 - d_1) = \rho g L, \quad (12.34)$$

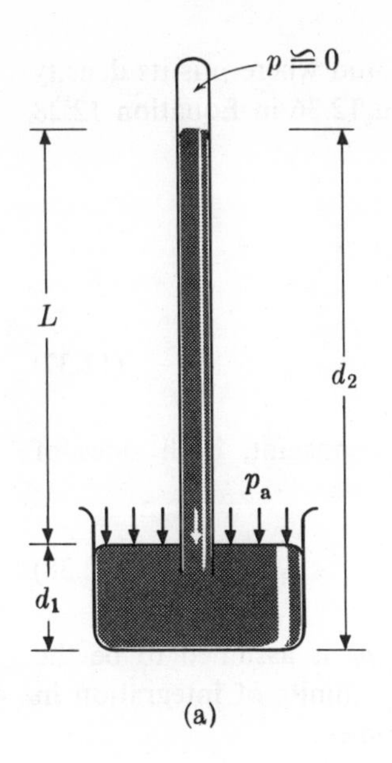

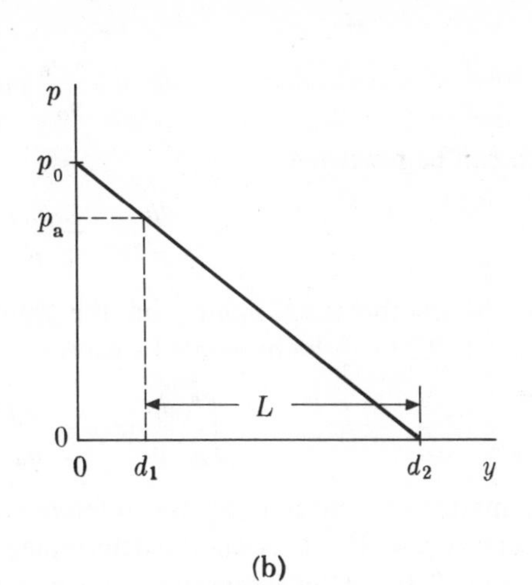

FIGURE 12.11 The mercury barometer. (a) At level d_1, the pressure in the liquid is p_a, both at the interface with the air and within the column. At level d_2, the pressure is approximately zero. The distance L and air pressure p_a are related by $p_a = \rho g L$. (b) A graph of pressure vs height in the mercury. This is analogous to Figure 12.10, with $p_s = 0$.

a formula for the air pressure p_a in terms of the known quantities ρ and g and the measurable length L of the mercury column in the tube. The standard atmosphere is defined by a mercury column length of exactly 76 cm at a place where $g = 980.665$ cm/sec^2 and at a temperature of 0 °C, for which the density of mercury is 13.595 gm/cm^3. These figures give

Mercury: 1 atm in 0.76 m

$$\begin{aligned} p_a &= (13.595 \text{ gm/cm}^3)(980.665 \text{ cm/sec}^2)(76.00 \text{ cm}) \\ &= 1.0132 \times 10^6 \text{ dyne/cm}^2 \\ &= 1.0132 \times 10^5 \text{ N/m}^2 = 1.0132 \text{ bars.} \end{aligned} \tag{12.35}$$

THE ATMOSPHERE

Unlike a liquid, a gas undergoes large changes of density as its pressure changes. At constant temperature, the density of a gas is proportional to its pressure.* We may write

$$\rho = \frac{\rho_0}{p_0} p, \tag{12.36}$$

A version of Boyle's law: $\rho \sim p$ if T = constant

* This statement is equivalent to Boyle's law, which states that the pressure of a fixed quantity of gas is inversely proportional to its volume if its temperature is constant. Boyle's law will be discussed in Section 13.4.

where ρ_0 is its density at some reference pressure p_0 and where ρ is its density at any other pressure p. Substituting from Equation 12.36 in Equation 12.28 gives

$$dp = \frac{-\rho_0}{p_0} pg\, dy,$$

which can be rewritten

$$\frac{dp}{p} = -\frac{\rho_0 g}{p_0}\, dy. \tag{12.37}$$

Since the combination $\rho_0 g/p_0$ on the right side is constant, both sides of Equation 12.37 can be integrated easily:

$$\int_{p_0}^{p} \frac{dp}{p} = -\frac{\rho_0 g}{p_0} \int_0^y dy. \tag{12.38}$$

As a matter of convenience, the reference pressure p_0 is assumed to be the pressure at $y = 0$. This choice determines the lower limits of integration in Equation 12.38. After integration, this equation becomes

$$\ln\left(\frac{p}{p_0}\right) = -\frac{\rho_0 g}{p_0}\, y,$$

which is equivalent to

$$\frac{p}{p_0} = e^{-(\rho_0 g/p_0)y}. \tag{12.39}$$

This result takes on a slightly simpler appearance if we define a "scale height" b by

$$b = \frac{p_0}{\rho_0 g}. \tag{12.40}$$

The law of atmospheres: Pressure varies exponentially with altitude if temperature is constant

Then the pressure p as a function of altitude y can be written

$$p = p_0 e^{-y/b}. \tag{12.41}$$

If the temperature of the air were constant and if the variation of g with altitude were negligible, air pressure would decrease exponentially, falling by a factor e (to 36.8 percent of its surface value) at an altitude equal to the scale height b.

■ EXAMPLE 2: What is the scale height of air if its pressure and density at sea level are $p_0 = 1.013 \times 10^5$ N/m^2 and $\rho_0 = 1.29$ kg/m^3? At what altitude is air pressure half of its sea-level value? Equation 12.40 gives

The approximate scale height of air

$$b = \frac{1.013 \times 10^5}{1.29 \times 9.8} = 8.01 \times 10^3 \text{ m} = 8 \text{ km}, \tag{12.42}$$

which is about 5 miles, or 26,000 ft. The "half-distance," in which p decreases by a factor of two, is

$$L_{1/2} = 0.693b \tag{12.43}$$

(since $e^{-0.693} = 0.5$). For the calculated value of b,

$$L_{1/2} = 5.55 \text{ km} = 3.45 \text{ miles} = 18{,}200 \text{ ft}.$$ ■

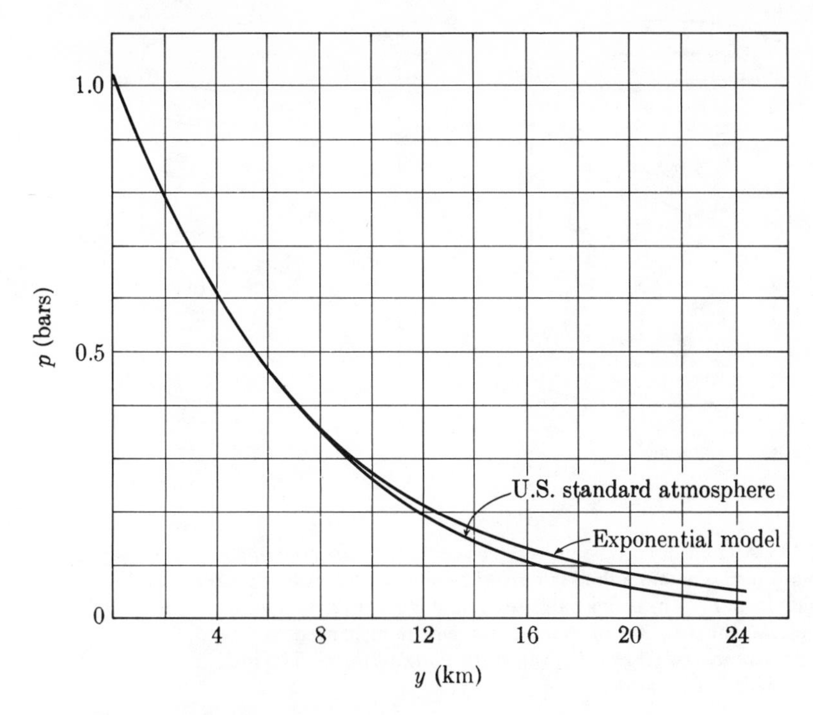

FIGURE 12.12 Atmospheric pressure vs altitude. A realistic model of the atmosphere, the U.S. standard atmosphere, is compared with the exponential model, as expressed by Equation 12.41 (with $b = 8$ km).

In fact, the temperature of the air varies quite significantly over vertical distances of a few kilometers. The variation of g is less important, but it has some significance if great heights are considered. In Figure 12.12, the simple exponential model of the atmosphere, as given by Equation 12.41, is compared graphically with a more accurate model, the so-called U.S. Standard Atmosphere. The exponential curve is qualitatively correct..

ARCHIMEDES' PRINCIPLE

Archimedes' principle: a law of buoyant force

The upward buoyant force on an object partially or wholly submerged in a fluid is equal to the weight of the displaced fluid. This useful principle is attributed to Archimedes, who lived in the third century B.C. To understand the principle, focus attention first on a particular part of a fluid in equilibrium, such as the triangular wedge shown in Figure 12.13(a). The sum of all forces acting on this wedge of fluid must be zero. Therefore, the sum of all the pressure forces acting on it must be directed upward and must be equal in magnitude to the downward-acting force of gravity on the wedge, that is, to the weight of the wedge of fluid. If this element of fluid is now replaced by a wedge of wood [Figure 12.13(b)], the rest of the fluid remains as it was, and the pressure acting on each part of the wedge is unchanged. The net upward force produced by the surrounding fluid is still the same, equal to the weight of fluid that occupied the volume of the wedge. This is called the buoyant force. The weight of the wood, F_g', might be less than the weight of the displaced fluid, F_g (if, for instance, the fluid is water).

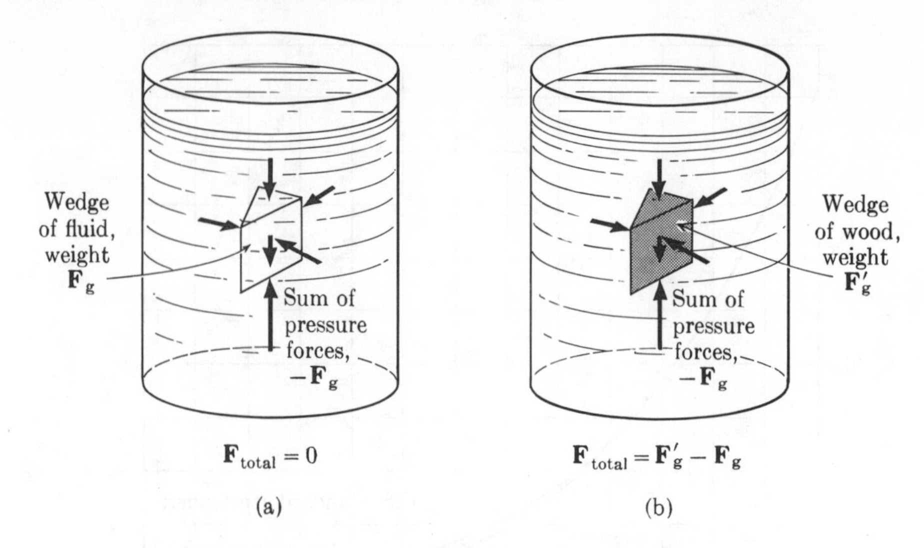

FIGURE 12.13 Archimedes' principle. (a) The sum of the forces acting on a wedge-shaped region within a fluid is zero. The pressure forces balance the gravitational force $\mathbf{F}_g$. (b) If this part of the fluid is replaced by a wedge of wood, whose weight is $\mathbf{F}'_g$, the pressure forces remain unchanged, and sum to a net upward force equal to $-\mathbf{F}_g$, called the buoyant force. The total force on the wedge of wood is $\mathbf{F}'_g - \mathbf{F}_g$.

Then the piece of wood experiences a net upward force, $F_g - F'_g$, and equilibrium no longer prevails.

Archimedes' principle is equally valid for a partially submerged object, such as the rod shown in Figure 12.14. Again the surrounding fluid must provide a net upward force equal to the weight of the displaced fluid, for that is the force which would be required to maintain equilibrium if the rod were removed and if the "hole" it left behind were filled with the same fluid.

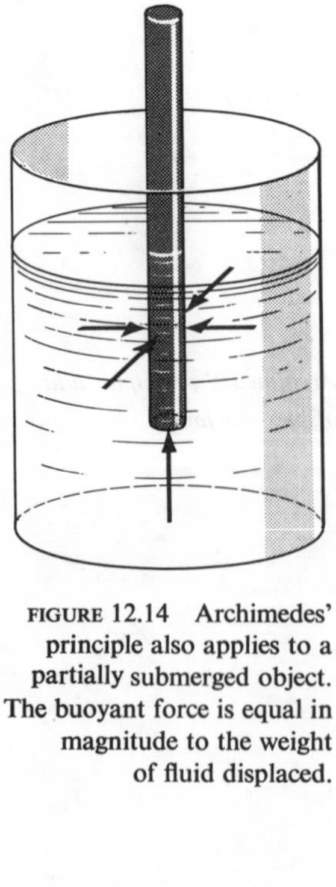

FIGURE 12.14 Archimedes' principle also applies to a partially submerged object. The buoyant force is equal in magnitude to the weight of fluid displaced.

A hot-air balloon (Figure 12.15) is a flying machine that gains its lift from Archimedes' principle. A typical two-man balloon displaces about 1,600 m^3 (56,500 ft^3) of air, whose mass on a nice spring day (T = 20 °C = 68 °F) is about 1,925 kg. The buoyant force on the balloon can therefore support this same total mass. If the hot air within the balloon has a mass of 1,600 kg, corresponding to a temperature of about 80 °C, everything else—fabric, ropes, gondola, burner, fuel, occupants, and baggage—can add up to 325 kg, or a little more than 700 lb.

12.4 Fluid flow

Hydrodynamics, the study of fluids in motion, is a subject about which multi-volume textbooks have been written, and in a single section we can only touch upon a few aspects of this vast subject. In particular, we want to show how mass conservation and energy conservation influence the steady flow of fluids.

Steady flow means that the way in which the fluid moves does not change with time. One element of fluid follows a certain path. Behind it, another element of fluid follows the same path. Such a path pursued by a small element of

FIGURE 12.15 A hot-air balloon. The heated air within the balloon has less mass than the air it displaces. In equilibrium, the mass of the heated air plus the mass of the balloon and its occupants is equal to the mass of displaced air. (Photograph courtesy of Don Piccard Balloons, Newport Beach, California.)

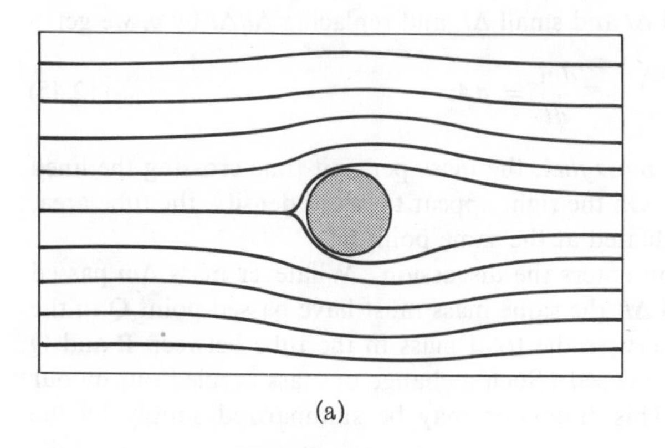

(a)

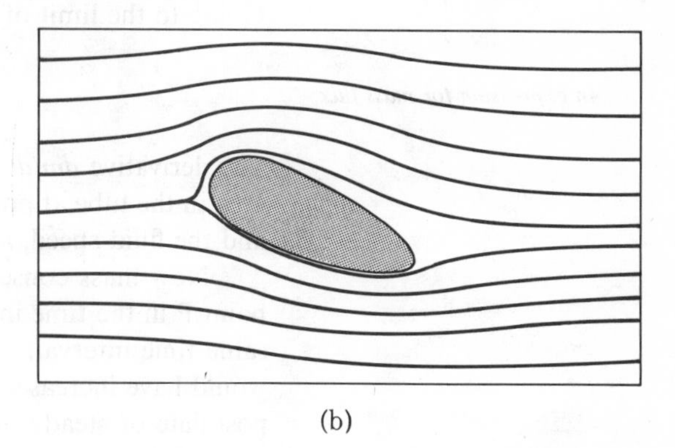

(b)

FIGURE 12.16 Streamlines of steady fluid flow.

fluid is called a streamline. In steady flow, streamlines are stationary. A diagram of streamlines shows at a glance the pattern of fluid flow (Figure 12.16).

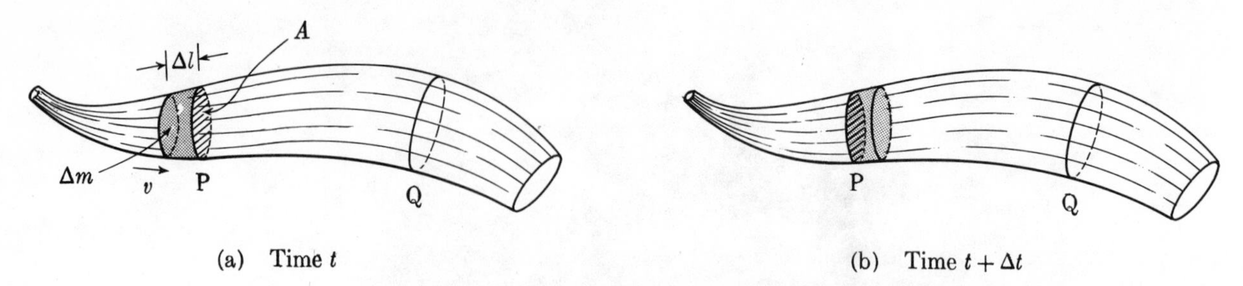

FIGURE 12.17 A tube of flow. (a) At time t, the shaded element of fluid has its front surface at location P. (b) At time $t + \Delta t$, the rear surface of this element of fluid is at P. A mass Δm has flowed past P. Since the flow is steady, the same mass of fluid, Δm, must have passed Q during this time.

THE EQUATION OF CONTINUITY

A tube of flow is bounded by streamlines

A set of streamlines may be drawn to enclose what is called a tube of flow (Figure 12.17). Such a tube of flow may be an actual physical tube enclosing the flow, or it may be imbedded in a larger pattern of flow, a part singled out for attention. Figure 12.17(a) is meant to be a snapshot of a tube of flow at a particular time t. The shaded portion of fluid in the tube has area A, length Δl, and mass

$$\Delta m = \rho A \, \Delta l, \tag{12.44}$$

where ρ is the density of the fluid. At some later time, $t + \Delta t$, this portion of fluid has moved forward a distance Δl [Figure 12.17(b)] so that its average speed in the interval Δt is $v = \Delta l/\Delta t$. Division of both sides of Equation 12.44 by Δt gives

$$\frac{\Delta m}{\Delta t} = \rho A \frac{\Delta l}{\Delta t}.$$

Going to the limit of small Δl and small Δt, and replacing $\Delta l/\Delta t$ by v, we get

An expression for mass flux

$$\frac{dm}{dt} = \rho A v. \tag{12.45}$$

The derivative dm/dt is the *mass flux*, the mass per unit time crossing the lined area in the tube at point P. On the right appear the fluid density, the tube area, and the fluid speed, all evaluated at the same point P.

Now mass conservation enters the discussion. Whatever mass Δm passed point P in the time interval Δt, the same mass must have passed point Q in the same time interval, for otherwise the total mass in the tube between P and Q would have increased or decreased. Such a change of mass is ruled out by our postulate of steady flow. This discussion may be summarized simply by the statement

$$\frac{dm}{dt} = \textit{constant}. \tag{12.46}$$

In steady flow, the mass flux is the same everywhere in a tube of flow. According to Equation 12.45, this in turn means

$$\rho A v = \textit{constant}. \qquad (12.47)$$

The equation of continuity for steady flow

Although each of these three factors may vary from one part of a tube of flow to another part, their product is constant. If two points labeled 1 and 2 are considered, Equation 12.47 is equivalent to

$$\rho_1 A_1 v_1 = \rho_2 A_2 v_2. \qquad (12.48)$$

For nearly incompressible liquids, the factors ρ_1 and ρ_2 are nearly equal; then Equation 12.48 simplifies to

$$A_1 v_1 = A_2 v_2 \qquad (\rho = \text{constant}). \qquad (12.49)$$

The product Av, with the dimension (length)^3/time, is the volume flux, that is, the volume of fluid passing a point in the tube per unit time. For incompressible fluids, both the mass flux and the volume flux are constant in a tube of flow. Equation 12.47 (or its equivalent, Equation 12.48) is known as the equation of continuity.

■ EXAMPLE 1: Steady flow of water is established in a funnel, as shown in Figure 12.18(a). If $A_1 = 5A_2$ and $A_2 = 5A_3$, what are the relative speeds of the water in the upper part of the funnel (v_1), at the outlet of the funnel (v_2), and in the stream just below the faucet (v_3). Since water is nearly incompressible, Equation 12.49 applies. The speed ratios are simply the inverse area ratios:

$$\frac{v_2}{v_1} = \frac{A_1}{A_2} = 5,$$

$$\frac{v_3}{v_1} = \frac{A_1}{A_3} = 25.$$

■

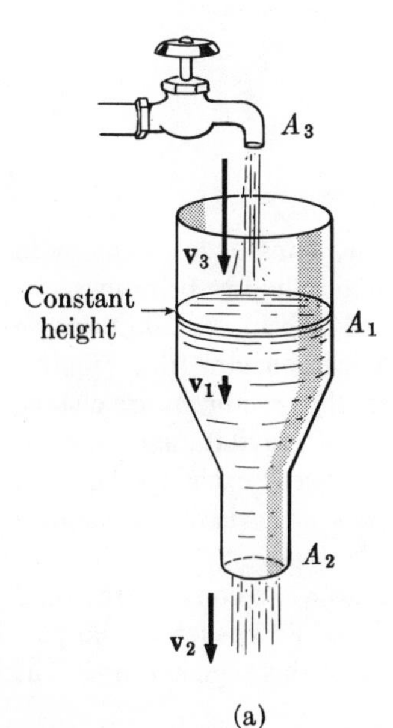

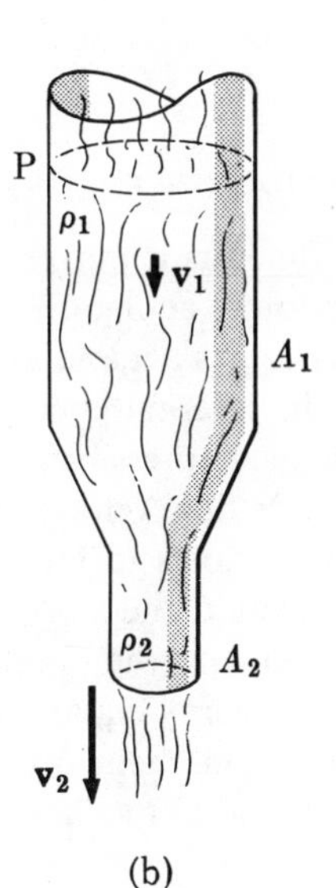

FIGURE 12.18 Examples of steady fluid flow. (a) Water in a funnel. (b) Compressible gas in a pipe.

■ EXAMPLE 2: A compressible gas flows steadily through a pipe of cross-sectional area A_1 and out of a narrower opening of area $A_2 = A_1/5$. If its density at a certain point P in the pipe is twice its density at the outlet and if its speed at P is $v_1 = 2$ m/sec, what is its speed v_2 at the outlet? From Equation 12.48,

$$v_2 = \frac{A_1}{A_2} \cdot \frac{\rho_1}{\rho_2} v_1 = 10v_1 = 20 \text{ m/sec}.$$

Lower density, like smaller area, increases the speed of flow. ■

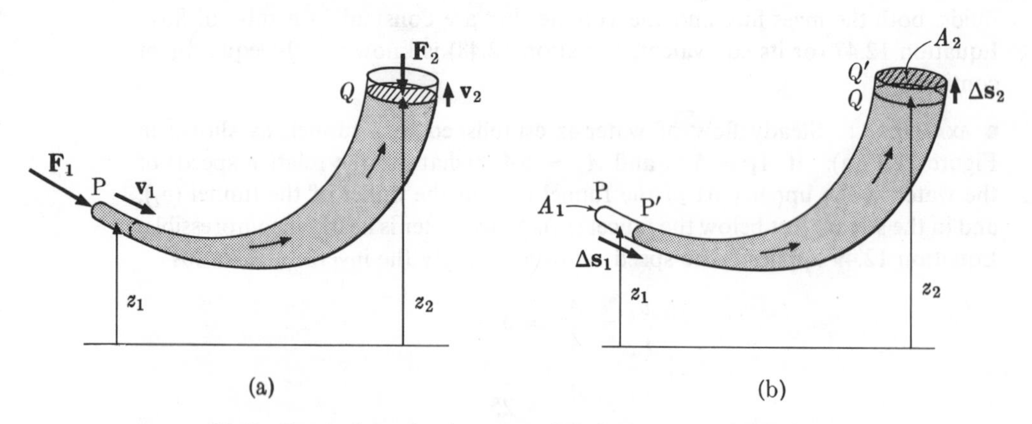

FIGURE 12.19 Steady flow of an incompressible fluid. As the front surface of an element of fluid moves from Q to Q′ and its rear surface moves from P to P′, pressure forces $\mathbf{F}_1$ and $\mathbf{F}_2$ do work, and the mechanical energy of the fluid changes. These energy considerations lead to Bernoulli's equation (Equation 12.59).

BERNOULLI'S EQUATION FOR INCOMPRESSIBLE FLUIDS

An element of fluid, as it moves along a tube of flow, gains or loses energy in response to the forces acting on it. We shall assume the fluid to be nonviscous so that no shear forces act and no energy is dissipated as heat. In addition, we postulate that the internal energy of the fluid remains constant (this requires that the fluid be incompressible). The energy changes that *can* occur are change of kinetic energy as flow velocity changes and change of gravitational potential energy as the fluid flows to higher or lower levels. Energy exchange between any portion of the fluid and a neighboring portion takes place via the mechanism of work associated with the pressure forces within the fluid.

To derive an energy equation for this idealized flow, we consider the fluid that at a particular moment occupies part of a tube of flow, such as the part between points P and Q in Figure 12.19(a). One end of this segment, at P, has cross-sectional area A_1 and is moving with speed v_1. The other end, at Q, has cross-sectional area A_2 and is moving with speed v_2. The pressure at P is p_1; the pressure at Q is p_2. During an interval of time Δt, the rear end of the fluid segment being considered moves ahead a distance Δs_1 from P to P′, and the forward end moves ahead a distance Δs_2 from Q to Q′ [Figure 12.19(b)]. The net work done by the adjoining portions of fluid is

Pressure forces do work on a fluid in motion

$$\Delta W = \mathbf{F}_1 \cdot \Delta \mathbf{s}_1 + \mathbf{F}_2 \cdot \Delta \mathbf{s}_2. \tag{12.50}$$

This equation utilizes the basic definition of work (Equation 10.22). We can take advantage of the facts that the pressure force $\mathbf{F}_1$ is parallel to $\Delta \mathbf{s}_1$ (whence $\mathbf{F}_1 \cdot \Delta \mathbf{s}_1 = F_1 \, \Delta s_1$) and that $\mathbf{F}_2$ is antiparallel to $\Delta \mathbf{s}_2$ (whence $\mathbf{F}_2 \cdot \Delta \mathbf{s}_2 = -F_2 \, \Delta s_2$). Noting also that $F_1 = p_1 A_1$ and $F_2 = p_2 A_2$, we write

$$\Delta W = p_1 A_1 \, \Delta s_1 - p_2 A_2 \, \Delta s_2. \tag{12.51}$$

The combination $A_1 \, \Delta s_1$ in this equation is ΔV_1, the volume of the tube of flow between P and P′. Similarly, $A_2 \, \Delta s_2 = \Delta V_2$, the volume of the tube of flow between Q and Q′. However, these two volumes are equal. Since the flow is steady, the mass leaving the region PP′ in time Δt is the same as the mass reaching the region QQ′ in the same time. Because of the postulated incompressibility of the fluid, equal masses of fluid occupy equal volumes. Therefore, we can write

$$A_1 \, \Delta s_1 = A_2 \, \Delta s_2 = \Delta V. \tag{12.52}$$

The expression for work given by Equation 12.51 then simplifies further:

$$\Delta W = (p_1 - p_2) \, \Delta V. \tag{12.53}$$

This net work done on the fluid segment must be equal to the change of kinetic energy plus the change of potential energy of the segment:

The fluid energy changes in response to the work

$$\Delta W = \Delta K + \Delta U. \tag{12.54}$$

Although some fluid has flowed into the region between P′ and Q during the time interval Δt and some has flowed out of it, the properties of the fluid between P′ and Q have not changed. At the moment pictured in Figure 12.18(b), the mass of fluid in this region and its distribution of pressure and velocity are the same as at the earlier moment pictured in Figure 12.18(a). Therefore, the change of energy of the whole fluid segment, which initially lay between P and Q and later lies between P′ and Q′, must be equal to the *difference* between the energy of the part in region QQ′ and the energy of the part in region PP′. If Δm designates the mass in either of these regions, the change of kinetic energy is

$$\Delta K = \tfrac{1}{2} \, \Delta m \cdot (v_2{}^2 - v_1{}^2). \tag{12.55}$$

If z measures the height of the tube of flow above any chosen base level, the change of potential energy is

$$\Delta U = \Delta m \, g \cdot (z_2 - z_1). \tag{12.56}$$

Using Equations 12.53, 12.55, and 12.56 in Equation 12.54 gives

$$(p_1 - p_2) \, \Delta V = \tfrac{1}{2} \, \Delta m \cdot (v_2{}^2 - v_1{}^2) + \Delta m \, g \cdot (z_2 - z_1). \tag{12.57}$$

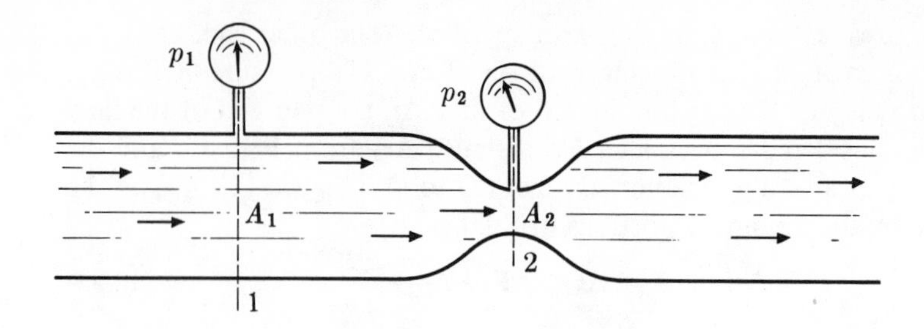

FIGURE 12.20 Where the pipe has a constriction, the flow velocity is higher and the pressure is lower than in the rest of the pipe. The decrease of pressure with increase of speed is sometimes called Bernoulli's principle.

The ratio $\Delta m/\Delta V$ is equal to ρ, the constant fluid density.* Therefore, Equation 12.57 is equivalent to

$$p_1 - p_2 = \tfrac{1}{2}\rho\cdot(v_2{}^2 - v_1{}^2) + \rho g\cdot(z_2 - z_1),$$

which, in turn, can be rewritten

$$p_1 + \tfrac{1}{2}\rho v_1{}^2 + \rho g z_1 = p_2 + \tfrac{1}{2}\rho v_2{}^2 + \rho g z_2. \tag{12.58}$$

Since the combination $p + \frac{1}{2}\rho v^2 + \rho g z$ has the same value at the arbitrarily chosen points P and Q, it must be the same everywhere along a tube of flow, and we can write

Bernoulli's equation, a consequence of energy conservation

$$p + \tfrac{1}{2}\rho v^2 + \rho g z = \textit{constant}. \tag{12.59}$$

This is *Bernoulli's equation*, valid for the steady flow of an incompressible non-viscous fluid. It dates from 1738, when Daniel Bernoulli, a Swiss mathematician (whom we would now call a theoretical physicist as well), published the first major work on hydrodynamics.

APPLICATIONS OF BERNOULLI'S EQUATION

1. *Static Equilibrium.* In the absence of flow, Equation 12.59 reduces to

$$p + \rho g z = \textit{constant},$$

equivalent to Equation 12.30, which governs the static equilibrium of an incompressible fluid. The earlier equation was derived by setting the total force on a fluid element equal to zero. Here the same result follows from energy considerations.

2. *Flow in a Horizontal Pipe.* Another useful simplification of Bernoulli's

* For an incompressible fluid, the statement $\rho = \Delta m/\Delta V$ is valid for any finite volume. Nevertheless, the validity of Equation 12.57 does depend on going to the limit $\Delta V \to 0$. In the derivation of Equation 12.57, we assumed that the speed v_1 and the area A_1 are constant over the region PP′ and that v_2 and A_2 are constant over the region QQ′. These assumptions are valid only in the limit that Δs_1 and Δs_2 approach zero.

equation occurs if horizontal flow is considered. Then the term ρgz can be discarded and the equation takes the form

$$p + \tfrac{1}{2}\rho v^2 = constant. \tag{12.60}$$

Bernoulli's equation for horizontal flow

Consider, for example, a pipe of nonuniform cross-sectional area (Figure 12.20). According to Equation 12.60, pressure decreases when speed increases. This statement is sometimes called *Bernoulli's principle.* The equation of continuity (Equation 12.49), in turn, relates speed and cross-sectional area. For the places labeled 1 and 2 in Figure 12.20, Bernoulli's equation may be written

Bernoulli's principle: Higher speed means lower pressure

$$p_1 + \tfrac{1}{2}\rho v_1{}^2 = p_2 + \tfrac{1}{2}\rho v_2{}^2. \tag{12.61}$$

From Equation 12.49,

$$v_2 = \left(\frac{A_1}{A_2}\right) v_1. \tag{12.62}$$

If the right side of Equation 12.62 is substituted for v_2 in Equation 12.61, the result is

$$p_1 - p_2 = \frac{1}{2}\rho\left[\left(\frac{A_1}{A_2}\right)^2 - 1\right] v_1{}^2. \tag{12.63}$$

This equation finds a use in a method of determining flow velocity by measuring pressure. Suppose that the pipe shown in Figure 12.20 has uniform cross-sectional area A_1 everywhere except in a short region of constriction, where its area is A_2. The areas A_1 and A_2 are known, and the pressure difference, $p_1 - p_2$, may be measured. Then the speed v_1 in the uniform part of the pipe may be calculated from the formula

$$v_1 = \sqrt{\frac{2(p_1 - p_2)}{\rho[(A_1/A_2)^2 - 1]}}, \tag{12.64}$$

which follows directly from Equation 12.63.

3. *Stagnation and Dynamic Pressure.* Even though a gas is compressible, it may sometimes exhibit approximately incompressible flow. Then Bernoulli's equation is approximately valid. The requirement for this condition is that the speed of flow be small compared with the speed of sound in the gas.

Figure 12.21 shows the streamlines of slow flow of air around a disk. At the central point of the disk, the speed of flow is zero: $v_1 = 0$. This is called a *stagnation point* of the flow. If the conditions of uniform flow well-removed from the disk are described by pressure p_0 and speed v_0, Bernoulli's equation yields the pressure p_1 at the stagnation point:

At a stagnation point, fluid velocity is zero and pressure is maximum

$$p_1 = p_0 + \tfrac{1}{2}\rho v_0{}^2. \tag{12.65}$$

(We continue to assume that potential energy changes are negligible.) Suppose, for example, that the speed of flow is one-tenth the speed of sound, $v_0 =$ 33 m/sec (74 mile/hr). Then the amount by which the pressure is increased at the stagnation point is

$$\Delta p = \tfrac{1}{2}\rho v_0{}^2 = \tfrac{1}{2}(1.29\ \text{kg/m}^3)(33\ \text{m/sec})^2 = 700\ \text{N/m}^2.$$

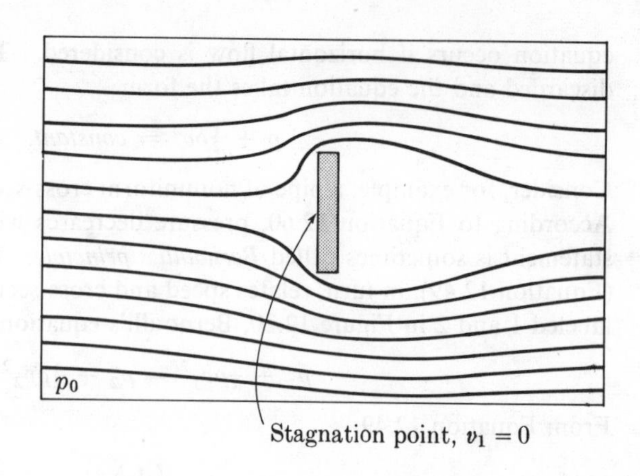

FIGURE 12.21 Flow of air around a disk. If the speed of flow is much less than the speed of sound, the density of the air changes little, and the approximation of incompressible flow is valid.

Dynamic pressure $= \frac{1}{2}\rho v^2$

The ratio of this pressure change to normal air pressure (10^5 N/m^2) is 7×10^{-3}, or 0.7 percent. The quantity $\frac{1}{2}\rho v^2$, which is kinetic energy per unit volume, is also known as the "dynamic pressure." It is comparable in magnitude to normal pressure p_0 in the so-called transonic region when the speed of flow is comparable to the speed of sound. In the supersonic region, the dynamic pressure exceeds the normal pressure. Even though Bernoulli's equation is no longer valid in these regions of high-speed gas flow, the concept of dynamic pressure remains useful.

4. *The Pitot Tube.* Equation 12.65 provides the basis for understanding the action of the pitot tube (Figure 12.22), which is used for measuring the airspeed

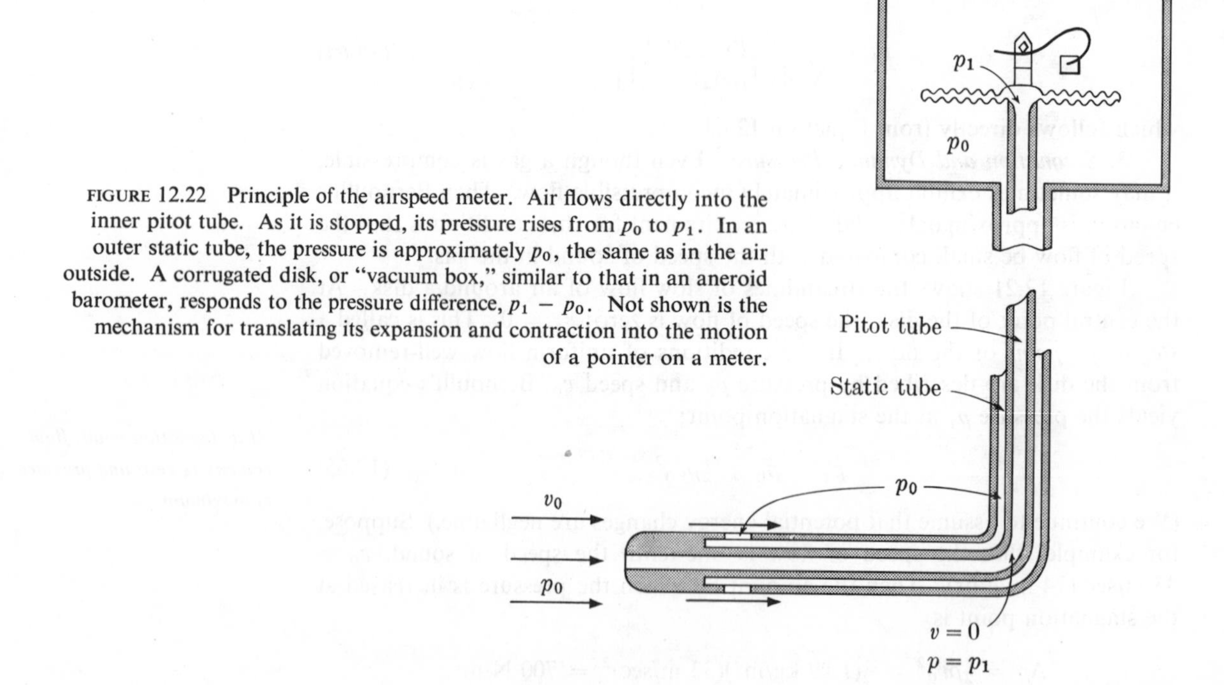

FIGURE 12.22 Principle of the airspeed meter. Air flows directly into the inner pitot tube. As it is stopped, its pressure rises from p_0 to p_1. In an outer static tube, the pressure is approximately p_0, the same as in the air outside. A corrugated disk, or "vacuum box," similar to that in an aneroid barometer, responds to the pressure difference, $p_1 - p_0$. Not shown is the mechanism for translating its expansion and contraction into the motion of a pointer on a meter.

of subsonic aircraft. Within the pitot tube, air is brought to rest (relative to the airplane). There its pressure is p_1. Air flowing around the outside of the pitot tube with little change of speed has pressure p_0. The pressure difference $p_1 - p_0$ is measured, and a meter is calibrated to read in proportion to the square root of this pressure difference. The relative speed of the air and the airplane is, from Equation 12.65,

Stagnation pressure minus free-stream pressure determines speed

$$v_0 = \sqrt{\frac{2(p_1 - p_0)}{\rho}}. \qquad (12.66)$$

The calibration of the airspeed meter is such that it reads correctly only when the air density is about 1.2 kg/m^3, corresponding to sea level at 20 °C. At high altitude, where the density ρ is less, Equation 12.65 shows that a given speed v_0 produces a smaller pressure difference $p_1 - p_0$. This means that the reading of the airspeed meter at high altitude is less than the actual airspeed. The correction amounts to about 2 percent per thousand feet of altitude. Fortunately for safe flying, the lift on an airplane wing is proportional to ρv_0^2, just as is the pressure difference $p_1 - p_0$. If an airplane should be landed at 70 mile/hr at sea level, it might require 80 mile/hr at a certain high-altitude field. The extra speed required is exactly the same as the error of the airspeed meter. The pilot approaches either field with the same indicated airspeed, 70 mile/hr. At the higher field, his actual speed is greater, just enough greater so that his wings develop the same lift as at the lower field.

12.5 Frictional air drag

Any fluid exerts a frictional force on an object moving through it. For very small objects moving slowly—a grain of dust sinking in water, for instance, or a tiny droplet of water falling through air—this frictional force is proportional to the speed of the object. For larger objects moving faster—airplanes, baseballs, hailstones, skydivers—the frictional force varies more nearly in proportion to the square of the speed. This drag force, as it is called, may be written

An approximate law of drag force

$$F_D = \tfrac{1}{2} C_D A \rho v^2. \qquad (12.67)$$

In this formula, v is the speed of the object, A is its cross-sectional area, ρ is the density of the fluid, and C_D is a dimensionless quantity called the *drag coefficient*. For objects of a particular shape, such as spheres, the drag coefficient is nearly a constant over a rather wide range of sizes and speeds. In what follows, we shall assume that C_D is exactly a constant.

Formula 12.67 is of obvious practical importance for air travel. In steady cruising flight, the drag force on an airplane is equal in magnitude and opposite in direction to the propulsive force. Since the drag force is proportional to v^2, a twofold increase of cruising speed requires a fourfold increase of propulsive force. The power expended (force times speed) is then increased eightfold. According to Formula 12.67, the drag force decreases as the air density decreases. This is one reason why jets fly at high altitudes.

Another example of interest is the vertical fall of a skydiver (Figure 12.23). Acting downward on him is the gravitational force $\mathbf{F}_g$; acting upward is the

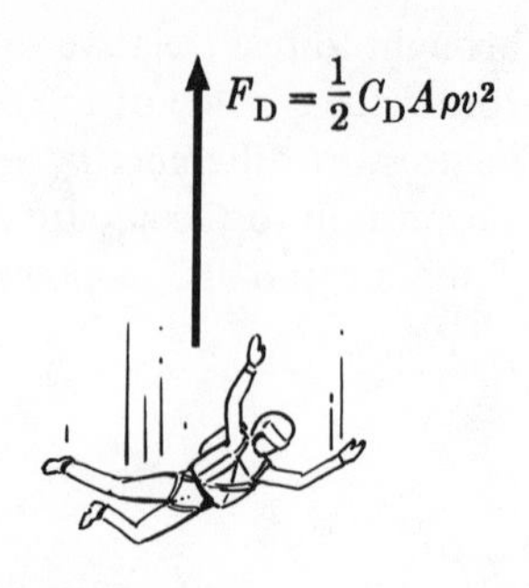

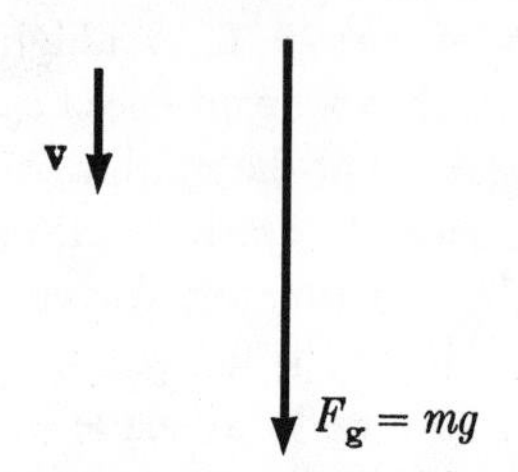

FIGURE 12.23 In "free fall," a skydiver is acted upon by a gravitational force and a force of frictional air drag. When he reaches terminal speed, these forces sum to zero.

drag force $\mathbf{F}_D$. If we measure x downward, the x component of force is $F_x = F_g - F_D$, or

$$F_x = mg - \tfrac{1}{2}C_D A \rho v^2. \tag{12.68}$$

The terminal speed v_T is the speed for which $F_x = 0$. It is

Terminal speed

$$v_T = \sqrt{\frac{2mg}{C_D A \rho}}. \tag{12.69}$$

Expressed in terms of v_T, Equation 12.68 takes on a simpler appearance:

$$F_x = mg\left(1 - \frac{v^2}{v_T^2}\right). \tag{12.70}$$

Since $F_x = ma_x$, the net downward acceleration is

Acceleration in downward fall

$$a_x = g\left(1 - \frac{v^2}{v_T^2}\right). \tag{12.71}$$

From this equation—which is a differential equation, since $a_x = dv/dt = d^2x/dt^2$—one can find the speed as a function of distance or of time, or the distance as a function of time. The quantity probably of greatest interest to the skydiver is speed as a function of distance, $v(x)$. This is also the easiest function to extract from Equation 12.71. We use the identity

$$a_x = \frac{1}{2}\frac{d(v^2)}{dx}. \tag{12.72}$$

This relationship is derived as follows:

$$a_x = \frac{dv}{dt} = \frac{dv}{dx}\cdot\frac{dx}{dt} = v\frac{dv}{dx} = \frac{1}{2}\frac{d(v^2)}{dx}.$$

Equation 12.71 can therefore be written

$$\frac{1}{2}\frac{d(v^2)}{dx} = g\left(1 - \frac{v^2}{v_T^2}\right),$$

which is equivalent to

$$\frac{d(v^2)}{1 - \dfrac{v^2}{v_T^2}} = 2g\,dx. \tag{12.73}$$

We make a change of variable, defining

$$u = \frac{v^2}{v_T^2}, \tag{12.74}$$

and integrate both sides of Equation 12.73:

$$\int \frac{du}{1 - u} = \frac{2g}{v_T^2}\int dx.$$

The integration gives

$$\ln(1 - u) = -\frac{2gx}{v_T^2}. \tag{12.75}$$

Here the constants of integration are chosen such that $u = v = 0$ when $x = 0$. On the left we reintroduce $u = v^2/v_T^2$, and on the right we define a characteristic distance x_c by

$$x_c = \frac{v_T^2}{2g} = \frac{m}{C_D A \rho}. \tag{12.76}$$

With these changes, Equation 12.75 becomes

$$\ln\left(1 - \frac{v^2}{v_T^2}\right) = -\frac{x}{x_c}.$$

From this equation follows an expression for the function $v(x)$. It is best expressed in terms of v^2:

Speed vs distance in a fall from rest

$$v^2 = v_T^2(1 - e^{-x/x_c}). \tag{12.77}$$

The two constants appearing on the right, v_T and x_c, are defined by Equations 12.69 and 12.76. Figure 12.24 shows graphs of v^2 vs x and v vs x. It is left as an exercise to estimate for a skydiver the quantities v_T and x_c, which determine the scales of these graphs.*

■ EXAMPLE: What is the terminal speed in air of a Ping-Pong ball of mass 2.5 gm and cross-sectional area 11.1 cm^2? How far must it fall before its downward acceleration is reduced to $0.5g$? An average value for the drag

* Other functions that may be derived from Equation 12.71 are $v(t) = v_T \tanh(gt/v_T)$ and $x(t) = 2x_c \ln[\cosh(gt/v_T)]$. Tanh and cosh designate the hyperbolic tangent and hyperbolic cosine functions.

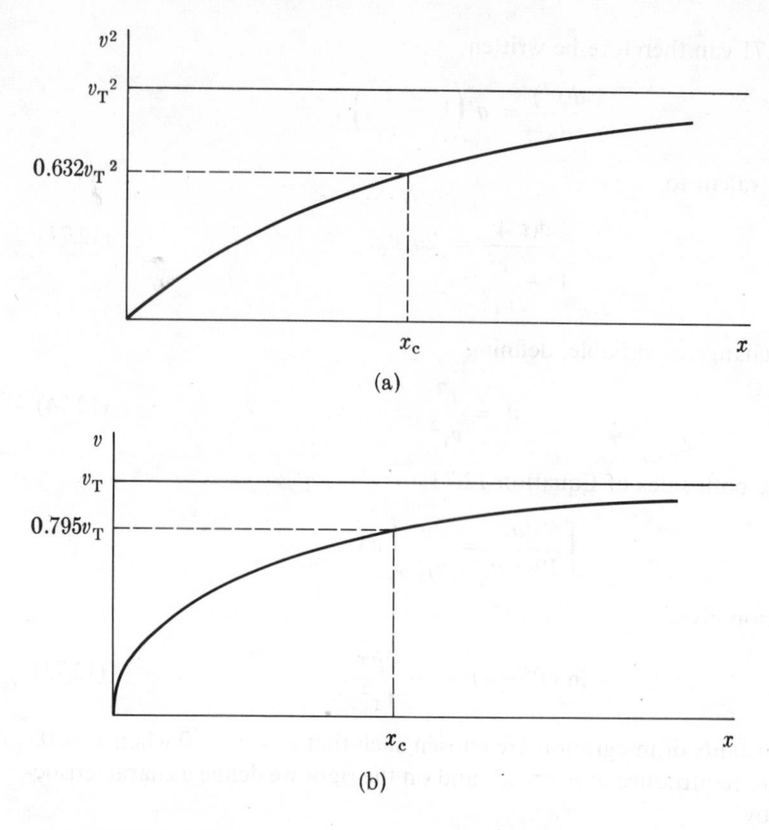

FIGURE 12.24 Graphs of (a) v^2 vs x and (b) v vs x for fall from rest, if the drag force is proportional to v^2 (Equation 12.67).

coefficient of a smooth sphere is $C_D = 0.45$, and for the density of air, we may use $\rho = 1.20 \times 10^{-3}$ gm/cm³. Using numbers expressed in cgs units in Equation 12.69, we have

$$v_T = \sqrt{\frac{(2)(2.5 \text{ gm})(980 \text{ cm/sec}^2)}{(0.45)(11.1 \text{ cm}^2)(1.20 \times 10^{-3} \text{ gm/cm}^3)}}$$

$$= \sqrt{81.7 \times 10^4 \text{ cm}^2/\text{sec}^2} = 904 \text{ cm/sec} = 9.04 \text{ m/sec}.$$

This is a speed of about 20 mile/hr. To answer the second question, we look back at Equation 12.71 and see that $a_x = 0.5g$ when $v^2 = 0.5v_T{}^2$. This occurs, according to Equation 12.77, when

$$e^{-x/x_c} = 0.5,$$

a condition satisfied by $x = 0.693x_c$. From Equation 12.76 it follows that

$$x_c = \frac{v_T{}^2}{2g} = \frac{81.7 \times 10^4 \text{ cm}^2/\text{sec}^2}{(2)(9.8 \times 10^2 \text{ cm/sec})}$$

$$= 417 \text{ cm} = 4.17 \text{ m}.$$

The distance of fall to reach an acceleration of $0.5g$ is, therefore,

$$x = 0.693x_c = 289 \text{ cm} = 2.89 \text{ m}.$$

This is a distance of 9.5 ft. ■

For a tiny sphere moving slowly through a fluid, the law of drag force has a form quite different from Equation 12.67. It is

Stokes's law: drag force for low speed and small size

$$F_D = 6\pi\mu r v. \tag{12.78}$$

This formula, called Stokes's law, states that the drag force is directly proportional to the radius r of the sphere and to its speed v. The quantity μ is called the *viscosity* of the fluid. It is expressed in kg/m sec or in gm/cm sec, the latter combination being called *poise*. Viscosity is a measure of the shear stress that can exist in a fluid (its exact definition can be found in more advanced texts). Viscosity is also a measure of the frictional force acting on a fluid flowing over a solid surface, and that is the reason for its appearance in Stokes's law.

How small or how slow must an object be in order that the drag force be given by Equation 12.78 instead of Equation 12.67? The answer is provided approximately by the numerical value of a parameter called the *Reynolds number*. The Reynolds number is defined by

The Reynolds number, a useful parameter for fluid flow

$$R = \frac{\rho v d}{\mu}, \tag{12.79}$$

where ρ is the density of the fluid, μ is its viscosity, v is the speed of flow past an object (or the speed of the object through the fluid), and d is the diameter of the object. For an irregularly shaped object, d can be given only approximately; for a sphere, it is well-defined. Different values of the Reynolds number R determine different regimes of flow in which different laws of drag force are applicable. These different regimes are summarized in Table 12.1.

TABLE 12.1 REGIMES OF FLUID FLOW

Reynolds Number	Law of Drag Force
0 to 10	Stokes's law, Equation 12.78: $F_D \sim v$
10 to 300	Transition region
3×10^2 to 3×10^5	Approximately constant drag coefficient, Equation 12.67: $F_D \sim v^2$
Greater than 3×10^5	Some variation of drag coefficient: F_D only roughly proportional to v^2

12.6 Two-body collisions

From billiard balls to elementary particles, two-body collisions make their appearance in many realms of physics. In this section we consider two examples: the perfectly elastic collision of equally massive particles and the completely inelastic collision of two bodies.

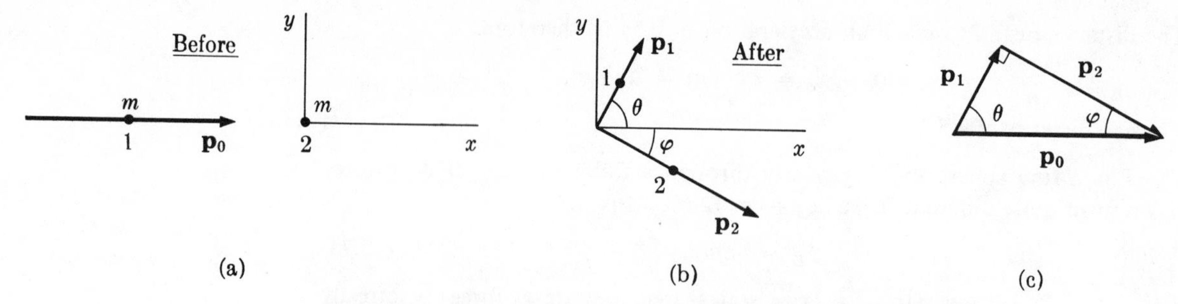

FIGURE 12.25 (a) A particle of mass m approaches a stationary particle of the same mass. (b) After their collision, the particles fly at angles θ and φ to the original direction of particle 1. (c) If energy as well as momentum is conserved, the final momentum vectors $\mathbf{p}_1$ and $\mathbf{p}_2$ are perpendicular, so that $\theta + \varphi = \frac{1}{2}\pi$.

ELASTIC COLLISION OF EQUALLY MASSIVE PARTICLES: LABORATORY FRAME

Consider two particles of equal mass m, the first moving with momentum p_0 toward the second, which is at rest [Figure 12.25(a)]. After their collision, particle 1 flies off with momentum $\mathbf{p}_1$ at angle θ to the initial flight direction, and particle 2 recoils with momentum $\mathbf{p}_2$ at angle φ to the initial flight direction [Figure 12.25(b)]. The most interesting aspect of such a collision, if it is elastic, is that $\mathbf{p}_1$ and $\mathbf{p}_2$ are always at right angles to one another. Let us prove this result. The equation of momentum conservation for this example is

$$\mathbf{p}_1 + \mathbf{p}_2 = \mathbf{p}_0. \tag{12.80}$$

Conservation laws for an elastic collision

Since an elastic collision is, by definition, one in which kinetic energy is conserved, the equation of energy conservation is

$$\frac{p_1^{\ 2}}{2m} + \frac{p_2^{\ 2}}{2m} = \frac{p_0^{\ 2}}{2m}. \tag{12.81}$$

For convenience, we use $p^2/2m$ rather than $\frac{1}{2}mv^2$ as the expression for kinetic energy. The factors of $2m$ cancel, leaving

$$p_1^{\ 2} + p_2^{\ 2} = p_0^{\ 2}. \tag{12.82}$$

Equation 12.80 can be represented by the arrow diagram of Figure 12.25(c). The arrows representing $\mathbf{p}_1$, $\mathbf{p}_2$, and $\mathbf{p}_0$ form the sides of a triangle. According to Equation 12.82, the sum of the squares of two sides of this triangle is equal to the square of the third side. This is the Pythagorean theorem. Therefore the triangle is a right triangle: $\mathbf{p}_1$ and $\mathbf{p}_2$ are perpendicular.

Also of interest is the change of kinetic energy of particle 1 in the collision. The geometry of Figure 12.25(c) shows that

$$p_1 = p_0 \cos\theta. \tag{12.83}$$

Squaring both sides of this equation and dividing by $2m$ leads to a relation for kinetic energies:

$$K_1 = K_0 \cos^2\theta. \tag{12.84}$$

For $\theta = 0$, particle 1 is undeflected and loses no energy ($K_1 = K_0$). For $\theta = 45$ deg, $\cos^2\theta = 0.5$, and $K_1 = 0.5K_0$. Then particle 1 emerges with half its initial energy. The other half is given to the recoiling particle. As θ approaches $\pi/2$, according to Equation 12.84, K_1 approaches zero. For an *almost* head-on collision, the incident particle is deflected through 90 deg but is left with negligible kinetic energy. For an *exactly* head-on collision, the angle of deflection actually has no meaning because particle 1 is stopped and particle 2 is propelled straight ahead ($\varphi = 0$), with $K_2 = K_0$. Note that $\theta + \varphi = \pi/2$. As the angle of deflection θ grows from 0 to $\frac{1}{2}\pi$, the angle of recoil φ diminishes from $\frac{1}{2}\pi$ to 0. The *maximum* possible value of both θ and φ is $\frac{1}{2}\pi$. This follows from the fact that they are interior angles of a right triangle [Figure 12.25(c)]. Finally, we note that a formula for the *change* of kinetic energy of particle 1 is

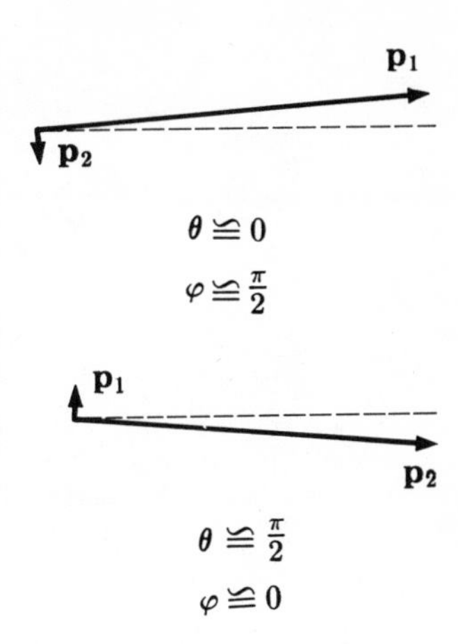

$$\Delta K = K_0 \sin^2\theta, \tag{12.85}$$

where $\Delta K = K_0 - K_1$. This is a simple consequence of Equation 12.84. How is ΔK related to the energy K_2 of the recoiling particle?

★ELASTIC COLLISION OF EQUALLY MASSIVE PARTICLES: CENTER-OF-MASS FRAME

Additional insight is gained by examining the elastic collision of equally massive particles in the center-of-mass frame of reference. Since the center of mass lies halfway between the two particles, its speed before the collision is

$$v_c = \tfrac{1}{2}v_0. \tag{12.86}$$

This must, of course, also be its speed after the collision. Figure 12.26 compares the before and after views of the collision in the laboratory frame and the center-of-mass frame. In the center-of-mass frame, the two particles approach each other with equal speed v, the same as the center-of-mass speed. Since the total momentum in this frame is zero, the particles must also separate with equal speed along a straight line. Energy conservation requires that this separation speed after the collision be the same as the approach speed v before the collision. Thus, as is evident in Figure 12.26, the description of the two-body elastic collision is particularly simple in the center-of-mass frame of reference.

In the center-of-mass frame, no kinetic energy is exchanged

Any particle velocity $\mathbf{v}_L$ in the laboratory frame is equal to the vector sum of the particle velocity $\mathbf{v}_p$ in the center-of-mass frame and the velocity $\mathbf{v}_c$ of the center of mass:

Velocity transformation between frames of reference

$$\mathbf{v}_L = \mathbf{v}_p + \mathbf{v}_c. \tag{12.87}$$

Applied to the example under discussion, this rule means that the transformation from the center-of-mass frame to the laboratory frame is achieved by adding to each velocity in the center-of-mass frame a velocity $\mathbf{v}_c$ of magnitude v, directed to the right. Consider, for instance, the lower left portion of Figure 12.26. Adding the velocity $\mathbf{v}_c$ to the velocity of particle 1 gives $\mathbf{v}_0$, directed to the right, with magnitude $2v$. Adding $\mathbf{v}_c$ to the velocity of particle 2 gives zero. In short, this transformation changes the lower left portion of the figure into the upper left portion of the figure.

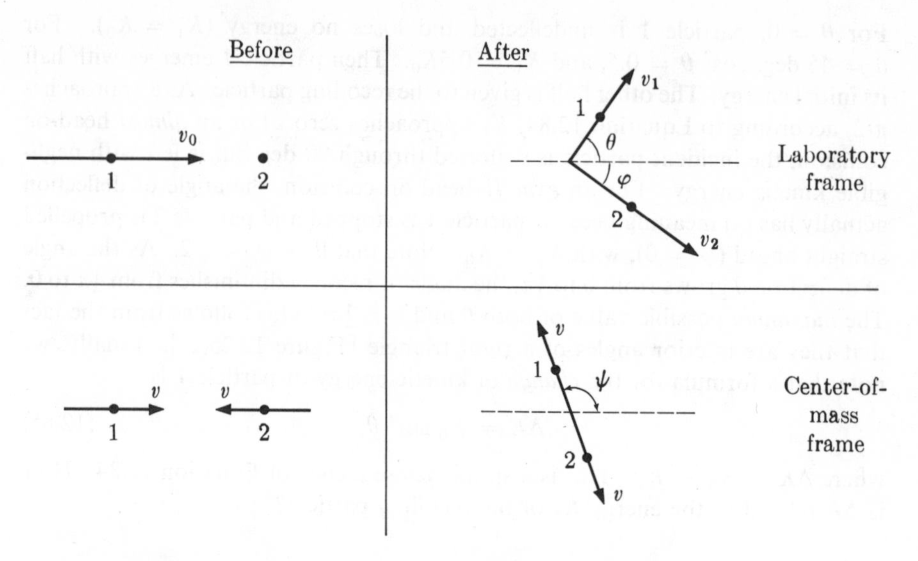

FIGURE 12.26 Elastic collision of equally massive particles. The upper diagrams (analogous to Figure 12.25) show before and after views of the collision in the laboratory frame of reference. The lower diagrams show before and after views of the same collision in the center-of-mass frame.

The same rule of vector addition converts the lower right portion of Figure 12.26 to the upper right portion. The geometry of this transformation is shown in Figure 12.27. Consider first the triangle ABC. The side AB, of length v, represents the final velocity of particle 1 in the center-of-mass frame. The side BC, also of length v, represents the center-of-mass velocity. The sum of these two velocities, represented by the side AC, is the final velocity $\mathbf{v}_1$ of particle 1 in the laboratory frame. Since ψ has been defined to be the deflection angle of particle 1 in the center-of-mass frame (Figure 12.26), the angle ABC

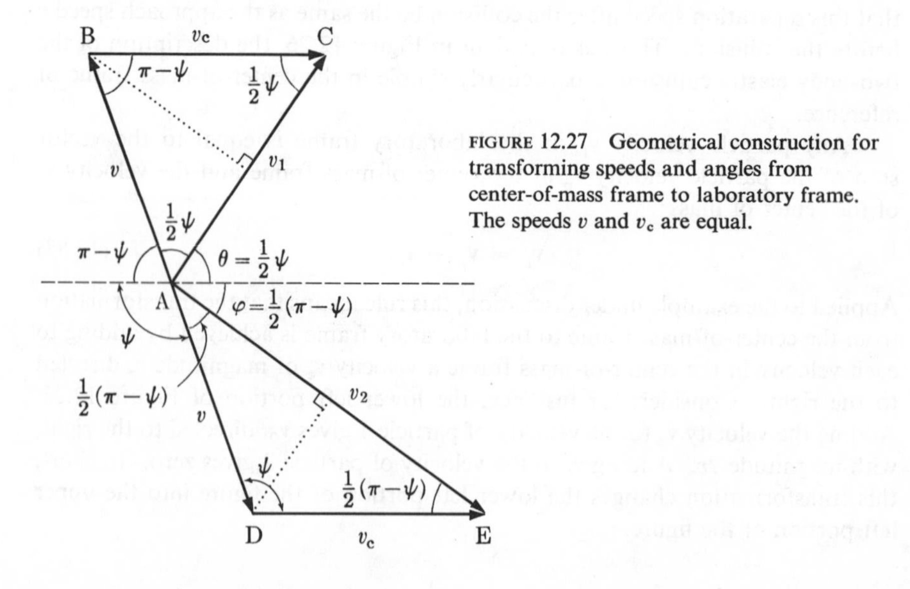

FIGURE 12.27 Geometrical construction for transforming speeds and angles from center-of-mass frame to laboratory frame. The speeds v and v_c are equal.

is equal to $\pi - \psi$. The other two angles of the isosceles triangle, BAC and BCA, are each equal to $\frac{1}{2}\psi$. Finally, the angle of side AC (or of $\mathbf{v}_1$) with respect to the horizontal direction is

$$\theta = \tfrac{1}{2}\psi. \tag{12.88}$$

The deflection angle of particle 1 in the laboratory frame (θ) is just half its deflection angle in the center-of-mass frame (ψ). Similar geometrical reasoning applied to triangle ADE leads to the result

$$\varphi = \tfrac{1}{2}(\pi - \psi). \tag{12.89}$$

The recoil angle of particle 2 in the laboratory frame (φ) is also half its recoil angle in the center-of-mass frame ($\pi - \psi$). The angle between final particle velocities is 180 deg in the center-of-mass frame. As deduced earlier, it is 90 deg in the laboratory frame.

Finally, we can relate the magnitudes of the final velocities in the two frames of reference. Each dotted line in Figure 12.27 divides an isosceles triangle into two right triangles. Consideration of these right triangles leads to the relations

$$v_1 = 2v \cos(\tfrac{1}{2}\psi),$$

$$v_2 = 2v \cos(\tfrac{1}{2}\pi - \tfrac{1}{2}\psi).$$

With the help of Equations 12.88 and 12.89, this pair of equations may also be written

$$v_1 = v_0 \cos\theta, \tag{12.90}$$

$$v_2 = v_0 \cos\varphi. \tag{12.91}$$

Here we have made use of the equality

$$v_0 = 2v. \tag{12.92}$$

COMPLETELY INELASTIC COLLISION OF TWO BODIES

We can call a collision "completely" inelastic if during the collision, the maximum possible fraction of the kinetic energy is transformed into other forms of energy. A ball of putty striking and sticking to another ball of putty is an example of such a collision; kinetic energy is transformed to heat. The collision of two elementary particles can also be completely inelastic if new particles are created and if the greatest possible transformation of kinetic energy to mass energy occurs.

Consider a collision of this type in which an object of mass m_1 and speed v_1 strikes an object of mass m_2 initially at rest (Figure 12.28). We suppose first that the increase of mass of the system is negligible, so the product of the collision is an object of mass $M = m_1 + m_2$. We call its speed v_f. The conservation of momentum, expressed by

$$m_1 v_1 = M v_f,$$

leads to the expression for the final speed,

$$v_f = \frac{m_1}{M} v_1. \tag{12.93}$$

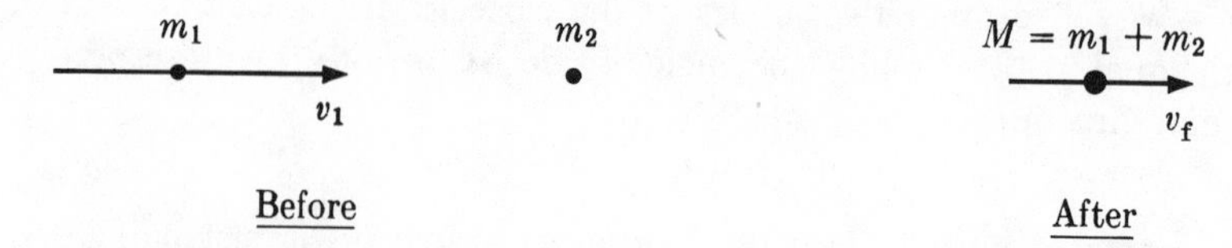

FIGURE 12.28 A completely inelastic collision. Momentum is conserved, but mechanical energy is not. (See also Figure 8.17).

This result was obtained previously (Equation 8.59). Now we are interested in the energy transformed. The change of kinetic energy is

Kinetic energy is lost; other forms of energy must gain

$$\Delta K = K_1 - K_f = \tfrac{1}{2}m_1v_1^2 - \tfrac{1}{2}Mv_f^2. \tag{12.94}$$

Substituting the right side of Equation 12.93 for v_f gives, after simplification,

$$\Delta K = \frac{m_1 m_2 v_1^2}{2M}. \tag{12.95}$$

This result may be expressed even more simply as

Available energy

$$\Delta K = \frac{m_2}{M} \cdot K_1. \tag{12.96}$$

The quantity ΔK is called the "available energy." It is that part of the initial kinetic energy that may be transformed into other kinds of energy. If most of the mass M belongs to the incident object (a freight train striking an automobile, an alpha particle striking an electron), the ratio m_2/M is much less than 1, and the available energy ΔK is much less than the initial kinetic energy K_1. Little of the energy can be transformed. If the reverse situation holds true (a snowball hitting a man, an electron hitting a nucleus), the ratio m_2/M is nearly equal to 1, and ΔK is nearly equal to K_1. Most of the initial kinetic energy is available for transformation. For the collision of equally massive particles, the ratio m_2/M is equal to $\frac{1}{2}$, and $\Delta K = \frac{1}{2}K_1$.

The easiest way to prove that the quantity ΔK is indeed the *maximum* energy available for transformation is to look at a collision in the center-of-mass frame of reference. Figure 12.29 shows the center-of-mass view of the collision whose laboratory view appears in Figure 12.28. From the general definition of the center-of-mass velocity (Equation 8.18), it follows that the center of mass moves to the right in the laboratory with velocity

$$\mathbf{v}_c = \frac{m_1}{M}\mathbf{v}_1. \tag{12.97}$$

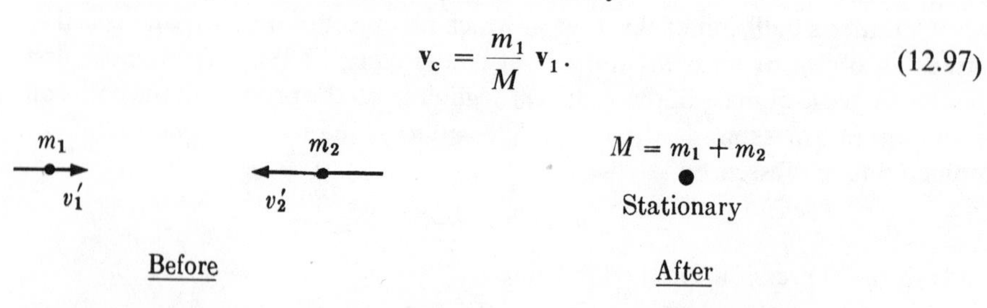

FIGURE 12.29 A completely inelastic collision in the center-of-mass frame. The kinetic energy after the collision is zero.

The velocities of the two objects in the center-of-mass frame are

$$\mathbf{v}_1' = \mathbf{v}_1 - \mathbf{v}_c = \frac{m_2}{M}\mathbf{v}_1, \tag{12.98}$$

$$\mathbf{v}_2' = 0 - \mathbf{v}_c = -\frac{m_1}{M}\mathbf{v}_1. \tag{12.99}$$

The total kinetic energy in this frame of reference is

$$K_c = K_1' + K_2' = \tfrac{1}{2}m_1 v_1'^2 + \tfrac{1}{2}m_2 v_2'^2. \tag{12.100}$$

Substituting from Equations 12.98 and 12.99 in Equation 12.100 gives

$$K_c = \tfrac{1}{2}m_1 v_1^2 \left(\frac{m_2}{M}\right). \tag{12.101}$$

In the center-of-mass frame, the total momentum is zero: $m_1\mathbf{v}_1' + m_2\mathbf{v}_2' = 0$. Therefore, the products of the collision in this frame need not move. The final kinetic energy may be zero. If it is, *all* of the initial kinetic energy is available energy, and

In the center-of-mass frame, all kinetic energy is available energy

$$\Delta K = K_c. \tag{12.102}$$

Equations 12.102 and 12.101 are indeed equivalent to Equation 12.96. The available energy is the center-of-mass kinetic energy.

In the derivation above, we set $M = m_1 + m_2$, an assumption of mass conservation. For the production of new mass in particle collisions, we should instead use $M = m_1 + m_2 + \Delta M$, where ΔM is the net increase in mass. However, since we use nonrelativistic formulas ($\mathbf{p} = m\mathbf{v}$ and $K = \frac{1}{2}mv^2$), our derivation is valid only at relatively low energies; then the new mass ΔM is small compared with the original mass, $m_1 + m_2$. Consequently, Equations 12.96 and 12.102 for available energy remain nearly correct for all non relativistic collisions even if new mass is created.

Near the speed of light these principles remain valid, but the formulas are changed. In particular, for very high energy, the available energy in a proton-proton collision ($m_1 = m_2 = \frac{1}{2}M$) is not half of the initial laboratory kinetic energy but less than half. In the 6-GeV accelerator in Berkeley, California, the available energy in such a collision is about 2 GeV. At the National Accelerator Laboratory, a 500-GeV proton striking another proton yields up only 29 GeV as available energy. Since accelerators are very costly devices, this "waste" of energy is unfortunate. It can be avoided if the colliding particles are both in motion and have equal and opposite velocity before the collision, as in Figure 12.29. Then the center-of-mass frame and the laboratory frame are the same, and *all* of the kinetic energy is available energy. To take advantage of this fact, several laboratories have developed "colliding beams" (see Figure 12.30), in which oppositely directed particles are focused into a narrow space where they undergo collision and can give up any fraction (or all) of their kinetic energy.*

Colliding beams increase the available energy

* See Gerard K. O'Neill, "Particle Storage Rings," *Scientific American*, November, 1966.

FIGURE 12.30 Colliding beams. Protons from the CERN 28-GeV accelerator (in Geneva, Switzerland) are guided magnetically into two intersecting storage rings. This photograph shows one of the intersection points. At the crossing point of the two beam tubes, protons can strike other protons nearly head on. The available energy in such a collision is more than 50 GeV. [Photograph courtesy of European Organization for Nuclear Research (CERN).]

The principles of mechanics invade every realm of science and engineering. In this chapter we have considered but a tiny fraction of the applications of mechanics. Yet such is the power of basic science, you have, in studying these chapters, gained understanding and control over a whole range of applications that have not been mentioned here and that you may not have even thought about; some of them no one has yet thought about.

Summary of ideas and definitions

FRICTION

Surface friction is, by definition, a force acting parallel to a surface of contact.

For static contact, the coefficient of static friction μ_s is defined as the ratio of the maximum frictional force to the normal force (Equation 12.4).

For sliding contact, the coefficient of kinetic friction μ_k is defined as the ratio of the frictional force to the normal force (Equation 12.10).

For objects of ordinary size and speed moving through air, the frictional force, or drag force, is proportional to the cross-sectional area of the object and to the square of its speed (Equation 12.67).

For a falling body, terminal speed is achieved when the drag force and the gravitational force are equal in magnitude (Equation 12.69).

The kinetic energy of a falling body is expressed in terms of an exponential function of the distance of fall (Equation 12.77).

For microscopic bodies moving through air, the drag force follows Stokes's law (Equation 12.78); it is proportional to the diameter of the body and to its speed.

STATICS

If a rigid body is in static equilibrium, the total force and the total torque acting on it must both be zero.

The gravitational force acting on the parts of a system near the earth are equivalent to a single force acting at the center of mass of the system.

In solving a statics problem, the choice of origin is arbitrary; a well-chosen origin may make the problem easier.

FLUIDS

In fluids, forces of tension and shear can often be ignored; only compressional forces, or pressure forces, then need to be considered.

Pressure is a scalar quantity defined as the compressional force per unit area in a fluid.

For a fluid in static equilibrium, the pressure as a function of height is governed by a simple differential equation,

$$\frac{dp}{dy} = -\rho g. \tag{12.29}$$

For an incompressible fluid in equilibrium, pressure is a linear function of height (Equation 12.30 or 12.32).

For a compressible fluid whose density is proportional to pressure, pressure is an exponential function of height (Equation 12.39 or 12.41).

One bar is defined as a pressure of 10^5 N/m^2, which is nearly equal to 1 atm.

Archimedes' principle: The upward buoyant force on an object partially or wholly submerged in a fluid is equal to the weight of the displaced fluid.

In steady flow, mass flux is constant; this gives the equation of continuity,

$$\rho A v = \textit{constant}. \tag{12.47}$$

In steady *incompressible* flow, the work done by pressure forces changes the mechanical energy (kinetic plus potential) of the fluid; this energy balance leads to Bernoulli's equation,

$$p + \tfrac{1}{2}\rho v^2 + \rho g z = \textit{constant}. \tag{12.59}$$

In horizontal incompressible flow, $p + \frac{1}{2}\rho v^2 = \textit{constant}$; then higher speed implies lower pressure, a relationship called Bernoulli's principle.

A point of zero velocity is called a stagnation point; it is a point of maximum pressure.

TWO-BODY COLLISIONS

An *elastic* collision is one in which total kinetic energy is conserved.

When a moving object strikes a stationary object of the same mass, the paths of the two objects after the collision are perpendicular.

In such a collision, the incident object can transfer any fraction of its energy to the recoiling object, depending on the angle of deflection (Equation 12.85).

In the center-of-mass frame, particle momenta are equal and opposite, both before and after the collision.

A completely *inelastic* collision is one in which the maximum possible fraction of the initial kinetic energy is transformed into other kinds of energy; this fraction is called the available energy.

If a moving particle of mass m_1 strikes a stationary particle of mass m_2, the ratio $m_2/(m_1 + m_2)$ gives the available fraction of the initial energy (Equation 12.96).

In the center-of-mass frame of reference, the total kinetic energy is available energy.

QUESTIONS

Section 12.1

Q12.1 (1) On a given surface, a sports car cannot exceed a certain acceleration, no matter what improvements are made in its engine. Why is this? (2) Suggest a kind of wheeled vehicle that might be free of this restriction.

Q12.2 Suppose that for a block on a plank the coefficient of kinetic friction were greater than the coefficient of static friction. Describe what would happen as the angle of the plank to the horizontal is gradually increased from zero. In particular, what happens when the angle reaches arc tan μ_s and arc tan μ_k?

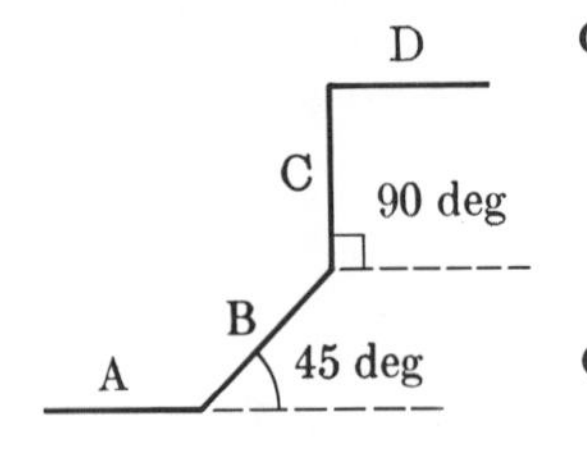

Q12.3 In the figure, A represents a valley floor, B and C represent the wall of a mesa (or bluff), and D represents the top of the mesa. (1) Which part of the profile is most likely to consist of loose material? Which part is most likely to consist of hard material? (2) For which part(s) of the profile can a coefficient of static friction be calculated? What is it?

Q12.4 Piles of loose material, such as sand or sawdust or wheat, are observed to display characteristic and reproducible "angles of repose" if the piles are fed slowly from the top. Discuss carefully what conclusions can be drawn from this fact. Does it prove that the maximum frictional force on a piece of the material is proportional to the perpendicular (normal) force on the piece (Equation 12.4)?

Q12.5 Name two effects that might cause a pile of loose material to have sides that slope at less than the critical angle θ_c.

Q12.6 If you wished to calculate the maximum horizontal component of force that a rolling tire could exert on a pavement without slipping, would you use the coefficient of static friction or the coefficient of kinetic friction? Why?

Q12.7 Suggest a definition of a coefficient of rolling friction. In general, why is such a coefficient likely to be much less than coefficients of static friction and kinetic friction?

Q12.8 If a block with small surface irregularities is removed from a metal plate and then polished and returned to the plate, its coefficient of friction will have *increased.* Why?

Q12.9 Microscopic theories of friction take into account surface *roughness* (analogous to stones and ruts impeding the progress of a car on a country road) and surface *adhesion* (analogous to soft tar impeding the progress of a car on a smooth road). In an outside reference read more about friction, and report on the relative importance of these two contributors to surface friction.

Section 12.2

Q12.10 The phrases "center of gravity" and "center of mass" are often used interchangeably. Why?

Q12.11 (1) Why does a body in a uniform gravitational field, no matter what its shape or distribution of mass, experience no torque about its center of mass? (2) What is an important physical consequence of this fact?

Q12.12 If the bottom of a ladder rests on a frictionless floor, it always slips, no matter what the coefficient of friction at its point of contact with the wall. Explain why.

Q12.13 A secretary has a small spring-balance postal scale that weighs envelopes up to 4 oz. With this scale, she finds it easy to determine the weight of a large envelope weighing 12 oz or more if its center of mass is at the center of the envelope. With a little more effort, she can succeed in finding the weight of the large envelope even if she does not know in advance the location of its center of mass. Explain her technique.

Section 12.3

Q12.14 Pressure is force per unit area. Why is pressure often measured in units of length, such as millimeters of mercury?

Q12.15 (1) Why are liquids nearly incompressible? (2) Explain why no substance can ever be truly incompressible.

Q12.16 What property of motor oil is measured by its "weight"—10-weight, 20-weight, 30-weight, etc.?

Q12.17 Some liquids can exert a small negative pressure. What does this mean?

Q12.18 In a coasting spacecraft, a small amount of water in a glass forms itself into a sphere instead of conforming to the shape of the glass. Why?

Q12.19 Why is mercury used in barometers? What would be the disadvantages of a water barometer?

Q12.20 Explain the action of a soda straw. Is it more difficult to suck through a long straw than a short one? Does the ease of drinking through a straw depend on the density of the liquid? Answer in terms of the basic ideas of fluid mechanics.

Q12.21 An altimeter is a device that translates pressure measurements into altitude readings. (1) If an altimeter is carried down a mine shaft, will it correctly record the depth? (2) If an altimeter is lowered into a lake, will it correctly record the depth?

Q12.22 Painted on the sides of many ships are numbers that indicate the vertical distance downward to the keel of the ship. Explain how these numbers can be used to determine the total weight carried by the ship.

Q12.23 (1) Why does a fat person float in water more easily than a thin person? (2) Why is it easier to float in salt water than in fresh water? (3) Why does a slender person who can barely float in fresh water find it helpful to inhale? (4) Could you float in strong drink (half alcohol and half water)?

Section 12.4

Q12.24 Water flows through a horizontal pipe of circular cross section and variable diameter. How does its speed depend on the diameter of the pipe?

Q12.25 Can air flow with constant speed through a pipe of circular cross section and variable diameter? If not, why not? If so, what other conclusions can you draw about the properties of the air in the pipe?

Q12.26 On a certain mythical planet the familiar laws of physics are all valid with one exception—the mass of any object depends on its height above the surface of the planet. (1) On this planet, is steady flow of a fluid possible in a pipe that is not horizontal? (2) Does the equation $dm/dt = \rho A v$ correctly describe the mass flux at any point? (3) Is the equation $\rho A v = constant$ valid?

Q12.27 Why does the stream of water flowing from a kitchen faucet become narrower as it falls?

Q12.28 Under what conditions might a falling stream of water form a cylinder of constant cross-sectional area? Would such flow be incompressible? Would the equation of continuity apply to it?

Q12.29 Assume that the flow depicted in Figure 12.16(b) is steady incompressible flow. (1) Where is the speed greatest and where is it least? (2) Where is the pressure greatest and where is it least?

Q12.30 In principle it is possible to build an airspeed meter that would indicate true airspeed at all altitudes. What information would have to be supplied to the meter in addition to the pressure difference in a pitot tube? How might this additional information be acquired?

Q12.31 What is cavitation? What causes it? Does it occur for both gases and liquids? Does it invalidate either the equation of continuity or Bernoulli's equation?

Section 12.5

Q12.32 If the force of frictional air drag on a body were independent of its speed, would the body reach a terminal speed? Discuss the rate of dissipation of heat energy by such a body as it falls through the air.

Q12.33 Is the net force acting on a skydiver (Equation 12.70) a conservative force? Explain why it is or is not.

Q12.34 At a given inclination of a wing to its direction of motion through the air (its "angle of attack"), the ratio of the lift force to the drag force on the wing is approximately independent of the speed of the wing. Why?

Q12.35 Pairs of spheres are dropped from rest in air. For each of the three following pairs, identify the one that reaches the floor first: (a) spheres of equal mass and unequal radius; (b) spheres of equal radius and unequal mass; and (c) spheres of equal density, unequal in both mass and radius.

Section 12.6

Q12.36 A neutron moderator is a substance used to reduce the kinetic energy of neutrons. Why is paraffin a good moderator? Why is lead a poor moderator?

Q12.37 A certain particle of mass m is observed to have the same kinetic energy K_1 in two different frames of reference. What is the minimum relative speed of these two frames of reference? What is their maximum relative speed?

Q12.38 Since colliding beams offer such an advantage in increasing available energy, why are not all accelerators built with colliding beams?

EXERCISES

Section 12.1

E12.1 A pebble on a plank is characterized by a coefficient of static friction $\mu_s = 0.8$ and a coefficient of kinetic friction $\mu_k = 0.6$. The plank is tilted until the originally stationary pebble starts to slide. (1) What is the angle of the plank to the horizontal? (2) What is the acceleration of the pebble down the plank? (3) Use two methods to find the speed of the pebble after it slides 1 m along the plank: (a) work from its acceleration; (b) work from energy considerations.

E12.2 A 60-kg man and a 20-kg child are seated side by side on the rear seat of a golf cart; they have no foot support and nothing to hold onto. When the cart starts with acceleration $a = 0.3g$, neither falls off. When it starts with acceleration $a = 0.4g$, the child falls off and the man does not. What quantitative conclusions can you draw about the coefficient of friction of either person with the seat?

E12.3 A man wishes to pile sand in a square of area l^2 in his yard without having any of the sand spill out into the surrounding area. Show that the maximum volume of sand that he can accommodate is $\frac{1}{6}l^3\mu_s$, where μ_s is the coefficient of static friction for sand resting on sand.

E12.4 An entertainer sets a table with fine china and then, with a flourish, yanks the tablecloth out from under the dishes. Assume that the dishes on the cloth have coefficients of static and kinetic friction satisfying the inequality $\mu_s > \mu_k$. (1) With what initial acceleration must the entertainer pull the cloth? With what acceleration must he continue to pull it once it is in motion? (2) What role, if any, does the coefficient of friction between the cloth and the table play in this performance? (3) Why is an inexperienced person likely to have only partial success with this trick, pulling the tablecloth part way out and then having the dishes start to move?

E12.5 The figure shows a block of mass m that is attached to a spring and that can slide over a horizontal surface. When the block is moving, the horizontal components of force acting on it are the restoring force of the spring, $F_s = -kx$, and a constant force of friction, $F_f = \pm\mu mg$, the sign depending on the direction of motion. The block is pulled aside to $x = -x_0$ and then released. Using energy considerations, show that it slides to $x = x_0 - (2\mu mg/k)$ before reversing direction.

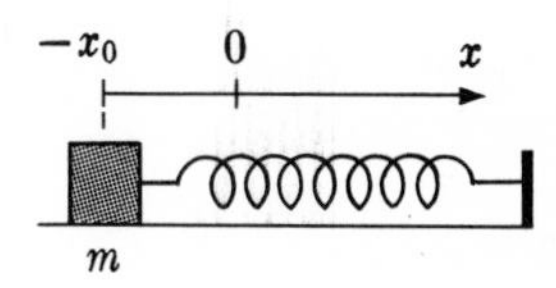

Section 12.2

E12.6 Four particles of equal mass m are arranged at the corners of a square of side l, as shown in the first figure. (1) Evaluate the gravitational torque on each particle with respect to the origin, and show that the sum of these torques is the same as if a single force of magnitude $4mg$ acted at the center of mass. (2) Repeat this demonstration using the center of mass as the reference point for calculating torque. (3) Now the square is rotated through 90 deg about the dashed line in the figure, giving the arrangement shown in the second figure. Repeat parts 1 and 2.

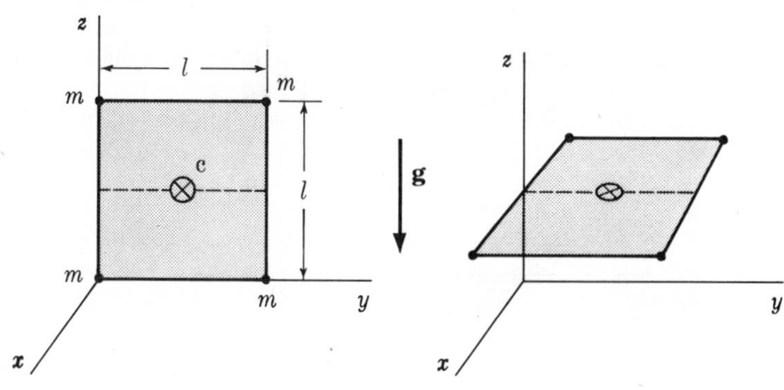

E12.7 Work from the conditions of static equilibrium of a rigid body to derive the formula for the mechanical advantage of a simple lever.

E12.8 Solve the door problem of Example 1 in Section 12.2, picking some point other than B in Figure 12.4 as the reference point for calculating torques.

E12.9 Solve the ladder problem of Example 2 in Section 12.2, picking some point other than the bottom of the ladder as the reference point for calculating torques.

E12.10 (1) Obtain an expression for the frictional force F_f exerted on the bottom of the ladder in Figure 12.5 as a function of the distance s that the man has climbed along the ladder. (2) If the coefficient of friction of the ladder on the floor is $\mu_s = 0.35$ and the angle of the ladder to the horizontal is $\theta = 60$ deg, how far can the man climb up the 3-m ladder before it slips?

E12.11 Solve the ladder problem of Example 2 in Section 12.2 if the mass of the ladder is $\frac{1}{4}M$ (M is the mass of the man).

E12.12 A man grasps a pole at one end with one hand and holds it horizontally. The mass of the pole is 1 kg and its length is 2 m. (1) What total force does the man exert on the pole? (2) What total torque does he exert on the pole? (3) If his action on the pole is approximated by two forces that are applied 10 cm apart and act in opposite directions, what are the magnitudes and directions of these two forces?

E12.13 An object of weight 2,000 N is suspended as shown in the figure. What are the magnitudes of the forces $\mathbf{F}_1$ and $\mathbf{F}_2$?

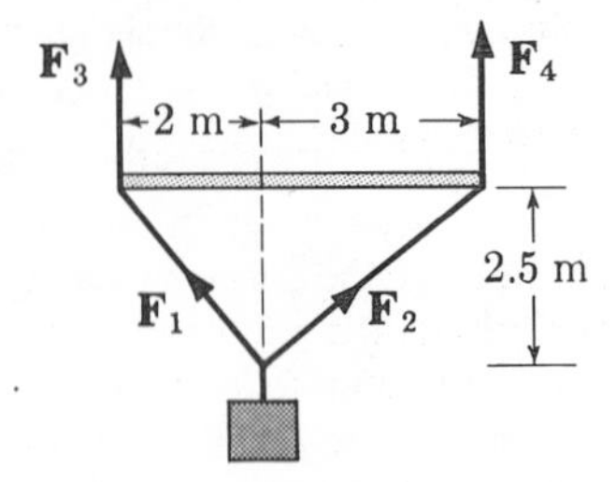

E12.14 The horizontal beam shown in the figure of the preceding exercise weighs 1,000 N; the ropes have negligible weight. What are the magnitudes of the forces $\mathbf{F}_3$ and $\mathbf{F}_4$?

E12.15 Timbers of length 2 m and 3.46 m that have the same mass per unit length are bolted together to make an "L." Resting on a horizontal surface, as shown in the figure, the structure is supported by vertical forces $\mathbf{F}_1$ and $\mathbf{F}_2$. What is the ratio F_1/F_2?

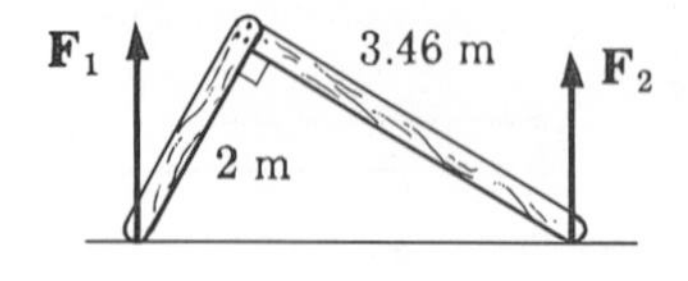

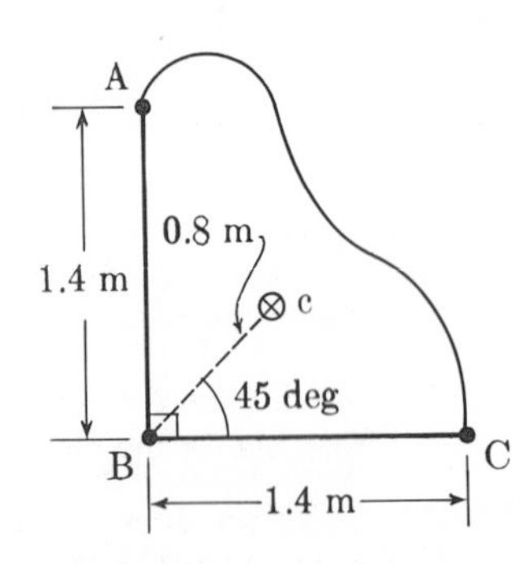

E12.16 Three men lift a grand piano at the points A, B, and C shown in the diagram. Also shown in the diagram is the location of the center of mass of the piano. If the weight of the piano is 3,500 N, what force does each man exert when the piano is held motionless off the floor?

E12.17 A flagpole is supported as shown in the figure. The mass of the pole (with its flag) is 20 kg; its center of mass is at the point shown in the figure. (1) Give the x and y components of the three forces $\mathbf{F}_g$, $\mathbf{F}_1$, and $\mathbf{F}_2$. (2) If the base of the pole is not attached to the building, what is the minimum coefficient of static friction of the pole against the building that will keep the pole from slipping?

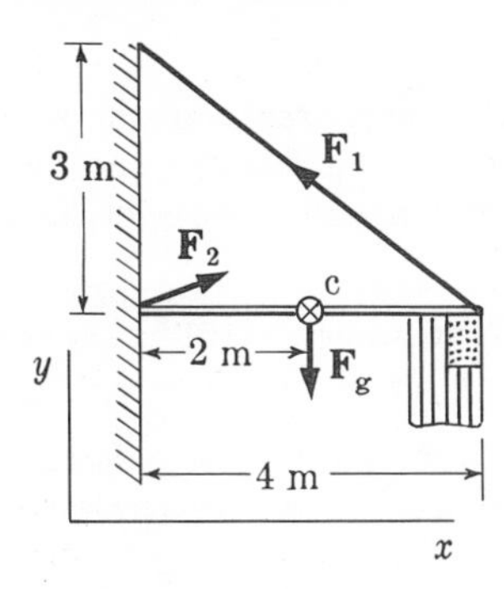

Section 12.3

E12.18 Water from a spring is carried through a pipe of length 100 m down a hill inclined at 20 deg to the horizontal. When the pipe is full but water is not flowing, what is the pressure at the lower end of the pipe?

E12.19 A piston pushes down on the top surface of a container of water with a pressure that increases gradually from 1 atm to 10 atm. (1) If the container is a circular cylinder of depth 1 m, what happens to the pressure at the bottom of the container? (2) If the depth of the container is 1 m and its area at the bottom is twice its area at the top, what happens to the pressure at the bottom?

E12.20 What is the theoretical maximum height to which water can be raised by a lift pump (a) from a well near sea level and (b) from a well at an elevation of 5,000 ft? (A lift pump "sucks" water from a well by reducing the air pressure in a pipe that extends into the water.)

E12.21 The water in an aquarium is 0.4 m deep and the length of the aquarium is 0.6 m. (1) What is the force exerted by the water on one side of the aquarium? (2) What is the oppositely directed force of the outside air on this side of the aquarium? (3) What is the total force of air and water on the side of the aquarium? Show that this total force would remain unchanged if the aquarium were transported to high altitude.

E12.22 (1) A man accustomed to living at sea level goes skiing at an elevation of 10,000 ft. By what factor is his oxygen intake per breath decreased? (2) If the pressure in a commercial airliner at 35,000 ft were to sink to the pressure of the outside air, how would the available oxygen per breath compare with that available at sea level?

E12.23 (1) Show that at altitudes less than a few kilometers, air pressure decreases approximately linearly with altitude according to the formula

$$p = p_0 - \alpha y,$$

where p_0 is sea-level pressure and y is altitude. Give an expression for α that includes the scale height b. (2) What is the percentage decrease of pressure per thousand feet of elevation at low altitudes?

E12.24 Within a few thousand feet of sea level, a certain altimeter is accurate to within 20 ft. With what accuracy, both absolute and relative, must it sense air pressure?

E12.25 (1) A piece of wood floats in water with 40 percent of its volume above the surface. What is the density of the wood? (2) Would this piece of wood float in gasoline? If so, approximately what percentage of its volume would be above the surface?

E12.26 The lower half of a pan contains water; its upper half contains oil of density 0.6 gm/cm^3. An object of density 0.9 gm/cm^3 is dropped into the pan. In equilibrium, what fraction of the volume of this object is below the oil-water interface?

E12.27 An object weighs 3 N in air and 2.5 N in water. (1) What is its density? (2) What is its weight when it is immersed in oil of density 600 kg/m^3?

E12.28 A 1-liter container open to the air in a laboratory room is found to weigh exactly 5 N. What is its weight under each of the following conditions? (1) The container is evacuated and then weighed in air. (2) Air at 1-atm pressure is readmitted to the container and it is weighed in a vacuum. (3) The container is evacuated and weighed in a vacuum.

E12.29 A certain hot-air balloon displaces 1,925 kg of air. At a particular moment, it contains 1,600 kg of heated air and the rest of its mass is 325 kg, so it experiences no net force. A little later, the mass of the air inside the balloon rises to 1,650 kg. Then what is the net force on the balloon (magnitude and direction)? What is the acceleration of the balloon? (NOTE: The air inside the balloon is part of the accelerating system.)

Section 12.4

E12.30 In steady flow, water flows from a pipe of three-quarter-inch diameter to a pipe of half-inch diameter. (1) By what factor does its speed change? (2) By what factor does its mass flux change?

E12.31 As steadily flowing air passes from a pipe of 1-ft diameter to a pipe of 6-in. diameter, its density increases by 20 percent. By what factor does its speed change? Does this answer require that the temperature of the air remain constant?

E12.32 The mechanical advantage of the simple hydraulic lift shown in the figure is $M.A. = A_2/A_1$, where A_1 and A_2 are the areas of the two cylinders. Derive this result in two ways: (1) use Pascal's principle and the relation of force to pressure—a static method; and (2) use the equation of continuity and the equality of input and output power—a dynamic method.

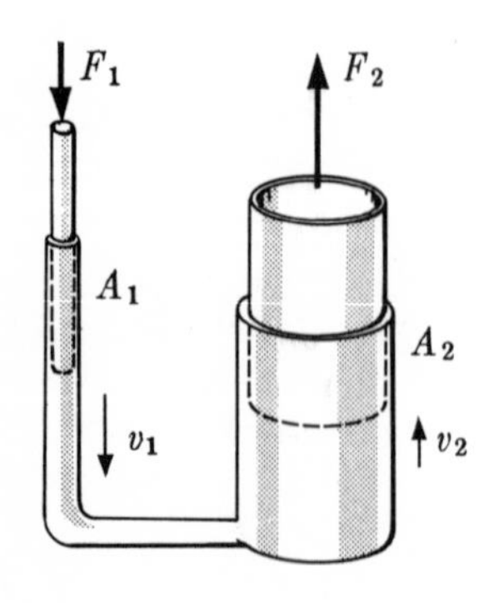

E12.33 The difference in pressure between two ends of a horizontal pipe carrying water is 1 atm. (1) What is the change of kinetic energy per unit volume (in J/m^3) of water going through the pipe? (2) What is the change of kinetic energy per unit mass (in J/kg) of water going through the pipe?

E12.34 (1) A liquid whose density is $\rho = 600$ kg/m^3 is flowing through the pipe shown in Figure 12.20. If the ratio of areas is $A_1/A_2 = 1.5$ and the difference of pressures is 0.05 atm, what is the speed v_1? (2) If the pressure in the unconstricted pipe is $p_1 = 2$ atm, what is the maximum possible pressure that this fluid could exert at any point in steady horizontal flow?

E12.35 What is the pressure at the stagnation point on a baseball approaching home plate at 30 m/sec? Express the answer in N/m^2 and in atmospheres.

E12.36 How fast must air at standard density and 1-atm pressure flow in order that its stagnation pressure be 2 atm? Express the answer in m/sec and in Mach number (ratio to the speed of sound).

E12.37 A water tower serving a small community is represented in the figure. At the upper surface of the water, the pressure is 1 atm and the speed of the water is negligible. (1) When no water is flowing, what is the pressure at the exit pipe? (2) By how much is the pressure at the exit pipe diminished if water flows there at a speed of 10 m/sec? (3) If the pressure at the exit pipe is to remain above 3 atm, what is the maximum permissible speed of flow through this pipe? (4) If the need for water may be as great as 10^3 kg/sec, what minimum diameter of the exit pipe is needed to maintain a pressure of 3 atm or more?

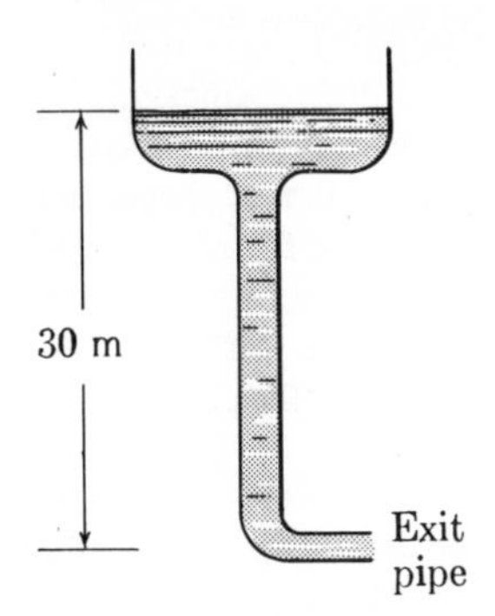

E12.38 In a child's toy rocket, air under pressure forces water from a small nozzle. Suppose that the water inside the rocket is at a pressure of 3 atm and has negligible speed. (1) Assuming the pressure at the nozzle to be 1 atm, calculate the exhaust speed of the water. (2) If the mass flow rate of water from the nozzle is 0.05 kg/sec, what mass can the rocket lift? (Rocket thrust is given by Equation 8.69.)

E12.39 (1) For steady incompressible flow in a horizontal pipe, show that the pressure is given by

$$p = p_1 - \frac{C}{A^2},$$

where p_1 is the stagnation pressure (Equation 12.65), A is the area of the pipe, which may be variable, and C is a positive constant. (2) How is C related to the mass flux and density of the fluid? (3) Sketch a graph of p vs A, and point out its significant features.

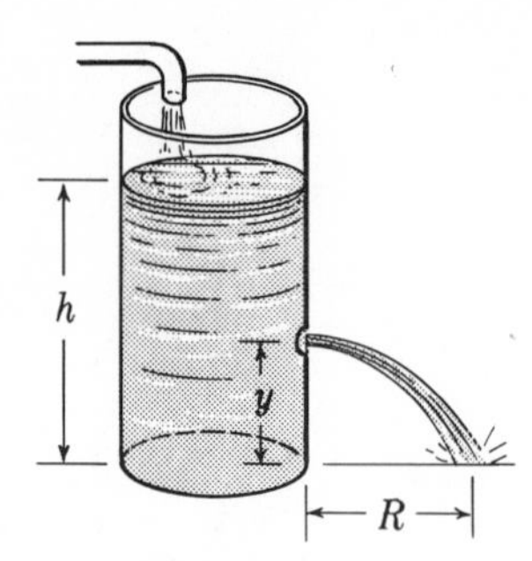

E12.40 A pattern of steady flow is established in a cylinder with a hole in its side, as shown in the figure. Prove that the speed of the liquid leaving the hole is $v \cong \sqrt{2g(h-y)}$. Does this speed change as air pressure changes?

E12.41 The jet of water emerging from the side of the cylinder shown in the preceding exercise covers a horizontal distance R before striking the ground. (1) Derive a formula for R. For an identical cylinder on the moon, where the acceleration of gravity is $\frac{1}{6}g$, what is R? (2) Show that the maximum range of the jet of water is $R_{\text{max}} = h$, achieved when $y = \frac{1}{2}h$.

E12.42 (1) At sea level, what pressure difference $p_1 - p_0$ in a pitot tube signals an airspeed of 60 m/sec? (2) What is the true airspeed of the airplane at an altitude of 10,000 ft if the same pressure difference exists in the pitot tube (thus causing the airspeed meter to read 60 m/sec)?

Section 12.5

E12.43 Verify that the drag coefficient C_D in Equation 12.67 is dimensionless.

E12.44 (1) Calculate the ratio of drag forces on a jet-driven airplane flying at 550 mile/hr at an altitude of 36,000 ft and on a propeller-driven airplane flying at 300 mile/hr at an altitude of 20,000 ft. Assume that both airplanes have the same frontal area and drag coefficient. (2) Making the same assumptions, calculate the ratio of drag forces on a supersonic transport flying at 1,500 mile/hr at 60,000 ft and on the subsonic jet described above.

E12.45 (1) For airplanes of the same frontal area and drag coefficient, how does the power required to move them through the air depend on speed? (2) Show that this relationship between power and speed is *not* obeyed by the three fixed-gear airplanes whose properties appear in the following table. Suggest a reason for the difference between the simple theoretical formula and the actual data. (3) Assuming the frontal areas of the four airplanes described in the table to be equal, find their *relative* drag coefficients.

Type of Airplane	Power Used in Cruising Flight (horsepower)	Cruising Speed (mile/hr)
Fixed landing gear	70	105
	105	125
	160	155
Retractable landing gear	125	160

E12.46 Show that the acceleration of an object falling from rest in air is given by

$$a_x(x) = ge^{-x/x_c},$$

where x is measured downward from the point where the fall starts (x_c is the characteristic distance given by Equation 12.76). Assume a constant drag coefficient.

E12.47 (1) Calculate roughly the terminal speed of a skydiver in air near sea level. Make reasonable estimates of his (or her) mass and cross-sectional area; use $C_D = 0.5$. (2) For this terminal speed, what is the characteristic distance x_c (Equation 12.76)?

E12.48 Expand Equation 12.77 in powers of x/x_c and show that for $x \ll x_c$, the speed of an object falling from rest is given approximately by

$$v \cong \sqrt{2gx}\left(1 - \frac{1}{4}\frac{x}{x_c}\right).$$

E12.49 Two spheres whose radii differ by a factor of 2 are observed to remain together as they fall, both of them reaching the same terminal speed. What is the relative density of the two spheres?

E12.50 (1) A tiny droplet falling in air experiences a drag force given by Equation 12.78. Obtain an expression for the droplet's terminal speed v_T. (2) Show by qualitative arguments that the droplet should approach this speed in a characteristic distance given by $x_c = v_T^2/g$.

E12.51 What is the physical dimension of the Reynolds number defined by Equation 12.79?

E12.52 The viscosity of air is $\mu = 1.8 \times 10^{-5}$ kg/m sec. Answer the following questions using data from Table 12.1. (1) For what range of speed of a ball of diameter 2 cm is the force of air drag given by Stokes's law? (2) For what range of speed is the drag force proportional to v^2?

E12.53 A pellet of diameter 2×10^{-4} m and mass 10^{-8} kg falls through water whose viscosity at room temperature is $\mu = 1.0 \times 10^{-3}$ kg/m sec. (1) Assuming the validity of Stokes's law, calculate the terminal speed of the pellet, find its Reynolds number at its terminal speed, and use Table 12.1 to verify the correctness of the initial assumption. (2) Give an order-of-magnitude estimate for the time needed for the pellet to reach terminal speed if it starts from rest.

Section 12.6

E12.54 Neutrons from a reactor enter a surrounding tank of water, where they collide repeatedly with protons (hydrogen nuclei). For sufficiently energetic neutrons, the protons may be approximated as being at rest before being struck. (1) Give the final energy of a 1-MeV neutron that strikes a proton and is deflected through (a) 30 deg, (b) 45 deg, or (c) 60 deg. (2) If, on the average, a neutron loses half of its kinetic energy in each collision with a proton, about how many collisions with protons are needed to degrade the energy of a neutron from 1 MeV to 1 keV? (NOTE: Neutrons also collide with oxygen nuclei in the water, but they lose less energy in such collisions.)

E12.55 A 10-MeV neutron strikes a proton at rest. If the proton recoils with an energy of 3 MeV, what angle does its path make with the incident direction of the neutron? Through what angle is the neutron deflected?

E12.56 Two billiard balls of equal mass collide head-on; their paths before and after the collision lie along the same line. Prove that in an elastic collision, the balls "change roles"—that is, the final velocity of each one is equal to the initial velocity of the other one.

E12.57 What is the available energy in each of the following collisions? (1) A 10-MeV neutron strikes a stationary carbon nucleus (mass of nucleus = 12 × mass of neutron). (2) A 10-MeV carbon nucleus strikes a stationary proton. (3) A 10-MeV neutron strikes a stationary proton. (4) Two 10-MeV protons collide head-on?

PROBLEMS

Using friction to help move boxes

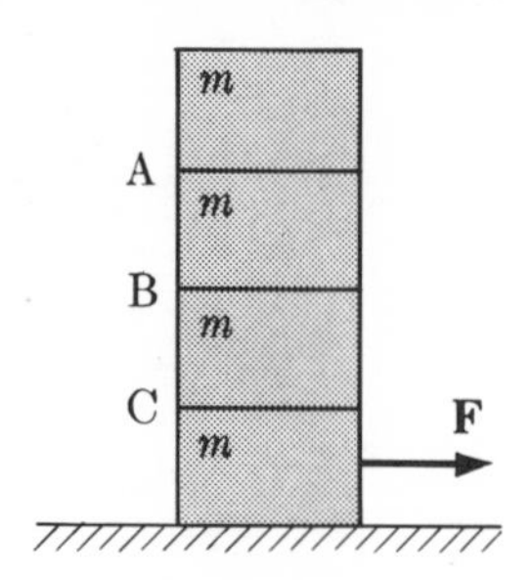

P12.1 Four identical boxes, each of mass m, are stacked on the floor, and a horizontal force $\mathbf{F}$ is applied to the lowest box (see the figure). Suppose that the coefficients of static and kinetic friction are equal ($\mu_s = \mu_k = \mu$) and are the same at all surfaces of contact, including the floor. (1) Prove that the maximum acceleration that the boxes can sustain without slipping is $a_{max} = \mu g$. (2) Show that if this acceleration is exceeded, the no-slip condition is violated simultaneously at all three surfaces A, B, and C. (3) To move the boxes without mishap, what is the minimum force that must be applied, and what is the maximum force that can be applied?

Minimum force to move a block

P12.2 A man applies a force $\mathbf{F}$ to a block of mass m at rest on a horizontal surface. The coefficient of static friction of the block on the surface is μ_s. Show that the least magnitude of force that will start the block moving is

$$F = \frac{\mu_s mg}{\sqrt{1 + \mu_s^2}}$$

and that the required angle of application of this force (see the figure) is $\theta = \arctan \mu_s$.

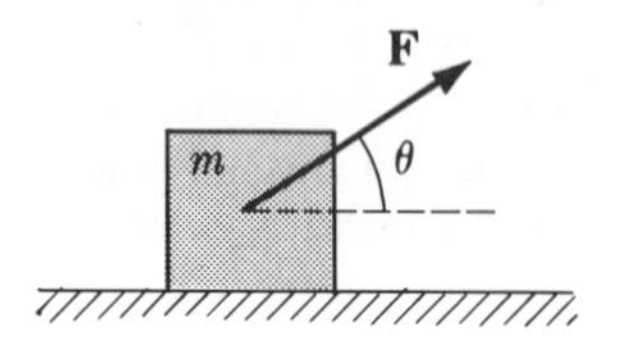

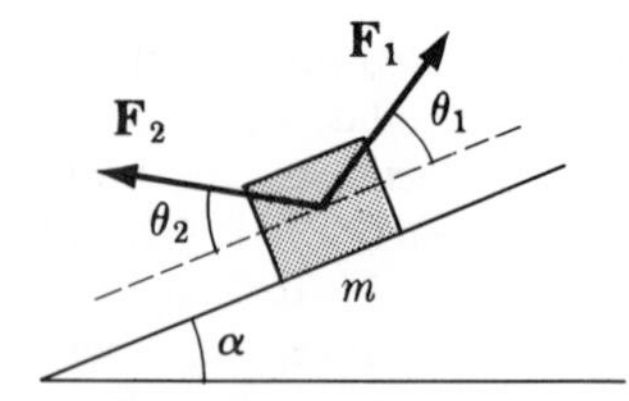

P12.3 As shown in the figure, a block of mass m rests on a plank inclined at an angle α to the horizontal; $\tan \alpha$ is less than the coefficient of static friction μ_s. As in the preceding problem, a man wishes to apply the least magnitude of force that will start the block moving. (1) Show that whether the block is started up the slope or down the slope, the force should be applied at the same angle to the slope, given by $\theta_1 = \theta_2 = \arctan \mu_s$. (2) Show that the required magnitudes of force are given by

$$F_{1,2} = \frac{mg(\mu_s \cos \alpha \pm \sin \alpha)}{\sqrt{1 + \mu_s^2}},$$

the plus sign in the numerator giving F_1 (the force required to start the block upward) and the minus sign in the numerator giving F_2 (the force required to start the block downward).

Minimum work to move a block

P12.4 In both of the preceding problems, the work to displace the block along the surface is minimized by applying the force vertically upward. (1) Prove this statement mathematically for a horizontal surface. What is the magnitude of the work for displacement s? (CAUTION: Not all minima are accompanied by a zero derivative.) (2) In terms of general energy considerations, explain why a vertically applied force also minimizes the work to move a block up an inclined plank.

P12.5 *Motion with a constant coefficient of kinetic friction*

A girl in her soapbox racer rolls down a slope of length 50 m inclined at an angle of 10 deg to the horizontal and then coasts on a horizontal surface. If the net frictional force is approximately independent of speed and is given by Equation 12.10, with $\mu_k = 0.05$, how far does she coast on the horizontal surface before coming to rest?

P12.6 *Friction of a linear air track*

A linear air track is tilted at 2 deg (0.035 radian) to the horizontal. Starting from rest, a car slides 2 m down the track, rebounds from a spring, and returns to within 20 cm of its starting point. If the car loses 5 percent of its kinetic energy in rebounding, what is the coefficient of kinetic friction of the car on the track? (NOTE: It is a satisfactory approximation to set the normal force of the track on the car equal to the weight of the car, mg.) *Optional:* If a linear air track is available to you in a laboratory, investigate its frictional properties in order to test Equation 12.10 and to obtain experimental values of μ_k. Is μ_k independent of speed?

P12.7 *Trick motorcyclist*

A motorcyclist at a carnival rides in a horizontal circle around the inside wall of a circular cylinder of radius r (see the figure). (1) If the frictional force preventing the tires from slipping down the wall is characterized by a coefficient of friction μ, what is the minimum speed v_{min} that the motorcyclist must maintain? (2) For $\mu = 0.75$ and $r = 5$ m, evaluate v_{min} in mile/hr. (3) At speed v, what should be the motorcyclist's angle of "bank" away from the vertical in order that the net force exerted by the wall is directed through the center of the motorcycle? (4) To maintain a horizontal circle, the front wheel of the motorcycle must be turned slightly. In which direction?

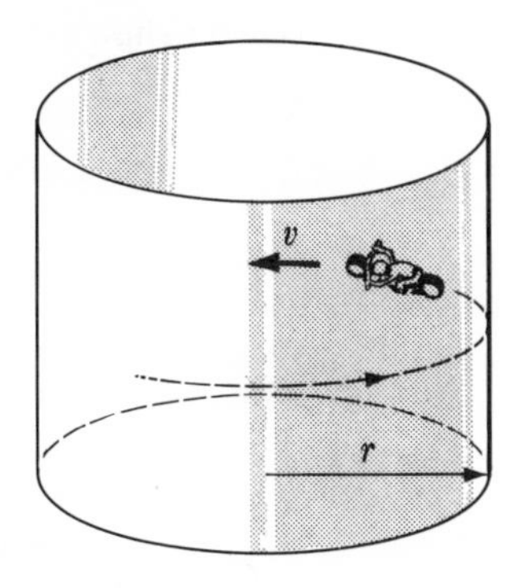

P12.8 *Problems of static equilibrium*

A man who weighs 1,000 N wishes to ride with the load shown in Exercise 12.13 as it is being raised at constant velocity. He can stand on the 2,000-N weight, or he can sit at any point along the 1,000-N beam. If all four ropes—the two running from the lower weight and the two running upward from the ends of the beam—are of equal strength, at what point should the man choose to ride in order to minimize the chance of breaking one of the ropes?

P12.9 Three beams of equal mass m and equal length l are arranged as shown. A lightweight cable joins the lower ends of two of the beams. Where the beams rest on supporting piers, friction is negligible. Find the compressional force F_c in the upper beam (it is a force that acts horizontally outward on the sloping beams) and the tensional force F_t in the cable (it is a force that acts horizontally inward on the sloping beams). (NOTE: The upper beam must also exert vertical components of force on the sloping beams.)

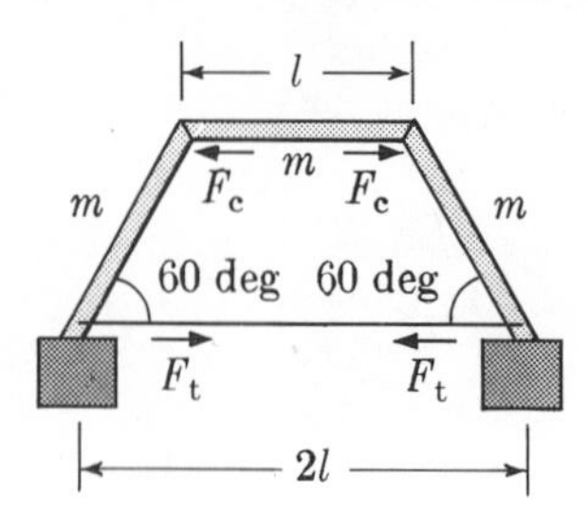

P12.10 The lower end of a ladder of length $2L$ rests on a floor, and its upper end rests on a smooth ledge a height L above the floor (see the figure). The ledge exerts force $\mathbf{F}'$ on the ladder, directed perpendicular to the ladder. The force of the floor on the ladder is $\mathbf{F} + \mathbf{N}$, and the force of gravity is $\mathbf{W}$, acting effectively at the center of mass of the ladder. (1) Prove that if the ladder does not slip, the ratio of frictional force to normal force at the floor is

$$\frac{F}{N} = \frac{\sin^2\theta\cos\theta}{1 - \cos^2\theta\sin\theta},$$

where θ is the angle of the ladder to the horizontal. (2) Show that if the coefficient of friction at the floor is $\mu_s = 0.50$, there is a range of angles for which the ladder will slip; for smaller and larger angles it will not slip. (It is not necessary to identify the angular range precisely; show only that it exists.)

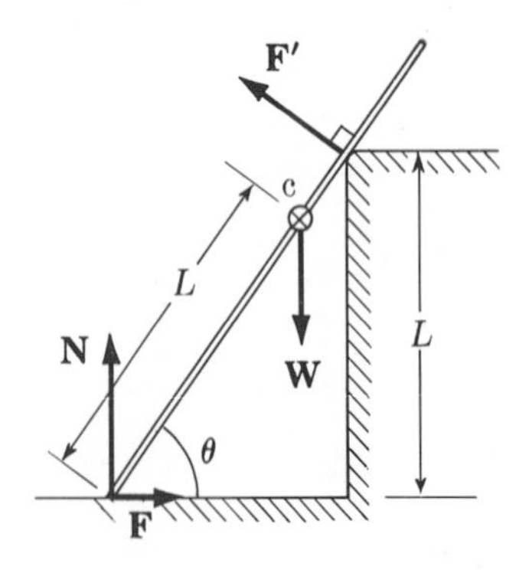

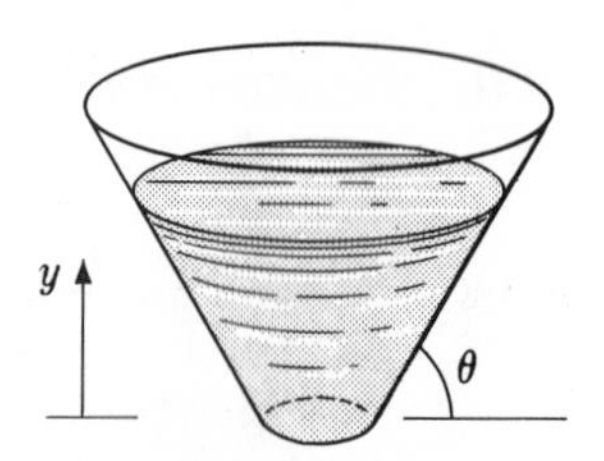

Pressure in a container with sloping sides

P12.11 Derive Equation 12.29 ($dp/dy = -\rho g$) for a conical container (see the figure). Take into account the facts that the area at height $y + \Delta y$ is slightly greater than the area at height y and that the walls exert a vertical component of force on the fluid. The effects of these two additions to the analysis of Section 12.3 cancel.

P12.12 An incompressible liquid of density ρ is stored in a tank on a small planet where the variation of gravitational acceleration with height cannot be neglected. Let the radius of the planet be R and let its gravitational acceleration be $g = g_0R^2/(R + y)^2$, where y is the height above ground. (1) Show that the pressure in the liquid is given by

Incompressible fluid on a small planet

$$p = p_0 - \rho g_0 y \frac{R}{R + y}.$$

(2) Sketch a graph of p vs y, and point out its features for $y \ll R$ and $y \gg R$. (3) Consider tanks of water of depth 30 m, one on earth and one on a planet of radius 100 m. What fractional error does Equation 12.30 give in calculations of the differences in pressure between the tops and bottoms of these tanks?

P12.13 To good approximation, the acceleration of gravity near the earth is given by

Incompressible fluid on earth

$$g = g_0\left(1 - \frac{2y}{R}\right),$$

where g_0 is the acceleration of gravity at the surface, R is the radius of the earth, and y is the height above the surface. (1) Using this formula for g, find an expression for the pressure p in an incompressible fluid as a function of y. Show that this expression is equivalent to the formula for p given in the preceding problem if $y \ll R$.

P12.14 (1) Show that the acceleration of gravity at depth d in the ocean is given by $g = g_0(R - d)/R$, where g_0 is the acceleration of gravity at sea level and R is the radius of the earth. (2) Approximating the ocean water as incompressible, with density ρ, derive the following formula for pressure as a function of depth:

$$p = p_0 + \rho g_0 d\left(1 - \frac{d}{2R}\right)$$

($p_0 = 1$ atm). Does Equation 12.30 slightly underestimate or slightly overestimate ocean pressure? (3) Evaluate the ocean pressure at a depth of 8 km. Is the difference between Equation 12.30 and the equation derived here significant?

P12.15 The density of a hypothetical gas is proportional to the square of its pressure. Obtain expressions for the density and the pressure of this gas as a function of altitude in a uniform gravitational field.

A hypothetical atmosphere

P12.16 The effect of the decrease of temperature with increasing altitude can be approximately taken into account by writing for the density of the atmosphere

An improved exponential model of the atmosphere

$$\rho = \frac{\rho_0}{p_0} p \left(1 + \frac{y}{a}\right)$$

(compare this with Equation 12.36); y is altitude and a is a characteristic distance that is considerably larger than the scale height b (Equation 12.40). (1) Show that this formula for density leads to the following expression for atmospheric pressure as a function of altitude:

$$p = p_0 e^{-y/b} e^{-y^2/2ab}.$$

(2) By comparing the graphs of the U.S. standard atmosphere and the exponential model in Figure 12.12 (or, better, by working with tabulated data on atmospheric pressure from an outside reference), show that a value of a of about 60 km substituted in the above equation leads to a substantial improvement over the simple exponential model.

Harmonic oscillation of a floating object

P12.17 A cylinder is weighted at one end so that it floats vertically upright in water. In equilibrium, it extends a distance s_0 below the water surface (see the figure). (1) Prove that if the cylinder is pushed down slightly and released, it will oscillate harmonically with angular frequency given by $\omega = \sqrt{g/s_0}$. (2) How does the angular frequency depend, if at all, on the density of the liquid in which the cylinder floats?

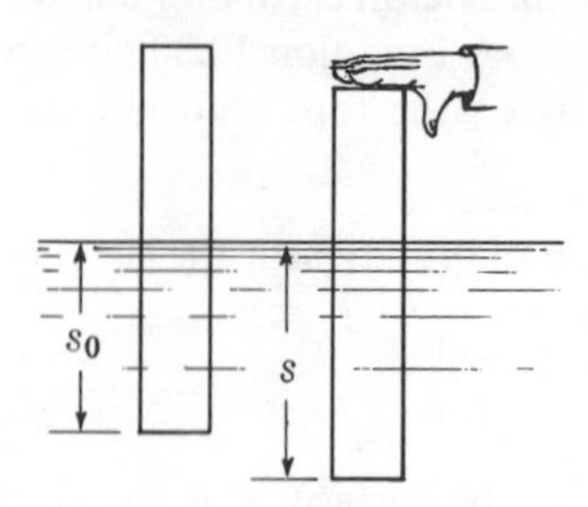

Water in free fall

P12.18 Water leaves a faucet of inside diameter D with speed v_0. Derive an expression for the diameter of the stream as a function of distance below the outlet of the faucet. Assume that the water does not break into droplets and that air friction on the stream is negligible.

P12.19 Water leaving a faucet passes through a screen and is broken into a myriad of tiny droplets. (1) Describe the fall of the water in terms of the mechanics of the individual droplets. Assume that the droplets are all the same size, leave the faucet with the same low speed, and approach the same terminal speed. (2) Describe the fall of the water in terms of its bulk properties averaged over many droplets. In particular, use the equation of continuity. (3) As a function of distance from the faucet, sketch qualitative graphs of the speed of the water, the density of a single droplet, the average density of many droplets, and the average spacing between droplets.

Litf force on a wing

P12.20 Tubes of flow of air above and below a wing surface are shown in the figure. (1) Obtain a formula for the net upward pressure on the wing as a function of the air density, ρ, the speed of the air relative to the wing, v_0, and the average areas of the tubes of flow above and below the wing, A_1 and A_2.

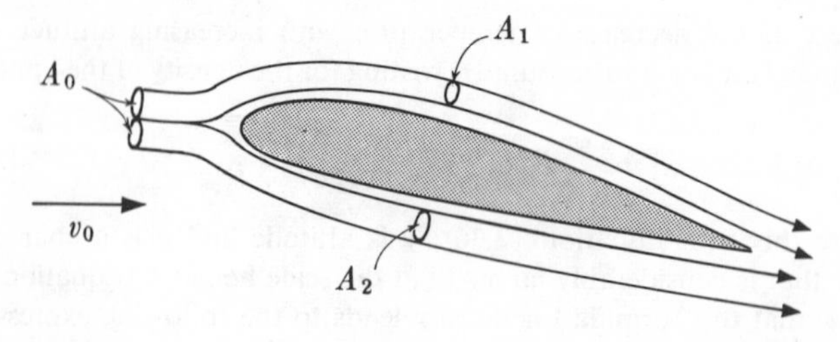

Note the implication that the force of lift on the wing is proportional to ρv_0^2 (provided the areas A_1 and A_2 are nearly independent of speed, which is a good approximation). (2) If an airplane flies at a speed of 50 m/sec at an altitude where $\rho = 1.0\ \text{kg/m}^3$ and if the areas are $A_1 \cong 0.8A_0$ and $A_2 \cong A_0$, what is the numerical value of the force of lift per unit area? Make an estimate of the weight and wing area of a light plane to see if the calculated number is reasonable. (3) The existence of lift implies that the wing must deflect air downward. Why?

P12.21 *Water-hammer effect*

A straight pipe of length l and cross-sectional area A joins a hot-water tank at one end of a house to a faucet at the other end of the house. With the faucet fully open, the water moves through the pipe with speed v. (1) By considering the rate of change of momentum of the water, derive an expression for the excess pressure at the faucet valve during the time Δt that the faucet is being turned off. (2) Evaluate this excess pressure in atmospheres for $l = 10$ m, $v = 5$ m/sec, and $\Delta t = 0.2$ sec. *Optional:* Find out what is done in home water systems to prevent this "water-hammer effect" from making an objectionable noise and/or damaging the pipes.

P12.22 *Force and energy in fluid flow*

A force F_1 is applied to a piston of area A_1 at one end of a pipe to establish a condition of steady incompressible flow in the pipe. At the other end of the pipe, the fluid exerts a force F_2 on a piston of area A_2 (see the figure). The piston and fluid at one end of the pipe move with speed v_1; at the other end, they move with speed v_2. (1) Calculate the input power at the first piston, the output power at the second piston, and the rate of change of the kinetic energy of the fluid between the pistons. Verify that energy is conserved. (2) Show that the force F_3 required to keep the pipe from moving is

$$F_3 = F_1\left(\frac{A_1 - A_2}{A_1}\right) + \tfrac{1}{2}\rho v_1^2\left(\frac{A_1^2 - A_2^2}{A_2}\right).$$

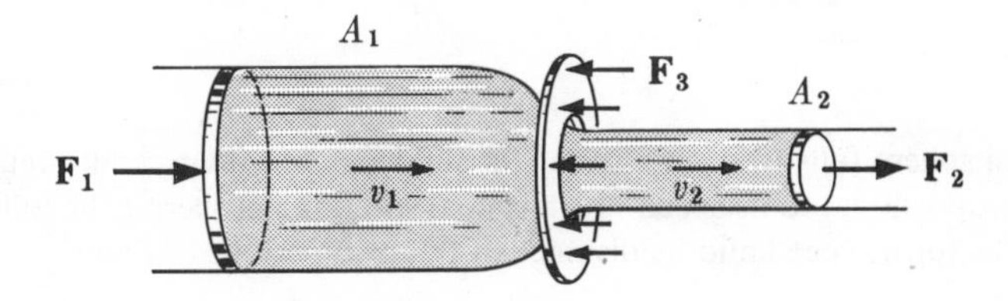

P12.23 *Approximate law of drag for a large parachute*

When it landed in 1971 with only two of its parachutes open, the Apollo 15 spacecraft was reported to have hit the water at 32 ft/sec. Had all three parachutes been open, its terminal speed would have been 28 ft/sec. (1) If the drag force on an Apollo parachute is proportional to v^n, where v is its speed, what is the exponent n? (2) A terminal speed of 32 ft/sec is equivalent to free fall without friction from what height?

P12.24 *Work done by a falling object*

An object experiencing the drag force given by Equation 12.67 falls from rest through a distance l. Find the work that it does on the air in two ways: (1) Find F_D as a function of x and perform the necessary integration over x to obtain the work. (2) Find the loss of mechanical energy of the object as it falls through the distance l.

Simple model of drag

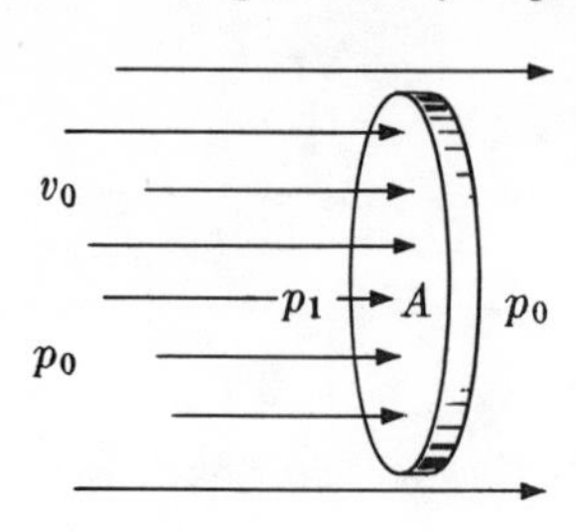

P12.25 To calculate the drag on a disk of area A in a simple way, assume that the air, which has pressure p_0 and speed v_0 far from the disk, is brought to rest at one surface of the disk and is undisturbed elsewhere. (See the accompanying figure. For a more realistic view of what happens, see Figure 12.21.) Assume that the pressure immediately behind the disk is p_0. What net force is exerted on the disk by the air? What is the drag coefficient according to this simplified theoretical calculation? *Optional:* From an outside reference, obtain information on the drag coefficient of a disk and compare accurate values with the approximate value you have calculated.

P12.26 As an alternative to the method suggested in the preceding problem, analyze the drag on a disk in the following way. Let the disk move through still air at speed v. Assume that as the disk advances a distance Δx, it gives speed v to all the air in its path in the cylinder of area A and height Δx (the shaded portion of the figure). The air gains kinetic energy, so the disk loses kinetic energy. From the rate of change of energy of the disk, a force can be calculated. Show that the force has the form of Equation 12.67. What is the drag coefficient calculated in this way? (This method is equivalent to the original approach of Newton.) *Optional:* Explain the relation of this method to the method used in the preceding problem.

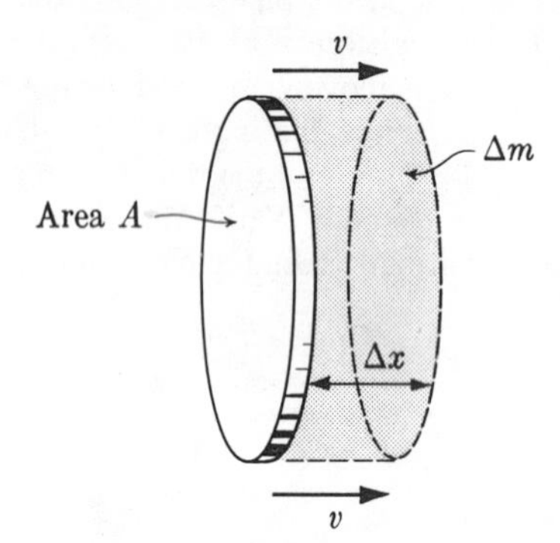

Motion influenced by Stokes's law of drag force

P12.27 A small sphere falls from rest under the influence of gravity and a retarding force proportional to its speed. Its terminal speed is v_T. Derive the following formulas for its speed and its distance of fall as functions of time:

$$v = v_T(1 - e^{-gt/v_T}),$$

$$x = v_T t - \frac{v_T^2}{g}(1 - e^{-gt/v_T}).$$

Average energy loss of neutrons striking protons

P12.28 For neutrons of moderate energy striking protons at rest, the number of deflected neutrons per unit angle is proportional to the sine of twice the scattering angle:

$$\frac{dN}{d\theta} \sim \sin 2\theta.$$

(1) What is the most probable angle of deflection? What are the least probable angles of deflection? (2) For neutrons of initial energy K_0, find the average energy loss in such collisions. (NOTE: This is a weighted average.) (3) In the center-of-mass frame, what is the form of $dN/d\psi$, the number of neutrons per unit angle of deflection?

P12.29 A body of mass m has momentum p_0 and kinetic energy K_0 in a frame of reference S_0. In a frame of reference S_1 whose velocity with respect to S_0 is $V\mathbf{i}$, what are the momentum and the energy of the body? Give answers (a) in general and (b) in particular if $\mathbf{p}_0 = p_0\mathbf{i}$.

Galilean transformation of momentum and energy

P12.30 A spacecraft approaches the moon with velocity $\mathbf{v}_1$ directed nearly opposite to the moon's velocity $\mathbf{v}_M$ (see the figure). (1) Show that the spacecraft may leave the vicinity of the moon with a speed v_2 nearly as great as $v_1 + 2v_M$. (SUGGESTION: Consider the encounter in the center-of-mass system of the moon and spacecraft.) (2) After an encounter of the kind shown in the figure, is the spacecraft more likely to follow an elliptical path around the earth or to escape from the earth? (3) Are there any limitations in principle or in practice to using successive encounters with moons and planets to give a spacecraft arbitrarily great energy? (Planning for future missions to Jupiter and beyond does include use of this technique to increase spacecraft energy.)

The moon as an energy source for space travel

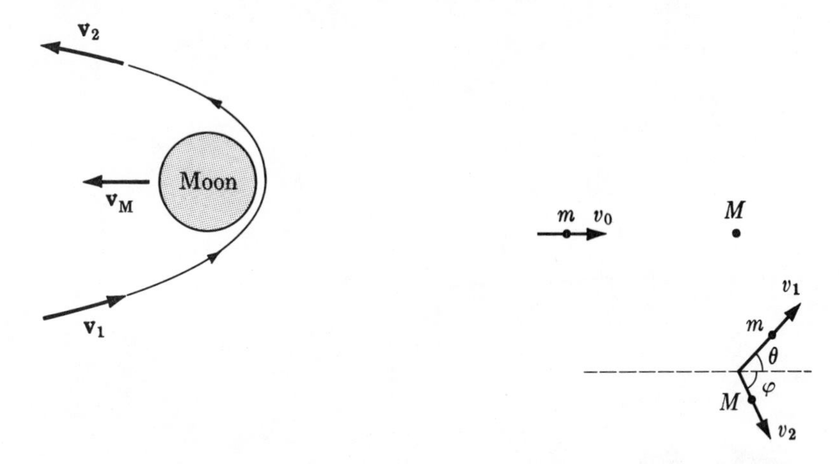

P12.31 A particle of mass m and initial speed v_0 collides elastically with a particle of mass M that is initially at rest. Paths of the particles after the collision are shown in the figure. (1) Display the equations of energy and momentum conservation that govern this process. (2) Show that after the collision, the speed of the particle of mass m is given by

$$v_1 = v_0 \frac{\alpha \cos \theta \pm \sqrt{1 - \alpha^2 \sin^2 \theta}}{1 + \alpha},$$

where $\alpha = m/M$. Show that this agrees with results derived in the text if $\alpha = 1$. (3) Consider a neutron striking a deuteron (for which $\alpha = \frac{1}{2}$). Why must the + sign be chosen in the formula above? What is the maximum deflection angle of the neutron? What is its maximum fractional loss of energy? What is its fractional loss of energy for $\theta = 45$ deg?

Elastic collision of bodies of unequal mass

Appendices

1	Units in the international system (SI)	A3
2	Physical quantities: symbols and SI units	A5
3	Numerical data	A8
4	Conversion factors	A13
5	Equations of electromagnetism for SI (mks) and Gaussian (cgs) units	A15
6	Mathematical formulas	A18
7	Trigonometric functions	A27
8	The exponential function	A29
9	The logarithmic function	A30

APPENDIX 1. Units in the international system (SI)

(This appendix is adapted from E. A. Mechtly, "The International System of Units," NASA Report SP-7012, 1969, available for 30 cents from the Superintendent of Documents, U.S. Government Printing Office, Washington, D.C. 20402.)

Basic units

Length — The METER (m) is the length equal to 1,650,763.73 wavelengths in vacuum of the radiation corresponding to the transition between the levels $2p_{10}$ and $5d_5$ of the krypton 86 atom.

Mass — The KILOGRAM (kg) is the mass of the international prototype of the kilogram (a particular cylinder of platinum-iridium alloy preserved in a vault in Sèvres, France).

Time — The SECOND (sec*) is the duration of 9,192,631,770 periods of the radiation corresponding to the transition between the two hyperfine levels of the ground state of the cesium 133 atom.

Current — The AMPERE (A) is that constant current which, if maintained in two straight parallel conductors of infinite length, of negligible circular cross section, and placed 1 m apart in vacuum, would produce between those conductors a force per unit length equal to 2×10^{-7} N/m.

Temperature — The KELVIN (K) is the fraction 1/273.16 of the thermodynamic temperature of the triple point of water.

Luminous intensity — The CANDELA (cd) is the luminous intensity, in the perpendicular direction, of a surface of 1/600,000 m^2 of a blackbody at the temperature of freezing platinum under a pressure of 101,325 N/m^2.

Some other important units

Force — The NEWTON (N) is that force which gives to a mass of 1 kg an acceleration of 1 m/sec^2.

Energy — The JOULE (J) is the work done when the point of application of 1 N is displaced a distance of 1 m in the direction of the force.

Power — The WATT (W) is the power which gives rise to the production of energy at the rate of 1 J/sec.

Charge — The COULOMB (C) is the charge transported in 1 sec by a current of 1 A.

Potential — The VOLT (V) is the difference of potential between two points of a conducting wire carrying a constant current of 1 A when the power dissipated between these points is equal to 1 W.

Resistance — The OHM (Ω) is the resistance between two points of a conductor

* The symbol s is also commonly used for the second.

	when a constant difference of potential of 1 V, applied between these two points, produces in this conductor a current of 1 A, this conductor not being the source of any electromotive force.
Capacitance	The FARAD (F) is the capacitance of a capacitor between the plates of which there appears a difference of potential of 1 V when it is charged by 1 C.
Inductance	The HENRY (H) is the inductance of a closed circuit in which an electromotive force of 1 V is produced when the current in the circuit varies uniformly at a rate of 1 A/sec.
Magnetic flux	The WEBER (Wb) is the magnetic flux which, linking a circuit of one turn, produces in it an electromotive force of 1 V as it is reduced to zero at a uniform rate in 1 sec.

APPENDIX 2. Physical quantities: symbols and SI units

Quantity	Common Symbol	Unit	Unit Expressed in Terms of Basic SI Units
Acceleration	**a**	m/sec^2	m/sec^2
Angle	θ, φ	radian	
Angular acceleration	α	radian /sec^2	sec^{-2}
Angular frequency	ω	radian/sec	sec^{-1}
Angular momentum	**L, J**	kg m^2/sec	kg m^2/sec
Spin	**S**		
Angular velocity	$\boldsymbol{\omega}$	radian/sec	sec^{-1}
Angular speed	ω		
Area	**S**, A	m^2	m^2
Atomic number	Z		
Capacitance	C	farad (F) (= C/V)	A^2 sec^4/kg m^2
Charge	q, e	coulomb (C)	A sec
Charge density			
Volume	ρ	C/m^3	A sec/m^3
Surface	σ	C/m^2	A sec/m^2
Line	τ	C/m	A sec/m
Conductivity	σ	1/Ω m	A^2 sec^3/kg m^3
Current	I	AMPERE	A
Current density	**J**	A/m^2	A/m^2
Density	ρ	kg/m^3	kg/m^3
Dielectric constant	κ_e		
Displacement	**s**	METER	m
Distance	d		
Length	l, L		
Electric dipole moment	**p**	C m	A sec m
Electric field	**E**	V/m	kg m/A sec^3
Electric flux	Φ_E	V m	kg m^3/A sec^3
Electromotive force	$\mathscr{V}$	volt (V)	kg m^2/A sec^3

Quantity	Common Symbol	Unit	Unit Expressed in Terms of Basic SI Units
Energy	E	joule (J)	kg m^2/sec^2
Internal energy	U		
Kinetic energy	K		
Potential energy	U		
Entropy	S	J/K (often kcal/K)	kg m^2/sec^2 K
Force	$\mathbf{F}$	newton (N)	kg m/sec^2
Frequency	ν	hertz (Hz)	sec^{-1}
Heat	Q	joule (J) (often cal or kcal)	kg m^2/sec^2
Inductance	L	henry (H)	kg m^2/A^2 sec^2
Magnetic dipole moment	μ	N m/T	A m^2
Magnetic field	$\mathbf{B}$	tesla (T) (= Wb/m^2)	kg/A sec^2
Magnetic flux	Φ_B	weber (Wb)	kg m^2/A sec^2
Mass	m, M	KILOGRAM	kg
Mass number	A		
Molar specific heat	C'	J/kmole K (often kcal/kmole K)	kg m^2/sec^2 kmole K
Molecular weight	$M.W.$	kg/kmole (= gm/mole) (= amu/molecule)	kg/kmole
Moment of inertia	I	kg m^2	kg m^2
Momentum	$\mathbf{p}$	kg m/sec	kg m/sec
Period	T	sec	sec
Permeability	κ_m		
Permeability constant	μ_0	N/A^2 (= H/m)	kg m/A^2 sec^2
Permittivity of space	ϵ_0	C^2/N m^2 (= F/m)	A^2 sec^4/kg m^3
Pole strength	P	N/T	A m
Potential Voltage	V	volt (V) (= J/C)	kg m^2/A sec^3
Power	P	watt (W) (= J/sec)	kg m^2/sec^3
Pressure	P, p	N/m^2	kg/m sec^2
Resistance	R	ohm (Ω) (= V/A)	kg m^2/A^2 sec^3

Quantity	Common Symbol	Unit	Unit Expressed in Terms of Basic SI Units
Specific heat (see also molar specific heat)	C	J/kg K (often kcal/kg K)	m^2/sec^2 K
Temperature	T	KELVIN	K
Time	t	SECOND	sec
Torque	$\mathbf{T}$	N m	kg m^2/sec^2
Velocity	$\mathbf{v}$	m/sec	m/sec
Speed	v		
Volume	V	m^3	m^3
Wave function	ψ	Usually $m^{-3/2}$ ($m^{-1/2}$ in one dimension)	$m^{-3/2}$
Wave number	k, $1/\lambda\!\!\!^{-}$	m^{-1}	m^{-1}
Wavelength	λ	m	m
Work	W	joule (J) (= N m)	kg m^2/sec^2

ALPHABETICAL LIST OF STANDARD ABBREVIATIONS OF UNITS

Abbreviation	Unit	Abbreviation	Unit
A	ampere	Hz	hertz
Å	angstrom	in	inch
A.U.	astronomical unit	J	joule
amu	atomic mass unit	K	kelvin
atm	atmosphere	kcal	kilocalorie
C	coulomb	kg	kilogram
°C	degree Celsius	kmole	kilomole
cal	calorie	lb	pound
cm	centimeter	m	meter
deg	degree (angle)	min	minute
esu	electrostatic unit	N	newton
eV	electron volt	°R	degree Rankine
F	faraday	rpm	revolutions per minute
°F	degree Fahrenheit	sec	second
fm	fermi, femtometer	T	tesla
ft	foot	V	volt
G	gauss	W	watt
gm	gram	Wb	weber
H	henry	μm	micrometer, micron
hr	hour	Ω	ohm

APPENDIX 3. Numerical data

For physical data, see, in addition to this appendix, appropriate tables in the text.

Table 3.1 Some of the More Important Elementary Particles
Table 9.2 Some Properties of Planetary Orbits Known to Kepler
Table 11.1 Principal Properties of the Sun's Nine Known Planets
Table 13.1 Liquefaction Temperatures of Some Common Gases
Table 13.2 Specific Heats of Some Common Materials at Room Temperature
Table 13.3 Heats of Fusion and Vaporization of Some Common Materials at Atmospheric Pressure
Table 15.1 Conductivities of Some Common Materials at Room Temperature
Table 15.2 Dielectric Constants of Some Materials
Table 16.1 Selected Permeabilities κ_m and Susceptibilities χ_m
Table 18.1 The Speed of Sound in Common Substances
Table 23.1 Photoelectric Threshold Frequency and Work Function for Several Metals
Table 24.3 Periodic Table of the Elements
Table 26.1 Alpha-Decay Properties of Several Nuclei

A. Physical constants

(This table is adapted from B. N. Taylor, W. H. Parker, and D. N. Langenberg, *The Fundamental Constants and Quantum Electrodynamics* [New York: Academic Press, 1969]. A good popular article on the fundamental constants, by the same authors, is to be found in the October, 1970, issue of *Scientific American*. The numbers recorded here have been truncated so that the uncertainty in each is at most ± 1 in the last digit.)

Quantity	Symbol	Value
Gravitational constant	G	6.67×10^{-11} N m^2/kg^2 (or m^3/kg sec^2)
Avogadro's number	N_0	6.0222×10^{23} particles/mole (or amu/gm)
Boltzmann's constant (microscopic gas constant)	k	1.3806×10^{-23} J/K 8.617×10^{-5} eV/K
	$\frac{1}{k}$	11,605 K/eV
Macroscopic gas constant	$R (= N_0 k)$	8.314 J/mole K 1.9872 kcal/kmole K
Quantum unit of charge	e	1.60219×10^{-19} C 4.8033×10^{-10} esu
Faraday constant (1 mole of electricity)	$F (= N_0 e)$	9.6487×10^{4} C/mole 2.8926×10^{14} esu/mole

Quantity	Symbol	Value
Permittivity of space	$\epsilon_0 \left(= \frac{1}{\mu_0 c^2}\right)$	8.85419×10^{-12} C^2/N m^2
	$4\pi\epsilon_0 \left(= \frac{4\pi}{\mu_0 c^2}\right)$	1.112650×10^{-10} C^2/N m^2
	$\frac{1}{4\pi\epsilon_0} \left(= \frac{\mu_0 c^2}{4\pi}\right)$	8.98755×10^{9} N m^2/C^2
Permeability constant	μ_0	$4\pi \times 10^{-7}$ N/A^2 *exact, by definition* or 1.256637×10^{-6} N/A^2
	$\frac{\mu_0}{4\pi}$	*exactly* 10^{-7} N/A^2
Speed of light	c	2.997925×10^{8} m/sec
Planck's constant	h	6.6262×10^{-34} J sec 4.1357×10^{-15} eV sec 4.1357×10^{-21} MeV sec
	$\hbar \left(= \frac{h}{2\pi}\right)$	1.05459×10^{-34} J sec 6.5822×10^{-16} eV sec 6.5822×10^{-22} MeV sec
Charge-to-mass ratio or electron	$\frac{e}{m_e}$	1.75880×10^{11} C/kg 5.2728×10^{17} esu/gm
Mass of electron	m_e	9.1096×10^{-31} kg 5.4859×10^{-4} amu
Mass of proton	m_p	1.67261×10^{-27} kg 1.0072766 amu $1836.11 m_e$
Mass of neutron	m_n	1.67492×10^{-27} kg 1.0086652 amu $1838.64 m_e$
Intrinsic energy of electron	$m_e c^2$	0.51100 MeV
Intrinsic energy of proton	$m_p c^2$	938.26 MeV
Intrinsic energy of neutron	$m_n c^2$	939.55 MeV
Rydberg constant for infinitely massive nucleus	$\mathscr{R}_\infty \left[= \left(\frac{1}{4\pi\epsilon_0}\right)^2 \frac{m_e e^4}{4\pi\hbar^3 c}\right]$	1.0973731×10^{7} m^{-1}
Rydberg constant for hydrogen 1	$\mathscr{R}_H$	1.0967758×10^{7} m^{-1}
Fine structure constant	$\alpha \left(= \frac{1}{4\pi\epsilon_0} \frac{e^2}{\hbar c}\right)$	7.29735×10^{-3} or 1/137.036

Quantity	Symbol	Value
Bohr radius	$a_0 \left(= \dfrac{4\pi\epsilon_0\hbar^2}{m_e e^2}\right)$	5.29177×10^{-11} m 0.529177 Å
Compton wavelength of the electron	$\lambda_C \left(= \dfrac{h}{m_e c}\right)$	2.42631×10^{-12} m
Reduced Compton wavelength of the electron	$\bar{\lambda}_C \left(= \dfrac{\hbar}{m_e c}\right)$	3.86159×10^{-13} m 386.159 fm
Bohr magneton	$\mu_B \left(= \dfrac{e\hbar}{2m_e}\right)$	9.2741×10^{-24} J/T
	Useful Combinations of Constants	
	$\dfrac{e^2}{4\pi\epsilon_0}$	2.3071×10^{-28} J m 14.400 eV Å 1.4400 MeV fm
	$\hbar c$	3.1616×10^{-26} J m 1.97329×10^3 eV Å 197.329 MeV fm
	$\dfrac{\hbar^2}{2m_e}$	6.1044×10^{-39} J m^2 3.8100 eV Å^2 3.8100×10^4 MeV fm^2
	$\dfrac{\hbar^2}{2m_p}$	3.3246×10^{-42} J m^2 2.0751×10^{-3} eV Å^2 20.751 MeV fm^2
	c^2	8.98755×10^{16} J/kg 9.3148×10^8 eV/amu 931.48 MeV/amu

B. Terrestrial data* (Footnote on page A12)

Quantity	Value
Acceleration of gravity at sea level (g)	9.80665 m/sec^2, standard reference value 9.7804 m/sec^2 at equator 9.8322 m/sec^2 at poles
Mass of earth (M_E)	5.98×10^{24} kg
Mass of earth times gravitational constant ($M_E G$)	3.9860×10^{14} N m^2/kg (or m^3/sec^2)
Radius of earth (R_E)	6.37×10^6 m, 6370 km, 3960 miles } approximate average value 6378.2 km at equator 6356.8 km at poles
Equatorial circumference of earth	4.008×10^7 m 24,902 miles

Quantity	Value
The Atmosphere	
Standard air pressure at sea level (760 mm of Hg)	1.013×10^5 N/m^2
Standard dry air density at sea level and 0 °C	1.293 kg/m^3
Typical moist air density at sea level and 20 °C	1.20 kg/m^3
Speed of sound in standard air at 0 °C	331 m/sec 740 mile/hr
Typical speed of sound in moist air at 20 °C	344 m/sec 770 mile/hr
Approximate composition of atmosphere, by number of molecules	N_2, 78 percent O_2, 21 percent Ar, 1 percent
Mean molecular weight of dry air	28.97
Specific heats of standard air	$C_p = 0.2403$ kcal/kg K $C_v = 0.1715$ kcal/kg K $C_p' = 3.503R$ $C_v' = 2.500R$
Ratio of specific heats of standard air (γ)	1.401

C. Densities of common materials at standard conditions of temperature and pressure

Substance	Density (gm/cm^3)	(kg/m^3)
Hydrogen (H_2)	8.99×10^{-5}	0.0899
Helium (He)	1.785×10^{-4}	0.1785
Nitrogen (N_2)	1.250×10^{-3}	1.250
Oxygen (O_2)	1.429×10^{-3}	1.429
Air	1.293×10^{-3}	1.293
Gasoline	~0.7 ~700	660–690
Alcohol (ethanol)	0.806	806
Water	1.000	1.000×10^3
Mercury	13.60	1.360×10^4
Aluminum	2.70	2.70×10^3
Iron	7.86	7.86×10^3
Copper	8.96	8.96×10^3
Lead	11.4	1.14×10^4

D. Astronomical data*

Quantity	Value
Distance from center of earth to center of moon	3.844×10^8 m 2.389×10^5 miles
Period of moon	27.32 days 2.360×10^6 sec
Mass of moon	7.35×10^{22} kg $0.0123M_E$
Radius of moon	1.738×10^6 m $0.2728R_E$
Acceleration of gravity at the surface of the moon	1.62 m/sec^2 $0.165g$
Distance from center of earth to center of sun (1 A.U.)	1.496×10^{11} m } average 9.30×10^7 miles } average 1.471×10^{11} m at perihelion 1.521×10^{11} m at aphelion
Mass of sun (M_S)	1.99×10^{30} kg $3.329 \times 10^5 M_E$
Mass of sun times gravitational constant ($M_S G$)	1.3272×10^{20} N m^2/kg (or m^3/sec^2)
Radius of sun (R_S)	6.960×10^8 m $109.2R_E$
Period of earth	365.26 days 3.156×10^7 sec
Average orbital speed of earth	2.98×10^4 m/sec
Average orbital acceleration of earth	5.93×10^{-3} m/sec^2 $6.05 \times 10^{-4}g$

* Reference: C. W. Allen, *Astrophysical Quantities*, second edition (London: The Athlone Press, University of London, 1963). Other useful references for physical data are the *Handbook of Chemistry and Physics* (Cleveland, Ohio: The Chemical Rubber Co.), frequently revised; and the *American Institute of Physics Handbook*, third edition (New York: McGraw-Hill Book Co., 1972).

APPENDIX 4. Conversion factors

For convenience in units arithmetic, this appendix lists conversion factors directly (such as 2.54 cm/in.) rather than equations (such as 1 in. = 2.54 cm). Any quantity can be multiplied or divided by appropriate conversion factors since each conversion factor is equivalent to unity.

Conversion factors preceded by a dot (●) are exact and serve to define one unit in terms of another. For example, the factor 0.3048 m/ft defines the foot as exactly 0.3048 m.

1. Length

● 10^2 cm/m
● 10^3 m/km

● 2.54 cm/in.
● 12 in./ft
● 5,280 ft/mile

● 0.3048 m/ft
● 1.609344×10^3 m/mile
● 1.609344 km/mile

1.49598×10^{11} m/A.U.
9.461×10^{15} m/light-year
3.084×10^{16} m/parsec

● 10^{-6} m/μm (or m/micron)
● 10^{-10} m/Å
● 10^{-15} m/fm

2. Volume

● 10^{-3} m^3/liter
● 10^3 cm^3/liter
0.94635 liter/quart
3.7854×10^{-3} m^3/gallon

3. Time

(The day is a mean solar day; the year is a sidereal year.)

● 3,600 sec/hr
● 8.64×10^4 sec/day
365.26 day/year
3.1558×10^7 sec/year

4. Speed

● 0.3048 (m/sec)/(ft/sec)
1.609×10^3 (m/sec)/(mile/sec)
0.4470 (m/sec)/(mile/hr)
1.609 (km/hr)/(mile/hr)

5. Acceleration

● 0.3048 (m/sec^2)/(ft/sec^2)

6. Angle

● 60 second of arc(″)/minute of arc(′)
● 60 minute of arc(′)/deg
● $180/\pi$ ($\cong$ 57.30) deg/radian
● 2π ($\cong$ 6.283) radian/revolution

7. Mass

● 10^3 gm/kg

453.59 gm/lb
0.45359 kg/lb
2.2046 lb/kg

1.66053×10^{-27} kg/amu
6.0222×10^{26} amu/kg
6.0222×10^{23} amu/gm

8. Density

● 10^3 (kg/m^3)/(gm/cm^3)
16.018 (kg/m^3)/(lb/ft^3)
1.6018×10^{-2} (gm/cm^3)/(lb/ft^3)

9. Force

● 10^5 dyne/N
● 10^{-5} N/dyne
4.4482 N/lbf
(1 lbf = weight of 1 pound at standard gravity [g = 9.80665 m/sec^2])

10. Pressure

- 0.1 $(N/m^2)/(dyne/cm^2)$
- 10^5 (N/m^2)/bar
- 1.01325×10^5 (N/m^2)/atm
- 1.01325×10^6 $(dyne/cm^2)$/atm
- 1.01325 bar/atm

133.32 (N/m^2)/mm of Hg (0 °C)
3.386×10^3 (N/m^2)/in. of Hg (0 °C)
6.895×10^3 $(N/m^2)/(lbf/in.^2$, or psi)

11. Energy

(For mass-to-energy conversion, see the values of c^2 at the end of Appendix 3A.)

- 10^7 erg/J
- 10^{-7} J/erg
- 4.184 J/cal
- 4,184 J/kcal
- 10^3 cal/kcal

(The kilocalorie [kcal] is also known as the food calorie, the large calorie, or the Calorie.)

1.60219×10^{-19} J/eV
1.60219×10^{-13} J/MeV

- 10^6 eV/MeV
- 3.60×10^6 J/kW hr

4.20×10^{12} J/kiloton
4.20×10^{15} J/megaton

0.04336 (eV/molecule)/(kcal/mole)
23.06 (kcal/mole)/(eV/molecule)

12. Power

- 746 W/horsepower

13. Temperature

- 1.00 F°/R°
- 1.00 C°/K
- 1.80 F°/C°
- 1.80 R°/K
- T (K) = T (°C) + 273.15
- T (°C) = [T (°F) − 32]/1.80
- T (K) = T (°R)/1.80

14. Electrical quantities

(Note that 2.9979 is well approximated by 3.00.)

Charge: 2.9979×10^9 esu/C
Current: 2.9979×10^9 (esu/sec)/A
Potential: 299.79 V/statvolt
Electric field: 2.9979×10^4 (V/m)/(statvolt/cm)

- Magnetic field: 10^4 G/T
- Magnetic flux: 10^8 G cm^2/Wb
- Pole strength: 10 cgs unit/michel

(cgs unit = $\sqrt{\text{erg cm}}$;
michel = Am)

APPENDIX 5. Equations of electromagnetism for SI (mks) and Gaussian (cgs) units

Magnetic poles are excluded from the equations that follow. Equation numbers match those of the text.

A. Equations that are the same for both sets of units

Description of Equation's Content	Equation	
Relation of current and charge	$I = \frac{dq}{dt}$	(15.5)
Relation of electric field and electric force	$\mathbf{F}_E = q'\mathbf{E}$	(15.18)
Definition of electric flux	$\Phi_E = \int \mathbf{E} \cdot d\mathbf{S}$	(15.30)
Definition of magnetic flux	$\Phi_B = \int \mathbf{B} \cdot d\mathbf{S}$	(16.19)
Solenoidal character of magnetic field	$\oint \mathbf{B} \cdot d\mathbf{S} = 0$	(16.23)
Definition of potential	$V = \frac{U}{q}$	(15.55)
Relations of potential and static electric field	$V_2 - V_1 = -\int_{\mathbf{r}_1}^{\mathbf{r}_2} \mathbf{E} \cdot d\mathbf{s}$	(15.48)
	$\mathbf{E} = -\nabla V$	(15.67)
Ohm's law	$\mathbf{J} = \sigma \mathbf{E}$	(15.84)
	$V = IR$	(15.88)
Power associated with current and potential difference	$P = IV$	(15.57)
Power in linear circuit	$P = I^2R$	(15.107)
	$P = \frac{V^2}{R}$	(15.108)
Definition of electric dipole moment	$\mathbf{p} = q\mathbf{l}$	(15.71)
Energy of electric dipole	$U = -\mathbf{p} \cdot \mathbf{E}$	(15.73)
Energy of magnetic dipole	$U = -\boldsymbol{\mu} \cdot \mathbf{B}$	(16.16)
Definition of capacitance	$C = \frac{q}{V}$	(15.109)
Energy stored in capacitor	$U = \frac{1}{2}CV^2$	(15.119)
Definition of inductance	$L = -\frac{\mathcal{V}}{\left(\frac{dI}{dt}\right)}$	(17.44)
Energy stored in inductor	$U = \frac{1}{2}LI^2$	(17.71)

B. Equations that are different for the two sets of units

Description of Equation's Content	Equation for SI Units	Equation for Gaussian Units
Coulomb's law	$\mathbf{F}_{12} = \frac{1}{4\pi\epsilon_0}\frac{q_1q_2\mathbf{i}_{12}}{r^2}$ (15.15)	$\mathbf{F}_{12} = \frac{q_1q_2\mathbf{i}_{12}}{r^2}$
Electric field of point charge	$\mathbf{E} = \frac{1}{4\pi\epsilon_0}\frac{q\mathbf{i}_r}{r^2}$ (15.20)	$\mathbf{E} = \frac{q\mathbf{i}_r}{r^2}$
Potential of point charge	$V = \frac{1}{4\pi\epsilon_0}\frac{q}{r}$ (15.60)	$V = \frac{q}{r}$
Gauss's law	$\oint \mathbf{E}\cdot d\mathbf{S} = \frac{q}{\epsilon_0}$ (15.36)	$\oint \mathbf{E}\cdot d\mathbf{S} = 4\pi q$
Electric field near a conductor	$E = \frac{\sigma}{\epsilon_0}$ (15.44)	$E = 4\pi\sigma$
Capacitance of a parallel plate capacitor	$C = \frac{\epsilon_0 A}{d}$ (15.112)	$C = \frac{A}{4\pi d}$
Energy density of electromagnetic field	$u = \frac{1}{2}\epsilon_0 E^2 + \frac{1}{2\mu_0}B^2$ (16.25)	$u = \frac{1}{8\pi}(E^2 + B^2)$
Magnetic force on a moving charge	$\mathbf{F_M} = q'\mathbf{v}\times\mathbf{B}$ (16.27)	$\mathbf{F_M} = \frac{q'}{c}\mathbf{v}\times\mathbf{B}$
Magnetic force on a current element	$d\mathbf{F} = I'\,d\mathbf{s}\times\mathbf{B}$ (16.43)	$d\mathbf{F} = \frac{I'}{c}\,d\mathbf{s}\times\mathbf{B}$
Magnetic field created by a moving charge	$\mathbf{B} = \frac{\mu_0}{4\pi}\frac{q\mathbf{v}\times\mathbf{i}_r}{r^2}$ (16.36)	$\mathbf{B} = \frac{q}{c}\frac{\mathbf{v}\times\mathbf{i}_r}{r^2}$
Magnetic field created by a current element	$d\mathbf{B} = \frac{\mu_0}{4\pi}\frac{I\,d\mathbf{s}\times\mathbf{i}_r}{r^2}$ (16.56)	$d\mathbf{B} = \frac{I}{c}\frac{d\mathbf{s}\times\mathbf{i}_r}{r^2}$
Radius of curvature of charge orbiting in magnetic field	$r = \frac{p_\perp}{q'B}$ (16.30)	$r = \frac{p_\perp c}{q'B}$
Magnetic moment of circling particle	$\boldsymbol{\mu} = \frac{q}{2m}\mathbf{L}$ (16.54)	$\boldsymbol{\mu} = \frac{q}{2mc}\mathbf{L}$
Magnetic moment of current loop	$\boldsymbol{\mu} = I\mathbf{A}$ (16.49)	$\boldsymbol{\mu} = \frac{I\mathbf{A}}{c}$
Force per unit length on parallel currents	$\frac{dF}{ds} = \frac{\mu_0}{2\pi}\frac{I_1I_2}{d}$ (16.74)	$\frac{dF}{ds} = \frac{2I_1I_2}{c^2d}$
Magnetic field of long straight wire	$B = \frac{\mu_0 I}{2\pi x}$ (16.61)	$B = \frac{2I}{cx}$
Magnetic field within a long solenoid	$B = \mu_0 nI$ (16.69)	$B = \frac{4\pi nI}{c}$
Law of electromagnetic induction	$\oint \mathbf{E}\cdot d\mathbf{s} = -\frac{d\Phi_B}{dt}$ (17.1)	$\oint \mathbf{E}\cdot d\mathbf{s} = -\frac{1}{c}\frac{d\Phi_B}{dt}$

Description of Equation's Content	Equation for SI Units	Equation for Gaussian Units
Ampère's law and law of magnetoelectric induction	$\oint \mathbf{B} \cdot d\mathbf{s} = \mu_0 I + \mu_0 \epsilon_0 \frac{d\Phi_E}{dt}$ (17.29)	$\oint \mathbf{B} \cdot d\mathbf{s} = \frac{4\pi I}{c} + \frac{1}{c}\frac{d\Phi_E}{dt}$
The speed of light	$c = \frac{1}{\sqrt{\mu_0 \epsilon_0}}$ (17.77)	No counterpart; c appears explicitly in the Gaussian equations

C. The differential form of Maxwell's equations

SI	Gaussian
$\nabla \cdot \mathbf{E} = \frac{\rho}{\epsilon_0}$	$\nabla \cdot \mathbf{E} = 4\pi\rho$
$\nabla \cdot \mathbf{B} = 0$	$\nabla \cdot \mathbf{B} = 0$
$\nabla \times \mathbf{E} = -\frac{\partial \mathbf{B}}{\partial t}$	$\nabla \times \mathbf{E} = -\frac{1}{c}\frac{\partial \mathbf{B}}{\partial t}$
$\nabla \times \mathbf{B} = \mu_0 \mathbf{J} + \mu_0 \epsilon_0 \frac{\partial \mathbf{E}}{\partial t}$	$\nabla \times \mathbf{B} = \frac{4\pi \mathbf{J}}{c} + \frac{1}{c}\frac{\partial \mathbf{E}}{\partial t}$

APPENDIX 6. Mathematical formulas

Some of the formulas below go beyond the immediate needs of this text in order to provide a reference source for other courses or for optional additional work an instructor may wish to assign. For a much more extensive compendium of formulas, see Herbert Dwight's *Tables of Integrals and Other Mathematical Data*, 4th edition (New York: The Macmillan Company, 1961). This excellent reference volume, modest in size and price, is a good investment. It will prove useful throughout one's student and professional careers.

A. Mathematical signs

$=$ is equal to
$\neq$ is not equal to
$\simeq$ is approximately equal to
$\equiv$ is identical to, is defined as
$>$ is greater than
$\geq$ is greater than or equal to
$\gg$ is much greater than
$<$ is less than
$\leq$ is less than or equal to
$\ll$ is much less than
$\sim$ is proportional to

B. Arithmetic: powers of 10

$$10^a 10^b = 10^{a+b}$$

$$10^a/10^b = 10^{a-b}$$

$$(10^a)^b = 10^{ab}$$

C. Algebra

FRACTIONS

$$a\left(\frac{b}{c}\right) = \frac{ab}{c}$$

$$\frac{\left(\frac{b}{c}\right)}{d} = \frac{b}{cd}$$

$$\left(\frac{a}{b}\right)\left(\frac{c}{d}\right) = \frac{ac}{bd}$$

$$\frac{\left(\frac{a}{b}\right)}{\left(\frac{c}{d}\right)} = \frac{ad}{bc}$$

$$\frac{a}{b} + \frac{c}{d} = \frac{ad + bc}{bd}$$

ROOTS OF A QUADRATIC EQUATION

If $ax^2 + bx + c = 0$ then $x = \dfrac{-b \pm \sqrt{b^2 - 4ac}}{2a}$.

If $x^2 + 2\beta x + \gamma = 0$ then $x = -\beta \pm \sqrt{\beta^2 - \gamma}$.

BINOMIAL EXPANSIONS

Factorial of an integer n: $n! = n(n-1)(n-2)\cdot\cdots\cdot 2\cdot 1$

Binomial coefficient for integers q and n: $\dbinom{q}{n} = \dfrac{q!}{n!\,(q-n)!}$

$(a \pm b)^2 = a^2 \pm 2ab + b^2$

$(a \pm b)^3 = a^3 \pm 3a^2b + 3ab^2 \pm b^3$

To evaluate $(a + b)^p$, write it as

$a^p(1 + x)^p$, where $x = b/a$, or as

$b^p(1 + x)^p$, where $x = a/b$.

$$(1 \pm x)^p = 1 \pm px + \frac{p(p-1)}{2!}x^2 \pm \frac{p(p-1)(p-2)}{3!}x^3 + \cdots.$$

This is a finite series if p is a positive integer. For other values of p, it is an infinite series that converges for $|x| < 1$.

Special cases:

$$p = -1: \quad \frac{1}{1 \pm x} = 1 \mp x + x^2 \mp x^3 + x^4 \mp \cdots$$

$$p = -2: \quad \frac{1}{(1 \pm x)^2} = 1 \mp 2x + 3x^2 \mp 4x^3 + 5x^4 \mp \cdots$$

$$p = \tfrac{1}{2}: \quad \sqrt{1 \pm x} = 1 \pm \tfrac{1}{2}x - \tfrac{1}{8}x^2 \pm \tfrac{1}{16}x^3 - \cdots$$

$$p = -\tfrac{1}{2}: \quad \frac{1}{\sqrt{1 \pm x}} = 1 \mp \tfrac{1}{2}x + \tfrac{3}{8}x^2 \mp \tfrac{5}{16}x^3 + \cdots$$

COMPLEX NUMBERS

$(a + ib) + (c + id) = (a + c) + i(b + d)$

$(a + ib)(c + id) = (ac - bd) + i(bc + ad)$

$$\frac{a + ib}{c + id} = \frac{(a + ib)(c - id)}{c^2 + d^2}$$

$|a + ib|^2 = (a + ib)^*(a + ib) = (a - ib)(a + ib) = a^2 + b^2$

$a + ib = re^{i\theta}$, where $r = \sqrt{a^2 + b^2}$, $\theta = \arctan(b/a)$

$e^{i\theta} = \cos\theta + i\sin\theta$

D. Trigonometry

DEFINITIONS OF TRIGONOMETRIC FUNCTIONS

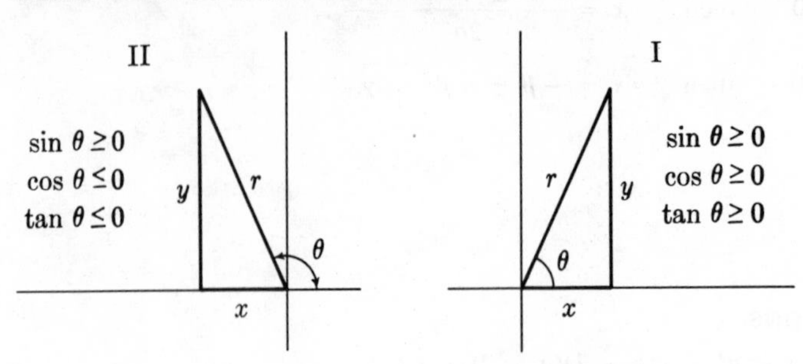

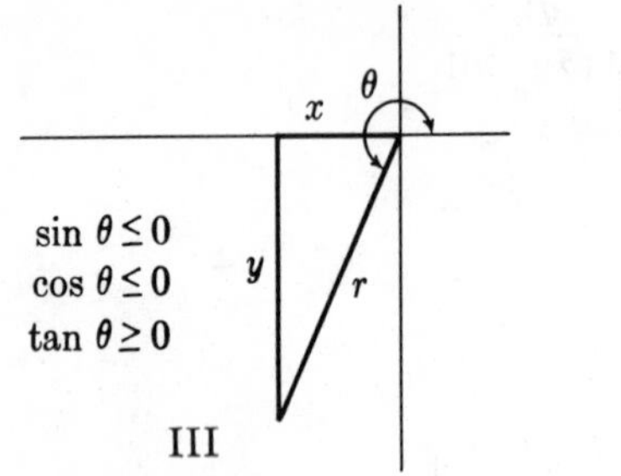

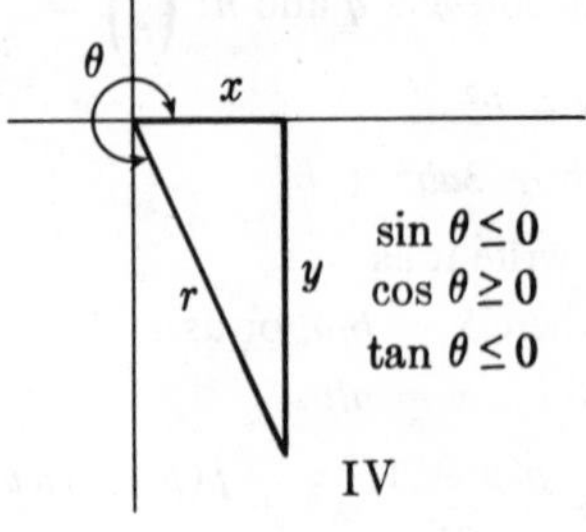

$$\sin\theta = \frac{y}{r} \qquad \csc\theta = \frac{1}{\sin\theta} = \frac{r}{y}$$

$$\cos\theta = \frac{x}{r} \qquad \sec\theta = \frac{1}{\cos\theta} = \frac{r}{x}$$

$$\tan\theta = \frac{\sin\theta}{\cos\theta} = \frac{y}{x} \qquad \text{ctn}\,\theta = \frac{1}{\tan\theta} = \frac{x}{y}$$

Inverse functions: If $u = \sin\theta$, then $\theta = \text{arc}\sin u$, sometimes written $\theta = \sin^{-1} u$. The other inverse functions are similarly designated: arc cos u, arc tan u, etc.

SIMPLE PROPERTIES

$$\sin(-\theta) = -\sin\theta \qquad \cos(-\theta) = \cos\theta$$

$$\sin\left(\theta \pm \frac{\pi}{2}\right) = \pm\cos\theta \qquad \cos\left(\theta \pm \frac{\pi}{2}\right) = \mp\sin\theta$$

$$\sin(\theta \pm \pi) = -\sin\theta \qquad \cos(\theta \pm \pi) = -\cos\theta$$

$$\tan(-\theta) = -\tan\theta$$

$$\tan\left(\theta \pm \frac{\pi}{2}\right) = -\frac{1}{\tan\theta} = -\text{ctn}\,\theta$$

$$\tan(\theta \pm \pi) = \tan\theta$$

VALUES FOR SPECIAL ANGLES

Function	Angle				
	0 deg	30 deg	45 deg	60 deg	90 deg
$\sin\theta$	0	$\frac{1}{2}$	$\frac{1}{\sqrt{2}} = 0.7071$	$\frac{\sqrt{3}}{2} = 0.8660$	1
$\cos\theta$	1	$\frac{\sqrt{3}}{2} = 0.8660$	$\frac{1}{\sqrt{2}} = 0.7071$	$\frac{1}{2}$	0
$\tan\theta$	0	$\frac{1}{\sqrt{3}} = 0.5774$	1	$\sqrt{3} = 1.7321$	∞

For other values and for graphs, see Appendix 7.

TRIGONOMETRIC FORMULAS

$$\sin^2\theta + \cos^2\theta = 1$$

$$\sec^2\theta - \tan^2\theta = 1$$

$$\csc^2\theta - \text{ctn}^2\theta = 1$$

$$\sin 2\theta = 2\sin\theta\cos\theta \qquad \sin\tfrac{1}{2}\theta = \sqrt{\frac{1-\cos\theta}{2}}$$

$$\cos 2\theta = \cos^2\theta - \sin^2\theta \qquad \cos\tfrac{1}{2}\theta = \sqrt{\frac{1+\cos\theta}{2}}$$

$$= 2\cos^2\theta - 1$$

$$= 1 - 2\sin^2\theta$$

$$\tan 2\theta = \frac{2\tan\theta}{1-\tan^2\theta} \qquad \tan\tfrac{1}{2}\theta = \sqrt{\frac{1-\cos\theta}{1+\cos\theta}}$$

$$\sin(A \pm B) = \sin A\cos B \pm \cos A\sin B$$

$$\cos(A \pm B) = \cos A\cos B \mp \sin A\sin B$$

$$\tan(A \pm B) = \frac{\tan A \pm \tan B}{1 \mp \tan A\tan B}$$

$$\sin A \pm \sin B = 2\sin[\tfrac{1}{2}(A \pm B)]\cos[\tfrac{1}{2}(A \mp B)]$$

$$\cos A + \cos B = 2\cos[\tfrac{1}{2}(A + B)]\cos[\tfrac{1}{2}(A - B)]$$

$$\cos A - \cos B = 2\sin[\tfrac{1}{2}(A + B)]\sin[\tfrac{1}{2}(B - A)]$$

$$\tan A \pm \tan B = \frac{\sin(A \pm B)}{\cos A\cos B}$$

$$\sin\theta + \sin 2\theta + \sin 3\theta + \cdots + \sin n\theta = \frac{\sin[\tfrac{1}{2}(n+1)\theta]\sin(\tfrac{1}{2}n\theta)}{\sin(\tfrac{1}{2}\theta)}$$

$$\cos\theta + \cos 2\theta + \cos 3\theta + \cdots + \cos n\theta = \frac{\cos[\tfrac{1}{2}(n+1)\theta]\sin(\tfrac{1}{2}n\theta)}{\sin(\tfrac{1}{2}\theta)}$$

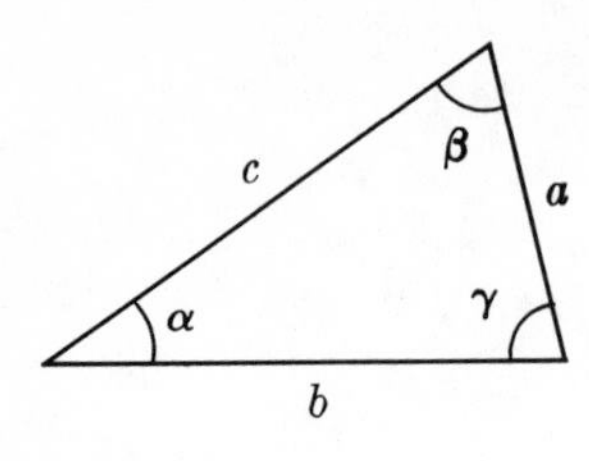

PROPERTIES OF A TRIANGLE

$\alpha + \beta + \gamma = \pi$

$a^2 = b^2 + c^2 - 2bc \cos \alpha$

$b^2 = c^2 + a^2 - 2ca \cos \beta$

$c^2 = a^2 + b^2 - 2ab \cos \gamma$

$$\frac{a}{\sin \alpha} = \frac{b}{\sin \beta} = \frac{c}{\sin \gamma}$$

For a right triangle $\left(\gamma = \frac{\pi}{2}\right)$, $a^2 + b^2 = c^2$

SERIES EXPANSIONS

$$\sin x = x - \frac{x^3}{3!} + \frac{x^5}{5!} - \frac{x^7}{7!} + \cdots$$

$$\cos x = 1 - \frac{x^2}{2!} + \frac{x^4}{4!} - \frac{x^6}{6!} + \cdots$$

These series converge for all x.

E. Exponential and logarithmic functions

For graphs and numerical values, see Appendices 8 and 9.

$e = 2.71828$ $\qquad$ $e^0 = 1$

$e^x e^y = e^{x+y}$ $\qquad$ $e^x/e^y = e^{x-y}$

$(e^x)^n = e^{nx}$ $\qquad$ $a^x = e^{x \ln a}$

$e^{\ln x} = x$ $\qquad$ $e^{-\ln x} = 1/x$

$e^{-t/\tau} = 0.5$ for $t = \tau \ln 2 = 0.6931\tau$

$e^{t/\tau} = 2$ for $t = \tau \ln 2 = 0.6931\tau$

$$e^{\pm x} = 1 \pm x + \frac{x^2}{2!} \pm \frac{x^3}{3!} + \frac{x^4}{4!} \pm \cdots.$$

Series converges for all x.

$\ln e = 1$ $\qquad$ $\ln 1 = 0$

$\ln (xy) = \ln x + \ln y$ $\qquad$ $\ln (x/y) = \ln x - \ln y$

$\ln (1/x) = -\ln x$ $\qquad$ $\ln (x^n) = n \ln x$

$\ln (e^x) = x$ $\qquad$ $\ln (a^x) = x \ln a$

$\ln a = 2.3026 \log_{10} a$ $\qquad$ $\log_{10} a = 0.43429 \ln a$

$$\ln (1 \pm x) = \pm x - \frac{x^2}{2} \pm \frac{x^3}{3} - \frac{x^4}{4} \pm \cdots$$

$$\ln \left(\frac{1 + x}{1 - x}\right) = 2\left(x + \frac{x^3}{3} + \frac{x^5}{5} + \frac{x^7}{7} + \cdots\right)$$

Series converge for $|x| < 1$.

F. Calculus

In what follows, f, g, and u are functions; a, b, and n are constants.

SOME RULES OF DIFFERENTIATION

$$\frac{d}{dx}(fg) = \frac{df}{dx}g + f\frac{dg}{dx}$$

$$\frac{d}{dx}\left(\frac{f}{g}\right) = \frac{\frac{df}{dx}g - f\frac{dg}{dx}}{g^2}$$

$$\frac{d}{dx}[f(u)] = \frac{df}{du}\cdot\frac{du}{dx}$$

LINEARITY PROPERTIES

$$\frac{d}{dx}(af + bg) = a\frac{df}{dx} + b\frac{dg}{dx}$$

$$\int (af + bg)\,dx = a\int f\,dx + b\int g\,dx$$

THE DEFINITE INTEGRAL

$$D = \int_a^b f(x)\,dx = I(x)\Big|_a^b = I(b) - I(a),$$

where I is the indefinite integral, $I(x) = \int f(x)\,dx$, or the antiderivative: $f(x) = dI(x)/dx$.

INTEGRATION BY PARTS

$$\int_a^b f(x)\frac{dg}{dx}\,dx = f(x)g(x)\Big|_a^b - \int_a^b \frac{df}{dx}g(x)\,dx$$

TAYLOR SERIES

If all derivatives of a function exist at a certain point, the function may be written as a power series about that point. Empirical functions may be similarly approximated.

$$f(x) = f(x_0) + (x - x_0)\left(\frac{df}{dx}\right)_{x_0} + \frac{(x - x_0)^2}{2!}\left(\frac{d^2f}{dx^2}\right)_{x_0} + \frac{(x - x_0)^3}{3!}\left(\frac{d^3f}{dx^3}\right)_{x_0} + \cdots.$$

All the derivatives are evaluated at $x = x_0$.

SOME DERIVATIVES (See also Table 5.2.)

$$\frac{d}{dx}(x^n) = nx^{n-1}$$

$$\frac{d}{dx}(\sin ax) = a\cos ax$$

$$\frac{d}{dx}(\cos ax) = -a\sin ax$$

$$\frac{d}{dx}(\tan ax) = a\sec^2 ax = \frac{a}{\cos^2 ax}$$

$$\frac{d}{dx}\left(\text{arc sin}\,\frac{x}{a}\right) = \frac{\pm 1}{\sqrt{a^2 - x^2}}$$

+ sign in 1st and 4th quadrants
− sign in 2nd and 3rd quadrants

$$\frac{d}{dx}\left(\text{arc cos}\,\frac{x}{a}\right) = \frac{\mp 1}{\sqrt{a^2 - x^2}}$$

− sign in 1st and 2nd quadrants
+ sign in 3rd and 4th quadrants

$$\frac{d}{dx}\left(\text{arc tan}\,\frac{x}{a}\right) = \frac{a}{a^2 + x^2}$$

$$\frac{d}{dx}(e^{ax}) = ae^{ax}$$

$$\frac{d}{dx}(\ln ax) = \frac{1}{x}$$

SOME INDEFINITE INTEGRALS (See also Table 5.7)

To each of the following integrals an arbitrary constant should be added.

$$\int x^n\,dx = \frac{x^{n+1}}{n+1}, \qquad n \neq -1$$

$$\int \frac{1}{x}\,dx = \ln |x|$$

$$\int (a + bx)^n\,dx = \frac{(a + bx)^{n+1}}{b(n+1)}, \qquad n \neq -1$$

$$\int \frac{dx}{a + bx} = \frac{1}{b} \ln |a + bx|$$

$$\int \frac{dx}{a^2 + x^2} = \frac{1}{a}\,\text{arc tan}\,\frac{x}{a}$$

$$\int \frac{dx}{a^2 - x^2} = \frac{1}{2a} \ln \left|\frac{a + x}{a - x}\right|$$

$$\int \sqrt{a + bx}\,dx = \frac{2}{3b}(a + bx)^{3/2}$$

$$\int \frac{dx}{\sqrt{a + bx}} = \frac{2}{b}\sqrt{a + bx}$$

$$\int \sqrt{x^2 + a^2}\,dx = \tfrac{1}{2}x\sqrt{x^2 + a^2} + \tfrac{1}{2}a^2 \ln (x + \sqrt{x^2 + a^2})$$

$$\int \frac{dx}{\sqrt{x^2 + a^2}} = \ln (x + \sqrt{x^2 + a^2})$$

$$\int \sqrt{x^2 - a^2}\,dx = \tfrac{1}{2}x\sqrt{x^2 - a^2} - \tfrac{1}{2}a^2 \ln |x + \sqrt{x^2 - a^2}|$$

$$\int \frac{dx}{\sqrt{x^2 - a^2}} = \ln |x + \sqrt{x^2 - a^2}|$$

$$\int \sqrt{a^2 - x^2}\, dx = \tfrac{1}{2}x\sqrt{a^2 - x^2} + \tfrac{1}{2}a^2 \arcsin \frac{x}{a}$$

$$\int \frac{dx}{\sqrt{a^2 - x^2}} = \arcsin \frac{x}{a}$$

$$\int \sin ax\, dx = -\frac{1}{a}\cos ax$$

$$\int \cos ax\, dx = \frac{1}{a}\sin ax$$

$$\int \tan ax\, dx = -\frac{1}{a}\ln|\cos ax|$$

$$\int \csc ax\, dx = \frac{1}{a}\ln|\tan \tfrac{1}{2}ax|$$

$$\int \sec ax\, dx = \frac{1}{2a}\ln\left(\frac{1 + \sin ax}{1 - \sin ax}\right)$$

$$\int \operatorname{ctn} ax\, dx = \frac{1}{a}\ln|\sin ax|$$

$$\int \arcsin \frac{x}{a}\, dx = x \arcsin \frac{x}{a} + \sqrt{a^2 - x^2}$$

$$\int \arccos \frac{x}{a}\, dx = x \arccos \frac{x}{a} - \sqrt{a^2 - x^2}$$

$$\int \arctan \frac{x}{a}\, dx = x \arctan \frac{x}{a} - \tfrac{1}{2}a \ln(a^2 + x^2)$$

$$\int e^{ax}\, dx = \frac{1}{a}e^{ax}$$

$$\int xe^{ax}\, dx = \frac{1}{a}\left(x - \frac{1}{a}\right)e^{ax}$$

$$\int \ln ax\, dx = x\ln ax - x$$

$$\int x \ln ax\, dx = \tfrac{1}{2}x^2 \ln ax - \tfrac{1}{4}x^2$$

G. Vectors

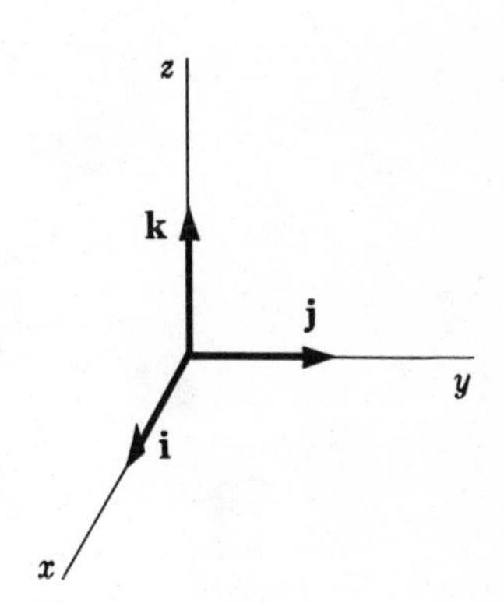

Unit vectors **i**, **j**, and **k** are parallel to the x, y, and z axes respectively.

Vector in terms of Cartesian components: $\mathbf{a} = a_x\mathbf{i} + a_y\mathbf{j} + a_z\mathbf{k}$

The position vector: $\mathbf{r} = x\mathbf{i} + y\mathbf{j} + z\mathbf{k}$

Magnitude of a vector: $|\mathbf{a}| = a = \sqrt{a_x^2 + a_y^2 + a_z^2}$

SCALAR PRODUCT

$\mathbf{a} \cdot \mathbf{b} = ab\cos\theta$; θ is the smaller angle between **a** and **b**.

$\mathbf{a} \cdot \mathbf{b} = a_xb_x + a_yb_y + a_zb_z$

$\mathbf{a} \cdot \mathbf{b} = \mathbf{b} \cdot \mathbf{a}$

VECTOR PRODUCT

$\mathbf{a} \times \mathbf{b} = (a_y b_z - a_z b_y)\mathbf{i} + (a_z b_x - a_x b_z)\mathbf{j} + (a_x b_y - a_y b_x)\mathbf{k}$

$|\mathbf{a} \times \mathbf{b}| = ab \sin \theta;$ θ is the smaller angle between **a** and **b**.

$\mathbf{a} \times \mathbf{b} = -\mathbf{b} \times \mathbf{a}$

PROPERTIES OF UNIT VECTORS

$\mathbf{i} \cdot \mathbf{i} = \mathbf{j} \cdot \mathbf{j} = \mathbf{k} \cdot \mathbf{k} = 1$

$\mathbf{i} \cdot \mathbf{j} = \mathbf{j} \cdot \mathbf{k} = \mathbf{k} \cdot \mathbf{i} = 0$

$\mathbf{i} \times \mathbf{j} = \mathbf{k}, \qquad \mathbf{j} \times \mathbf{k} = \mathbf{i}, \qquad \mathbf{k} \times \mathbf{i} = \mathbf{j}$

$\mathbf{j} \times \mathbf{i} = -\mathbf{k}, \qquad \mathbf{k} \times \mathbf{j} = -\mathbf{i}, \qquad \mathbf{i} \times \mathbf{k} = -\mathbf{j}$

H. Vector calculus

F and **G** are vector functions; f is a scalar function; a and b are numerical constants.

DERIVATIVES

$$\frac{d\mathbf{F}}{dt} = \frac{dF_x}{dt}\mathbf{i} + \frac{dF_y}{dt}\mathbf{j} + \frac{dF_z}{dt}\mathbf{k}$$

$$\frac{d}{dt}(f\mathbf{F}) = \frac{df}{dt}\mathbf{F} + f\frac{d\mathbf{F}}{dt}$$

$$\frac{d}{dt}(a\mathbf{F} + b\mathbf{G}) = a\frac{d\mathbf{F}}{dt} + b\frac{d\mathbf{G}}{dt}$$

$$\frac{d}{dt}(\mathbf{F} \cdot \mathbf{G}) = \frac{d\mathbf{F}}{dt} \cdot \mathbf{G} + \mathbf{F} \cdot \frac{d\mathbf{G}}{dt}$$

$$\frac{d}{dt}(\mathbf{F} \times \mathbf{G}) = \frac{d\mathbf{F}}{dt} \times \mathbf{G} + \mathbf{F} \times \frac{d\mathbf{G}}{dt}$$

INTEGRALS

$$\int \mathbf{F}(t)\, dt = \left[\int F_x(t)\, dt\right]\mathbf{i} + \left[\int F_y(t)\, dt\right]\mathbf{j} + \left[\int F_z(t)\, dt\right]\mathbf{k}$$

Line integral: $\int \mathbf{F} \cdot d\mathbf{s} = \int F_{\parallel}\, ds$, where $F_{\parallel}$ is the component of **F** parallel to the designated path of integration at each point.

Surface integral: $\int \mathbf{F} \cdot d\mathbf{S} = \int F_{\perp}\, dS$, where $F_{\perp}$ is the component of **F** perpendicular to the designated surface of integration (or parallel to the vector $d\mathbf{S}$) at each point.

VECTOR OPERATIONS NEEDED IN MORE ADVANCED WORK

The *gradient* of a scalar function is a vector function:

$$\nabla f = \frac{\partial f}{\partial x}\mathbf{i} + \frac{\partial f}{\partial y}\mathbf{j} + \frac{\partial f}{\partial z}\mathbf{k}$$

The *divergence* of a vector function is a scalar function:

$$\nabla \cdot \mathbf{F} = \frac{\partial F_x}{\partial x} + \frac{\partial F_y}{\partial y} + \frac{\partial F_z}{\partial z}$$

The *curl* of a vector function as an axial vector function:

$$\nabla \times \mathbf{F} = \left(\frac{\partial F_z}{\partial y} - \frac{\partial F_y}{\partial z}\right)\mathbf{i} + \left(\frac{\partial F_x}{\partial z} - \frac{\partial F_z}{\partial x}\right)\mathbf{j} + \left(\frac{\partial F_y}{\partial x} - \frac{\partial F_x}{\partial y}\right)\mathbf{k}$$

APPENDIX 7. Trigonometric functions

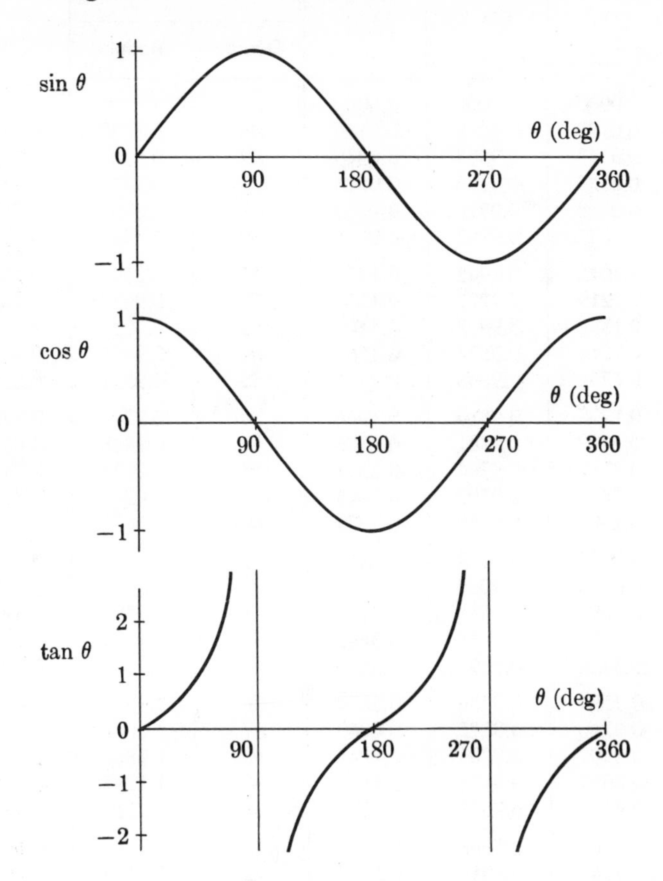

The table on the next page gives $\sin\theta$, $\cos\theta$, and $\tan\theta$ in the first quadrant ($0 \leq \theta \leq 90$ deg).

For the second quadrant, measure backward from 180 deg, and use

$$\sin(\pi - \theta) = \sin\theta$$
$$\cos(\pi - \theta) = -\cos\theta$$
$$\tan(\pi - \theta) = -\tan\theta$$

For the third quadrant, measure forward from 180 deg, and use

$$\sin(\pi + \theta) = -\sin\theta$$
$$\cos(\pi + \theta) = -\cos\theta$$
$$\tan(\pi + \theta) = \tan\theta$$

For the fourth quadrant, measure backward from 0 deg (or 360 deg), and use

$$\sin(-\theta) = -\sin\theta$$
$$\cos(-\theta) = \cos\theta$$
$$\tan(-\theta) = -\tan\theta$$

Other trigonometric functions are defined by

$$\csc\theta = \frac{1}{\sin\theta} \qquad \sec\theta = \frac{1}{\cos\theta} \qquad \text{ctn}\,\theta = \frac{1}{\tan\theta}$$

Angle θ		$\sin\theta$	$\cos\theta$	$\tan\theta$	Angle θ		$\sin\theta$	$\cos\theta$	$\tan\theta$
Degree	Radian				Degree	Radian			
0	0.0000	0.0000	1.0000	0.0000					
1	0.0175	0.0175	0.9998	0.0175	46	0.8029	0.7193	0.6947	1.0355
2	0.0349	0.0349	0.9994	0.0349	47	0.8203	0.7314	0.6820	1.0724
3	0.0524	0.0523	0.9986	0.0524	48	0.8378	0.7431	0.6691	1.1106
4	0.0698	0.0698	0.9976	0.0699	49	0.8552	0.7547	0.6561	1.1504
5	0.0873	0.0872	0.9962	0.0875	50	0.8727	0.7660	0.6428	1.1918
6	0.1047	0.1045	0.9945	0.1051	51	0.8901	0.7771	0.6293	1.2349
7	0.1222	0.1219	0.9925	0.1228	52	0.9076	0.7880	0.6157	1.2799
8	0.1396	0.1392	0.9903	0.1405	53	0.9250	0.7986	0.6018	1.3270
9	0.1571	0.1564	0.9877	0.1584	54	0.9425	0.8090	0.5878	1.3764
10	0.1745	0.1736	0.9848	0.1763	55	0.9599	0.8192	0.5736	1.4281
11	0.1920	0.1908	0.9816	0.1944	56	0.9774	0.8290	0.5592	1.4826
12	0.2094	0.2079	0.9781	0.2126	57	0.9948	0.8387	0.5446	1.5399
13	0.2269	0.2250	0.9744	0.2309	58	1.0123	0.8480	0.5299	1.6003
14	0.2443	0.2419	0.9703	0.2493	59	1.0297	0.8572	0.5150	1.6643
15	0.2618	0.2588	0.9659	0.2679	60	1.0472	0.8660	0.5000	1.7321
16	0.2793	0.2756	0.9613	0.2867	61	1.0647	0.8746	0.4848	1.8040
17	0.2967	0.2924	0.9563	0.3057	62	1.0821	0.8829	0.4695	1.8807
18	0.3142	0.3090	0.9511	0.3249	63	1.0996	0.8910	0.4540	1.9626
19	0.3316	0.3256	0.9455	0.3443	64	1.1170	0.8988	0.4384	2.0503
20	0.3491	0.3420	0.9397	0.3640	65	1.1345	0.9063	0.4226	2.1445
21	0.3665	0.3584	0.9336	0.3839	66	1.1519	0.9135	0.4067	2.2460
22	0.3840	0.3746	0.9272	0.4040	67	1.1694	0.9205	0.3907	2.3559
23	0.4014	0.3907	0.9205	0.4245	68	1.1868	0.9272	0.3746	2.4751
24	0.4189	0.4067	0.9135	0.4452	69	1.2043	0.9336	0.3584	2.6051
25	0.4363	0.4226	0.9063	0.4663	70	1.2217	0.9397	0.3420	2.7475
26	0.4538	0.4384	0.8988	0.4877	71	1.2392	0.9455	0.3256	2.9042
27	0.4712	0.4540	0.8910	0.5095	72	1.2566	0.9511	0.3090	3.0777
28	0.4887	0.4695	0.8829	0.5317	73	1.2741	0.9563	0.2924	3.2709
29	0.5061	0.4848	0.8746	0.5543	74	1.2915	0.9613	0.2756	3.4874
30	0.5236	0.5000	0.8660	0.5774	75	1.3090	0.9659	0.2588	3.7321
31	0.5411	0.5150	0.8572	0.6009	76	1.3265	0.9703	0.2419	4.0108
32	0.5585	0.5299	0.8480	0.6249	77	1.3439	0.9744	0.2250	4.3315
33	0.5760	0.5446	0.8387	0.6494	78	1.3614	0.9781	0.2079	4.7046
34	0.5934	0.5592	0.8290	0.6745	79	1.3788	0.9816	0.1908	5.1446
35	0.6109	0.5736	0.8192	0.7002	80	1.3963	0.9848	0.1736	5.6713
36	0.6283	0.5878	0.8090	0.7265	81	1.4137	0.9877	0.1564	6.314
37	0.6458	0.6018	0.7986	0.7536	82	1.4312	0.9903	0.1392	7.115
38	0.6632	0.6157	0.7880	0.7813	83	1.4486	0.9925	0.1219	8.144
39	0.6807	0.6293	0.7771	0.8098	84	1.4661	0.9945	0.1045	9.514
40	0.6981	0.6428	0.7660	0.8391	85	1.4835	0.9962	0.0872	11.430
41	0.7156	0.6561	0.7547	0.8693	86	1.5010	0.9976	0.0698	14.301
42	0.7330	0.6691	0.7431	0.9004	87	1.5184	0.9986	0.0523	19.081
43	0.7505	0.6820	0.7314	0.9325	88	1.5359	0.9994	0.0349	28.636
44	0.7679	0.6947	0.7193	0.9657	89	1.5533	0.9998	0.0175	57.290
45	0.7854	0.7071	0.7071	1.0000	90	1.5708	1.0000	0.0000	∞

APPENDIX 8. The exponential function

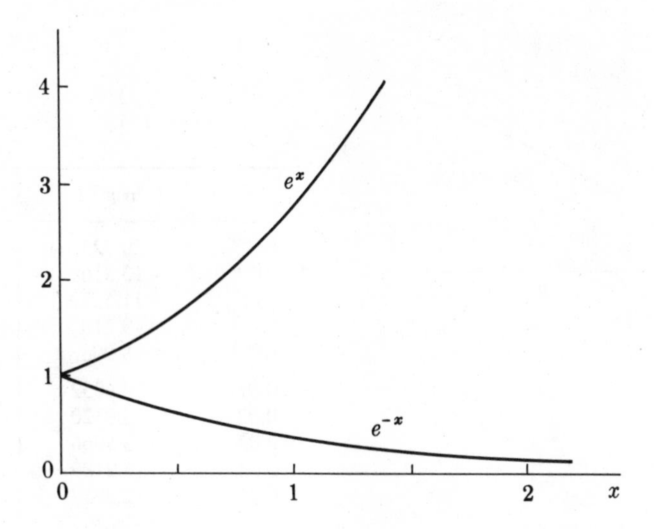

x	e^x	e^{-x}	x	e^x	e^{-x}
0	1.0000	1.0000	2.6	13.464	0.07427
0.1	1.1052	0.9048	2.8	16.445	0.06081
0.2	1.2214	0.8187	3.0	20.086	0.04979
0.3	1.3499	0.7408	3.2	24.533	0.04076
0.4	1.4918	0.6703	3.4	29.964	0.03337
0.5	1.6487	0.6065	3.6	36.598	0.02732
0.6	1.8221	0.5488	3.8	44.701	0.02237
0.7	2.0138	0.4966	4.0	54.598	0.01832
0.8	2.2255	0.4493	4.2	66.686	0.01500
0.9	2.4596	0.4066	4.4	81.451	0.01228
1.0	2.7183	0.3679	4.6	99.484	0.01005
1.1	3.0042	0.3329	4.8	121.51	0.00823
1.2	3.3201	0.3012	5.0	148.41	0.00674
1.3	3.6693	0.2725	5.5	244.69	0.00409
1.4	4.0552	0.2466	6.0	403.43	0.00248
1.5	4.4817	0.2231	6.5	665.14	0.00150
1.6	4.9530	0.2019	7.0	1096.6	0.00091
1.7	5.4739	0.1827	7.5	1808.0	0.00055
1.8	6.0496	0.1653	8.0	2981.0	0.00034
1.9	6.6859	0.1496	8.5	4914.8	0.00020
2.0	7.3891	0.1353	9.0	8103.1	0.00012
2.2	9.025	0.11080	9.5	13,360.	0.00007
2.4	11.023	0.09072	10.0	22,026.	0.00005

APPENDIX 9. The logarithmic function

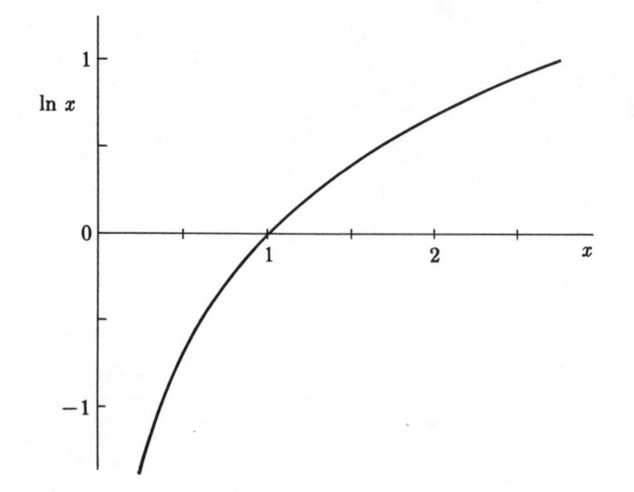

x	$\ln x$	x	$\ln x$
10^{-9}	−20.723	1.05	0.0488
10^{-6}	−13.816	1.10	0.0953
10^{-5}	−11.513	1.15	0.1398
10^{-4}	−9.210	1.20	0.1823
10^{-3}	−6.908	1.25	0.2231
0.01	−4.6052	1.30	0.2624
0.02	−3.9120	1.35	0.3001
0.03	−3.5066	1.40	0.3365
0.04	−3.2189	1.45	0.3716
0.05	−2.9957	1.50	0.4055
0.06	−2.8134	1.55	0.4383
0.07	−2.6593	1.60	0.4700
0.08	−2.5257	1.65	0.5008
0.09	−2.4079	1.70	0.5306
0.10	−2.30259	1.75	0.5596
0.12	−2.1203	1.80	0.5878
0.14	−1.9661	1.85	0.6152
0.16	−1.8326	1.90	0.6419
0.18	−1.7148	1.95	0.6678
0.20	−1.6094	2.00	0.69315
0.22	−1.5141	2.1	0.7419
0.24	−1.4271	2.2	0.7885
0.26	−1.3471	2.3	0.8329
0.28	−1.2730	2.4	0.8755
0.30	−1.2040	2.5	0.9163
0.35	−1.0498	2.6	0.9555
0.40	−0.9163	2.7	0.9933
0.45	−0.7985	2.8	1.0296
0.50	−0.6931	2.9	1.0647
		3.0	1.0986
0.55	−0.5978	3.1	1.1314
0.60	−0.5108	3.2	1.1632
0.65	−0.4308	3.3	1.1939
0.70	−0.3567	3.4	1.2238
0.75	−0.2877	3.5	1.2528
0.80	−0.2231	3.6	1.2809
0.85	−0.1625	3.7	1.3083
0.90	−0.1054	3.8	1.3350
0.95	−0.0513	3.9	1.3610
1.00	0.0000	4.0	1.3863

x	$\ln x$	x	$\ln x$
4.2	1.4351	16	2.773
4.4	1.4816	17	2.833
4.6	1.5261	18	2.890
4.8	1.5686	19	2.944
5.0	1.6094	20	2.996
5.2	1.6487	22	3.091
5.4	1.6864	24	3.178
5.6	1.7228	26	3.258
5.8	1.7579	28	3.332
6.0	1.7918	30	3.401
6.2	1.8245	32	3.466
6.4	1.8563	34	3.526
6.6	1.8871	36	3.584
6.8	1.9169	38	3.638
7.0	1.9459	40	3.689
7.2	1.9741	42	3.738
7.4	2.0015	44	3.784
7.6	2.0281	46	3.829
7.8	2.0541	48	3.871
8.0	2.0794	50	3.912
8.2	2.1041	55	4.007
8.4	2.1282	60	4.094
8.6	2.1518	65	4.174
8.8	2.1748	70	4.248
9.0	2.1972	75	4.317
9.2	2.2192	80	4.382
9.4	2.2407	85	4.443
9.6	2.2618	90	4.500
9.8	2.2824	95	4.554
10.0	2.30259	100	4.605
10.5	2.3514	200	5.298
11.0	2.3979	300	5.704
11.5	2.4423	400	5.991
12.0	2.4849	500	6.215
12.5	2.5257	600	6.397
13.0	2.5649	10^3	6.908
13.5	2.6027	10^4	9.210
14.0	2.6391	10^5	11.513
14.5	2.6741	10^6	13.816
15.0	2.7081	10^9	20.723

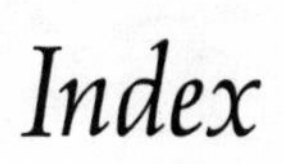

INDEX

Italic letters following page numbers are used with the following meanings: *f*, figure; *n*, footnote; and *t*, table. Parentheses following a page number enclose the number of a question, exercise, or problem; such an end-of-chapter item is referenced only if it contains factual information that does not occur elsewhere in the book.

Absolute temperature, 549, 552
Absolute zero, 551–52, 558, 647
Acceleration: *see also* Centripetal acceleration, Motion
  absolute, 248
  in air, 514
  average, 115, 185*f*
  of Earth, 247
  instantaneous, 115, 185
  of Mars, 248
  of sun, 248
Acceleration of gravity, 224–25, 228, 448, 457, 463*t*, A10*t*
Accelerators, 51, 52*f*, 60–61, 69(P3.4)
Action and reaction, 280
Action at a distance, 303
ADAMS, JOHN C. (1819–1892), 472
Adiabatic change, 640–42, 645, 648–49
Air, properties of, A11*t*
Air drag, 513–17
  laws of, 513, 517
Air pressure, 500–503, A11*t*
Airspeed meter, 512–13
Air track, 213, 221*f*
ALDRIN, EDWIN E., JR. (1930–), 480 (E11.40)
Algebra, A18–A19
  commutative law, 94
AMONTONS, GUILLAUME (1663–1705), 557
Ampere (A), 33, A3
amu: *see* Atomic mass unit
ANDERSON, CARL D. (1905–), 47, 48*f*
Angle of repose, 491
Angles, 106–109
Angstrom (Å), 24*t*
Angular acceleration, 341
Angular frequency, 136, 222
Angular momentum, 33–35, 316–58
  of circular motion, 320–21
  definition of, 316–17
  law of change, 337, 415
  orbital, 35, 78, 79*f*, 321, 325–27, 335
  of a particle, 316–17
  of precessing wheel, 346–47
  quantization, 321
  significance of, 357–58
  spin, 35, 325–27, 329, 336
  of straight-line motion, 319–20
  of systems, 322–27
  unit, 319
  vector nature, 316–18
Angular-momentum conservation, 78–79, 85, 347–58, 461, 465
Angular speed, 107, 189
Angular velocity, 332, 340–41
Anharmonic oscillation, 410
Antibaryons, 73
Anticommutative property, 180–81
Antiderivative, 121, 124
Antilambda particle, 56*f*
Antineutron, 49, 56
Antiparticles, 54–57
Antiproton, 47, 49, 56, 77*f*
Aphelion, 355–56, 461
  of Earth, 461
Apogee, 354–55
ARCHIMEDES (*c.* 287–212 B.C.), 396–97, 503
Archimedes' principle, 503–504
Area under curve, 125–33
Areas: *see* Law of areas
ARISTOTLE (384–322 B.C.), 206, 464
Arithmetic, 161–63, A18
  geometrical, 161–62
  laws of, 162
Arrow diagrams, 163–68
Arrow of time, 650–54
Astronomical data, A12*t*
Astronomical unit (A.U.), 24*t*
Atmosphere, 500–503
  exponential model, 502, 503*f*
  improved model, 539–40(P12.16)
  properties of, A11*t*

scale height, 502
U.S. standard, 503*f*
Atomic mass, 554
Atomic mass unit (amu), 553–54
Atomic weight, 553–54
Available energy
in particle collisions, 76, 522–23
in thermodynamics, 622, 636–37, 652
Averages, 258(P7.2)
AVOGADRO, AMEDEO (1776–1856), 554–55
Avogadro's hypothesis, 555, 571
Avogadro's number, 554, A8*t*
Axial vector: *see* Vectors

Balloon, hot-air, 504, 505*f*
Bank, angle of, 234–35, 262(P7.20)
Bar (pressure unit), 500
BARDEEN, JOHN (1908–), 9*f*
Barometer, 500, 501*f*
Baryon family, 57
conservation, 72–73
Base of logarithms, 141
Beads on a wire, 582–84
BERNOULLI, DANIEL (1700–1782), 510, 565, 566*f*, 624
Bernoulli's equation, 508–510
applications, 510–13
Bernoulli's principle, 510*f*, 511
Beta decay, 48–49
Bevatron, 47
Binary stars, 82–83
Binomial coefficients, 612, A19
Binomial expansion, A19
Block and tackle, 432(E10.34)
BODE, JOHANN E. (1747–1826), 487 (P11.25)
Bode's law, 487(P11.25)
BOHR, NIELS (1885–1962), 35, 46
Boiling, 576
temperature of, 552*t*
BOLTZMANN, LUDWIG (1844–1906), 558, 625
Boltzmann's constant, 558, 619, A8*t*
Boundary conditions, 122, 124; *see also* Initial conditions
BOYLE, ROBERT (1627–1691), 556
Boyle's law, 501, 557, 639
BRAHE, TYCHO DE (1546–1601), 59, 355
BRATTAIN, WALTER H. (1902–), inside rear cover
British system of units, 18, 30
British thermal unit (Btu), 31
Brownian motion, 584–86
BROWN, ROBERT (1773–1858), 584
Bubble chamber, 64, 65*f*
Bubble-chamber photographs, 4*f*, 48*f*, 56*f*, 58*f*, 77*f*
Bulk energy: *see* Energy
Buoyant force, 503–504

Calculus, 101–107, 112–14, 121–34, A23–A26
fundamental theorem, 132
Caloric, 564, 624
Calorie, 31, 561–62
Calorimetry, 575
Candela (cd), 19, A3
Carbon 12, 553
Carnot cycle, 644–47
Carnot engine, 644–47, 649–50
efficiency, 647
CARNOT, SADI (1796–1832), 622–23, 625, 632
Cart and horse paradox, 279–80
Cartesian coordinates, 98
Catenary, 155(E5.56)
Cathode rays, 45–46
CAVENDISH, HENRY (1731–1810), 469–70
Celsius scale, 552–53
Center of mass, 270–77
and angular momentum, 324–25
and momentum conservation, 292–95
and Newton's second law, 281–82
relation to momentum, 274
and total kinetic energy, 390–91
of two-body system, 271, 274
Center-of-mass
acceleration, 272
frame, 519–21
velocity, 272, 390–91
Central force, 233, 351–56, 465
Centripetal acceleration, 189, 233, 261(P7.14), 453
Centripetal force, 233
cgs system of units: *see* Gaussian units
Chain, accelerating, 285–87
Chain hoist, 396*f*, 397
Chain rule, 105
Chance, 64–65; *see also* Probability
Change, spontaneous, 618; *see also* Adiabatic change, Isothermal change
Chaos and order, 74
Charge, 32–33

of particles, 54, 55*t*
quantum unit, 33, A8*t*
Charge conservation, 71–72
Charged particles, motion of, 239–44
Charge multiplets, 57
CHARLES, JACQUES A. C. (1746–1823), 557, 566
Charles's law, 557
Circular motion: *see* Motion
Classical physics, 12
CLAUSIUS, RUDOLPH J. E. (1822–1888), 565, 625, 652
Cloud chamber, 63
Cloud-chamber photograph, 48*f*
Cockcroft-Walton accelerator, 61
Coefficient of friction
kinetic, 491
static, 490, 495–96
Coefficient of thermal expansion
linear, 599–600(E13.7)
volume, 601(E13.23)
Coins and probability, 609–615
Colliding beams, 523, 524*f*
Collisions
elastic, 294–95, 518–20, 565
inelastic, 291–92, 521–23
molecular, 565–69, 585, 591, 618
of two bodies, 291–92, 294–95, 517–23
Combining volumes, law of, 554–55
Comets, 460
Complex numbers, A19
Components, vector, 172–78, 186
Compression, 496–97
Compton wavelength, 68(E3.2), A10*t*
Concepts, quantitative, 16–18, 215, 266
primitive and derived, 17
CONDUITT, JOHN (1688–1737), 467
Conic sections, 459–60
Conservation laws, 70–85; *see also* Charge conservation, Energy conservation, etc.
Conservation of work, 395–98
Conservative force, 401, 404, 416–22
Constants, A8–A10*t*
Continuity, equation of, 506–508
Conversion factors, 22, A13–A14
Coordinate systems, 97–99
Copernican revolution, 248
COPERNICUS, NIKOLAUS (1473–1543), 65, 464
Cosine function, 108, 135–37, 222–23, A20–A22, A27–A28*ft*
Cosmic radiation, 51, 53*f*
Coulomb (C), 33, A3
Couple, 338, 339*f*
Creation of matter, 49
Critical angle, 490–91
Curl, A26
Cyclical processes, 644–50
Cyclotron, 61
Cyclotron frequency, 243
Cylinder: *see* Moment of inertia
Cylindrical coordinates, 99

Decay of particles, 57, 58*f*, 287; *see also* Kaon decay, Muon decay, etc.
one-body (absence of), 77–78
three-body, 77, 287, 289*f*
two-body, 76, 78–79, 287
Deformation, 210
Degrees of freedom, 582–84, 586–87
of diatomic molecules, 588
frozen, 591–93
of monatomic gas, 590
nuclear, 592–93
rotational, 582, 586–87
translational, 582–84, 586
Densities of materials, A11*t*
Derivative
definition of, 101
partial, 417–18
relation to slope, 112–13
Derivatives, 103–106, 184–87, A23–A24; *see also* Differentiation
Detectors, particle, 61–64
Diatomic gas, 588–89
DICKE, ROBERT H. (1916–), 231, 232*f*
Difference of specific heats, 590, 638
Differential equations, 124–25, 218–19, 406, 408
Differentiation: *see also* Derivatives
chain rule, 105
linearity, 104
of vectors, 184–87, A26
Diffusion, 571
Dimensional analysis, 42
Dimensional consistency, 20–22
Dimensionless physics, 36
Dimensionless quantities, 19
Dimensions, 19–21
DIRAC, PAUL A. M. (1902–), 47, 54, 56, 97
Disk: *see* Moment of inertia
Disorder, 614–15

Disordered energy: *see* Energy
Displacement vector, 171–72, 174–75
Dissipation of energy, 580, 622
Divergence, A26
Downhill rule of decay, 75
Drag: *see* Air drag, Friction
Drag coefficient, 513
  of sphere, 516
Dynamic pressure, 512

Earth
  acceleration of surface, 247, 457
  orbit of, 355*t*, 461, 463*t*
  orbital acceleration, 189
  properties of, 463*t*, A10–A12*t*
  shape of, 356, 454–57
Earth-moon system, 327
Eccentricity of ellipse, 460
Effective one-body problem, 300–302
Efficiency
  of Carnot engine, 647
  of gasoline engine, 649–50
  maximum of heat engine, 634, 650
EINSTEIN, ALBERT (1879–1955), 47, 57*n*, 585
Electric charge: *see* Charge
Electric field, 239
Electric force, 239–40
Electromagnetism, equations, A15–A17*t*
Electron, 45–46
  properties of, A8–A10*t*
  stability, 72
Electron family, 57
Electron-family conservation, 74–75
Electron volt (eV), 32
Electrostatic unit (esu), 33
Elementary particles, 44–65; *see also* specific particles and properties
  decay, 57–58, 287
  properties of, 54–58, 55*t*
  significance of, 64–65
Ellipse, 458–60
  equation of, 460, 479(E11.33)
Energy, 30–32, 376–424; *see also* Available energy, Heat, Internal energy, Kinetic energy, Mass energy, Potential energy, Work
  bulk, 560, 578–81
  definition of, 392
  in fluid flow, 508–510
  in inverse-square force field, 453
  manifold form of, 377*f*, 424
  mechanical, 400, 405–412, 562
  ordered and disordered, 560, 578–81, 636
  rotational, 412–16
  significance of, 393
  thermal, 592–93
  units, 31, 32*t*
Energy conservation, 30–31, 75–76, 85, 394, 400, 403–412, 562, 578–81, 624
  significance of, 423–24
Energy diagrams, 405–411, 452*f*
Energy transfer, 392–93, 560, 579–81
Entropy
  additive property, 619
  Boltzmann definition, 619, 643
  Clausius definition, 625–27
  development of concept, 625
  relation of different definitions, 627–29
  unit, 619
Entropy change, 621, 625–26
  in free expansion, 643
  in heat flow, 630
  in isothermal change, 638, 640
  in mixing, 631
EÖTVÖS, ROLAND VON (1848–1919), 231
Equilibrium
  of fluids, 496–504, 510
  of rigid bodies, 492–96
  thermal, 550, 584, 626
    law of, 555–56
Equipartition theorem, 582–86, 621
Erg, 31
Error, 143
Error bars, 144–45
Escape, 384, 412, 452–53
Escape speed, 412, 451
Eta particle, 55*t*
Ether, 246
EUCLID (lived *c.* 300 B.C.), 93
Exchange force, 50
Exhaust: *see* Rocket
Expectation value, 568
Experiment and theory, 7–8
Explosion of shell, 288–90
Exponential decay, 139–40
Exponential function, 137–40, A22, A29*ft*
Exponential growth, 140
Extensive variable, 597(Q13.23)
External force: *see* Forces

Fahrenheit scale, 552–53
Family-number conservation laws, 57, 72–75

Farad (F), A4
Fermi (fm), 24*t*
FERMI, ENRICO (1901–1954), 48–49
First law of thermodynamics, 578–81, 624
  for adiabatic expansion, 640
  applied to heat engine, 633
  applied to refrigerator, 635
  applied to universe, 652
  for changing volume, 638
  at constant volume, 637
  for isothermal expansion, 638
Fluid flow, 504–513
  regimes of, 517
  steady, 505–513
Fluids, 496–513
  in equilibrium, 496–504
  incompressible, 499–513
  in motion, 504–513
Foci of ellipse, 459*f*, 460
Food calorie, 31
Force, 209–212; *see also* Conservative force, Electric force, Gravitational force, Magnetic force
  average, 268–69
  constant, 217–18
  of molecular collision, 569
  time-dependent, 218–19
  unit, 209
  vector nature, 210–11
Force constant, 220, 408
Forces
  conservative: *see* Conservative force
  equal and opposite, 277–80, 303
  external, 281, 403
  internal, 281–83, 389, 403
  in modern physics, 303
  of type 1, 398, 401
  of type 2, 398, 400, 402, 421
Frames of reference, 99–100, 208*f*, 244–48
  accelerated, 208–209, 245–46
  inertial, 208–209, 244–47
FRANKLIN, BENJAMIN (1706–1790), 32
Free expansion, 642–44
Free fall, 117–18, 225, 229, 406–407
  with air friction, 514–17
Frequency, 136
Friction
  and energy dissipation, 580, 622
  fluid, 513–17
  kinetic, 491
  sliding, 491
  static, 490
  surface, 489–92
Functions, 110–13, A23, A26
  exponential, 137–40, A22, A29*ft*
  linear, 110–12
  logarithmic, 141–42, A22, A30–A31*ft*
  quadratic, 112–14
  trigonometric, 108, 134–37, 220–23, A20–A22, A27–A28*ft*
Fundamental theorem of integral calculus, 132

g force, 262(P7.20)
Galaxy, 5*f*, 33
Galilean relativity, 244–46
Galilean transformation, 200(P6.11), 264 (P7.30)
GALILEO GALILEI (1564–1642), 11, 59, 114, 158(P5.11), 206, 226, 244, 246, 464
Gamma rays, 48*f*
  detection of, 64
Gas constant, 557–58, A8*t*
Gas thermometer, 548–51
Gases: *see also entries under* Ideal
  behavior of, 637–44
  kinetic theory of, 564–72
  non-ideal, 569
Gasoline engine, 647–50
GAUSS, KARL FRIEDRICH (1777–1855), 94
Gaussian units, 18
  equations of electromagnetism in, A15–A17*t*
GAY-LUSSAC, JOSEPH L. (1778–1850), 554, 557, 566
Geiger counter, 61–64
GEIGER, HANS W. (1882–1945), 61
Geometry, 93–94
GILBERT, WILLIAM (1544–1603), 11, 464
GLASER, DONALD A. (1926–), 65*f*
GOUDSMIT, SAMUEL A. (1902–), 35
Gradient, 401, 418, A26
Graphs, 111–14, 144
Gravitation, 443–72; *see also entries under* Gravitational, Gravity
  law of universal, 444, 469
Gravitational constant, 444–45, 448, 469–70, A8*t*
Gravitational force, 224, 401, 410, 444
  mass symmetry of, 446*f*
  superposition, 445
  on a system, 492
  weakness of, 445

Gravitational mass: *see* Mass
Gravitational potential energy: *see* Potential energy
Gravitational radius, 477(E11.15)
Gravitational torque
  in nonuniform field, 456, 457*f*
  in uniform field, 492–93
Graviton, 49
Gravity: *see also* Gravitation *and entries under* Gravitational
  of the earth, 448–57
  historic role, 444
  inside the earth, 449–52
  in the large, 410–12
  near the earth, 401–403, 405–407
  outside the earth, 449–53
  of a sphere, 471
  of a spherical shell, 446–48

Half life, 139–40
HALLEY, EDMUND (1656–1742), 467
Heat, 560–64, 579–80
  compared with internal energy, 560
  related to temperature change, 573
  relation to work, 560
  unit, 561
Heat death, 653
Heat engines, 622, 632–34, 644–50
  Carnot, 644–47
  diesel, 650
  efficiency of, 633–34
  gasoline, 647–50
  and second law, 622
Heat flow, 580, 621–22, 629–30
Heat of fusion, 576
Heat of vaporization, 576
Helical motion, 244
HELMHOLTZ, HERMANN L. F. VON (1821–1894), 563*n*, 564
Henry (H), A4
HERSCHEL, SIR WILLIAM (1738–1822), 472
Homogeneity: *see also* Uniformity
  of space, 80, 84–85
  of time, 82, 85
HOOKE, ROBERT (1635–1703), 467
Hooke's law, 210, 220
Hoop, 414–15; *see also* Moment of inertia
HUYGHENS, CHRISTIAAN (1629–1695), 465
Hydrodynamics, 504–513
Hyperbola, 459

Ideal diatomic gas, 588–89
Ideal gases, 637–44
Ideal monatomic gas, 576–78, 590
Ideal-gas law, 556–60, 569, 638
Impulse, 268–70
  molecular, 568
Inertial frames: *see* Frames of reference
Inertial mass: *see* Mass
Initial conditions, 122, 124, 218, 222
Instability of particles, 57
Integrals: *see also* Integration, Line integral
  definite, 125–34, A23
  indefinite, 121–24, 123*t*, 132–33, A24–A25
  relationship of two kinds, 131–34
Integrand, 121
Integration, 121–34; *see also* Calculus, Integrals
  additive property, 127
  linearity, 122, 128
  by parts, A23
  of vectors, A26
Intensive variable, 597(Q13.23)
Internal energy, 389, 391, 393, 637, 560–64, 578–81
  compared with heat, 560
  of ideal diatomic gas, 588
  of ideal gas, 637
  of ideal monatomic gas, 577
Internal force: *see* Forces
International system of units: *see* SI units
Invariance, 82–84
  of laws of motion, 246
Invariance principles, 196(E6.22)
Inverse-square force, 420–22, 453
Inversion of coordinates, 183
Ionization, 62–63
Isolated systems, 207, 299, 338, 350, 619
Isothermal change, 638–40, 645
  relation to free expansion, 643
Isotropy of space, 80, 85, 357–58; *see also* Uniformity

Joule (J), 19, A3
JOULE, JAMES P. (1818–1889), 562, 564–65, 624
Jupiter, 355*t*, 463*t*

K particle: *see* Kaon
Kaon, 51, 55*t*
  decay, 58*f*, 76, 77*f*, 289*f*, 350–51

Kelvin (K), 549, A3
KELVIN, LORD (WILLIAM THOMSON) (1824–1907), 625
Kelvin scale, 548–53
KEPLER, JOHANNES (1571–1630), 11, 351, 353–56, 462
Kepler's first law, 458–61, 467
Kepler's laws, 458–62, 465–67
  compared with Newton's laws, 458
Kepler's second law, 461, 465; *see also* Law of areas
Kepler's third law, 461–62, 466–67
Kilogram (kg), 28, A3
Kilomole, 554
Kinematics
  one-dimensional, 100–103, 110–12, 114–20, 124–25, 130–31, 133–34
  rotational, 106–108
  vector, 184–87
Kinetic energy
  definition, 30
  of fluid, 509
  internal, 391
  molecular, 570, 591–92
  in particle decay, 75
  relation to work, 380–82, 385, 387–88, 399
  rotational, 391, 413–14, 416*t*
  of system, 390–92
  total, 390–92
  translational, 390–91, 570, 591
Kinetic theory, 564–72
  basic assumptions of, 565
KRÖNIG, AUGUST K. (1822–1879), 565

Lambda particle, 51, 55*t*
  decay, 58*f*, 71, 79*f*
LAPLACE, PIERRE S. (1749–1827), 472, 624
Latent heats, 576
Latitude, 455
Law of areas, 351–56, 461, 465
Length, 23–24
  standard, 24, A3
  units, 24*t*
Lever, 395–97
  mechanical advantage of, 396
LEVERRIER, URBAIN J. J. (1811–1877), 472
Limiting process, 101, 103, 129–30
Line integral, 386, 419–20
Linear function, 110–12
Linearity properties, 104, 122, 128, 170, 181, A23
Liquefaction temperatures, 552*t*
Logarithmic function, 141–42, A22, A30–A31*ft*
Logarithms, 141–42, A30–A31*t*
  Napierian, 141
  natural, 141
LOVELL, JAMES A., JR. (1928–), 480 (E11.40)
LOWELL, PERCIVAL (1855–1916), 472

Mach's principle, 248
Magnetic force, 240
Magnitude of vector, 165, A25
Mars, 354–55, 463*t*
  acceleration of, 248
Mass
  definition, 215
  equality of gravitational and inertial, 231
  gravitational, 29–30, 215, 445
    relation to charge, 230
  inertial, 29, 213–15, 224, 230
  introduction to concept, 28–30
  relation to weight, 224, 229
  scalar nature of, 214
  standard, 28, A3
Mass conservation, 214, 506–507
Mass energy, 31, 75
Mass flux, 506–507
Mass spectrometer, 264(P7.29)
Mathematical formulas, A18–A26
Mathematical truth, 92–95
Mathematics, 92–146
  axioms, 95, 156(P5.1, P5.2)
  definition of, 94
  relation to physics, 8–9, 95–97, 145–46
MAUPERTUIS, PIERRE L. M. DE (1698–1759), 455
Maxwell-Boltzmann distribution, 616–17
Maxwell's equations, A17
MAYER, JULIUS R. (1814–1878), 562, 564
Mean life, 139
Mechanical advantage, 396–97
Mechanical energy: *see* Energy
Mechanical equivalent of heat, 562–64
Melting, 576
  temperatures, 576*t*
Mercury, 355, 356*f*, 463*t*
Mesons, 57
Meter (m), 24, A3
Mixed product, 181–82
Mixing, 630–32

mks system of units: *see* SI units
Molar latent heats, 576
Molar specific heat, 574, 590
  of hydrogen, 593*f*
  of ideal monatomic gas, 577
Mole, 554
Molecular speed
  average, 617–18
  mean-square, 569–71, 585, 616–17
  root-mean-square (rms), 571–72, 618
Molecular weight, 554
Moment of inertia, 328–36, 413
  of cylinder, 360(Q9.10)
  of disk, 333, 372(P9.12)
  of half-cylinder, 371(P9.9)
  of hoop, 328–29
  of rigid body, 330
  of rod, 331, 367(E9.26)
  of sphere, 333
  unit, 328
Momentum, 266–304
  components, 291–92
  definition of, 267
  in modern physics, 303–304
  unit, 267
Momentum conservation, 76–78, 84, 175–76, 287–99
  and center of mass, 292–95
  relation to Newton's third law, 303
  significance of, 302–304
  in systems not isolated, 291–92
Monatomic gas, 576–78, 590
Moon, 470–71
  properties of, A12*t*
Moons of Jupiter and Saturn, 468, 470
Motion: *see also* Kinematics, Rotation
  with constant acceleration, 116–20, 217–18, 225, 239–40, 406–407
  with constant velocity, 115–16
  in electric and magnetic fields, 239–44
  near the earth, 224–29
  in one dimension, 383–85
  oscillatory: *see* Oscillation, Oscillator
  projectile, 236–39
  relative, 167
  simple harmonic: *see also* Oscillator
    inside the earth, 449–50
  in three dimensions, 386–87
  in two dimensions, 232–44
  uniform circular, 187–90, 233–36, 242–43, 320–21
  with variable acceleration, 218–24
Mu-e-gamma puzzle, 74
Mu meson: *see* Muon
Muon, 51, 55*t*
  decay, 58*f*, 74
Muon-electron puzzle, 64
Muon family, 57
Muon-family conservation, 74–75

National Accelerator, 60*f*, 61
Natural motion, 207
Neptune, 463*t*
  discovery of, 472
Neutrinos, 49, 53, 55*t*, 75, 351
Neutron, 47, 55*t*
  decay, 74
  mass, A9*t*
  stabilization, 57–58
Newton (N), 19, 209, A3
NEWTON, SIR ISAAC (1642–1727), 11, 244, 356, 446, 454–56, 460, 462, 464–65, 467–71, 653
Newtonian synthesis, 464–72
Newton's first law, 206–209, 464
  for systems, 282
Newton's second law, 83, 212–13
  applications of, 215–29
  for systems, 275–77, 281–82
  in terms of impulse, 268
  in terms of momentum, 267
Newton's third law, 83–84, 277–80, 445, 468–69
  applications of, 284
  connection to first and second laws, 282–83
  and momentum conservation, 303
  original version of, 280
Noncommutative algebra, 94–95
Nonequilibrium process, 642–44
Non-Euclidean geometry, 93–94
Nucleus, atomic, 46–47
Null vector, 192(Q6.7)
Numerics, 161–63

Ocean pressure, 539(P12.14)
Ohm (Ω), A3
Omega particle, 53, 55*t*
Operational definition, 18
Orbital angular momentum: *see* Angular momentum
Ordered energy: *see* Energy
Orthogonal systems, 99
Oscillation

anharmonic, 410
small, 409–410
Oscillator, simple harmonic, 136–37, 219–23, 226–28, 404–405, 407–410; *see also* Motion
amplitude, 223, 408
angular frequency, 222
period, 220, 222–23
phase constant, 222
potential energy, 404, 408*f*

Parabola, 113, 218*f*, 237–38, 459
Partial derivatives, 417–18, 421
Partial pressures, 607(P13.8)
Particles: *see* Elementary particles *and see* specific particles (Electron, Muon, etc.)
Pascal's principle, 500
Path independence, 416–21
PAULI, WOLFGANG (1900–1958), 49
Pendulum, 136–37, 226–29, 422–23
period, 228–29, 423
potential energy, 423
seconds, 228
Perigee, 354–55
Perihelion, 355–56
of Earth, 461
Period: *see* Oscillator, Pendulum
Perpetual motion, 622–24
Perturbations, 356, 472
Phase change, 575–76
Photographic emulsion, tracks in, 50*f*, 53*f*, 289*f*
Photon, 47–48, 55*t*
Physics
definition of, 3
evolution of, 11–12
nature of, 3–12
theories of, 6–7
Physics research activity, 140, 155(E5.54)
PLANCK, MAX K. E. L. (1858–1947), 35
Planck's constant, 35–36, A9*t*
Planetary orbits, 355*t*, 356*f*, 460, 462, 463*t*, 465–67
Planets, 354–56, 463*t*
Plasma, 592
Pluto, 463*t*, 472
Pi meson: *see* Pion
Pion, 51, 55*t*
decay, 50*f*, 58*f*, 71, 74, 78*f*
Pitot tube, 512–13
Poise, 517
Polar coordinates, 98
Polar vector: *see* Vectors
Position vector, 170–72, 174–75, 177
for uniform circular motion, 187
Positron, 47, 49, 55*t*, 57
Potential energy, 398–411, 417–23
arbitrary zero of, 401
for constant force, 418–19
definition of, 399, 416–17
of fluid, 509
in general, 417
gravitational, 402, 405, 410–11, 450–52
for inverse-square force, 421
in one dimension, 398–411
of pendulum, 423
relation to work, 400, 417
of simple harmonic oscillator, 404, 408*f*
Taylor series expansion, 410
POWELL, CECIL F. (1903–1969), 51
Power, 394–95
rotational, 415
unit, 394, A3
Power plants, 634
Precession, 344–47
of equinoxes, 456
Pressure
atmospheric, 500, A11*t*
definition of, 496
fluid, 496–503
of gas, 566–70
in ocean, 539(P12.14)
relation to work, 509
standard, 501, A11*t*
Pressure-volume relationships
adiabatic, 642
graphical, 639*f*, 645*f*, 648*f*
isothermal, 639
Principal axes, 334, 335*f*
Principia, 11, 467, 469*f*, 653
Probability
a priori, 620, 627, 643
for coins, 609–615
of energy, 627–28
and experimental error, 142–45
in kinetic theory, 567
molecular, 615–19
of position, 615–16, 643
in random events, 609–621
relation to disorder, 614
relative, 620
of speed, 616

Projectile motion, 236–39
Projection, 108–109, 172–73
Propulsion of vehicles, 280; *see also* Rocket
Proton, 46
 properties of, 55*t*, A8–A10*t*
 stability of, 73
Pseudoscalar, 183
Publication, physics, 140, 155(E5.54)
PV diagram: *see* Pressure-volume relationships
Pyrometer, 548

Quadratic equation, roots of, A19
Quadratic function, 112–14
Quantum energies, 592
 effect on degrees of freedom, 587, 591
Quarks, 72*n*

Radian, 106, 134
Radioactive decay, 139–40
Radiometer, 597(Q13.27)
Radius of curvature in magnetic field, 243–44
Radius of gyration, 368(E9.39)
Randomness, postulate of, 611
Range of projectile, 238–39
Rankine scale, 552–53
Ratio of specific heats, 571, 641–42
Reaction (force), 280
Real axis, 161–62
Recoil definition of mass, 29–30
Reduced mass, 300
Refrigerators, 634–36
Relative acceleration, 299
Relative motion, 167
Relative velocity, 299
Relativity of time, 653
Resolution of a vector, 172
Resonances, 53–54
Reynolds number, 517
Right-handed coordinates, 98–99
Right-hand rule, 34, 180, 317*f*
Rigid body, 389–90, 412–15, 492–96
 rotation of, 329–30
Rocket
 equations, 297, 299
 exhaust, 581
 propulsion, 295–99
 thrust, 299
Rotation: *see also* Motion
 with constant angular acceleration, 341–43
 with constant angular velocity, 340–41
 of coordinates, 177–78, 183, 196
 about fixed axis, 415
 kinematics of, 106–108
 and translation compared, 340*t*, 416*t*
Rotational energy: *see* Energy
RUMFORD, COUNT (BENJAMIN THOMPSON) (1753–1814), 562
RUSSELL, BERTRAND (1872–1970), 80
RUTHERFORD, SIR ERNEST (1871–1937), 46, 61

Satellites, 235–36, 317*f*, 354–56, 421–22, 455
 in circular orbits, 453
 communications, 236, 354–55
 energy of, 441(P10.21)
 low-altitude, 235
  period of, 235
Saturn, 355*t*, 463*t*
Scalar product, 179–80, A25
 commutative property, 179
Scalars, 17, 161–63, 183
Scale height of air, 502
Science, nature of, 96; *see also* Physics
Second (sec), 24–26, A3
Second law of thermodynamics
 applied to heat engine, 633
 applied to refrigerator, 635
 applied to the universe, 652
 and one-way time, 650–54
 significance of, 621, 623, 636, 650
 in terms of
  disorder, 615, 619
  entropy, 619
  heat engines, 622
  heat flow, 622
  perpetual motion, 623
  probability, 615, 618
Seconds pendulum, 228
Shear, 496–97, 517
Shell, exploding, 288–90
Sigma particle, 51, 55*t*
Simple harmonic oscillator: *see* Oscillator
Simple pendulum: *see* Pendulum
Simplicity, faith in, 8
Sine function, 108, 134–37, 220–23, A20–A22, A27–A28*ft*
SI units, 18, A3–A7*t*
 equations of electromagnetism in, A15–A17*t*
Slope, 112–13

Small oscillations, 409–410
Soddy, Frederick (1877–1956), 46
Sound, 571
Space: *see* Homogeneity, Isotropy, Uniformity
Spark chamber, 63
Spark-chamber photograph, 63*f*
Specific heat, 572–75; *see also* Difference of specific heats, Ratio of specific heats
  of common materials, 574*t*
  at constant pressure, 589
  at constant volume, 577, 588
  of ideal diatomic gas, 588–89
  of ideal monotamic gas, 577
  at low temperature, 592
  molar, 574
  unit of, 572
  of water, 573
Specific impulse, 314(P8.15)
Speed, 27–28, 100–102; *see also* Molecular speed
  average, 100
  definition of, 27
  instantaneous, 27, 101
Speed of light, 27, 36, A9*t*
Speed of sound, 571
Sphere: *see* Gravity, Moment of inertia
Spherical coordinates, 98
Spherical pendulum, 263(P7.22)
Spherical shell, gravity of, 445–48
Spin, 54, 55*t*, 78, 79*f*; *see also* Angular momentum
Spin flip, 327
Spring, 210–11
Square-root-of-n rule, 143–44
Stability of particles, 57
Stagnation, 511
Standard conditions, A11
Standards, 17, A3
Stanford Linear Accelerator, 52*f*
State of a system, 656(Q14.7)
Statics
  of fluids, 496–504, 510
  of rigid bodies, 492–96
Statistical fluctuations, 143, 585
Statistical mechanics, 546*n*, 625
Stevin, Simon (1548–1620), 11
Stokes, George G. (1819–1903), 39 (E2.11)
Stokes's law, 517
  motion influenced by, 542(P12.27)
Strange particles, 51
Streamlines, 505–506, 511, 512*f*
Summation notation, 128
Sun
  acceleration of, 248
  properties of, A12
Superposition, 445, 446*f*
Symmetries, 80–85, 304, 357–58, 425
Symmetry axis, 332–34
Systems, 270, 275–77, 281–83, 322–27, 388–92, 492–93; *see also* Isolated systems
  in one dimension, 284–87
  two-body, 271, 274–76, 294–95, 299–302, 391–92, 517–23

Tangent function, 108, A20–A21, A27–A28*ft*
Taylor series, 410, A23
Technology
  relation to physics, 8–10
  relation to second law of thermodynamics, 637
Telescope, 5*f*
Temperature, 547–53
  absolute, 549, 552
  definition
    macroscopic, 548, 551*f*
    submicroscopic, 570
  negative, 559*n*
  related to energy, 593*t*
  scalar nature of, 551
  scales, 549, 552–53
  thermodynamic, 647
  units of, 549, 552–53
Tension, 260(P7.11), 496
Terminal speed, 225–26, 514–15, 516*f*
Terrestrial data, A10–A11*t*
Theories of physics, 6–7
Theory and experiment, 7–8
Thermal energy: *see* Energy
Thermal equilibrium: *see* Equilibrium
Thermal pollution, 634
Thermistor, 605(P13.1)
Thermodynamics
  history of, 546, 562, 564, 624
  laws of: *see* First, Second, and Zeroth laws
  significance of, 546–47
Thermodynamic temperature scale, 647
Thermometer
  calibration of, 551

gas, 548–51
and thermal equilibrium, 556
THOMSON, SIR JOSEPH J. (1856–1940), 45–46, 61
Three-body decay: *see* Decay of particles
Three-vectors, 164
Thrust, 299, 402
Time, 24–27, 244, 653
and relativity, 653
standard of, 24–25, A3
uniformity of, 424
Time-reversal invariance, 651, 653
TITIUS, JOHANN D. (1729–1796), 487 (P11.25)
Torque, 337–47, 415, 416*t*
definition of, 337
on the earth, 456, 457*t*
external, 348
gravitational, 492–93
internal, 348, 357
relation to angular acceleration, 341
unit of, 337
vector nature of, 337, 339*f*
Torsion balance, 470*f*
Transformation
of components, 176–78
coordinate, 150(E5.5)
Galilean, 200(P6.11)
Translational and rotational motion compared, 340*t*, 416*t*; *see also* Motion
Triangle, properties of, A22
Trigonometric functions, 108–109, 134–37, A20–A22, A27–A28*ft*
Triple point of water, 549–51
Triple product, 181–83
Tube of flow, 506
Turning points, 405, 406*f*, 408*f*
Two-body decay: *see* Decay of particles
Two-body systems: *see* Systems

UHLENBECK, GEORGE E. (1900–), 35
Unavailable energy, 622
Uncertainty, experimental, 142–45
Uniform circular motion: *see* Motion
Uniformity
of space, 80, 82–85, 304
of time, 424
Units, 18–19, A3–A4, A5–A7*t*; *see also* Gaussian units, SI units
conversion of, A13–A14
natural, 35–36
Units arithmetic, 22–23
Units consistency, 20–23
Unit vectors, 172–73, 179, A25–A26
Universe
lifetime of, 27
mass of, 30
as a physical system, 652–53
size of, 6*t*
Uranus, 463*t*, 472

Van Allen radiation belt, 243
Van de Graaf accelerator, 61
Vector arithmetic, 163–70
addition, 164–65
associative law, 165, 170
commutative law, 165
distributive law, 169
linearity property, 170
multiplication by numeric, 168–70, 174
subtraction, 166
Vector calculus, 184–87, 190, A26
Vector product, 180–82, A26
anticommutative property, 180–81
linearity property, 181
Vector quantity, 17
Vectorial consistency, 178–79, 183
Vectors, 96, 161–90; *see also* Unit vectors
axial, 183–84, 316*n*
components, 172–78
magnitude of, 165, A25
polar, 183–84
properties of, A25–A26
Vehicle propulsion, 280
Velocity
average, 100, 114, 159(P5.17), 184
instantaneous, 101, 114, 185
for uniform circular motion, 188
Venus, 355, 463*t*
Viscosity, 517
Vise, 433(E10.36)
Volt (V), A3

Water hammer, 541(P12.21)
Watt (W), 31–32, 394, A3
Weber (Wb), A3
Weight, 224, 449, 450*f*
relation to mass, 224, 229
Width of probability curve, 613
WILSON, ROBERT R. (1914–), inside rear cover
Wind triangle, 167–68

Work
  in Carnot cycle, 646
  definition of, 377, 383, 386
  done by a constant force, 376–79
  done by gravity, 382, 384
  as energy transfer, 392–93, 579–80
  on an expanding gas, 589, 638
  in a fluid, 509
  as an integral over distance, 383
  as an integral over time, 387
  as a line integral, 386
  in one dimension, 379–85
  related to heat, 560
  related to kinetic energy, 380–82, 385, 387–88, 399
  related to potential energy, 400, 417
  significance of, 381
  for a system, 388–89
WREN, SIR CHRISTOPHER (1632–1723), 467

Xi particle, 51, 55*t*
  decay, 58*f*

YUKAWA, HIDEKI (1907–   ), 50

Zero of potential energy, 401, 421
Zeroth law of thermodynamics, 555–56, 581–84, 621
Zero vector, 192(Q6.7)